Institute of Physics Publishing

11th IAEA Technical Meeting on H-Mode Physics and Transport Barriers 2007

Journal of Physics: Conference Series Vol. 123

September 26-28, 2007
Tsukuba, Japan

Printed from e-media with permission by:

Curran Associates, Inc.
57 Morehouse Lane
Red Hook, NY 12571
www.proceedings.com

ISBN: 978-1-60560-467-1

Some format issues inherent in the e-media version may also appear in this print version.

Institute of Physics Publishing
Dirac House, Temple Back
Bristol BS1 6BE UK

Institute of Physics Publishing

11th IAEA Technical Meeting on H-Mode Physics and Transport Barriers
2007

TABLE OF CONTENTS

11th IAEA Technical Meeting on H-mode Physics and Transport Barriers

Tomonori Takizuka 2008 *J. Phys.: Conf. Ser.* **123** 011001 (1p) doi: 10.1088/1742-6596/123/1/011001

Tomonori Takizuka

This volume of *Journal of Physics: Conference Series* contains papers based on invited talks and contributed posters presented at the 11th IAEA Technical Meeting on H-mode Physics and Transport Barriers. This meeting was held at the Tsukuba International Congress Center in Tsukuba, Japan, on 26–28 September 2007, and was organized jointly by the Japan Atomic Energy Agency and the University of Tsukuba. The previous ten meetings in this series were held in San Diego (USA) 1987, Gut Ising (Germany) 1989, Abingdon (UK) 1991, Naka (Japan) 1993, Princeton (USA) 1995, Kloster Seeon (Germany) 1997, Oxford (UK) 1999, Toki (Japan) 2001, San Diego (USA) 2003, and St Petersburg (Russia) 2005.

The purpose of the eleventh meeting was to present and discuss new results on H-mode (edge transport barrier, ETB) and internal transport barrier, ITB, experiments, theory and modeling in magnetic fusion research. It was expected that contributions give new and improved insights into the physics mechanisms behind high confinement modes of H-mode and ITBs. Ultimately, this research should lead to improved projections for ITER.

As has been the tradition at the recent meetings of this series, the program was subdivided into six topics. The topics selected for the eleventh meeting were:

- H-mode transition and the pedestal-width
- Dynamics in ETB: ELM threshold, non-linear evolution and suppression, etc
- Transport relations of various quantities including turbulence in plasmas with ITB: rotation physics is especially highlighted
- Transport barriers in non-axisymmetric magnetic fields
- Theory and simulation on transport barriers
- Projections of transport barrier physics to ITER

For each topic there was an invited talk presenting an overview of the topic, based on contributions to the meeting and on recently published external results. The six invited talks were:

- A Leonard (GA, USA): Progress in characterization of the H-mode pedestal and L-H transition
- N Oyama (JAEA, Japan): Progress and issues in physics understanding of dynamics, mitigation and control of ELMs

- J Rice (MIT, USA): Spontaneous rotation and momentum transport in tokamak plasmas
- K Ida (NIFS, Japan): Transport barriers in non-axisymmetric magnetic fields
- F Jenko (IPP, Germany): Transport barriers: Recent progress in theory and simulation
- T Hoang (CEA, France): Internal transport barriers: Projection to ITER

Every talk satisfied the objective of the meeting. A discussion period followed each invited talk in order to expand physics understandings, projection capabilities, and the direction of research around the topic. Short talks were presented by contributing speakers in addition to questions, answers, comments and discussion among the participants. For each topic there was an associated poster session for contributed papers, and lively discussion took place in front of every poster. Through the meeting six invited papers and 77 contributed papers were presented in total. The final session of the meeting was devoted to summaries; R Groebner, T S Hahm and K Ida of the IAC summarized the fruits of topics 1 and 2, 3 and 5, and 4 and 6, respectively.

I would like to thank Dr A Malaquias, the IAEA Scientific Secretary, for his continuous support and useful suggestions on the arrangements of the meeting. I am very grateful to the IAC members for their cooperation in selecting topics and invited speakers, and for their important advices on the meeting strategy and proceedings publication. I also wish to express my gratitude to LOC colleagues for their hard work organizing the meeting. Young students of the University of Tsukuba helped us during the meeting. Financial and personel support from JAEA and the University of Tsukuba were essential. Finally I would like to acknowledge the participants of the meeting and the referees for the present proceedings. All of the above contributions contributed to the success of the meeting.

Tomonori Takizuka Editor

International Advisory Committee

T Takizuka (Japan Atomic Energy Agency, Japan: Chair)
R J Groebner (General Atomics, USA)
T S Hahm (Princeton Plasma Physics Laboratory, USA)
A E Hubbard (MIT Plasma Science and Fusion Center, USA)
K Ida (National Institute for Fusion Science, Japan)
S V Lebedev (Ioffe Institute, Russia)
G Saibene (EFDA CSU Garching, Germany)
W Suttrop (Max-Plank-Institut für Plasmaphysik, Germany)
Additional information about this meeting (H-mode-TM-11) is available in its homepage http://www-jt60.naka.jaea.go.jp/h-mode-tm-11/.

List of Participants

N Aiba (Japan Atomic Energy Agency, Japan)
T Akiyama (National Institute for Fusion Science, Japan)
N Asakura (Japan Atomic Energy Agency, Japan)
L G Askinazi (Ioffe Institute, Russia)
M N A Beurskens (EURATOM/UKAEA Fusion Association, UK)
J D Callen (University of Wisconsin, USA)
T Cho (University of Tsukuba, Japan)
P C DeVries (EURATOM/UKAEA Fusion Association, UK)
X T Ding (Southwestern Institute of Physics, China)

E J Doyle (University of California, Los Angels, USA)
A Fukuyama (Kyoto University, Japan)
P Gohil (General Atomics, USA)
R J Groebner (General Atomics, USA)
T S Hahm (Princeton Plasma Physics Laboratory, USA)
N Hayashi (Japan Atomic Energy Agency, Japan)
Y Higashiyama (Nagoya University, Japan)
Y Higashizono (University of Tsukuba, Japan)
M Hirata (University of Tsukuba, Japan)
G T Hoang (Association Euratom-CEA sur la Fusion Controle, France)
G M D Hogeweij (FOM-Institute for Plasma Physics Rijnhuizen, The Netherlands)
M Honda (Japan Atomic Energy Agency, Japan)
L D Horton (Max-Plank-Institut für Plasmaphysik, Germany)
W A Houlberg (ITER Organization)
A E Hubbard (MIT Plasma Science and Fusion Center, USA)
J W Hughes (MIT Plasma Science and Fusion Center, USA)
M Ichimura (University of Tsukuba, Japan)
K Ida (National Institute for Fusion Science, Japan)
T Ido (National Institute for Fusion Science, Japan)
T Imai (University of Tsukuba, Japan)
F Imbeaux (Association Euratom-CEA sur la Fusion Controle, France)
A Itakura (University of Tsukuba, Japan)
K Itoh (National Institute for Fusion Science, Japan)
S-I Itoh (Kyushu University, Japan)
F Jenko (Max-Plank-Institut für Plasmaphysik, Germany)
D Kalupin (Institut für Plasmaphysik, Forschungszentrum Jülich GmbH, Germany)
Y Kamada (Japan Atomic Energy Agency, Japan)
N Kasuya (National Institute for Fusion Science, Japan)
I Katanuma (University of Tsukuba, Japan)
M Kimura (Kyushu University, Japan)
A Kirk (EURATOM/UKAEA Fusion Association, UK)
S Kitajima (Tohoku University, Japan)
S Kobayashi (Kyoto University, Japan)
T Kobuchi (Tohoku University, Japan)
J Kohagura (University of Tsukuba, Japan)
P T Lang (Max-Plank-Institut für Plasmaphysik, Germany)
S V Lebedev (Ioffe Institute, Russia)
A W Leonard (General Atomics, USA)
J Q Li (Kyoto University, Japan)
A Malaquias (International Atomic Energy Agency)
Y R Martin (Centre de Recherches en Physique des Plasmas, EPFL, Switzerland)
C J McDevitt (University of California, San Diego, USA)
D C McDonald (EURATOM/UKAEA Fusion Association, UK)
H Meyer (EURATOM/UKAEA Fusion Association, UK)
C A Michael (National Institute for Fusion Science, Japan)
K Miki (Kyushu University, Japan)

R Minami (University of Tsukuba, Japan)
T Minami (National Institute for Fusion Science, Japan)
Y Miyata (University of Tsukuba, Japan)
N Miyato (Japan Atomic Energy Agency, Japan)
Y Motegi (University of Tsukuba, Japan)
V Mukhovatov (ITER Organization)
S Murakami (Kyoto University, Japan)
Y Nagashima (Kyushu University, Japan)
Y Nakashima (University of Tsukuba, Japan)
T Numakura (University of Tsukuba, Japan)
S Ohshima (National Institute for Fusion Science, Japan)
T Oishi (National Institute for Fusion Science, Japan)
T Onjun (Sirindhorn International Institute of Technology, Thailand)
T H Osborne (GENERAL Atomics, USA)
N Oyama (Japan Atomic Energy Agency, Japan)
T Ozeki (Japan Atomic Energy Agency, Japan)
V Parail (EURATOM/UKAEA Fusion Association, UK)
A Polevoi (ITER Organization, France)
J E Rice (MIT Plasma Science and Fusion Center, USA)
F Ryter (Max-Plank-Institut für Plasmaphysik, Germany)
H Saimaru (University of Tsukuba, Japan)
R Sakamoto (National Institute for Fusion Science, Japan)
Y Sakamoto (Japan Atomic Energy Agency, Japan)
M Sasaki (University of Tokyo, Japan)
Y Shi (Institute of Plasma Physics, Chinese Academy of Science, China)
A Shimizu (National Institute for Fusion Science, Japan)
T Shimozuma (National Institute for Fusion Science, Japan)
P B Snyder (General Atomics, USA)
C Suzuki (National Institute for Fusion Science, Japan)
H Takahashi (National Institute for Fusion Science, Japan)
Y Takahashi (Nagoya University, Japan)
Y Takeiri (National Institute for Fusion Science, Japan)
H Takenaga (Japan Atomic Energy Agency, Japan)
M Takeuchi (Nagoya University, Japan)
T Takizuka (Japan Atomic Energy Agency, Japan)
N Tamura (National Institute for Fusion Science, Japan)
K Tanaka (National Institute for Fusion Science, Japan)
S Tokuda (Japan Atomic Energy Agency, Japan)
S Tokunaga (Kyushu University, Japan)
G Turri (Centre de Recherches en Physique des Plasmas, EPFL, Switzerland)
H Urano (Japan Atomic Energy Agency, Japan)
H Utoh (Tohok University, Japan)
K Uzawa (Kyoto University, Japan)
M Valovic (EURATOM/UKAEA Fusion Association, UK)
L Vermare (Max-Plank-Institut für Plasmaphysik, Germany)
F Watanabe (Nagoya University, Japan)

M Yagi (Kyushu University, Japan)
Y Yamaguchi (University of Tsukuba, Japan)
K Yamazaki (Nagoya University, Japan)
M Yokoyama (National Institute for Fusion Science, Japan)
M Yoshida (Japan Atomic Energy Agency, Japan)
M Yoshinuma (National Institute for Fusion Science, Japan)

Journal of Physics: Conference Series **123** (2008) 012001 doi:10.1088/1742-6596/123/1/012001

Progress in characterization of the H-mode pedestal

A W Leonard

General Atomics, PO Box 85608, San Diego, CA 92186-5608, USA

E-mail: Leonard@fusion.gat.com

Abstract. The H-mode pedestal remains a critical subject for magnetic fusion energy research. Recent experimental efforts to characterize the edge pedestal, for tests of pedestal models now under development, are summarized in this report. The pedestal transport is characterized by the profiles and energy and particle fluxes through both the electron and ion channels. The flux measurements are made difficult by the short scale lengths of the pedestal and the difficulty of measuring the neutral ionization source within the pedestal. Recent results on pedestal scaling with a number of parameters are also summarized. Measurements of the pedestal scaling with the ion gyroradius, global beta, input power, shape, critical gradients, neutral density, aspect ratio, rotation, toroidal ripple and the toroidal field direction are presented. These measurements provide a strong test for emerging pedestal models as well as improve confidence of extrapolating existing experiments towards future devices. Recent work on the scaling of the H-mode power threshold is also presented.

1. Introduction

The H-mode pedestal, a narrow region of reduced transport near the plasma edge, remains a critical issue for magnetic fusion energy research. For the core plasma the edge pedestal serves as a boundary condition for the central density and temperature profiles. With improved models of core plasma transport, the H-mode edge pedestal conditions now represent the largest uncertainty in predicting the performance of future burning plasma experiments [1,2]. The pedestal region also represents a boundary condition for the scrape off layer (SOL) plasma. Models of the SOL, important for predicting the heat flux onto the plasma facing component, require the poloidal distribution of heat and particle flux across the separatrix. A complete and validated pedestal model will be required to predict the details of these fluxes. Finally, the plasma conditions required for accessing the H-mode regime are an important aspect for the design of future devices. Access to H-mode is critical for a burning plasma tokamak such as ITER and thus any reduction in uncertainty could result in improved design flexibility and increased confidence in achieving performance goals. All of these issues indicate the high priority of developing a pedestal model with predictive capability.

Development of the peeling-ballooning model [3-5] for the MHD stability limit of the pedestal has provided a powerful constraint on predictions for the achievable pedestal height in a given configuration. The peeling-ballooning model describes a limit for the edge pressure gradient and current that build in the pedestal region during H-mode. The edge gradients build in the pedestal until the MHD limit is reached and an edge localized mode (ELM) typically results. After an ELM the gradients build again until the next ELM. While these models make certain assumptions, such as exclusion of the plasma very near the separatrix, the peeling-ballooning MHD limit has been tested in

Journal of Physics: Conference Series **123** (2008) 012001 doi:10.1088/1742-6596/123/1/012001

a number of configurations and conditions with generally good agreement within experimental uncertainty. An example of this agreement taken from reference [4] is shown in figure 1. In this case pedestal electron density and temperature profiles were selected just before an ELM in a set of discharges from DIII-D with a wide density range, but at constant plasma current, field, shape and pedestal width. The stability limit of these discharges was estimated with the ELITE code [5] and a series of magnetic equilibria with increasing pedestal temperature for the experimental configuration, density and pedestal width. In each case examined the pedestal temperature at which the equilibrium becomes unstable was in agreement with the observed pedestal temperature. The pedestal width is an important aspect of the analysis as numerical studies have found the critical pressure gradient to decrease with pedestal width as $p' \propto \Delta^{-1/4}$, where Δ is the width of a hyperbolic tangent function used to parameterize the pedestal pressure profile [5]. With the success of the peeling-ballooning model the calculation of pedestal height in future devices then relies upon a predictive capability for the width of the pedestal steep pressure gradient region.

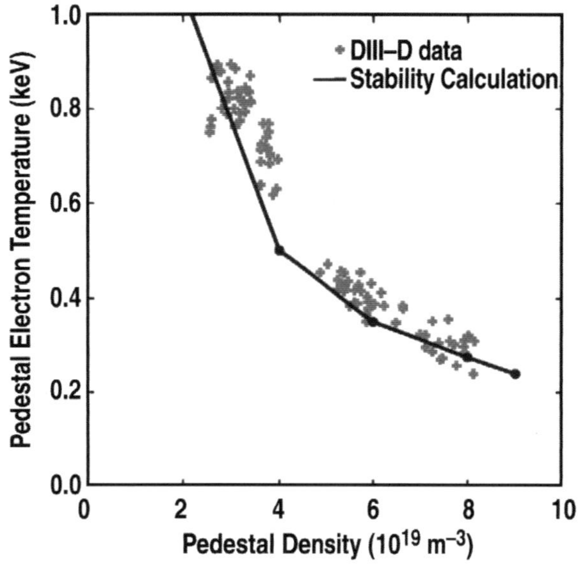

Figure 1. Comparison of pedestal stability limit calculated with ELITE to DIII-D data as a function of pedestal density. (Reprinted with permission from Snyder P B 2004 *Plasma Phys. Control. Fusion* **46** A131. Copyright 2004, Institute of Physics.)

In order to improve predictive capability for the pedestal width, and thus its height, a multi-device pedestal database was established by the ITPA Pedestal Topical Group [6]. This database contains pedestal measurements from many of the tokamaks in the world fusion research community. There are several shortcomings when examining the scaling of this database, primarily that the tokamaks all have different measurement characteristics and capabilities for the pedestal. Several devices could not measure both the pedestal electron density and temperature profiles simultaneously with the pedestal ion temperature. In these cases analysis of the database required setting equal the ion and electron pedestal profiles. Also the pedestal is narrow and its width cannot be resolved in many cases so only the pedestal height can be examined. Given these shortcomings, several studies examined the ITPA database for scaling of the pedestal top pressure [7-9]. While these studies included different analysis techniques, such as utilizing stability constraints, all found significant uncertainty when projecting their respective scaling results to ITER.

The difficulty in producing a scaling relationship from a pedestal database is illustrated in a study of the ITPA database using theory-based models to examine the pedestal width [10]. Because only the pedestal pressure top was generally available the width was obtained from an analytic MHD stability constraint. The derived widths were then fit to a number of proposed theories for the pedestal width. The theory based models examined included the width being set by: i) magnetic and shear flow stabilization, $\Delta \propto \rho S$, ii) flow shear stabilization, $\Delta \propto \sqrt{\rho R q}$, iii) diamagnetic stabilization, $\Delta \propto$

Journal of Physics: Conference Series **123** (2008) 012001 doi:10.1088/1742-6596/123/1/012001

$\rho^{2/3}R^{1/3}$, iv) neutral penetration, $\Delta \propto 1/n_{ped}^{3/2}$, v) ion orbit loss, $\Delta \propto \sqrt{\varepsilon\rho_\theta}$, and vi) normalized poloidal pressure, $\Delta \propto \sqrt{\beta_\theta}R$, where Δ is the pedestal width, ρ is the ion gyroradius, S is the magnetic shear at the pedestal top, R is the major radius, q the safety factor, n_{ped} the pedestal density, ε the aspect ratio, ρ_θ the ion poloidal gyro-radius, and β_p the pedestal pressure normalized by the poloidal field. As can be seen from a summary of the different model fits to the data in Table 1, most of the model fits to the data are of equal quality. Of particular concern in this database exercise is that the models have a very different dependence on ion gyroradius, ranging from linear dependence to independent of gyroradius, with a similar quality fit to the data. The ion gyroradius is the one parameter that current experiments cannot match the expected values in a burning plasma class tokamak. The range of gyroradius dependence implies an uncertainty of a factor of 2, or greater, when projecting the pedestal height to a tokamak such as ITER.

Table 1. Root mean square error (RMSE) in fit of the ITPA database to different models of the pedestal width.

Model	Width	RMSE (%)
Magnetic and flow shear	$\Delta \propto \rho S$	32
Shear flow	$\Delta \propto \sqrt{\rho R q}$	30.8
Diamagnetic stabilization	$\Delta \propto \rho^{2/3}R^{1/3}$	33.7
Neutral penetration	$\Delta \propto 1/n_{ped}^{3/2}$	53.4
Ion orbit loss	$\Delta \propto \sqrt{\varepsilon\rho_\theta}$	34.4
Poloidal pressure	$\Delta \propto \sqrt{\beta_\theta}R$	32.9

The difficulty in distinguishing simple models by use of the database arises not only from the measurements as described earlier, but also from the assumptions made by these models. The pedestal location, just inside the separatrix, is a region of varying shape and magnetic shear that may depart significantly from analytic approximations. Also the analytic approximations for neoclassical ion transport may not be valid due to the radial electric field and separatrix geometry. In addition, most models of pedestal transport assume equilibrium, while as described above the pedestal is continually evolving from one ELM to the next violating this assumption of steady state. Finally the pedestal particle source is not poloidally uniform due to edge and divertor recycling. This non-uniform particle source could result in 2D transport issues for which models will have to account.

All of the complicating factors outlined above motivate a more careful and dedicated approach to developing a validated pedestal model that can be used with confidence to predict pedestal performance in future tokamaks. To directly test newer pedestal transport models that are under development documentation is required of the profiles of all relevant parameters, including both the electrons and ions and their fluctuations as well as the energy and particle flux through both channels. A second valuable approach is to measure changes in pedestal parameters across single parameter scans. Dedicated experimental scans can bring out trends that may be masked by incomplete data sets and correlated parameters in typical database studies. The trends obtained in dedicated experimental scans can be used to guide and validate model development as well as for developing empirical scaling relations for pedestal prediction in future tokamaks.

This report summarizes the current status, and recent progress in experimental characterization of the H-mode pedestal and criteria for H-mode access. In particular, areas that require further work to facilitate progress in edge pedestal model development will be highlighted. In section 2, progress characterizing transport through the pedestal will be described. These measurements will be important to test more complete models of the pedestal that are now under development. Section 3 will survey recent experiments that seek to illuminate the underlying physics controlling the pedestal height through single parameter scans in individual devices and cross device comparisons. The consistent trends observed in these experiments can serve to guide and validate pedestal model development. In section 5 some recent progress in H-mode power threshold scaling will be summarized. Finally section 6 will summarize progress in pedestal research and highlight high priority topics required for further progress.

2. Characterization of pedestal transport

New code models of pedestal transport now under development will capture many aspects of physics relevant to transport in the edge pedestal region. The code TGLF [11,12] uses quasi-linear approximations to full gyrokinetic simulation of driftwave turbulence that can be active in the pedestal region. In addition the code uses realistic separatrix geometry and includes other pedestal effects such as collisions and a description of neoclassical ion transport. The TGLF code is now at the stage of computing growth rates of instabilities for measured pedestal profiles. The code GEM [13,14] is modeling full turbulence in the edge and finding that electromagnetic turbulence must be included along with electrostatic in order to accurately model edge transport. An ambitious approach is being taken by the codes TEMPEST [15] and XGC-1 [16]. These codes are implementing a full gyrokinetic simulation of the pedestal and SOL region to encompass all the relevant physics. The TEMPEST code is taking a continuum approach while the XGC-1 is using a particle-in-cell (PIC) method to describe transport. Both codes will employ self-consistent neoclassical transport that will take into account the self-generated radial electric field in the edge pedestal. The codes will also model realistic recycling neutral particles to accurately model the particle source. Already these codes are showing the potential of neoclassical transport to drive a pedestal radial electric field and determine much of the pedestal characteristics [15,17]. An accurate model for turbulent electron transport must still be added before the simulations can be directly compared to experiment.

All of the pedestal modeling codes described above need accurate pedestal characterization to guide their development and validate their results. The most basic characterization of a pedestal transport model is the ion and electron profiles of density and temperature and the radial energy and particle flux profile through these two species. Several SOL and divertor fluid modeling codes, such as SOLPS [18], UEDGE [19] and EDGE2D [20], encompass the pedestal region and can be used to characterize transport in the pedestal. However, these codes are typically utilized in a predictive mode with preset transport characteristics. Interpretive transport analysis, such as the example shown in figure 2 [21], is generally more useful for determining energy and particle source terms, and then calculating the corresponding fluxes through the continuity equation. The interior energy source, which flows radially through the pedestal, is typically straightforward to determine from ohmic and auxiliary heating analysis. However within the pedestal significant radial energy flux can be transferred from the ions to the electrons due to a stronger gradient in the pedestal electron temperature. This is illustrated in figure 2(a) and can result in significant changes to the radial energy flux profiles. An accurate determination of the separate energy fluxes through the electrons (Q_e) and ions (Q_i) then requires that both of their respective profiles be resolved through the entire pedestal region.

Time dependence is another critical issue for characterising pedestal transport as the profiles evolve from one ELM to the next. This is particularly true for pedestal diagnostics that can vary in time response from the almost instantaneous Thomson scattering measurement, to much slower ion temperature measurements from CER. For some applications, such as stability analysis, profiles just before an ELM are required, while for others, such as implications for core confinement, averaging over the ELM cycle is more appropriate. For transport analysis evolution of the profiles between ELMs is required. While the example in figure 2 is averaged over the ELM cycle, further analysis of the same discharge [22] has indicated that the change in stored energy, $\mathrm{d}W/\mathrm{d}t$, and particles, $\mathrm{d}n/\mathrm{d}t$, can significantly change the energy and particle flux in the pedestal from that inferred from time averaged profiles.

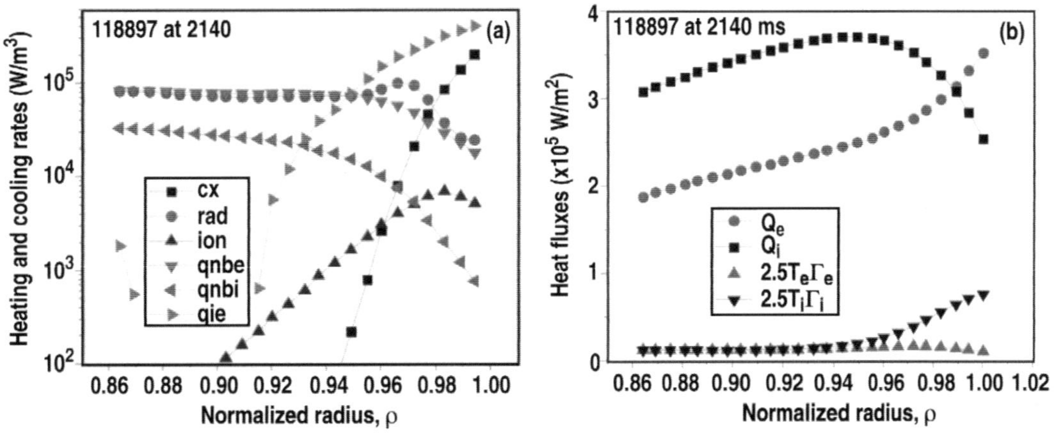

Figure 2. (a) Heating and cooling rates during ELM-free H-mode in the pedestal region of DIII-D (cx = charge-exchange+elastic scattering cooling, rad = radiation cooling, qnbe = neutral beam heating of electrons, qnbi = neutral beam heating of ions, qie = collisional energy exchange from ions to electrons). (b) Total electron (Q_e) heat flux, ion heat flux (Q_i) and convective heat fluxes for both ions and electrons during ELM-free H-mode in the pedestal region. (Reprinted with permission from Stacey W M 2007 *Phys. Plasmas* **14** 012501. Copyright 2007, American Institute of Physics.)

Determination of the pedestal particle flux is more problematic. This is because a large fraction of pedestal particle flux arises from neutral ionization within the pedestal region itself due to the SOL and divertor recycling source. This effect can be seen in figure 2(b) where the convective ion energy flux ($2.5 T_i \Gamma_i$) greatly increases within the pedestal region. In addition, as shown in figure 2(a), the edge recycling neutrals can cause significant charge-exchange loss to the energy transported radially through the ion channel. While the neutral ionization and density in figure 2 is determined from a more sophisticated 2-point SOL and divertor model, than is used in typical transport analysis, a significant uncertainty still exists. A more accurate determination of the 2D profile of neutral density within the pedestal is required to address these uncertainties.

The uncertainty in determining the profile of neutral ionization and density within the pedestal arises from several factors. Because the direct measurement of neutral density within a plasma is difficult, neutral ionization is usually determined by spectroscopic measurements with the neutral density inferred from these measurements using the measured pedestal densities and temperatures. These spectroscopic measurements are difficult firstly because the line of sight measurements require inversion of a hollow profile where the pedestal emission is much less than in the surrounding SOL and divertor. Secondly, interpretation of the most common measurements of ionization, D_α emission, is subject to large uncertainties due to the strong dependence of the emission rate on the background plasma density and temperature. While ionization measurements based on Ly_α emission is less dependent on the background plasma [23], implementing a 2D Ly_α diagnostic is technically challenging and expensive, and thus may not be widely deployable. A 2D measurement of the ionization profile is required because of the very strong local neutral source in the divertor compared to the main chamber. This can lead to a very different width of the ionization profile depending on the assumed location of the dominant recycling when the flux expansion from the midplane to the divertor can be a factor of 5.

One approach to determining the neutral density and ionization profile is through interpretive modeling. With this concept one first obtains a neutral particle flux from all of the plasma facing surfaces equal to the ion flux profile onto these surfaces. The ion flux profile to all surfaces can be determined by measuring the flux profile in the divertor and selected main chamber locations with fixed Langmuir probes. The complete ion flux profile can then be interpolated using the decay length of the plasma density and temperature in the SOL and assuming sonic plasma flow to surfaces [24]. Initial analysis of this kind has indicated that the total main chamber ion flux is the same order as the divertor ion flux [24-26]. The inferred neutral flux from plasma facing surfaces can then be converted to an ionization profile through a Monte Carlo simulation, which launches the same profile of neutral flux into a reconstructed plasma background that matches the experimental conditions. The reconstructed plasma background can be obtained by a fluid model of the pedestal and boundary plasma, such as UEDGE, SOLPS, EDGE2D or OSM [27], constrained by as many experimental measurements as possible. The resulting 2D profile of ionization can then be checked for consistency with profiles of D_α recycling emission. The uncertainties with this approach arise from the accuracy of the reconstructed plasma background and the assumption of saturated walls where the ion flux matches the neutral flux.

An example of this type of interpretive modeling is shown in figure 3 [28]. The 2D D_α taken at a time in between ELMs, figure 3(a), clearly shows how the emission is peaked outside the separatrix, particularly in the divertor. The poloidal profile of neutrals crossing the separatrix, figure 3(b), obtained from a Monte Carlo simulation, indicates the divertor region represents the largest source of ionization fueling the pedestal. There is still significant development work needed however. In the example given, the Monte-Carlo wall neutral flux has not yet been made consistent with the wall ion flux profile. In addition a sensitivity study is needed to determine the dependence of the inferred ionization profile on uncertainties in the reconstructed plasma background. Finally time dependence of the ionization profile should be addressed. Particularly it has been shown that during an ELM the wall ion flux can be very large and with a very different profile than the quiescent period between ELMs [26]. The ion radial flux during ELMs typically represents 20%-40% of the total radial ion flux across the separatrix [26,29] and will need to be taken into account.

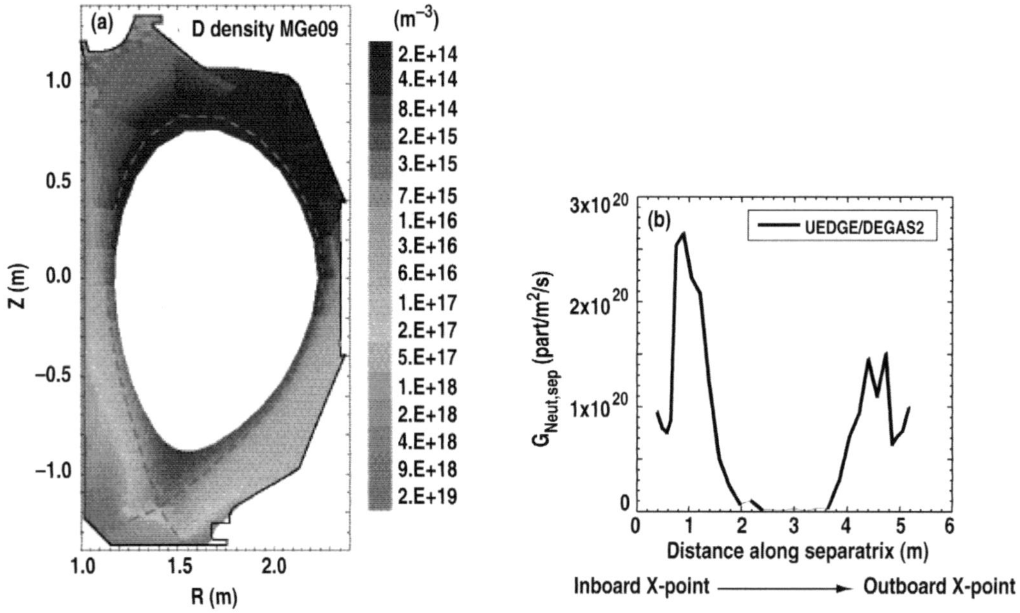

Figure 3. (a) The neutral deuterium density as simulated by UEDGE constrained by experimental measurements. (b) The Monte-Carlo (DEGAS) calculated neutral flux across the separatrix in H-mode as a function of poloidal angle. (Groth M *et al.* 2005 *J. Nucl. Mater.* **337-339** 425)

Other profiles are also important for testing pedestal transport models. The fluctuation characteristics within the pedestal are important for validating the physics within models of the pedestal. The pedestal arises in a region where the naturally occurring fluctuation-driven transport is reduced. Earlier data has shown that the region of edge fluctuation suppression in H-mode corresponds roughly to the steep gradient region in the pedestal [30-32]. Further work is required to determine if the fluctuation suppression corresponds to the region of steep temperature gradient. Also measurements of the residual fluctuations in the pedestal will be valuable for testing models of pedestal transport that sets the electron and ion temperature gradients.

The magnetic shear is another important characteristic of the pedestal that is needed to test models. The magnetic shear in the pedestal is important firstly because it is an important parameter determining the stability limit of the peeling-ballooning model. A lower magnetic shear stabilizes the pressure driven modes that lead to an ELM and limit the growth of the pedestal pressure. The shear is also an important parameter for pedestal transport. Theoretical considerations suggest it may play a role in setting the pedestal width [33]. Variation in the pedestal magnetic shear arises because of the edge bootstrap current that is driven by the steep gradients in the pedestal [34]. Initial measurements of the pedestal bootstrap current find it consistent with a neoclassical collisional bootstrap current model [35,36]. Most of these measurements are taken during long ELM-free periods for better averaging. Additional measurements are needed to characterize the pedestal bootstrap current and resulting magnetic shear in the presence of ELMs that may dissipate some of the edge current.

3. Experimental scaling of the pedestal height and width

Another important approach to experimental characterization of the pedestal is empirical scaling of controlling parameters. This approach can illuminate the important physical processes controlling the pedestal and reveal effects that must be included in any pedestal model. Single parameter scans can provide an important test for pedestal models to exhibit the same trends. Also empirical scaling can be used in its own right to predict pedestal characteristics in future tokamaks and lend confidence to model predictions for those devices.

In contrast to multi-machine pedestal database studies, with their significant uncertainties in scaling, recent experiments dedicated to varying single parameters and cross-machine comparisons are now yielding consistent trends in pedestal characteristics. With the advance of pedestal peeling-ballooning stability models these trends can now be analyzed to separate the contribution of MHD stability from pedestal width changes to the pedestal height.

In the following section recent pedestal scaling experiments will be summarized. The parameters varied in these studies include the ion gyroradius, injected power, global beta, plasma shape, neutral density, critical gradient scale lengths and toroidal ripple. The scaling results will be examined in terms of the important physical processes that should be included in any accurate pedestal model.

3.1. Ion gyroradius

The ion gyroradius, $\rho*$, is an important scaling parameter because it is the one dimensionless parameter of a burning plasma scale tokamak that cannot be reproduced in current experiments. Projections to future larger tokamaks such as ITER rely upon extrapolation of experimental scaling of $\rho*$ from current tokamaks and theoretical models based upon validated physics. As mentioned in the introduction a number of theories of pedestal width, and thus height, predict different dependencies upon $\rho*$. For theories based upon pressure driven flow shear suppression of edge turbulence the $\rho*$ dependence could range from $\rho^{1/2}$ to ρ^1, depending upon details of the dominant turbulence and suppression mechanism [23,37,38]. For theories based upon ion orbit loss setting up a radial electric field shear to suppress turbulence the width should scale as the poloidal ion gyroradius, ρ_θ [39], or possibly the toroidal ion gyroradius in the case of orbit squeezing due to the self-consistent radial electric field [40]. Other theoretical considerations [41] would even suggest no dependence of the pedestal width on ion gyroradius. These different $\rho*$ dependencies project to a factor of two or more uncertainty in the pedestal height of ITER.

An early experiment in DIII-D found no change to the pedestal width, or height, when ρ* was varied by a density scan [42]. However other parameters such as pedestal collisionality, ν*, were not kept constant. Because the pedestal height may vary with other parameters, such as shape and plasma beta, that affect the pedestal stability limit, it is important to carry out experimental scans of ρ* with all other dimensionless parameters held fixed. Another examination of ρ* scaling was carried out in a cross-machine comparison between JET and DIII-D [43]. The pedestal widths as a function of ρ* just before an ELM is shown in figure 4. In this comparison when β, ν* and ρ* were matched at the top of the pedestal the normalized width of the transport barrier, as parameterized by a hyperbolic tangent fit to the pedestal electron temperature, Δ_T, was similar in both devices. Other comparison experiments have also found the transport barrier width to remain constant when keeping the dimensionless parameters fixed [44,45]. For the comparison in figure 4 the pedestal density was higher in DIII-D in order to match the dimensionless parameters. This resulted in a narrower pedestal density width in DIII-D, consistent with a shorter neutral penetration length. The scaling with density will be described in section 3.4. In DIII-D ρ* was scanned from the matched condition to one with ρ* larger by a factor of 1.6, while only a 15% variation in ρ* was achieved in JET. Over this range of ρ* no consistent change in the transport barrier width was observed, as shown in figure 4.

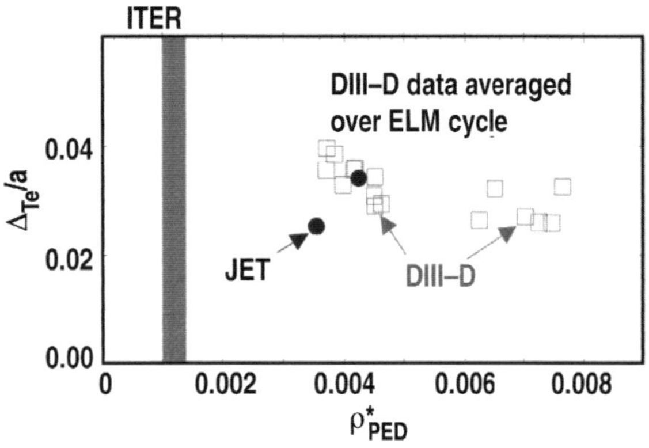

Figure 4. The electron temperature pedestal width normalized by the minor radius in JET and DIII-D as a function of the toroidal ion gyroradius also normalized by the minor radius. Other dimensionless parameters are kept constant including shape, collisionality and beta.

The implications of the observed lack of ρ* dependence for the pedestal width are significant. This scaling would imply a wider, and higher, pedestal in ITER than would otherwise be expected, resulting in higher confinement and significantly better performance. In addition many theories of H-mode transport suppression would have to be revised to accommodate this scaling. Obviously future experiments across devices are of high priority to address this issue.

3.2. Input power and global beta

In a transport model where gradients are driven by fluxes it is typical to scale the pressure with power flowing through the region. With this motivation an examination of the ITPA pedestal database [9] found W_{ped} the pedestal energy as defined by the pedestal top pressure times the plasma volume, to increase with power as in the following expression:

$$W_{ped} = 0.00807\, I_p^{1.41}\, R^{1.37} P^{0.50}\, n_e^{-0.15}\, B_t^{0.32}\, m^{0.2}\, F_q^{1.61}\, \kappa_a^{1.21}\ , \qquad (1)$$

where I_p is the plasma current R is the major radius, P the thermal loss power, n_e the density, B_t the toroidal field, m the atomic mass, κ_a the elongation and $F_q = q_{95}/q_{cyl}$ where $q_{cyl} = 5\,\kappa a^2 B_t/RI_p$. Another ITPA study [46] examined carefully selected profiles from ASDEX-Upgrade, JET, JT-60U and DIII-D in ELMing H-mode and found W_{ped} to increase with input power consistent with the earlier study. Further examination of the dataset revealed the normalized poloidal beta, β_{pol}, at the top of the pedestal was also strongly correlated with the total, or global, β_{pol}, as shown in figure 5. For this dataset the pedestal β_{pol} was averaged over the ELM cycle to examine the implications for core plasma confinement.

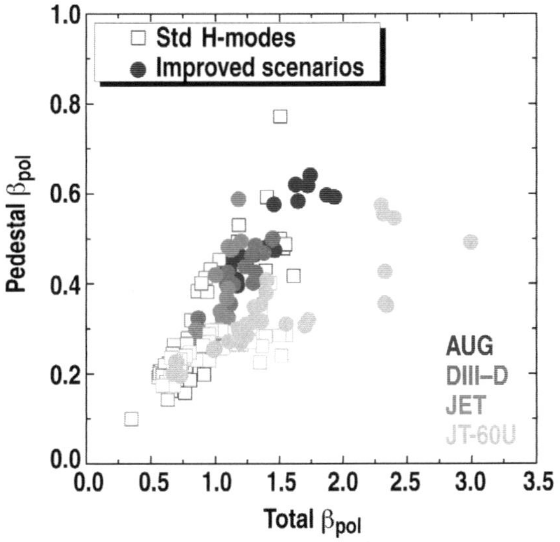

Figure 5. Pedestal β_{pol} versus global β_{pol} for H-mode discharges in AUG, DIII-D, JET and JT-60U. Standard H-mode discharges are shown with open symbols and discharges in advanced regimes are indicated by closed symbols. (Reprinted with permission from Maggi C F 2007 *Nucl. Fusion* **47** 535. Copyright 2007, Institute of Physics.)

In the peeling-ballooning model the pedestal pressure gradient, and thus the top of the pedestal, is limited by the MHD stability limit. This limit is dependent on a number of factors including, pedestal width, shape, plasma current, toroidal field density and temperature. The influence of these factors can be seen in the dependencies of the scaling in equation 1. In addition at least part of the power dependence can be attributed to an increase in the pedestal stability limit. Experiment and stability modeling has documented that higher poloidal beta, and the resulting Shafranov shift, due to higher heating power can significantly raise the pedestal gradient stability limit [47]. This effect is stronger in more highly shaped plasmas and may add to the shaping dependence.

In addition to an increase in the pedestal pressure gradient, the pedestal width has also been found to increase at higher input power and/or global beta [46,48,49]. This results in an even greater increase in the pedestal pressure with β than would be expected from the stability increase alone. The width increase is most apparent in the pedestal ion temperature, but the electron temperature width is increased as well. The underlying physical cause for this increase in width is not yet established, but several candidate mechanisms are possible. First, the increase in stability limit at higher β could allow for a higher ion temperature, or greater ρ^*. This could lead to a greater width of the pedestal transport suppression region as described by several theories. Another possibility is that simply the greater energy flux through the pedestal at higher beta could increase the width of transport suppression as described in reference [38]. Or alternatively the increase in pedestal pressure gradient at high β could modify the magnetic shear profile in the edge increasing the spatial region over which conditions for H-mode transport suppression are met. It is important to resolve which of these mechanisms are responsible for the increase in width with β and/or power as they have different implications for scaling to larger tokamaks, particularly for advanced scenarios that aim to operate at high β.

3.3. Plasma shape

It has long been recognized that plasma shape can significantly affect the height of the H-mode pedestal in a tokamak. Increasing triangularity has yielded higher global confinement in ELMing H-mode, particularly at higher density, in a number of tokamaks [50-52]. This has generally been understood as higher triangularity allowing for a higher pedestal pressure through a higher MHD stability limit in the peeling-ballooning model [4,5]. Higher moments of the shape, such as squareness and location of a secondary separatrix, can also significantly affect pedestal stability [49]. While a few stability studies have parameterized the stability limit with respect to basic parameters of shape much more work is required to quantify the stability effects of other aspects of shape such as the separate influence of upper and lower triangularity, higher moments of the shape, and location of the secondary X-point.

An examination of two plasmas with different squareness in DIII-D, but otherwise similar conditions revealed that the influence of shaping on pedestal structure can extend beyond changes to the MHD stability limit [49]. The two shapes shown in figure 6(a), with otherwise constant conditions, produced the pedestal profiles just before an ELM as shown in figure 6(b). This example shows how a modest change in shape can produce a significant increase in pedestal pressure. Stability analysis of these shapes using pedestal profiles of the same width, as parameterized by a hyperbolic tangent function, revealed that only half of the measured increase in pedestal top pressure could be attributed to the improved stability limit of the less square shape. The other half of the increase of the pedestal is presumably from an increase in the pedestal width. Because of the limited spatial resolution of the diagnostics in the narrow pedestal the uncertainty in the measurement of the pedestal width is typically much larger than that of the pedestal height. This type of analysis uses the validated stability model to constrain the pedestal measurements and thus obtain a better assessment of variation in the pedestal width for what would otherwise be a difficult measurement.

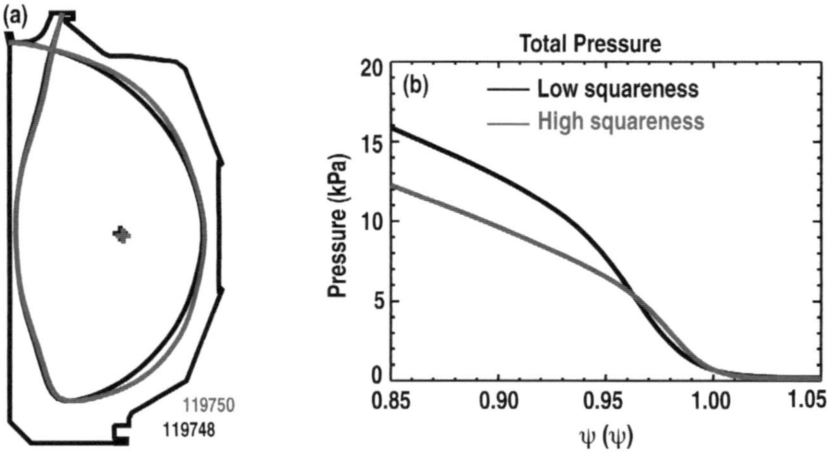

Figure 6. Modest change in the discharge shape squareness (a) produces a significant change in the pedestal pressure profile. (Reprinted with permission from Leonard A W 2007 *Nucl. Fusion* **47** 552. Copyright 2007, Institute of Physics.)

The issue of pedestal width variation with shape is very similar to that for global β as described in section 3.2. The increase in width with stronger shaping could be due to an increasing ρ* as the improved stability limit allows higher pressure. Or, the effect of the shape could broaden the region where the conditions are met for H-mode suppression of turbulent transport. Understanding these mechanisms will be required to improve confidence in models and scaling results to future larger tokamaks.

3.4. Neutral density in the pedestal

The particle ionization source within the pedestal itself due to recycling neutrals can also affect the structure of the pedestal. It is generally expected that the pedestal density profile and its gradient result from the convolution of this particle source with the underlying transport. In addition theoretical considerations suggest that the particle flux from this source could be one of the factors driving the H-mode transport barrier formation [38]. An additional motivation for studying the pedestal density profile separately is its role in obtaining burning plasma conditions in the core plasma. Because neutral penetration of recycling neutrals will be much less in a large burning plasma tokamak than in current devices central fueling through pellet injection is envisioned. The role of the edge pedestal in setting the core density profile will be an important factor in obtaining burning plasma conditions.

As described earlier in section 2, an accurate measurement of the 2D ionization profile in the pedestal is not generally available in current experiments. To address the issues of neutrals a number of scaling experiments and cross-machine comparisons have been carried out and compared to reduced models of edge recycling and particle transport. An earlier scaling study in DIII-D found the pedestal density width to scale inversely with the pedestal density height [42,53]. This is in rough agreement with a simple model of neutral penetration and particle continuity in the plasma edge where the pedestal ionization source is balanced by outward particle flux. This study also found that the pedestal temperature width was similar to the density width in most instances. A follow-on experiment constructed a dimensionless parameter comparison between DIII-D and Alcator C-Mod (C-Mod) [44]. Because of the higher field and density in C-Mod a simple neutral penetration model would expect a smaller normalized pedestal width in C-Mod. However, as shown in figure 7 where the DIII-D density and temperature profiles were normalized to match the C-Mod dimensionless parameters, the normalized density width was very similar in both devices. Other studies have produced mixed results, with ASDEX-Upgrade [54] finding a constant pedestal density width over a range of densities while JET [55] and MAST have found the density width to scale as expected from a more detailed neutrals model [56].

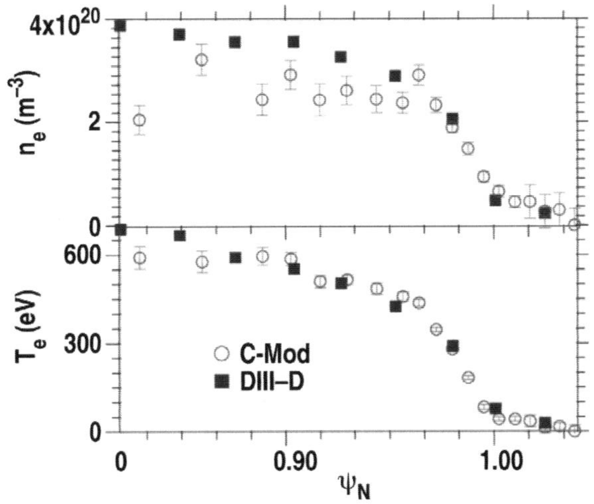

Figure 7. Dimensionless comparison of pedestal profiles in C-Mod and DIII-D. The DIII-D profiles are normalized to match the dimensionless parameters of ion gyroradius, collisionality and β. (Reprinted with permission from Mossessian D A 2003 *Phys. Plasmas* **10** 689. Copyright 2003, American Institute of Physics.)

Several competing effects could explain some of the ambiguous results of pedestal density width scaling. For pedestal fueling dominated by the X-point region, as suggested in some DIII-D studies [28], the flux expansion is larger resulting in a shorter normalized neutral penetration length than if the recycling is poloidally symmetric. It has been postulated that the higher density and metal surfaces of C-Mod could lead to more poloidally symmetric pedestal fueling compared to the lower density and carbons surfaces of DIII-D and lead to similar normalized density profiles. Other effects that must be taken into account are the kinetic details of neutral transport and ionization. A 1-D kinetic neutrals model [57] found that while the density width should be inversely proportional to the pedestal density at DIII-D conditions, over the range of the higher C-Mod pedestal density the pedestal density width could be independent of density [58]. Finally simple 1-D pedestal density scaling studies have indicated that either the neutral penetration length or the transport barrier width could set the pedestal density width depending on the regime and relative scale lengths of the two processes [59]. Clearly a detailed kinetic neutrals model coupled with experimentally constrained recycling distribution will be required to assess and model pedestal fueling and its relationship to the edge transport barrier in setting the pedestal density profile. This also implies that an accurate model of SOL particle transport and recycling will be needed to predict the pedestal density profile of future tokamaks.

3.5. Pedestal transport scaling

The pedestal density and temperature profiles may also be analyzed to determine the underlying transport controlling the pedestal gradients. In ASDEX-Upgrade the Thomson scattering diagnostic collected data for electron density and temperature in an H-mode pedestal during the quiescent period between ELMs [60-62]. An example of data throughout the ELM cycle from the Thomson diagnostic is shown figure 8. The pedestal electron temperature plotted versus the electron density on a log scale reveals a linear relationship, $\eta_e = d(\ell n T_e)/d(\ell n n_e) = 2$, or the electron density scale length is twice the temperature scale length. This relationship holds throughout the transport suppression region of the pedestal, and throughout the ELM cycle as the profiles evolve. This constant ratio of the temperature and density scale lengths is indicative of drift wave turbulence dominating electron thermal transport. The η_e ratio is also seen to generally increase with stronger shaping as theoretical considerations would suggest [62]. A similar scale length ratio has also been observed in DIII-D [63] indicating some universality of this transport mechanism in the pedestal. The scale lengths of the ion temperature and density were also examined in ASDEX-Upgrade [64]. In this case $\eta_i = d(\ell n T_i)/d(\ell n n) = 2$ indicating that ion temperature gradient driven drift modes were not likely responsible for setting the ion temperature profile. Further modeling of these discharges with the fluid code SOLPS indicated that neoclassical ion transport would be sufficient to match the measured ion temperature profile.

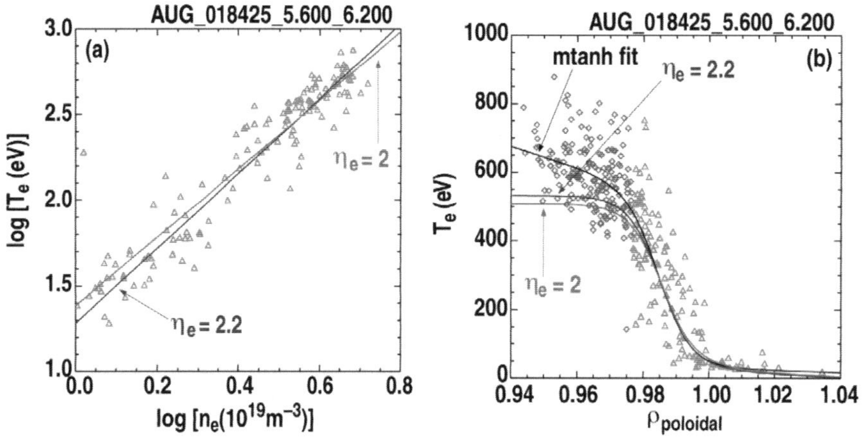

Figure 8. (a) Linear fit of log(electron temperature) versus log (electron density) in the pedestal region (pink triangles) of a Type I ELMy H-mode. The best fit gives η_e=2.2, with η_e=2.0 shown for comparison. (b) Profiles of the edge electron temperature profile; in black a hyperbolic tangential function fit to T_e, in blue the assumption of η_e=2.2, and in red the assumption of η_e=2.0. (Reprinted with permission from Horton L D 2005 *Nucl. Fusion* **45** 856. Copyright 2005, Institute of Physics.)

Pedestal transport can also be characterized by examining trends in the pedestal pressure gradient. In C-Mod a database of pedestal profiles in EDA H-mode was obtained for a range of plasma current, toroidal field and density [65,66]. EDA H-mode in C-Mod is an ELM-free regime where the typical density rise of H-mode is limited by a quasi-coherent mode (QCM). While there is no consistent variation in the width of the pedestal over the range of parameters examined, clear trends in the pedestal pressure height and gradient are evident. In particular the pedestal pressure and the pressure gradient consistently scale with I_p^2. Due to the invariance of the pedestal width across this dataset the pressure gradient exhibited the same I_p^2 scaling. This scaling for the pressure gradient is what would be expected for an ideal MHD limit, though these pedestals were computed to be stable to high-n and intermediate-n peeling-ballooning modes. Similar dependencies of the pedestal pressure gradient on I_p have also been observed in ASDEX-Upgrade [61] and JET [67]. These results are consistent with transport being governed by electromagnetic fluid drift turbulence [14,37]. It is possible that both the electrostatic temperature gradient modes as well as electromagnetic turbulence could be active simultaneously in setting the profiles in the edge pedestal.

The profile scaling presented above represents a strong constraint, or crosscheck, of simulation and modeling of pedestal transport. Further work is required to extend this type of analysis to a wider range of parameters, particularly collisionality. The time dependence of the profiles between ELMs must also be examined more closely to determine if the gradients saturate due to this transport with the pedestal width growing until the more global edge MHD limit is approached. Though this analysis highlights the physical mechanisms for transport in the pedestal that models must eventually reproduce, it does not address scaling of the pedestal width needed for projection to future tokamaks.

3.6. Toroidal field drift direction

The toroidal field direction has been shown to affect the pedestal structure. This effect may not be unexpected as it is well documented that the H-mode power threshold is considerably reduced when the toroidal field direction is such that the ∇B drift direction is towards the dominant divertor X-point. In DIII-D with the ∇B drift towards the X-point the pedestal pressure has been reported higher with less frequent ELMs than for toroidal field in the opposite direction [68]. The higher pressure is primarily a result of higher pedestal density. A similar result has been documented in JET [69]. The dependence of the pedestal height and structure on magnetic field direction is, at this time, mostly anecdotal. Additional experiments and analysis could offer insight into processes that particularly control the pedestal width.

Though scaling of the pedestal with ∇B drift direction does not significantly aid in pedestal prediction for future burning plasma tokamaks, all are expected to be in the favorable ∇B configuration to ensure H-mode access, these results do offer insight to the processes controlling the pedestal width. The ∇B drift direction is known to influence flows in the scrape off layer that could couple flow shear into the pedestal region [70]. On the other hand the toroidal field direction also affects ion orbits that are lost from the pedestal region and potentially alter the region of transport suppression. These are the same issues that arise when examining the H-mode power threshold dependence on toroidal field direction as will be discussed in section 4.

3.7. Toroidal field ripple

Toroidal field ripple may also play a role in determining the pedestal profiles. A dimensionless comparison between JET and JT-60U failed to achieve a match of pedestal conditions [71] that was speculated to be due to differences in field ripple. This was also a dimensional comparison with the same discharge size and fields employed in both devices. The attempt to match parameters resulted in a pedestal pressure in JT-60U that was up to 30% lower than in JET, as shown in figure 9, and consequently significantly lower global confinement. The larger toroidal field ripple, ~1.2%, in the edge of JT-60U in the matched shape compared to JET, ~0.1%, became a leading candidate to explain the differences in what should otherwise be identical plasmas. The toroidal field ripple causes a loss of heating beam ions in the edge of JT-60U and results in a toroidal rotation of the plasma counter to the plasma current direction. The JT-60U pedestal could be affected by change to the toroidal rotation profile or perhaps more directly by the beam ion loss altering the radial electric field shear in the edge.

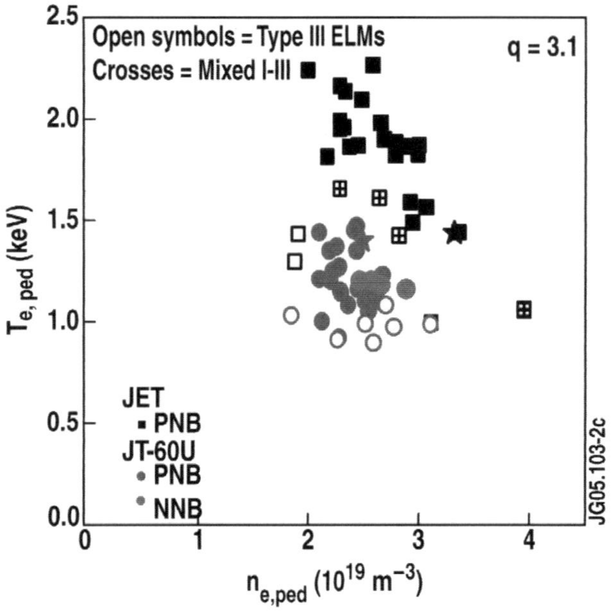

Figure 9. Comparison of JET and JT-60U pedestal T_e versus n_e (top of the pedestal values), for all similarity discharges at q_{95} = 3.1. The black squares are for JET data, the circles for JT-60U data; red with positive NBI injection and green (large circles) with negative NBI. The stars mark the pedestal values corresponding to the best dimensionless matching points. (Reprinted with permission from Saibene G 2007 *Nucl. Fusion* **47** 969. Copyright 2007, Institute of Physics.)

To reduce the toroidal field ripple in JT-60U ferritic steel tiles were installed in the vacuum vessel [72]. These tiles reduced the separatrix averaged field ripple by almost half, from 1.7% to 1.0%, in the large bore configuration and from 0.35% to 0.17% in the smaller bore configuration. In the larger, higher ripple configuration, the reduced ripple significantly increased the pedestal pressure, ~20%, and global confinement, ~10%. The increase in pedestal pressure was largely a result of an increase in the pedestal width, particularly evident in the ion temperature as the profile of electron temperature was not fully resolved. The improvement in confinement was also accompanied by a reduction in the counter toroidal rotation with respect to the plasma current. For the small bore configuration where the field ripple was well below 1% there was little change to the confinement or rotation with addition of ferritic inserts. The improvements in confinement and pedestal pressure were correlated with both less counter rotation and reduced ion orbit loss.

Complementary experiments at JET have examined the effect of increasing the toroidal field ripple on pedestal and ELM characteristics [73]. A scan of average ripple was performed in JET from its nominal value of 0.08% up to 1% by powering alternating toroidal field coils to differing current levels. Preliminary results indicate that without gas puffing even a modest increase in ripple to 0.3% caused a reduction of pedestal pressure and global confinement of ~15%. The decrease in pedestal pressure was mostly a result of a drop in pedestal density with more rapid ELMs. The pedestal electron and ion temperatures increased, but not enough to overcome the density decrease. The decrease in pedestal pressure and confinement became less with additional gas puffing to raise the density. The JET ripple experiments also found a correlation of increased ripple and reduced pedestal parameters with a shift of toroidal rotation in the direction counter to the plasma current and injected neutral beam momentum. Further work will be required from both JT-60U and JET to determine the respective roles of ion orbit loss and toroidal rotation profile in setting the pedestal structure.

The recent pedestal scaling experiments are quite illuminating, particularly when coupled with the MHD constraints of the peeling-ballooning model. Experiments now showing consistent trends with shape, power, β, rotation, toroidal ripple and gradients associated with different transport mechanisms will provide a potent test for newly developed models of the pedestal. While most models of the pedestal width are expected to exhibit a significant ion-gyroradius scaling, such scaling has not yet been unambiguously observed in experiment. On the other hand plasma configurations that lead to improved edge stability, such as strong shaping and higher global beta also result in a wider pedestal. This is suggestive of interplay between stability and transport where increased stability results in a higher pressure and greater ion-gyroradius and in turn increases the pedestal width and height. Ion orbit loss may still be playing a role in setting the pedestal width and might be responsible for the observations of pedestal scaling with toroidal ripple and ∇B drift direction. The observation of critical gradients is an important test for identifying the important transport mechanisms operating in the pedestal. Finally the neutral density is certainly playing a role in setting the pedestal density profile, but its effects on the transport barrier setting the pedestal temperature profiles is mostly unknown.

4. H-mode access conditions

The conditions and parameters required to achieve access to H-mode have important implications for both design of future tokamaks and physics understanding of the pedestal structure. Future tokamaks, such as ITER, will require the high confinement of H-mode to achieve burning plasma conditions. Because of the high cost of heating systems it will be important to have an accurate prediction of the required auxiliary heating power required to achieve the H-mode transition. In addition an understanding of the conditions required for H-mode access adds insight to the processes that control the pedestal structure, particularly the width of the H-mode transport suppression region.

The ITPA Confinement Database and Modeling topical group has had an ongoing database effort to parameterize the H-mode power threshold [74]. Analysis of data from six of the larger tokamaks produced a power threshold scaling similar to previous studies with increasing power threshold at higher density and field. This scaling indicates that for ITER operating at its full field and a low density of $0.5\times10^{20}\,\mathrm{m}^{-3}$, 55 MW of additional heating should be sufficient to achieve H-mode. However at a higher density of $1.0\times10^{20}\,\mathrm{m}^{-3}$ the expected required auxiliary power rises to 95 MW. Significant uncertainties remain in that analysis from single devices produces a stronger density dependence than the combined scaling. This may lead to a higher power threshold than the scaling indicates at the higher densities of ITER. Also a number of machines find a minimum density for the power threshold scaling before the power threshold begins increasing. This lower density limit may increase with toroidal field and would indicate a higher minimum density and power threshold in ITER. Also the minimum power above the threshold required to achieve good confinement has been examined, but no firm conclusion can yet be drawn for this aspect of H-mode operation.

The correlation of H-mode power threshold and plasma rotation has come under increased study recently. It has long been observed that the H-mode power threshold is significantly lower when the $B\times\nabla B$ drift direction is towards the dominant X-point [75-76]. The power threshold can be even lower when the magnetic configuration is very near to a balanced double-null, as shown by results from ASDEX-Upgrade and MAST [77]. A correlation between the radial electric field of H-mode and SOL plasma flows has been examined theoretically by Rozhansky [78]. Examining both these issues together plasma flows were measured in the C-Mod SOL for USN and LSN configurations while the toroidal field and plasma current were kept constant [70]. In the USN configuration the ∇B drift direction was downward away from the X-point and resulted in an H-mode power threshold about a factor of two higher than the LSN configuration. When the plasma current and toroidal field were reversed in direction the LSN configuration exhibited the higher power threshold indicating the significance of the ∇B drift direction. In all configurations plasma in the SOL was carried by parallel flow from the outboard midplane around the periphery of the plasma to the inboard divertor. The toroidal component of these flows was always in the direction of the plasma current for the favorable ∇B drift direction and counter to the current in the unfavorable ∇B drift direction. The core rotation in ohmic plasmas is counter to the plasma current in all configurations, with the greatest counter rotation for the unfavorable ∇B drift direction, consistent with the edge toroidal rotation providing a boundary condition for the bulk plasma rotation. In H-mode as the stored energy increases the core rotation shifts toward the plasma current direction. The C-mod data suggests that in the favorable ∇B drift direction the SOL flows couple to plasma inside the separatrix and shift the toroidal rotation towards that more favorable for H-mode access. The edge electron temperature at the H-mode transition was also found to be higher in the unfavorable configuration. This lends support to the concept that the transition is triggered by shear in the $E \times B$ flows rather than the gradients in the temperature and density profiles directly.

In DIII-D the H-mode power threshold was examined by varying the central torque with co and counter neutral beam injection [79]. The power threshold was found to be significantly lower with more counter injection torque. The reduction in power was most significant with the toroidal field in the unfavorable ∇B drift direction. Also the edge fluctuation levels in L-mode were found to be lower with the counter injection compared to co injection at the same power level. This is in contrast to the C-Mod results where lower power thresholds were associated with core rotation more in the co direction with the plasma current.

The leading paradigm for the H-mode transition is still that when the shear in the edge radial electric field, or $E \times B$ flow shear, exceeds a threshold value turbulent transport is suppressed leading to higher edge pressure gradients and in turn greater flow shear. While the edge pressure may be responsible for part of this shear, toroidal momentum transport may also play an important role in establishing the sheared flow required for the H-mode transition. Still unresolved is the role ion orbit loss or neoclassical transport may play in establishing the $E \times B$ flow shear observed. Experiments are needed to separate the effects of ion orbit loss from direct momentum transport in establishing the required shear flow for the H-mode transition.

5. Summary and conclusions

There has been significant recent progress in the experimental characterization and modeling of pedestal structure. A number of models under development will be capable of taking into account a number of physical effects of the pedestal region that are important, including shaping, full gyrokinetic turbulence, self-consistent neoclassical transport, collisionality and neutral particle sources. Detailed experimental testing of these codes will be required to provide confidence in using them for prediction of future burning plasma tokamak performance.

The pedestal models should be tested through two exercises: 1) prediction of experimentally measured profiles, including fluxes and fluctuations, and 2) reproduction of experimentally observed scaling trends. Obtaining adequate experimental measurements for these tests provides significant challenges that must be overcome. One of the more challenging of these measurements is that of the particle source profile. Knowledge of the particle source, derived from the neutral density profile, is one of the fundamental characteristics of a transport system that must be determined to test any model of that system. There are reasonable prospects for measuring the 2D neutral density profile, but it requires additional resources through extending coverage of existing diagnostics and extensive interpretive modeling. Measurement of the neutral density profile will also provide other benefits, such as determination of charge-exchange losses. Other diagnostic improvements that are required are more straightforward but will require significant additional resources. These improvements include measuring the profiles of all species in each tokamak with sufficient spatial resolution to resolve the pedestal and sufficient time resolution to follow evolution of the profiles between ELMs. Finally a few other diagnostics that have been primarily developed for the core, particularly fluctuation and current profile diagnostics, will also have to be adapted for the pedestal and its spatial and temporal resolution issues.

While the pedestal scaling results have been very illuminating several areas still require further, careful experimental measurements. Of very high priority is the scaling of the pedestal with the ion gyroradius. This is one parameter of future burning plasma tokamaks that cannot be matched in current experiments. Because of this, predictions of the pedestal in future tokamaks rely on extrapolation of experimental trends and validation of the ion gyroradius physics in the predictive model. Also the ion gyroradius is fundamental to theories that rely on $E \times B$ flow shear stabilization of turbulent transport to establish the width of the pedestal. Models under development are expected to exhibit this scaling. If the lack of ion-gyroradius dependence observed in a few initial experiments is confirmed by further experiment and modeling this result will have significant implications for scaling current tokamak performance to future larger tokamaks such as ITER. Because of its implications further experiments to study gyroradius scaling should be given a higher priority. These studies should also be coordinated across devices of varying size to increase the range of gyroradius values examined.

There are also other mechanisms that may play a role in the pedestal structure that still need to be examined. Unlike current devices particle recycling will contribute very little to the particle source of the pedestal in ITER because of the high density and longer path lengths. Because of this central fueling with pellets will provide most of the ion flux through the pedestal. Efforts should be made to test the effects of such ion flux on the pedestal structure, particularly the density profile. The pedestal width was found to increase with power and/or global beta, but their relative importance has not yet been determined. While these two parameters are typically very well correlated they might be separated by comparing discharges in two different confinement regimes while other parameters are held constant. Finally pedestal collisionality, which has implications for turbulent transport, should also be examined. The challenge of isolating this parameter experimentally is that the pedestal bootstrap current, and the resulting magnetic shear, is strongly correlated with the collisionality over typical parameters of current devices. A scan of low collisionality, though, where the bootstrap current is not so dependent may be beneficial.

The H-mode power threshold requirement is also a very important issue in the context of pedestal research. The power requirement to achieve H-mode is an important design issue for burning plasma tokamaks and reducing uncertainty can result in significant cost savings. A better understanding of the threshold conditions is also likely to illuminate mechanisms controlling the pedestal width. Recent results indicate a coupling of toroidal momentum to shear in the edge electric field. Ion orbit losses may also be playing a role. These results coupled with further L-H transition database work could result in reduced uncertainty in predicting the requirements for achieving H-mode.

Acknowledgment
This work was supported in part by the U.S. Department of Energy under Cooperative Agreement DE-FC02-04ER54698.

References

[1] Hubbard A E 2000 *Plasma Phys. Control. Fusion* **42** A15
[2] Hatae T, Sugihara M and Hubbard A E 2001 *Nucl. Fusion* **41** 285
[3] Wilson H R, Snyder P B, Huysmans G T A and Miller R L 2002 *Phys. Plasmas* **9** 1277
[4] Snyder P B, Wilson H R, Ferron J R, Lao L L, Leonard A W, Mossessian D, Murakami M, Osborne T H, Turnbull A D and Xu X Q 2004 *Nucl. Fusion* **44** 320
[5] Snyder P B, Wilson H R, Osborne T H and Leonard A W 2004 *Plasma Phys. Control. Fusion* **46** A131
[6] Hatae T 2001 *Nucl. Fusion* **41** 285
[7] Osborne T H *et al.* 2002 *Proc 19th IAEA Fusion Energy Conference (Lyon)* IAEA-CN-94/CT-3
[8] Sugihara M, Mukhovatov V, Polevoi A and Shimada M 2003 *Plasma Phys. Control. Fusion* **45** L55
[9] Cordey J G for the ITPA H-Mode Database Working Group and the ITPA Pedestal Database Working Group 2003 *Nucl. Fusion* **43** 670
[10] Onjun T, Bateman G, Kritz A H and Hammett G 2002 *Phys. Plasmas* **9** 5018
[11] Staebler G M, Kinsey J E and Waltz R E 2005 *Phys. Plasmas* **12** 102508
[12] Staebler G M, Kinsey J E and Waltz R E 2006 *Proc. 21st IAEA Fusion Energy Conference (Chengdu)* IAEA-CN-149/TH/1-2
[13] Scott B 2000 *Phys. Plasmas* **7** 1845
[14] Scott B 2005 *Phys. Plasmas* **12** 102307
[15] Xu X Q *et al.* 2007 *Nucl. Fusion* **47** 809
[16] Chang C S, Seunghoe Ku and Weitzner H 2004 *Phys. Plasmas* **11** 2649
[17] Chang C S *et al.* 2006 *Proc. 21st IAEA Fusion Energy Conference (Chengdu)* IAEA-CN-149/TH/P6-14
[18] Coster D P *et al.* 2002 *Proc. 19th IAEA Fusion Energy Conference (Lyon)* IAEA-CN-94/THP2/13
[19] Rognlien T, Milovich J, Rensink M and Porter G 1992 *J. Nucl. Mater.* **196-198** 347
[20] Simonini R, Corrigan G, Radford G, Spence J and Taroni A 1994 *Contrib. Plasma Phys.* **34** 368
[21] Stacey W M and Groebner R J 2007 *Phys. Plasmas* **14** 012501
[22] Stacey W M submitted to *Phys. Plasmas*

[23] Degeling A W, Weisen H, Zabolotsky A, Duval B P, Pitts R A, Wischmeier M, Lavanchy P, Marmillod Ph and Pochn G 2004 *Rev. Sci. Instrum.* **75** 4139
[24] Whyte D G, Lipschultz B L, Stangeby P C, Boedo J, Rudakov D L, Watkins J G and West W P 2005 *Plasma Phys. Control. Fusion* **47** 1579
[25] Lipschultz B, Whyte D and LaBombard B 2005 *Plasma Phys. Control. Fusion* **47** 1559
[26] Leonard A W, Boedo J A, Groth M, Lipschultz B L, Porter G D, Rudakov D L and Whyte D G 2007 *J. Nucl. Mater.* **363-365** 1066
[27] Stangeby P C 2002 "The Plasma Boundary of Magnetic Fusion Devices", Institute of Physics, Bristol
[28] Groth M *et al.* 2005 *J. Nucl. Mater.* **337-339** 425
[29] Porter G D, Casper T A and Moller J M 2001 *Phys. Plasmas* **8** 5140
[30] Moyer R A, Tynan G R, Holland C and Burin M J 2001 *Phys. Rev. Lett.* **87** 135001
[31] Kurzan B *et al.* 2007 *Plasma Phys. Control. Fusion* **49** 825
[32] Shirmer J *et al.* 2007 *Plasma Phys. Control. Fusion* **49** 1019
[33] Sugihara M, Igitkhanov Yu, Janeschitz G, Hubbard A E, Kamada Y, Lingertat J, Osborne T H and Suttrop W 2000 *Nucl. Fusion* **40** 1743
[34] Sauter O, Angioni C and Lin-Liu Y R 1999 *Phys. Plasmas* **6** 2834
[35] Wade M R, Murakami M and Politzer P A 2004 *Phys. Rev. Lett.* **92** 235005
[36] Thomas D M, Leonard A W, Groebner R J, Osborne T H, Casper T A, Snyder P B and Lao L L 2005 *Phys. Plasmas* **12** 056123
[37] Rogers B N and Drake J F 1999 *Phys. Plasmas* **6** 2797
[38] Hinton F L and Staebler G M 1993 *Phys. Fluids B* **5** 1281
[39] Shaing K C 1992 *Phys. Fluids B* **4** 290
[40] Chang C S, Seunghoe Ku and Weitzner H 2004 *Phys. Plasmas* **11** 2649
[41] Osborne T H, Groebner R J, Lao L L, Leonard A W, Maingi R, Miller R L, Porter G D, Thomas D M and Waltz R E 1998 *Plasma Phys. Control. Fusion* **40** 845
[42] Groebner R J, Mahdavi M A, Leonard A W, Osborne T H, Porter G D, Colchin R J and Owen L W 2002 *Phys. Plasmas* **9** 2134
[43] Fenstermacher M E *et al.* 2005 *Nucl Fusion* **45** 1493
[44] Mossessian D A, Groebner R J, Moyer R A, Osborne T H, Hughes J W, Greenwald M, Hubbard A and Rhodes T L 2003 *Phys. Plasmas* **10** 689
[45] Kallenbach A *et al.* 2002 *Nucl. Fusion* **45** 1493
[46] Maggi C F *et al.* 2007 *Nucl. Fusion* **47** 535
[47] Snyder P B *et al.* 2007 *Nucl. Fusion* **47** 961
[48] Kamada Y, Takenaga H, Isayama A, Hatae T, Urano H, Kubo H, Takizuka T and Miura Y 2002 *Plasma Phys. Control. Fusion* **44** A279
[49] Leonard A W, Casper T A, Groebner R J, Osborne T H, Snyder P B and Thomas D M 2007 *Nucl. Fusion* **47** 552
[50] Saibene G *et al.* 2002 *Plasma Phys. Control. Fusion* **44** 1769
[51] Kamada Y and the JT-60 Team 1999 *Plasma Phys. Control. Fusion* **41** B77
[52] Stober J, Gruber O, Kallenbach A, Mertens V, Ryter F, Stäbler A, Suttrop W, Treutterer W and the ASDEX Upgrade Team 2000 *Plasma Phys. Control. Fusion* **42** A211
[53] Mahdavi M A, Maingi R, Groebner R J, Leonard A W, Osborne T H and Porter G 2003 *Phys. Plasmas* **10** 3984
[54] Nunes I *et al.* 2004 *Proc. 31st EPS Conference on Plasma Physics (London)* Vol 28B (ECA) P4.137
[55] Kallenbach A *et al.* 2004 *Plasma Phys. Control. Fusion* **46** 431
[56] Kirk A, Counsell G F, Arends E, Meyer H, Taylor D, Valovic M, Walsh M, Wilson H and the MAST Team 2004 *Plasma Phys. Control. Fusion* **46** A187
[57] Hughes J W, LaBombard B, Mossessian D A, Hubbard A E, Terry J, Biewer T and the Alcator C-Mod Team 2006 *Phys. Plasmas* **13** 056103
[58] Hughes J W, LaBombard B, Terry J, Hubbard A and Lipschultz B 2007 *Nucl Fusion* **47** 1057
[59] Owen L W *et al.* 2005 *J. Nucl. Mater.* **337-339** 410
[60] Horton L D *et al.* 2005 *Nucl. Fusion* **45** 856
[61] Neuhauser J *et al.* 2002 *Plasma Phys. Control. Fusion* **44** 855
[62] Kurzan B *et al.* 2007 *Proc. 33rd EPS Conference on Plasma Physics (Warsaw)*
[63] Groebner R J, Leonard A W, Luce T C, Fenstermacher M E, Jackson G L, Osborne T H, Thomas D M and Wade M R 2006 *Plasma Phys. Control. Fusion* **48** A109

[64] Wolfrum E *et al.* 2007 *Proc. 33rd EPS Conference on Plasma Physics (Warsaw)* Vol 28B
[65] Hughes J W *et al.* 2007 *Fusion Sci. Technol.* 51 317
[66] Hubbard A E *et al.* 2007 *Phys. Plasmas* **14** 056109
[67] Loarte A *et al.* 2004 *Phys. Plasmas* **11** 2668
[68] Petrie T W *et al.* 2003 *J. Nucl. Mater.* 313-316 834
[69] Loarte A *et al.* 2005 *Proc. 32nd EPS Conference on Plasma Physics (Taragona)* Vol 29C, P-2.007
[70] LaBombard B *et al.* 2005 *Phys. Plasmas* **12** 056111
[71] Saibene G *et al.* 2007 *Nucl. Fusion* **47** 969
[72] Urano H, Oyama N, Kamiya K, Koide Y, Takenaga H, Takizuka T, Yoshida M, Kamada Y and the JT-60 Team 2007 *Nucl. Fusion* **47** 706
[73] Sartori R *et al.* 2007 this conference
[74] Martin Y R *et al.* 2007 this conference
[75] Wagner F *et al.* 1985 *Nucl. Fusion* **25** 1490
[76] Carlstrom T N *et al.* 1994 *Plasma Phys. Control. Fusion* **36** A147
[77] Meyer H *et al.* 2006 *Nucl. Fusion* **46** 64
[78] Rozhansky V *et al.* 2007 *J. Nucl. Mater.* **363-365** 664
[79] Gohil P, McKee G R, Wang G, Ferron J R, Kinsey J E, Petty C C and Schlossberg D J 2007 this conference

Journal of Physics: Conference Series **123** (2008) 012002 doi:10.1088/1742-6596/123/1/012002

Progress and issues in understanding the physics of ELM dynamics, ELM mitigation, and ELM control

N Oyama

Japan Atomic Energy Agency, Naka, Ibaraki-ken, 311-0193 Japan

oyama.naoyuki@jaea.go.jp

Abstract. Recent experimental progress in understanding the dynamics of type I ELM, small/no ELM regimes to achieve ELM mitigation and active ELM controls is reviewed. As for the type I ELM dynamics, the smaller growth rate of the ELM precursor relative to the Alfvén frequency, the importance of ELM filaments to evaluate the ELM heat load, the evolution of pedestal pressure in the recovery phase, and the effects of edge toroidal rotation and toroidal field ripple on ELM energy loss have been observed in many devices. In low collisionality (ν_e^*) small/no ELM regimes, the type V ELM has been obtained with one or two filaments in $\nu_e^*<1$ condition. Small normalized ELM energy loss less than 1% has been achieved in the grassy ELM regime in non-rotating plasmas. The highest pedestal pressure has been achieved with smaller edge toroidal rotation counter to the plasma current in the QH-mode. ELM control/suppression by pellet pacing and external magnetic field perturbation has been demonstrated, and so that a design activity of ELM control coils for ITER has started. Various effects of the edge toroidal rotation upon ELM characteristics have been found such as ELM energy loss (ELM frequency) in the type I ELM regime and the grassy ELM regime, changes in edge harmonic oscillation and achievable pedestal pressure in the QH-mode regime, and a screening effect in evaluation of island formation by ELM control coils.

1. Introduction

High-confinement mode (H-mode) plasma with edge localized mode (ELM), the so-called ELMy H-mode, has been considered as part of the standard operating scenario with Q = 10 in ITER. Although ELMs play an important role in the particle control needed to achieve stationary H-mode plasmas, instantaneous heat load transferred to divertor target plates due to ELM has been considered as an important and urgent issue in ITER [1, 2]. Therefore, extensive studies of ELM dynamics [3, 4] and models/theories of ELMs [5] have been made for a better prediction of ELM size in ITER. In addition to this research into the type I ELM regime, small/no ELM regimes as an alternative regime [6, 4] and various active ELM control methods are also investigated to establish small ELM heat load operation in ITER. Nevertheless, further investigations to gain understanding of the physics of ELMs are required to address the remaining issues of ELM heat load in ITER.

The acceptable level of the ELM heat load is determined by the heat load limit of 0.5 MJ/m^2 for both of the tile materials, carbon and tungsten, which are candidates for the plasma facing components (PFCs) of the ITER divertor [7]. The value has been reduced to half of the previous assessment reported in [1, 2]. In addition to this, the observation of an asymmetry of the divertor heat load between inner and outer divertor target plates (P_{OUT}/P_{IN} = 1/2) [8] and the reassessment of ELM wetted area (1.3 m^2 for inner divertor), based on the expected SOL width in the outer midplane, of ~5 mm [9] have cased a considerable reduction in the acceptable ELM energy loss ($\Delta W_{\rm ELM}$) for ITER from 5 MJ to 1 MJ, which corresponds to about 1% of the pedestal stored energy ($W_{\rm ped}$) [10]. In addition to this severe requirement upon ELM size, higher pedestal pressure ($p^{\rm ped}$) is required to achieve $Q = 10$ plasma in ITER, because all transport models predict higher Q values at higher pedestal temperatures with fixed plasma density [11]. Therefore, simultaneous achievement of high $p^{\rm ped}$ and small $\Delta W_{\rm ELM}$ in the ITER relevant low collisionality condition should be achieved to establish the ITER standard operating scenario.

Studies of ELM physics have been recognized as an important research area contributing not only to ITER, but also to common MHD phenomena in a torus system. The ELM-like burst at the plasma edge has been observed in helical devices such as W7-AS [12] TJ-II [13] and LHD [14, 15]. The characteristics of ELM-like oscillation and its relation to edge MHD modes were investigated in LHD [16]. Looking at ELMs from a broader viewpoint, an explosive event similar to ELMs with filamentary structures can be found in space plasmas such as a solar flare [17]. Both ELMs and solar flares exhibit interesting non-linear dynamics in magnetized plasmas.

In this paper, recent progress and remaining issues in understanding the physics of ELMs are reviewed in terms of the dynamics of type I ELM, ELM mitigation by small ELM regimes or steady ELM free regimes, and active ELM control methods. Section 2 summarizes experimental observations of type I ELM dynamics such as ELM precursors, edge stabilities determining ELM threshold, filament structures in the collapse phase, recoveries of $p^{\rm ped}$, effects of the toroidal rotation ($V_{\rm T}$) and the toroidal field (TF) ripple on the $\Delta W_{\rm ELM}$. Section 3 describes progress in small/no ELM regimes such as the expansion of the operational regime of type V ELM, the investigation of ELM frequency dependence on $V_{\rm T}$ in the grassy ELM regime, and the mechanism of ELM suppression and the role of $V_{\rm T}$ in the QH-mode regime. Active ELM controls using pellet injection and external magnetic field perturbations are described in section 4. Finally, section 5 discusses the importance of $V_{\rm T}$ in determining ELM characteristics.

2. Dynamics of type I ELM

The series of collapses and recoveries of the edge pressure and the current cased by ELMs plays a fundamental role in the control of heat and particles to achieve a quasi-steady H-mode in ELMy H-mode plasmas. The dynamics of type I ELM consists of several different phases, such as a precursor phase as a linear growth phase of MHD instability, a collapse phase as a non-linear evolution of MHD instability, and a recovery phase where the lost edge pressure and current are recovered for the next ELM. Since the $\Delta W_{\rm ELM}$, the rate of increase in the plasma stored energy (dW/dt) determined by inter-ELM transport, and the evolution of the edge bootstrap current in the recovery phase determine the ELM frequency, the whole process of ELM dynamics must be understood to achieve the better prediction of ELM characteristics in ITER. In this section, the recent progress in edge stability analyses determining the $\Delta W_{\rm ELM}$ and experimental observations of fast type I ELM dynamics measured with fast, local and multi-dimensional edge plasma diagnostics are described.

2.1. Edge stability and precursor phase

The edge stability is strongly affected by the plasma shape. The plasma elongation (κ) and the plasma triangularity (δ) are well known as important parameters. In addition to these parameters, squareness (ζ) [18] and sharpness (σ) [19, 20] have recently been found to change the edge stability. The pedestal performance and ELM characteristics during ζ scan around a shape similar to ITER ($\zeta = 0.35$) while keeping the δ fixed has been investigated in DIII-D [21]. As ζ increased from 0.32 to 0.405, the $p^{\rm ped}$ decreased from 14 kPa to 9 kPa, which is consistent with the stability analysis of the measured profiles. Since the $\Delta W_{\rm ELM}$ also decreased by the same ratio as the $p^{\rm ped}$, the normalized ELM energy loss ($\Delta W_{\rm ELM}/W_{\rm ped}$) was almost constant over the ζ scan.

The radial profile of the linear eigenfunction of edge MHD modes such as the peeling-ballooning mode reveals the region expected to be affected by ELM [22-24]. It is important to understand what parameters can change the radial profile of the linear eigenfunction to predict the ELM size in a future device. Although the edge stability is mainly determined by the edge pressure gradient and the edge current density, the effect of the pressure profile and the current density profile inside the pedestal has been found by the MHD stability analysis using the MARG2D code [25]. With a shape similar to ITER, the radial profile of the edge MHD mode is expanded by the spreading of the envelope of the edge ballooning mode structure due to the increase in the pressure gradient inside the pedestal and by the decrease in the toroidal mode number of the most unstable mode due to the increase in the current density inside the pedestal.

When the edge plasma condition reaches a stability limit, the ELM precursor is often observed in density, temperature and magnetic fluctuations indicating a linear growth phase of edge MHD modes. The study of ELM precursor is important to reveal the MHD modes which trigger ELMs. In JT-60U, clear density and temperature precursors are observed 100-350 μs before the onset of ELMs [26]. One of the important parameters to characterize the ELM precursor is the growth rate, as shown in figure 1. A systematic study in JT-60U showed that the growth rate normalized to the Alfvén frequency (γ/ω_A) is about 0.1%. A similar density precursor was also observed in MAST. The edge interferometer exhibited a growth of density fluctuation with $\gamma/\omega_A \sim 0.5\%$ as shown in figure 5(d) in [27]. A precursor in magnetic fluctuation with the frequency of 50-100 kHz was observed in low collisionality edge plasmas ($0.2 < \nu_e^* < 1$) with large ELMs in Alcator C-Mod. The mode amplitude grows at the rate $\gamma/\omega_A \sim 1\%$ as shown in figure 3(a) in [28]. All three observations indicate a small growth rate of the ELM precursor, γ/ω_A = 0.1-1%.

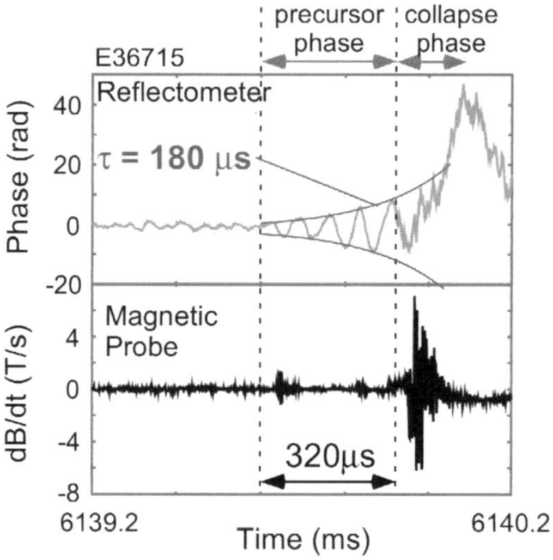

Figure 1. Time evolution of the density precursor measured with the O-mode reflectometer (upper) together with the magnetic fluctuation (lower) in JT-60U. Normalized growth rate of the amplitude of the density fluctuation to the Alfvén frequency was $\gamma/\omega_A \sim 0.1\%$. No detectable magnetic precursor was observed in this case.

The origin of the ELM precursor has not been identified experimentally. One candidate is the rotating seed filaments formed at the plasma edge. In MAST, the density precursor could be measured with an edge interferometer before the filaments are visible in the fast camera [29]. However, such density perturbation seems to continue to exist after the filaments are visible in the camera. Therefore, the density precursor may be a growing structure which develops into the filaments. Further investigation using more sensitive edge plasma diagnostics should reveal the direct relation among the ELM precursor, seed filaments and edge MHD modes.

2.2. Collapse phase and ELM filaments

During the collapse phase after the precursor phase, pedestal particles and energy are quickly released to the scrape-off-layer (SOL) plasma due to ELM. The duration of the collapse phase has been reported to be ~200 μs in a number of devices, and various diagnostics indicate that the collapse of the pedestal structure is localized to the outer midplane [3]. Recent experiments making fast and local measurements in the SOL/divertor plasma region reveal that filamentary structures during the collapse phase as shown in figure 2 can be considered

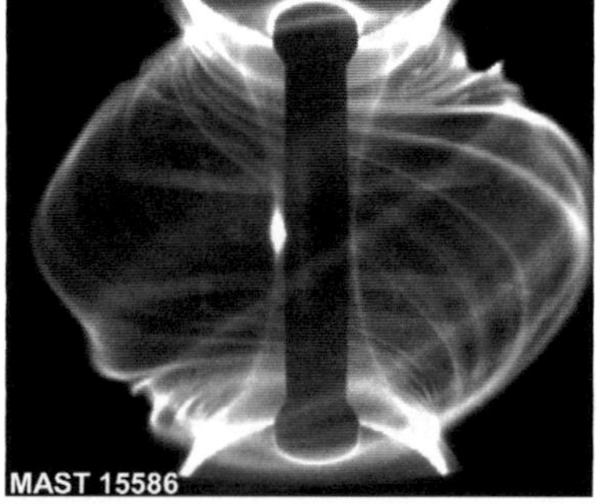

Figure 2. Fast visible image obtained during type I ELM in MAST ($\Delta W_{ELM} \sim$ 1.4 kJ and $W_{ped} \sim$ 11.8 kJ).

as a common feature of type I ELM in most tokamaks (Recent publications: Alcator C-Mod [28], ASDEX Upgrade [30, 31], DIII-D [24], JET [32, 33], JT-60U [34, 35], MAST [27, 29] NSTX [36, 37]). A comparison of the spatial and the temporal structure of ELM filaments in various devices is given in [38]. Therefore, the importance of ELM filaments in evaluating the ELM heat load is highlighted in this section.

The propagation of ELM filaments in the SOL region has been measured with various diagnostics such as fast camera (infrared and visible imaging), Thomson scattering, probes, and beam emission spectroscopy (BES). Among these, one of the important experimental results is the evaluation of the filament energy content from Thomson scattering, which shows that each filament contains up to 2.5 % of the ΔW_{ELM} in MAST [39] and JET [40]. Therefore, apparently some amount of ELM energy loss (e.g. 25% of ΔW_{ELM} in the case of 10 filaments) is carried by filaments both to divertor plates and to the first wall depending on the radial and toroidal velocity of the filaments.

The inter-machine comparisons between MAST and ASDEX Upgrade in terms of the radial transport of ELM filaments and e-folding lengths of the ion saturation current have also been performed [41]. The filaments rotating toroidally and poloidally with velocities close to that of the pedestal are detached from the edge plasma. Then, the velocity decreases as the filaments propagate radially. In both devices, the e-folding lengths of the ion saturation current of filaments exhibit a weak dependence on $\Delta W_{ELM}/W_{ped}$.

In contrast to the above observation, the fraction of ELM energy loss arriving at the divertor plates exhibits a strong dependence on the ELM size, as observed in JET [32, 33]. As the $\Delta W_{ELM}/W_{ped}$ increases, the ELM heat load transferred to the divertor plates decreases as shown in figure 3, which implies that the ELM heat load carried to the first wall by filaments becomes larger. In the case of large ELMs with $\Delta W_{ELM}/W_{ped} \sim 0.28$, more than 50% of ΔW_{ELM} should accordingly be deposited on the first wall. Therefore, to obtain a reliable prediction of the respective ELM heat loads on the divertor plates and the first wall in ITER, further investigations of the energy transfer mechanism of ELM heat load and its parameter dependence including precise evaluation of the energy content in each filament are important.

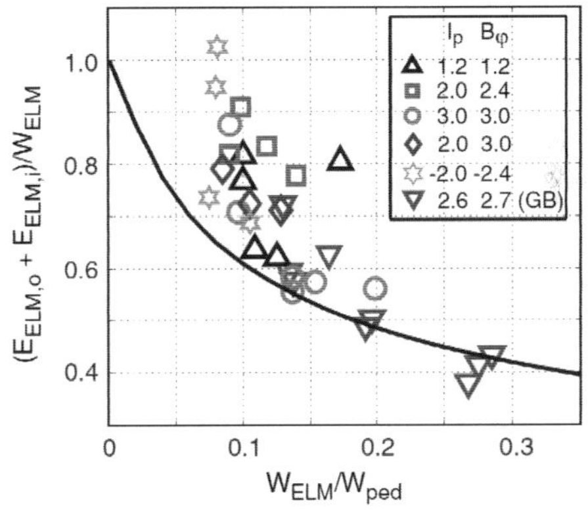

Figure 3. Divertor ELM energy fraction versus normalized ELM energy loss for diagnostic optimized configuration-L discharges in JET [33]. $E_{ELM,o}$ and $E_{ELM,i}$ are total energy load toward outer and inner divertor region, respectively.

2.3. Recovery phase

During the collapse of the pedestal structure, the edge E_r shear structure, which plays an important role in suppressing edge turbulences, is also destroyed. Several ms after the collapse, the edge E_r shear structure forms again. Then, the p^{ped} starts to recover together with the further increase in the edge E_r shear structure toward next ELM [22]. The temporal and spatial evolutions of pedestal parameters in the recovery phase have been investigated in DIII-D [42]. Although the ion and electron pedestal temperatures tend to saturate during long ELM cycles, saturation of the electron density is typically not observed. Looking at the behavior of the electron pedestal pressure (p_e^{ped}) and its gradient (∇p_e), most of the lost p_e^{ped} and ∇p_e are recovered within the first 10-20 ms (~20% of an ELM cycle). Then, both p_e^{ped} and ∇p_e gradually evolve toward the next ELM. The behavior of the total stored energy exhibits the same trend as the time evolution of p_e^{ped}.

The thermal transport properties in DIII-D plasmas have also been estimated from the experimental temperature and density profiles, and calculated conductive flux profiles in the edge plasma [43]. The inferred thermal diffusivities, both χ_i and χ_e, increase with time between ELMs. Observed change in the thermal diffusivities seems to be consistent with the observed change in the time scale of the recovery phase in T_i and T_e.

In contrast to the case of DIII-D, both pedestal parameters and the total stored energy typically linearly recover toward the next ELM in JT-60U typically [44], even if the edge electron heat diffusivity during the inter-ELM phase is close to the level of the neoclassical ion heat diffusivity [45]. The different time scale in the recovery phase can be understood by a peeling-ballooning model as shown in figure 4. In DIII-D, there first is fast recovery of the pressure gradient (∇p) corresponding to the movement to the right inside the stable region, and after this, edge current density gradually recovers together with small recovery of ∇p toward the peeling-ballooning stability limit. On the other hand, in JT-60U, chiefly ∇p is lost and then this recovers to near the ballooning stability limit. In order to confirm this kind of model, the direct measurement of the change in the edge current density as well as the edge pressure gradient over the ELM cycle will be required.

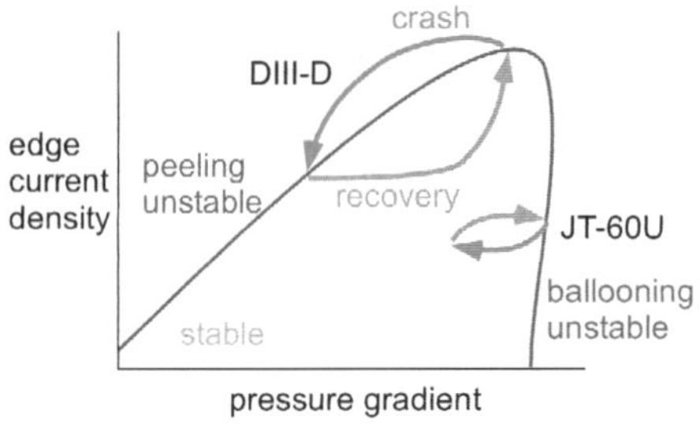

Figure 4. Schematic image of ELM cycle based on peeling-ballooning instabilities.

2.4. Effects of toroidal field ripple and toroidal rotation on ELM energy loss

The effects of the V_T on type I ELM characteristics have been investigated in JT-60U [46]. As counter V_T and/or fast ion losses due to TF ripple increased, $\Delta W_{ELM}/W_{ped}$ decreased while the edge ν_e^* remained fixed. An increase in $\Delta W_{ELM}/W_{ped}$ was also observed in the JT-60U/JET dimensionless identity experiments together with the reduction of fast ion losses and consequent smaller counter V_T when perpendicular NBIs were replaced with tangential NBIs with negative-ion sources [47]. In order to distinguish between effects of the V_T and the TF ripple on the ELM characteristics and the pedestal performance, dedicated ripple experiments have been performed in JT-60U [48, 49].

In order to reduce the ripple amplitude (δ_{ripple}), ferritic steel tiles are installed inside the vacuum vessel in JT-60U. In this way, the δ_{ripple} at the outer midplane can be reduced from ~1.7% without FSTs to ~1% with FSTs, in large plasma configurations with plasma volume (V_P) of ~75 m^3. Figure

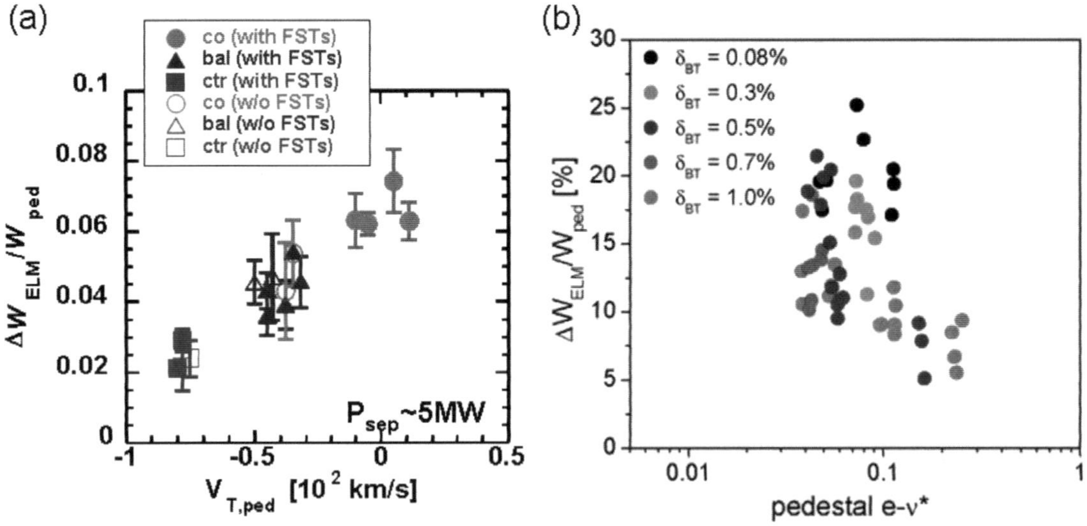

Figure 5. (a) Normalized ELM energy loss as a function of toroidal rotation speed at the top of pedestal in JT-60U. Two different ripple amplitude (with and without FSTs) are compared in large plasma configuration (V_p ~ 75 m^3). Data are from [48, 49]. (b) Collisionality dependence of normalized ELM energy loss in JET ripple experiment [50].

5(a) compares the dependence of the $\Delta W_{ELM}/W_{ped}$ on the V_T with and without FSTs [48, 49]. As counter V_T increased, the $\Delta W_{ELM}/W_{ped}$ decreased with the same trend appearing both with and without FSTs. Although the p^{ped} changes slightly in the course of the V_T scan, the change in thc ΔW_{ELM} is much larger than that in the W_{ped}. This result indicates that it is V_T not δ_{ripple} that plays an important role in determining $\Delta W_{ELM}/W_{ped}$. In addition to this, it also indicates that not the torque input but rather the resultant V_T is important, because the same $\Delta W_{ELM}/W_{ped}$ is obtained in plasmas without FSTs heated by co-NBIs and in plasmas with FSTs heated by balanced-NBIs.

In the usual operation of JET, the 32 TF coils produce very low δ_{ripple} of 0.08%. The TF system can also be configured in such a way as to feed different currents to the odd and even set of coils. In this operation mode, the δ_{ripple} can actively be changed by selecting the appropriate current differential between each set of coils. As the δ_{ripple} increases, the amplitude of type I ELMs appears to be smaller in terms of the D_α emission from the divertor. To quantify this observation, figure 5(b) shows the collisionality dependence of the $\Delta W_{ELM}/W_{ped}$ during the δ_{ripple} scan [50] Even with similar ν_e^*, the $\Delta W_{ELM}/W_{ped}$ can be reduced as the δ_{ripple} increases. It also can be seen that it is difficult to separate the effect of δ_{ripple} and V_T in JET plasmas, because V_T which is in the same direction as the plasma current (co-V_T) is also reduced as the δ_{ripple} increases.

Since analysis shows that the MHD stability of the two devices is similar and probably cannot explain the observed difference in ELMy H-mode performance [51], the V_T and the δ_{ripple} have been focused upon as the major difference between JET and JT-60U. Therefore, new inter-machine experiments on the V_T and the δ_{ripple} have been performed in both devices using plasma with matched shapes [52]. Similar to the dedicated ripple experiment in JT-60U shown in figure 5(a), larger co-V_T seems to increase the ELM energy loss and also reduce the ELM frequency. Especially, the ELM frequency normalized to the loss power through the separatrix (P_{sep}) seems to be correlated with V_T in both devices. As co-V_T increases, smaller ELM frequency is observed, indicating larger ΔW_{ELM}.

The $\Delta W_{ELM}/W_{ped}$ can be reduced by changing V_T. However, the minimum level of $\Delta W_{ELM}/W_{ped}$ is still larger than 2 % even in plasmas chiefly rotating in the counter direction as shown in figure 5(a). Further reduction of $\Delta W_{ELM}/W_{ped}$ is required to achieve the acceptable level for ITER.

3. Small/no ELM regimes

Several small ELM or no ELM (steady ELM free) regimes with good confinement properties, such as EDA, grassy ELM, HRS, QH-mode, type II and type V ELM, have been obtained in various devices, as summarized and categorized in [6]. All these regimes show considerable reduction of the ELM heat load onto divertor target plates in contrast to conventional type I ELM. The $\Delta W_{ELM}/W_{ped}$ in these regimes has been evaluated as less than ~5% as shown in figure 6. Since the values in some regimes are determined by the resolution limit of the diagnostics for the ΔW_{ELM} measurement, the level of $\Delta W_{ELM}/W_{ped}$ in all these regimes seems to be acceptable in ITER.

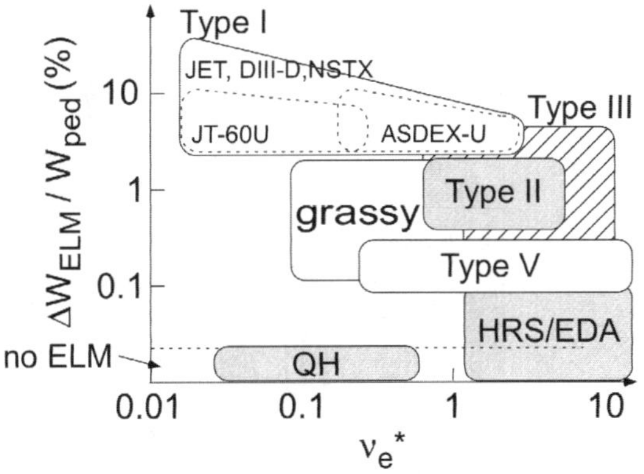

Figure 6. Comparison of operational space in normalized ELM energy loss ($\Delta W_{ELM}/W_{ped}$) and edge collisionality. Some regimes marked by shade indicate that edge fluctuations play important role in ELM mitigation/suppression.

Another important parameter to discuss the applicability of these alternative regimes to ITER is the edge ν_e^*. Although the operational space shown in figure 6 indicates that some regimes seem to be separated by $\nu_e^* \sim 1$, in NSTX the operational space of type V ELM regime can be expanded toward low ν_e^* regime ($\nu_e^* < 1$) by increasing the plasma triangularity up to $\delta \sim 0.75$ [53]. Therefore, the type V ELM regime is now considered as a different regime from the HRS regime in JFT-2M and the EDA regime in Alcator C-Mod, even though the type V regime also exhibits the characteristics of the enhanced recycling. The existence of edge fluctuations is considered as the reason for the enhanced recycling in HRS [54] or EDA [55] regimes. The fact that no edge fluctuations are observed in the type V ELM regime is another big difference between the type V ELM regime and the HRS/EDA regimes.

In this section, the recent progress on some small ELM or no ELM regimes, such as type V ELM, grassy ELM and QH-mode regimes, obtained with low ν_e^* plasmas are summarized.

3.1. Type V ELM regime

A single null configuration has been found to be a required condition to obtain the type V ELM regime in NSTX [56]. Since ITER cannot operate using the double null configuration and Δ_{sep} (the distance between the separatrix and the flux surface through the upper X-point at the outer midplane) should be kept at 4 cm, the fact that type V ELM can be obtained with a larger gap makes this favorable for ITER. In low δ plasmas, the type V ELM regime can be obtained when $\Delta_{sep} \geq 1.5$ cm. In high δ plasmas, a slight movement in the plasma configuration, where the upper X-point is pulled upward by ~4 cm from a balanced double null configuration, changes the ELM characteristic from the type I ELM to the type V ELM. This small change corresponds to $\Delta_{sep} \sim 0.5$ cm [53].

The spatial structure of the type V ELM has been investigated using various diagnostics such as fast visible camera, ultra-soft X-ray array, interferometers, magnetic probes and gas-puff imaging [53, 37, 56]. Dedicated measurements using these diagnostics reveal the following characteristics of the type V ELM. The Type V ELM appears as single or sometimes double filaments aligned with the magnetic field. The filament first appears on the closed field lines near the top of the H-mode pedestal. There is a clear electromagnetic $n = 1$ precursor rotating counter to the plasma current before the type V ELM crash. Then, the filament propagates toroidally in the direction counter to the plasma current. The filament drifts radially outward as it moves toroidally, eventually appearing on visible camera images of the entire plasma and the divertor region. Figure 7 compares the fast visible camera images of the type V ELM and the type I ELM. Similar behavior of some filaments was observed in the type I ELM shown in figure 2 and figure 7(b), while a single filament along the open field line corresponding

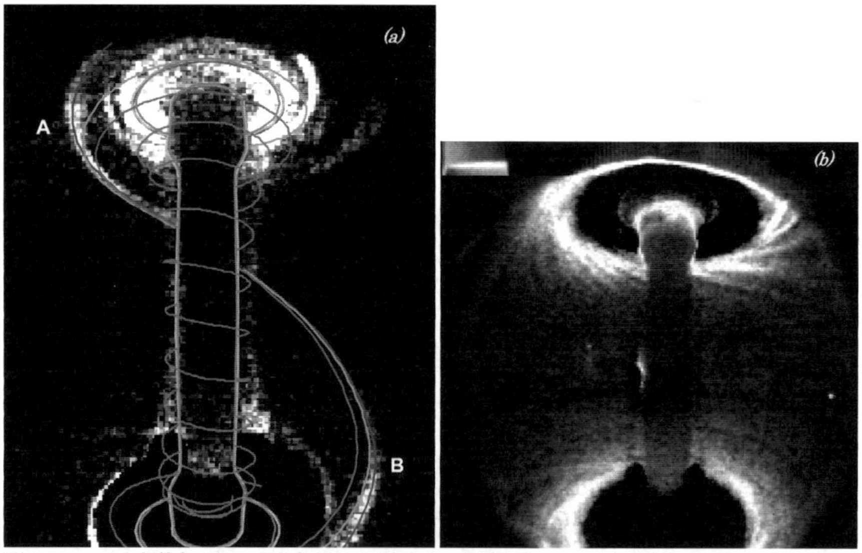

Figure 7. Visible camera image of (a) single filament taken in type V ELM and (b) many independent filaments taken in type I ELM. Type V ELM is obtained in lower single null configuration, which has a second X-point inside a vacuum vessel. Background frame is subtracted in both figures [53].

to the normalized poloidal flux of 1.49 (far outside the second X-point) was clearly observed in type V ELM. These differences between the type I ELM and the type V ELM indicate that the number of filaments is also important in determining the ELM amplitude and to distinguish ELM types. However, the identification of the underlying instability responsible for the onset of the type V ELM requires further investigation.

3.2. Grassy ELM regime

The grassy ELM regime discovered in JT-60U is a small ELM regime which combines tolerable ELM energy losses with low pedestal collisionality and no degradation of pedestal pressure. To enter the grassy ELM regime in JT-60U, several important parameters have been found such as high safety factor (q_{95}), high δ, and high poloidal beta (β_p) [57]. The importance of these parameters in reproducing grassy ELM has been confirmed in JET [58] and ASDEX Upgrade [59]. In addition to these parameters, V_T has been found to influence the appearance of the grassy ELM in JT-60U, ELM frequency gradually increasing together with a reduction in the ELM amplitude as counter V_T increases [23]. Since the V_T in ITER plasma is expected to be small (e.g. ASTRA code predicts 0.26 kHz at the pedestal [60]), systematic study of the effect of V_T on the appearance of the grassy ELM and its parameter dependence has been performed in JT-60U [61].

The ELM frequency observed for different V_T and β_p are summarized in figure 8(a). In order to change the toroidal rotation frequency while keeping a given β_p fixed, combinations of tangential co-, ctr- and balanced-NBIs and perpendicular-NBIs were varied. The ELM frequency clearly increases with increase in counter V_T independently of β_p. It is also found that there is a threshold value of β_p in configuration with lower δ ($\delta < 0.5$). When the toroidal plasma rotation frequency becomes higher than ~1 kHz in the co-direction, on the other hand, type I ELM with a frequency of ~20 Hz is observed.

Figures 8(b)-(d) compare the time evolution of the stored energy in the plasmas marked in figure 8(a), measured with a diamagnetic loop. In contrast to the type I ELMs shown in figure 8(d), there is no clear reduction of the plasma stored energy in the other cases with higher ELM frequency, because the ΔW_{ELM} of the grassy ELM is smaller than the resolution limit of the diamagnetic loop measurement (~5-10 kJ). The ΔW_{ELM} calculated from the change in the temperature profile measured with an ECE radiometer (no detectable density loss has been observed) is several kJ, even in 200 kHz grassy ELM with co-V_T. Since this value corresponds to 0.1-0.8% of W_{ped}, the $\Delta W_{ELM}/W_{ped}$ of the grassy ELM obtained in plasmas with small V_T is at an acceptable level for ITER.

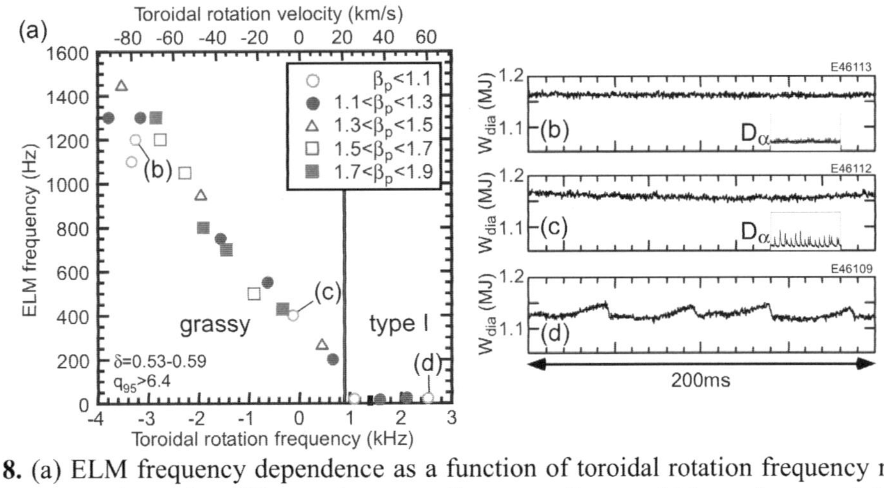

Figure 8. (a) ELM frequency dependence as a function of toroidal rotation frequency measured at the top of T_i pedestal for different β_p ranges. Corresponding toroidal rotation speed for typical plasma condition is also shown at the upper horizontal axis. (b)-(d) Time traces of the plasma stored energy correspond to the data marked with (b)-(d) in (a). Time traces of divertor D_α signal in grassy ELM are also plotted in (b) and (c). Data are from [61].

The β_p achievable in co-rotating plasmas over the range of rotation frequency is limited, because the higher heating power resulting from the perpendicular NBIs provides larger counter V_T due to ripple induced fast ion losses. It is expected that the grassy ELM will be obtained in plasmas with higher β_p even at larger co-V_T than 1 kHz, because grassy ELM is obtained in JET and ASDEX Upgrade with higher co-V_T than that in JT-60U. Further survey of sufficient conditions to achieve grassy ELM and the understanding of the mechanism of change in the ELM type (or ELM affected area) with change in V_T are important to establish an operation scenario for the grassy ELM in ITER.

3.3. QH-mode regime

The QH-mode regime, where confinement levels similar to those of standard ELMy H-mode plasmas can be sustained without any ELMs, were originally observed in DIII-D [62] and then produced in AUG [63], JT-60U [64] and JET [65]. In these QH-mode plasmas, counter-NBI heating and a large clearance between the plasma separatrix and the first wall (i.e. gap) are typically required to obtain QH-mode plasmas. The edge harmonic oscillation (EHO), which is considered to enhance the particle fluxes, is usually observed near the separatrix at the onset of the QH-mode plasma in all devices. The effect of the V_T and the direction of the torque input have been investigated in JT-60U [23]. Since partial QH-mode phases are observed during co- and balanced-NBI heated plasmas with slightly counter or almost zero V_T, a dedicated experiment to confirm the threshold of the V_T for entrance into the QH-mode regime as well as the physical mechanism of suppression of ELMs has been conducted.

After the reorientation of two neutral beam sources to allow co-, counter, and balanced NBI, that provides fine control of the plasma rotation in DIII-D, the investigation of effects of V_T on QH-mode plasmas was performed [66, 67]. As is shown in figure 9, the pedestal density in the QH-mode can be increased by reducing the counter V_T after changing the fraction of the co-NBI power while keeping the total NBI power fixed. With the lower torque (125730 and 125731) when the counter V_T is reduced to less than ~20 km s^{-1}, ELMs return as the pedestal density increases. In QH-mode plasmas, enhanced edge particle transport caused by the presence of the EHO has been observed. During experiments varying the rotation speed, the EHO becomes less robust at the lower input torque. These experimental results indicate that changes in the EHO caused by the reduction of the counter V_T are connected to the decrease in the edge particle transport.

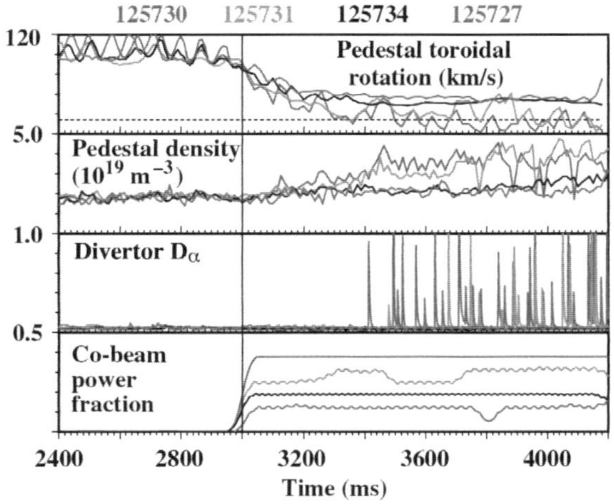

Figure 9. Comparison of waveforms in rotation scan experiments in DIII-D. Discharges with less than 1/4 co NBI power fraction remain QH-mode (125734 and 125729), while those with larger co NBI power fractions and with smaller counter V_T ~ 20 km s^{-1} redevelop ELMs [66].

A previous stability analysis using the ELITE code indicates that the typical QH-mode lies near a current limited stability boundary at high edge current density and high ∇p [68]. A peeling-ballooning stability analysis spanning the transition from QH-mode to ELMy H-mode has been performed [67, 69]. The operational points in ELMy H-mode are closer to or slightly beyond the calculated stability boundary, while those in QH-mode are just below the peeling boundary. Therefore, the ELMs reappear together with the change in EHO characteristics when the pedestal operating point crosses the stability boundary. However, the cause for the gradual change in the edge ∇p and the current density as counter V_T is reduced is not yet understood.

These experimental results suggest that the operational space of QH-mode can be expanded into a better plasma configuration having a higher stability limit. This kind of plasma optimization has also been performed in DIII-D [67]. The ELITE code has been used as a guide to optimize the plasma shape to achieve the widest possible stable operating region, and the new shape with different flux surfaces near the centerpost improves the overall edge stability. Using the new shape, the p^{ped} in QH-mode plasmas can be increased by a factor of two by reducing the input torque (by reducing counter V_T). Thus, the total plasma performance is also improved even though the input power is constant. These results demonstrate that the edge p^{ped} can be controlled by changing the input torque at constant input power while keeping the QH-mode phase. Making use of this optimization, record pedestal densities in the QH-mode reaching 50 % of the Greenwald density limit (n_{GW}) can be achieved. As the counter V_T decreases, the EHO characteristics are changed so that the MHD oscillations detected by the Mirnov loops develop a broadband character with many of the dominant toroidal mode numbers, rotating in the direction of the plasma current. Further investigation to understand how these changes influence change in the transport will be required.

4. Active ELM control

In order to suppress or control the type I ELM, a variety of techniques have been studied in various devices as follows. Pellet pacing, this is, forced ELM triggering by continuous pellet injection, has been demonstrated in ASDEX Upgrade [70] and is planned for JET [71] and DIII-D [72]. In contrast to a pellet injection, no prompt ELM triggering has been achieved using a pulsed gas jet [73]. Edge current modification by vertical plasma displacement has been demonstrated in COMPASS-D [74], and reproduced in TCV [75] and ASDEX Upgrade [76] with a vertical position control system. Edge modulated ECH which produced a local pressure gradient was found in ASDEX Upgrade [77]. Recently, a quasi-stationary ELM-free H-mode regime was obtained during third-harmonic X-mode ECH in TCV [78]. Although the physical mechanism of the suppression of ELMs is not discussed in [78], the ELM-free period seems to be related not to the duration of the third-harmonic X-mode ECH but to the duration of the modulated ECH phase [79]. The achievement of steady H-mode with small ELMs has been achieved in JFT-2M with high m resonant modes ($m > 10$) [80] and control of type III ELM frequency has been demonstrated in COMPASS-D [74]. The complete suppression of type I ELM by the $n = 3$ RMP producing edge stochastic magnetic field boundary, where the particle loss is enhanced, has been demonstrated in DIII-D with high ν_e^* plasmas [81] and then expanded to ITER relevant low ν_e^* [82]. In JET, $n = 1$ error field correction coils (EFCC) can reduce the ELM amplitude together with higher ELM frequency [83]. Because of these successful ELM control using external coils, an installation of such control coils has been planned in MAST and ASDEX Upgrade. Technique to optimize ΔW_{ELM} and p^{ped} with gas puffing (D_2, N_2, Ne) for advanced tokamak operation is also being studied in JET [84].

In this section, the recent results of pellet pacing (planned in ITER) and external magnetic field perturbation for ELM suppression (reviewed for ITER) are summarized.

4.1. Pellet pacing

Since only a pellet injector for large size fueling pellets has been installed in JET, a demonstration of pellet pacing cannot be performed. However, the initial ELM trigger process initiated by the pellet has been investigated to confirm the concept of the pellet pacing in a large scale device [85]. The time evolution of D_α signal in the pellet ablation monitor and magnetic fluctuation indicate the timing of ELM triggering, when only a small fraction (less than 1%) of the fueling size pellet is ablated and deposited. More information on the threshold condition for pellet produced perturbations, which trigger ELMs, will be investigated in JET with a new pellet injection system optimized for the pellet pacing.

In order to design a pellet injector optimized for ELM control and to predict the operation space of pellet pacing in ITER, the minimum penetration depth required for the forced ELM triggering is one of the most important parameters. Systematic analysis of the delay of ELM onset from the time the pellet crosses the separatrix has been performed in ASDEX Upgrade, and the result is summarized in figure 10 [86]. The seed location where the ablating pellet triggers an ELM lies with 99% probability in a narrow range extending ±1.2 cm from the most probable position (peak of the probability function of the seed location) in the middle of the pedestal. The location is independent of the pellet mass and velocity. However, we have to confirm whether smaller pellets, which can only penetrate the middle of the pedestal, can trigger an ELM or not. In addition to this, the minimum penetration depth obtained in ASDEX Upgrade should be compared with other devices to find a reliable operational space for pellet pacing in ITER.

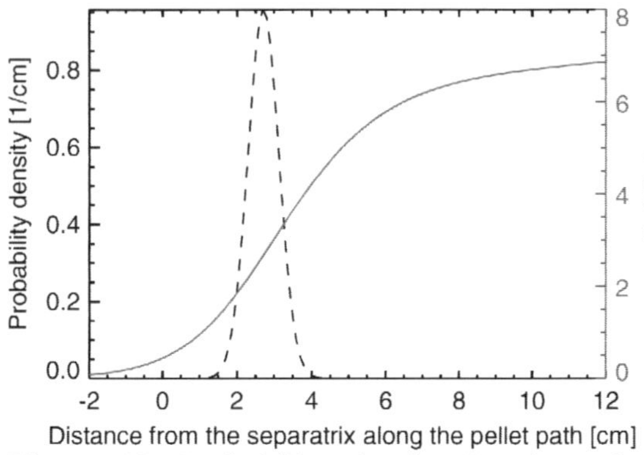

Figure 10. Probability density function of the seed location (dashed line) and electron plasma pressure profile as a function of the distance along the pellet path measured from the position of the separatrix. The pedestal top is located at about 6 cm [86].

The operation space for the pellet pacing in ITER is determined by several parameters such as the pellet frequency required to reduce the ΔW_{ELM} to the acceptable level, the pellet size, the pumping speed, the particle confinement time, the fueling efficiency, and the minimum penetration depth for ELM triggering. The new criterion of the acceptable ELM energy loss described in the introduction requires a pellet frequency higher than 16 Hz, which is 10 times higher than the natural ELM frequency. An estimation of the operational space of pellet pacing in ITER has been performed assuming appropriate values of above mentioned parameters [87]. A combination of a pellet injector at the high field side (HFS) and one at the low field side (LFS) can create a reasonable operational space for pellet pacing at 16-22 Hz as shown in figure 11. The operational space is limited by the total pumped flux $S_{tot\ max}$ of 120 Pa m^3 s^{-1} in the analysis. The uncertainties of the particle confinement time in each operational scenario define the possible operational range as shown by horizontal dashed lines.

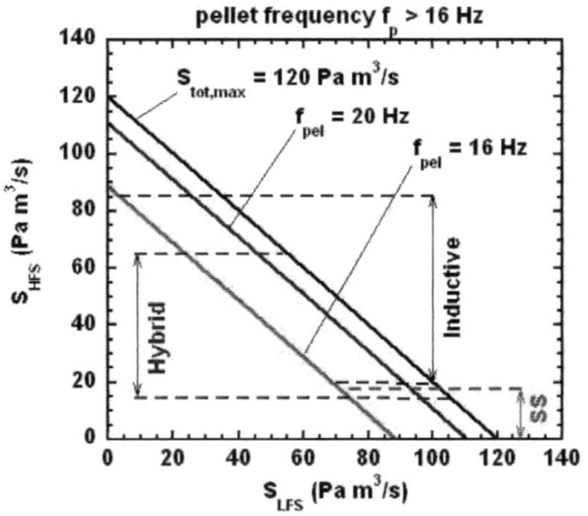

Figure 11. Operational space of pellet pacing in ITER. Horizontal and vertical axis are fueled particle flux from HFS and LFS pellet injectors, respectively. Horizontal dashed lines correspond to the estimate of the uncertainties of the knowledge of core particle transport for inductive, hybrid and steady state (SS) ITER scenarios [87].

In the analysis shown in figure 11, a pellet penetration to the top of pedestal is adopted as a conservative required condition for the forced ELM triggering. This requires a pellet size of $d > 4$ mm. However, if seed perturbation at the middle of pedestal is enough for the ELM triggering, a smaller pellet ($d = 3$ mm in [88]) can be used for pellet pacing at LFS. In order to accurately determine the operational space in ITER, further studies are required to reduce these uncertainties.

4.2. External magnetic field perturbation

In plasmas in DIII-D with ITER relevant ν_e^*, type I ELMs are completely eliminated with radial magnetic field perturbations (δb_r) at the $q = m/n = 11/3$ surface exceeding $\delta b_r/B_T = 2.6\times10^{-4}$, where B_T is the toroidal magnetic field on the plasma axis and the value is evaluated in a vacuum. The typical edge plasma response to RMP in ITER relevant ν_e^* plasmas is a significant change in the global particle balance. The pedestal density gradient is reduced, while the pedestal temperature gradient increases modestly. Consequently, the electron pressure gradient across the pedestal region is reduced [89]. A strong relation between the size of the resulting pressure gradient and the suppression of the type I ELM has been found experimentally. The level of the change in pedestal profile varies with several parameters such as the RMP coil current, the RMP coil configuration, the edge safety factor, the NBI heating power, and the gas fueling rate.

These experimental observations of ELM suppression with RMP can be explained by the peeling-ballooning model [66]. Similar to ELM suppression in the QH-mode regime, the pedestal parameters can be kept either close to or well below the peeling-ballooning stability boundary as shown in figure 12. These results indicate that the edge ergodic layer produced by the RMP plays a role similar to the EHO in the QH-mode, driving substantial transport (especially particle transport) and providing steady H-mode plasmas without ELMs.

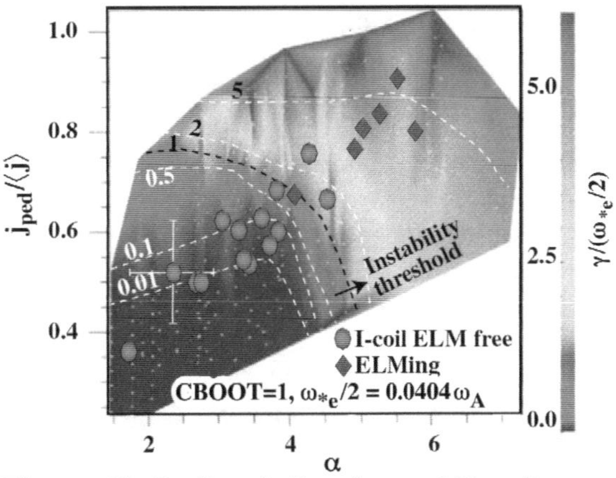

Figure 12. Peeling–ballooning stability diagram for a series of RMP ELM suppressed discharges (circles) compared with phases of similar discharges with ELMs (diamonds) [66].

After the modification of divertor configuration in DIII-D, complete ELM suppression with an ITER similar shape (ISS) was also demonstrated [90]. In the ISS configuration, a larger RMP coil current producing $\delta b_r/B_T = 3.4\times10^{-4}$ is required for complete ELM suppression than the low δ configuration. In addition to this, the resonant window in q_{95} is narrower in the ISS configuration which has higher magnetic shear. Nevertheless, the operational boundary between the ELMy phase and ELM suppression phase can be explained by the peeling-ballooning model [69], the same as in the case of low δ shown in figure 12. From the comparison between the region of island overlap and the change in the ∇p required for complete ELM suppression, a physics based required condition of the mode spectrum should be obtained in future research. For more accurate assessment, the effects of edge bootstrap currents on the q profile and the screening effect of the V_T on the resonant magnetic field should be considered.

In JET, active ELM control using a static external perturbation field with $n = 1$, generated by four EFCCs located far from the plasma, also has been demonstrated [83]. During the application of the $n = 1$ perturbation field, the ELM frequency increased by a factor of 4 and the amplitude of the D_α signal decreased as shown in figure 13. The ΔW_{ELM} normalized to the total stored energy, $\Delta W_{ELM}/W_{total}$, dropped to values below 2%. Transport analyses shows no or only moderate

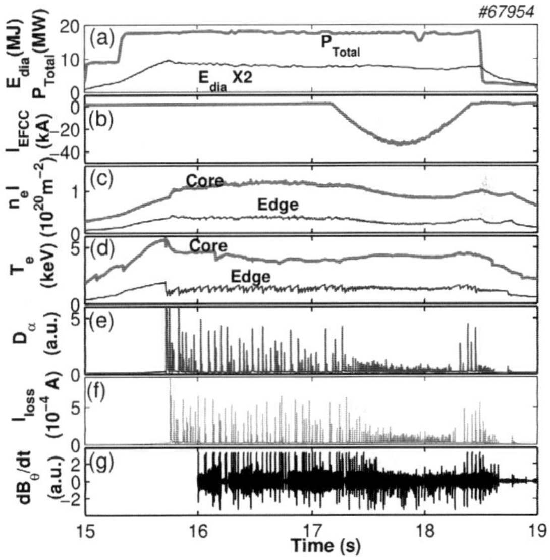

Figure 13. Waveforms of typical ELM control using EFCC in JET. (a) total input power and the stored energy, (b) EFCC current, (c) the line-integrated electron densities in the central chord (upper trace) and the pedestal chord (lower trace), (d) the electron temperature at the plasma center (upper trace) and at the pedestal top (lower trace), (e) the D_α signal measured at the outer divertor, (f) the fast ion loss current and (g) magnetic fluctuation [83].

(up to 20%) degradation of the energy confinement time during the ELM mitigation phase. The core and edge density are lowered by the density pump out, as shown in figure 13(c). It is noted that there is an operational window of the EFCC current amount which avoids a locked mode and achieves ELM mitigation [91]. The fact that the window seems to be wider in the case of $n = 2$ perturbation field indicates that a higher toroidal mode number is advantageous for ELM control without destabilizing the locked mode.

The effect of density pump out is a common characteristic in plasmas under the ELM control using external magnetic field perturbation. So far, the recovery of plasma density has not been achieved. However, a plasma operating contour (POPCON) plots for ITER in volume averaged density and temperature indicates that an averaged density of ~0.6 n_{GW} is required to achieve $Q = 10$ (figure 93 in [92]). Therefore, it is important to investigate how to achieve controllability of the plasma density with external magnetic field perturbation.

Based on these successful ELM controls using external magnetic field perturbation demonstrated in DIII-D and JET, a number of possible designs of external and in-vessel coils generating the RMP for the ELM control in ITER have been analyzed for the reference scenarios (H-mode, Hybrid and Steady-State) taking into account physical, technical and spatial constraints [93]. The evaluation of RMP coils designs has been made to achieve an RMP spectrum that produces enough edge ergodisation and minimum central perturbation at minimum current. One possible candidate for the coil design for ITER comprises port-plug coils in the midplane port with 150 kAt, which could be reasonably adapted to various scenarios.

5. Importance of V_T for ELM studies

As described in the previous section, the edge V_T clearly affects ELM characteristics. In the type I ELM regime, the $\Delta W_{ELM}/W_{ped}$ can be reduced with smaller co-V_T or larger counter V_T, the ELM frequency varying with the V_T accordingly. In the grassy ELM regime, grassy ELM frequency rapidly increases while maintaining similar pedestal performance as counter V_T increases. In the QH-mode regime, EHO characteristics and p^{ped} seem to be changed by V_T.

Not only in the experiments, but also in the design of the ELM control coils, the importance of V_T can be found. As described in the previous section, the level of central perturbation due to the external magnetic field perturbation is one of the important aspects of the optimization of the design. In such assessment, the screening effect of the V_T should be taken into account. The possible screening of RMP is estimated using both analytical and numerical results from the JOREK code [94]. In the estimations of the degree of screening of the RMP by the V_T, there was a strong reduction of the magnetic field perturbation amplitude towards the plasma center, but relatively little change in the pedestal region where the resistive time and the toroidal rotation frequency have usually much smaller values than the plasma center. In other words, if the toroidal rotation frequency at the pedestal is larger than the current estimation of ~0.3 kHz in ITER, a higher coil current will be needed to generate an edge ergodic region with the same size.

All these results indicate the importance of a reliable prediction of the toroidal rotation velocity and/or frequency in ITER with TF ripple which produces an effective torque at the edge region. This effect on plasmas in JET and JT-60U with TF ripple has been modeled using the ASCOT code [95]. The analysis shows that the observed negative plasma rotation near the plasma edge requires extra negative torque, which might come from ripple losses of thermal ions, to satisfy the toroidal force balance. It is important to understand the physical mechanism determining the torque source profile with TF ripple and momentum transport properties for better prediction/modeling.

The physical mechanism of change in the ELM characteristics and the pedestal performance with change in V_T is another issue. Two possible mechanisms can be considered here. One is that the V_T affects the plasma transport and/or the edge fluctuation, which in turn causes the change in the edge ∇p. In this mechanism, the edge radial electric field and its shear may play an important role. The other is that the V_T profile directly affects the edge MHD stability, although experimentally relevant V_T shear only has a stabilizing effect in the case of larger toroidal mode number ($n > 20$) [5]. Recently, a new concept, edge localized resistive wall mode, has been proposed in the QH-mode [66]. To address this issue, dedicated and simultaneous measurements of plasma density, temperature, rotation and current profiles using high resolution edge diagnostics are required.

6. Conclusions

The recent improvement of edge plasma diagnostics provides a better understanding of the physics in the ELM dynamics. In precursor phase as a linear growth phase in MHD instability, a small growth rate of the type I ELM precursor, $\gamma / \omega_A = 0.1$-1%, is observed. A phenomenon commonly observed during the collapse phase is the propagation of ELM filaments in the SOL region, which is important to evaluate the ELM heat load. However, the mechanism of the loss of particles and heat during the short-lived (~200 μs) collapse phase remains to be understood. After the ELM filament is ejected from the last closed flux surface, further investigation is required to answer the question whether there is a direct path from the pedestal to the divertor or not. The different evolutions of pedestal parameters observed in DIII-D and JT-60U in the recovery phase can be understood qualitatively introducing different operational point in a stability boundary determined by peeling-ballooning model. Direct measurement of the change in the edge current density during the ELM cycle will be required to validate this model quantitatively. Although the V_T and the TF ripple clearly affect the $\Delta W_{ELM}/W_{ped}$, the mechanism by which the $\Delta W_{ELM}/W_{ped}$ can be reduced by making co-V_T smaller or counter-V_T larger is not yet understood. Moreover, the unacceptable level of the $\Delta W_{ELM}/W_{ped}$ for the type I ELM regime in cases including counter-rotating plasmas is still an important remaining issue for ITER.

The establishment of plasma operations in small ELM or no ELM regimes is especially important in ITER because ITER cannot accept the ΔW_{ELM} of type I ELMs. Therefore, further expansion of the operational space of various small ELM or no ELM regimes through multi-machine experiments and improved edge diagnostics is strongly encouraged. The progress achieved in low ν_e^* small ELM or no ELM regimes is as follows. A type V ELM regime in NSTX characterized by small ELM amplitude with one or two filaments can be obtained in lower collisionality ($\nu_e^* < 1$). The ELM frequency in the grassy ELM regime increases as the V_T changes in the counter direction. A small $\Delta W_{ELM}/W_{ped} < 1\%$ is achieved with the grassy ELM in plasmas with small or no V_T. In QH-mode plasmas in DIII-D, higher p^{ped} together with changes in EHO characteristics are achieved at smaller counter-V_T. The edge operational conditions can be kept just below the peeling boundary because of the existence of EHO/broadband fluctuations, which is considered as a physical mechanism of ELM suppression in the QH-mode regime.

Systematic studies of pellet pacing and the external magnetic field perturbation have progressed in developing ELM control methods applicable to ITER. The middle of the pedestal has been confirmed as a minimum penetration depth for forced ELM triggering. In JET, it has been found that only a small fraction of the fueling size pellet is enough for ELM triggering. Complete ELM suppression in DIII-D low ν_e^* plasmas with ITER similar shape was demonstrated. Because particle loss is increased by RMP, the pedestal parameters can be kept either close to or well below the peeling-ballooning stability boundary. This is why the ELMs are suppressed in RMP plasmas. ELM control with EFCC creating $n = 1$ perturbation also was demonstrated in JET. Based on the results, both pellet pacing and external magnetic field perturbation seem to be effective in ITER. However, the narrow operational margin imposed by the uncertainties due to the lack of the knowledge concerning the forced ELM triggering cased by pellet pacing and the density pump-out effect with external magnetic field perturbation are remaining issues to achieve $Q = 10$ plasmas in ITER. New experiments using newly installed ELM control methods in various devices will contribute to establish active ELM control in the next-step device.

Acknowledgments

The author would like to thank the following persons for helpful discussions and contributions to this work: Drs. N. Aiba, N. Asakura, M. Bécoulet, M. Beurskens, K. Burrell, T.E. Evans, W. Fundamenski, R. Groebner, L.D. Horton, Y. Kamada, K. Kamiya, A. Kirk, G. Kocsis, B. Kurzan, P. Lang, A.

Leonard, Y. Liang, A. Loarte, R. Maingi, V. Mukhovatov, T. Osborne, V. Parail, R.A. Pitts, A. Polevoi, G. Saibene, Y. Sakamoto, R. Sartori, P. Snyder, M. Sugihara, J.L. Terry, T. Takizuka, H. Urano, F. Watanabe, M. Yoshida. This work was partly supported by the Grant-in-Aid for Scientific Research (S) 17106013 and for Young Scientists (A) 18686076, Japan Society for the Promotion of Science.

References

[1] Federici G, Loarte A, Strohmayer G 2003 *Plasma Phys. Control. Fusion* **45** 1523
[2] Federici G 2006 *Phys. Scr.* **T124** 1
[3] Leonard A W *et al.* 2006 *Plasma Phys. Control. Fusion* **48** A149
[4] Kamiya K *et al.* 2007 *Plasma Phys. Control. Fusion* **49** S43
[5] Wilson H R, Cowley S C, Kirk A and Snyder P B 2006 *Plasma Phys. Control. Fusion* **48** A71
[6] Oyama N *et al.* 2006 *Plasma Phys. Control. Fusion* **48** A171
[7] Zhitlukhin A *et al.* 2007 *J. Nucl. Mater.* **363-365** 301
[8] Eich T, Kallenbach A, Pitts R A, Jachmich S, Fuchs J C, Herrmann A, Neuhauser J, ASDEX Upgrade Team and JET-EFDA Contributors 2007 *J. Nucl. Mater.* **363-365** 989
[9] Lipschultz B *et al.* 2007 *Nucl. Fusion* **47** 1189
[10] Polevoi A *et al.* 2007 *"Comment on Requirements for ELM Mitigation in ITER"* In discussion session of the workshop in topic 2
[11] Mukhovatov V, Shimada M, Chudnovskiy A N, Costley A E, Gribov Y, Federici G, Kardaun O, Kukushkin A S, Polevoi A, Pustovitov V D *et al.* 2003 *Plasma Phys. Control. Fusion* **45** A235
[12] Weller A, Anton M, Geiger J, Hirsch M, Jaenicke R, Werner A, Nührenberg C, Sallander E and Spong D A 2001 *Phys. Plasmas* **8** 931
[13] Jiménez J A, de la Luna E, García-Cortés I and Shchepetov S V 2006 *Plasma Phys. Control. Fusion* **48** 515
[14] Toi K *et al.* 2006 *Plasma Phys. Control. Fusion* **48** A295
[15] Morita S *et al.* 2006 *Plasma Phys. Control. Fusion* **48** A269
[16] Watanabe F *et al.* 2007 "ELM-like oscillation observed in the Large-Helical-Device Plasmas with/without L-H Transition" *this conference P2-14*
[17] Fundamenski W, Naulin V, Neukirch T, Garcia O E and Juul Rasmussen J 2007 *Plasma Phys. Control. Fusion* **49** R43
[18] Ferron J R *et al.* 2000 *Phys. Plasmas* **7** 1976
[19] Aiba N, Tokuda S, Takizuka T, Kurita G and Ozeki T 2007 *Nucl. Fusion* **47** 297
[20] Saarelma S, Hender T C, Kirk A, Meyer H, Wilson H R and MAST Team 2007 *Plasma Phys. Control. Fusion* **49** 31
[21] Leonard A W, Casper T A, Groebner R J, Osborne T H, Snyder P B and Thomas D M 2007 *Nucl. Fusion* **47** 552
[22] Wade M R, Burrell K H, Hogan J T, Leonard A W, Osborne T H, Snyder P B and Coster D 2005 *Phys. Plasmas* **12** 056120
[23] Oyama N *et al.* 2005 *Nucl. Fusion* **45** 871
[24] Fenstermacher M E *et al.* 2005 *Nucl. Fusion* **45** 1493
[25] Aiba N *et al.* 2007 "Effect of equilibrium properties on the structure of the edge MHD modes in tokamaks" *this conference P2-01*
[26] Oyama N *et al.* 2004 *Nucl. Fusion* **44** 582
[27] Kirk A *et al.* 2007 *Plasma Phys. Control. Fusion* **49** 1259
[28] Terry J L, Cziegler I, Hubbard A E, Snipes J A, Hughes J W, Greenwald M J, LaBombard B, Lin Y, Phillips P and Wukitch S 2007 *J. Nucl. Mater.* **363-365** 994
[29] Scannell R, Kirk A, Ben Ayed N, Carolan P G, Cunningham G, McCone J, Prunty S L and Walsh M J 2007 *Plasma Phys. Control. Fusion* **49** 1431
[30] Herrmann A *et al.* 2007 *J. Nucl. Mater.* **363-365** 528
[31] Kurzan B, Horton L D, Murmann H, Neuhauser J, Suttrop W and the ASDEX Upgrade Team 2007 *Plasma Phys. Control. Fusion* **49** 825
[32] Eich T, Andrew P, Herrmann A, Fundamenski W, Loarte A, Pitts R A and JET-EFDA contributors 2007 *Plasma Phys. Control. Fusion* **49** 573
[33] Pitts R A, Andrew P, Arnoux G, Eich T, Fundamenski W, Huber A, Silva C, Tskhakaya D and JET EFDA Contributors 2007 *Nucl. Fusion* **47** 1437

[34] Asakura N *et al.* 2006 *Proc. 21st Int. Conf. on Fusion Energy (Chengdu, 2006)* (Vienna: IAEA) CD-ROM file EX/9-2

[35] Asakura N *et al.* 2007 "ELM propagation in the low- and high-field-side Scrape-off Layer of the JT-60U tokamak" *this conference P2-02*

[36] Maingi R *et al.* 2006 *Phys. Plasmas* **13** 092510

[37] Maqueda R J, Maingi R, Tritz K, Lee K C, Bush C E, Fredrickson E D, Menard J E, Roquemore A L, Sabbagh S A and Zweben S J 2007 *J. Nucl. Mater.* **363-365** 1000

[38] Kirk A *et al.* 2007 "Comparison of the spatial and temporal structure of type-I ELMs" *this conference P2-05*

[39] Kirk A, Koch B, Scannell R, Wilson H R, Counsell G, Dowling J, Herrmann A, Martin R, Walsh M and the MAST team 2006 *Phys. Rev. Lett.* **96** 185001

[40] Beurskens M N A *et al.* 2007 "ELM dynamics in the JET edge pedestal" *13th ITPA pedestal topical group meeting (Naka, 2007)*

[41] Kirk A *et al.* 2007 "Comparison of the spatial structure of type I ELMs in ASDEX Upgrade and MAST" *this conference P2-06*

[42] Groebner R J *et al.* 2007 "Temporal and spatial evolution of the H-mode pedestal in DIII-D" *this conference P1-03*

[43] Stacey W M 2006 *Phys. Plasmas* **13** 112506

[44] Yoshida M *et al.* 2006 *Plasma Phys. Control. Fusion* **48** A209

[45] Urano H, Takizuka T, Kamada Y, Oyama N Takenaga H and Miura Y 2005 *Phys. Rev. Lett.* **95** 035003

[46] Kamiya K, Urano H, Koide Y, Takizuka T, Oyama N, Kamada Y and the JT-60 Team 2006 *Plasma Phys. Control. Fusion* **48** A131

[47] Saibene G, Oyama N, Lönnroth J, Andrew Y, de la Luna E *et al.* 2007 *Nucl. Fusion* **47** 969

[48] Urano H, Oyama N, Kamiya K, Koide Y, Takenaga H, Takizuka T, Yoshida M, Kamada Y and the JT-60 Team 2007 *Nucl. Fusion* **47** 706

[49] Kamiya K, Urano H, Oyama N and Kamada Y 2007 *Plasma and Fusion Research* **2** 005

[50] Saibene G *et al.* 2007 "Toroidal field ripple effects on H-modes in JET and implications for ITER" *34th EPS Conf. on Plasma Physics (Warsaw, Poland, 2–6 July 2007)*

[51] Lönnroth J-S, Parail V, Hynönen V, Johnson T, Kiviniemi T, Oyama N, Beurskens M *et al.* 2007 *Plasma Phys. Control. Fusion* **49** 273

[52] Oyama N *et al.* 2007 "Effect of toroidal field ripple and toroidal rotation on H-mode performance and ELM characteristics in JET/JT-60U similarity experiments" *this conference P2-11*

[53] Maingi R *et al.* 2006 *Phys. Plasmas* **13** 092510

[54] Kamiya K, Oyama N, Ido T, Bakhtiari M and JFT-2M Group 2006 *Phys. Plasmas* **13** 032507

[55] Terry J L *et al.* 2005 *Nucl. Fusion* **45** 1321

[56] Maingi R *et al.* 2005 *Nucl. Fusion* **45** 264

[57] Kamada Y *et al.* 2000 *Plasma Phys. Control. Fusion* **42** A247

[58] Saibene G *et al.* 2005 *Nucl. Fusion* **45** 297

[59] Stober J *et al.* 2005 *Nucl. Fusion* **45** 1213

[60] Polevoi A *et al.* 2002 J. Plasma Fusion Res. Ser. **5** 82

[61] Oyama N, Kamada Y, Isayama A, Urano H, Koide Y, Sakamoto Y, Takechi M, Asakura N and the JT-60 Team 2007 *Plasma Phys. Control. Fusion* **49** 249

[62] Burrell K H *et al.* 2002 *Plasma Phys. Control. Fusion* **44** A253

[63] Suttrop W *et al.* 2003 *Plasma Phys. Control. Fusion* **45** 1399

[64] Sakamoto Y, Shirai H, Fujita T, Ide S, Takizuka T, Oyama N and Kamada Y 2004 *Plasma Phys. Control. Fusion* **46** A299

[65] Suttrop W *et al.* 2005 *Nucl. Fusion* **45** 721

[66] Snyder P B *et al.* 2007 *Nucl. Fusion* **47** 961

[67] Burrell K H *et al.* 2007 *GENERAL ATOMICS REPORT* "Quiescent H-mode experiments with co- plus counter-neutral beam injection" GA-C25837

[68] West W P *et al.* 2005 *Nucl. Fusion* **45** 1708

[69] Osborne T H *et al.* 2007 "Edge Stability of Stationary ELM-Suppressed Regimes on DIII-D" *this conference P2-10*

[70] Lang P T *et al.* 2003 *Nucl. Fusion* **43** 1110

[71] Geraud A *et al.* 2006 "The JET high frequency pellet injector project" *24th SOFT Conference (Warsow, 2006) P4-H-201*

Journal of Physics: Conference Series **123** (2008) 012002 doi:10.1088/1742-6596/123/1/012002

[72] Baylor L R *et al.* 2006 *Bull. Am. Phys. Soc.* **51** 115
[73] Lang P T *et al.* 2005 *Plasma Phys. Control. Fusion* **47** 1495
[74] Fielding S J *et al.* 2001 *Proc. 28th EPS Conf. on Contr. Fusion and Plasma Phys. (Madeira)*, ECA 25A 1825
[75] Degeling A W, Martin Y R, Lister J B, Villard L, Dokouka V N, Lukash V E and Khayrutdinov R R 2003 *Plasma Phys. Control. Fusion* **45** 1637
[76] Martin Y *et al.* 2004 *Proc. 31st EPS Conf. on Plasma Phys. (London)*, ECA 28G P4-133
[77] Horton L D *et al.* 2004 *Plasma Phys. Control. Fusion* **46** B511
[78] Porte L *et al.* 2007 *Nucl. Fusion* **47** 952
[79] Turri G *et al.* 2007 "The role of MHD in the sustainment of electron internal transport barriers and H-mode in TCV" *this conference P6-12*
[80] Shoji T *et al.* 1992 *J. Nucl. Mater.* **196-198** 296
[81] Evans T E *et al.* 2004 *Phys. Rev. Lett.* **92** 235003
[82] Evans T E *et al.* 2005 *Nucl. Fusion* **45** 595
[83] Liang Y *et al.* 2007 *Phys. Rev. Lett.* **98** 265004
[84] Beurskens M N A *et al.* 2007 "Optimized ELM penetration and pedestal pressure for JET AT scenarios" *this conference P2-04*
[85] Lang P T, Alper B, Buttery R, Gal K, Hobirk J, Neuhauser J, Stamp M and JET-EFDA contributors 2007 *Nucl. Fusion* **47** 754
[86] Kocsis G, Kálvin S, Lang P T, Maraschek M, Neuhauser J, Schneider W, Szepesi T and the ASDEX Upgrade Team 2007 *Nucl. Fusion* **47** 1166
[87] Polevoi A R *et al.* 2007 "Assessment of independent fuelling and ELM pace making by pellet injection in ITER" *34th EPS Conf. on Plasma Physics (Warsaw, Poland, 2–6 July 2007)*
[88] Baylor L R, Parks P B, Jernigan T C, Caughman J B, Combs S K, Foust C R, Houlberg W A, Maruyama S and Rasmussen D A 2007 *Nucl. Fusion* **47** 443
[89] Evans T E *et al.* 2006 *Phys. Plasmas* **13** 056121
[90] Evans T E *et al.* 2007 "ELM control during RMP H-modes in DIII-D with ITER similar shapes" *Stochasticity in Fusion Plasmas, Jülich, Germany*; to appear in *Nucl. Fusion*
[91] Liang Y *et al.* 2007 *12th ITPA pedestal topical group meeting (Garching, 2007)*
[92] Progress in the ITER Physics Basis 2007 *Nucl. Fusion* **47** S18
[93] Bécoulet M *et al.* 2007 "Numerical study of the resonant magnetic perturbations for type I edge localized modes control in ITER" *Stochasticity in Fusion Plasmas, Jülich, Germany*; to appear in *Nucl. Fusion*
[94] Bécoulet M *et al.* 2006 et al in Fusion Energy 2006 (Proc. 21st Int. Conf. Chengdu, 2006) (Vienna: IAEA) CD-ROM file IT/P1-29
[95] Parail V *et al.* 2007 "Predictive modelling of ripple-induced effects in ELMy H-mode plasmas" *this conference P2-12*

Journal of Physics: Conference Series **123** (2008) 012003 doi:10.1088/1742-6596/123/1/012003

Spontaneous rotation and momentum transport in tokamak plasmas

J E Rice

MIT Plasma Science and Fusion Science Center, Cambridge, MA 02139, USA

E-mail: rice@psfc.mit.edu

Abstract. Recently there has been widespread attention paid to rotation and momentum transport in tokamak plasmas. Of particular interest is spontaneous (intrinsic) toroidal rotation in plasmas without external momentum input. The strong co-current spontaneous rotation in enhanced confinement regimes, with ion thermal Mach numbers up to 0.3, may allow for resistive wall mode suppression in high-pressure ITER discharges, without requiring the use of neutral beam injection. Spontaneous rotation in L-mode discharges exhibits a complex dependence on plasma parameters and magnetic configuration compared to the relatively simple scaling of Alfven Mach number ($M_A = V_\phi/C_A$, where C_A is the Alfven speed) $M_A \sim \beta_N$ observed in enhanced confinement plasmas. There is currently no comprehensive, quantitative explanation of this phenomenon. An accurate prediction of the expected rotation velocity profile from whatever neutral beam injection is available on ITER requires a detailed understanding of momentum transport. There have been extensive investigations into correlations between energy and momentum diffusivities, and whether there are systematic trends of the Prandtl number with plasma parameters. Of late, there has been vigorous theoretical activity regarding a possible momentum pinch that could help enhance the rotation in the plasma interior. There has been a renewed interest in poloidal rotation, especially in ITB discharges, which is generally found to be at odds with the predictions of neo-classical theory. This calls into question the common practice of the determination of E_r from toroidal rotation measurements with the assumption of neo-classical poloidal rotation.

1. Introduction

Plasma rotation and velocity shear play important roles in the transition to high confinement mode (H-mode) [1-6], in the formation of internal transport barriers (ITBs) [7-11] and in suppression of resistive wall modes (RWMs) [12,13] in tokamak discharges. Rotation is usually externally provided by the momentum input from neutral beam injection in the current generation of tokamaks. This may not be available in future reactor-grade devices due to the large machine sizes, high densities and the limitations of beam current. It has been estimated that an Alfven Mach number ($M_A = V_\phi/C_A$, where $C_A \sim B^2/\mu_0 n_e m_{ave}$) $M_A = 0.02$ will be required for RWM stabilization in certain ITER operational scenarii, [14,15], depending on the velocity profile and normalized pressure, β_N. This corresponds to a rotation speed of 200 km/s (30 kRad/s) in a particular ITER case with $B_T = 5.2$ T and $n_e = 6.7 \times 10^{19}/m^3$, and it is unknown whether this level of rotation will be generated from neutral beams. There are currently two approaches to this problem. The first is to utilize the best estimates of anticipated neutral beam injection for ITER and to calculate the expected rotation velocity profile assuming a particular form for the momentum diffusivity, χ_ϕ. The usual assumption is that momentum diffusivity is equal to the ion thermal conductivity, χ_i, which has been observed in several experiments. However, since the expected beam power is possibly marginal for RWM suppression, a more accurate knowledge of χ_ϕ is desirable. There also exists evidence for an inward momentum pinch in some experiments that could enhance the internal rotation for a given input torque. In either case, a better understanding of momentum transport is necessary for an accurate evaluation of the rotation velocity in ITER from neutral beam injection. The other approach is to take advantage of the spontaneous (intrinsic) rotation observed in many tokamaks without external momentum input to provide the necessary velocity. This relies on the accuracy of extrapolation of the current observations to the ITER plasma and machine parameters. To this end a database of spontaneous rotation has been recently assembled [16]. In section 2 observations of spontaneous rotation in a wide variety of operational scenarii will be summarized and the results of the database will be presented. In section 3 the current understanding of momentum transport will be reviewed.

2. Spontaneous rotation

Spontaneous rotation (in the absence of external momentum input) in low confinement mode (L-mode) Ohmic plasmas is often directed counter-current, and depends very sensitively on the magnetic configuration [17-19] and in a complicated fashion on other parameters, such as the plasma current, electron density and ion temperature [18-21]. While the study of spontaneous rotation in L-mode discharges is of interest in its own right (e.g. for its relation to the H-mode power threshold [18,19]), these plasmas will not be considered here, since most ignition scenarii in future devices require H-mode confinement. The spontaneous toroidal rotation in H-mode (and in enhanced confinement regimes in Tore Supra) is generally in the co-current direction [18,22-37] and has been observed on many devices and produced with a wide variety of techniques (ICRF, Ohmic, ECH), demonstrating its fundamental nature.

Substantial rotation velocities have been obtained in ion cyclotron range of frequencies (ICRF) heated plasmas on JET [23,33], Alcator C-Mod [24,25,30,31,34,18] and Tore Supra [26,29,32]. Co-current spontaneous rotation during H-mode operation was first observed in JET ICRF heated plasmas [23], where a strong correlation was found between the local angular momentum density and the ion pressure for low recycling discharges (figure 10 of reference [23]). Intrinsic rotation velocities as high as 60 km/s (20 kRad/s) were measured using passive x-ray spectroscopy, viewing the plasma at r/a ~ 0.35. Spontaneous rotation in ICRF heated H-mode plasmas has been studied extensively in Alcator C-Mod [38], with velocities up to 130 km/s (200 kRad/s, thermal ion Mach number M_i = 0.3) measured in discharges with no direct momentum input, also utilizing passive x-ray spectroscopy. The main C-Mod results are summarized as follows: the rotation during H-mode is in the co-current direction, changing direction, but remaining co-current when the current direction is reversed. The time histories of two discharges, with forward and reversed current, are shown in figure 1. Both plasmas enter H-mode after 0.7 s with noticeable increases in the stored energy; at the same time, the rotation switches from slightly counter-current in both cases to substantially co-current. The time history of the rotation velocity tracks the time evolution of the plasma stored energy, W_p, and not the electron density or ion temperature independently. It was not possible to change the magnetic field and plasma current directions separately. The change in the rotation velocity between L- and H-mode increases with the change in the plasma stored energy during ICRF heating and decreases with the magnitude of the plasma current, which is demonstrated in figure 2. This appears to be a common feature of spontaneous rotation in enhanced confinement regimes. The intrinsic rotation is seen to propagate in from the plasma boundary following the H-mode transition, with a momentum confinement time similar to the energy confinement time [30,17], and highly anomalous compared to neo-classical theory. This is demonstrated in the top of figure 3, which shows the velocity profile evolution at three different radii for an EDA [39] H-mode (transition time 1.11 s) plasma. The solid curves are from a simple momentum diffusion model with a diffusivity of 0.05 m^2/s. The rotation velocity profiles are flat in EDA H-mode and

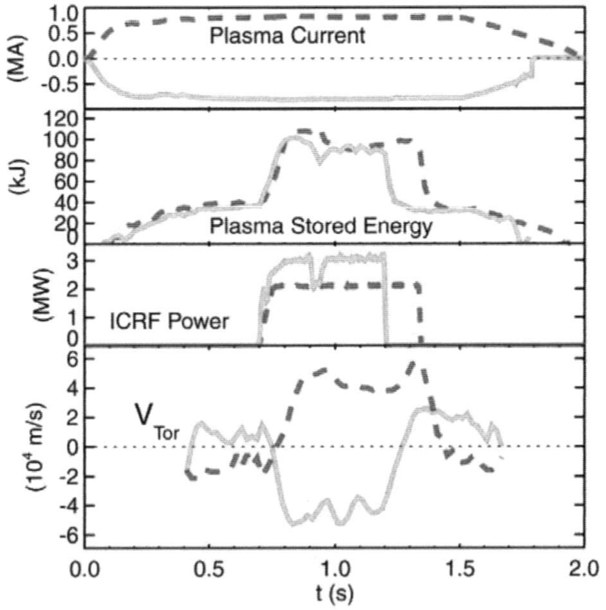

Figure 1. Time histories of the plasma current, stored energy, ICRF power and central rotation velocity for forward (red dashed, 1051202012) and reversed (green, 1060519022) current Alcator C-Mod H-mode discharges.

centrally peaked in ELM-free H-mode (bottom of figure 3), implying the presence of an inward momentum pinch in the latter case of ~ 10 m/s (see next section). In Tore Supra, strong co-current toroidal rotation, with velocities reaching 80 km/s (35 kRad/s), has similarly been observed in ICRF heated enhanced confinement regimes [26,29,32], also using passive x-ray spectroscopy. A scaling similar to the C-Mod results, with the change in the rotation velocity increasing with the change in the stored energy normalized to the plasma current, has been observed (figure 3 of reference [32]). In this case the slope of a line through the data points is a little over a factor of two less than in C-Mod, and this presumably contains some machine size scaling information. The best correlation of the intrinsic rotation velocity in Tore Supra is with the ion pressure [26].

Co-current rotation has been seen in Ohmic H-mode discharges in COMPASS-D [22], Alcator C-Mod [25,27,28], DIII-D [35,36] and TCV. Spontaneous rotation in Ohmic H-mode plasmas has been studied extensively in C-Mod [27,28], and the observations are essentially identical to those in ICRF H-modes in terms of the scaling with W_p/I_p (see figure 2), and in the overall rotation time histories, as shown in figure 4. This is suggestive of a common driving mechanism not involving energetic ions [34]. The rotation velocity profiles during Ohmic H-modes in C-Mod and DIII-D are flat.

Similarly, co-current rotation has been observed in the outer regions (*r/a* ~ 0.8) of electron cyclotron heated (ECH) H-mode plasmas on DIII-D [35,36], on JT-60U [37] with a combination of lower hybrid (LH) waves and ECH, and in TCV with core ECH [40]. In DIII-D ECH H-modes, rotation velocities as high as 30 km/s (20 kRad/s) at *r/a* = 0.8 have been measured. The profile shape depends on the ECH resonance location, with hollow profiles occurring under on-axis heating; no velocity increase in the plasma center over the L-mode background value is seen. Off-axis ECH leads to relatively flat (see figure 5)

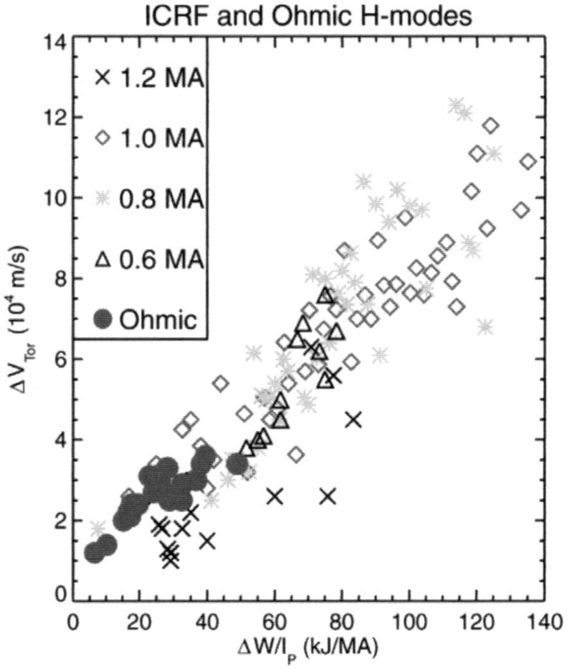

Figure 2. The change in the central rotation velocity between L- and H-mode as a function of the change in the stored energy divided by the current, for ICRF and Ohmic C-Mod [30] plasmas.

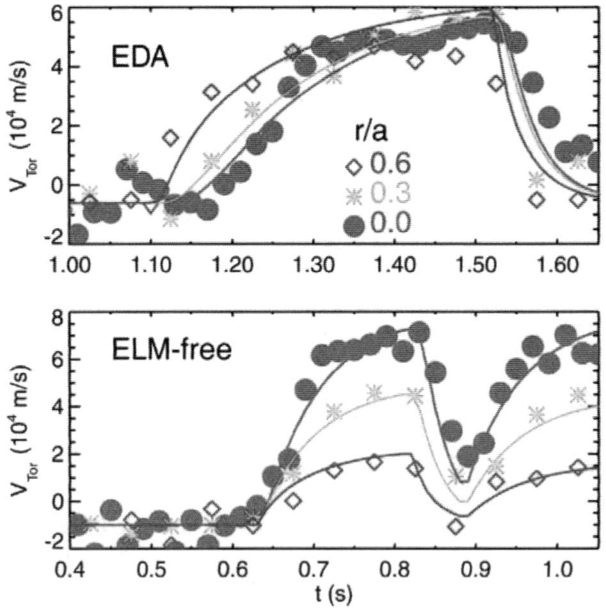

Figure 3. The rotation velocity profile evolution at three different radii for EDA (top, 1020830023) and ELM-free (bottom, 1021105017) H-mode C-Mod [18] discharges.

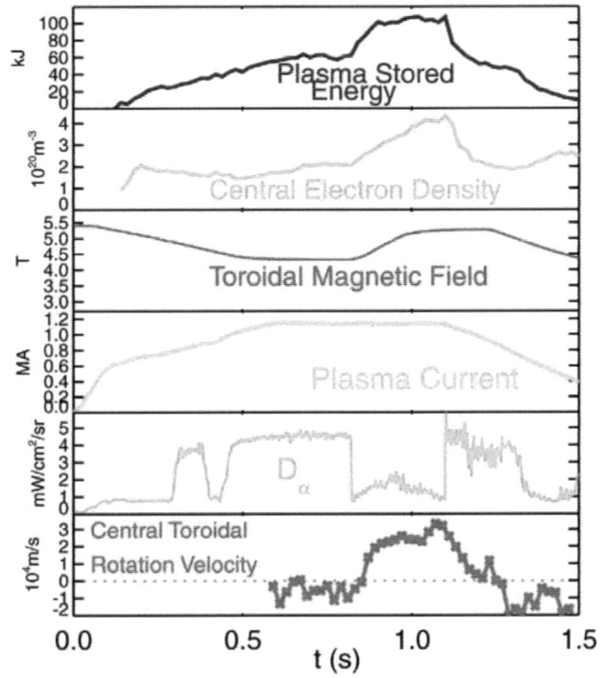

Figure 4. The time histories of plasma parameters for a C-Mod Ohmic H-mode discharge (990521014) [28], with the transition time at 0.82 s.

velocity profiles. Regardless of the resonance location, the maximum co-current velocity increase is near r/a = 0.8. A scaling of the rotation velocity near r/a = 0.8 as a function of the plasma stored energy normalized to the plasma current in DIII-D ECH and Ohmic H-modes shows a very similar scaling to those seen in C-Mod and Tore Supra (see figure 6 top), suggesting a fundamental connection between observations of intrinsic rotation, regardless of the production technique. A slightly better fit to the DIII-D data was found multiplying W_p/I_p by $T_e(0)/T_i(0)$, as shown in figure 6 bottom. The slope of the fit is lower than that found in C-Mod and Tore Supra. Strong co-current intrinsic rotation, with velocities as high as 120 km/s (35 kRad/s), has been observed in JT-60U plasmas with a combination of LH and ECH [37], utilizing the very beginning of the heating neutral beam pulse. These discharges were at low electron density ($n_e = 4 \times 10^{18}/m^3$ and $T_e >> T_i$), had electron temperature ITBs and exhibited slightly peaked velocity profiles. Similar to the scalings described above, there was a strong correlation between the rotation velocity and the electron pressure. Co-current spontaneous rotation has also been observed in TCV ECH H-mode [40] plasmas, with velocities up to 35 km/s (40 kRad/s) at r/a ~ 0.6 - 0.7.

Co-current spontaneous rotation has also been revealed in DIII-D neutral beam heated H-mode plasmas [36]. Using careful tuning of co- and counter-current directed neutral beam injection, H-mode plasmas have been produced which have near zero input torque profiles. These plasmas exhibit a net co-current toroidal rotation consistent with the scalings of figure 6.

A common feature of all of these observations, regardless of the heating method, is a strong correlation between the toroidal rotation velocity and the plasma pressure or stored energy. In experiments that can operate with a large range of plasma current, the rotation velocity is found to be inversely proportional to I_p. The coefficient of this scaling is different on different devices, and probably includes some machine size scaling information. In order to unify the various observations

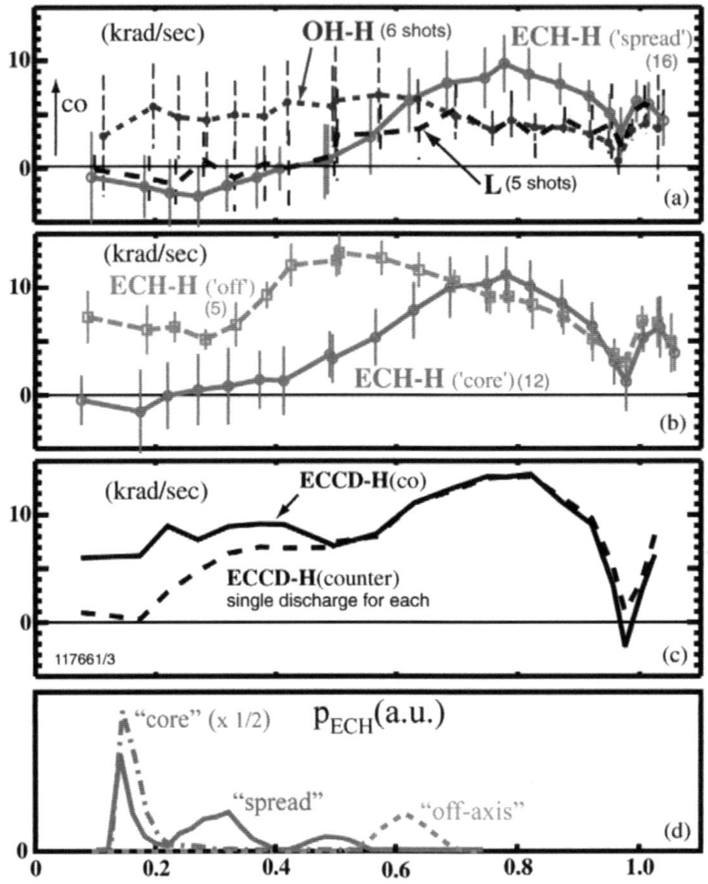

Figure 5. Rotation profiles for a.) L-mode, Ohmic H-mode and ECH spread, b.) ECH core and off-axis, c.) ECCD co- and counter discharges, and d.) ECH power deposition profiles in DIII-D [36]. X-axis is *r/a*.

from the different tokamaks, with an eye towards extrapolation to future devices and to assist in a better theoretical understanding of the process(es) driving spontaneous rotation, the approach of utilizing dimensionless parameters has been followed [16]. For the normalized velocities, the ion thermal and Alfven Mach numbers have been considered. A parameter that captures the ratio of the stored energy to the plasma current, and includes some size scaling is the normalized pressure, β_N. The scaling of M_A as a function of β_N is shown in figure 7 [16], depicting co-current toroidal rotation in H-mode in C-Mod, DIII-D, JET, JT-60U and TCV and enhanced confinement regimes Tore Supra. (The top axis is $\beta_N/40\pi$, which is truly dimensionless.) This scaling seems to unify the results fairly well and extrapolates to $M_A \sim 3\%$ for ITER with β_N of 2.6, high enough for RWM suppression. Whether this scaling indicates that an MHD process is driving spontaneous rotation is not known. In contrast, there is no correlation of Mach number with normalized gyro-radius, ρ^* (= 1.02×10^{-4} sqrt(μ(AMU)T(eV))/B(T)a(m)), as shown in figure 8; points from various machines are arranged in vertical strips. It is an open question if this implies that diamagnetic or certain turbulence effects are not involved in generating the spontaneous rotation. Similarly, there is no correlation found with collisionality [16]. A proper regression analysis has been performed with dimensionless parameters and the result is shown in figure 9 [16]; the best fit has $M_A = 0.65\ \beta_T^{1.4}\ q^{*2.3}$ (β_T is the ratio of the kinetic and magnetic pressures, and $q^* = 2\pi\kappa a^2 B/\mu_0 R I_p$) which has a stronger plasma current dependence than $1/I_p$. This scaling also bodes well for ITER with regard to RWM suppression. In line with this dimensionless scaling approach, a series of dimensionless similarity experiments has been undertaken between C-Mod and DIII-D [36]. In these experiments, several different values of the dimensionless parameters q, β, ρ^* and ν^* have been matched in discharges on both devices and the thermal Mach numbers were also found to match. The scaling from a regression analysis using dimensional machine parameters [16] is shown in figure 10. Here ΔV is the change in toroidal rotation velocity between L- and H-mode, B is the toroidal magnetic field, $\Delta \langle P \rangle$ is the change in the average kinetic pressure between L- and H-mode, I_p is the plasma current, R is the major radius and C is a constant. An extrapolation to ITER predicts rotation velocities in the 100 km/s range.

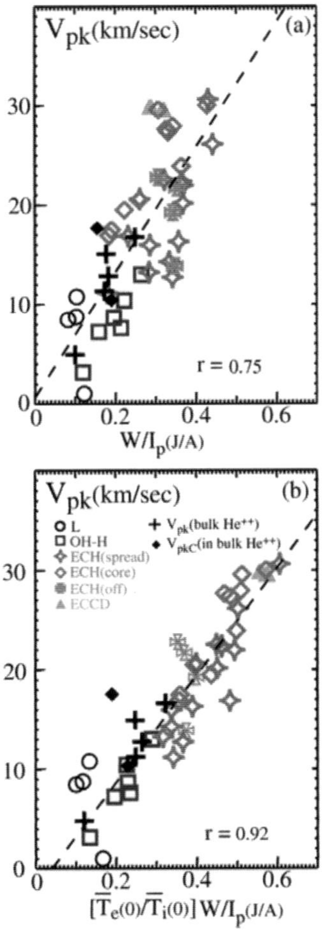

Figure 6. Rotation in DIII-D plasmas [36] as a function of stored energy normalized to the plasma current (top) and with a factor $T_e(0)/T_i(0)$ (bottom).

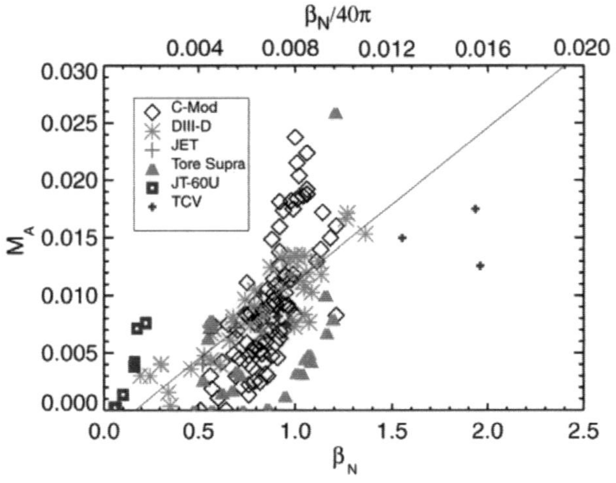

Figure 7. Alfven Mach number as a function of β_N for six devices [16].

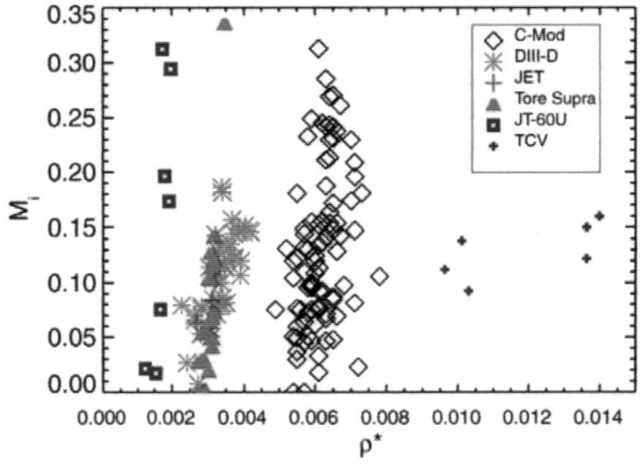

Figure 8. Ion thermal Mach number as a function of ρ^* [16] for six devices. $M_i = V_\phi/v_{th}$.

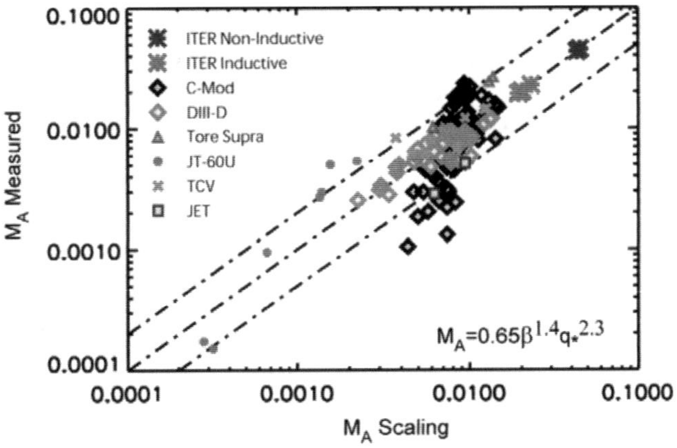

Figure 9. The measured Alfven Mach number as a function of dimensionless scaling $M_A = 0.65\ \beta_T^{1.4} q^{*2.3}$ [16].

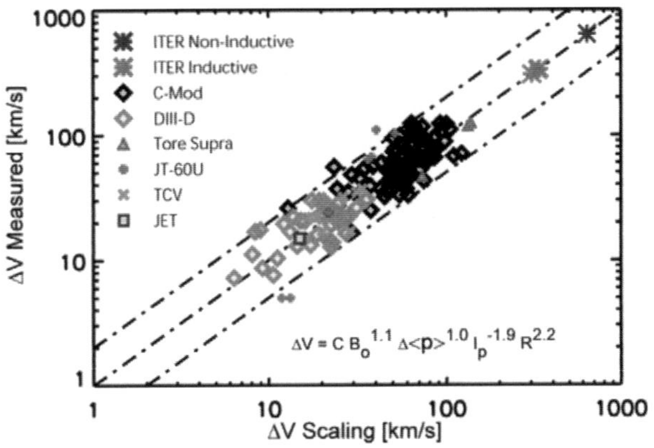

Figure 10. The measured rotation velocity as a function of the machine parameter scaling [16].

Several approaches toward an explanation of the observed intrinsic/spontaneous rotation have been undertaken. An obvious starting point is a comparison with the predictions of neo-classical theory. Using a moment formulation [41] in conjunction with radial force balance, the impurity toroidal rotation has been determined as a function of plasma parameters [42], with the toroidal rotation velocity proportional to the pressure gradient and radial electric field, and inversely proportional to the poloidal magnetic field. This captures some of the general features of the observed intrinsic rotation velocity scaling shown in figure 2: the plasma pressure or stored energy is widely found to increase with the pedestal pressure gradient, and the plasma current is directly related to the poloidal magnetic field. Unfortunately, the radial electric field is rarely measured directly (in practice it is inferred from the routinely available rotation velocity [43]) and remains largely an unknown, yet dominant, parameter. Furthermore, the standard neo-classical theory may not be valid at the edge of some H-mode discharges because the normal ordering, that the ion gyro-radius is small compared to the ion temperature gradient scale length, is violated. The sub-neo-classical theory [44,45] properly treats this situation, and has the added benefit of providing E_r from an ambipolarity constraint. The predicted toroidal rotation velocity is in reasonable quantitative agreement with observations on C-Mod [46] under certain operating conditions [34]. However, a fundamental problem with neo-classical and sub-neo-classical theory is that the momentum diffusivity [47] (viscosity) is orders of magnitude

Journal of Physics: Conference Series **123** (2008) 012003 doi:10.1088/1742-6596/123/1/012003

smaller than that observed in experiments [31 and references therein] (see next section).

Since the early observations of intrinsic rotation were in ICRF heated discharges, it is natural to consider models based on energetic ion orbit shift mechanisms [48-52]. However, the similarity of the intrinsic rotation observed in ICRF, ECH and purely Ohmic plasmas suggests that it is not due to ICRF wave or fast ion orbit effects. The prediction of reversal of the rotation direction with high magnetic field side off-axis ICRF absorption [49,51] has not been observed in the C-Mod experiments [34]. Energetic ion orbit effects have also been ruled out as a source of rotation in JET [23] and Tore Supra [29].

An alternative approach to explain the spontaneous generation of rotation is based on turbulence [53]. Fluctuation induced toroidal stress [54] can give rise to toroidal rotation where the direction of the rotation depends upon the mode frequency spectrum, and may explain the reversal of the observed rotation in going from L- to H-mode. A quantitative comparison with experimental results is not yet possible since the calculated magnitude of the rotation is a function of the level of the turbulence, which is not routinely measured. A comparison of predicted velocity profile shapes in C-Mod plasmas has produced mixed results [34]. A similar model [55] of electrostatic modes driven by the ion pressure gradient is in qualitative agreement with many of the features observed in the C-Mod experiment: direction of rotation in L- and H-mode, scaling with W_P/I_P in H-mode and the drop in the rotation observed in ITB plasmas. However, there are no quantitative predictions about the magnitude of the rotation or the size of the momentum diffusivity for further comparison with the experimental results. Furthermore, these turbulence-based theories do not provide a scaling of the turbulence levels with plasma parameters that would allow a direct comparison. A recent model of turbulence driven transport of toroidal momentum [56], which includes a momentum convection related to the particle pinch and an off-diagnoal fluctuation-induced torque density enabled by the symmetry breaking provided by radially sheared poloidal ***ExB*** flow, looks promising. It has recently been suggested [57] that the blob transport at the edge of C-Mod plasmas could possibly drive the rotation if the blobs exit the plasma at some toroidal angle. Whether the total momentum of blobs leaving the plasma edge is enough to spin the plasma in the opposite direction remains an open question for future work.

As discussed above, it is difficult to make quantitative comparisons of the measured toroidal rotation with the predictions of neo-classical theory, largely because the radial electric field is very rarely independently measured. The expression for the neo-classical poloidal rotation velocity [42] is relatively simple, proportional to the ion temperature gradient with a coefficient depending on the collisionality regime. Recent measurements of poloidal rotation velocities are at odds with neo-classical predictions, quantitatively and even in direction. In JET ITB plasmas [58], the measured poloidal rotation velocity profiles in the vicinity of the ITB foot substantially disagree with the calculated neo-classical values. Similarly, the measured poloidal rotation velocity profiles of H-mode plasmas in DIII-D [59] can differ in sign and by an order of magnitude quantitatively when compared to neo-classical calculations. This certainly calls into question the common practice of determining E_r from the measured toroidal rotation velocity profile and using the calculated neo-classical poloidal profile in the absence of direct measurements of the latter. One possible contribution to poloidal rotation could come from poloidal viscous stresses due to drift waves [60]. The generation of poloidal rotation from ITG and TEM turbulence has been simulated in JET plasmas [61].

3. Momentum transport

In a majority of tokamak plasmas, the toroidal rotation is generated externally by neutral beam injection. Momentum transport may then be characterized [61-74] with various levels of sophistication. The global momentum confinement time, τ_ϕ, can be determined from the ratio of the total plasma angular momentum to the input torque. This type of analysis has been performed on many devices [63,65,66,69,71,73,61] and it has been generally found that τ_ϕ is very close to the energy confinement time, with some systematic trends in the ratio between τ_ϕ and τ_E as a function of electron density in JET [73,61], as shown in figure 11. By measuring the steady state rotation velocity profiles, either from the associated beam diagnostics or by passive techniques, and calculating the input torque profiles from the beam injection, momentum transport coefficient profiles may be evaluated. Momentum diffusivity profiles have been obtained on many devices [64-67,72,73,61]; momentum confinement is generally found to be anomalous [62-74,31,61], with a diffusivity, χ_ϕ, similar to the ion thermal conductivity, χ_i, but much larger than the neo-classical [47,75-77,44] diffusivity (viscosity). Prandtl (χ_ϕ/χ_i) numbers between 0.1 and 2 have been observed, depending on radial location and plasma parameters; some systematic trends have been seen in this ratio with plasma parameters [72-74,61]. Evidence for a momentum pinch (off diagonal transport) has been clearly demonstrated using off-axis beam injection [67,70], utilizing co-counter beam modulation [68] and from the time evolution of spontaneous rotation, as shown in the bottom of figure 3 [31]. Analysis of steady state velocity profiles does not allow a unique determination of the pinch velocity; transient analysis from beam modulation experiments is necessary [68,74]. An inward pinch velocity of ~ 10 m/s has been inferred from JT-60U measurements [74] in this way, as shown in figure 12.

It is generally believed that the source of anomalous momentum transport is turbulence. It has been found that for diffusive momentum transport driven by ITG turbulence [78], χ_ϕ is of order χ_i, similar to what has been observed in most experiments. Viscous stress due to ITG modes has been considered in reference [79]. Anomalous viscosity from Reynolds stress has been calculated from a non-linear electro-magnetic gyro-kinetic equation [80] and the effects of fluctuation-induced stress have been included in a neo-classical quasi-linear treatment of momentum confinement [81]. Anomalous toroidal viscosity has been included in transport codes such as GYRO [82], GLF23 [83] and LINART [84]. Recently, there have been explanations put forward for a toroidal momentum pinch, due to the Coriolis drift effect [85] and from symmetry breaking due to magnetic field curvature [86].

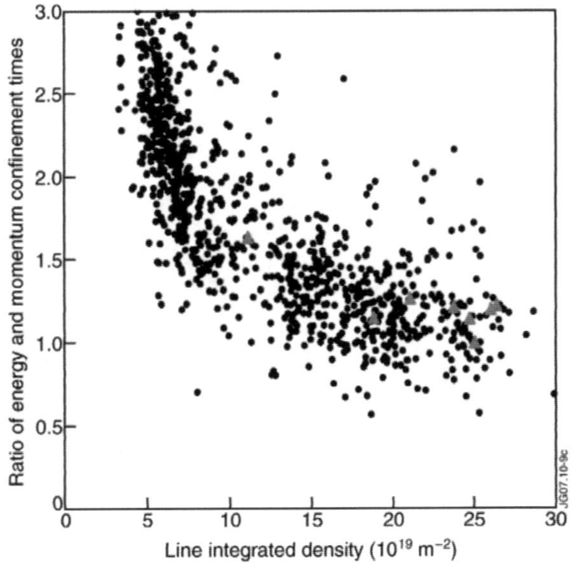

Figure 11. The ratio of the energy and momentum confinement times as a function of density for JET discharges [73].

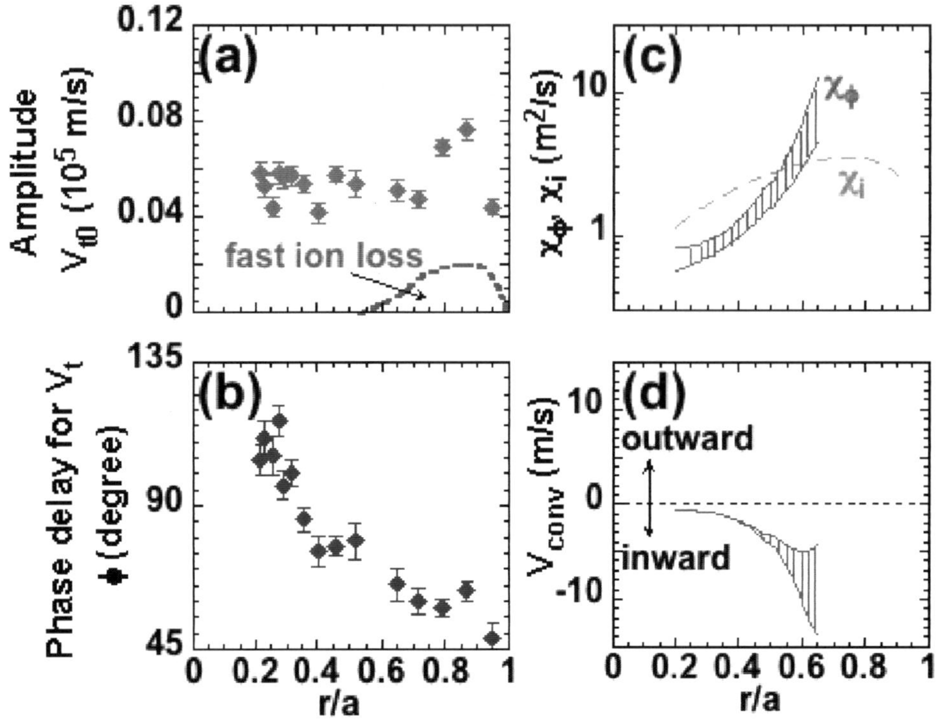

Figure 12. Profiles of the modulated amplitude (a), phase delay (b), momentum diffusivity (c) and momentum convection velocity (d) obtained from transient momentum balance analysis in JT-60U [74].

4. Conclusions

Spontaneous (in the absence of external momentum input) co-current toroidal rotation has been observed on many tokamaks in H-mode (and in enhanced confinement regimes in Tore Supra [26]), produced by a variety of different methods: ICRH, ECH, Ohmic heating, ECH with lower hybrid heating and balanced neutral beam injection. A common feature of these observations is that the rotation velocity increases with the plasma pressure or stored energy. A simple dimensionless scaling, with the Alfven Mach number proportional to the normalized pressure ($M_A = \beta_N/40\pi$), unifies the results. No dependence is found with normalized gyro-radius or collisionality. A more rigorous regression yields $M_A = 0.65\ \beta_T^{1.4}\ q^{*2.3}$ for dimensionless parameters and $\Delta V = C\ B^{1.1}\ \Delta\langle P\rangle^{1.0}\ I_p^{-1.9}\ R^{2.2}$ for machine variables. All of these scalings predict rotation velocities in ITER high enough for RWM suppression without external momentum input. At present there is no comprehensive, quantitative explanation for spontaneous rotation. Some recent observations of poloidal rotation are inconsistent with the predictions of neo-classical theory.

Momentum transport has been found to be closely coupled to energy transport, and with momentum diffusivities much higher than neo-classical values. These results are consistent with diffusive momentum transport driven by ITG turbulence. Evidence for an inward momentum pinch has also been observed, which could be due to the Coriolis drift effect or from symmetry breaking due to magnetic field curvature

Acknowledgments

Many thanks to J. deGrassie, P. deVries, A. Ince-Cushman and M. Yoshida. Work supported at C-Mod by DoE Contract No. DE-FC02-99ER54512.

References

[1] ASDEX Team 1989 *Nucl. Fusion* **29** 1959
[2] Shaing K C and Crume E C 1989 *Phys. Rev. Lett.* **63** 2369
[3] Biglari H *et al.* 1990 *Phys. Fluids* **B2** 1
[4] Groebner R J *et al.* 1990 *Phys. Rev. Lett.* **64** 3015
[5] Ida K *et al.* 1990 *Phys. Rev. Lett.* **65** 1364
[6] Terry P W 2000 *Rev. Mod. Phys.* **72** 109
[7] Hahm T S 1994 *Phys. Plasmas* **1** 2940
[8] Burrell K 1997 *Phys. Plasmas* **4** 1499
[9] Synakowski E J 1998 *Plasma Phys. Contr. Fusion* **40** 581
[10] Wolf R C 2003 *Plasma Phys. Contr. Fusion* **45** R1
[11] Connor J W *et al.* 2004 *Nucl. Fusion* **44** R1
[12] Strait E J *et al.* 1994 *Phys. Rev. Lett.* **74** 2483
[13] Zheng L J , Kotschenreuther M and Chu M S 2005 *Phys. Rev. Lett.* **95** 255003
[14] Polevoi A R *et al.* 2002 in *Fusion Energy (Proc. 19th Int. Conf. Lyon, 2002) (Vienna: IAEA)*
[15] Liu Yueqiang *et al.* 2004 *Nucl. Fusion* **44** 232
[16] Rice J E *et al.* 2007 *Nucl. Fusion* **47** 1618
[17] Rice J E *et al.* 1997 *Nucl. Fusion* **37** 421
[18] Rice J E *et al.* 2004 *Nucl. Fusion* **44** 379
[19] LaBombard B *et al.* 2004 *Nucl. Fusion* 44 1047
[20] Scarabosio A *et al.* 2006 *Plasma Phys. Contr. Fusion* **48** 663
[21] Bortolon A *et al.* 2006 *Phys. Rev. Lett.* **97** 235003
[22] Coffey I H *et al.* 1996 *Proc. 11th Colloquium on UV and X-ray Spectroscopy of Astrophysical and Laboratory Plasmas, Nagoya, Japan, 1995, p.431, Frontiers Science Series No.15 (Editors: K. Yamashita and T. Watanabe), Universal Academy Press, Tokyo, Japan, 1996*
[23] Eriksson L G *et al.* 1997 *Plasma Phys. Contr. Fusion* **39** 27
[24] Rice J E *et al.* 1998 *Nucl. Fusion* **38** 75
[25] Rice J E *et al.* 1999 *Nucl. Fusion* **39** 1175
[26] Hoang G T *et al.* 2000 *Nucl. Fusion* **40** 913
[27] Hutchinson I H *et al.* 2000 *Phys. Rev. Lett.* **84** 3330
[28] Rice J E *et al.* 2000 *Phys. Plasmas* **7** 1825
[29] Eriksson L G *et al.* 2001 *Nucl. Fusion* **41** 91
[30] Rice J E *et al.* 2001 *Nucl. Fusion* **41** 277
[31] Lee W D *et al.* 2003 *Phys. Rev. Lett.* **91** 205003
[32] Assas S *et al.* 2003 'Toroidal plasma rotation in ICRF heated Tore Supra discharges', 30th European Physical Society Conference on Plasma Physics and Controlled Fusion, St. Petersburg, Russia, 7-11 July 2003, ECA Vol. **27A** P-1.138
[33] Noterdaeme J M *et al.* 2003 *Nucl. Fusion* **43** 274
[34] Rice J E *et al.* 2004 *Phys. Plasmas* **11** 2427
[35] deGrassie J S *et al.* 2004 *Phys. Plasmas* **11** 4323
[36] deGrassie J S *et al.* 2007 *Phys. Plasmas* **14** 056115
[37] Sakamoto Y *et al.* 2006 *Plasma Phys. Contr. Fusion* **48** A63
[38] Rice J E *et al.* 2007 *Fusion Sci. Tech.* **51** 288
[39] Greenwald M *et al.* 1997 *Nucl. Fusion* **37** 793
[40] Porte L *et al.* 2007 *Nucl. Fusion* **47** 952
[41] Hirshman S P and Sigmar D J 1981 *Nucl. Fusion* **21** 1079
[42] Kim, Y B *et al.* 1991 *Phys. Fluids* **B3** 2050
[43] Ida K 1998 *Plasma Phys. Contr. Fusion* **40** 1429
[44] Rogister A L 1999 *Phys. Plasmas* **6** 200
[45] Claasen H A *et al.* 2000 *Phys. Plasmas* **7** 3699
[46] Rogister A L *et al.* 2002 *Nucl. Fusion* **42** 1144

[47] Hinton F L and Wong S K 1985 *Phys. Fluids* **28** 3082
[48] Chang C S *et al.* 1999 *Phys. Plasmas* **6** 1969
[49] Perkins F W *et al.* 2001 *Phys. Plasmas* **8** 2181
[50] Chan V S *et al.* 2002 *Phys. Plasmas* **9** 501
[51] Eriksson L G and Porcelli F 2002 *Nucl. Fusion* **42** 959
[52] Zheng L J *et al.* 2006 *Fusion Energy (Proc. 21st Int. Conf. Cheng Du, 2006) (Vienna: IAEA) TH/P3-14*
[53] Diamond P H *et al.* 1996 in *Plasma Physics and Controlled Nuclear Fusion Research 1994 (Proc. 15th Int. Conf. Seville, 1994), Vol. 3, p. 323, IAEA, Vienna (1996)*
[54] Shaing K C 2001 *Phys. Rev. Lett.* **86** 640
[55] Coppi B 2002 *Nucl. Fusion* **42** 1
[56] Gurcan O D *et al.* 2007 *Phys. Plasmas* **14** 042306
[57] Myra J R *et al.* 2006 *'Blobs, Momentum Transport and Tokamak Rotation', Proc. 19th US TTF Meeting, Myrtle Beach, S.C., Apr. 2006*
[58] Crombe K *et al.* 2005 *Phys. Rev. Lett.* **95** 155003
[59] Solomon W M *et al.* 2006 *Phys. Plasmas* **13** 056116
[60] Staebler G M 2004 *Phys. Plasmas* **11** 1064
[61] Tala T *et al.* 2007 *Nucl. Fusion* **47** 1012
[62] Suckewer S *et al.* 1981 *Nucl. Fusion* **21** 1301
[63] Burrell K H *et al.* 1988 *Nucl. Fusion* **28** 3
[64] Scott S D *et al.* 1990 *Phys. Rev. Lett* **64** 531
[65] Kallenbach A *et al.* 1991 *Plasma Phys. Contr. Fusion* **33** 595
[66] Asakura N *et al.* 1993 *Nucl. Fusion* **33** 1165
[67] Nagashima K *et al.* 1994 *Nucl. Fusion* **34** 449
[68] Ida K *et al.* 1995 *Phys. Rev Lett.* **74** 1990
[69] Zastrow K-D *et al.* 1998 *Nucl. Fusion* **38** 257
[70] Gohil P *et al.* 1998 *Nucl. Fusion* **38** 432
[71] deGrassie J S *et al.* 2003 *Nucl. Fusion* **43** 142
[72] Nishijima D *et al.* 2005 *Plasma Phys. Contr. Fusion* **47** 89
[73] deVries P C *et al.* 2006 *Plasma Phys. Contr. Fusion* **48** 1693
[74] Yoshida M *et al.* 2007 *Nucl. Fusion* **47** 856
[75] Connor J W *et al.* 1987 *Plasma Phys. Control. Fusion* **29** 919
[76] Stacey W M and Sigmar D J 1985 *Phys. Fluids* **28** 2800
[77] Sugama H and Horton W, 1997 *Phys. Plasmas* **4** 2215
[78] Mattor N and Diamond P H 1988 *Phys. Fluids* **31** 1180
[79] Staebler G M and Dominguez R R 1993 *Nucl. Fusion* **33** 77
[80] Sugama H and Horton W 1998 *Phys. Plasmas* **5** 2560
[81] Shaing K C 2001 *Phys. Plasmas* **8** 193
[82] Candy J and Waltz R E 2003 *J. Comp. Phys.* **186** 545
[83] Waltz R E *et al.* 1997 *Phys. Plasmas* **4** 2482
[84] Peeters A G *et al.* 2006 *Plasma Phys. Control. Fusion* **48** B413
[85] Peeters A G *et al.* 2007 *Phys. Rev. Lett.* **98** 265003
[86] Hahm T S *et al.* 2007 *Phys. Plasmas* **14** 072302

Journal of Physics: Conference Series **123** (2008) 012004 doi:10.1088/1742-6596/123/1/012004

Transport barriers in non-axisymmetric magnetic fields

K Ida

National Institute for Fusion Sciences, Toki, Gifu 509-5292, Japan

E-mail : ida@nifs.ac.jp

Abstract. A transport barrier which is characterized by the steeper temperature or/and density gradient has been observed both in the core region (internal transport barrier : ITB) and edge region (edge transport barrier : ETB) in the plasmas with non-axisymmetric magnetic fields compared with the gyro-Bohm type L-mode transport. A reduction of thermal conductivity and particle diffusivity in these plasmas compared with the L-mode levels is observed and the parameter dependence of thermal conductivity and particle diffusivity changes significantly from that in L-mode. A negative or at least a weak dependence of the thermal diffusivity on temperature is observed in the heat transport barrier, while a strong temperature dependence (temperature to the power of 1.5) is normally observed in the L-mode region. In this paper, transport barriers that have been observed in non-axisymmetric magnetic field configuration are discussed from the view point of the parameter dependence of transport coefficients, not the shape of temperature or density profiles. The transition between the L-mode and the improved mode with a transport barrier is also discussed based on the idea of a bifurcation of multi-state transport (discrete transport states with different relations of heat flux to temperature gradient that are self-consistent with the different turbulence states realized).

1. Introduction

The physics mechanism of transport barriers in confinement systems with non-axisymmetric magnetic fields has been studied since the high confinement mode (H-mode) and internal transport barrier (ITB) were experimentally observed in stellarators and heliotron devices. In the H-mode, steep pressure gradients and $E \times B$ flows localized near the plasma edge were observed in Wendelstein-7AS[1, 2, 3], CHS[4, 5, 6], LHD[7], Heliotron-J[8], TJ-II stellarator[9], H-1 heliac[10] and Tohoku-heliac[11, 12, 13]. The H-modes in these devices have been found to be sensitive to the edge rotation transform (iota). Clear edge iota windows for the H-mode transition were observed in Wendelstein 7-AS and Heliotron-J, which has not been observed in the H-mode in a tokamak. Although there are some differences in the trigger mechanism of the H-mode, the role of the $E \times B$ shear flow on the suppression of turbulence and the reduction of transport is similar to that in a tokamak H-mode. The details of the characteristics of an H-mode in the confinement system with a non-axisymmetric magnetic field has been discussed[14].

In non-axisymmetric confinement devices, both the electron and ion ripple losses are sensitive to the radial electric field and electron and ion radial fluxes are functions of the radial electric field in the plasma. Since the electron radial flux is equal to the ion radial flux in the steady state condition, the radial electric field which gives identical ion and electron radial flux is called a "root" of ambipolar condition ($\Gamma_i = \Gamma_e$). The radial electric field with positive (negative) value is called electron (ion) root, because the electron (ion) flux exceeds the ion (electron) flux at zero radial electric field. Because the magnitude of the radial electric field is much larger in the electron root, the reduction of neoclassical transport due to the radial electric field is expected in the electron root rather than in the ion root. Therefore, the transition of the radial electric field is triggered by the neo-classical non-ambipolar ion and electron fluxes, and the radial electric field becomes positive (in the electron root) when the plasma collisionality is low enough, while it is negative (in the ion root) at higher collisionality[15, 16]. Associated with a transition of the radial electric field from negative (ion root) to large and positive (electron root), the reduction of the electron thermal diffusivity is observed in CHS[17], Wendelstein-7AS[18] and LHD plasmas[19, 20, 21, 22]. The coupling between the radial electric field shear and the turbulence levels has been investigated to understand the anoumalous transport in the TJ-II stellarator [23].

Recently zonal flows have been identified experimentally using a heavy ion beam probe (HIBP) and probes and the coupling between turbulence and zonal flows has been investigated in CHS[24, 25] and in the H-1 Heliac[26, 27] as well as in tokamaks[28, 29, 30, 31, 32]. The physics mechanism for the formation of edge and internal transport barriers has been studied. In this paper, the physics mechanism to determine the mean $E \times B$ flow and its shear, zonal flows and spontaneous toroidal flow is described. The transport characteristics, especially the dependence of the heat flux on the temperature and temperature gradient, are discussed in comparison with that in tokamaks. The electron internal transport barrier associated with the transition of the mean $E \times B$ flow and other transport improvements (ion transport and particle transport) are also described.

2. Physical mechanism determining spontaneous, mean, zonal and GAM flows

Transport of the plasma in the optimized non-axisymmetric magnetic field system is governed by the turbulence as it is in the plasma in a tokamak, because the transport due to collisional processes (neoclassical transport) has been reduced to a level below

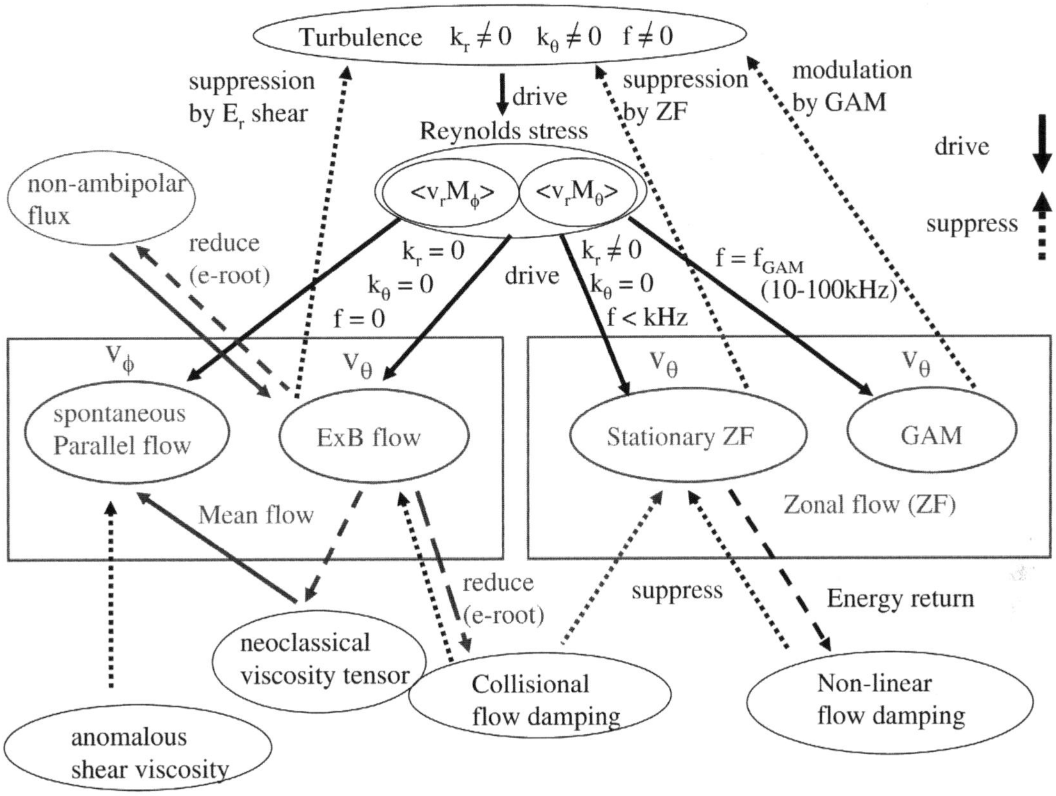

Figure 1. Physical mechanism determining spontaneous toroidal flow, mean poloidal flow, stationary zonal flow and geodesic acoustic modes (GAMs) in non-axisymmetric magnetic field systems.

the transport due to turbulence (anomalous transport). Figure 1 shows the physical mechanism determining the spontaneous toroidal flow, the mean poloidal flow, the stationary zonal flow and geodesic acoustic modes (GAMs) in a non-axisymmetric magnetic field system. Turbulence in the plasma produces a mean flow in the direction of $E \times B$ (nearly the poloidal direction) and in the direction parallel to the magnetic field (nearly the toroidal direction) through the Reynolds stress. These mean flows tend to be localized at the region of the transport barrier, where the temperature an/or density gradient are significantly large. The mean flow in the direction parallel to the magnetic field is called the spontaneous toroidal rotation, while the mean flow in the direction of $E \times B$ is called the poloidal spin-up.

These mean flows are damped by the parallel and perpendicular viscosities [33]. The parallel viscosity is a damping mechanism through collisions between trapped particles (banana particles in a tokamak, ripple trapped particles in a non-axisymmetric magnetic system) and passing particles. This parallel viscosity is the dominant damping process in the poloildal direction. The perpendicular viscosity is the damping mechanism through momentum exchange due to turbulence in the region where a velocity shear exists. Therefore this viscosity is called shear viscosity or momentum diffusivity in momentum transport analysis. This perpendicular viscosity is a dominant damping process in the toroidal direction in a tokamak. In non-axisymmetric magnetic systems both parallel and perpendicular viscosities are important for determining the toroidal rotation.

Non-ambipolar radial flux and the neoclassical viscosity tensor are additional mechanisms driving these mean flows. Non-ambipolar radial flux produces a radial electric field and sustains the $E \times B$ mean flow. The non-ambipolar radial flux due to ion and electron ripple loss is a strong driving term of the $E \times B$ flow especially in non-axisymmetric magnetic field systems. In general (except for at the magnetic axis) the magnetic field is oblique to the symmetry direction (toroidal in tokamak and helical in non-axisymmetric system) because of the rotational transform of magnetic field. Because the plasma flow along the symmetry direction has minimum viscosity, the plasma tends to flow along the direction of minimum viscosity (symmetry direction). The neoclassical viscosity tensor (called poloidal viscosity in tokamak) acts as the forces to change the direction of plasma flow towards the symmetry direction. The neoclassical viscosity tensor is significantly large and sometimes larger than the shear viscosity and therefore the mean flow parallel to the magnetic field can be driven by the $E \times B$ flow. Because of the existence of these additional mechanisms driving these mean flows, the generation of the mean flows does not require the existence of turbulence.

On the other hand, zonal flows ($E \times B$ flows with finite frequency and radial wave number) are always generated by turbulence and these flows should vanish if the turbulence completely disappears. Since the zonal flows have no poloidal wave number these flow suppress the turbulence and reduce the radial transport by the $E \times B$ shear effect. The zonal flows have finite frequency and radial wave number and there is no net poloidal momentum when the momentum of the zonal flow is averaged in time and space. This is in contrast to the mean flows which have finite momentum by exhausting the momentum out of the plasma through the plasma periphery. The magnitude of the zonal flows are determined by the driving term due to turbulence and damping due to collisional flow damping. The other damping mechanism is non-linear flow damping which suppresses the growth of the zonal flow by consuming the free energy used to produce a zonal flow. In a non-axisymmetric system, bifurcation of the radial electric field occurs because the E_r dependence of the ion radial flux is different than that of the electron radial flux. As the collisionality is decreased below the critical value, the radial electric field of the plasma changes from small negative (ion-root) to large positive (electron-root) and the radial flux also decreases significantly. The collisional flow damping is sensitive to the radial electric field and the mean $E \times B$ flow in a non-axisymmetric system. The collisional flow damping becomes small in the electron root plasma where the large E_r reduces ripple loss and collional damping. Therefore a larger zonal flow is expected in the electron root in a non-axisymmetric system.

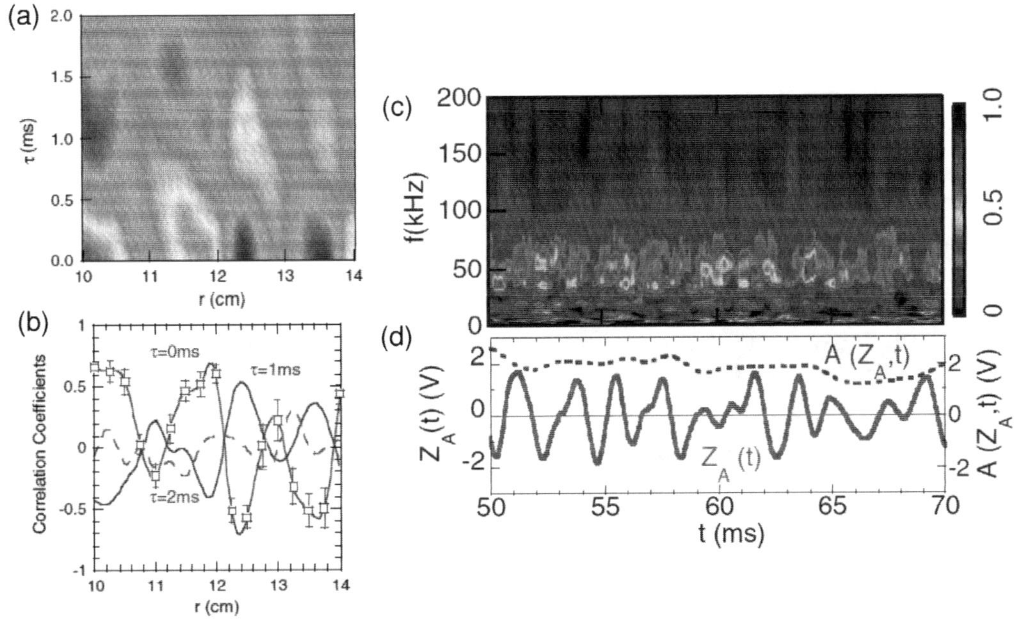

Figure 2. (a) Spatio-temporal structure of stationary zonal flows evaluated from the correlation function in CHS. (b) Spatial correlation with the time lag of $\tau = 0$, 1 and 2 ms.[34] (c) Evolutions of wavelet spectrum of electric field fluctuations as a function of frequency, $W(E_A; f, t)$. The unit of the color bar is V^2/kHz. (d) Evolution of the zonal flow, $Z_A(t)$. The waveform of the zonal flow is the electric field numerically filtered in the frequency range around ~ 0.5 kHz. The blue dashed line indicates the evolution of the zonal flow amplitude using wavelet analysis, $A(Z_A; t)$ [35].

3. Coupling between fluctuations and stationary zonal flows and GAMs

Identification of zonal flows in the high temperature core was accomplished in CHS using twin HIBPs located at two different toroidal positions separated by 90 degrees [24]. This zonal flow has a toroidal symmetry of n = 0 with finite radial wavenumbers and has a broad frequency spectrum from $\sim$ 0.5 to $\sim$ 1 kHz which is in contrast to the zonal flow with a sharp frequency peak at the GAM frequency (10 - 20 kHz) described later. This zonal flow is called stationary zonal flow, however, the flow pattern changes within ms and is not in a steady state because of the coupling between the zonal flow and turbulence. The change in the flow pattern of the stationary zonal flow is clearly demonstrated in the spatio-temporal pattern of the radial correlation function between the electric field at r = 12 cm and at another location varying from 10 to 14 cm. This observation shows that zonal flows with a finite radial wavelength of $\sim$ 1 cm do exist, and the structure of a zonal flow changes completely in $\sim$ 2 ms.

The linkage between a zonal flow and turbulence is examined with wavelet analysis to extract the relation between the intermittent characteristics of the turbulence and the zonal flow phase (or direction). Figure 2(c) shows examples of the evolution of the wavelet power spectrum of electric field fluctuations, $W(E_A; f, t)$. The wavelet analysis

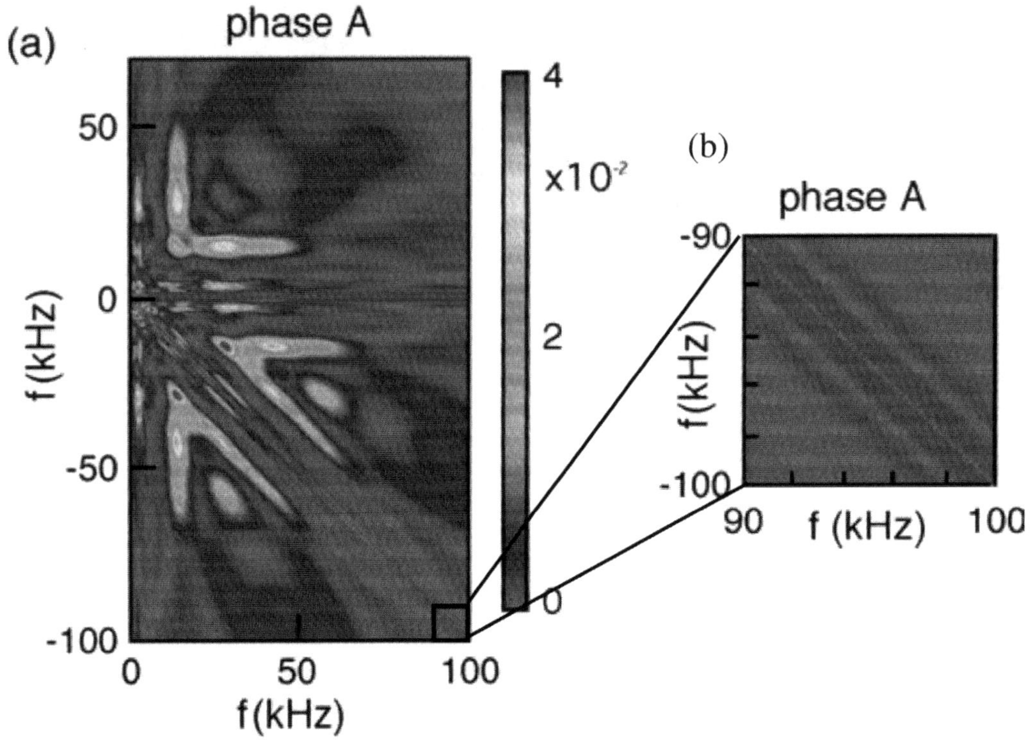

Figure 3. (a) Diagrams of wavelet squared bicoherence near the positive peak (maxima) of the zonal flow and (b) an expanded view of the diagram[25].

successfully reveals the intermittent nature of turbulence manifested in the electric field fluctuations. Figure 2(d) shows the zonal flow component filtered out from the electric field fluctuation. Here the positive sign means that the resultant $E \times B$ -drift is in the ion diamagnetic direction to which the bulk rotation orients itself.

It has been confirmed qualitatively that the turbulence power is modulated by the phase of the zonal flow and that the fluctuation spectra clearly differ from each other at the maxima and minima of the zonal flow $Z_A(t)$ The electric field fluctuations measured with an HIBP have three characteristic frequency ranges for the fluctuation spectrum: zonal flows in the range of less than $\sim$ 1.5 kHz, the coherent modes at 19 and 38 kHz which are conjectured as GAMs and broad band turbulence fluctuations in the range around 50 kHz. Wavelet bicoherence analysis is used to evaluate the dependence of the coupling constant on the phase of the zonal flow with the help of a conditional average. As seen in figure 3, a clear couplings between fluctuations and a zonal flow (f $<$ 1.5kHz) has been found near the positive peak (maxima) of the zonal flow. The strong couplings appear along the lines of $f_1 + f_2 \pm 0.5$kHz, which is clearer in the expanded view in the frequency range (90 - 100 kHz) of the diagram. Couplings between fluctuations and a GAM zonal flow ($\sim$ 19kHz) are also observed in the medium frequency range

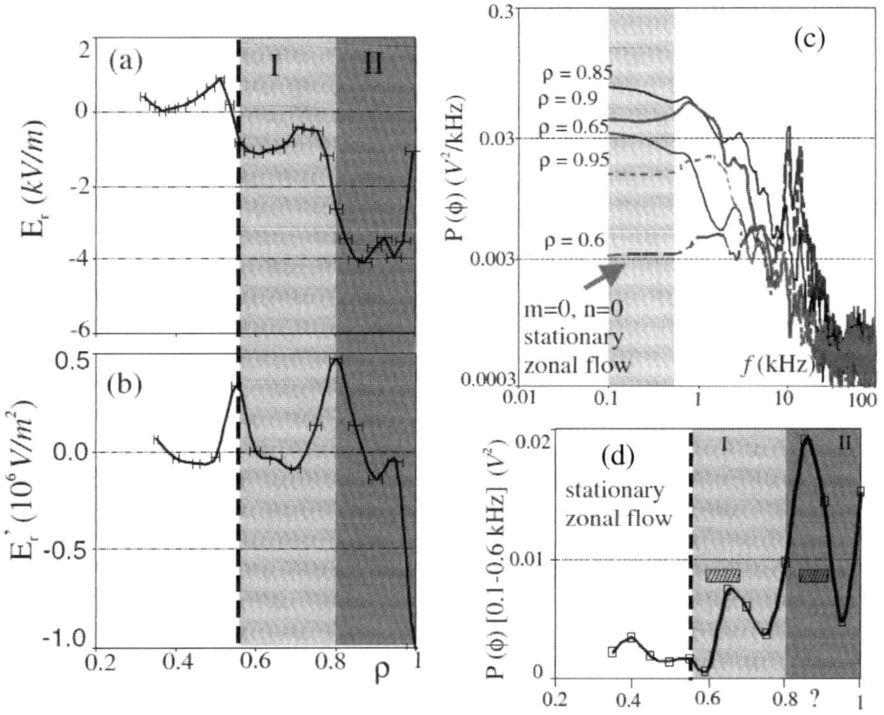

Figure 4. Radial profiles of (a) radial electric field and (b) shear in the radial electric field, (c) power spectra of the plasma potential in different radial regions, (d) radial profile of the spectral power density of stationary zonal flows (0.1-0.6kHz) of H-mode plasmas in the H-1 heliac. Hatched boxes indicate radial positions of the E_r maxima in regions I and II[36].

(30 - 60kHz) of the turbulence frequency. These observations clearly demonstrate the coupling between fluctuations and stationary zonal flows and between fluctuations and GAMs.

4. Relation between mean $E \times B$ flow and zonal flow

It is well known that the $E \times B$ flow shear suppresses turbulence in the plasma. This E_r suppression effect causes strong coupling between the turbulence and the mean $E \times B$ flow shear. Because the turbulence in the plasma couples with both the stationary zonal flow and mean $E \times B$ flow shear, the stationary zonal flow is expected to be related to the $E \times B$ flow through turbulence in the plasma. In the H-1 heliac, the contribution of a zonal flow to the spatial modulation of the radial electric field, E_r, profile was investigated. Figure 4 shows the radial profile of the radial electric field, its shear and spectral power density of stationary zonal flows measured with Langmuir probes. As seen in figure 4(c), plasma potential fluctuations with low frequency (0.1-0.6kHz) and with zero toroidal (n) and poloidal (m) mode numbers are dominant in the H-mode and localized near the plasma edge. There are three E_r regions observed: a slightly positive E_r inside the top of the transport barrier, a negative E_r region (I) and a more negative E_r region (II). The zonal flow maximum spatially coincides with the maximum in negative E_r, where the E_r shear becomes small. In general, the E_r shear tends to reduce the intensity of the zonal flow by the reduction of turbulence (driving terms) due to the shear suppression effect, while the E_r itself may enhance the intensity of the zonal flow by reducing the collisional flow damping (damping term). Therefore both the small E_r shear at the peak of E_r and the large E_r contribute to increase the zonal flow intensity as seen in figure 4. However, from these experiments it is not clear whether the E_r or E_r shear contribute to the enhancement of the stationary zonal flow observed.

Figure 5(a) shows the intensity of the zonal flow, $|E_{ZF}|^2$, and the normalized fluctuation amplitude, $|eE_r/\nabla T_0|^2$ during the period of the electron internal transport barrier (e-ITB) and in the case without the ITB. A large intensity of the zonal flow and a small fluctuation amplitude are observed in the plasma with an electron internal transport barrier, where large E_r and large E_r shear exists inside ($\rho < 0.25$) and at the boundary ($\rho \sim 0.3$) of the ITB, respectively. This is in contrast to the small intensity of the zonal flow and the large fluctuation amplitudes which are observed in the plasma without an ITB. This experimental result clearly demonstrates the role of E_r in collisional damping rather than the role of E_r shear in turbulence suppression. The data in figure 5(a) are measured at $\rho = 0.2$ where the E_r shear is too weak (see the radial profile of the plasma potential) to suppress turbulence even in the ITB. This experimental result clearly demonstrates that the energy of the fluctuations is preferentially transferred into zonal flows when the collisional flow damping becomes weak in the electron ITB.

In order to study the effects of E_r and E_r shear on transport, the radial profiles of thermal diffusivity are calculated with a one-dimensional transport code based on the nonlinear current-diffusive interchange mode using typical T_e, T_i, n_e profiles. In this code, the E_r effect on the collisional damping rate and E_r shear effect on the reduction of thermal diffusivity are included. The radial profiles of E_r in figure 5(b) are mainly determined by neoclassical non-ambipolar fluxes. The reduction of turbulent transport in the entire region of strong positive E_r (not just the region of strong E_r shear) is quantitatively demonstrated by use of transport code analysis. The transport coefficient by bare fluctuations (without Zonal flow effects) does not show a noticeable reduction for $r < r_{inv}$, where r_{inv} is the minor radius at the boundary between electron root (large positive E_r) and ion root (small negative E_-) owing to the increased temperature. By including the screening by Zonal flow thermal diffusivity is predicted to be quenched in the core of plasmas with electron ITB.

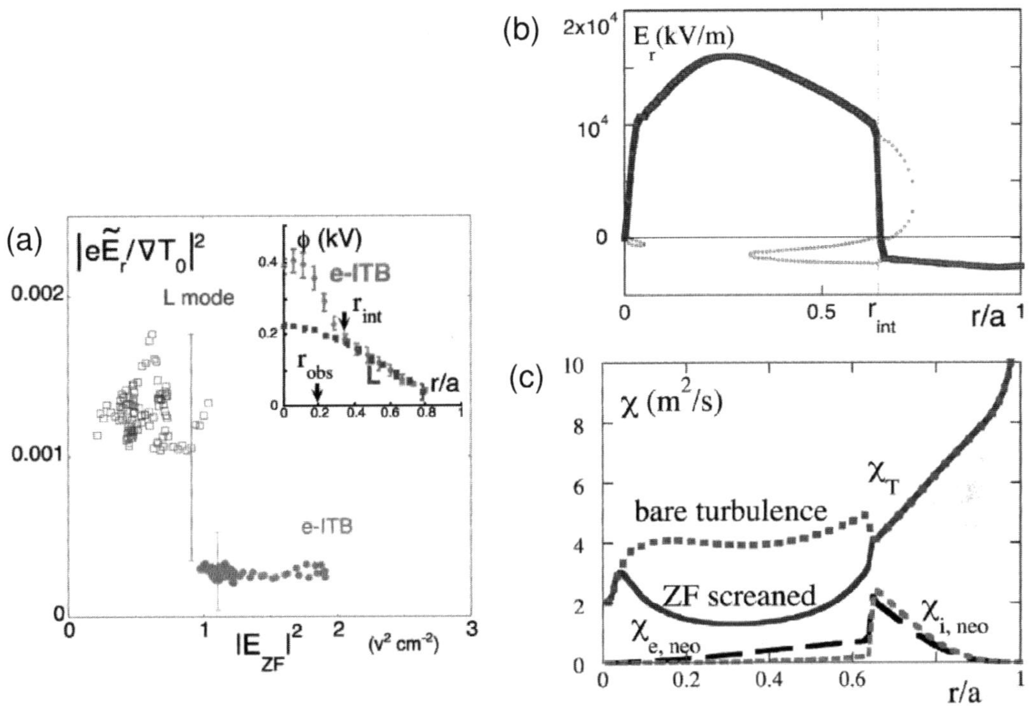

Figure 5. (a) Simultaneously observed zonal flow and microturbulence intensities during the electron internal transport barrier (e-ITB) phase (solid circle) and the L-phase (open circle) in CHS. Vertical and horizontal axes indicate fluctuation amplitude normalized by temperature gradients and the magnitude of the zonal flow, respectively. Transport code analysis of LHD plasma with an electron ITB. Radial electric field profile (solid line) is shown with solutions of local charge neutrality condition (dots) in (b). Profiles of neoclassical thermal conductivity and turbulent conductivity (c). Predictions based on the bare microturbulence (dotted line) and the one with screening by zonal flows (ZFs : solid line) are shown[37].

5. Temperature and temperature gradient dependence of transport in axisymmetric and non-axisymmetric systems

Figure 6 shows the radial profiles of electron temperature in the plasmas with an electron ITB for the ECH power scan and the radial profiles of thermal diffusivity at the highest ECH powers in JT-60U tokamak and the LHD heliotron. The electron temperature profile becomes peaked after the formation of the electron ITB and the central electron temperature strongly depends on the ECH power. As seen in figure 6(a)(b), the electron temperature profile observed in the electron ITB in LHD is peaked at the plasma axis, while the electron temperature profile observed in JT-60U has a flattened region near the

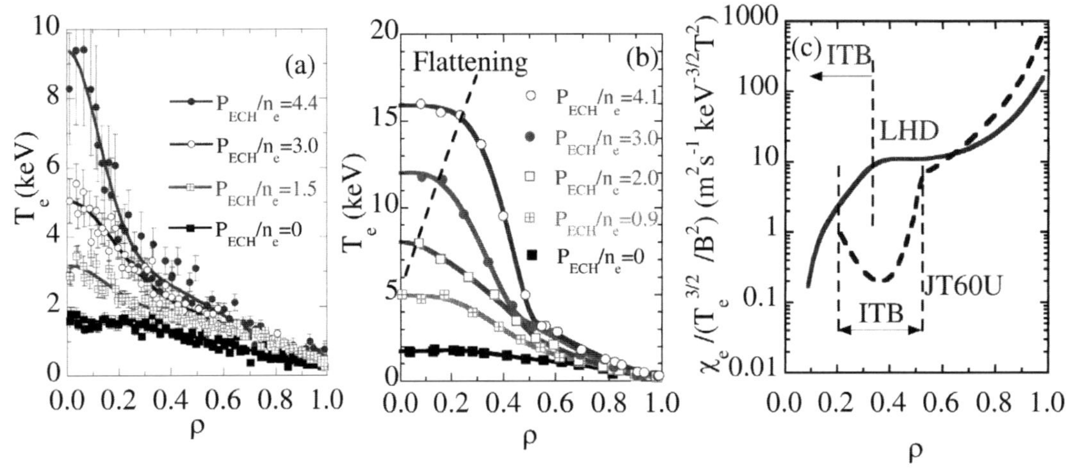

Figure 6. Radial profiles of electron temperature in (a) LHD and (b) JT60U plasmas for various heating power of ECH, P_{ECH}/n_e. (c) electron thermal diffusivity normalized by $T_e^{3/2}/B^2$[38].

plasma center. Although the mechanism causing the flattening is not well understood, the flattening shows that the transport at the plasma core is worse than the L-mode level and the transport barrier is localized in the narrow region.

In order to compare the radial structure of heat transport between tokamak and heliotron plasmas with an electron ITB, the electron thermal diffusivity is normalized by the gyro-Bohm scaling of $T_e^{3/2}/B^2$ as shown in figure 6(c). The normalized electron heat conductivity in the outer plasma is nearly identical for JT-60U and LHD. The normalized thermal diffusivity decreased towards the plasma center and reaches low levels close to 0.1 $m^2s^{-1}/(keV^{3/2}T^{-2})$ in the LHD electron ITB. This is in contrast to the radial profile of the normalized electron thermal diffusivity observed in JT60U, where the minimum normalized thermal diffusivity is located at one-third of plasma minor radius ($r/a = 0.35$). It is interesting that the reduction of thermal diffusivity in a tokamak is localized in the narrow region with the $E \times B$ shear, where the reduction of thermal diffusivty in a heliotron extends to the plasma core where E_r shear is small but E_r is large enough. These observation are consistent to the predictions of transport modeling including the zonal flow screening effect as seen in figure 5 The difference in the radial profile of the normalized electron thermal diffusivity between LHD and JT-60U could be explained by the role of E_r on the collisional flow damping of the zonal flow which has a strong contribution towards determining the thermal diffusivity.

The temperature dependence of the thermal diffusivity $\alpha = (T_e/\chi_e)(d\chi_e/dT_e)$ is an important parameter for studying the plasma with an ITB. In the L-mode, the parameter α is positive and typically it is 1.5, which is predicted from the gyro-reduced

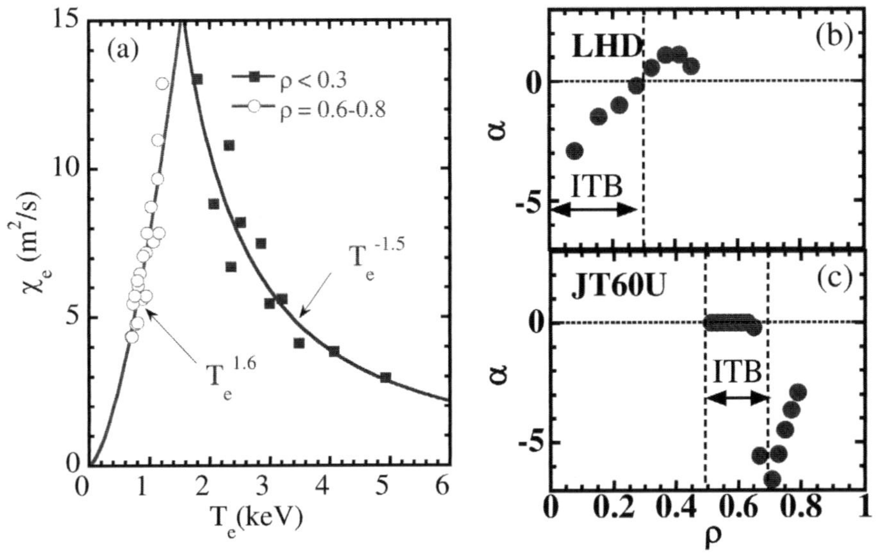

Figure 7. (a) Electron thermal diffusivity as a function of electron temperature inside the ITB region in LHD and radial profile of the temperature dependence derived from the cold pulse propagation experiments in (b)LHD and (c)JT-60U[22, 39].

Bohm scaling and is also consistent with the power degradation of the global energy confinement in LHD of $\tau_E \propto (P/n)^{-0.6}$ [40]. If the parameter α stays positive, the formation of an ITB would never occur, because a spontaneous increase of electron temperature during the formation of an ITB requires a negative α. Therefore α would be the most reasonable parameter to confirm whether the plasma is in the L-mode regime or the ITB regime, especially near the boundary between L-mode and ITB mode in space and in time, and therefore negative α can be a definition of an ITB plasma. The temperature dependence is investigated by power balance analysis and cold pulse experiments.

As shown in figure 7, the temperature dependence parameter of the electron thermal diffusivity in the core region becomes negative (-1.5) associated with the formation of an ITB, while the temperature dependence parameter outside the ITB region remains positive, $\alpha = 1.6$, which is consistent with the temperature dependence ($\alpha = 1.5$) of the gyro-reduced Bohm transport. Pulse propagation analysis of the cold pulse experiment also shows the negative α inside the ITB and a positive α outside the ITB in LHD as shown in figure 7(b). The change of sign of the temperature dependence of the

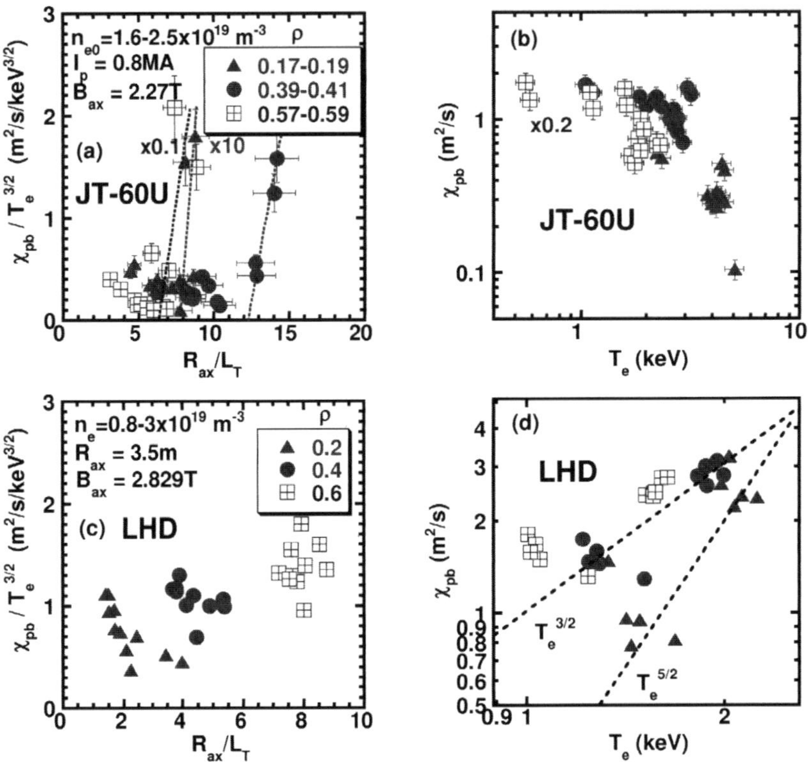

Figure 8. (a) ∇T_e and (b) T_e dependence of χ_e at different normalized radii in the JT-60U NBI plasmas and (c)(d)those in LHD NBI plasmas. The χ_e is normalized by the gyro-Bohm T_e dependence and ∇T_e is normalized by (R/T_e), where R is the major radius and $L_T = \nabla T_e/T_e$ in (a) and (c)[39].

electron thermal diffusivity is a key to the formation of the ITB in LHD. In contrast, as seen in figure 7(c), the radial profile of the temperature dependence factor, α, in a tokamak is quite different from that observed in LHD. The temperature dependence is zero inside the ITB and negative outside the ITB in JT-60U. This data imply that the temperature dependence alone can not explain the formation of an ITB in a tokamak. In order to cause the spontaneous transition to the ITB plasma, the temperature gradient dependence should play a role inside the ITB in the JT-60U tokamak, because the transport reduction of $\alpha = 0$ is not enough to cause the transition as described above.

Figure 8 shows the ∇T_e and T_e dependence of χ_e (obtained from the static analysis i.e. the power balance analysis) in JT-60U and LHD. In figure 8(a) and (c), the thermal diffusivity is normalized by $T_e^{3/2}$, which is predicted by Gyro-Bohm transport, in order to eliminate the T_e dependence of χ_e and make the ∇T_e dependence clear. The ∇T_e is normalized by R/T_e, Because the ratio of major radius to scale length of temperature is expected to give the threshold of turbulence in the critical temperature gradient transport model.

As seen in figure 8(a)(c), there is no clear critical gradient scale length observed in LHD, although an existence of the critical gradient scale length is suggested in JT-60U. In contrast, A clear T_e dependence of χ_e is observed in LHD, while there is no clear T_e dependence of χ_e in JT-60U. The dependence of heat transport on T_e and ∇T_e changes significantly depending on the scale length of the temperature in JT-60U. The thermal diffusivity χ_e shows a Gyro-Bohm type T_e dependence when the temperature gradient is small enough. When the temperature gradient exceeds a critical value, it depends mostly on the scale length in a tokamak. This implies that there is a strong nonlinear mechanism in transport. One of the candidates for the mechanism causing strong non-linearity is non-linear flow damping of the zonal flow in a tokamak plasmas. In the LHD heliotron, the thermal diffusivity χ_e shows a clear T_e dependence and it suggests that the nonlinear mechanism in transport is weaker. The weak non-linearity might be due to the stronger collisional flow damping of the zonal flow than that in tokamaks. More study on driving and damping mechanisms of zonal flows is required to clarify the mechanism causing the differences in the T_e and ∇T_e dependence of heat transport between heliotrons and tokamaks.

In LHD heliotrons, detailed experiment to investigate the T_e and ∇T_e dependence of heat transport was performed using dynamic transport analysis of the discharges with repetitive pellet injection. After the pellet injection there are clear phases characterized by the sign of the time derivative of the T_e gradients as indicated by phases I, II, III, IV. Both T_e and the T_e gradient increase in time as the electron density decreases in phases II and IV, which is a normal characteristic of heat transport in plasma. However, in phases I and III, the T_e gradients decrease even with the increase of T_e. Figure 9(a) shows the normalized heat as a function of T_e and the T_e gradient in the weak T_e dependence branch (phase II) and in the strong T_e dependence branch (phase IV) after the pellets (12 pellets in series) are injected into the plasmas with different densities. The experimental data in phase II shows that the normalized heat flux gradually increases as the T_e and T_e gradient are increased, while that in phase IV shows a sharp increase of the normalized heat flux. It is noted that all the experimental data points of the 12 events are connected and located along curves (data points in phase II are on a weak T_e dependence curve and data points in phase IV on a strong T_e dependence curve). The data points in the transition phase (Phase III) are scattered between the weak and the strong T_e dependence curves. When the plasma is on one of the branches, both T_e and the T_e gradient are uniquely determined for the given normalized heat flux, while they are not uniquely determined during the transition phase.

In order to investigate quantitatively the dependence of the normalized heat flux on the T_e and T_e gradient dependence, the normalized heat flux is given by $Q_e/n_e \propto T_e^{\alpha}(-\nabla T_e)^{\beta}$, where α and β are the T_e and T_e gradient dependence parameters, respectively. The difference between the Q_e/n_e measured and Q_e/n_e calculated with the parameters (α, β) are investigated in a wide range of Q_e/n_e for these two branches. Figure 9(b) and 9(c) show the contours of the square of differences between the normalized heat flux measured and that calculated, $\chi^2 = [Q_e/n_e - cT_e^{\alpha}(-\nabla T_e)^{\beta}]^2$,

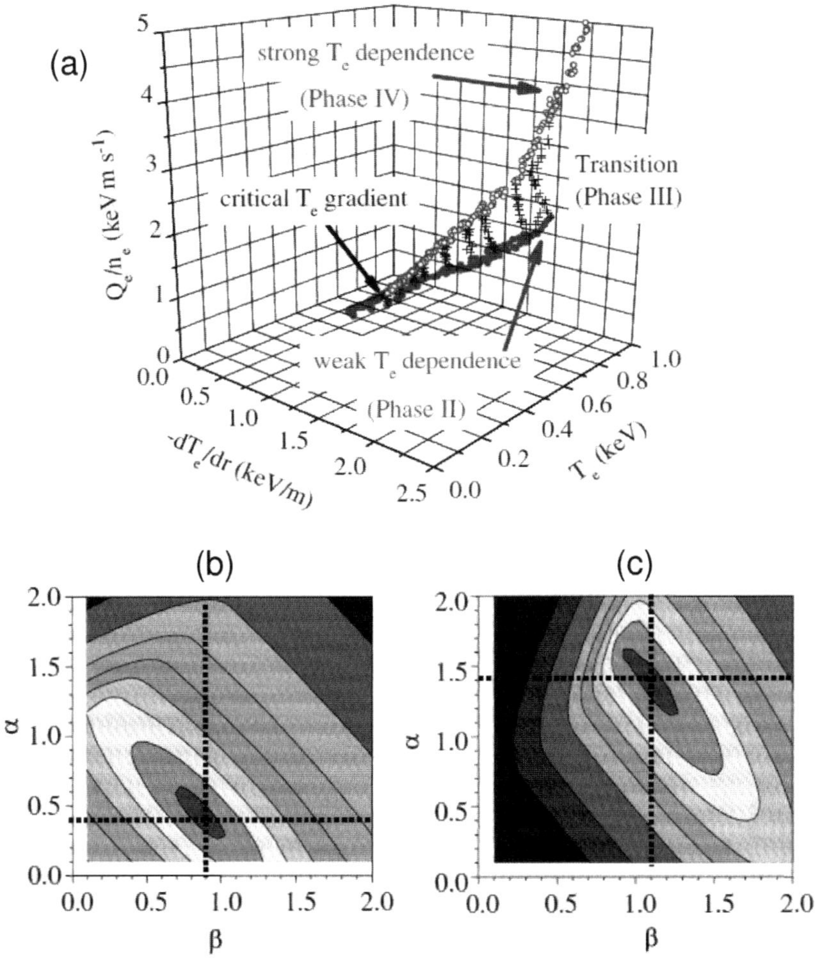

Figure 9. (a) Heat flux normalized by electron density as a function of electron temperature T_e and temperature gradient in the weak T_e dependence branch (phase II), in the transition period (phase III), and in the strong T_e dependence branch (phase IV) at $\rho = 0.65$. The contours of the difference between the normalized heat flux measured and that calculated with the model of $cT_e^{\alpha}T_e^{\beta}$ with various values of (α, β) for (b) the weak T_e dependence branch (phase II) and (c) the strong T_e dependence branch (phase IV) at $\rho = 0.65$[41].

for various values of (α, β) in the weak and strong Te dependence branches. The T_e dependence parameter α is 0.44 (-0.23, +0.25) for the weak T_e dependence transport branch and 1.4 (-0.4, +0.3) for the strong T_e dependence transport branch, while the T_e gradient dependence parameter, β is close to unity ; 0.86 (-0.25, +0.25) and 1.1 (-0.2, +0.27) for the weak and the strong T_e dependence branches, respectively. This strong T_e dependence is consistent with the observation in tokamak when the T_e gradient is

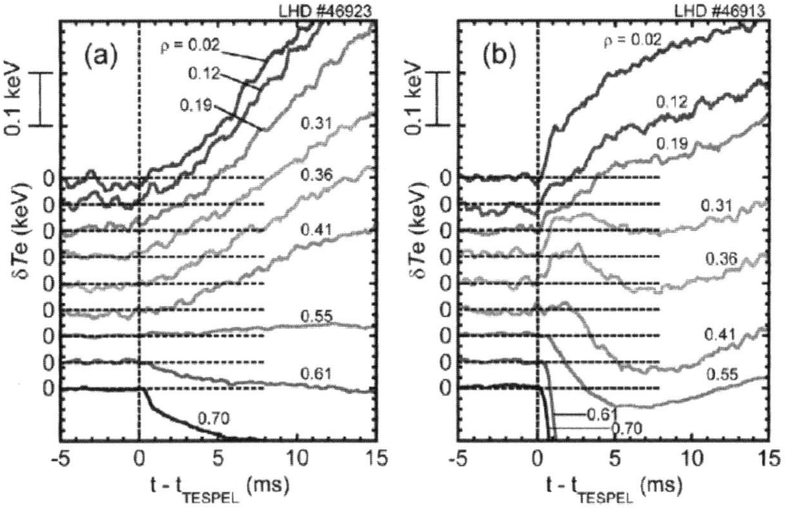

Figure 10. Temporal evolutions of the perturbed electron temperature with (a) the $\sim$ 420 μm diameter TESPEL injection, (b) the larger ($\sim$ 670 μm diameter) TESPEL injection into the plasma with a line-averaged electron density of 0.5 $\times$ 10^{19} m^3 [46].

below the threshold [42]. The error bar is estimated from the range of α and β values for $\chi^2 < 1.3\chi^2_{min}$. In this experiment, the T_e increases as the T_e gradient is increased. The co-linearity between T_e and the T_e gradient causes large errors in the estimated α and β values. The sum of α and β is evaluated more accurately and it is 1.3 (-0.12, +0.12) for the weak T_e dependence branch and 2.5 (-0.24, +0.26) for the strong T_e dependence branch.

The strong (T_e, ∇T_e) dependence of $\alpha + \beta = 2.3$ observed in phase VI is consistent with the gyro-Bohm like dependence of $\alpha + \beta = 2.5$ in the collisionless regime. However the weak (T_e, ∇T_e) dependence of $\alpha + \beta = 1.3$ observed in phase II is much weaker than the gyro-Bohm dependence and it can be understood by taking account of the influence of a zonal flow. When zonal flows coexist with drift wave turbulence, the turbulent transport coefficient is strongly influenced by the damping rate of the zonal flow and deviates from the gyro-Bohm dependence[43]. The (T_e, ∇T_e) dependence $\alpha + \beta$ depends on fluctuation levels and, when the collisional flow damping of a zonal flow is the dominant process, it can be 1.5, which is consistent with the (T_e, ∇T_e) dependence in phase II. These observation suggests that the collisional flow damping of a zonal flow in a non-axisymmetric system and is consistent with the fact that zonal flows are enhanced in the electron root plasma, where the collision flow damping becomes small.[24]

6. Non-local transport in non-axisymmetric system

The non-linearity of transport described above is due to the non-linearity of the driving and damping of turbulence and plasma flow. Since the scale lengths of the turbulence ($\sim$

poloidal gyro radius) and zonal flows ($\sim$ tens of poloidal gyro radius) are different, the coupling between the turbulence and zonal flows is a candidate to cause the non-linearity of transport. Therefore the observation of nonlocal T_e rise in a helical device will give new insight into the phenomenon of nonlocal T_e rise, since helical systems have quite a different magnetic configuration (normally negative magnetic shear), which can affect the turbulence. The characteristics of the nonlocal T_e rise invoked by the rapid edge cooling in helical plasmas have been investigated in LHD[44, 45, 46]. In LHD, a variety of time responses of the nonlocal Te rise are obtained by changing the plasma parameters or the TESPEL size. Figure 10(a) shows a typical example of an abrupt increase in core Te just after the edge cooling. In this example, the $\sim$ 420 μm diameter TESPEL is injected into the LHD plasma at a density of 0.5×10^{19} m$^{?3}$ and as a consequence, the TESPEL reaches $\rho \sim 0.8$. As can be seen in figure 10(b), when the larger TESPEL ($\sim$ 670 μm diameter) is injected into the LHD plasma, the time derivative of the core T_e just after the edge cooling becomes much larger, compared with the case of figure 10(a). In this case, the line-averaged electron density and the TESPEL penetration depth are almost the same as those in the case of figure 10(a).

It is interesting that non-local transport appears both in LHD and tokamaks, although the T_e and ∇T_e dependences of heat transport have significant differences between a heliotron and a tokamak. One of the candidate for the physics mechanism causing the non-local transport is energy transfer between high k_r turbulence determining the transport and low k_r zonal flows, because energy transfer between separated regions is required to explain propagation much faster than the diffusion of turbulence. It is important to investigate how the zonal flow changes during the non-local transport phenomena in order to understand turbulent transport.

7. Energy transfer between flows and turbulence and generation of parallel flow

Experimentally, direct measurements have been made of the radial-perpendicular component and radial-parallel component of the Reynolds stress in the boundary region in the TJ-II stellarator. As seen in figure 11(a), the perpendicular phase velocity of fluctuations (perpendicular to the magnetic field) is computed by means of the two point correlation technique using two floating potential signals measured by probes poloidally separated about 0.3 cm. The resulting radial profile of v_θ is radially flat for a mean plasma density below the threshold value (0.55×10^{-19}m^{-3}), whereas above this critical density the perpendicular phase velocity reverses and the naturally occurring velocity shear layer appears in the proximity of the LCFS.

From the radial derivative of the mean parallel velocity and the radial-parallel component of the Reynolds stress, the radial-parallel contribution to the production of turbulent kinetic energy P was experimentally evaluated in TJ-II as seen in figure 11(b). The production term combines the velocities cross correlation $< v_r M_\parallel >$ (momentum flux) with the mean velocity gradient ($\partial M_\parallel / \partial r$) and gives a measure of the amount of

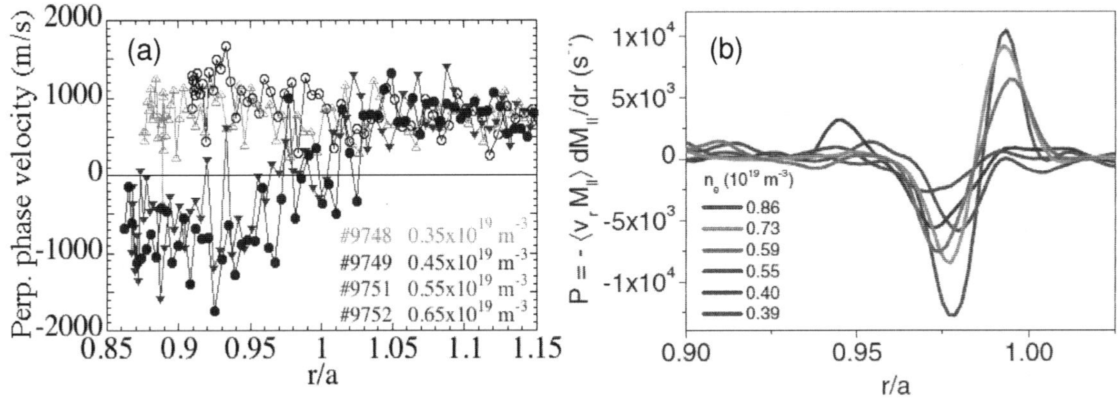

Figure 11. (a) Radial profile of perpendicular phase velocity for various densities near the threshold density for spin-up. (b)Radial profile of production term (P) at different plasma densities in TJ-II[23, 47].

energy per unit mass and unit time that is transferred locally between the mean flow and fluctuations. In low density plasmas, the production term P as well as gradients in $(\partial M_{\|}/\partial r)$ are small and, in some cases, within experimental error bars. That means that, at low density, there is not any significant energy transfer between flows and turbulence. As the density approaches the threshold value to trigger $E \times B$ perpendicular flows the results show a different behavior. The production term clearly increases in the region where sheared flows are developed, reaching values up to $10^4 s^{-1}$. Furthermore, two different signs are found in P thus implying that the turbulence can act as an energy sink ($P > 0$) for the mean flow or energy source ($P < 0$) near the shear layer [20]. This result suggests that parallel turbulent forces are relevant in momentum dynamics during the development of sheared flows in the proximity of the LCFS.

There are $E \times B$ flows and parallel flows driven by non-ambipolar flux in non-axi-symmetric systems. These flows become more significant in the plasmas where an internal barrier appears associated with the transition from ion root (small negative E_r to electron root (large positive E_r). Figure 12(a) and (b) show the time evolution of the poloidal and toroidal flows in the NBI plasma with and without a 2nd ECH pulse. During the 2nd ECH pulse, the electron collisionality is low enough for the plasma to be in the electron root (positive electric field) as predicted by neoclassical theory. The positive radial electric field is mainly made up of the poloidal flow in the direction parallel to the $< E_r \times B_\phi >$ drift. The large electric field of 10-20 kV/m and the large poloidal flow of 10-20 km/s are produced near the plasma center ($\rho = 0.3$) [figure 11 (a)], where the sharp temperature gradient is observed. The neutral beam is injected to the plasma tangentially in the co-direction. Here the "co" direction (positive flow velocity) is defined as parallel to the equivalent toroidal plasma current, which would produce the average transform actually produced by the external coil current. The "counter" direction (negative flow velocity) is defined as anti-parallel to the equivalent

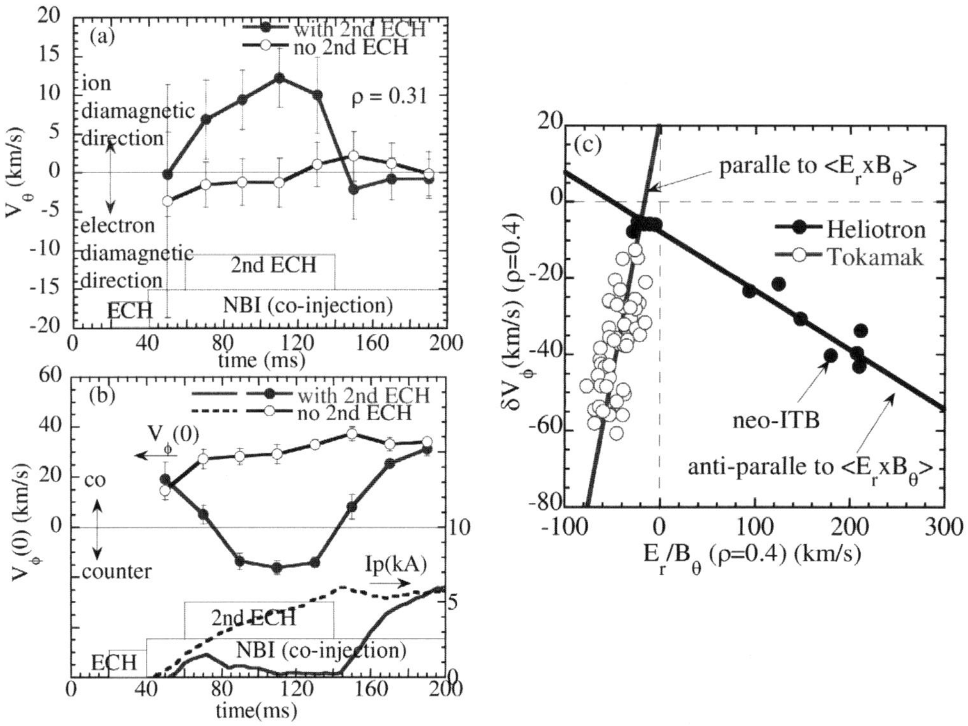

Figure 12. Time evolution of (a) central electron temperatures and densities, (b) poloidal flow velocity at $\rho = 0.38$[48], and (c) spontaneous toroidal flow velocity as a function of E_r/B_θ at $\rho = 0.4$ for tokamak and helical plasmas. The solid lines are best fits for the measured data and $\delta V_\phi = 0.15\ (E_r/B_\theta)$ - 7.9 for a Heliotron plasma and $\delta V_\phi = 1.34\ (E_r/B_\theta) + 22.3$ for a tokamak plasma[49].

toroidal plasma current. As seen in figure 11(b), the plasma rotates parallel to the NBI when there is no 2nd ECH. This is simply because the toroidal momentum from the NBI causes the plasma toroidal flow. However, when the 2nd ECH is turned on, the toroidal flow velocity decreases and finally the plasma rotates anti-parallel to the NBI. These measurements clearly show the driven toroidal flow associated with a large poloidal flow (and large positive radial electric field) during the 2nd ECH pulse. The driven toroidal flow reaches 5x10^{4}m/s at the plasma center and it is large enough to overcome the toroidal flow driven by a tangentially injected neutral beam. It should be emphasized that the direction of the toroidal flow is anti-parallel to the direction of the $< E_r \times B_\theta >$ drift. This is in contrast to the spontaneous flow in a tokamak, where the direction of the toroidal flow is parallel to the direction of the $< E_r \times B_\theta >$ drift, because the toroidal viscosity is nearly zero due to the toroidal symmetry.

The significant difference between tokamaks and a heliotron devices is the relationship between the poloidal field direction and the direction of dominant symmetry as seen in figure 12(c). The pitch angle of dominant symmetry is even larger than that of the averaged poloidal field in a Heliotron device, while it is zero due to the toroidal symmetry in a tokamaks. The formula for toroidal flow qualitatively explains the reversal of the sign of v_ϕ between CHS and tokamak plasmas, because $\alpha < \beta$ in Heliotron and $\alpha > \beta = 0$ in tokamak, where α is the pitch angle of the magnetic field and β is the pitch angle of the minimum ∇B direction . The spontaneous toroidal flow observed in the plasma with an electron ITB in CHS is anti-parallel to the $< E_r \times B_\theta >$ drift direction. This is in contrast to the spontaneous toroidal flow in the direction parallel to the direction of the $< E_r \times B_\theta >$ drifts in tokamak plasmas, which is clearly demonstrate in figure 11(c). Here the spontaneous toroidal flow is given by the difference of magnitude of toroidal flow velocity between co-injection and counter-injection with a similar magnitude of momentum input and the radial electric field is given by the temperature gradient. In general, the toroidal flow in Heliotron plasmas is much smaller than that in a tokamak due to toroidal viscosity [33] which does not exist in tokamak plasmas. However, the spontaneous flow observed in CHS is comparable to that observed in the JFT-2M tokamak. This is because the radial electric field in CHS plasma with an internal transport barrier is much larger than that in the L-mode plasma in the JFT-2M tokamak. It should be noted that the ratio of spontaneous flow velocity to the radial electric field normalized by the poloidal field in the CHS Heliotron is 0.15, which is much smaller than that observed in a tokamak (1.34) by nearly an order of magnitude because of the parallel viscosity in the toroidal direction.

8. Ion internal transport barrier

Although the transition from L-mode to the high ion temperature mode plasma, which is characterized by the peaked ion temperature profile, has been achieved, the ion temperature gradient normalized R/L_{Ti} was relatively small (< 10) compared with that observed in tokamaks. The high T_i mode is characterized by a high central ion temperature and a low central ion thermal diffusivity, associated with a peaked electron density produced by NB fuelling with low wall recycling[50, 51].

More recently a clear ion internal transport barrier was observed associated with the formation of an electron internal transport barrier in the electron root[52]. As seen in figure13, a sharp temperature gradient of 12 keV/m appears at $\rho = 0.66$, which is much further outside of the electron transport barrier ($\rho < 0.3$). The ion temperature at the ion transport barrier is 0.3 - 0.4 keV and the normalized ion temperature gradient R/L_{Ti} exceeds 30, which is comparable to the normalized electron gradient R/L_{Te} observed inside the electron ITB[19]. The large separation of ITB locations between the ion transport barrier and the electron transport barrier is an interesting characteristic in non-axi-symmetric magnetic field systems, which is not observed in tokamaks. (The locations of electron ITBs and ion ITBs are close to each other[54]).

The high ion temperature plasma obtained by NBI heating (no additional ECH) in LHD has a moderate ion temperature gradient of $R/L_{Ti} = 10$ -15 as seen in figure 13(b). In this discharge, there is no electron ITB observed ($T_e(0) < T_i(0)$). The radial

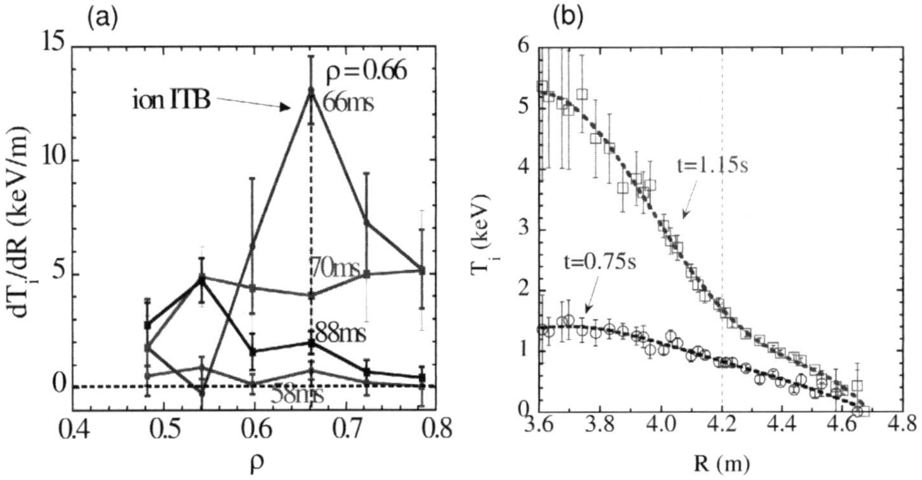

Figure 13. Radial profiles of (a) ion temperature gradient of an ion transport barrier plasma in CHS[53] and of (b) ion temperature of high T_i discharges in LHD. In figure (a) the peak ion temperature gradient of 12keV/m gives the the large normalized ion temperature gradient R/L_{Ti} of 30 - 40.

profile of ion temperature observed in LHD is similar to that observed in the high T_i mode without ECH heating[50, 51]. It should be noted that the change in ∇T_i appears at R = 4.2m, which suggest the formation of a weak internal transport barrier. Further study of ion transport for this high T_i discharge should be done. One of the interesting features of this discharge is impurity exhaust. The large error bar in the ion temperature near the axis is due to the low intensity of the charge exchange line measured. The low intensity is mainly due to the hollow carbon density rather than the beam attenuation. It should be noted that the simultaneous achievement of good energy confinement and poor impurity confinement is a preferable transport characteristic in the transport of non-axisymmetric systems.

9. High density operation

Accessibility to high density operation is one of the important features of non-axisymmetric systems. In Wendelstein 7-AS, a high density regime with good energy confinement and low impurity concentration was achieved by building up the density rapidly with strong gas puffing at the very start of the discharge during NBI initiation[55, 56]. As discussed in the previous section, poor impurity confinement is achieved simultaneously with good energy confinement. In non-axisymmetric systems, simultaneous achievement of high density and poor impurity confinement has been also demonstrated experimentally. In LHD, a highly peaked density profile was created by injection of multiple pellets with a low recycling configuration[57, 58, 59]. The central densities achieved in Wendelstein 7-AS and LHD are 4-5$\times10^{20}$m^{-3} which is more than

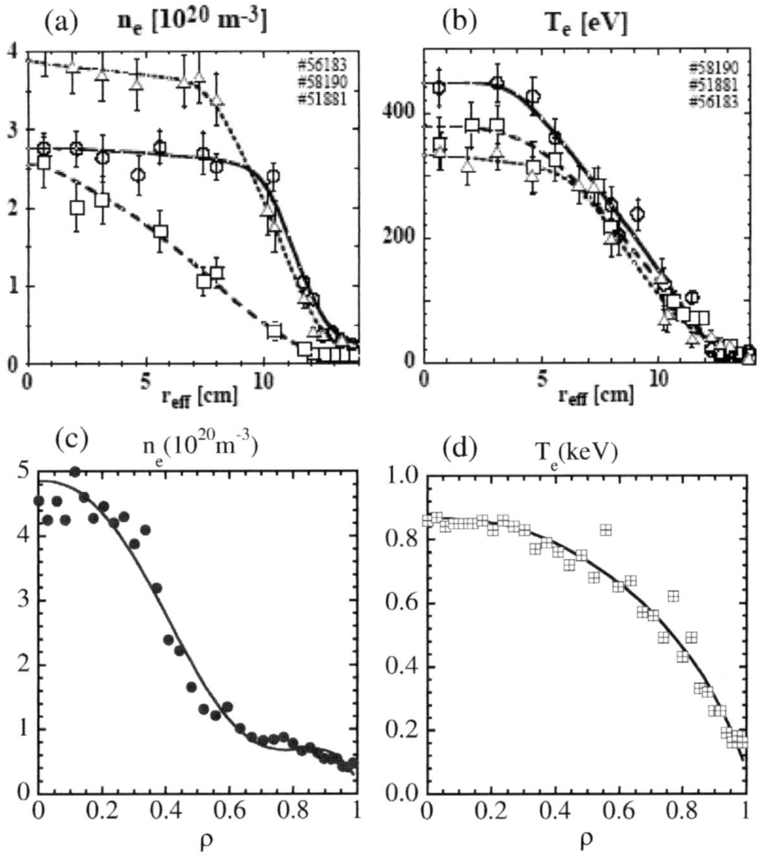

Figure 14. Radial profiles of (a) electron density and (b) electron temperature in the normal confinement (NC), and attached and detached high density H-mode (HDH) in Wendelstein 7-AS (B= 2.5T)[56] and radial profiles of (c) electron density and (d) electron temperature in the super dense core plasma in LHD (B=2.64T)[58].

twice that of a tokamak with similar size and magnetic field strength. By replacing the plasma current in tokamak with the equivalent current in a helical system, the densities in helical systems exceed the Greenwald limit by a factor of 2.

Figure 14(a) and (b) show radial profiles of electron density and temperature in the normal confinement (NC), and the attached and detached high density H-mode (HDH) in Wendelstein 7-AS (B = 2.5T). Although the density profile of the normal confinement (NC) is peaked at the magnetic axis, the density profile of the high density H-mode (HDH) is broad. The significant difference between NC and HDH discharges is energy confinement and impurity confinement time. The energy confinement was up to twice that of standard scaling, while the impurity confinement time was reduced and approaching the energy confinement time by the reduction of the inward pinch velocity which normally exists in the NC discharge. A super dense core (SDC) plasma develops naturally in LHD as a peaked, high density profile is generated by multiple pellet injections from the outside mid-plane. As shown in figure 14(c)(d), A core region

Journal of Physics: Conference Series **123** (2008) 012004 doi:10.1088/1742-6596/123/1/012004
with electron densities $\sim 4.6 \times 10^{20}\text{m}^{-3}$ and temperatures ~ 0.85 keV is maintained by an internal diffusion barrier (IDB) located at normalized minor radius $0.3 < \rho < 0.5$ Unlike the tokamak internal transport barrier (ITB) discharges, there is no increase in the temperature gradient, possibly because the density becomes extremely large in LHD. The radial width of the IDB is ~ 0.1m ($\rho \sim 0.2$). It should be pointed out that there is no impurity accumulation causing the termination of the discharge observed in LHD. The density is limited not by radiation collapse but by MHD instabilities trigged by the increase of pressure in the IDB region[60].

10. Summary

Transport barriers in confinement systems with non-axisymmetric magnetic fields are summarized focusing on the physics mechanisms determining the turbulence level in the plasma. There are differences in transport barrier characteristics between plasmas with axisymmetric magnetic fields (tokamak plasmas) and plasmas with non-axisymmetric magnetic fields (helical plasmas), although in both cases the turbulence in the plasma has a dominant contribution to the transport.

(1) The radial-perpendicular component of the Reynolds stress due to turbulence in the plasma produces a zonal flow with finite k_r and zero k_θ k_ϕ as well as a mean flow ($k_r = 0$). There are two types of zonal flows one is a stationary zonal flow with the frequency below $1/\tan_{damping}$ and the other is a zonal flow excited at the GAM frequency. The stationary zonal flow, which is considered to be more effective for the suppression of turbulence, is driven by turbulence and damped by collisional flow damping and non-liner flow damping.

(2) The internal transport barrier in a helical plasma extends to the plasma core where there is no strong E_r shear, while that in a tokamak tends to be localized in the narrow region where the strong E_r shear exists. The mechanism causing these differences is considered to be the E_r effect on the reduction of collisional flow damping of the zonal flow. In a helical plasma, the radial electric field is expected to reduce the collisional flow damping, which is much larger than in a tokamak because of the ripple (if $E_r = 0$), enhance stationary zonal flows and hence reduce the turbulence level. On the other hand, E_r does not contribute to the collisional flow damping and E_r shear can be a dominant process to reduce the turbulence in a tokamak.

(3) In general, the heat flux normalized by density depends on the temperature, T, and the temperature gradient, ∇T. Therefore the ratio of normalized heat flux to ∇T (thermal diffusivity) should have some dependence on T and on the scale length of the temperature $T/\nabla T$. The thermal diffusivity in a helical plasma has a clear temperature dependence but has no clear scale length dependence. This is in contrast to tokamaks where there is a clear threshold of scale length of T above which a significant enhancement of the thermal diffusivity exists and the T dependence is often masked by the ∇T dependence. This difference can be also explained by the differences in the damping mechanism of zonal flows. The T and ∇T dependence of heat flux measured

in a helical plasma is consistent with that expected by zonal flow theory assuming that collisional flow damping is the dominant term. The existence of a threshold of scale length in tokamak plasmas suggests that the non-linear flow damping of zonal flows is dominant

(4) Non-local transport observed in helical plasmas is similar to that observed in tokamaks, although the T and $T/\nabla T$ dependence of transport is quite different between helical plasma and tokamak plasmas. This fact implies that the physics mechanism to transfer the turbulence energy from one location to another location is not plasma current but the fluctuations with small k_r (stationary zonal flow is one of the candidate). The coupling between turbulence(high-k)-turbulence(low-k) and turbulence-flow, that should cause the non-local transport is quite crucial for understanding transport both in tokamak and helical plasma.

(5) In helical plasmas the mean $E \times B$ and spontaneous parallel flow are driven by the radial-parallel component of the Reynolds stress due to turbulence and neoclassical non-ambipolar flux due to collisional processes. The viscosity tensor acts as damping and driving terms and the relation between $E \times B$ and spontaneous parallel flow is strongly affected by the magnitude of the viscosity in the polodal and toroidal directions. There is a difference in the direction and magnitude of the parallel flow between helical and tokamak plasmas. For the given radial electric field, the spontaneous parallel flow is small and anti-parallel to the direction of the $< E_r \times B_\theta >$ drift in a helical plasma, while in a tokamak it is large and parallel to the direction of $< E_r \times B_\theta >$. The difference of the direction is due to the differences in symmetry, helical symmetry or toroidal symmetry.

(6) The simultaneous achievements of good energy confinement and poor impurity confinement and high density and poor impurity confinement are observed in a helical plasma. A peaked ion temperature with moderate gradient (R/L_{Ti} = 10-15) is achieved in NBI heated plasma by reducing the recycling. By applying ECH a strong ion temperature gradient (R/L_{Ti} = 30-40) is obtained associated with the formation of an electron ITB in the electron root. The ion ITB and electron ITB are located in different regions of the helical plasma, which is in contrast to that the proximity of the electron ITB and ion ITB in a tokamak plasma. The density achieved in a helical system is twice of that in a tokamak with similar size and magnetic field (equivalent Greenwald factor of 2). Even with quite high density operation, the impurity is not a problem affecting the plasma performance.

Acknowledgments

The author would like to thank Drs. A.Fujisawa and K.Itoh (NIFS) for fruitful discussions in organizing this review paper. The author also would like to thank Drs. C.Hidalgo (CIMAT), M.Hirsch (Max-Planck Insitute), M.G.Shats (Australian National University) for providing the data and useful discussions on the interpretation of the experimental data from their devices.

References

[1] Erckmann V, Wagner F, Baldzuhn J, Brakel R, Burhenn R, Gasparino U, Grigull P, Hartfuss H J, Hofmann J V, Jaenicke R, *et al.* 1993 *Phys. Rev. Lett.* **70** 2086

[2] Wagner F, Baldzuhn J, Brakel R, Burhenn R, Erckmann V, Estrada T, Grigull P, Hartfuss H J, Herre G, Hirsch M, *et al.* 1994 *Plasma Phys. Control. Fusion* **36** A61

[3] Hirsch M, Baldzuhn J, Ehmler H, Holzhauer E and Wagner F 2006 *Plasma Phys. Control. Fusion* **48** S155

[4] Toi K, Ohdachi S, Yamamoto S, Sakakibara S, Narihara K, Tanaka K, Morita S, Morisaki T, Goto M, Takagi S, *et al.* 2005 *Phys. Plasmas* **12** 020701

[5] Okamura S, Minami T, Oishi T, SUzuki C, Ida K, Isobe M, Yoshimura Y, Nagaoka K, Toi K, Fujisawa A, *et al.* 2004 *Plasma Phys. Control. Fusion* **46** A113

[6] Minami T, Okamura S, Akiyama T, Isobe M, Oishi T, Kado S, Ida K, Fujisawa A, Nakano H, Nakamura K, *et al.* 2006 *Plasma and Fusion Research* **1** 047

[7] Toi K, Ohdachi S, Morisaki T, Sakakibara S, Morita S, Minami T, Tanaka K, Yamada H, Okamura S, Nishimura K, *et al.* 1996 *Plasma Phys. Control. Fusion* **38** 1289

[8] Sano F, Mizuuchi T, Nagasaki K, *et al.* 2004 *Fusion Sci. Technol.* **46** 288

[9] Alejaldre C, *et al.* 2004 Proc. 20th Int. Conf. on Plasma Physics and Controlled Nuclear Fusion Research 2004 (Villamoura, Portugal, 2004) (Vienna: IAEA) OV/4-3

[10] Shats M G and Rudakov D L 1997 *Phys. Rev. Lett.* **79** 2690

[11] Kitajima S, Takahashi H, Tanaka Y, Utoh H, Sasao M, Takayama M, Nishimura K, Inagaki S and Yokoyama M, *et al.* 2006 *Nucl. Fusion* **46** 200

[12] Takahashi H, *et al.* 2006 *Plasma Phys. Control. Fusion* **48** 39

[13] H.Utoh, Nishimura K, Inagaki S, Takahashi H, Tanaka Y, Takenaga M, Ogawa M, Shinde J, Iwazaki K, Okamoto A, *et al.* 2006 *Fusion Sci. Technol.* **50** 434

[14] Wagner F, Hirsch M, Hartfuss H-J, Laqua H P and Maassberg H 2006 *Plasma Phys. Control. Fusion* **48** A217

[15] Idei H, Ida K, Sanuki H, Yamada H, Iguchi H, Kubo S, Akiyama R, Arimoto H, Fujiwara M, Hosokawa M, *et al.* 1993 *Phys. Rev. Lett.* **71** 2220

[16] Ida K, Funaba H, Kado S, Narihara K, Tanaka K, Takeiri Y, Nakamura Y, Ohyabu N, Yamazaki K, Yokoyama M, *et al.* 2001 *Phys. Rev. Lett.* **86** 5297

[17] Fujisawa A, Iguchi H, Minami T, *et al.* 1999 *Phys. Rev. Lett.* **82** 2669

[18] Stroth U, Itoh K, Itoh S.-I, Hartfuss H and Laqua H 2001 *Phys. Rev. Lett.* **86** 5910

[19] Ida K, Shimozuma T, Funaba H, Narihara K, Kubo S, Murakami S, Wakasa A, Yokoyama M, Takeiri Y, Watanabe K Y, *et al.* 2003 *Phys. Rev. Lett.* **91** 085003

[20] Shimozuma T, Kubo S, Idei H, Yoshimura Y, Notake T, Ida K, Ohyabu N, Yamada I, Narihara K, Inagaki S, *et al.* 2003 *Plasma Phys. Control. Fusion.* **45** 1183

[21] Takeiri Y, Shimozuma T, Kubo S, Morita S, Osakabe M, Kaneko O, Tsumori K, Oka Y, Ikeda K, Nagaoka K, *et al.* 2003 *Phys. Plasmas* **10** 1788

[22] Ida K, Inagaki S, Shimozuma T, Tamura N, Funaba H, Narihara K, Kubo S, Murakami S, Wakasa A, Yokoyama M, *et al.* 2004 *Phys. Plasmas* **11** 2551

[23] Hidalgo C, Pedrosa M A, Garcia L and Ware A 2004 *Phys. Rev. E* **70** 067402

[24] Fujisawa A, Itoh K, Iguchi H, Matsuoka K, Okamura S, Shimizu A, Minami T, Yoshimura Y, Nagaoka K, Takahashi C, *et al.* 2004 *Phys. Rev. Lett.* **93** 165002

[25] Fujisawa A, Shimizu A, Nakano H, Ohshima S, Itoh K, Nagashima Y, Itoh S-I, Iguchi H, Yoshimura Y, Minami T, *et al.* 2007 *Plasma Phys. Control. Fusion* **49** 211

[26] Shats M G, Solomon W M and Xia H 2003 *Phys. Rev. Lett.* **90** 125002

[27] Punzmann H and Shats M G 2004 *Phys. Rev. Lett.* **93** 125003

[28] Mckee G R, Fonck R J, Jakubowski M, Burrell K H, Hallatschek K, Moyer R A, Nevins W, Rudakov D L and Xu X 2003 *Plasma Phys. Control. Fusion* **45** A477

[29] Conway G D, Scott B, Schirmer J, Reich M, Kendl A, the ASDEX Upgrade Team 2005 *Plasma*

Phys. Control. Fusion **47** 1165

[30] Hamada Y, Nishizawa A, Ido T, Watari T, Kojima M, Kawasumi Y, Narihara K, Toi K and JIPPT-IIU Group 2005 *Nucl. Fusion* **45** 81

[31] Melnikov A V, Vershkov V A, Eliseev L G, Grashin S A, Gudozhnik A V, Krupnik L I, Lysenko S E, Mavrin V A, Perfilov S V, Shelukhin D A, *et al.* 2006 *Plasma Phys. Control. Fusion* **48** S87

[32] Ido T, Miura Y, Kamiya K, Hamada Y, Hoshino K, Fujisawa A, Itoh K, Itoh S-I, Nishizawa A, Ogawa H, *et al.* 2006 *Plasma Phys. Control. Fusion* **48** S41

[33] Ida K, Yamada H, Iguchi H, Itoh K and CHS group 1991 *Phys. Rev. Lett.* **67** 58

[34] Fujisawa A, Ido T, Shimizu A, Okamura S, Matsuoka K, Iguchi H, Hamada Y, Nakano H, Ohshima S, Itoh K, *et al.* 2007 *Nucl. Fusion* **47** S718

[35] Fujisawa A, Shimizu A, Nakano H, Ohshima S, Itoh K, Nagashima Y, Itoh S-I, Iguchi H, Yoshimura Y, Minami T, *et al.* 2007 *J. Phys. Soc. Jpn.* **76** 033501

[36] Xia H, Shats M G and Punzmann H 2006 *Phys. Rev. Lett.* **97** 225003

[37] Itoh K, Toda S, Fujisawa A, Itoh S-I, Yagi M, Fukuyama A, Diamond P H and Ida K 2007 *Phys. Plasmas* **14** 020702

[38] Ida K, Fujita T, Fukuda T, Sakamoto Y, Ide S, Toi K, Inagaki S, Shimozuma T, Kubo S, Idei H, *et al.* 2004 *Plasma Phys. Control. Fusion* **46** A45

[39] Inagaki S, Takenaga H, Ida K, Isayama A, Tamura N, Takizuka T, Shimozuma T, Kamada Y, Kubo S, Miura Y, *et al.* 2006 *Nucl. Fusion* **46** 133

[40] Yamada H, Watanabe K Y, Sakakibara S, Murakami S, Osakabe M, Kaneko O, Narihara K, Tanaka K, Ida K, Minami T, *et al.* 2000 *Phys. Rev. Lett.* **84** 1216

[41] Ida K, Inagaki S, Sakamoto R, Tanaka K, Funaba H, Takeiri Y, Ikeda K, Michael C, Tokuzawa T, Yamada H, *et al.* 2006 *Phys. Rev. Lett.* **96** 125006

[42] Imbeaux F. Tyter F and Garbet X 2001 *Plasma Phys. Control. Fusion* **43** 1503

[43] Diamond P H, Itoh S-I, Itoh K and Hahm T S 2005 *Plasma Phys. Control. Fusion* **47** R35

[44] Tamura N, Inagaki S, Ida K, Shimozuma T, Kubo S, Tokuzawa T, Tanaka K, Neudatchin S V, Itoh K, Kalinina D, *et al.* 2005 *Phys. Plasmas* **12** 110705

[45] Inagaki S, Tamura N, Tokuzawa T, Ida K, Itoh K, Neudatchin S V, Tanaka K, Nagayama Y, Kawahata K, Yakovlev M, *et al.* 2006 *Plasma Phys. Control. Fusion* **48** A251

[46] Tamura N, Inagaki S, Tanaka K, Michael C, Tokuzawa T, Shimozuma T, Kubo S, Sakamoto R, Ida K, Itoh K, *et al.* 2007 *Nucl. Fusion* **47** 449

[47] Goncalves B, Hidalgo C, Pedrosa M A, Orozco R O, Sanchez E and Silval C 2006 *Phys. Rev. Lett.* **96** 145001

[48] Ida K, Minami T, Yoshimura Y, Fujisawa A, Suzuki C, Okamura S, Nishimura S, Isobe M, Iguchi H, Itoh K, *et al.* 2001 *Phys. Rev. Lett.* **86** 3040

[49] Ida K, Minami T, Yoshinmura Y, Fujisawa A, Suzuki C, Yokoyama M, Murakami S, Okamura S, Nishimura S, Isobe M, *et al.* 2002 *Plasma Phys. Control. Fusion* **44** 361

[50] Ida K, Kondo K, Nagasaki K, Hamada T, Hidekuma S, Sano F, Zushi H, Mizuuchi T, Okada H, Besshou S, *et al.* 1996 *Phys. Rev. Lett.* **76** 1268

[51] Ida K, Osakabe M, Tanaka K, Minami T, Nishinura S, Okamura S, Fujisawa A, Yamazaki K, Yoshimura Y, Kubo S, *et al.* 1999 *Nucl. Fusion* **39** 1649

[52] Minami T, Fujisawa A, Iguchi H, Liang Y, Ida K, Nishimura S, Yokoyama M, Murakami S, Yoshimura Y, Isobe M, *et al.* 2004 *Nucl. Fusion* **44** 342

[53] Okamura S, *et al.* 2006 Proc. 21th Int. Conf. on Plasma Physics and Controlled Nuclear Fusion Research 2004, (Fusion Energy Conference, Chengdu, Chinal, 2006) (Vienna: IAEA). EX/5-5Rb

[54] Fujita T 2002 *Plasma Phys. Control. Fusion* **44** A19

[55] McCormick K, Grigull P, Burhenn R, Brakel R, Ehmler H, Feng Y, Gadelmeier F, Giannone L, Hildebrandt D, Hirsch M, *et al.* 2002 *Phys. Rev. Lett.* **89** 015001

[56] Hirsch M 2007 submitted to *Plasma Phys. Control. Fusion*

[57] Ohyabu N, Morisaki T, Masuzaki S, Sakamoto R, Kobayashi M, Miyazawa J, Shoji M, Komori A and Motojima O 2006 *Phys. Rev. Lett.* **97** 055022

[58] Ohyabu N, Morisaki T, Masuzaki S, Sakamoto R, Kobayashi M, Miyazawa J, Shoji M, Funaba H, Harris J H, Hirooka Y, *et al.* 2006 *Plasma Phys. Control. Fusion* **48** B383

[59] Morisaki T, Ohyabu N, Masuzaki S, Kobayashi M, Sakamoto R, Miyazawa J, Gunaba H, Ida K, Ikeda K, Kaneko O, *et al.* 2007 *Phys. Plasmas* **14** 056113

[60] Ohdachi A, *et al.* 2007 submitted to *Plasma and Fusion Research*

The structure, evolution and role of the radial edge electric field in H-mode and L-mode on MAST

H Meyer, C Bunting, P G Carolan, N J Conway, M R Dunstan, A Kirk, R Scannell, D Temple, M Walsh, and the MAST and NBI Teams

EURATOM/UKAEA Fusion Association, Culham Science Centre, Abingdon, Oxfordshire, OX14 3DB, UK

E-mail: Hendrik.Meyer@ukaea.org.uk

Abstract. The first measurements of the structure of the edge radial electric field, E_r, in a spherical tokamak (MAST) are presented. Using active Doppler spectroscopy on He^+ with 120 lines of sight E_r profiles are calculated from the leading terms of the radial momentum balance. A spatial resolution up to $\Delta r \approx 1.5$ mm with a typical time resolution of $\Delta t = 5$ ms can be achieved. In L-mode the field is largely determined by the diamagnetic term of the force balance, and fields of only a few kV/m are observed. The measured impurity flow is mostly parallel to **B**, and is greatly affected by MHD, such as sawteeth or mode locking of tearing modes, or error fields. In H-mode a strong perpendicular flow evolves with poloidal and toroidal velocities up to $v_{\phi,\theta}^{\mathrm{He}^+} \approx -20\,\mathrm{km/s}$, and a deep negative electric field well $E_r^{\mathrm{min}} \gtrsim -15\,\mathrm{kV/m}$ develops. The profile form is dominated by the diamagnetic term.

1. Introduction

Plasmas confined by a toroidal magnetic field configuration exhibit a regime with a sudden improvement of confinement, the so called H-mode [1–6]. In this regime a transport barrier develops at the plasma edge (ETB), where the turbulent (anomalous) transport perpendicular to the magnetic field is largely suppressed. The physics of the formation of the ETB is still not fully understood.

It was soon recognised that strong perpendicular flows develop during H-mode [2], and that a negative radial electric field is generated [2, 7–9], causing a sheared $\mathbf{E} \times \mathbf{B}$ flow. The common theoretical picture is that this sheared flow is responsible for the suppression of the turbulent transport [10, 11]. The mechanisms leading to the generation of the electric field in H-mode, however, are less clear. One theoretical picture is that macroscopic flows generated by the turbulence itself, zonal flows, lead to the strong E_r [12], and indeed evidence for such flows has been found experimentally [13]. Other theoretical work and experimental evidence points towards the flow in the scrape-off-layer (SOL) as an important factor in establishing E_r [14, 15]. Here, the field is generated by viscous transport of momentum from the SOL into the edge region, and the field is mainly neoclassical [15].

Only on a few devices is the electric field in the edge measured to allow comparison with theory [7, 9, 16]. Here, usually the edge flows are measured either by active or passive Doppler spectroscopy [7, 9, 17] or by Doppler reflectometry [18]. In smaller devices sometimes also reciprocating Langmuir probes are used. The time resolution of these measurements, however, is seldom high enough to resolve the L/H transition, and the spatial resolution is often too coarse to resolve the detailed field structure.

In this paper we present the first profile measurements of E_r in a spherical tokamak, using active Doppler spectroscopy on He^+. On the Mega-Ampére-Spherical-Tokamak (MAST) an existing system for measuring the toroidal velocity at the plasma edge at 17 spatial points ($\Delta R = 6$ mm) was recently upgraded to a higher spatial resolution up-to $\Delta R \gtrsim 1.5$ mm. In addition, poloidal chords were added. In section 2 we will introduce the new diagnostic (ECELESTE) and discuss the main sources of errors in the evaluation of E_r. The structure of the field in L-mode is treated in section 3 and in H-mode in section 4. Finally we will discuss and summarise the results (section 5)

2. Edge Doppler spectroscopy on MAST (ECELESTE)

The radial electric field in a magnetically confined plasma can be calculated from the force balance of any charged species α

$$m_\alpha n_\alpha \left\{ \frac{\partial \mathbf{u}_\alpha}{\partial t} + (\mathbf{u}\cdot\nabla)\,\mathbf{u} \right\} = -\nabla\cdot \overleftrightarrow{P}_\alpha + n_\alpha q_\alpha (\mathbf{E} + \mathbf{u}_\alpha \times \mathbf{B}) + \mathbf{F}_\alpha - m_\alpha \mathbf{u}_\alpha S_\alpha \tag{1}$$

Here, m_α is the mass, n_α the density, $\mathbf{u}_\alpha$ is the fluid velocity, $\overleftrightarrow{P}_\alpha = p_\alpha \overleftrightarrow{I} + \overleftrightarrow{\pi}_\alpha$ the pressure tensor (p_α: scalar pressure, $\overleftrightarrow{\pi}_\alpha$: viscosity tensor), q_α the charge, $\mathbf{F}_\alpha$ the friction, and S_α the sources. Neglecting inertia, viscosity, friction, and sources [9] E_r is approximated by the leading terms in equation (1)

$$E_r \approx \frac{k_B T_\alpha}{q_\alpha}\left(\frac{1}{L_{T_\alpha}} + \frac{1}{L_{n_\alpha}}\right) + u_{\phi,\alpha} B_\theta - u_{\theta,\alpha} B_\phi, \tag{2}$$

with $L_f = \partial \ln f/\partial r$ the gradient length of f, and $k_B \approx 1.38\cdot 10^{-23}$ the Boltzmann constant. The first term on the right side is the diamagnetic contribution to E_r, and the two other terms are the contribution from the Lorentz force.

Using Doppler spectroscopy on He^+, atomic data with the electron temperature, T_e, and density, n_e, and the magnetic equilibrium reconstruction all of the terms on the right hand side of equation (2) can be calculated. It is also worth noting that in the diamagnetic term only the density gradient length is not directly measured, but has to be calculated from the measured intensity with the help of the atomic excitation data using T_e and n_e profiles measured by a different diagnostic.

In a conventional aspect ratio tokamak it is usually enough to measure the poloidal velocity, since the toroidal magnetic field is much higher than the poloidal field, and the poloidal and toroidal flows are similar [9]. In the low-field-side (LFS) edge of a spherical tokamak, however, both field components are of similar magnitude, and hence, both velocity components have to be measured.

Until recently only toroidal measurements in 17 spatial locations with a radial resolution of $\Delta R =$ 6 mm (The spot size of the optics in the plasma was $\Delta R_s = 12$ mm, with a hexagonal packing of the fibres leading to a nominal resolution of $\Delta R = 6$ mm due to the imaging properties) were available on MAST and only *differences* in E_r could be estimated with the help of neoclassical theory [19]. This system has now been upgraded to 120 lines of sight. The 120 slits are connected to a patch panel where they can be connected to an individual selection out of 64 toroidal chords, 64 poloidal chords, one radial chord, and one chord illuminated by a Zn-lamp. The latter two chords are used to determine the absolute velocity. Details of the spectrometer can be found in [17]. The data are stored on the storage part of the frame transfer CCD throughout the discharge and only read-out after the shot. This allows a fast time resolution of up to $\Delta t \approx 110\,\mu$s if the measurement is restricted to 10 chords to avoid smearing between chords (see section 2.2). Here, we concentrate on measurements with high spatial resolution. In the normal chord layout, used for the measurements throughout the paper, 58 toroidal chords and 59 poloidal chords as well as the Zn-lamp and the radial chord were connected. In this set-up a time resolution of $\Delta t = 4.85$ ms has to be used.

The toroidal and poloidal view optics have a spot size of $d_t = 3$ mm, and $d_p = 2.8$ mm respectively. The fibres at the image plane have a hexagonal packing in 6 rows and 64 columns of 180 μm fibres for the toroidal system, and 2 rows and 64 columns of 400 μm fibres for the poloidal system. The overlap between the 64 radial points due to the hexagonal packing allows for a maximum spatial resolution of

Journal of Physics: Conference Series **123** (2008) 012005 doi:10.1088/1742-6596/123/1/012005

$\Delta r \approx 1.5$ mm after spatial deconvolution to be done in future. Both optics have a narrow field of view of about 10 cm. The poloidal in-vessel optic is fixed, whereas the field of view of the toroidal ex-vessel optics can be moved manually between 1.2 m $\leq R_{\text{mid}} \leq$ 1.5 m. For the measurement of E_r both systems have to cover the same radial range. The poloidal and toroidal views are orthogonal to each other, but tilted by 6° with respect to the horizontal mid-plane. Note, that the field line angle at the LFS mid-plane is about 30° on MAST, hence about 36° with respect to the toroidal axis of the diagnostic.

The lines of sight of both systems pass through the plume of a thermal He jet usually injected within 10 cm of the plasma edge at the LFS mid-plane. This non-perturbing He jet not only provides about 10 times higher signal levels than the background emission of He^+, but also a good localisation. In the most simple way this localisation can be estimated by the $\cos^5\gamma$ angular distribution (γ: angle) of the plume, and the curvature radius of the plasma R_c. The length of the chord at the distance x from the nozzle containing about 90% of the density of the plume is $g(x) = 8|x|e/15$ ($e \approx 2.71\ldots$), giving a height of the segment of a circle with radius R and base g is $\Delta R(x) = R_c - \sqrt{R_c^2 - g^2(x)/4}$. For the geometry on MAST this gives a spatial resolution of 2 mm $\lesssim \Delta R_t \lesssim$ 6 mm for the toroidal views (R_c = 1.4 m) and 5 mm $\lesssim \Delta R_p \lesssim$ 17 mm for the poloidal views $R_c \approx 0.5$ m over the 10 cm field of view. Again a full spatial deconvolution can be used to increase the spatial resolution to $\Delta R \approx 1.5$ mm. More work is underway to characterise the He plume experimentally and theoretically, and to deconvolve the data.

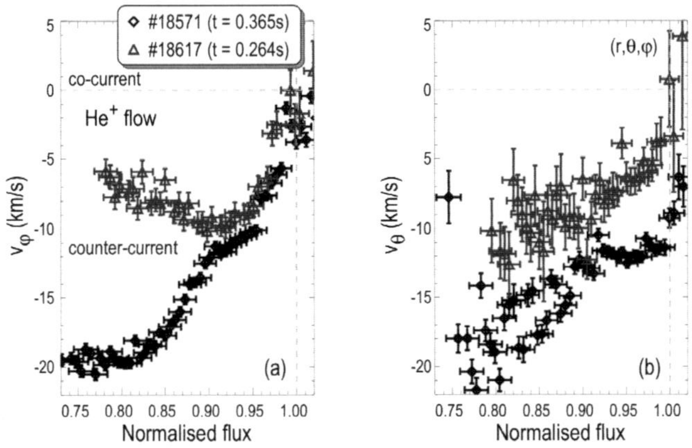

Figure 1. Example of (a) toroidal and (b) poloidal He^+flow profiles in H-mode for discharges with different plasma current #18571 (black,◇, I_p = 0.7 MA), and #18617 (blue, △, I_p = 1.1 MA). The minor radius r, poloidal angle θ and toroidal angle ϕ define a right handed coordinate system with ϕ in the co-current direction, and r outward from the plasma centre.

Example profiles of (a) the toroidal and (b) the poloidal He^+ flow in ELM free H-mode are shown in figure 1 as function of normalised poloidal flux $\psi_N = (\psi - \psi_s)/(\psi_0 - \psi_S)$, with $\psi = \Psi_{\text{pol}}/2\pi$ the poloidal flux function, and ψ_0, ψ_S the flux at the magnetic axis and the last closed flux surface (LCFS) respectively. The error bar of the velocity comes from the uncertainty of the fit (see section 2.1) and the wavelength calibration. The error bar on the radial position comes from the calculation above in the absence of a deconvolution. The resulting E_r from these measurements will be discussed in section 4.

2.1. Data analysis

The data presented in this paper are restricted to measurements with the He puffing. Each of the 120 line shapes are fitted by a Gaussian convoluted with the detector pixel size for each time slice. The apparatus function measured with the spectral line of a cold, low pressure Zn spectral lamp is also very close to a Gaussian shape. Therefore, the Doppler width is simply $\Delta\lambda_D = \sqrt{(\Delta\lambda_{\text{fit}})^2 - (\Delta\lambda_A)^2}$, where $\Delta\lambda_{\text{fit}}$ and $\Delta\lambda_A$ are the fitted line width of the He^+ line and the fitted width from the apparatus function respectively. From the fits we get the He^+ intensity, the relative He^+ velocity, and the He^+ temperature values for each chord. The absolute velocity is calculated using the recorded Zn spectrum at $\lambda_{\text{ZnI}} = 468.1034$ nm close to the measured He^+ line at $\lambda_{He^+} = 468.5673$ nm with its known spectral offset, and the relative positions from the spectral calibration of each chord also with the Zn lamp. The data are checked for consistency against the data from radial chord assuming a negligible radial velocity.

To evaluate the diamagnetic term, $E_r^{\mathrm{dia}} = k_B T_{\mathrm{He}^+}(L_T^{-1} + L_n^{-1})$ of the radial force balance (equation (2)) the experimental profiles are fitted with suitable functions, since the gradient length of T_{He^+} and n_{He^+} enter into E_r^{dia}. The T_e and n_e profiles from the multi point edge Thomson scattering (TS) [20] are fitted using a modified hyperbolic tangent function with the fitting parameters **p**[21]

$$f(x;\mathbf{p}) = \frac{p_1 - p_2}{2}\left[\frac{\left(1 + p_3\frac{p_4 - x}{2p_5}\right)\exp\left\{\frac{p_4 - x}{2p_5}\right\} - \exp\left\{-\frac{p_4 - x}{2p_5}\right\}}{\exp\left\{\frac{p_4 - x}{2p_5}\right\} + \exp\left\{-\frac{p_4 - x}{2p_5}\right\}} + 1\right] + p_2. \tag{3}$$

The logarithm of He$^+$ intensity as well as the He$^+$ temperature from the higher intensity toroidal chords are fitted using Legendre polynomials:

$$g_n(x;\mathbf{p}) = \sum_{i=0}^{n} p_i P_i(x) \text{ and } P_n(x) = \frac{1}{2^n n!}\frac{d^n}{dx^n}\left[(x^2 - 1)^n\right] \tag{4}$$

The orthogonal Legendre polynomials are quite stiff for the fitting, since the fitting almost represents a Fourier-Legendre-Series expansion. The He$^+$ temperature is not in equilibrium with the bulk plasma, since the ionisation into He^{2+} is faster than the thermalisation time, hence the He$^+$ temperature can't be used to determine the bulk ion temperature.

Without a sophisticated inversion algorithm, data from the innermost and the outermost chords are currently rejected. The intensity of the outermost chords is too low and suffers from smearing as discussed in section 2.2 and possibly reflections. On the innermost chords the intrinsic He$^+$ emission becomes comparable to the emission from the local gas puff and therefore requires careful inversion. The relative He$^+$ density is calculated from ADAS data [22] using the fitted T_e and n_e profiles. The Lorentz term, $E_r^{\mathrm{Lor}} = u_{\phi,\mathrm{He}^+}B_\theta - u_{\theta,\mathrm{He}^+}B_\phi$ is calculated without a fit to the profile of the experimental velocity data. The components of the magnetic field are taken from an equilibrium reconstruction (EFIT).

2.2. *Sources of systematic errors*

The main source of errors is the relative uncertainty between the radial calibrations of the profiles from the TS and the ECELESTE diagnostic. A relative shift between the two measurements of the order of $\Delta R \approx 1$ cm is enough to change the profiles significantly in strong H-mode. Great care was taken in the radial calibration of both diagnostics, by back-illuminating the fibres and measuring their position inside the vessel. The data of ECELESTE was also compared to other diagnostics with high accuracy of the radial calibration, giving an uncertainty of $\sigma_R^{\mathrm{EC}} \approx 3$ mm. The uncertainty of the position of the TS profile, however, is of the order of $\sigma_R^{\mathrm{TS}} \approx 5$ mm. An outward shift of the ECELESTE measurement with respect to the TS measurement leads to a reduction of the diamagnetic contribution by several km/s, whereas an inward shift increases the diamagnetic contribution. However, an outward shift leads also to a relative He$^+$ density profile which peaks outside the LCFS, which is unlikely at electron temperatures of typically 20 eV $\lesssim T_e^{\mathrm{LCFS}} \lesssim$ 30 eV. No relative shifts between the two measurements were employed.

A further error source comes from the lack of an optical shutter. Therefore, the CCD is exposed to light whilst it is clocked vertically. Since the 120 slits are arranged as an array of 12 rows and 10 columns, a given measurement picks up a fraction of light from all the 12 measurements vertically stacked on the CCD. This unwanted pick-up may amount to 22% due to the ratio of vertical transfer time $\tau_v \approx 0.3\,\mu$s, exposure time $\Delta t_{\mathrm{exp}} = 4.9$ ms, and binning on the CCD. The chord positions on the CCD were chosen in a way to reduce this pick-up, and the strong chords usually only have a pick-up of 2%. In the absence of transient events the data can be corrected.

A problem, however, arises for the radial chord, which doesn't pass through the He puff, and hence has a much reduced intensity. Analysis of this chord has to be restricted to time slices before the He puff, and preferably during L-mode phases of the discharge, which have low rotation speeds. A wrongly calculated velocity offset leads to a shift of the E_r profile by roughly $0.3 \times \Delta v_{\mathrm{offset}}$. The error on E_r was calculated by using a relative shift between the TS profiles and the ECELESTE profiles of $\Delta R = \pm 2.5$ mm to show the uncertainty and using the statistical errors of the velocity measurement. No error was added for the fitting error of the profiles.

3. The radial electric field in L-mode and during the L/H transition

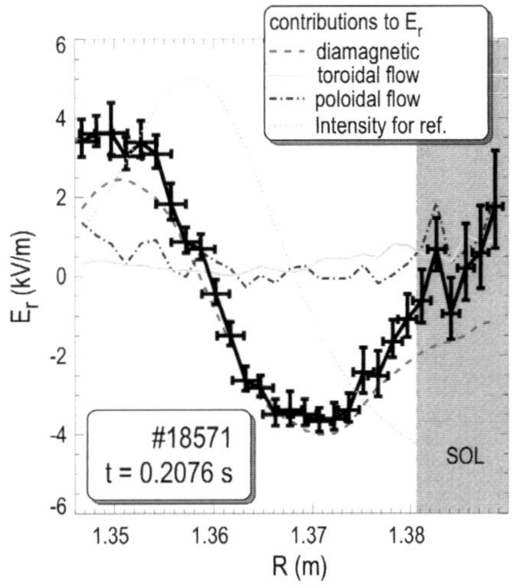

Figure 2. Typical radial electric field in L-mode (black,——). The contributions from the diamagnetic term (red, - - - -), and the individual components from the Lorentz term of u_ϕ (cyan,— · · —) and u_θ (blue, — · —), as well as the intensity profile for reference (green, · · · · · ·) are also shown.

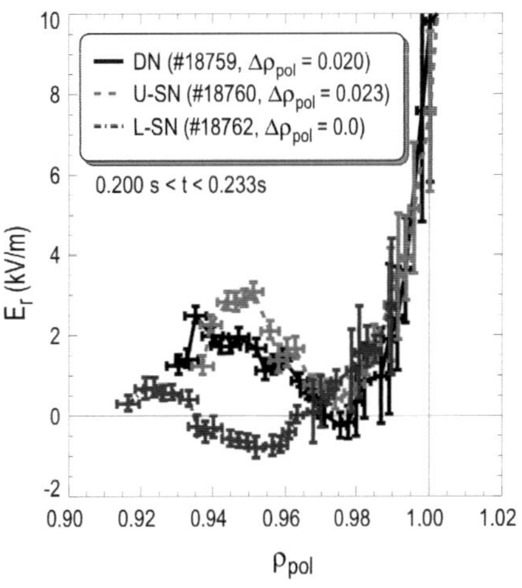

Figure 3. Comparison of the electric field in Ohmic discharges in DN ($\Delta R^{out}_{sep} = 0$ mm, black, ——), U-SN ($\Delta R^{out}_{sep} = 10$ mm, red, - - - -) and L-SN ($\Delta R^{out}_{sep} = -10$ mm blue, — · —). $\delta R^{out}_{sep} = R^{l}_{X} - R^{u}_{X}$, with $R^{l,u}_{X}$ the outer radius of the flux surface passing through the lower and upper X-point respectively.

The radial electric field in L-mode on MAST is very shallow and close to zero. An example of an L-mode E_r field at a single time point ($\delta t = 5$ ms, $t = 0.208$ s) is shown in figure 2 (black, ——) for a beam heated double null (DN) discharge immediately before a relatively sharp L/H transition (#18571, $I_p = 0.7$ MA, $\bar{n}_e = 2 \times 10^{19}$ m^{-3}, $P_{NBI} = 1.8$ MW). There are some unavoidable H-mode dithers during the measurement period, which may lead to slightly deeper profile than in an L-mode further away from H-mode. The TS profile, however, was measured during a true L-mode phase. The radial force balance of He^+ (equation (2)) is dominated by the diamagnetic term (red, - - - -), which determines E_r almost entirely. Any He^+ flow measured in L-mode is mainly parallel to **B**. Measurements in L-mode are particularly influenced by MHD affecting the edge plasma flow. Changes to mainly $u_\phi^{He^+}$ of the order of a few km/s have been observed during internal reconnection events, sawtooth crashes, and mode locking or growth of locked modes. These effects can even lead to positive E_r and will be the subject of further investigations. In the data presented here such MHD is absent.

Changes in the flow perpendicular to **B**, however, do occur, for example when the magnetic configuration is changed from DN to single null (SN) [16, 19, 23]. A comparison of the fields in similar Ohmic discharges with different magnetic configurations is shown in figure 3. The profiles are plotted against the normalised poloidal radius $\rho_{pol} = \sqrt{\psi_N}$. A problem in these set of discharges was the different behaviour of low frequency MHD ($f < 5$ kHz) susceptible to mode locking. The data are averaged between $0.2\text{ s} \leq t \leq 0.233$ s where the MHD was believed to be absent. The TS profiles were assembled from the individual time slices and fitted as one, because of the considerable fluctuation of the edge

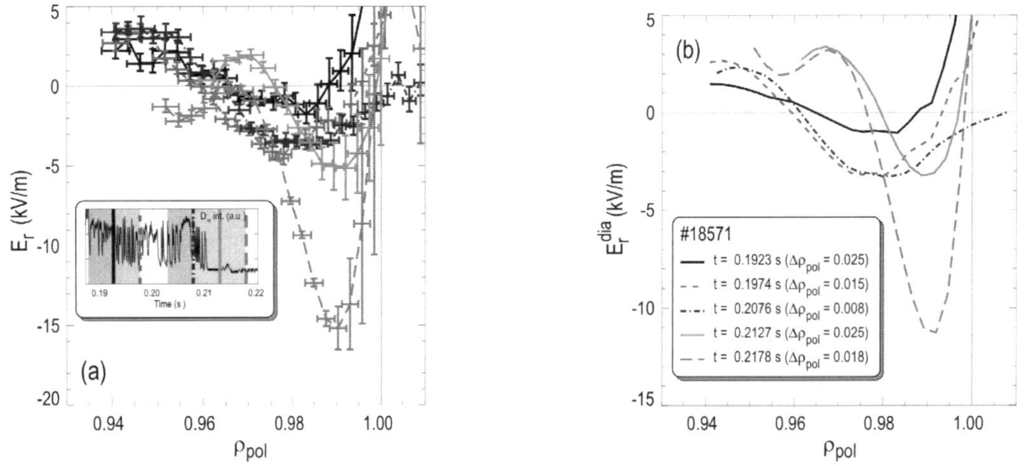

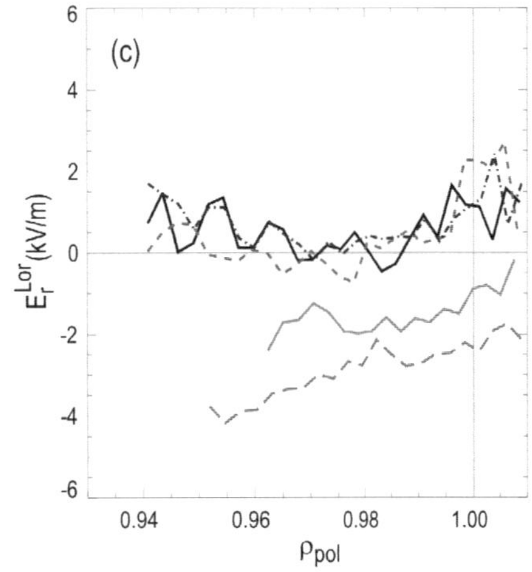

Figure 4. Evolution of E_r and its components through a fast L/H transition of #18571, (a) shows E_r, (b) the diamagnetic part, and (c) the Lorentz part. The different lines mark the different time points. The intensity of the D_α emission is shown in (a) with the vertical lines marking the end points of the ECELESTE measurement periods indicated by the shaded areas. Note that the profile at $t = 0.208$ s is the same as shown in figure 2.

plasma. In previous experiments on MAST and ASDEX Upgrade a $\Delta E_r \approx -1$ kV/m more negative E_r was observed in the DN compared to L-SN [19, 23]. The negative shear region of E_r, however, was very similar. In this data set the field in L-SN is the most negative with a broad E_r well, but the negative E_r shear seems to be strongest in DN with the narrowest E_r well. Furthermore, the strong change in $\partial E_r/\partial R$ observed on ASDEX Upgrade between upper SN (U-SN) and L-SN is not present in these discharges [23]. This difference between MAST and ASDEX Upgrade was also seen in B2SOLPS5.0 modelling [15]. Further experiments in more MHD stable plasmas are planned to investigate the subtle differences of E_r in the different configurations and compare them with theoretical predictions.

Of particular interest is the evolution of the radial electric field through an L/H transition. In figure 4 the evolution of E_r and its components through a relatively sharp L/H transition is shown. As the L/H transition is approached, the electric field well becomes deeper (figure 4a). Initially, during the dithery phase, the field is mainly determined by the diamagnetic term (figure 4b), and the Lorentz part is small and slightly positive (figure 4c). After the L/H transition at $t_{L/H} = 0.208$ s the Lorentz part becomes negative as a perpendicular He^+ flow develops, and a deep E_r well forms. The shear of E_r, however, dominated by the diamagnetic term of the He^+ force balance. The contribution of the Lorentz term to the minimum of $E_r^{min} \approx -14$ kV/m is in this case $E_r^{Lor} \approx -3$ kV/m.

Contour plots of (a) E_r, (b) E_r^{dia}, and (c) E_r^{Lor} through a very slow L/H transition are shown in figure 5.

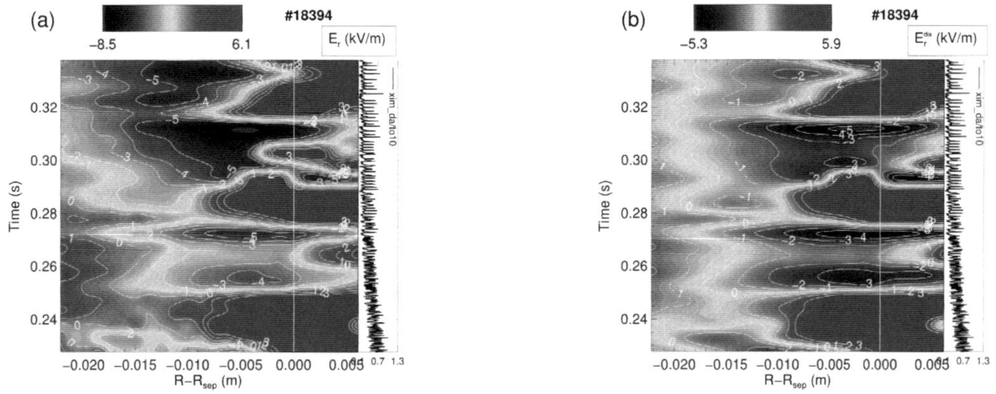

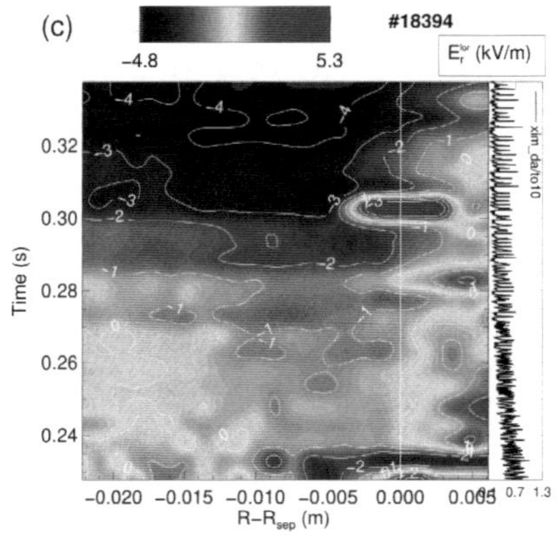

Figure 5. Contour plot of the evolution of E_r and its components through a slow L/H transition in #18394, (a) shows E_r, (b) the diamagnetic part, and (c) the Lorentz part. The intensity of the D_α emission is shown plotted on the right side of each graph for reference.

These plots show more clearly that as the H-mode progresses and deepens the Lorentz part , and hence the perpendicular velocity of the He^+ flow becomes increasingly more important, although the well is formed by the diamagnetic part. Care has to be taken when interpreting the diamagnetic part though, since the TS profile is an instantaneous snap shot ($\Delta t_{TS} = 7$ ns which sometimes is exactly on an ELM, whilst the ECELESTE data is averaged over $\Delta t = 5$ ms. This is the case at the times where the diamagnetic part is reduced around $t = 0.280$ s and $t = 0.315$ s. The ECELESTE measurement averages over several ELMs. Interestingly no strong shear is observed in the Lorentz part of the E_r on MAST.

4. The radial electric field in H-mode

As already shown in figure 4 a deep negative electric field well evolves in H-mode. The structure is mostly given by the diamagnetic part, as can be seen from figure 6, showing a low current $I_p = 0.7$ MA H-mode (#18571), and figure 7, showing a high current $I_p = 1.1$ MA H-mode (#18617). The Lorentz part contributes strongly to the minimum. The FWHM of the field well is only between 1 cm to 2 cm, and is narrower in the high current case (see also figure 8a). Also the pedestal width is narrower in the high current case than in the low current case. Both, the low and high current discharges have similar injected power of $P_{NBI} = 1.8$ MW (The initial input power in the high current discharge is $P_{NBI} = 3.2$ MW until $t = 0.215$ s), but different line averaged densities of $\bar{n}_e = 3.6 \times 10^{19}$ m^{-3} and $\bar{n}_e = 4.6 \times 10^{19}$ m^{-3} respectively at the time of the measurements. The densities in the low current discharge reaches $\bar{n}_e = 4.3 \times 10^{19}$ m^{-3} at the end of the H-mode and the E_r well broadens and deepens in time. Obviously, there is a difference between the two discharges in the power flowing over the LCFS, P_{loss}, due to the difference in I_p. In the $I_p = 1.1$ MA discharge $P_{loss} = 2.6$ MW, whereas in the $I_p = 0.7$ MA discharge $P_{loss} = 1.9$ MW.

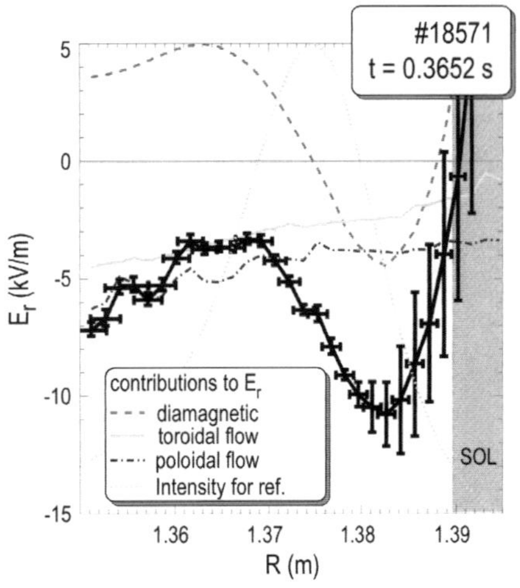

Figure 6. Typical radial electric field in a low current H-mode #18571 (black,——). The contributions from the diamagnetic term (red, - - - -), and the individual components from the Lorentz term of u_ϕ (cyan,— · · —) and u_θ (blue, — · —), as well as the intensity profile for reference (green, · · · · · ·) are also shown.

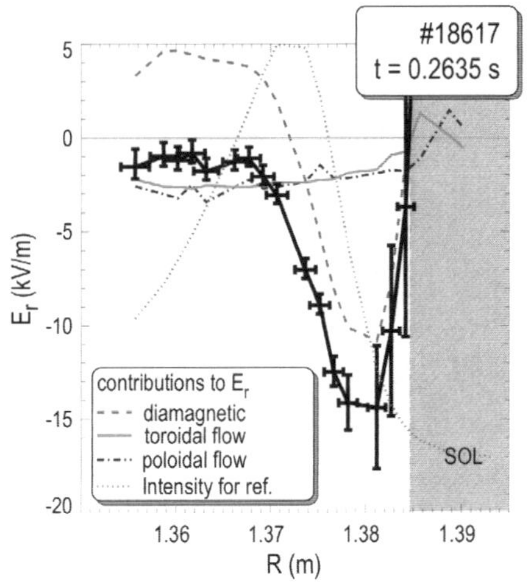

Figure 7. Typical radial electric field in a high current H-mode #18617 (black,——). The contributions from the diamagnetic term (red, - - - -), and the individual components from the Lorentz term of u_ϕ (cyan,— · · —) and u_θ (blue, — · —), as well as the intensity profile for reference (green, · · · · · ·) are also shown.

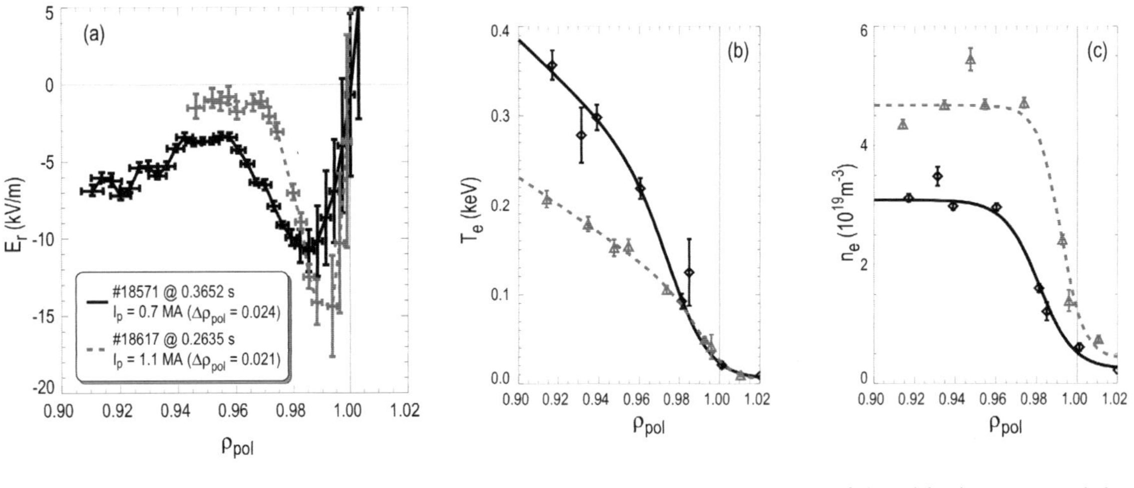

Figure 8. Comparison of (a) E_r, (b) T_e, and (c) n_e between the low current (#18571, black ——) and the high current discharge (#18617, red - - - -).

A further difference between the discharges is the fuelling location. The $I_p = 0.7$ MA discharge is fuelled from the high-field-side (HFS) with controllable valves at the top and bottom of the centre column, and the $I_p = 1.1$ MA discharge is continuously fuelled from the HFS mid-plane. These different fuelling locations generally lead to a different distribution of the pedestal pressure between T_e and n_e as can be seen from figure 8 (b,c). Nevertheless, pedestal pressure itself is relatively similar in both discharges. It is also noteworthy that the difference in T_e is not caused by a steeper temperature gradient, but rather by the wider transport barrier. The electron pressure gradient is larger in the low current discharge (see figure 9), although the negative electric field shear $\nabla E_r \approx -800\,\mathrm{kV/m^2}$ is 2 times smaller than in the high current discharge ($\nabla E_r \approx -2000\ \mathrm{kV/m^2}$) (see figure 9)

Because of the uncertainty in the absolute separatrix position by about $\Delta R \approx 1$ cm all the profiles have been shifted with respect to each other so that $E_r = 0\ \mathrm{kV/m}$ at the separatrix, $\rho_{pol} = 1$. In the SOL the plasma potential is usually positive $\phi_{pl} \propto T_e$ and increases from the outside, due to the requirement of a constant potential, $\phi_{target} = 0$ V along the earthed conducting target plates. Inside the LCFS the potential drops to give a negative electric field confining the ions. Therefore, ϕ_{pl} usually has a maximum at the LCFS, thus motivating the shift of the profiles. To arrive at a firm conclusion, whether the differences in E_r are due to the different plasma current, or the different fuelling location, will be the subject of future work.

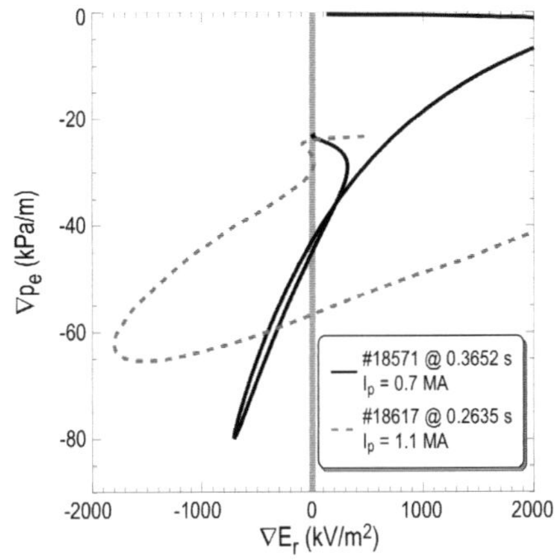

Figure 9. Electron pressure gradient against the radial electric field gradient for the low current (#18571, black ——) and the high current discharge (#18617, red - - - -) as in figure 8

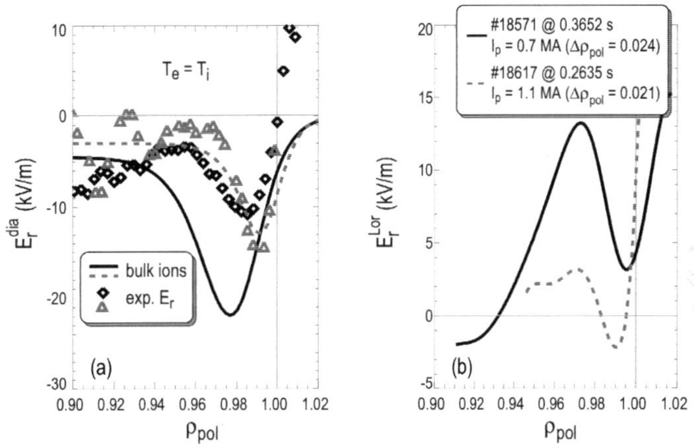

Figure 10. Contributions to the radial force balance for the bulk ions (a) diamagnetic part, and (b) Lorentz part for the low current (#18571, black ——) and the high current discharge (#18617, red - - - -) as in figure 8. Also shown in (a) is the measured E_r ($\Diamond$,$\triangle$).

The transport barrier seems to be related to the negative shear region of the E_r well rather than the strong positive shear approaching the LCFS as demonstrated in figure 9. Here, the pressure gradient is plotted against the electric field gradient. Both are derived from fits to the data using the functions given in equation 3 for the pressure, and equation 4 for E_r. Clearly the strongest gradient is in the region of the negative shear. In figure 10 the contributions of the force balance for the bulk ions are shown. Here, $T_e = T_i$ is assumed. The strong diamagnetic contribution (figure 10a) leads to an mostly positive Lorentz contribution (figure 10b). This is consistent with a co-current toroidal rotation of the bulk ions. Note, that the position of the strongest pressure gradient of the bulk ions is further inward than the strongest pressure gradient for the He^+ ions.

5. Summary and conclusions

For the first time in a spherical tokamak the edge radial electric field structure has been measured. Using a new active edge Doppler spectroscopy diagnostic, the field inside the first 10 cm of the plasma is calculated from the impurity ion momentum balance on up to 60 points with a spatial resolution of a few mm.

The L-mode field on MAST is rather shallow and the impurity flow which is mostly parallel to **B** plays a small role in the force balance. At the L/H transition the plasma spins up perpendicular to **B** and a strong negative electric field well develops in H-mode. The depths of the E_r well is of the order of $-15\ \mathrm{kV/m} \lesssim E_r$ and has a width of $1\ \mathrm{cm} \lesssim \Delta R \lesssim 2$ cm amounting to a negative shear up to

$\nabla E_r \approx -2\,\mathrm{MV/m^2}$. The He^+ flow, though important for the absolute value of E_r, contributes little to the overall shear of E_r. For the bulk ions the diamagnetic term (estimated from the electron profiles) is even larger than for He^+, and exceeds the values of E_r. Hence, the perpendicular bulk ion flow is likely to be in the opposite direction compared to He^+ flow consistent with a toroidal rotation generated by co-current neutral beam injection rotation. It is not surprising that due to the strong gradients in fully developed H-mode the diamagnetic term of the force balance is important. At the L/H transition this might be different as experimental evidence suggests [7]. The perpendicular flow could build up on a faster time scale than the pressure gradient, and therefore become more important for the transition physics. Faster measurements using the new diagnostic will be used in future to investigate this aspect on MAST.

The width of the H-mode E_r well reflects roughly the width of the transport barrier. The inner region of negative shear of the E_r well coincides with the position of the strongest pressure gradient, although the positive shear across the last closed flux surface is higher. This suggests that the negative shear is more important for the suppression of the turbulence. The pedestal temperature gradient and indeed the pressure gradient of the electrons on MAST is independent of the strong shear of the electric field, showing that once the shear is high enough, and the turbulence is largely suppressed other effects seem to govern ∇p_e. Insight into this threshold behaviour might be gained in future by investigations using the new ELM control coils on MAST, and turbulence measurements using a new beam emission spectroscopy system.

These first results show, that the new instrument is well suited to investigate the intimate relationship between the edge transport barrier and the radial electric field. Detailed comparison of the field structure with theory will allow a deeper understanding of what physics generates E_r, and what leads to the L/H transition. In particular the question what determines the width of the E_r well is important, because it may lead to better confinement regimes on MAST, and an understanding of the width of the pedestal.

Acknowledgements
This work was funded jointly by the United Kingdom Engineering and Physical Sciences Research Council and by the European Communities under the Contract of Association between EURATOM and UKAEA. The views and opinions expressed herein do not necessarily reflect those of the European Commission.

References

[1] Wagner F, Becker G, Behringer K *et al.* 1982 *Phys. Rev. Lett.* **49** 1408
[2] ASDEX-Team 1989 *Nucl. Fusion* **29** 1959–2040
[3] Burrell K H, Carlstrom T N, Doyle E J *et al.* 1992 *Plasma Phys. Control. Fusion* **34** 1859–1869
[4] Akers R J, Counsell G F, Sykes A, Appel L C, Arends E R, Byrom C, Carolan P G, Conway N J, Cunningham G, Dnestrowskij A, Dnestrowskij Y N, Field A R, Fielding S J, Gryaznevich M, Helander P, Kirk A, Korsholm S, Martin R, Meyer H, Nightingale M P S, Roach C M, Shevchenko V, Tournianski M R, Walsh M J, Warrick C D, Team T M and Team T N 2002 *Phys. Rev. Lett.* **88** 035002
[5] Maingi R *et al.* 2002 *Phys. Rev. Lett.* **88** 35003
[6] ITER-Team 1999 *ITER Physics Basis* vol 39 (Nucl. Fusion) chap 2, pp 2175–2249
[7] Burrell K H 1999 *Phys. of Plasmas* **6** 4418–4435 invited contribution to the APS-meeting 1999 in Atlanta
[8] Ida K 1998 *Plasma Phys. Control. Fusion* **40** 1429–1488
[9] Meyer H, Carolan P G, Conway N J, Field A R, Fielding S J and Helander P 2000 *Czechoslovak Journal of Physics* **50** 1451
[10] Hahm T S and Burrell K H 1995 *Physics of Plasmas* **2** 1648–1651
[11] Terry P W 2000 *Rev. Mod. Phys.* **72** 109–165
[12] Diamond P H, Itoh S, Itoh K and Hahm T S 2005 *Plasma Phys. Control. Fusion* **46** R35–R161
[13] Burrell K H 2006 *Plasma Phys. Control. Fusion* **48** A347–A363
[14] LaBombard B *et al.* Poloidal transport asymmetries, edge plasma flows and toroidal rotation in Alcator C-MOD 16th Int. Conf. on Plasma Surface Interactions in Controlled Fusion Devices (Portland, USA 24–28 May 2004) P2-57, to be published in J. Nucl. Mat.
[15] Rozhansky V, Kaveeva E, Voskoboynikov S, Counsell G, Kirk A, Meyer H, Coster D, Conway D, Schirmer J, Schneider R and ASDEX Upgrade team 2006 *Plasma Phys. Control. Fusion* **48** 1425–1435
[16] Schirmer J, Conway G D, Zohm H, Suttrop W and the ASDEX Upgrade Team 2006 *Nucl. Fusion* **46** S780–S791

[17] Carolan P G, Conway N J, Field A R, Jones P B and Meyer H 2001 *Rev. Sci. Instr.* **72** 881–887
[18] Conway G D, Schirmer J, Klenge S, Suttrop W, Holzhauer E and the ASDEX Upgrade Team 2004 *Plasma Phys. Control. Fusion* **46** 951–970
[19] Meyer H, Carolan P G, Conway N J, Counsell G F, Cunningham G, Field A R, Kirk A, McClements K G, Price M, Taylor D and the MAST team 2005 *Plasma Phys. Control. Fusion* **47** 843–867
[20] Scannell R, Walsh M J, Carolan P G, Conway N J, Darke A C, Dunstan M R, Hare D and Prunty S L 2006 *Rev. Sci. Instrum* **77** 10E510
[21] Groebner R J and Carlstrom T N 1998 *Plasma Phys. Control. Fusion* **40** 673–677
[22] Summers H P and von Hellermann M 1993 *Atomic and plasma-material interaction processes in controlled thermonuclear fusion* ed Janev R K and Drawin H W (P. O. Box 103, 1000 AC Amsterdam, The Netherlands: Elsevier Science Publishers B. V.)
[23] Meyer H, Carolan P G, Conway G D, Cunningham G, Horton L D, Kirk A, Maingi R, Ryter F, Saarelma S, Schirmer J, Suttrop W, Wilson H R and the MAST, ASDEX Upgrade and NSTX teams 2006 *Nucl. Fusion* **46** 64–72

Journal of Physics: Conference Series **123** (2008) 012006 doi:10.1088/1742-6596/123/1/012006

Characteristics of high density edge transport barrier with reheat mode on CHS

T Minami, S Okamura, T Akiyama, K Ida, T Oishi, M Isobe, H Nakano, A Fujisawa, K Nagaoka, M Yoshinuma, C Suzuki, Y Yoshimura, K Toi, S Ohshima, M Takeuchi, H Iguchi, S Nishimura, A Shimizu, K Matsuoka and C Takahashi[1]

National Institute for Fusion Science, Oroshi-cho 322-6, Toki 509-5292, Japan

E-mail: minami@nifs.ac.jp

Abstract. Edge transport barrier (ETB) formation and a reheat mode have been simultaneously realized on the Compact Helical System (CHS). The new mode is induced by neutral particle reduction in the edge region, which is caused by shutting off fueling after a strong gas-puffing. When both the reheat mode and the ETB are simultaneously realized, the density reduction is suppressed by the ETB in the peripheral region, and the temperature continues to increase by the reheat mode. This mode provides an enhanced confinement in the high density region ($\bar{n}_e \sim 1.2 \times 10^{20}\mathrm{m}^{-3}$) compared to the ETB formation without the reheat mode, because of a large suppression of an anomalous transport, which is confirmed with fluctuation measurements in the edge region.

1. Introduction

The realization of improved confinement for a high density plasma is important from the aspect of a helical device as a fusion reactor for the helical device, because high density operation is favorable for the helical devices compared to the tokamak devices[1]. However, it is difficult in the high density range to achieve an enhanced confinment. For example, an internal transport barrier of the helical devices that is formed by an ambipolar diffusion of neoclassical particle fluxes requires a low electron density ($5 \times 10^{18}\mathrm{m}^{-3}$) [1]. On the contrary, the formation of the edge transport barrier (ETB) and a reheat improved confinement mode occurred in a high density range compared to the helical ITB [2, 3, 4, 5, 6].

The reheat mode is a confinement improvement mode similar to an improved L-mode that is observed in tokamak experiments. This mode is initiated by shutting off fueling after strong gas-puffing[5, 6]. During the reheat mode, the density profile shape is peaked, and the temperature in the edge region considerably increases transiently. The electron temperature in the peripheral region increases resulting from a suppression of neutral particle density causing the charge exchange loss. The reheat mode is also similar to the high T_i mode[8] of the helical plasma, which is caused by the radial electric field that is created by an ion root transition. However, the reheat mode has a problem: the peripheral density continues to decrease after the fueling stops, and the stored energy rapidly goes down.

Meanwhile, the ETB improves the peripheral particle confinement by barrier formation. The ETB of CHS has a clear spontaneous drop of H_α emission followed by the increase of line-averaged electron density at the L-H transition. The quick density doubling ($< 100\mu$ sec) has been observed accompanying the density gradient increase in the edge region. The NBI power that is required for the barrier formation in the limiter-plasmas is similar to the tokamak divertor H-mode scaling [9, 2]. However, there are several different points from the tokamak experiments. In CHS experiments, the ETB is ELM free, therefore degradation by ELMs has not been observed[2]. In tokamak experiments, the clear pedestal structure has been observed, while the structure of the barrier is unclear in CHS experiments. In the previous CHS experiments, the H-factor, which is the ratio of the experimental confinement time to the calculation from the stellarator confinement scaling, increases during $\sim$ 20 ms after the transition, however, it gradually decreases subsequently with a density increase, because the confinement improvement is degraded. If the performance degradation of the ETB does not occurred with the density increase, the enhanced confinement is achieved in the high density range.

Accordingly, the simultaneous realization of both modes are favorable for the enhanced confinement in the high density range. However, it requires an extending the ETB performance in the higher density range. In addition,there is the issue of whether the ETB and the reheat mode are compatible modes or not. Recently, the edge transport barrier (ETB) during the reheat mode has been observed on CHS. The results have been preliminarily reported in [7].

This paper describes the detailed characteristics of the high density edge transport barrier formation during the reheat mode. Secondly, we show the structure of the improved region with the YAG thomson scattering measurements. Thirdly, we show that the results of the fluctuation measurement with a YAG phase contrast interferometer and discuss the anomalous transport. Finally, we summarize this paper.

2. Global behavior of high density edge transport barrier formation during reheat mode

Experiments are performed on CHS, which is a medium-sized heliotron/torsatron-type device with a periodicity of (l,m) = (2,8). The major and averaged minor radii are 1.0 and 0.2m, respectively. Because CHS is equipped with two co-NBIs (maximum power of each NBI is 0.8MW) and two gyrotrons (53GHz, 106GHz), we can study plasma physical characteristics for a wide variety of input powers and heating methods. When the reheat mode with the ETB is observed, the peripheral temperature increases while maintaining the plasma density in the edge region by the ETB formation.

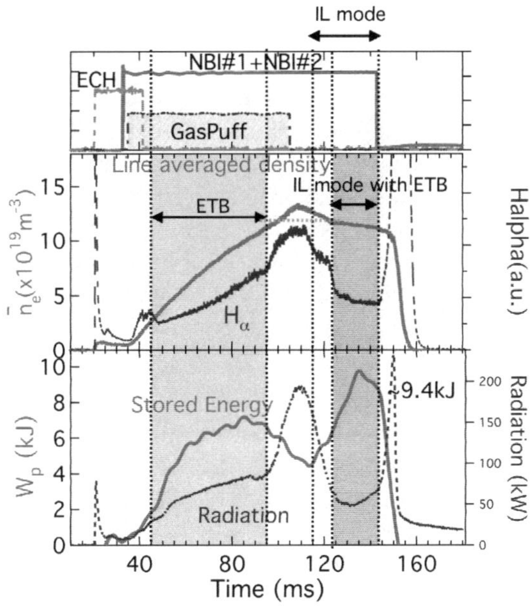

Figure 1. Global behavior of ETB plasma during reheat mode: line average density, H_α signal , radiation signal, and the stored energy W_p are plotted with the injection timings of NBI and ECH. Blue zone indicates period of first ETB formation without reheat mode. Red zone indicates period of ETB formation during reheat mode.

Figure 1 shows the global behavior of the reheat mode discharge with an ETB. Although high field strength (B_T = 1.86T) is favorable for the reheat mode, the formation of the ETB under a high magnetic field condition is difficult, because the NBI power threshold of the ETB formation depends on the

Journal of Physics: Conference Series **123** (2008) 012006 doi:10.1088/1742-6596/123/1/012006

magnetic field strength[2]. When the field strength increases, the NBI power threshold becomes larger. When the vacuum magnetic axis location (R_{ax}) shifts outwards, the threshold NBI power decreases. Accordingly, the experiment is carried out for the magnetic configuration of R_{ax} = 94.9 cm, which is larger than that of the standard configuration (R_{ax} = 92.1 cm).

As shown in figure1, the two co-NBIs (total power is 1.6 MW) are injected to the target plasma that is produced by a 54.5 GHz gyrotron. The ETB is formed below an upper density limit that is related to the NBI power threshold: the power threshold is determined by the heating power normalized with the electron density (P/n_e) [2]. On the other hand, because the higher plasma density is required for a reheat mode, the plasma density is increased by gas-puffing until the ETB formation disappears once.

As shown in figure 1, the ETB is formed at 45 ms by the strong gas-puff, then the ETB disappears and the plasma returns to the L-mode again at 95 ms resulting from the electron density exceeding the upper limit, which is determined by the density dependence of the power threshold ($P_{threshold} \propto n_e^{0.4}$). The stored energy is saturated and decreases after the transition with the density increase. After the L-mode transition, the stored energy further decreases. Then, the plasma density, as shown in the middle of figure 1, decreased after the gas-puff stop at 105ms. The onset of the reheat mode is denoted by the increased plasma stored energy from 115 ms due to the temperature increase in the peripheral region. The plasma radiation continues to reduce resulting from the reduction of peripheral neutral density. When the density decreased below the upper limit ($\bar{n}_e \sim 1.2 \times 10^{20}\mathrm{m}^{-3}$), the ETB is reformed in the edge region, which is denoted by the H_α reduction at 123 ms. And then, the density reduction is suppressed, as show in figure 1, and accordingly, the particle transport in the edge region is reduced. It is noted that this ETB occurred in a high density range, where enhanced improvement by ETB formation with gas-puffing have not been observed. In the reheat mode without the ETB, the density in the edge region is rapidly reduced. In contrast, when the ETB is formed in the reheat mode, the density reduction is supressed. Consequently, plasma confinement is improved by the synergistic effect of the simultaneous realization of both the reheat mode and the ETB. As a result of the continuous increase of the temperature by the reheat mode, the stored energy increased up to ~ 9.4kJ, which is larger than that of the first ETB phase (~ 7 kJ). After that, the increase of the temperature is saturated, the density decreases gradually due to no fueling, thus the stored energy is slightly reduced (134 ms). The confinement is degraded by the radiation increase in the edge region.

An H-factor based on ISS04 CHS/Heliotron/ATF[10] scaling is estimated, in which the effect of dW/dt is taken into account. In the CHS experiments, the H-factor of the NBI plasma without the ETB is below one. When the ETB is formed during the reheat mode, the H-factor increases up to ~ 1.3, which is larger than that during the L-mode just before the reheat mode. Thus, improved confinement is realized in the high density range ($\bar{n}_e \sim 1 \times 10^{20}\mathrm{m}^{-3}$) by the simultaneous realization of both the ETB and the reheat mode. The H-factor value of ~ 1.3 might be underestimated, because the CHS L-mode confinement is degraded by the outward shift[11]. The experiments are performed for the configuration of R_{ax} = 94.9 cm, while the ISS04 scaling is derived from the data in the standard configuration of the R_{ax} = 92.1 cm.

3. Pedestal structure during the high density edge transport barrier

Detailed plasma profiles on the ETB transition during the reheat mode were measured with the mulitpoint YAG laser Thomson scattering system. Figure 2 shows density, temperature, and pressure profiles for the typical discharge with the ETB formation during the reheat mode, respectively. The 4 profiles for different timing that represent the characteristics are plotted: the profiles before the fueling stops (100 ms), before the reheat mode (110 ms), during the reheat mode without the ETB (120 ms), and during the reheat mode with the ETB (130 ms) are plotted in figure 2.

The electron temperature and density in the L-mode continue to decrease during the first 10 ms after the timing of the gas-puff stop. However, the electron temperature increases inversely by the reheat mode from 110 ms, although the electron density continues to decrease in the edge region. Because the central density continues to increase, the density profile gradually becomes peaked during the reheat

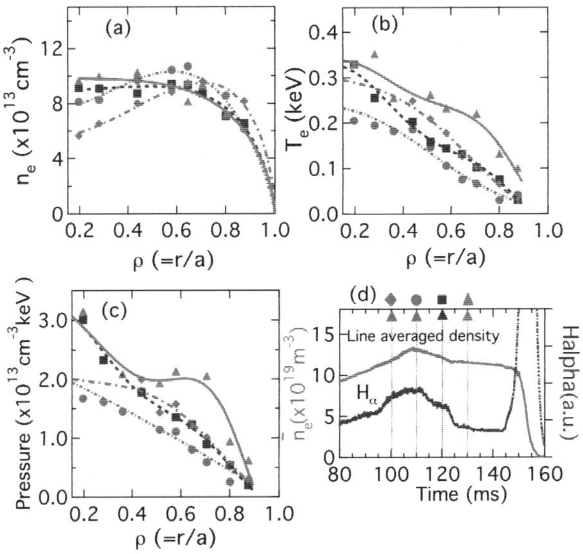

Figure 2. Density (a), Temperature (b) and pressure (c) profiles. Profiles before the gas-puff stop (100 ms), before the reheat mode (110 ms), during the reheat mode without the ETB (120 ms), and during the reheat mode with the ETB (130 ms) are plotted. Timings of the measurements are denoted (d).

mode. In contrast, because the edge temperature increases during the reheat mode, an inflection point appears in the temperature profile at $\rho \sim 0.5$, and the pedestal structure is formed, as is shown in figure 2(b). The estimated penetration length of the neutral particles to the plasma from the plasma surface is approximately in terms of $\Delta\rho \sim 0.4$, thus the region of the temperature increase almost coincides with the region of reduced neutral particles.

From the onset (123 ms) of the ETB formation, the reduction of the density is suppressed and the density profile shape is maintained during the ETB formation. This result shows that the diffusion of the peripheral plasma was blocked by the transport barrier formation. In addition, the electron temperature continues to increase by the reheat mode in the edge region. The previous results of the ETB experiments on the CHS show that the edge density gradient increases with the confinement improvement just after the L-H transition. However, the plasma confinement is degraded ~ 20 ms after the transition, because the edge electron temperature decreases with the edge density continuous increase[2], though the H_α signal is maintained at a small level. In contrast, the plasma pressure in the edge region at 140 ms increases up to three times larger than that of the first ETB. Consequently, the pressure gradient increase, and the pedestal structure is also formed in the pressure profile, as shown in figure 2 (c), thus the increase of the stored energy is mainly caused by the plasma pressure increase in the edge region. The deposited NBI power to the ETB plasma during the reheat mode is almost the same as that in the final timing of the first ETB (~ 90 ms), because the densities in both timings are almost the same. Therefore, global plasma confinement is improved by the ETB formation during the reheat mode.

4. Fluctuations during the high density edge transport barrier

Figure 3 shows the results of fluctuation measurement with a YAG laser phase contrast interferometer (YAG PCI)[12]. The fluctuations are integrated along the YAG laser path passing through the mid-plane of the plasma cross-section. The wave-number direction of the fluctuation component is poloidal (k_θ), because the YAG PCI can measure the perpendicular component to the laser path. In the ETB phase without the reheat mode, the fluctuation is reduced a few micro seconds before the L-H transition. After then, the fluctuation gradually increases with the density increase, and the increase of the stored energy is saturated because of the confinement degradation, as shown in figure 3.

In contrast, when the plasma enter the reheat mode that is denoted by the stored energy re-ascending, the reduction of the fluctuation is enhanced accompanying the H-factor increase. When the ETB is

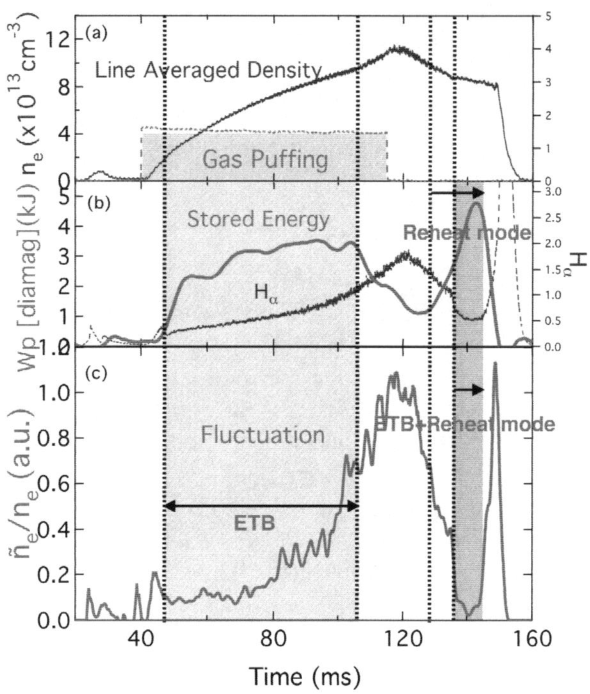

Figure 3. Result of fluctuation measurement with a YAG laser phase contrast method. (a) Density and gas-puffing (b) Stored energy and H_α. (c) Fluctuations, Blue zone indicates period of first ETB formation without reheat mode. Red zone indicates period of ETB formation during reheat mode.

formed during the reheat mode, the density reduction is reduced (figure 3 (a)) and the fluctuations sharply drop to the level of the onset of the first ETB phase, though the plasma density during the reheat mode is considerably larger than that of the first ETB phase. The reduction of the fluctuations is maintained during the ETB phase. The same tendency of the fluctuations in the ETB plasma during the reheat mode is confirmed with the other measurements such as a HCN scattering measurement and beam emission spectroscopy[9, 13].

The improved plasma confinement by the ETB during the reheat mode is related to the reduction of anomalous transport. As show in figure 1, the H_α signal behavior shows that the neutral particle density of the ETB phase during the reheat mode is lower than that of the first ETB phase. In addition, the confinement in the first ETB phase is gradually degraded resulting from the radiation increase. These results suggest that the expansion of the ETB formation range to the high density is related to the neutral particles in the edge region. The viscosity of the edge plasma might decrease due to the neutral particle reduction, then the anomalous transport is suppressed by the $E \times B$ shear flow in the edge region.

5. Summary

We have demonstrated the simultaneous realization of both a reheat mode and an edge transport barrier (ETB) on CHS. The profile measurements show that a pedestal-like structure is created in the temperature and the pressure profiles in the edge region. Consequently, the peripheral plasma pressure becomes larger than that of the ETB without the reheat mode. The ETB during the reheat mode shows the compatibility of both modes in the high density range ($\bar{n}_e \sim 1.2 \times 10^{20}\mathrm{m}^{-3}$), where the confinement improvement by the ETB without the reheat mode has not been observed. The new mode provides enhanced confinement in the high density region due to the reduction of anomalous transport that is confirmed by the fluctuation measurement with the YAG PCI.

References

[1] Minami T *et al.* 2004 *Nucl. Fusion* **44** 3421

[2] Minami T *et al.* 2006 *J. Plasma Fusion Res.* **1** 0.32
[3] Okamura S *et al.* 2004 *Plasma Phys. Control. Fusion* **46** A113
[4] Okamura S *et al.* 2003 *J. Plasma Fusion Res.* **79** 977
[5] Morita S *et al.* 1992 *Proc. 14th Int. Conf. Plasma Physics and Controlled Nuclear Fusion Research*, Würzburg, Germany
[6] Isobe M *et al.* 2006 *Fusion Science and Technology* **50** 229
[7] Minami T *et al.* 2006 *J. Plasma Fusion Res.* **1** 0.47
[8] Ida K *et al.* 1999 *Nucl. Fusion* **39** 1649
[9] Akiyama T *et al.* 2006 *Plasma Phys. Control. Fusion* **48** 1683
[10] Yamada H *et al.* 2004 *Proc. of 31th EPS Conf. on Contr. on Fusion and Plasma Phys.* ECA Vol. **28G** P-5 099
[11] Okamura S *et al.* 1999 *Nucl. Fusion* **39** 1337
[12] Matsuo K, Kado S Uchida N, Kawakubo M, Iguchi H, Okamurar S and Matsuoka K 2004 *Jpn. J. Appl. Phys.* **43** 2721
[13] Oishi T *et al.* 2007 "Observation of the high density edge transport barrier in CHS using beam emission spectroscopy" *this conference P1-11*

Journal of Physics: Conference Series **123** (2008) 012007 doi:10.1088/1742-6596/123/1/012007

Observation of the high-density edge transport barrier in CHS using beam emission spectroscopy

T Oishi[1,a], S Kado[2,b], M Yoshinuma[1], K Ida[1], T Minami[1], T Akiyama[1], M Isobe[1], C Suzuki[1], K Nagaoka[1], S Okamura[1] and CHS group[1]

[1] National Institute for Fusion Science, Toki 509-5292, Japan
[2] High Temperature Plasma Center, The University of Tokyo, 5-1-5 Kashiwanoha, Kashiwa, Chiba 277-8568, Japan

E-mail: t-oishi@mail.nucl.nagoya-u.ac.jp

Abstract. The formation of an edge transport barrier (ETB) has been observed in the high-density region ($n_e \sim 1.0 \times 10^{20}$ m^{-3}) of the Compact Helical System (CHS). The high-density ETB is observed in so-called "reheat mode" discharge, an improved confinement mode in which the temperature increases due to the change of density profile by controlling gas-puff fueling. In the present study, we investigated the behaviour of the density profile and the density fluctuations in the edge region accompanied by the formation of the high-density ETB using beam emission spectroscopy (BES), and we compared the behaviour with that of the normal ETB transition of CHS. BES has been implemented in CHS to simultaneously measure both local density fluctuations and gradients. It revealed that the density, density gradient, and inverse value of the density scale length increase in the edge region at both the normal ETB transition and the high-density ETB transition. The turbulence of the frequency at less than 20 kHz is suppressed at the formation of the high-density ETB.

1. Introduction

Studies of helical devices to confine high-density plasmas have been progressing as represented by the high-density H-mode in W7-AS [1], the super dense core maintained by an internal diffusion barrier in the Large Helical Device [2], and so on. The density operation region applied in these phenomenon reaches up to the order of 10^{20} m^{-3}. For realization in a fusion reactor, an improvement of temperature will be needed. Various types of improved confinement have been investigated in the compact helical system (CHS), which is a low-aspect-ratio, middle-sized heliotron (major radius = 1 m, minor radius = 0.2 m, toroidal period number = 8, polarity = 2), including the edge transport barrier (ETB) discharge [3-8] and the so-called "reheat mode" [9,10]. In the ETB discharge, a steep edge density gradient is formed due to improved particle confinement. In the "reheat mode", the temperature increases due to the change of density profile by controlling gas-puff fueling. This mode is similar to the "improved L-mode" [11] or "improved Ohmic confinement" [12] in tokamaks.

Recently, a new improved confinement mode by simultaneous realization of both ETB discharge and the improved L-mode was discovered. In this case, the formation of an ETB in the high-density region ($n_e \sim 1.0 \times 10^{20}$ m^{-3}) accompanied by an increase in edge electron temperature is observed.

[a] Present affiliation: Nagoya University, Chikusa-ku, Nagoya 464-8603 Japan
[b] Present affiliation: The University of Tokyo, 7-3-1 Hongo, Bunkyo-ku, Tokyo 113-8656 Japan

We call this new mode "high-density ETB" [13]. In the present study, we investigated the behaviour of the density profile and the density fluctuations in the edge region accompanied by the formation of the high-density ETB using beam emission spectroscopy (BES) [14]. The BES method, which is widely used to measure long-wavelength plasma density fluctuations, detects emissions from the collisionally excited neutral beam atoms (denoted as "beam emission") [15]. BES is a local diagnostic method that can measure multiple locations in plasmas simultaneously. It has been applied in several tokamak experiments and has provided important results, such as 2-D imaging measurement of turbulence [16], supression of turbulence at the edge region in the H-mode discharge [17], and so on. On the other hand, it has not been applied in helical systems before the case of CHS because the complicated magnetic configuration makes the spatial resolution worse. In order to overcome this difficulty, we designed the optics which can obtain optimized radial spatial resolution, which made a BES measurement in CHS successive. The diagnostic system and effort to improve the spatial resolution will be introduced briefly in the section 3.

2. Global characteristics of the ETB transition and the high-density ETB transition in CHS

A typical heating condition and temporal evolution of plasma parameters for the discharge including both the ETB transition and the high-density ETB transition are shown in figure 1. Plasma was initiated by electron cyclotron heating (ECH) and was further heated by two neutral beam injection (NBI) systems. Additional gas was puffed to increase density. The magnetic configuration in which vacuum magnetic axis position R_{ax} was 94.9 cm from the torus centre and toroidal magnetic field strength at the magnetic axis B_{ax} was 1.86 T was applied. B_{ax} = 1.86 T is about twice as large and R_{ax} = 94.9 cm is outward-shifted, compared to the standard configuration of CHS in which B_{ax} = 0.95 T and R_{ax} = 92.1 cm. In the improved L-mode ("reheat mode") discharges, higher W_p values have been obtained in the high magnetic field operation [10]. However, the threshold value of the heating power of NBI normalized by electron density (P/n_e) to form ETB in CHS increases in the high magnetic field. On the other hand, the threshold decreases when using an outward-shifted magnetic axis [5,6]. Therefore, the combination of high magnetic field and an outward-shifted magnetic axis is applied to achieve the high-density ETB, which is the simultaneous realization of the reheat mode and ETB. The empirical rule of the dependence of threshold value of P/n_e on the strength of magnetic field and the position of the magnetic axis is summarized in references 5 and 6. The magnetic well, the rotational transform in the edge region, the orbit of trapped particles, the plasma viscosity and the distance of the last closed flux surface from the wall could be controlled by the magnetic axis shift. However, it has not been clarified which parameter determines the dependence of threshold on the position of the magnetic axis yet [5].

When P/n_e exceeds a certain threshold, a sudden drop in the H_α intensity signal is observed (49 msec). This transition phenomenon represented by the transition of the H_α signal level has been called "ETB transition" in CHS. An increase of the edge density, its gradient, and the stored energy has been observed at the ETB transition. After the ETB transition, the electron density keeps on increasing due to gas puffing. After 60 msec, the edge electron temperature decreases while the density continues to increase, which results in the stored energy leveling off. When P/n_e becomes lower than the threshold of the ETB transition because of the increased density, a back transition to the L-mode occurs (96 msec). After the gas puffing is shut off (105 msec) in the high-density L-mode plasma, the improved L-mode (reheat mode) appears (107 msec), where the edge electron temperature as well as the stored energy increases. On the other hand, the electron density decreases because the gas puffing is shut off and P/n_e increases. When P/n_e exceeds the threshold condition of the ETB transition again, a sudden drop in the H_α intensity signal is observed (113 msec). We call this second transition of the H_α signal the "high-density ETB transition".

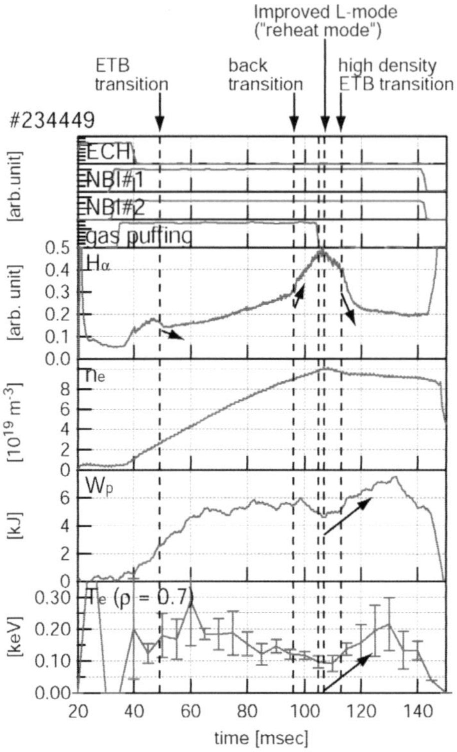

Figure 1. A typical experimental condition and temporal evolution of plasma parameters for the discharge with the ETB transition. Heating condition, H_α intensity, line-averaged density, stored energy, and edge electron temperature are shown together.

3. BES diagnostics

Figure 2 is a schematic drawing of BES systems in CHS. The magnetic field configuration, the line of NBI as the probe beam, the sightlines of BES, and the detection system are shown in this figure. H_α emission from the neutral hydrogen atomic beam for heating was measured as the beam emission in the present case. The electron impact excitation and ion impact excitation are comparable in the rate coefficient for the typical beam energy in CHS, 25–32 keV [18].

The spatial channels of BES consist of 16 optical fibres with an object lens that focuses on the position along the minor radius of CHS. The spatial pitch between sightlines Δx is about 1 cm, which yields the Nyquist wavenumber, $k_N = \pi / \Delta x$, of 3.1 rad cm^{-1}. Using the ramor radius evaluated from the electron temperature, $\rho_s = (2m_iT_e / e)^{1/2} / B$, of 4.6 mm in the standard parameter in CHS of $T_e = 1$ keV and $B = 1$ T, the wavenumber range $k\rho_s < 1.4$ is measureable. Since the observable region is the intersection of the beam line and the sightline for each fibre channel, local values and their correlations are available. In the present study, the normalized radius ρ was determined from the magnetic flux surfaces calculated using the three-dimensional equilibrium code VMEC [19] for $\beta =$ 0.5%, which yielded the observable region of BES.

We have to consider that multiple magnetic field surfaces are contained in a sightline, which is the so-called "line integral effect". In order to reduce this line integral effect, the optics are designed such that the sightlines are approximately tangent to the local magnetic field. In our case, a typical width of each observable region $\Delta\rho$ with considering the line integral effect is approximately ± 0.15 in the normalized minor radius in the region of interest in the present paper, while the separation between the sightlines is about 0.1 in the normalized minor radius. However, the actual spatial resolution will be better than that of the present evaluation, because the signal from the beam edge is weak [7]. In fact, the information obtained from each sightline is separated from that of other sightlines, as stated later in this paper, which supports the hypothesis that the crosstalk of the emission-weighted signal between channels is negligible. The line integral effect should be evaluated quantitatively in the future, taking into consideration the beam and the plasma parameters profiles.

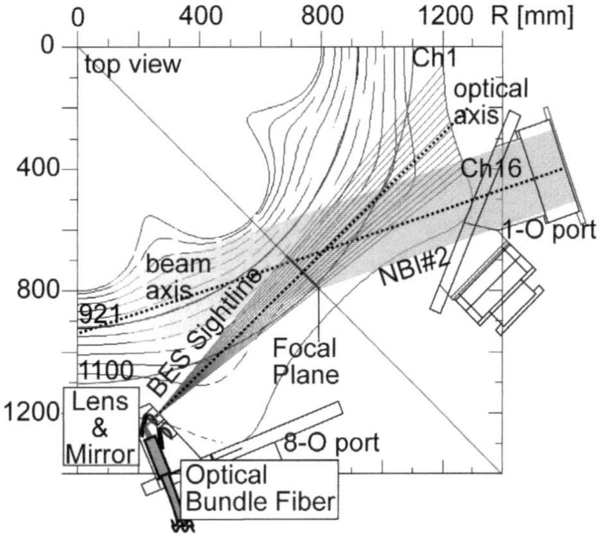

Figure 2. Schematic drawing of the BES system on the CHS device.

Beam emission can be distinguished from bulk plasma emission by taking advantage of the Doppler shift. Optical band-pass interference filters having a pass band of 0.5 or 1.0 nm in full width at half maximum (FWHM) are employed to transmit the wavelength range, including only the Doppler-shifted full energy (E) beam emission and eliminating the emission from the bulk plasma, impurity lines, and other beam components such as second ($E/2$) or third ($E/3$) energy fractions. The temporal resolution of BES measurement is 1 μs, which corresponds to the sampling rate of isolation digitizers of 1MHz. However, measurable frequency of fluctuations is limited below 100 kHz by the low-pass filter of avalanche photo diodes [14].

Provided that beam attenuation is negligible, the beam emission intensity is proportional to the plasma density. Under such conditions, we can regard the fluctuations of the signals as density fluctuations directly. Furthermore, BES can provide the density gradient ∇n and the inverse density scale length L_n^{-1} defined by $\nabla n/n$, where n indicates density. ∇n and n are obtained from the difference and average between the intensity of signals measured in adjacent viewing chords or channels, respectively.

4. Pedestal dynamics of the high-density ETB transition

4.1. Temporal evolution of the density profiles at two types of ETB transitions

BES diagnostics give temporal evolutions of local densities near the edge region. Figures 3(a) and (b) show the BES signals for $\rho = 0.65 \sim 1.03$ together with the H_α signal around the timing of the ETB transition and the high-density ETB transition, respectively. As shown in figure 3(a), the density

Journal of Physics: Conference Series **123** (2008) 012007 doi:10.1088/1742-6596/123/1/012007

inside $\rho = 0.87$ increases, while that outside $\rho = 0.96$ decreases at the ETB transition, indicating that particle confinement was improved and outward flux decreased because of the formation of the ETB for particles between these two positions. On the other hand, as shown in figure 3(b), the density inside $\rho = 0.76$ increases, while that outside $\rho = 0.87$ decreases at the high-density ETB transition.

We can compare the profile of density, density gradient, and L_n^{-1} between the ETB transition and the high-density ETB transition. Figure 4(a) shows the radial profile of the intensity of BES signal at 2 msec before (solid line) and 8 msec after (dotted line) the ETB transition. We chose the density profile 8 msec after transition as representative of the ETB phase because the density profile is changed by the transition and it becomes almost stable about 8 msec after the transition. Variation between these two traces is shown in figure 4(b), which indicates that the density around $\rho = 0.7\sim0.8$ increases due to the formation of ETB. Figures 4(c) and (d) show the radial profile of the density gradient before/after the transition and their variation, respectively. The density gradient at $\rho = 0.9$ is steepened due to the transition, and the gradient around $\rho = 0.9\sim1.1$ slightly increases. Figures 4(e) and (f) show the radial profile of L_n^{-1} before/after the transition and their variation, respectively. L_n^{-1} around $\rho > 0.9$ slightly increases. On the other hand, the profiles of density, density gradient, and L_n^{-1} before/after transition and their variation due to the high-density ETB transition are shown in figure 5. Figures 5(a) and (b) indicate that the density around $\rho = 0.6\sim0.7$ increases due to the transition. The density gradient keeps the profile that has the peak value at $\rho = 0.9$, even though the transition occurs as shown in figure 5(c). The position in which the density gradient varies most significantly is $\rho = 0.8$, as shown in figure 5(d). Figures 5(e) and (f) show that L_n^{-1} around $\rho > 0.8$ increases.

The results shown in figures 3, 4, and 5 show that the density, the density gradient, and the inverse value of the density scale length increase in the edge region at both the normal ETB transition and the high-density ETB transition. Now, there remains a possibility that the spatial region in which the density profile varies in the high-density ETB transition is slightly different from that in the ETB transition. Namely, the density inside $\rho = 0.87$ increases at the normal ETB transition while that inside $\rho = 0.76$ increases at the high-density ETB transition, as shown in figure 3.

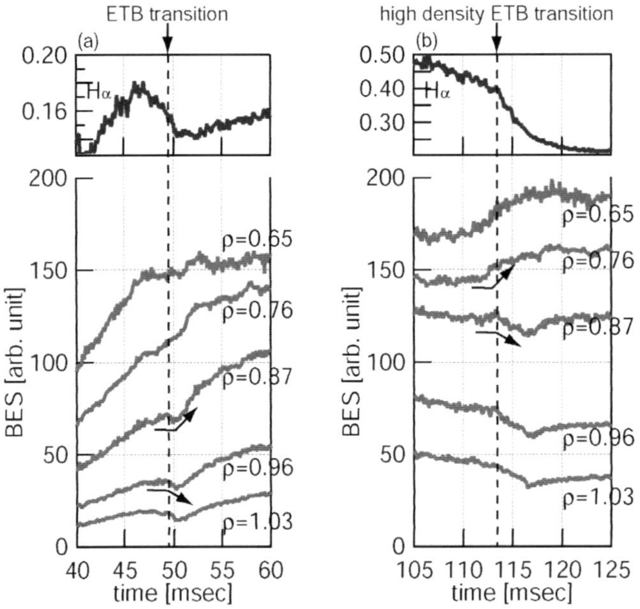

Figure 3. Temporal evolutions of BES signals for ρ = 0.65 ~ 1.03 together with the H_α signal around the timing of (a) the ETB transition and (b) the high-density ETB transition.

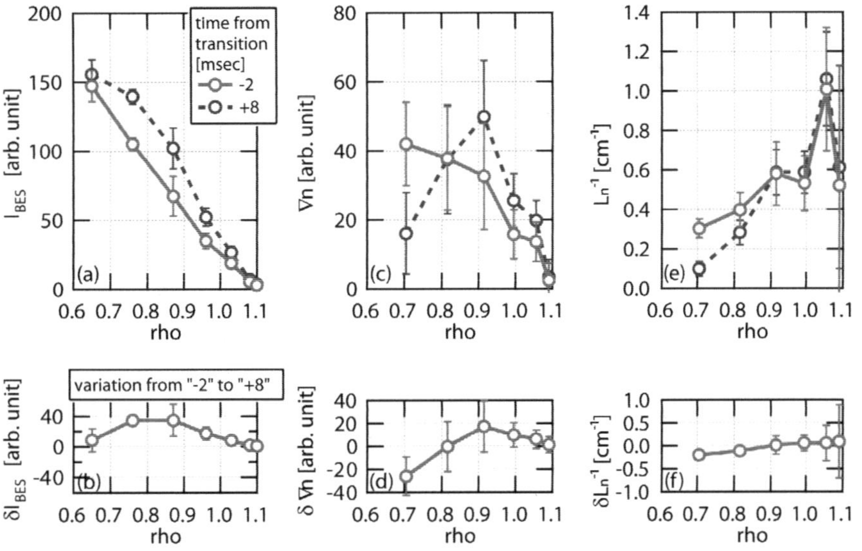

Figure 4. Variation of density profiles at the ETB transition. (a) Profile of the intensity of BES signal at 2 msec before (solid line) and 8 msec after (dotted line) the ETB transition. They are obtained by averaging the raw data of BES for 1 msec, which corresponds to 1000 data points. Error bars are the standard deviation of the raw data. (b) Variation between the two traces in (a). (c) Profile of the density gradient before/after the transition and (d) their variation. (e) Profile of L_n^{-1} before/after the transition and (f) their variation.

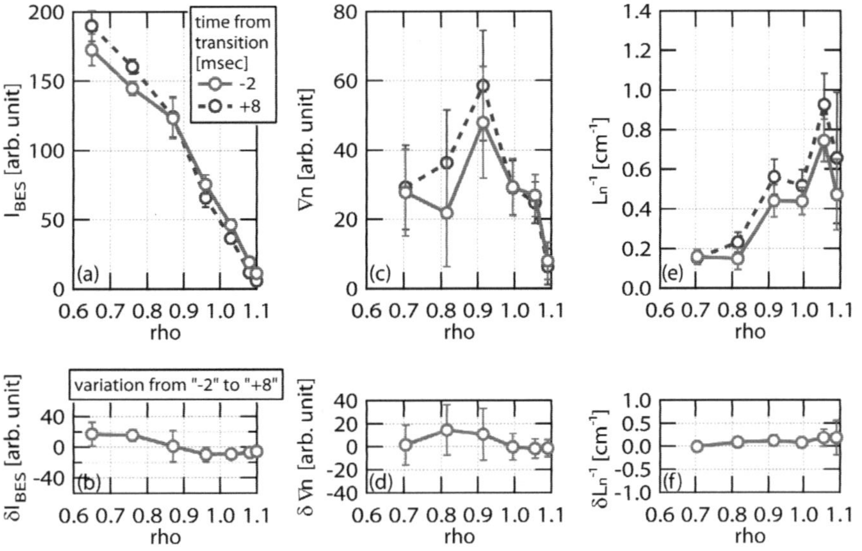

Figure 5. Variation of density profiles at the high-density ETB transition. (a) Profile of the intensity of BES signal at 2 msec before (solid line) and 8 msec after (dotted line) the high-density ETB transition. They are obtained by averaging the raw data of BES for 1 msec, which corresponds to 1000 data points. Error bars are the standard deviation of the raw data. (b) Variation between the two traces in (a). (c) Profile of the density gradient before/after the transition and (d) their variation. (e) Profile of L_n^{-1} before/after the transition and (f) their variation.

To examine the validity of this difference in the variation of the density profile between two types of ETBs more quantitatively, we should check the following three items as future studies. First is the effect of gas-puffing, which is shut off before the high-density ETB is formed. The change of source of fuelling might vary the density profile and the position where the density profile varies at the transition. Second, there is a study showing that impurity increases at the high-density ETB transition, which might affect the intensity of the BES signal. Third, the attenuation of the neutral beam should be evaluated. In the high-density case around 120 msec, a rough estimate shows that the beam is about 40% attenuated at $\rho = 0.65$, which is the innermost position in the region of interest in the present study. Namely, values in the inner region in figures 3(b) and 5(a) are underestimated because the intensity of the signal is proportional to both the plasma density and the beam density.

4.2. Density fluctuations at the high-density ETB transition
The characteristics of fluctuations in the high-density ETB discharge were investigated by means of BES. Figure 6 shows the power spectra of density fluctuations in the edge region averaged over 10 msec in the L-mode phase before the high-density transition (dotted line), and the high-density ETB phase (solid line). Density fluctuation of the frequency at less than 20 kHz, which is dominant in the spectra in the L-mode phase in the edge region, disappears in the high-density ETB phase. One might see peaks around 55 kHz and 81 kHz at $\rho = 0.96$ and around 60 kHz at $\rho = 0.65$ in the spectra of improved L-mode phase, but they have no repeatability, so they would not have a relationship with an improved L-mode or high-density ETB transition.

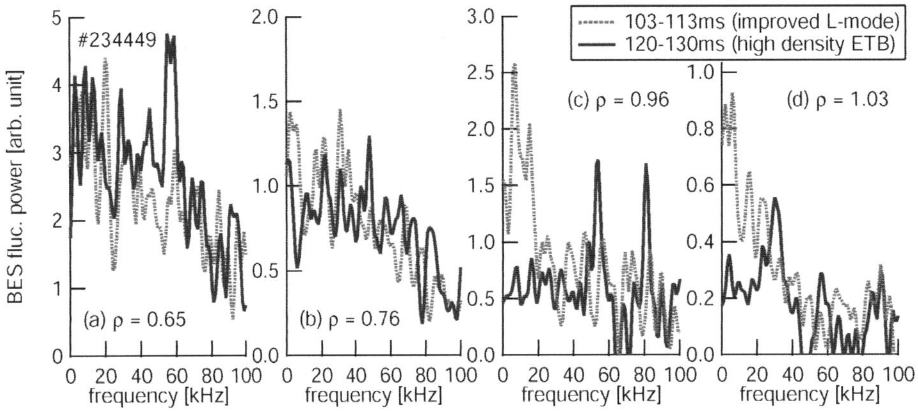

Figure 6. Power spectra of density fluctuations in the edge region averaged over 10 msec in the L-mode phase before the high-density transition (dotted line), and the high-density ETB phase (solid line).

Figures 7 and 8 show the profile of fluctuations before and after transition for two types of ETB. Power of fluctuation $|\tilde{n}|^2$ and relative fluctuation level $|\tilde{n}|/n$ at the frequencies below 20 kHz are integrated for 10 msec in the improved L-mode and high-density ETB phase for figure 7 and in the L-mode and ETB phase for figure 8. At the high-density ETB transition, both $|\tilde{n}|^2$ and $|\tilde{n}|/n$ decrease. On the other hand, $|\tilde{n}|^2$ increases while $|\tilde{n}|/n$ decreases at the normal ETB transition. It is common that $|\tilde{n}|/n$ decreases at the transition where $\rho > 0.9$ in both cases.

Figure 9 shows the temporal evolution of the power of density fluctuations at $\rho = 0.96$ integrated over 0–20 kHz. The H_α signal is shown. The fluctuation power decreases rapidly at the high-density ETB transition. This suppression of the fluctuation is also observed using the HCN laser scattering measurement and YAG laser phase contrast interferometer [20].

Figure 10 shows the power of density fluctuation integrated over 0-20 kHz plotted against the density gradient ∇n measured using BES. Density fluctuation increases with the edge density gradient in the initial ETB phase. The fluctuation power increases at the back transition, and it decreases suddenly at the high-density ETB transition, while the change of density gradient is not so large at either the back transition or the high-density ETB transition.

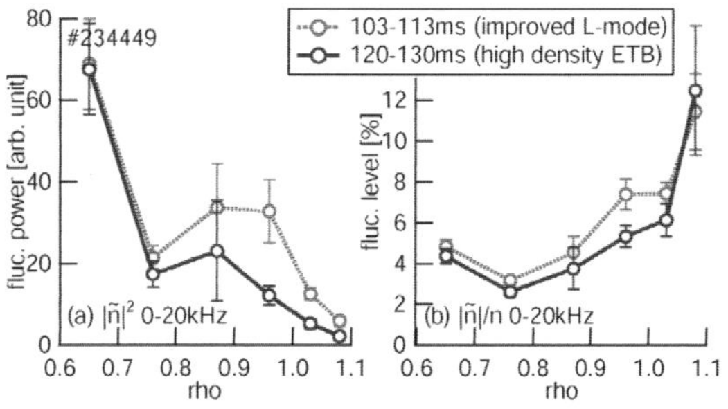

Figure 7. Profiles of power of fluctuation $|\tilde{n}|^2$ and relative fluctuation level $|\tilde{n}|/n$. Values at the frequencies below 20 kHz are integrated for 10 msec in the improved L-mode and high-density ETB phase.

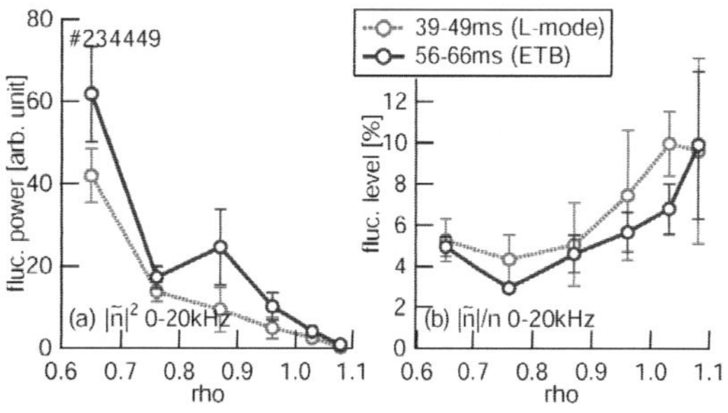

Figure 8. Profiles of power of fluctuation $|\tilde{n}|^2$ and relative fluctuation level $|\tilde{n}|/n$. Values at the frequencies below 20 kHz are integrated for 10 msec in the L-mode and ETB phase.

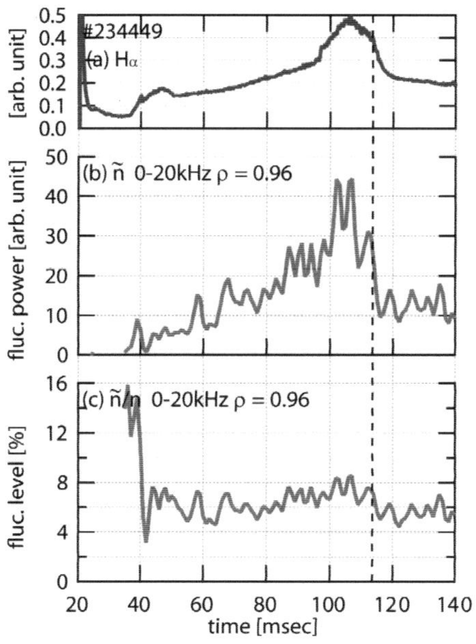

Figure 9. Temporal evolution of the power and fluctuation level of density fluctuations at $\rho = 0.96$ integrated over 0–20 kHz and the H_α signal.

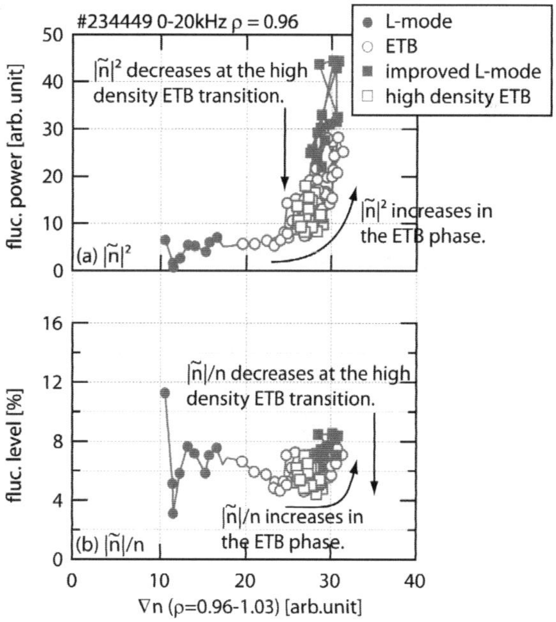

Figure 10. Power and fluctuation level of density fluctuation integrated over 0-20 kHz plotted against the density gradient ∇n.

5. Summary

Formation of the edge transport barrier (ETB) has been observed in the high-density region ($n_e \sim 1.0 \times 10^{20}$ m^{-3}) of the CHS. We investigated the behaviour of the density profile and the density fluctuations in the edge region accompanied by the formation of the high-density ETB using BES and compared the behaviour with that of the normal ETB transition of CHS.

The particle confinement is improved, and the density, ∇n, and L_n^{-1} increased in the edge region both at the ETB transition and at the high-density ETB transition.

Density fluctuation of the frequency at less than 20 kHz decreases rapidly at the high-density ETB transition in the edge region, which suggests the correlation between the fluctuation and the confinement property. The spatial regions where the relative fluctuation level decreases are similar between the ETB transition and the high-density ETB transition.

Acknowledgments

The work was supported in part by the National Institute for Fusion Science Collaborative Research Program (NIFS02KZPD003) directed by the second author.

References

[1] McCormick K *et al.* 2002 *Phys. Rev. Lett.* **89** 015001
[2] Ohyabu N *et al.* 2006 *Phys. Rev. Lett.* **97** 055002
[3] Okamura S *et al.* 2003 *J. Plasma Fusion Res.* **79** 977
[4] Okamura S *et al.* 2004 *Plasma Phys. Controlled Fusion* **46** A113
[5] Minami T *et al.* 2006 *Plasma Fusion Res.* **1** 032
[6] Akiyama T *et al.* 2006 *Plasma Phys. Control. Fusion* **48** 1683
[7] Oishi T *et al.* 2006 *Nucl. Fusion* **46** 317
[8] Takeuchi M *et al.* 2006 *Plasma Phys. Control. Fusion* **48** A277
[9] Morita S *et al.* 1993 *Proc. 14th Int. Conf. Plasma Physics and Controlled Nuclear Fusion Research 1992 (Wurzberg)* vol. **2** (Vienna: IAEA) p. 155
[10] Isobe M *et al.* 2006 *Fus. Sci. Tech.* **50** 229
[11] Mori M *et al.* 1988 *Nucl. Fusion* **28** 1892
[12] Soldner F X *et al.* 1988 *Phys. Rev. Lett.* **61** 1105
[13] Minami T *et al.* 2006 *Plasma Fusion Res.* **1** 047
[14] Oishi T, Tanaka S, Kado S, Yoshinuma M, Ida K, Okamura S and CHS group 2004 *Rev. Sci. Instrum.* **75** 4118
[15] Fonck R J, Duperrex P A and Paul S F 1990 *Rev. Sci. Instrum.* **61** 3487
[16] Jakubowski M, Fonck R, Kim J S and McKee G 1999 *Rev. Sci. Instrum.* **70** 874
[17] McKee G *et al.* 1999 *Rev. Sci. Instrum.* **70** 913
[18] Janev R K, Reiter D and Samm U 2003 *FZJ Report* Juel-4105
[19] Hirshman S P, Rij W I van and Merkel P 1986 *Comput. Phys. Commun.* **43** 143
[20] Minami T *et al.* 2007 "Characteristics of high-density edge transport barrier with reheat mode on CHS" *presented in this conference* P1-10

Journal of Physics: Conference Series **123** (2008) 012008 doi:10.1088/1742-6596/123/1/012008

Effect of equilibrium properties on the structure of the edge MHD modes in tokamaks

N Aiba, N Hayashi, T Takizuka, S Tokuda and T Ozeki

Japan Atomic Energy Agency, Naka, Ibaraki, 311-0193 Japan

E-mail: aiba.nobuyuki@jaea.go.jp

Abstract. Effects of the pressure profile and the current density profile inside the top of the pedestal and that of the plasma shape on the expansion of the structure of the unstable edge MHD mode are investigated numerically. The structure of the edge MHD mode is expanded by spreading the envelope of the edge ballooning mode structure due to the increase of the pressure gradient inside the top of the pedestal, and by decreasing the toroidal mode number of the most unstable mode due to the increase of the current density inside the top of the pedestal. In strongly shaped plasmas, the pressure gradient inside the top of the pedestal can approach to the ballooning mode stability boundary and the current density increases enough to reduce the toroidal mode number of the most unstable mode. These increases of the pressure gradient and the current density destabilize the edge MHD mode and expand the structure of this mode.

1. Introduction

For realizing a economical fusion reactor, H-mode operation is suitable due to a good confinement property near the plasma edge[1]. However, in the H-mode tokamak plasmas, edge localized modes (ELMs) sometimes occur, and constrain the maximum pressure gradient in the edge pedestal[2]. Since the so-called 'Type-I ELM' leads a large heat load on the divertor and reduces a life time of the divertor, the quantitative analysis about the amount of the energy loss due to the Type-I ELM is important for ITER and future devices.

Recent theoretical and experimental researches have shown that the Type-I ELM is triggered by ideal MHD modes (a peeling mode, a ballooning mode, and a coupled peeling-ballooning mode), and some numerical codes realize to analyze the linear stability of these MHD modes[3, 4, 5, 6, 7]. From the experimental analysis with such linear stability codes, the width of the eigenfunction of the edge MHD mode is sometimes related to the amount of the ELM energy loss[1, 4].

By comparing some experimental results, it has been found the amount of the energy loss by the Type-I ELM in JT-60U without Ferritic steel tiles tends to be smaller than that in JET and DIII-D[8]. The reason of such a difference of the ELM energy loss between JT-60U and JET/DIII-D needs to be revealed for reducing the amount of the ELM energy loss. In comparison between these devices, we can see that some equilibrium properties which have impacts on the edge MHD stability are different. From the viewpoint of the maximum pressure gradient at the pedestal, the difference of the shape of the plasma cross-section is crucial. In JET/DIII-D, the plasma shape can be optimized by changing shaping parameters (a ellipticity κ, a triangularity δ, a squareness[9], a sharpness[7] etc.) in wide range, and a pressure gradient near the pedestal can be improved due to stabilizing edge MHD modes. On the other hand, in JT-60U, since the

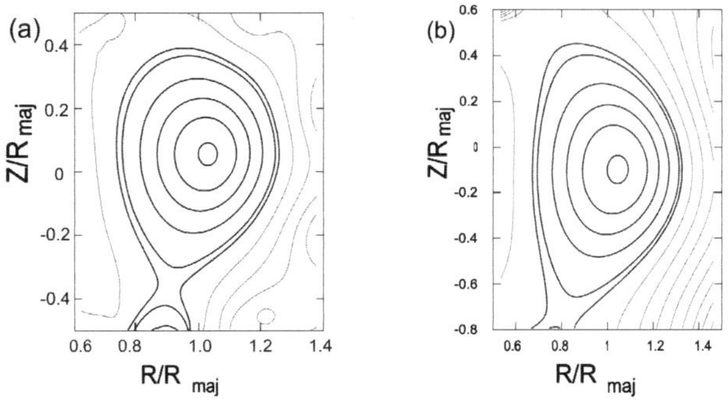

Figure 1. Shapes of the plasma cross-section of (a) JT-60U shape equilibrium and (b) ITER-like shape equilibrium.

optimization of the plasma shape is restricted by the layout of the magnetic coils, the maximum pressure gradient near the pedestal tends to be smaller than those in JET/DIII-D.

Moreover, we can see that there is a difference in the pressure profile not only near the pedestal but also inside the top of the pedestal among JT-60U, JET, and DIII-D. In particular, the pressure profile in JT-60U is more gradual than those in JET/DIII-D. The effect of the difference of the pressure profile inside the top of the pedestal on the amount of the ELM energy loss has been discussed numerically with the integrated simulation code TOPICS-IB, and Hayashi et al. have clarified that the pressure profile inside the top of the pedestal is important for the amount of the ELM energy loss due to the expansion of the width of the eigenfunction of the unstable MHD mode[10].

In this paper, we investigate numerically that the effect of the equilibrium properties on the width of the eigenfunction of the edge MHD modes by the MARG2D stability code[11, 12]. In particular, we pay attention to the effect of the pressure profile and the current density profile inside the top of the pedestal and the plasma shape on the expansion of the structure of the unstable edge MHD mode.

2. Stability analysis in different shape equilibria

2.1. Equilibria

In this paper, we investigate numerically the stability of edge MHD modes in the equilibria which have two kinds of the shape of the plasma cross-section as shown in figure 1. Figure 1(a) is the JT-60U (shot No. is 43075 and 5.5 sec.) like shape whose $A = 3.91$, $\kappa_{up} = 1.28$, $\kappa_{low} = 1.62$, $\delta_{up} = 0.16$, $\delta_{low} = 0.35$, and figure 1(b) is the ITER-like shape which will be realized in JT-60SA[13] whose $A = 3.12$, $\kappa_{up} = 1.66$, $\kappa_{low} = 2.03$, $\delta_{up} = 0.37$, $\delta_{low} = 0.60$. Here A is an aspect ratio, a subscript *up* (*low*) expresses the upper (lower) value. The plasma current is about 1.08MA (JT-60U) and 2.20MA (ITER like), which is slightly adjusted for fixing the edge safety factor as about $q_a = 6.4$, and the toroidal magnetic field is 3.11 (JT-60U) and 2.46 (ITER-like). The electron density n_e and electron temperature T_e profiles are determined artificially, the ion temperature T_i profile is assumed as $T_e = T_i$, and the effective charge Z_{eff} is given as 2.3. The current density profile, including a ohmic current, a bootstrap current, and additional (NBCD etc.) currents, is calculated by the ACCOME code[14].

Figures 2 and 3 show the profiles of (a) n_e and T_e, (b) the plasma pressure p, and (c) the parallel current density $\langle \boldsymbol{j} \cdot \boldsymbol{B} \rangle / \langle \boldsymbol{B}^2 \rangle$ and the safety factor q in the JT-60U shape (figure 2) and the ITER-like shape (figure 3) equilibria, respectively. To investigate the effect of the pressure gradient inside the top of the pedestal on the structure of the edge MHD modes, the T_e $(= T_i)$

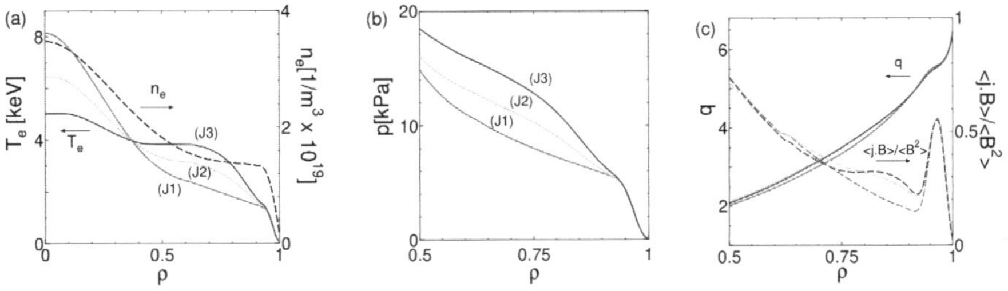

Figure 2. Profiles of (a) n_e and T_e, (b) p, and (c) q and $\langle \boldsymbol{j} \cdot \boldsymbol{B} \rangle / \langle \boldsymbol{B}^2 \rangle$ in JT-60U shape equilibria. The red, green, and blue lines show the profiles in cases (J1), (J2), and (J3), respectively.

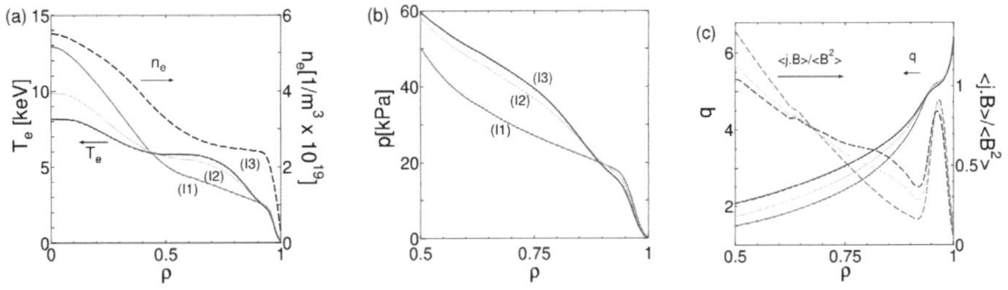

Figure 3. Profiles of (a) n_e and T_e, (b) p, and (c) q and $\langle \boldsymbol{j} \cdot \boldsymbol{B} \rangle / \langle \boldsymbol{B}^2 \rangle$ in ITER-like shape equilibria. The red, green, and blue lines show the profiles in cases (I1), (I2), and (I3), respectively.

profile is changed as shown in figure 2(a) (J1)-(J3) and figure 3(a) (I1)-(I3), but the n_e profile is unchanged. The profiles of T_e and T_i near the pedestal are changed linearly to identify the marginally stable condition, and these profiles in the plasma core region ($\rho \leq 0.5$) is adjusted to reduce the change of the poloidal beta value β_j in each equilibrium, where ρ is the minor radius defined as $\rho = \sqrt{V(\psi_N)/V_{total}}$, $V(\psi_N)$ is the volume at each magnetic surface, V_{total} is the volume of the plasma, and ψ_N is the poloidal magnetic flux normalized as $\psi_N = 0$ at the magnetic axis and $\psi_N = 1$ at the surface. As the result, β_j of each equilibrium is (J1) 0.76, (J2) 0.79, (J3) 0.82 in the JT-60U shape case, and (I1) 0.78, (I2) 0.80, (I3) 0.83 in the ITER-like shape case. The position of the top of the pedestal is fixed as $\rho_{ped} = 0.948$ in each equilibrium. Since the pressure gradient inside the top of the pedestal becomes larger, the bootstrap current at this region also increases as shown in figure 2 (c) and figure 3 (c), and the current density profile in the plasma core region ($\rho \leq 0.5$) is adjusted to maintain the q value at the magnetic axis $q_0 > 1.0$.

2.2. JT-60U shape case

We analyze numerically the ideal MHD stability of the equilibria shown in the previous subsection. The range of the toroidal mode number n of the analyzed MHD mode is from 1 to 20. Figure 4 shows the result of the stability analysis in the JT-60 shape case on the (n, λ_0) plane and the (n, γ_{IM}) plane, where λ_0 is the eigenvalue of the eigenvalue problem associated with the two-dimensional Newcomb equation[11]

$$\mathcal{N}\boldsymbol{Y} = -\lambda_0 \mathcal{R}\boldsymbol{Y}, \tag{1}$$

$\mathcal{N}$ is the Newcomb operator, $\mathcal{R}$ is the diagonal operator whose components are $\mathcal{R}_{m,m} \propto (m-nq)^2$, m is the poloidal mode number, and γ_{IM} is the growth rate of the MHD mode under the

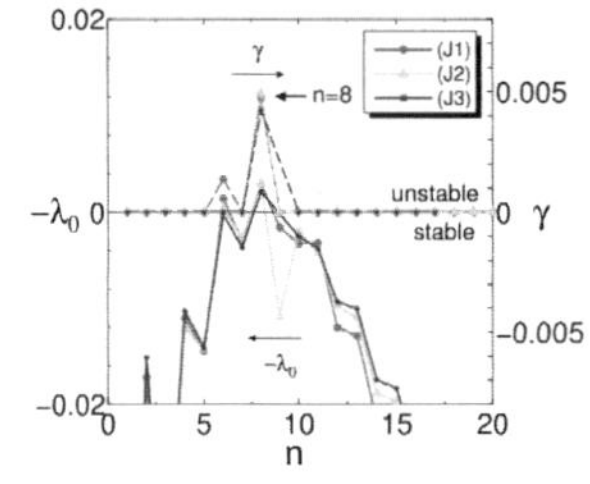

Figure 4. Dependence of the eigenvalue λ_0 (solid line) and the growth rate γ_{IM} (broken line) on n in JT-60U shape equilibria. The red, green, and blue lines show λ_0 and γ_{IM} in cases (J1), (J2), and (J3), respectively.

incompressible assumption normalized with the toroidal Alfvén frequency at the axis. Both λ_0 and γ_{IM} are calculated by the MARG2D code[11, 12]. The eigenvalue problem equation (1) realizes to identify explicitly whether the plasma is stable or unstable by determining the sign of the eigenvalue λ_0, and the λ_0 dependence on n roughly shows whether the plasma tends to be stable or unstable as n increases. In the case shown in figure 4, the most unstable mode is the $n = 8$ peeling-ballooning mode in each equilibrium, and the dependence of the edge MHD stability on n are similar to each other.

The structures of the poloidal Fourier harmonics of the eigenfunction $\boldsymbol{Y}$ are shown in figure 5 ((a) the real part and (b) the imaginary part), and the displacements in the radial direction ξ_r at $\theta = 0$ are also shown in figure 5(c), where the red, green, and blue lines show the eigenfunctions in cases (J1), (J2), and (J3), respectively. These figures indicate that the structure of the eigenfunction extends slightly to the plasma core region as the pressure gradient inside the top of the pedestal increases. To discuss a change of the width of the edge MHD mode, we define the width of $\xi_r(\theta = 0)$ as $w(\xi_r(\theta = 0)) = 1.0 - \rho(\xi_r(\theta = 0))$. Since the width of ξ_r is mainly determined by the ballooning mode structure in the plasma, $\xi_r(\theta = 0)$ is normalized as the top of the edge ballooning mode envelope is unity as shown in figure 5 (c). Afterward, ξ_r means $\xi_r(\theta = 0)$. Figure 5 (d) shows the dependence of w/w_{org} at $\xi_r = 0.3$ (red line), 0.2 (green line), and 0.1 (blue line) on the normalized pressure gradient at $\psi_N = 0.87$ ($\rho \simeq 0.87$), α_{87}, where w_{org} is the w value of case (J1), α is defined as $\alpha = -(\mu_0/2\pi^2)(dp/d\psi)(dV/d\psi)(VR/2\pi)^{0.5}$, μ_0 is the permeability in the vacuum, ψ is the poloidal magnetic flux, R is the major radius, and the subscript 87 expresses the value at $\psi_N = 0.87$. Since the ballooning mode becomes unstable first near $\psi_N = 0.87$ when the pressure gradient inside the top of the pedestal increases, we pay attention to the dependence of w on α_{87}. As shown in this figure, the width at $\xi_r = 0.3$ is almost unchanged in spite of that α_{87} increases from 0.20 to 0.68, but the w/w_{org} value becomes slightly larger as ξ_r decreases. This means that the tail of the edge ballooning mode envelope extends to the plasma core region as the pressure gradient inside the top of the pedestal increases, but the amount of variation of w/w_{org} is less than 20%.

2.3. ITER-like shape case

In this subsection, the edge MHD stability in the ITER-like shape equilibrium is investigated. The n dependence of λ_0 shown in figure 6 indicates that the n number of the unstable MHD mode tends to decrease as the pressure gradient inside the top of the pedestal increases, and as the result, the n number of the most unstable mode decreases from $n = 9$ (case (I1) and case (I2)) to $n = 6$ (case (I3)). Figure 7 shows the structures of the poloidal Fourier harmonics of (a) Re($\boldsymbol{Y}$) and (b) Im($\boldsymbol{Y}$) and (c) the structure of ξ_r, where red, green, and blue lines show the structures of the most unstable mode in cases (I1), (I2), (I3), and the purple line shows that of

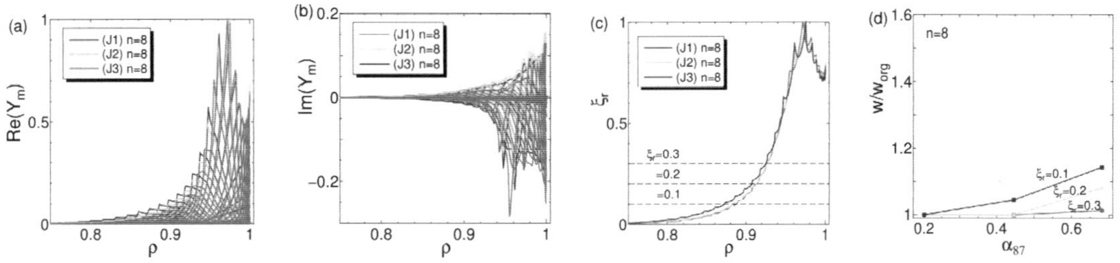

Figure 5. Profiles of the poloidal Fourier harmonics of (a) Re($\boldsymbol{Y}$) and (b) Im($\boldsymbol{Y}$), where $\boldsymbol{Y}$ is the eigenfunction of the eigenvalue problem equation (1), and (c) the displacement at $\theta = 0$, $\xi_r(\theta = 0)$ in JT-60U shape equilibria. The red, green, and blue lines show the results in cases (J1), (J2), and (J3), respectively. (d) Dependence of the width $w(\xi_r)$ normalized with $w_{org} = (w$ of (J1)) on α_{87}. The red, green, and blue lines show w/w_{org} at $\xi_r = 0.3$, 0.2, and 0.1, respectively.

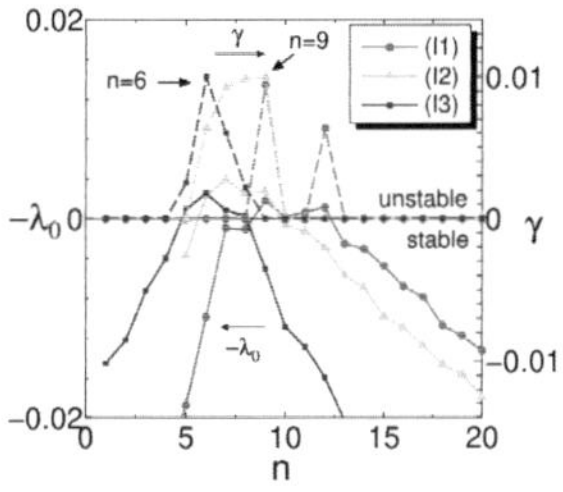

Figure 6. Dependence of λ_0 (solid line) and γ_{IM} (broken line) on n in ITER-like shape equilibria. The red, green, and blue lines show λ_0 and γ_{IM} of (I1), (I2), and (I3), respectively.

the $n = 9$ mode in case (I3). In this ITER-like shape case, the extension of ξ_r to the plasma core region is larger than that in the JT-60U shape case as shown in figure 7 even when the n number of the mode is unchanged as $n = 9$. Moreover, in case (I3), the decrease of the n number of the most unstable mode from 9 to 6 expands the structure of ξ_r. These results can be described with w normalized with $w_{org} = w$ of (I1) as shown in figure 7 (d), where the red, green blue lines shows w/w_{org} at $\xi_r = 0.3$, 0.2, 0.1, respectively. In particular, at $\xi_r = 0.1$, the width of ξ_r becomes about twice as α_{87} increases from 0.42 to 1.49.

3. Effect of the pressure profile and the current profile inside the top of the pedestal

In the previous section, not only the pressure gradient but also the current density profile inside the top of the pedestal becomes larger due to increasing the bootstrap current density. In this section, we identify separately the effect of the pressure gradient and the current density on the structure of the edge MHD mode.

3.1. Effect of the pressure profile

First, we investigate the effect of the pressure gradient. In this study, we change the pressure profile but fix the current density profile as shown in figures 8 (a), (b) (JT-60U shape) and figures 8 (c), (d) (ITER-like shape), respectively. The p and $\langle \boldsymbol{j} \cdot \boldsymbol{B} \rangle / \langle \boldsymbol{B}^2 \rangle$ profiles in case (Jp1), and those in case (Ip1) are same as those in case (J1) in figure 2 and those in case (I1) in figure 3.

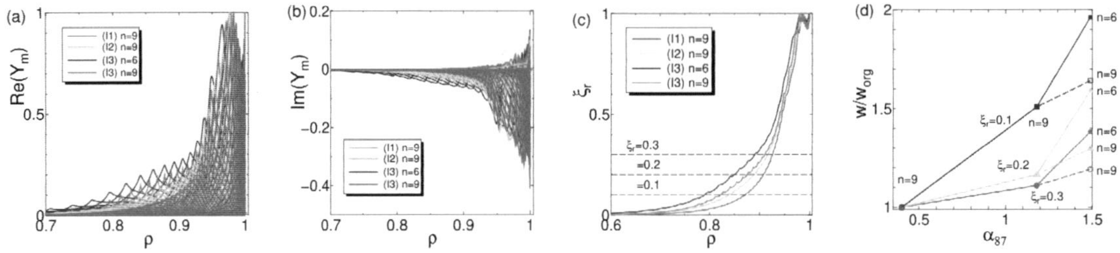

Figure 7. Profiles of the poloidal Fourier harmonics of (a) Re($\boldsymbol{Y}$) and (b) Im($\boldsymbol{Y}$), and (c) ξ_r in ITER-like shape equilibria. The red, green, and blue lines show the profile of the most unstable mode in cases (I1), (I2), (I3), and the purple line shows that of $n = 9$ mode in case (I3), respectively. (d) Dependence of the width $w(\xi_r)$ normalized with $w_{org} = (w$ of (I1)) on α_{87}. The red, green, and blue lines show w/w_{org} at $\xi_r = 0.3$, 0.2, and 0.1, respectively.

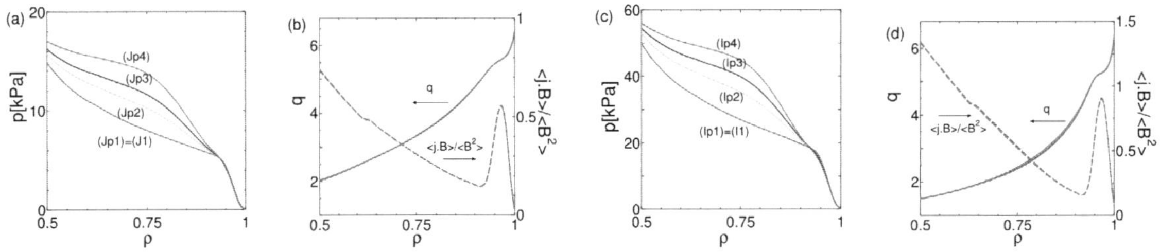

Figure 8. Profiles of (a) p, (b) q and $\langle \boldsymbol{j} \cdot \boldsymbol{B} \rangle / \langle \boldsymbol{B}^2 \rangle$ in JT-60U shape equilibria, and those of (c) p, (d) q and $\langle \boldsymbol{j} \cdot \boldsymbol{B} \rangle / \langle \boldsymbol{B}^2 \rangle$ in ITER-like shape equilibria. The red, green, blue, and purple lines show the profiles in cases (Jp1), (Jp2), (Jp3), (Jp4) in figures (a) and (b), and those in cases (Ip1), (Ip2), (Ip3), (Ip4) in figures (c) and (d), respectively.

The results of the stability analysis in JT-60U shape equilibria in figure 9 (a) show that the n dependence of the edge MHD stability changes little and the n number of the most unstable mode is same as $n = 8$ though the pressure gradient inside the top of the pedestal increases from $\alpha_{87} =$ (Jp1) 0.20 to (Jp2) 0.43, (Jp3) 0.59, and (Jp4) 0.80, where red, green, blue, and purple lines express the results in cases (Jp1), (Jp2), (Jp3), and (Jp4), respectively. This result is almost same as that in subsection 2.2. On the other hand, in ITER-like shape case, the result of the stability analysis shown in figure 9 (b) indicates the n number of the most unstable mode is also unchanged as $n = 9$ though the α_{87} value increases as (Ip1) 0.41, (Ip2) 0.91, (Ip3) 1.22, and (Ip4) 1.60. This result is different from that in subsection 2.3 (both profiles of p and $\langle \boldsymbol{j} \cdot \boldsymbol{B} \rangle / \langle \boldsymbol{B}^2 \rangle$ are changed), and in comparison between this result and that discussed in subsection 2.3, we reveal that the increase of the pressure gradient inside the top of the pedestal is not important for changing the n number of the most unstable mode.

Next, we compare the structure of ξ_r in figure 10. As shown in figure 10 (a), the expansion of ξ_r is relatively small in JT-60U shape case, and this result is similar to that shown in figure 5. On the other hand, the extension of ξ_r in ITER-like shape case shown in figure 10 (b) goes deeper to the plasma core region as α_{87} increases than that in JT-60U shape case. To discuss the expansion of the width due to the increase of the pressure gradient, the α_{87} dependence of w/w_{org} at $\xi_r = 0.3$, 0.2, and 0.1 are shown in figures 10 (c) (JT-60U shape case) and (d) (ITER-like shape case), where w_{org} is the w value in case (Jp1) in figure 10 (c) and that in case (Ip1) in figure 10 (d). Here the solid line in figure 10 (c) expresses w/w_{org} of ξ_r shown in

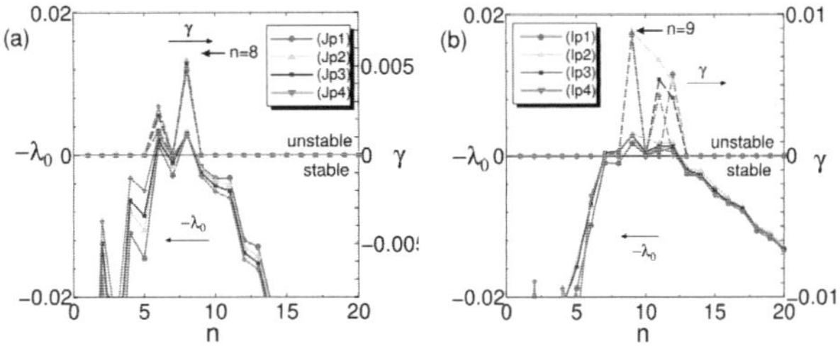

Figure 9. Dependence of λ_0 (solid line) and γ_{IM} (broken line) on n in (a) JT-60U shape equilibria and (b) ITER-like shape equilibria. The red, green, blue, and purple lines show λ_0 and γ_{IM} in cases (Jp1), (Jp2), (Jp3), (Jp4) in figure (a), and those in cases (Ip1), (Ip2), (Ip3), (Ip4) in figure (b), respectively. In these cases, the p profile is changed, but the $\langle \boldsymbol{j} \cdot \boldsymbol{B} \rangle / \langle \boldsymbol{B}^2 \rangle$ profile is fixed as shown in figure 8.

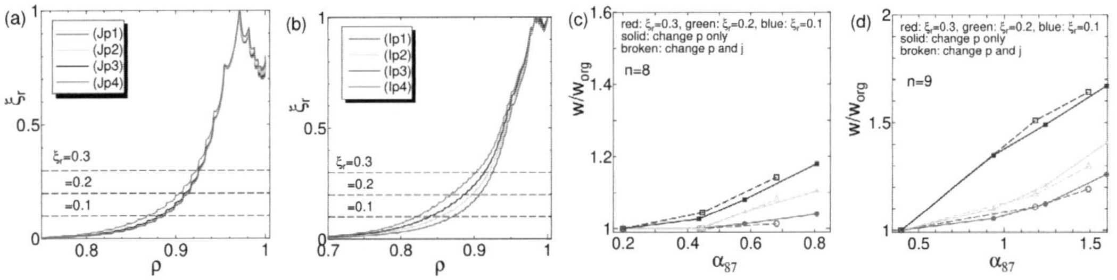

Figure 10. Profiles of ξ_r in (a) JT-60U shape equilibria and (b) ITER-like shape equilibria. The red, green, blue, and purple line show the profiles in cases (Jp1), (Jp2), (Jp3), (Jp4) in figure (a), and those in cases (Ip1), (Ip2), (Ip3), (Ip4) in figure (b), respectively. Dependence of w/w_{org} in (c) JT-60U shape equilibria and (d) ITER-like shape equilibria, where w_{org} is the w value in case (Jp1) in figure (c) and that in case (Ip1) in figure (d), respectively. The red, green, and blue lines show w/w_{org} at $\xi_r = 0.3$, 0.2, and 0.1.

figure 10 (a), and that in figure 10 (d) expresses w/w_{org} of ξ_r shown in figure 10 (b); in these cases, p profile is changed but $\langle \boldsymbol{j} \cdot \boldsymbol{B} \rangle / \langle \boldsymbol{B}^2 \rangle$ profile is unchanged. The broken line in figure 10 (c) expresses w/w_{org} of ξ_r whose $n = 8$ shown in figure 5 (c), and that in figure 10 (d) expresses w/w_{org} of ξ_r whose $n = 9$ shown in figure 7 (c); in these cases, both p and $\langle \boldsymbol{j} \cdot \boldsymbol{B} \rangle / \langle \boldsymbol{B}^2 \rangle$ profiles are changed. In comparison between these solid and broken lines, we find that the width of ξ_r, whose n number is unchanged as $n = 8$ in JT-60U shape case and $n = 9$ in ITER-like shape case, expands due mainly to the increase of the pressure gradient inside the top of the pedestal. Particularly, in ITER-like shape case, though the n number of the most unstable mode changes from 9 to 6 when both p and $\langle \boldsymbol{j} \cdot \boldsymbol{B} \rangle / \langle \boldsymbol{B}^2 \rangle$ profiles are changed as discussed in subsection 2.3, the α_{87} dependence of w/w_{org} of the $n = 9$ mode at $\xi_r = 0.3$, 0.2, 0.1 can be represented without the increase of the current density inside the top of the pedestal as shown in figure 10 (d).

3.2. Effect of the current density profile

In this subsection, we investigate the effect of the increase of the current density inside the top of the pedestal on the expansion of ξ_r. As shown in figure 10(c), since the increase of the pressure gradient can explain approximately the expansion of ξ_r in JT-60U shape case, we discuss the current density effect only in ITER-like shape case. In this study, we change the reference

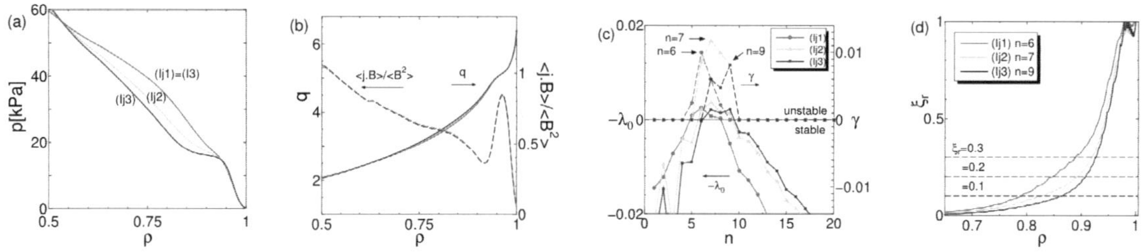

Figure 11. Profiles of (a) p and (b) q and $\langle \boldsymbol{j} \cdot \boldsymbol{B} \rangle / \langle \boldsymbol{B}^2 \rangle$ in ITER-like shape equilibria. (c) Dependence of λ_0 (solid line) and γ_{IM} (broken line) on n in ITER-like shape equilibria. (d) Profiles of ξ_r. The red, green, and blue lines show the results in cases (Ij1), (Ij2), and (Ij3), respectively. In these cases, the p profile is changed, but the $\langle \boldsymbol{j} \cdot \boldsymbol{B} \rangle / \langle \boldsymbol{B}^2 \rangle$ profile is fixed as shown in figures (a) and (b).

current density profile from (I1) to (I3) in figure 3. Figure 11 shows the profiles of (a) p and (b) q and $\langle \boldsymbol{j} \cdot \boldsymbol{B} \rangle / \langle \boldsymbol{B}^2 \rangle$, respectively. As already mentioned, the p and $\langle \boldsymbol{j} \cdot \boldsymbol{B} \rangle / \langle \boldsymbol{B}^2 \rangle$ profiles of (Ij1) in figure 11 is referred to those of (I3) in figure 3. The α_{87} value is 0.40 (Ij1), 0.76 (Ij2), and 1.49 (Ij3), respectively. By comparing the stability analysis results in these equilibria and those in the equilibria shown in figures 8 (c) and (d), we can discuss the effect of the current density inside the top of the pedestal on the expansion of ξ_r.

The results of the stability analysis in figure 11 (c) show that the n number of the unstable mode tends to increase as the pressure gradient inside the top of the pedestal decreases, and as the result, the n number of the most unstable mode changes from $n = 6$ in case (Ij1) to 7 in case (Ij2), and 9 in case (Ij3). As shown in figure 11 (d), ξ_r extends to the plasma core region as the pressure gradient inside the top of the pedestal increases. This expansion of ξ_r is thought to be caused by both increasing the pressure gradient inside the top of the pedestal and decreasing the n number of the most unstable mode. As shown in figure 11 (c), the increase of the bootstrap current density is necessary for decreasing n of the most unstable mode, and in combination with such a increase of the current density, the increase of the pressure gradient can help to decrease the n number of the mode.

4. Effect of the Plasma Shape

In this section, we discuss the reason that the plasma shape affects the expansion of the displacement of the unstable mode due to the increase of the pressure gradient and the current density inside the top of the pedestal. Figures 12 (a) and (b) show the stability diagram of the ideal MHD mode in JT-60U shape case on the (s, α) plane at (a) $\psi_N = 0.87$ ($\rho \simeq 0.87$) and (b) $\psi_N = 0.97$ ($\rho \simeq 0.97$) when the p profile is changed but the $\langle \boldsymbol{j} \cdot \boldsymbol{B} \rangle / \langle \boldsymbol{B}^2 \rangle$ profile is fixed discussed in subsection 3.1, where s is the magnetic shear defined as $s = 2V/q(dq/dV)$. The black line is the stability boundary of the $n = \infty$ ballooning mode, the red, green, blue and purple lines show those of the peeling/kink mode and the peeling-ballooning mode, and the red, green, blue, and purple points show the (s, α) point of the equilibria in cases (Jp1), (Jp2), (Jp3), and (Jp4), respectively. These equilibria access to the second stability region of ballooning mode at the pedestal ($\psi_N = 0.97$), but still exist in the first stability region inside the top of the pedestal ($\psi_N = 0.87$). In this case, the maximum pressure gradient at the pedestal ($\psi_N = 0.97$) $max(\alpha_{97})$ and the stable region against ideal MHD modes are almost unchanged though α_{87} increases about four times. These results and the result that the change of ξ_r is small as shown in figure 10 (a) imply that the increase of the pressure gradient inside the top of the pedestal affects little on the edge MHD stability in JT-60U shape case.

Journal of Physics: Conference Series **123** (2008) 012008 doi:10.1088/1742-6596/123/1/012008

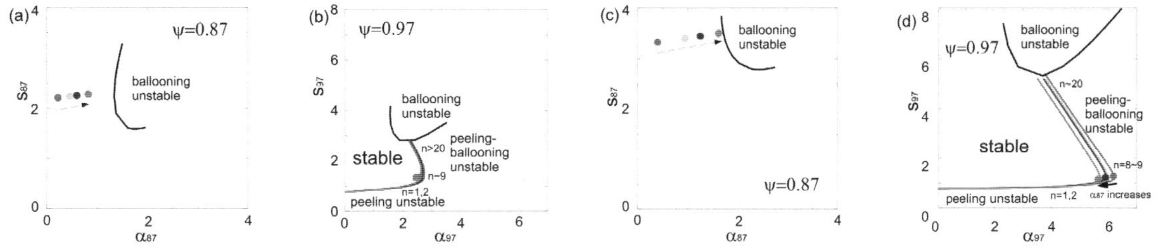

Figure 12. (a), (b) Stability diagram of the ideal MHD mode in JT-60U shape case discussed in subsection 3.1 on the (s, α) plane at (a) $\psi_N = 0.87$ and (b) $\psi_N = 0.97$. (c), (d) Stability diagram of the ideal MHD mode in ITER-like shape case discussed in subsection 3.1 on the (s, α) plane at (c) $\psi_N = 0.87$ and (d) $\psi_N = 0.97$. In both cases, the p profile is changed but the $\langle \boldsymbol{j} \cdot \boldsymbol{B} \rangle / \langle \boldsymbol{B}^2 \rangle$ profile is fixed. The black line is the stability boundary of the $n = \infty$ ballooning mode. The red, green, blue and purple lines show the stability boundaries of the peeling/kink mode and the peeling-ballooning mode, and the red, green, blue, and purple points show the (s, α) points of the equilibria in cases (Jp1), (Jp2), (Jp3), and (Jp4) in figures (a) and (b), and those of the equilibria in cases (Ip1), (Ip2), (Ip3), and (Ip4) in figures (c) and (d), respectively.

On the other hand, in ITER-like shape case shown in figures 12 (c) and (d), the $max(\alpha_{97})$ decreases about 10% and the stable region against ideal MHD modes becomes narrower as the pressure gradient inside the top of the pedestal increases about four times. Moreover, in figure 12 (c), the α_{87} value approaches to the stability boundary of the ballooning mode. These results and the result that the change of ξ_r is visible as shown in figure 10 (b) imply that the increase of the pressure gradient inside the top of the pedestal changes the edge MHD stability due to approaching the ballooning mode stability boundary inside the top of the pedestal. From these results, we find the reason of the difference in the expansion of the structure of ξ_r in the different shape equilibria as follows. In a strongly shaped plasma, the pressure gradient and the current density near the pedestal can increase due to stabilizing the ideal ballooning mode and the peeling-ballooning mode. However, since the shaping effect becomes weaker in the plasma core region, the pressure gradient inside the top of the pedestal can approach to the stability boundary of the ideal ballooning mode as shown in figure 12 (c). This approach changes the edge MHD stability, and expands the structure of the edge MHD mode.

Finally, we investigate the reason why the n number of the most unstable mode changes when both the p and $\langle \boldsymbol{j} \cdot \boldsymbol{B} \rangle / \langle \boldsymbol{B}^2 \rangle$ profiles are changed in ITER-like shape case as discussed in subsection 2.3. Figure 13 shows the (s, α) stability diagram of the equilibria shown in figure 3 at (a) $\psi_N = 0.87$ and (b) $\psi_N = 0.97$. In this case, the (s_{87}, α_{87}) point of the equilibria approaches to the ballooning mode stability boundary, but the trajectory of (s_{87}, α_{87}) does not pass across the ballooning mode stability boundary due to accessing the second stability region unlike in the fixed $\langle \boldsymbol{j} \cdot \boldsymbol{B} \rangle / \langle \boldsymbol{B}^2 \rangle$ case. At $\psi_N = 0.97$ shown in figure 13 (b), the stability boundaries of the peeling/kink mode and the peeling-ballooning mode straiten the stable region and the $max(\alpha_{97})$ decreases over 30%, because the relatively low-n ($n = 4 \sim 6$) peeling-ballooning modes and the low-n ($n = 1 \sim 3$) kink-ballooning modes are destabilized due to the increase of the current density inside the top of the pedestal.

In comparison between figures 12 (c) and (d) and figure 13, the difference of the edge MHD stability with/without the increase of the current density inside the top of the pedestal is in evidence. When the current density profile is fixed, since only the pressure driven component in the plasma potential energy becomes larger, the expansion of the eigenfunction is caused by making the pressure driven ballooning mode to be unstable as shown in figure 12 (c). When the

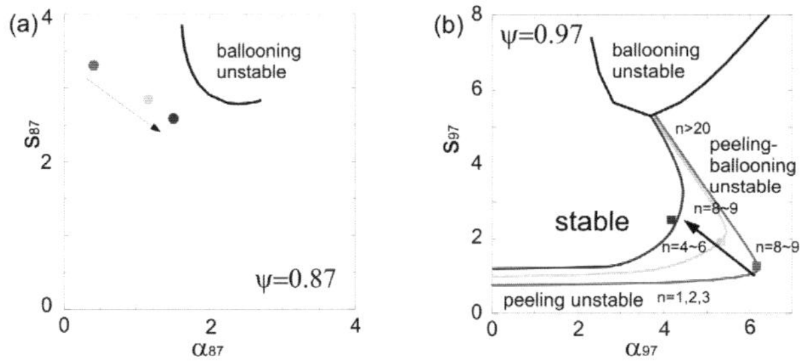

Figure 13. Stability diagram of the ideal MHD mode on the (s, α) plane at (a) $\psi_N = 0.87$ and (b) $\psi_N = 0.97$ when both the p and $\langle \boldsymbol{j} \cdot \boldsymbol{B} \rangle / \langle \boldsymbol{B}^2 \rangle$ profile are changed in ITER-like shape case discussed in subsection 2.3. The black line is the stability boundary of the $n = \infty$ ballooning mode. The red, green, blue and purple lines show the stability boundaries of the peeling/kink mode and the peeling-ballooning mode, and the red, green, blue, and purple points show the (s, α) point of the equilibria (I1), (I2), and (I3), respectively.

current density inside the top of the pedestal increases shown in figure 13, on the other hand, since not only the pressure driven component but also the current driven component become larger, the peeling-ballooning mode tends to be unstable and the maximum pressure gradient at the pedestal decreases. The reason of this destabilization is thought as the coupling between the peeling-ballooning mode localized near the pedestal and that localized inside the top of the pedestal, and the structure of the peeling-ballooning mode extends to the plasma core region as shown in figure 7.

5. Summary

In this paper, effects of the pressure profile and the current density profile inside the top of the pedestal and that of the plasma shape on the expansion of the structure of the unstable edge MHD mode are investigated numerically. The numerical results of the stability analyses showed that the pressure and the current density profiles inside the top of the pedestal is important for determining the structure of the eigenfunction of the edge MHD mode, and the expansion of the displacement of the unstable mode is enhanced in a strongly shaped plasma. From the results under the fixed current density profile assumption, it is revealed that the increase of the pressure gradient is essential for the expansion of the edge MHD mode due to spreading the envelope of the edge ballooning mode. On the other hand, the increase of the bootstrap current density is necessary for decreasing the toroidal mode number of the most unstable mode. Since the decrease of the toroidal mode number generally expands the structure of the edge MHD mode, both the pressure profile and the current density profile inside the top of the pedestal has an impact on the structure of the edge MHD mode. The expansion of the edge MHD mode is induced by making the ballooning mode or the peeling-ballooning mode to be unstable inside the top of the pedestal. The ideal ballooning mode is strongly stabilized by the plasma shaping but this stabilizing effect of the plasma shaping becomes weak in the plasma core region. In a strongly shaped equilibrium, the pressure gradient and the current density inside the top of the pedestal can increase largely and destabilize the edge MHD mode due to approaching to the ballooning mode stability boundary. This destabilization expands the structure of the edge MHD mode.

The numerical results in this paper showed that not only the pressure gradient but also the current density inside the top of the pedestal is important for discussing the structure of the edge MHD mode. This result implies that the global safety factor profile, in other words, the total plasma current, also affects the structure of the edge MHD mode by changing the ballooning mode and the kink/peeling mode stability on the (s, α) plane. This will be discussed in near future.

Acknowledgments

The authors would like to thank Dr. M. Kikuchi for his continuous encouragements and support. This research was partly supported by Japan Society for the Promotion of Science, grant-in-aid for science research (B), 18360448.

References

[1] Doyle E J *et al.* 2007 *Nucl. Fusion* **47** S72
[2] Connor J W, Hastie R J, Wilson H R and Miller R L 1998 *Phys. Plasmas* **5** 2687
[3] Wilson H R, Snyder P B, Miller R L and Huysmans G T A 2002 *Phys. Plasmas* **9** 1277
[4] Snyder P B, Wilson H R, Ferron J R, Lao L L, Leonard A W, Osborne T H, Turnbull A D, Mossessian D, Murakami M and Xu X Q 2002 *Phys. Plasmas* **9** 2037
[5] Becoulet M *et al.* 2003 *Plasma Phys. Control. Fusion* **45** A93
[6] Medvedev S Y, Martynov A A, Martin Y R, Sauter O and Villard L 2006 *Plasma Phys. Control. Fusion* **48** 927
[7] Aiba N, Tokuda S, Takizuka T, Kurita G and Ozeki T 2007 *Nucl. Fusion* **47** 297
[8] Oyama N, Asakura N, Chankin A V, Oikawa T, Sugihara M, Takenaga H, Itami K, Miura Y, Kamada Y, Shinohara K and Team J 2004 *Nucl. Fusion* **44** 582
[9] Ferron J R *et al.* 2000 *Phys. Plasmas* **7** 1976
[10] Hayashi N, Takizuka T, Aiba N, Ozeki T and Oyama N 2008 *Contrib. Plasma Phys.* **1** 196
[11] Tokuda S and Watanabe T 1999 *Phys. Plasmas* **6** 3012
[12] Aiba N, Tokuda S, Fujita T, Ozeki T, Chu M S, Snyder P B and Wilson H R 2007 *J. Plasma Fusion Res.* **2** 010
[13] Kikuchi M, JA-EU satellite tokamak working group and JT-60SA design team 2006 *Proc. 21st IAEA Fusion Energy Conf.* (Chengdu, 2006) (Vienna: IAEA) CDROM file FT/2-5 and http://www-naweb.iaea.org/napc/physics/FEC/FEC2006/html/index.html
[14] Tani K, Azumi M and Devoto R S J 1992 *J. Comput. Phys.* **98** 332

Journal of Physics: Conference Series **123** (2008) 012009 doi:10.1088/1742-6596/123/1/012009

ELM propagation in the low- and high-field-side Scrape-off Layer of the JT-60U tokamak

N Asakura[1], H Kawashima[1], N Ohno[2], G Matsunaga[1], T Nakano[1] and N Oyama[1]

[1] Japan Atomic Energy Agency, Naka Fusion Institute, Naka 311-0193, Japan

[2] EcoTopia Science Research Institute, Nagoya University, Nagoya 464-8603, Japan

E-mail: asakura.nobuyuki@jaea.go.jp

Abstract. In JT-60U, the filament structure of Edge Localized Modes (ELMs) is measured at three poloidal locations, and their temporal evolution is investigated at the inner (high magnetic-field-side, HFS) SOL as well as outer (low magnetic-field-side, LFS) SOL. At the inner SOL, filament structure with 7-8 temporal multi-peaks in ion-saturation probe signals and other diagnostics is, for the first time, measured close to the sepatratrix. The delay of the first peak after the start of MHD activity is faster than the characteristic time of the parallel convection from the outer midplane, and the Mach numbers of the plasma parallel flow reach ion sonic levels. These results show that ELM filaments extend from the outer to the inner plasma edge, and that a part of the filaments are ejected into the inner SOL. The toroidal mode numbers (n) and poloidal width of the filaments (δz) are evaluated from the interval of the multi-peaks and duration of each peak, yielding $n \sim 18$-44 and $\delta z \sim 2$-6 cm on the HFS SOL, while n and δz are comparable on the LFS SOL. After the multi-peaks appear, flow reversal of the SOL plasma toward is observed over a wide region of the inner SOL.

1. Introduction

Experimental data on edge and SOL plasma dynamics during ELMs has recently progressed by use of time-resolving diagnostics such as fast visible and infrared TV cameras, Langmuir probes and Beam-Emission-Spectroscopy [1-8]. In particular, evolution of ELM filaments has been measured, which is of great interest understanding heat and particle transport from the plasma edge to Plasma Facing Components (PFC) such as the divertor and the first wall [3, 9]. In JT-60U, time evolution of the ELM plasma has been measured with reciprocating Mach probes at different poloidal locations (i.e., outer midplane, divertor null-point, and above the inner baffle) [10, 11], and recently the sampling system was improved in order to measure the fast response of the plasma up to 500 kHz and those were synchronized with the principal fast diagnostics such as D_α emission and magnetic pickup coils. In this paper, the filament structure and evolution are investigated to compare the characteristics on the outer (low magnetic-field-side, LFS) and inner (high magnetic-field-side, HFS) SOLs. Especially noteworthy are measurements of filament structures and their transport at the inner SOL. Plasma parameters and diagnostics are presented in section 2. Evolution of the inner and outer SOL plasmas is presented in section 3. Filament structures at the outer and inner SOLs are discussed in sections 4 and 5, respectively. A summary is presented in section 6.

2. Experiments and diagnostics

ELM plasma propagation is investigated in Type-I ELMy H-mode discharges for the plasma current (I_p) of 1 MA, toroidal magnetic field (B_t) of 1.86 T, plasma triangularity (δ) of 0.32, elongation (κ) of 1.4, safety factor at 0.95 of the minor radius (q_{95}) of 3.3, neutral beam power (P_{NB}) of 5.5 MW, averaged electron density ($\bar{n}_e$) of 1.9 - 2.1×10^{19} m^{-3}, and the plasma volume (V_p) of 80 m^3. Here, P_{NB} is 1.5 times larger than the threshold power scaling necessary for L-H transition [12]. The ELM frequency is the range of 25 - 45 Hz.

Three Mach probes are simultaneously inserted into the different poloidal locations of the SOL, but their position relative to the flux surfaces at the LFS midplane are different since the insertion speed and the flux expansion of the magnetic field lines are differ at each location. 10-15 ELM events are measured at different radial positions during the inward movement of the reciprocating Mach probe (over ~ 0.5 s), and the SOL plasma conditions are not disturbed by the probe.

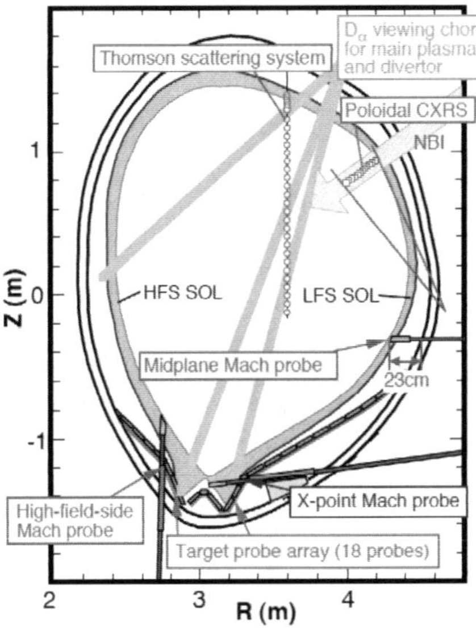

Figure 1. Main diagnostics for the edge and SOL measurements. Radial viewing chords of Thomson Scattering system, poloidal array of Charge eXchange Recombination Spectroscopy are plotted. Radial resolution of toroidal CXRS is about 6 cm. Three reciprocating Mach probes measure radial profiles of ion saturation currents, electron temperature and floating potential at the outer midplane, X-point and above the inner baffle. 18 target probes are installed on the inner and outer divertor plates. Three viewing chords of the D_α brightness measurement for the main and divertor are shown.

Figure 1 shows the plasma configuration and location of edge diagnostics such as the Thomson scattering system, charge-exchange recombination spectroscopy, and three reciprocating Mach probes. The magnetic flux surface corresponding to corresponding to 3 cm outside the separatrix at the LFS midplane (Δr^{mid} = 3 cm) is also shown, and the magnetic field lines between the HFS and LFS SOLs are connected within Δr^{mid} = 8 cm. Figure 2 shows radial profiles of ion and electron temperatures, electron density and toroidal rotation velocity at the plasma edge, which are measured between ELMs. The electron temperature, density and toroidal component of the parallel plasma flow in the SOL are measured by the Mach probes with the voltage sweep of 1 ms, and these profiles are obtained in 0.5 s. Here, data during transient period (~ 5 ms) in each ELM event are excluded, and the radial coordinate is mapping to the radius at the LFS midplane. Electron and ion temperatures, electron density and toroidal rotation velocity at the pedestal top (T_e^{ped}, T_i^{ped}, n_e^{ped} and v_t^{ped}) are 0.8 - 0.9 keV, 1.0 - 1.2 keV, 1.4 - 1.6×10^{19} m^{-3}and ~ 28 km/s, respectively. Thus, the loss fraction of the pedestal stored energy, ($\Delta W_{ELM}/W^{ped}$), due to Type-I ELMs is evaluated to be 0.09 - 0.12, where $W^{ped} \equiv (3/2)n_e^{ped}(T_e^{ped} + T_i^{ped})/V_p^{ped}$, and the plasma volume inside the pedestal (V_p^{ped}) is 70 m^3.

In the SOL, an in-out asymmetry of electron temperature is observed, while electron densities near the separatrix at the HFS and LFS SOLs are comparable. These result in a lower plasma pressure at the HFS SOL than at the LFS SOL. During the steady-state period, the plasma flow in the SOL is generally observed from the LFS SOL to the HFS divertor through the SOL opposite the X-point region (here, the top) [13]. The toroidal component of the SOL flow is evaluated as $v_t^{SOL} = M_{//}C_s^{SOL}(B_t/B_{//})$, where the Mach number of the SOL flow ($M_{//}$) is calculated from the ratio of ion saturation currents at the ion- and electron-sides of the Mach probe using Hutchinson's formula [14]: $M_{//} = 0.4\ \ln[j_s^{\,i}/j_s^{\,e}]$. C_s^{SOL} is the plasma sound velocity in the SOL, assuming $T_i = 2T_e$. Here, $M_{//}$ near the separatrix is relatively low (0.2 - 0.3) in the main SOL, i.e., at the LFS midplane and above the HFS baffle, and $v_t^{SOL} \sim 10$ km/s is larger than that measured just inside the separatrix ($v_t^{edge} \sim 3$ km/s), but lower than v_t^{ped}. On the other hand, in the LFS divertor, a subsonic level of $M_{//} \sim 0.45$ near the divertor X-point is measured, which corresponds to v_t^{SOL} = -20 to -30 km/s and is produced by reduction of the plasma pressure in the magnetic-sheath of the divertor. The direction is towards the divertor, which may cause the friction force to the edge plasma, provided that v_t^{edge} near the X-point is similar to the measurement at the LFS edge.

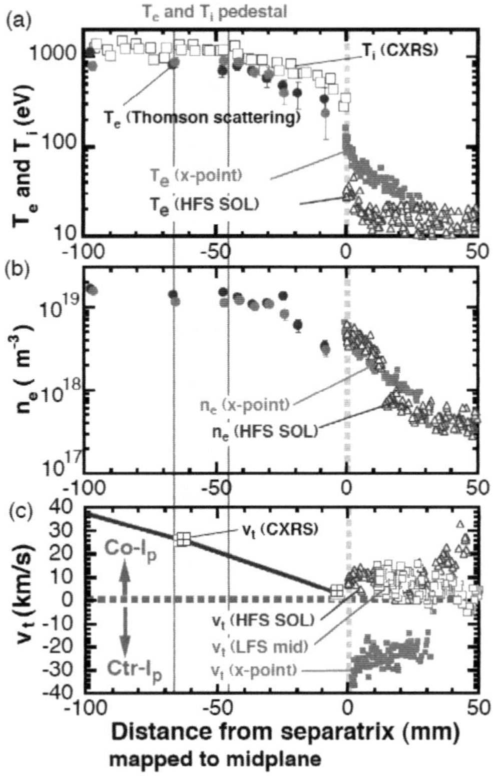

Figure 2. Radial profiles of (a) ion and electron temperatures, (b) electron density, (c) toroidal rotation velocity at the plasma edge. Electron temperature, density and toroidal component of the parallel plasma flow measured with the Mach probes are shown, where radial location is mapping to the radius at the LFS midplane. Vertical lines show edge pedestal (Δr = -67 to -47 mm) of the ion and electron temperature profiles.

3. Fast evolution of the SOL plasma during ELM

Figure 3 shows an example of one ELM event, i.e. rapid changes in magnetic fluctuations ($\tilde{B}_p$), j_s at the upstream (electron drift) side of the midplane probe (34 cm below the midplane) and the divertor strike-point (j_s^{mid} and j_s^{div}, respectively), and D_α brightness viewing the main plasma and the divertors. The corresponding midplane radial locations (Δr^{mid}) for the midplane Mach probe, and the target probe are 1.0 cm and 0.2 cm, respectively. The amplitude of $\tilde{B}_p$ rapidly increases after t = 5509.946 ms as seen as a vertical line (t_0^{MHD}) and this defined as the start for the time delay for various signals. The D_α signals, both in the main plasma and divertor, generally start to increase immediately after t_0^{MHD} as shown in figure 3 (e). The large $\tilde{B}_p$ turbulence is accompanied by plasma ejected from the edge region [11,15]. During the early period of the large $\tilde{B}_p$ turbulence (about 0.1 ms after t_0^{MHD}), large multi-peaks of j_s^{mid} appear both at the upstream and downstream sides of the Mach probe. Characteristics of the large multi-peaks of j_s^{mid} are investigated in the following sections. Then, the base-level of j_s^{mid} is increased on a longer time scale (~ 0.5 ms), and the maximum base-level and the delay are presented as $j_s^{mid}(base)$ and $\tau^{mid}(base)$, respectively.

Journal of Physics: Conference Series **123** (2008) 012009 doi:10.1088/1742-6596/123/1/012009

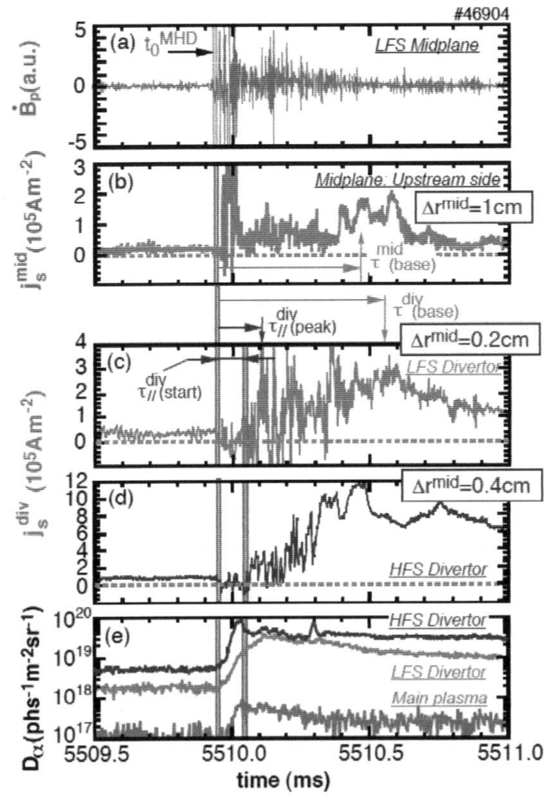

Figure 3. (a) Magnetic fluctuations at the LFS midplane, (b) j_s at upstream side of the midplane Mach probe for the midplane distance (Δr^{mid}) of 1 cm, (c) j_s at the LFS divertor for midplane distance of 0.2 cm, (d) j_s at the HFS divertor for midplane distance of 0.4 cm, (e) D_α brightness viewing the main plasma and the HFS and LFS divertors. Delays of the maximum base-level of the midplane j_s, start of the divertor j_s base-line increase and the first divertor j_s peak are represented by τ^{mid}*(base)*, $\tau_{//}^{div}$*(peak)* and $\tau_{//}^{div}$*(start)*, respectively.

Figure 3(c) shows that j_s^{div} at the LFS strike-point starts to increase after t = 5510.052 ms, i.e., after the early period of the large $\tilde{B}_p$ turbulence, then large multi-peaks appear in j_s^{div}. The first large j_s^{div} peak appears at t = 5510.105 ms during the increase in the j_s^{div} base-level. The plasma convection time from the LFS midplane to the LFS divertor is calculated as $\tau_{//}^{SOL\text{-}LFS} = L_s^{LFS\text{-}div}/C_s^{ped}$ = 105 μs, where $L_s^{LFS\text{-}div}$ (= 30 m) is the connection length from the LFS midplane to the divertor and C_s^{ped} is the plasma sound velocity for the pedestal plasma, i.e. $\sqrt{[(T_i^{ped}+T_e^{ped})/m_i]} = 2.8\times10^5$ m/s for T_i^{ped} = 1.2 keV and T_e^{ped} = 1.0 keV. Time delays of the start and the first peak in j_s^{div} enhancement ($\Delta t = t - t_0^{MHD}$ = 106 and 159 μs) are comparable to and larger than $\tau_{//}^{SOL\text{-}LFS}$, respectively. These delays can be explained by the parallel convection of the ELM plasma, as shown in $\tau_{//}^{div}$*(start)* and $\tau_{//}^{div}$*(peak)* in figure 3 (c).

Here, D_α signals at the HFS and LFS divertors increases simultaneously during the early period of the large $\tilde{B}_p$ turbulence. At the same time, a large increase in heat flux density is measured both at the HFS and LFS strike-points (q^{div} ~ 150 and 170 MW/m^2, respectively) with the infrared TV camera at a relatively slow frame rate of 250 μs. Figure 3 (d) shows that time delays of the start and the first peak in the HFS j_s^{div} enhancement are comparable to $\tau_{//}^{div}$*(start)* and $\tau_{//}^{div}$*(peak)* at the LFS divertor. The HFS results can not be explained by sonic convective transport of the ELM plasma from the LFS SOL to the HFS divertor; transport mechanism of the ELM filaments in the HFS SOL is discussed in section 5.

4. ELM filament propagation at LFS midplane

4.1. Time evolution of multi-peaks

Figure 4 shows the expanded time evolution of $\tilde{B}_p$, D_α brightness at the main plasma, j_s^{mid} at the upstream and downstream sides of the Mach probe as a function of the time delay from t_0^{MHD}, which is shown in figure 3. Multiple large positive peaks are observed from the two sides of the Mach probe, which indicates that exhausted ELM filaments extend to the Mach probe location along magnetic field lines. Three peaks in the upstream j_s^{mid} are observed at Δt = 9, 30, 53 μs. The second peak (5.2-5.7×10^5 Am^{-2}) is larger than the first peak (2.5×10^5 Am^{-2}), but the maximum value of the second peak is not clearly identified due to large fluctuations. Three peaks in the downstream j_s^{mid} are observed at Δt = 9, 28, 43 μs simultaneously except for the third peak, which is delayed by 6 μs. Here, the first peak (6×10^5 Am^{-2}) is larger than the other peaks (3.2×10^5, 2.2×10^5 Am^{-2}). After appearance of the multi-peaks, the upstream and downstream j_s^{mid} values reach 0.5-0.8×10^5 Am^{-2}.

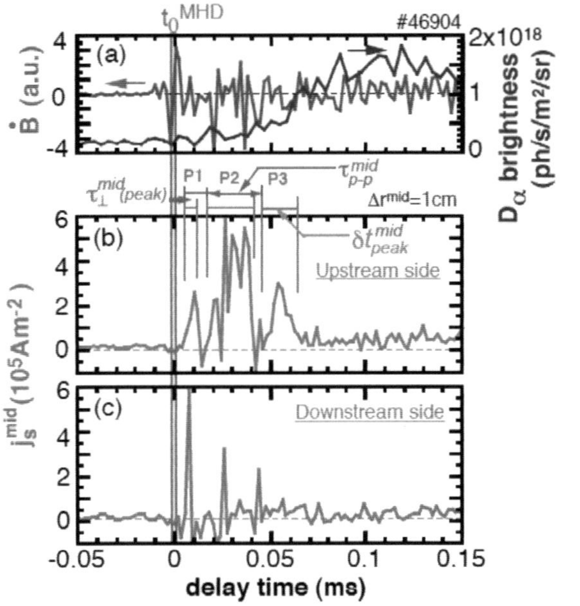

Figure 4. Enlarged time evolutions of (a) magnetic fluctuations, $\tilde{B}_p$, and D_α brightness in the main plasma, (b) ion saturation currents, j_s^{mid}, at the upstream side, and (c) at the downstream side of the LFS midplane Mach probe as a function of time lag from the start of the magnetic turbulence for the same ELM event as shown in figure 2. Delay of the first j_s^{mid} peak, separation and duration of three peaks are shown by $\tau_\perp^{mid}$*(peak)*, $\tau_{p\text{-}p}^{mid}$ and δt_{pk}^{mid}.

Characteristics of the first j_s^{mid} peak are presented in reference 16. The upstream j_s^{mid} is typically larger than the downstream j_s^{mid} near the separatrix. On the other hand, in the far SOL (Δr^{mid} > 3 cm), they are comparable at the outer flux surfaces. Radial propagation of the first ELM filament is investigated to quantify the fast propagation of the filaments. Here, the averaged value of the *first* j_s^{mid} *peaks* at the upstream and downstream sides and the delay of the upstream j_s^{mid} peak are defined as j_s^{mid}*(peak)* and $\tau_\perp^{mid}$*(peak)*, respectively. In this paper, we focus on evaluation of the spatial size and toroidal mode numbers from the duration ($\delta t^{mid}{}_{peak}$) and separation ($\tau^{mid}{}_{p\text{-}p}$) of the multiple peaks.

4.2. Radial propagation of filament

Radial distributions of j_s^{mid}*(peak)* and j_s^{mid}*(base)* are shown in figure 5(a). These distributions are deduced from four similar ELMy H-mode discharges, and the j_s^{mid}*(peak)* profile is *an envelope of* j_s^{mid} *peaks* rather than a profile appearing at one moment. Here, field lines at Δr^{mid} > 13 cm are connected to the upper LFS first wall, while those between 3.3 cm < Δr^{mid} < 13 cm are connected to the LFS baffle. Values of the first j_s^{mid} peaks are varied from 1×10^5 to 5.5×10^5 A/m^2 near the separatrix. The enhancement factor of j_s^{mid}*(peak)* compared the j_s^{mid} *base-level between ELMs*, j_s^{mid}*(ss)*, is 10 - 50 over a wide SOL region. Particularly, in the outer flux surfaces (Δr^{mid} > 2 cm), enhancement of radial propagation of j_s^{mid}*(peak)* is observed; its e-folding length (λ_{peak}^{mid}) of 7.5 cm is 2.5 times larger than those between ELMs (λ_{SS}^{mid} = 3.1 cm) and of j_s^{mid}*(base)* (λ_{base}^{mid} = 3.5 cm).

The radial propagation velocity of j_s^{mid}*(peak)* is evaluated as the average velocity from the separatrix to the Mach probe location, i.e., $v_\perp^{mid}$*(peak)* = $\Delta r^{mid}/\tau_\perp^{mid}$*(peak)*. Figures 5(b) and (c) show distributions of $\tau_\perp^{mid}$*(peak)* and $v_\perp^{mid}$*(peak)* as functions of Δr^{mid}. Since $\tau_\perp^{mid}$*(peak)* tends to increase with Δr^{mid} for Δr^{mid} < 5 cm, $v_\perp^{mid}$*(peak)* ranges between 0.4 and 1.5 km/s. The characteristic length of the ELM radial propagation, $L_{\perp}^{ELM}$, is defined as the radial distance, where the first j_s^{mid} peak moves during the parallel convection transport time to the divertor, i.e., $v_{\perp}^{mid}(peak) \cdot \tau_{//}^{SOL\text{-}LFS}$, resulting in $L_{\perp}^{ELM} = 4$ - 15 cm. It is found that the fast-radial-velocity ELM filaments reach the LFS baffle and the first wall. For $\Delta r^{mid} > 5$ cm, $\tau_{\perp}^{mid}(peak)$ ranges between 40 - 90 μs, indicating a large $v_{\perp}^{mid}(peak)$ of 1.5 - 2.9 km/s. As a result, $v_{\perp}^{mid}(peak)$ for the first j_s^{mid} peak tends to increase with Δr^{mid}.

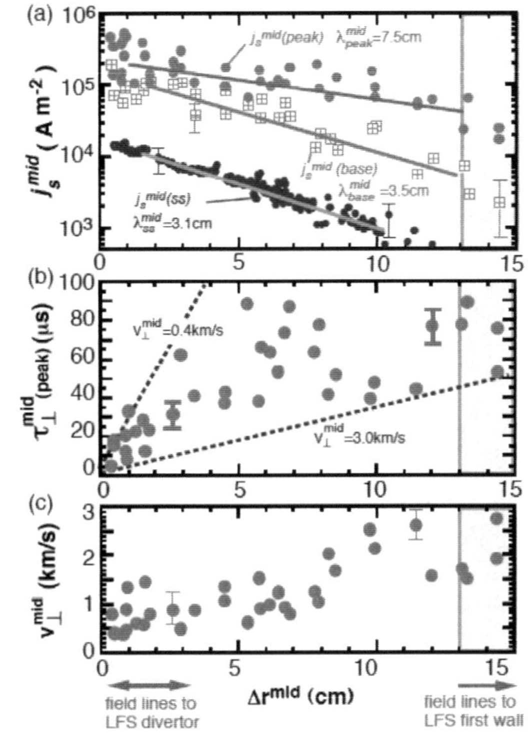

Figure 5. Radial distributions of the first j_s^{mid} peaks (large circles), maximum base-level of j_s^{mid} during ELM (squares), and j_s^{mid} between ELMs (small circles). Representative e-folding lengths are shown for three distributions. Radial distributions of (b) time delay of the first j_s^{mid} peak, $\tau_{\perp}^{mid}(peak)$, (c) averaged propagation velocity, $v_{\perp}^{mid}(peak)$.

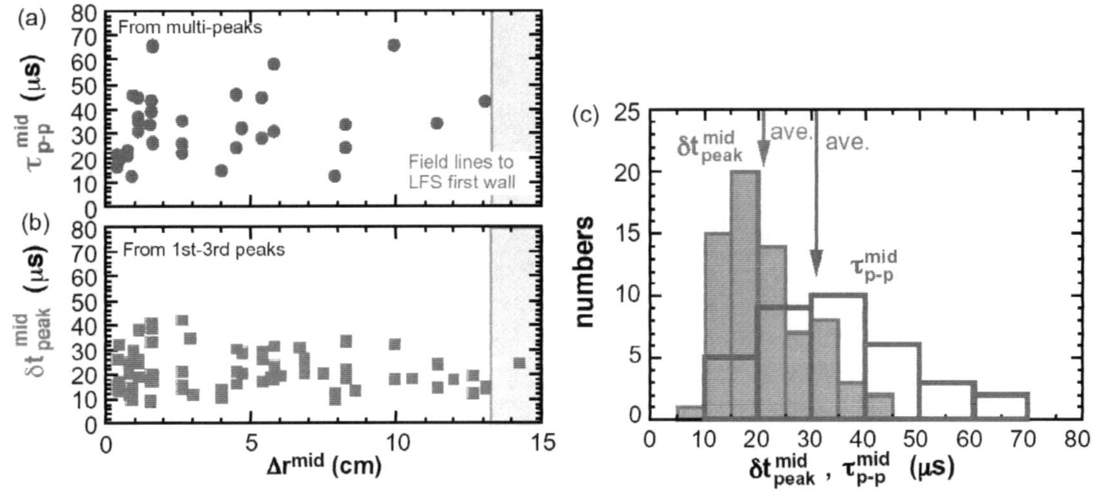

Figure 6. Radial distributions of (a) peak to peak separation, $\tau_{p\text{-}p}^{mid}$, (b) peak duration, δt_{pk}^{mid}, for the first to forth j_s^{mid} peaks for 20 ELM events. (c) Probability distributions of $\tau_{p\text{-}p}^{mid}$ and δt_{pk}^{mid}.

Journal of Physics: Conference Series **123** (2008) 012009 doi:10.1088/1742-6596/123/1/012009

4.3. Evaluation of LFS filament structure

Filament size and mode numbers are investigated from the database of $v_{\perp}^{mid}(peak)$, the j_s^{mid} peak duration (δt_{pk}^{mid}) and separation of the multi-peaks ($\tau_{p\text{-}p}^{mid}$). Values of δt_{pk}^{mid} and $\tau_{p\text{-}p}^{mid}$ are defined as the full width of the peak and as the separation time between the beginning of adjacent peaks in the time evolution of j_s^{mid}, respectively. For the three j_s^{mid} peaks in figure 4 (b), δt_{pk}^{mid} = 10, 23 and 18 μs and $\tau_{p\text{-}p}^{mid}$ = 12 and 25 μs. Radial profiles of $\tau_{p\text{-}p}^{mid}$ and δt_{pk}^{mid} and their probability distributions are shown in figure 6, using the database from the first to the fourth $j_s^{mid}(peak)$ in 20 ELM events from two similar ELMy H-mode discharges. It is found that average values of $\tau_{p\text{-}p}^{mid}$ and δt_{pk}^{mid} are 31 μs and 21 μs, and 72% of $\tau_{p\text{-}p}^{mid}$ and 70% of δt_{pk}^{mid} are in a range of 20 - 50 μs and 10 - 25 μs, respectively.

The ELM plasma filaments extend large distances along the magnetic field lines in SOL, and they are observed to move in the radial and toroidal directions. First, values of δt_{pk}^{mid} are closely related to the structure and dynamics of the plasma filament. Radial and poloidal scales of the plasma filament can be evaluated from the measurements of the radial and toroidal velocities, respectively. The radial scale of the plasma filament is evaluated as $\delta r_{pk}^{mid} = \delta t_{pk}^{mid} \cdot v_{\perp}^{mid}$, provided that the plasma filaments cross the Mach probe in the radial direction perpendicular to the field lines with the radial distribution of $v_{\perp}^{mid}(peak)$ as shown in figure 4. Figure 5 (c) suggests that radial scale of δr_{pk}^{mid} increases since $v_{\perp}^{mid}(peak)$ increases with radius, and δr_{pk}^{mid} increases from 0.5 - 4 cm near separatrix ($\Delta r^{mid} < 5$ cm) to 1 - 8 cm in the far SOL.

On the other hand, the poloidal scale of the plasma filament is evaluated as $\delta z_{pk}^{mid} = \delta t_{pk}^{mid} \cdot v_t^{ped} \cdot \tan(\alpha)$, provided that the plasma filaments extending along the magnetic field line pitch angle of α move in the toroidal direction with the rotation velocity at the pedestal (v_t^{ped}) and they cross the Mach probe [17]. Here, the poloidal rotation velocity of the pedestal top is also measured (4 - 8km/s in the electron diamagnetic direction) and the corresponding rotation frequency (~ 0.8 kHz) is small (about a half of the toroidal one), thus the effect of the poloidal velocity is not considered. From the experimental parameters of $v_t^{ped} \sim 28$ km/s as shown in figure 2 and $\alpha = 8°$, one finds δz_{pk}^{mid} = 2 - 5 cm for δt_{pk}^{mid} = 10 - 25 μs. The toroidal mode numbers (n) are deduced from $\tau_{p\text{-}p}^{mid}$ = 20 - 50 μs as shown in figure 6 (a) and (c) as $n = 2\pi R^{LFS}/(v_t^{ped} \cdot \tau_{p\text{-}p}^{mid})$, where the major radii at the LFS Mach probe (R^{LFS}) is 4.3 m, resulting in n = 18 - 44. These values of n are too large compared to those found in other experiments (n = 10 - 20) [18], which are consistent with peeling-ballooning model.

5. ELM filament propagation at HFS SOL

5.1. Time evolution of multi-peaks

Effects of plasma drifts and fast SOL flow on the divertor plasma and impurity transport have been recently studied in many tokamaks [13,19-22]. Plasma transport in the HFS SOL plays an important role on particle control and impurity transport in the L-mode plasma. Influence of the transient plasma transport produced by the ELM is also of great interest. Thus, filament structure and the dynamics in the HFS SOL are presented here using the fast measurement from the Mach probe.

Figure 7 shows the time evolution of D_{α} brightness at the HFS and LFS divertors, $\tilde{B}_p$, j_s *at* the upstream and downstream sides of the HFS Mach probe ($j_s^{HFS\text{-}up}$ and $j_s^{HFS\text{-}down}$) and j_s at the HFS strike point ($j_s^{div\text{-}HFS}$) during an ELM event, which is different time as shown in figure 3. The local distance from the separatrix to the Mach probe is 1.1 cm, which corresponds to just outside of the separatrix when mapped along the flux surface to the LFS midplane (Δr^{mid} = 0.3 cm). It is found that multi-peaks appear only in $j_s^{HFS\text{-}up}$ with the delay of 53 μs after t_0^{MHD}. The large peak observed on the LFS is not seen in $j_s^{HFS\text{-}down}$, although $j_s^{HFS\text{-}down}$ does increase after Δt = 53 μs. During the j_s^{HFS} increase, j_s at the HFS strike-point ($j_s^{div\text{-}HFS}$) also increases but again a large peak is not observed.

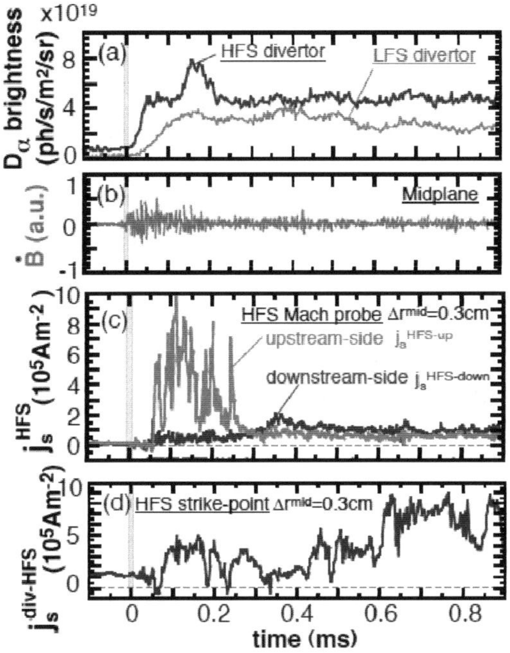

Figure 7. Time evolutions of (a) D_α brightness in the LFS and HFS divertors, (b) magnetic fluctuations, $\tilde{B}_p$, (c) ion saturation currents, j_s, at the up-stream and down-stream sides of the HFS Mach probe (Δr^{mid} = 0.3 cm), (e) j_s at the HFS divertor strike-point (Δr^{mid} = 0.3 cm), as a function of delay from the start of large $\tilde{B}_p$ turbulence.

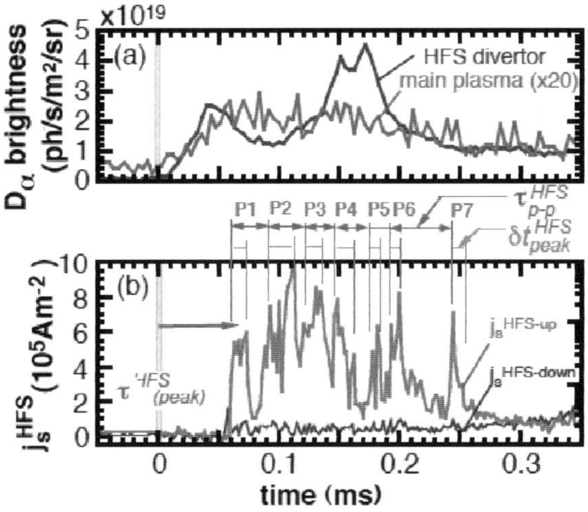

Figure 8. Enlarged time evolutions of (a) D_α brightness in the main plasma and the HFS divertor, (b) j_s^{HFS} at upstream and downstream sides of the HFS Mach probe as shown in figure 7. Delay of the first j_s^{HFS} peak, separation and duration of three peaks are shown by $\tau^{HFS}(peak)$, $\tau_{p\text{-}p}^{HFS}$ and δt_{pk}^{HFS}.

Figure 8 shows an expanded time evolution of the D_α brightness at the main plasma and HFS divertor, $j_s^{HFS\text{-}up}$ and $j_s^{HFS\text{-}down}$, which were shown in figure 7. During the early period of the j_s^{HFS} increase (Δt = 50 - 250 μs), about seven large peaks appear in $j_s^{HFS\text{-}up}$ at Δt = 66, 110, 130, 148, 180, 202 and 242 μs. The peak duration (δt_{pk}^{mid}) is between 10 and 22 μs, which is comparable to those measured at the LFS midplane. The separation of the maxima ($\tau_{p\text{-}p}^{mid}$) is between 25 and 55 μs, which is larger than that at the LFS midplane. The peak values of $j_s^{HFS\text{-}up}$ ($j_s^{HFS\text{-}up}(peak)$) are between 5.6×10^5 and 9.8×10^5 Am^{-2}, and they are significantly larger than $j_s^{HFS\text{-}down}$ (~ 1.2×10^5 Am^{-2}). Large enhancement in $j_s^{HFS\text{-}up}$ shows that the fast parallel plasma flow towards the HFS divertor is associated with the filaments, and the corresponding Mach number reaches close to the sonic level (0.7 to 1). Delays of the first to forth $j_s^{HFS\text{-}up}$ peaks (Δt = 68 - 145 μs) measured are smaller than the characteristic time of the convective transport from the LFS midplane to the HFS Mach probe, i.e., $\tau_{//}^{LFS\text{-}HFS} = L_s^{LFS\text{-}HFS}/C_s^{ped}$ = 186 μs, where the distance from the LFS midplane to the HFS Mach probe along the field line, $L_s^{LFS\text{-}HFS}$ is 55 m. On the other hand, the delay of the first peak is only slightly larger than the characteristic convection time from the HFS midplane to the HFS Mach probe: $\tau_{//}^{\text{HFS-HFS}} = L_s^{\text{HFS-HFS}}/C_s^{\text{ped}} = 46\ \mu\text{s}$ ($L_s^{\text{HFS-HFS}} = 12$ m). Thus, provided that ELM filaments extend to the HFS edge region along the field lines, and they are ejected into the SOL, those short delays can be explained as convection from the HFS midplane to the probe.

5.2. Radial distribution of HFS filaments

Figure 9 (a) shows the radial distributions of the large $j_s^{\text{HFS-up}}$ peaks ($j_s^{\text{HFS}}(peak)$), the maximum base-level after appearance of the large peaks ($j_s^{\text{HFS}}(base)$) and $j_s^{\text{HFS-up}}$ between ELMs ($j_s^{\text{HFS}}(ss)$) as functions of the LFS midplane radius. Enhancement factors of $j_s^{\text{HFS}}(peak)$ are 20 - 50, which are comparable to those at the LFS midplane (10 - 50). Very large enhancement (30 - 50) of $j_s^{\text{HFS}}(peak)$ is often observed near the separatrix ($\Delta r^{\text{mid}} < 0.6$ cm). On the other hand, in the outer flux surfaces, such large $j_s^{\text{HFS}}(peak)$ values are not observed and e-folding length of $j_s^{\text{HFS}}(peak)$ mapping to the LFS midplane is $\lambda_{\text{peak}}^{\text{HFS}} =$ 3.2 cm, which is about a half that at the LFS midplane ($\lambda_{\text{peak}}^{\text{mid}} = 7.5$ cm).

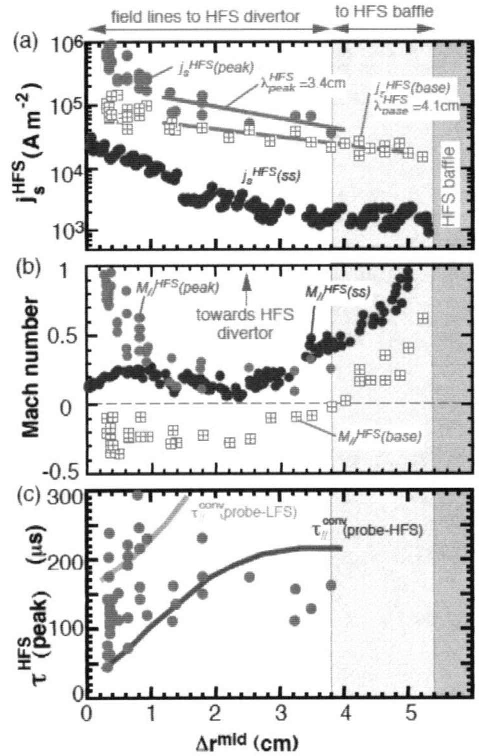

Figure 9. Radial distributions of (a) j_s^{HFS} peaks (red circles), maximum base-level (squares) during ELMs, and j_s^{HFS} between ELMs (blue circles) measured at upstream-side of the HFS Mach probe, (b) Mach numbers corresponding to $j_s^{\text{HFS}}(peak)$, $j_s^{\text{HFS}}(base)$, $j_s^{\text{HFS}}(ss)$, (c) delay of the j_s^{HFS} peaks. All profiles are mapped to the LFS midplane radius. Parallel transport times from LFS midplane to the HFS probe, and from HFS midplane to the HFS probe are shown.

The Mach numbers for $j_s^{\text{HFS}}(peak)$, $j_s^{\text{HFS}}(base)$ and $j_s^{\text{HFS}}(ss)$ are shown in figure 9 (b): $M_{//}^{\text{HFS}}(peak)$, $M_{//}^{\text{HFS}}(base)$ and $M_{//}^{\text{HFS}}(ss)$, respectively. $M_{//}^{\text{HFS}}(peak)$ increases from 0.2 - 0.3 away from the separatrix to near unity as close to the separatrix, i.e., a sonic level of the fast SOL flow is observed near the separatrix ($\Delta r^{\text{mid}} < 0.8$ cm). The delay of $j_s^{\text{HFS}}(peak)$, i.e., $\tau^{\text{HFS}}(peak)$, is plotted in figure 9 (c). Characteristic times of the convective transport from the LFS midplane, i.e., $\tau_{//}^{\text{conv}}(probe\text{-}LFS)$, and from the HFS midplane, i.e., $\tau_{//}^{\text{conv}}(probe\text{-}HFS)$, are also shown assuming a constant C_s^{ped} of 2.9×10^5 m/s, using $M_{//}^{\text{HFS}}(peak)$ data shown in figure 9 (b) and connection length varying with the radial position of the HFS probe. $\tau^{\text{HFS}}(peak)$ is comparable or larger than $\tau_{//}^{\text{conv}}(probe\text{-}HFS)$, except in the far SOL ($\Delta r^{\text{mid}} > 2$ cm). These facts suggest that the ELM filaments are ejected into the HFS SOL during the early period of ELM, and that the radial extent of the filament is mostly limited to flux surfaces in the narrow region with $\Delta r^{\text{mid}} < 1$ cm.

It is also found that the flow reversal of the SOL plasma occurs over a wide SOL region ($\Delta r^{mid} < 4$ cm) after the appearance of the multi-peaks. As a result, the plasma and impurity transport at the HFS SOL and divertor is influenced by two transient convective fluxes during the ELM event: one is the large, short convective flow from the upstream SOL, and another is the flow reversal over a wide SOL region, both of which can play an important role in deposition profiles of impurities.

5.3. Evaluation of LFS filament structure

Filament size and mode number are investigated from the database of the $j_s^{HFS\text{-}up}$ peak duration (δt_{pk}^{HFS}) and separation of the multi-peaks ($\tau_{p\text{-}p}^{HFS}$). Figure 10 (a) shows the radial distributions of $\tau_{p\text{-}p}^{HFS}$ and δt_{pk}^{HFS} for the $j_s^{HFS\text{-}up}$ peaks as a function of corresponding LFS midplane radius, Δr^{mid}. Values of δt_{pk}^{mid} are independent of Δr^{mid}. Multi-filaments at the HFS appear near the separatrix, where the radial distance is less than 3 cm corresponding the LFS midplane flux surfaces of $\Delta r^{mid} < 1$ cm. In section 4.3, the radial scale of the filament near the LFS separatrix (δr_{pk}^{mid}) is 0.5-4 cm, so relative small radial size of filaments may appear at the HFS SOL.

Next, $\tau_{p\text{-}p}^{HFS}$ can be determined near the separatrix and scattered between 18 and 64 μs. Probability distributions of δt_{pk}^{mid} and $\tau_{p\text{-}p}^{HFS}$ are shown in figure 10 (b), and 76% of δt_{pk}^{HFS} and 81% of $\tau_{p\text{-}p}^{HFS}$ are in a range of 10 - 30 μs and 20 - 50 μs, respectively. Average values are $\delta t_{pk}^{HFS} = 24$ μs and $\tau_{p\text{-}p}^{HFS} = 38$ μs, which are slightly larger than those for the LFS midplane ($\delta t_{pk}^{mid} = 21$ μs and $\tau_{p\text{-}p}^{mid} = 31$ μs). The poloidal scale of the plasma filament at the HFS SOL is evaluated from δt_{pk}^{HFS} by $\delta z_{pk}^{HFS} = \delta t_p^{HFS} \cdot v_t^{ped} \cdot \tan(\beta)$, provided that the plasma filament is formed along the local field line pitch angle of $\beta = 6°$ at the Mach probe, and that the toroidal rotation at the HFS SOL corresponds to $v_t^{ped}(HFS) = v_t^{ped} \times (R^{HFS}/R^{LFS})$, where $v_t^{ped} \sim 28$ km/s, the major radii at the HFS and LFS Mach probes are $R^{HFS} = 2.7$ m and $R^{LFS} = 4.3$ m. Again, effect of the poloidal velocity is not considered as is described in section 4.3 due to small poloidal rotation frequency (~ 0.8 kHz, about a half of the toroidal one at the LFS midplane). Then, δz_{pk}^{HFS} is 2 - 6 cm for $\delta t_{pk}^{HFS} = 10$ - 30 μs. The toroidal mode number is estimated to be $n = 18$ - 44 from separation time of $\tau_{p\text{-}p}^{HFS} = 20$ - 50 μs. As a result, averaged filament size and mode numbers are comparable to those at the LFS SOL, suggesting the filament extends to the HFS SOL at least when it is ejected into the SOL.

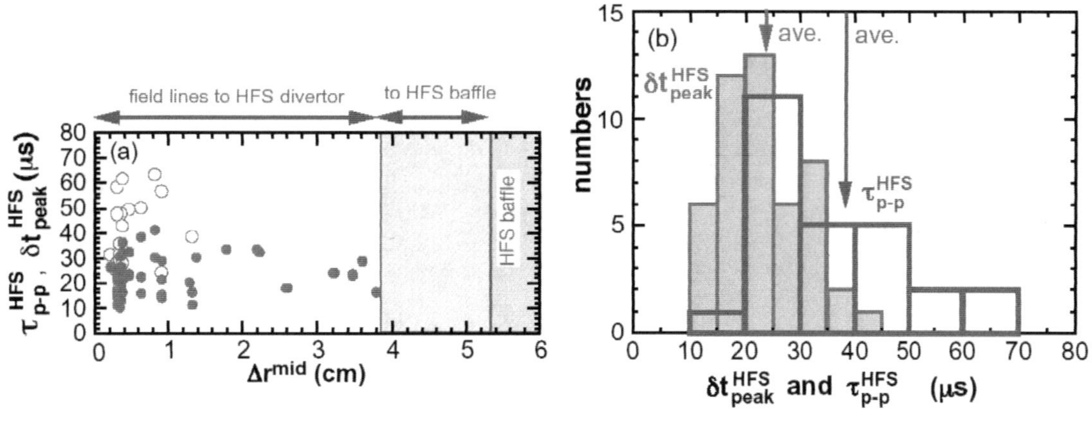

Figure 10. Radial distributions of (a) peak to peak separation, $\tau_{p\text{-}p}^{HFS}$ (open circles), and peak duration, δt_{pk}^{HFS} (closed circles), (b) Probability distributions of $\tau_{p\text{-}p}^{HFS}$ and δt_{pk}^{HFS}.

6. Summary

Time and spatial structures of Type-1 ELM filaments were investigated both at the High-Field-Side and Low-Field-Side SOLs, using reciprocating Mach probes with fast sampling rate of 500 kHz and synchronizing with the other principal fast diagnostics. Large multi-peaks of the ion saturation current

(j_s^{mid} of a few 10^5 to 10^6 Am^{-2}) appear during the early period of the ELM events (< 100 μs after start of the large magnetic activity produced by ELM) at both the HFS SOL and the LFS midplane, which shows that plasma filaments propagate to the first wall and the divertor plates.

At the LFS midplane, the propagation time of the large j_s^{mid} peaks to the Mach probe, i.e. $\tau_\perp^{mid}$ (= 40-90 μs) is faster than *parallel convection to the divertor* (~ 100 μs). Radial propagation of the filaments were enhanced with the radial velocity of $v_\perp^{mid}$ = 0.5-3 km/s (0.5-1.5 km/s near separatrix). Fast filament fluxes will cause local heat and particle loadings.

Several filaments appear near separatrix (Δr^{mid} < 0.4 cm) of the HFS SOL. At the same time, convective transport with $M_{//}$ = 0.5-1 was observed with short propagation time to the HFS Mach probe above the baffle (τ^{HFS} = 53 μs), which is faster than parallel convection transport time from the LFS midplane (~ 186 μs). Filaments thus may extend to the HFS plasma edge and they are ejected into the HFS SOL, producing fast propagation of the filaments to the HFS SOL and divertor. This convection flux causes large neutral desorption at the HFS divertor and "*flow reversal*" over a wide radial range of the HFS SOL.

For HFS and LFS SOLs, a large part of the peak-duration (δt_{peak} = 10 - 30 μs) and the peak-separation ($\tau_{p\text{-}p}$ = 20 - 50 μs) are independent of the radial and poloidal location. Radial and poloidal sizes were evaluated to be δr ~ 0.5-4 cm and δz ~ 2-6 cm, and toroidal mode number of n = 18 - 44 was estimated assuming a toroidal rotation velocity at the pedestal of ~ 28 km/s. However, n > 20 is too high to understand by conventional MHD (peeling-ballooning) model, and thus further study is necessary.

References

[1] Kirk A *et al.* 2004 *Phys. Rev. Lett.* **92** 245002
[2] Herrmann A *et al.* 2004 *Plasma Phys. Control. Fusion* **46** 971
[3] Eich T *et al.* 2005 *Plasma Phys. Control. Fusion* **47** 815
[4] Kirk A *et al.* 2005 *Plasma Phys. Control. Fusion* **47** 315
[5] Boedo J A *et al.* 2005 *J. Nucl. Matter* **337-339** 771
[6] Endler M *et al.* 2005 *Plasma Phys. Control. Fusion* **47** 219
[7] Boedo J A *et al.* 2005 *Phys. Plasmas* **12** 072516
[8] Pitts R A, Fundamenski W, Erents S K *et al.* 2006 *Nucl. Fusion* **46** 82
[9] Eich T *et al.* 2003 *Phys. Rev. Lett.* **91** 195003
[10] Asakura N, Sakurai S, Naito O *et al.* 2002 *Plasma Phys. Control. Fusion* **44** A313
[11] Asakura N, Takechi M, Oyama N *et al.* 2005 *J. Nucl. Matter.* **337-339** 712
[12] Ryter F and H-mode Threshold Database Group 2002 *Plasma Phys. Control. Fusion* **44** A415
[13] Asakura N and ITPA SOL and divertor Topical Group 2007 *J. Nucl. Mater.* **363-365** 41
[14] Hutchinson I H 1988 *Phys. Rev.* **A 37** 4358
[15] Oyama N *et al.* 2004 *Nucl. Fusion* **44** 582.
[16] Asakura N, Kawashima H, Ohno N *et al.* submitted to *Nucl. Fusion* 2007.
[17] Kirk A *et al.* 2005 *Plasma Phys. Control. Fusion* **47** 995
[18] Kirk A *et al.* 2007 "Comparison of the spatial and temporal structure of type-I ELMs" *this conference*
[19] Asakura N *et al.* 2004 *Nucl. Fusion* **44** 503
[20] LaBombard B, Rice J E, Hubbard A E *et al.* 2004 *Nucl. Fusion* **44** 1047
[21] Fundamenski W, Andrew P, Erents K *et al.* 2005 *J. Nucl. Mater.* **337-339** 305
[22] Pitts R *et al.* 2005 *J. Nucl. Mater.* **337-339** 146

Journal of Physics: Conference Series **123** (2008) 012010 doi:10.1088/1742-6596/123/1/012010

Radial electric field evolution in various operational modes in the TUMAN-3M tokamak

L G Askinazi[1], V A Kornev[1], S V Krikunov[1], L I Krupnik[2], S V Lebedev[1], A I Smirnov[1], M Tendler[3], A S Tukachinsky[1], M I Vildjunas[1] and N A Zhubr[1]

[1] Ioffe Institute, St Petersburg 194021, Russia
[2] IPP, Kharkov Institute of Physics and Technology, 61108 Kharkov, Ukraine
[3] Alfven Lab., Euratom NFR Association, Royal Institute of Technology, Stockholm, Sweden

E-mail: leonid.askinazi@mail.ioffe.ru

Abstract. Radial electric field evolution has been studied on the TUAMN-3M tokamak in different modes of operation: ohmic and NBI heating, L- and H-modes, with and without strong MHD activity. Peripheral radial electric field was measured using Langmuire probes, which were inserted up to 2cm inside LCFS, while core plasma potential evolution was measured using HIBP diagnostic. It was found, that in presence of strong MHD activity radial electric field in a vicinity of the island changed sign from negative to positive and could reach up to 4kV/m. Central plasma potential exhibited a positive perturbation of ~700V during the MHD burst. This positive radial electric field might lead to H-mode termination, both in ohmic and NBI heating cases. Possible mechanism of the positive E_r generation, namely the electron losses along ergodized magnetic field lines in the presence of MHD-island, is discussed. The same mechanism might be responsible for the positive potential spikes during a saw-tooth crash, also observed using HIBP. Another phenomenon observed using HIBP was quasi-coherent potential oscillations with the frequency close to one of the GAM. Possible location of these oscillations in the core region $r/a \sim 0.33$ is discussed.

1. Introduction

Radial electric field is known to play a key role in various important processes in tokamak plasma. Among them are confinement mode transitions, toroidal and poloidal rotation of the plasma, MHD island rotation. Besides, quasi-stationary or oscillating radial electric fields is exited by zonal flows and its higher frequency branch – the Geodesic Acoustic Mode (GAM), which, in turn, plays a role of mediator between turbulence and transport.

Spatial structure and temporal dynamics of the radial electric field in the small tokamak TUMAN-3M ($R = 0.55$m, $a = 0.22$m, $I_p^{max} = 0.16$MA, $B_t^{max} = 0.8$T, $n_e < 5\times10^{19}$m^{-3}, $T_e < 500$eV, $T_i < 200$eV, cylindrical $q_{lim} \sim 2.3$ - 2.6) was studied in different modes of plasma heating and confinement, namely, in the ohmic L- and H-modes and NBI, with and without low frequency MHD oscillations. Central plasma potential was measured by heavy ion beam probe (HIBP) [1], which gives a rare possibility of direct measurement of hot dense plasma potential. Due to the geometrical limitations, only a region of $0 < r < 16$cm (i.e. $0 < r/a < 0.73$) is typically covered by the HIBP. Peripheral radial electric field was measured using Langmuire probes [2]. Thanks to the proper design and materials used (and of course, moderate electron temperature and density of the TUMAN-3M peripheral plasma), the probes may have been inserted up to 2cm ($r/a = 0.91$) inside LCFS, thus allowing E_r measurement in the peripheral region of the core plasma. So, range $0.73 < r/a < 0.91$ is still unreachable for both Langmuire probes and HIBP, and obtaining information on E_r structure in this region is a challenge for future experiments. Useful information on plasma rotation in this region, qualitatively corresponding to E_r evolution measured with probes and HIBP has been obtained using microwave reflectometry [3].

2. Radial electric field and plasma potential evolution in the presence of rotating magnetic island

Earlier, radial electric field evolution in a vicinity of the rotating MHD island created by a resistive tearing mode has been measured in the TUMAN-3M tokamak using Langmuire probes [2] and reflectometry [3]. The probes were inserted from high filed side and located close to the island separatrix, staying at least 1cm away from it even when the island had the largest size. It has been found [2] that when the island is small and its rotation velocity is high, measured E_r is negative and close to the theoretical predictions [4]. This theory is based on neoclassical description of plasma flow near the island and takes into account anomalous viscosity. The theory predicts, in particular, that in order to maintain the common rotation speed of the island and the ambient plasma, the should exist radial electric field $E_r^{island} = - <B_t>\omega r_s/m$ in a vicinity of the island ($<B_t>$ is magnetic surface averaged toroidal field, ω, r_s and m are the island rotation frequency, resonance surface radius and poloidal mode number, respectively). Later, as the island grows and rotation decelerates, measured E_r changes sign and becomes positive (directed to the plasma boundary). It should be stressed, that there exists not only a quantitative disagreement between E_r and the island rotation velocity values, but also the positive direction of E_r contradicts to the cw direction of the island's poloidal rotation. The most probable cause of positive E_r build up is thought to be electron losses due to stochastical perturbation of the magnetic field lines in a vicinity of the island [5]. Measurements of the HXR emission from the limiter shows strong increase in electron losses during the MHD bursts [6]. In the recent experiments, these results were supplemented with the HIBP measurements of the central plasma potential evolution performed in similar experimental conditions. Figure 1 presents the results of peripheral E_r evolution measurements by probes discussed in detail in [2]. Initially, radial electric field is negative $E_r \sim$ -2kV/m, which is in good agreement with the theoretical formula $E_r^{island} = <B_t>\omega r_s/m$, see figure 1.

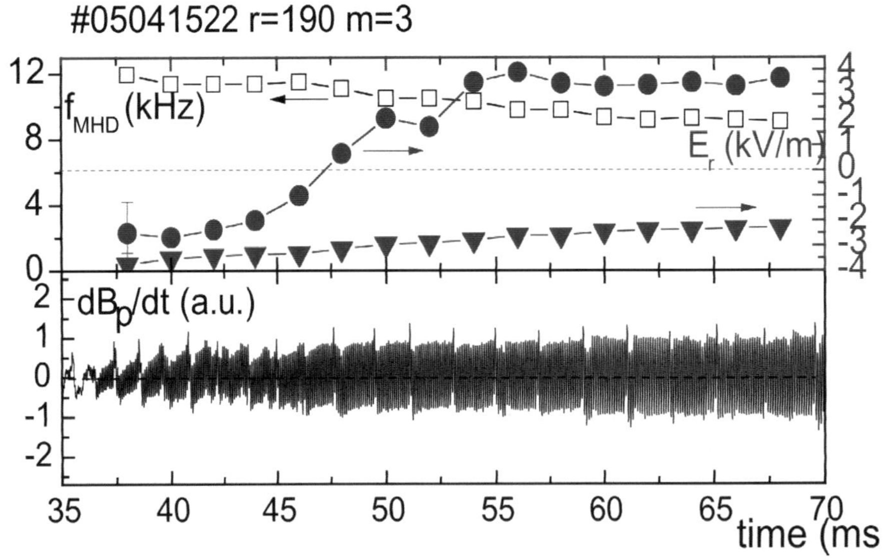

Figure 1. Top: MHD mode frequency evolution (black open squares), measured radial electric field $E_r = - \nabla\Phi_{float}$ near the island (closed blue circles) and theoretical value $E_r^{island} = - <B_t>\omega r_s/m$ from [4] (closed blue triangles); bottom: magnetic probe signal dB_p/dt (a.u.).

Formation of positive $E_r \sim 4$kV/m is clearly seen after $t \sim 55$ms. Figure 2 displays the HIBP results of central plasma potential evolution in a shot with relatively short burst of MHD oscillations (*m*/*n* = 2/1 mixed with 3/1). In these measurement, the initial location of the sample volume was at $r = 12$ cm, changing slowly to $r = 10$cm due to the B_t non-constancy. Hence, the point of potential measurement was located deeper in the core plasma than the magnetic islands resided near $r = 17$-19cm. The location of the island was obtained from the analysis of the MHD-caused perturbations on the chord signals of microwave interferometer, see [7] for details. It was observed, that core plasma potential had exhibited strong positive perturbation up to $\Delta\Phi = 600$-700V correlated with the MHD burst, so the change in average radial electric field between the point of HIBP measurement and the wall is $\Delta\Phi/\Delta r \sim 0.7$kV/0.11m = 6.4kV/m. This value agrees well with the change in peripheral radial electric field $\Delta E_r \sim 6$kV/m measured by probes. It should be stressed that in this experiment HIBP measured plasma potential at the distance of 10-12cm from the wall; this is significantly deeper than the peripheral the ergodic layer created by the MHD oscillations where positive radial electric field is generated. The positive radial electric field generated in the presence of magnetic island may have a dramatic effect on plasma confinement. In particular, if the MHD island perturbation develops during H-mode phase of the discharge, it may lead to H-mode termination and backward confinement transition. This situation is discussed in more detail in the next section of the paper.

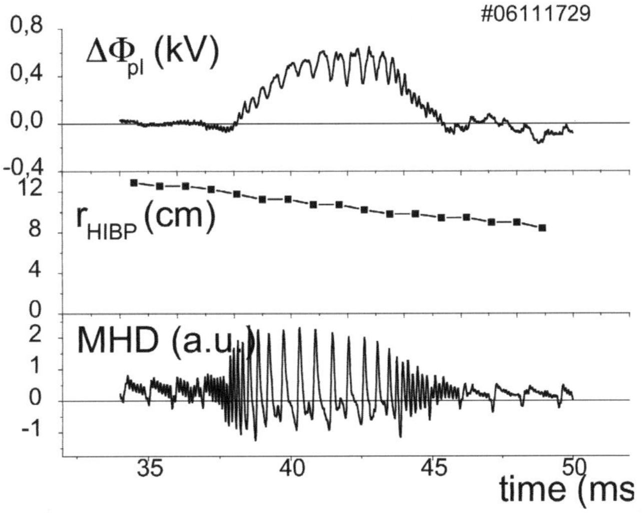

Figure 2. Positive perturbation of core plasma potential caused by the MHD burst.

As for the mechanism of positive E_r generation by the MHD island, the most probable candidate is thought to be the loss of electrons along ergodized magnetic field lines. Such an ergodic layer should exist near the island's separatrix. To sustain quasi-neutrality, this electron out flux must be compensated by an ion current, which, in turn, requires a positive radial electric field generation. This situation has a lot in common with plasma biasing experiments performed on a number of tokamaks (and in the TUMAN-3m, in particular [8]), when plasma responses with the transverse conductivity current to the externally exited biased electrode current. The idea of electron losses due to the magnetic lines ergodization was used to describe the E_r evolution during the Dinamic Ergodic Divertor (DED) operation on the TEXTOR tokamak [9]. A quantitative theoretical model of this mechanism of E_r generation in the presence of MHD island may be found in [10]. This theory gives E_r values close to ones measured on the TUMAN-3M tokamak. Moreover, this theory takes into account transverse inward transport of toroidal momentum and as a result, is capable of explaining deep penetration of E_r created at the plasma edge. The direct experimental evidence to electron losses was obtained in the measurements of HXR emission from the limiter. The significant increase in the emission in the

Journal of Physics: Conference Series **123** (2008) 012010 doi:10.1088/1742-6596/123/1/012010

presence of the MHD island was found [6]. The oscillating fraction of the HSR emission is observed to be in phase with magnetic probe signal, indicating an effect of local field line topology on the electron losses. So, it supports the hypothesis of the increased longitudinal electron losses caused by magnetic field lines ergodization by the presence of the MHD island.

3. Plasma potential evolution in counter-NBI heated plasma

Central plasma potential studies were carried out in a scenario with ohmic L-H transition followed (with a delay of ~ 10ms) by counter-NBI heating pulse. Co- or counter- injection of the neutral beam in the TUMAN-3M tokamak is selected by switching the plasma current direction, with the injection direction being constant (pointed reversely with respect to the toroidal field). Generally speaking, co-NBI is much more favorable from the point of view of fast ion confinement, and this holds for the TUMAN-3M experiments also [11]. Unfortunately, co-NBI heating on the TUMAN-3M is not compatible with the HIBP diagnostic set up due to the wrong direction of toroidal displacement of the primary and secondary ions trajectories, which prevent secondary beam from passing through the exit port. The NBI application on the TUMAN-3M is discussed in [11] in more detail. Measurements of the electric potential were performed by probing the core plasma with 80keV K^+ ions. Figure 3 (upper box) displays the plasma potential evolution measured with HIBP in two shots: #06111730, ohmic H-mode plus counter-NBI (red lines) and #06111724, ohmic H-mode only (blue lines). For a comparison, the potential evolution measured in ohmic L-mode shot is also displayed in the same box (black line). It is clearly seen that potential tends to be more negative after the L-H transition, though it changes rather slowly during ~ 8ms after the transition. The location of the point of potential measurement (shown by black squares in a separate box in the same figure) was nearby $r = 6$ - 8cm, $r/a = 0.25$ - 0.33, that is well inside the edge transport barrier located near $r = 20$cm. A slow drift of the sample volume is caused by the temporally non-constant B_t. A relatively slow evolution of the central potential corresponds to a gradual change in plasma density profile after the edge transport barrier formation. After the start of NBI heating, central plasma potential gradually becomes even more negative (by ~ 200V) than it was in ohmic-H-mode phase. Numerical modeling of the counter- NBI injection

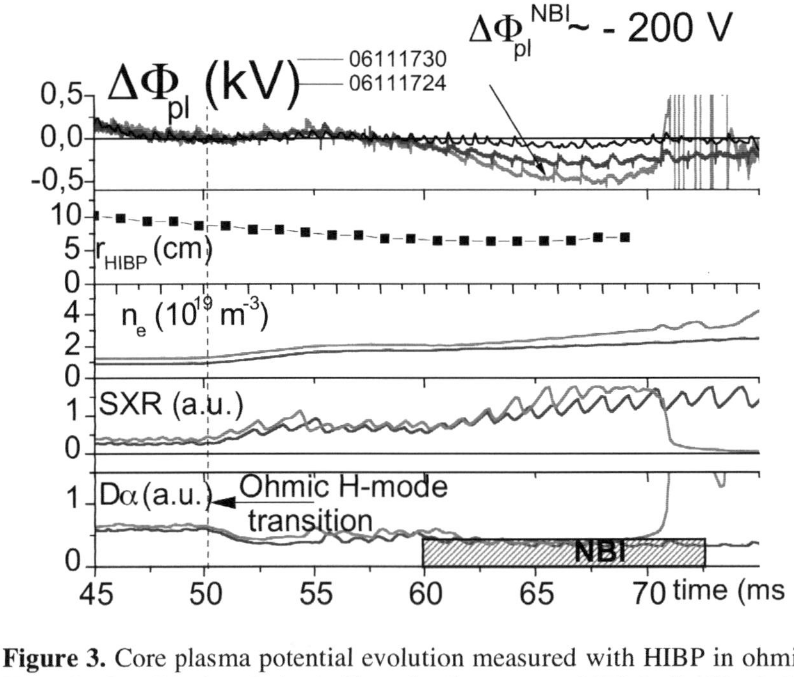

Figure 3. Core plasma potential evolution measured with HIBP in ohmic H-mode shot (blue) and ohmic H-mode plus counter-NBI (red). Ohmic L-mode shot is also displayed (black).

scenario in the TUMAN-3M performed using ASTRA code shows that due to the high orbit loss term the efficiency of direct power and momentum transfer is low [11]. Hence, the negative potential perturbation caused by the NBI may be a result of the orbit losses of fast ions created by the neutral beam ionization, with minor additional effect of toroidal momentum and plasma pressure input from the injection. In some shots a burst of MHD oscillations was excited in the H-mode stage of the discharge at the front of NBI heating pulse, see figure 4. In this case, similarly to the ohmic heating situation, the MHD burst was accompanied by a positive perturbation (by ~ 700V) of the central plasma potential. As may be concluded from figure 4, the MHD burst and positive perturbation of plasma potential were followed by H-L transition, which was featured by an increase in D_α emission, electron density and SXR radiation decay after $t \sim 61$ms. The average radial electric field needed to create such a perturbation of the potential may be estimated as $\Delta\Phi/\Delta r \sim 0.7$kV/0.17m = 4.1kV/m. This radial electric field perturbation resulted, apparently, in decrease of the peripheral radial electric field shear, which, in turn, led to turbulent transport reduction termination, and the low mode of confinement recovery. However, the experimental data available at present are not enough to quantitatively support this mechanism; more sophisticated multi-point measurement at the edge is needed to evidence it. Earlier, a similar effect of H-mode degradation due to the MHD activity burst was observed in pure ohimcally heated TUMAN-3M plasma [2]. A positive perturbation of the peripheral radial electric field $\Delta E_r \sim 5$kV/m was measured by the Langmuire probes in that regime.

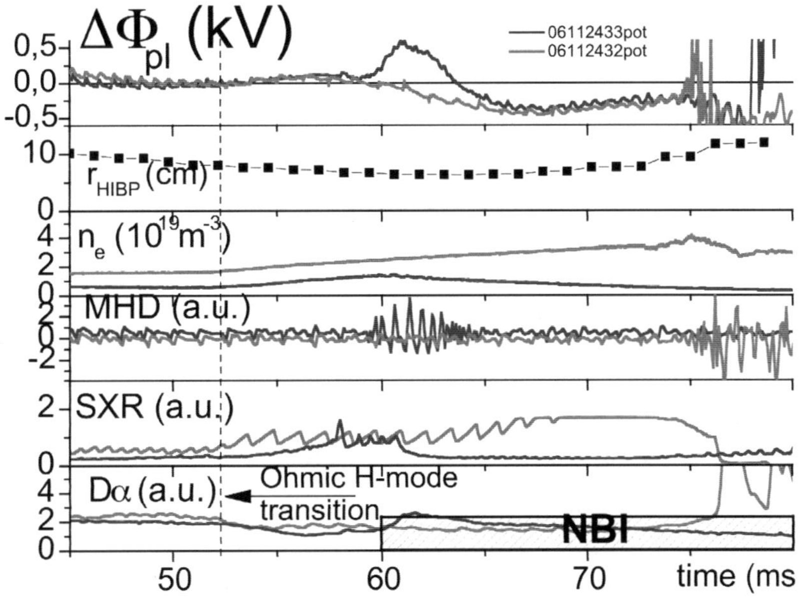

Figure 4. Core plasma potential evolution measured with HIBP in ohmic H-mode plus counter-NBI with (blue) and without (red) MHD burst.

4. Observation of Geodesic Acoustic Mode in core plasma

Another interesting phenomenon observed with the HIBP on the TUMAN-3M is quasi-coherent oscillations of plasma potential. These oscillations exist in the TUMAN-3M plasma at the initial stage of the discharge, while the current profile is not stable, usually disappearing after the current density profile becomes stationary, but sometimes surviving to the end of the shot. These oscillations has frequency around ~ 30kHz and relative amplitude up to $\delta\Phi/\Phi \sim 0.3$. At the same time, local density signal (which is simply proportional to the intensity of the secondary beam of the HIBP) does exhibit only a very low level oscillations ($\delta n/n \leq 0.05$), if any. No oscillations near that frequency were seen

Journal of Physics: Conference Series **123** (2008) 012010 doi:10.1088/1742-6596/123/1/012010

on any other diagnostic signals, such as Mirnov coils, interferometer and SXR chords. For example, spectral range of low frequency MHD oscillations in the TUMAN-3M have higher boundary of approximately 12 - 13kHz. The frequency of the potential oscillations is close to the geodesic acoustic mode scaling (GAM) $f_{GAM} \sim 1/(2\pi R)(2T_e/m_i)^{1/2} \sim 30$kHz, provided that $T_e \sim 100$eV. Another feature, namely $\delta\Phi/\Phi >> \delta n/n$ is also typical for the GAM, especially when observed in equatorial plane [12].

In the GAM studies, the sample volume was located close to the equatorial plane of the tokamak in a region of $r_{sv} \sim 8$cm, that corresponds to $r/a \sim 0.33$, according to numerical calculations of probing beam trajectories. It is not very usual for GAMs, which, as a rule, are located at the very edge of the plasma column [13]. Therefore, the central location of the GAM in our experiment had to be cross-checked. Two important issues should be considered in this connection. First, it would be useful to obtain independent experimental evidence to the central location of the sample during the GAM studies. Such evidence may be found in a shot with strong saw-tooth oscillations co-existing with GAM oscillations of the plasma potential, figure 5.

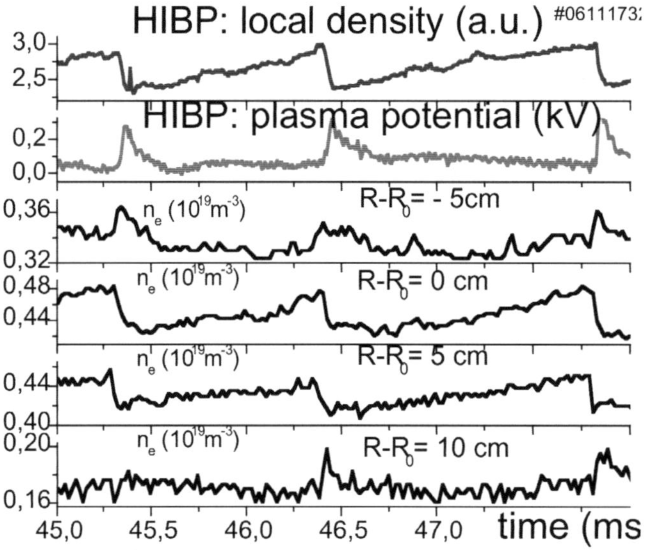

Figure 5. HIBP signals (density and potential) and signals of 4 central chords of microwave interferometer, showing coexistence of GAM and saw-teeth.

In this shot, the GAM frequency is clearly seen in HIBP potential signal and its spectrum, figure 6 and figure 7. At the same time, the saw-tooth oscillations were simultaneously registered by HIBP density channel and by the multi-chord microwave interferometer. Analyzing phase relations between the HIBP density channel and interferometer chord signals, it was found that the HIBP sample volume resided somewhere in a region $r < 8$cm. So, the sample volume location around $r_{sv} \sim 8$cm found from the calculation of the HIBP trajectories does look reliable. Second issue is of more general nature and is connected with the fact that HIBP diagnostic actually measure a potential difference between sample volume location r_{sv} and the wall located at r_w. This potential difference is caused by a radial electric field located (or spread) anywhere in a range $r_{sv} < r < r_w$. This radial electric field may have an arbitrary spatial structure $E_r(r)$, provided that integral of $E_r(r)$ over the range $r_{sv} < r < r_w$ is kept constant. In other words, local single point measurement of plasma potential does not mean local measurement of radial electric field and give only E_r averaged over $r_{sv} < r < r_w$ interval. Therefore, additional information is required to draw a conclusion on of E_r structure from single point plasma potential measurement. In case of GAMs, such information may be found in zonal structure of this mode. Being a higher frequency branch of zonal flows, GAM is thought to consist of two or more narrow adjacent (in radial direction) spatial regions with alternating direction of poloidal velocity and

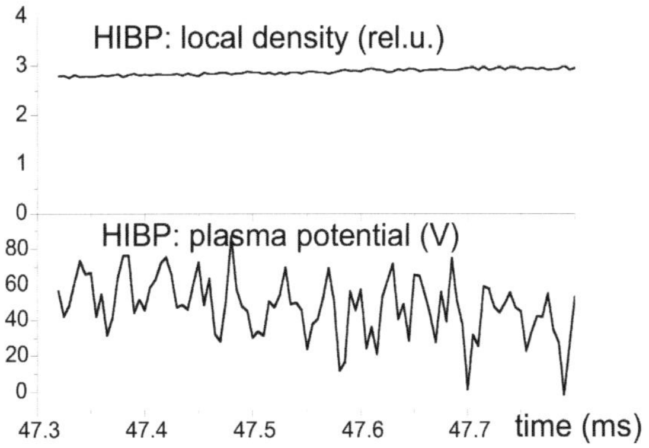

Figure 6. Zoomed HIBP signals for the same shot as in figure 5; potential and density oscillation spectra.

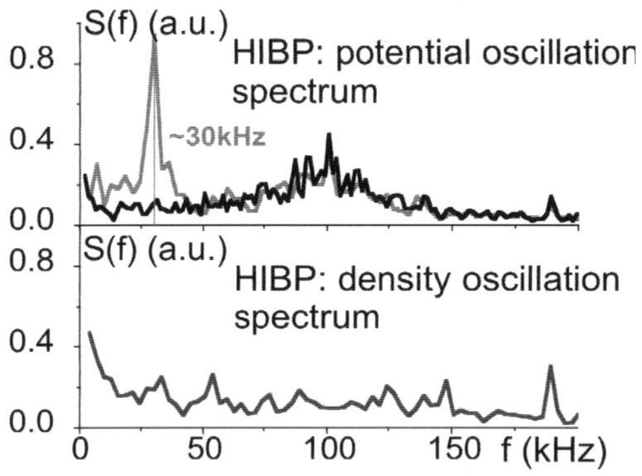

Figure 7. Potential and density oscillation spectra for the shot presented in figure 5. Black line indicates a noise level of the electronics.

radial electric fields. Hence, radial electric field averaged over the region occupied by GAM should be small: $\int E_r(r)dr \sim 0$. If the GAM was localized inside $r_{sv} < r < r_w$ interval, it would hardly be seen by the HIBP. And vise versa, if HIBP register non-zero GAM oscillation, it means that sample volume location r_{sv} is somewhere inside the radial correlation length of GAM. In other words, in case of the presence of GAM oscillations in the HIBP signal one may expect that the location of the investigated object coincides with the sample volume location. For our studies, it means that GAM is very probably localized in core plasma in a vicinity of $r = 8$cm

Another interesting observation about this shot is strong (~ 200V) positive perturbation of central plasma potential in tact with saw-tooth crash, clearly seen in figure 5. It may be caused by the loss of fast electrons as a result of the crash, in a way similar to that discussed above for the MHD island.

5. Conclusion

Strong positive perturbation of the core plasma potential was registered by the HIBP during the burst of peripheral MHDs with low m, n. If such a burst takes place in the H-mode (both ohmic and counter-NBI heated), the positive potential perturbation leads to H-mode termination. The most probable mechanism of the positive field build-up during MHD burst is though to be a loss of fast electrons along partly disturbed magnetic field lines near the island's separatrix [5, 6]. This mechanism is similar to the ergodic divertor's action on the TEXTOR [9], where radial electric field modification by the electron loses was also discussed. A quantitative analysis of the subject may be found in [10]. Similar mechanism may be responsible for a positive perturbation of central plasma potential registered in the saw-tooth crashes. The positive radial electric field caused by MHD activity is capable of H-mode termination, most probably through the canceling the H-mode "natural" negative radial electric field.

In a scenario with counter-NBI it was found using HIBP that, due to the NBI effect (most probably, orbit loss with some heating and momentum impact), core plasma potential plasma gradually became more negative (by ~ 200V).

The GAM with $\delta\Phi/\Phi \sim 0.3$ and $\delta\Phi/\Phi >> \delta n/n \sim 0.05$ where observed with HIBP in a core region of the TUMAN-3M $r/a \sim 0.33$ in the current ramp phase. Further studies are planned to reveal a possible connection between the GAM properties (localization, plasma condition dependence) and plasma confinement in the TUMAN-3M.

Acknowledgments
This work was jointly supported by the Federal Agency for Science and Innovations MES RF: State Contract # 02.518.11.7023, by the Grant of President of Russian Federation "Leading Scientific School" # 1550.2008.2 and by the Russian Foundation for Basic Research: grants # 07-02-00276, # 07-02-01311 and # 06-02-16785.

References

[1] Askinazi L G, Kornev V A, Lebedev S V*et al.* 2004 *Rev. Sci. Inst.* **75** 3517
[2] AskinaziL G, Golant V E, Kornev V A *et al.* 2006 *Plasma Phys. Contr. Fusion* **48** A85
[3] Bulanin V V, Askinazi L G, Lebedev S V *et al.* 2006 *Plasma Phys. Contr. Fusion* **48** A101
[4] Kaveeva E and Rozhansky V 2003 *Proc. 30th EPS Conference on Contr. Fusion and Plasma Phys. (St. Petersburg)* vol 27A P-3.150
[5] Bulanin V V, Dyachenko V V, Esipov L A *et al.* 2003 *Proc. Int. workshop on strong microwaves in plasmas (Nizhniy Novgorod)* vol 1 398
[6] Bulanin V V, Askinazi L G, Chugunov I N *et al.* 2006 *Proc. 33rd EPS Conference on Plasma Phys. (Rome)* vol 30I P-4.106
[7] Kornev V A, Askinazi L G, Vil'dzhyunas M I *et al.* 2005 *Plasma Phys. Reports* **31** 803
[8] Askinazi L G, Golant V E, Lebedev S V *et al.* 1992 *Nucl. Fusion* **32** 271
[9] Finken K H, Abdullaev S S, Jakubowski M W *et al.* 2007 *Phys. Rev. Lett.* **98** 065001
[10] Kaveeva E, Rozhansky V and Tendler M 2008 *Nucl. Fusion* **48** 075003
[11] Lebedev S V, Askinazi L G, Barsukov A G *et al.* 2006 *Proc. 21st IAEA Fusion Energy Conference (Chengdu)* EX/P3-15
[12] Melnikov A V, Vershkov V A, EliseevL G *et al.* 2006 *Plasma Phys. Control. Fusion* **48** S87
[13] Ido T, Miura Y, Kamiya K *et al.* 2006 *Plasma Phys. Control. Fusion* **48** S41

Comparison of the spatial and temporal structure of type-I ELMs

A Kirk[1], N Asakura[2], J A Boedo[3], M Beurskens[1], G F Counsell[1], T Eich[4], W Fundamenski[1], A Herrmann[4], Y Kamada[2], A W Leonard[5], S Lisgo[1], A Loarte[6], N Oyama[2], R A Pitts[7], A Schmid[4] and H R Wilson[8]
for the Pedestal and Edge Physics ITPA Topical Group

1 Euratom/UKAEA Fusion Association, Culham Science Centre, Abingdon, UK
2 Naka Fusion Research Establishment, Japan Atomic Energy Research Institute, Ibaraki, Japan
3 Center for Energy Research Eng. University of California, Building Unit II, 9500 Gilman Dr, La Jolla, CA 92093-0417, USA
4 Max-Planck-Institut fur Plasmaphysik, EURATOM Association, Garching, Germany
5 General Atomics, PO Box 85608, San Diego, CA, 92186-5608, USA
6 EFDA Close Support Unit—Garching, 2 Boltzmannstrasse, Garching, Germany
7 Centre de Recherches en Physique des Plasmas, Association Euratom Confedereration Suisse, Ecole Polytechnique, CH-1015 Lausanne, Switzerland
8 University of York, Heslington, York YO10 5DD UK

E-mail: Andrew.kirk@ukaea.org.uk

Abstract: A comparison of the spatial and temporal evolution of the filamentary structures observed during type I ELMs is presented from a variety of diagnostics and machines. There is evidence that these filaments can be detected inside the LCFS prior to ELMs. The filaments do not have a circular cross section instead they are elongated in the perpendicular (poloidal) direction and this size appears to increase linearly with the minor radius of the machine. The filaments start off rotating toroidally/poloidally with velocities close to that of the pedestal. This velocity then decreases as the filaments propagate radially. By comparing the results from all measurements and from comparison with models it is most likely that the filaments have at least their initial radial velocity when they are far out into the SOL and before they have interacted with the nearest limiter surface. There is a general consensus that the dominant loss mechanism in the separated filaments is through parallel transport and that the transport to the wall is through the radial propagation of these filaments. Measurements of the filament energy content show that each filament contains up to 2.5 % of the energy released by the ELM at the time it separates from the LCFS, assuming $T_i = T_e$. The parallel flux e-folding length measured on DIII-D, AUG and MAST has a weaker scaling with normalised ELM size than appears to be necessary to explain the deficit in the ELM energy arriving in the divertor on JET, assuming a purely exponential decay of the filament energy with time.

1. Introduction

Extrapolations for the amount of energy released by Type I Edge Localised Modes (ELMs) in ITER, based on data from existing tokamaks, indicate that the largest ELMs could not be tolerated regularly because of the damage they would cause [1]. However, there is considerable uncertainty associated with such predictions because of the lack of understanding of all of the processes involved. If the understanding of at least some aspects of the ELM event can be improved, it would mean that models could be further constrained and a reduction in the uncertainty of the predictions for ITER obtained.

Filament structures have been observed during ELMs in a wide range of Tokamaks using a variety of diagnostics (see [2] and references therein). In reference [3] a compilation of the radial size and propagation of these filaments was produced. It was found that the radial velocity varied widely across devices as well as in a single device, highlighting the fact that although it is clear that filament structures exist during the ELM, it is still unclear which parameters determine their size and how they propagate. Similar to reference [4], it is proposed in this paper that the ELM evolution can be broken down into three distinct phases: 1) the linear growth stage of the mode resulting in localised structures inside the Last Closed Flux Surface (LCFS), 2) the rapid growth and formation of filaments that extend outside the LCFS and 3) the separation of these filaments from the core and their radial motion towards the first wall/limiters. It is in the third stage of this process, when there is a clear gap between the plasma in the filament and the core plasma that energy and particles are transported to the walls.

The layout of this paper is as follows: in section 2 the evidence for precursor activity related to the growth of the filaments will be discussed. In section 3 and 4 a review of the size and motion of the filaments will be presented. In section 5 the energy content of the filaments will be discussed and in section 6 a study of the dependence of the power e-folding length on plasma parameters will be presented.

2. The linear growth stage and pre-cursor activity

The peeling-ballooning mode theory is now widely accepted as the explanation for the observed pressure limit prior to an ELM [5]. This model predicts that Type I ELMs in tokamaks will onset due to an instability with a toroidal mode number in the range 3-40 [6]. Several Type I ELM precursors, consistent with predictions of this model, are observed using a range of techniques as illustrated below.

On ASDEX Upgrade an ELM precursor signal is detected in the electron temperature from Electron Cyclotron Emission (ECE) channels resonant within the pedestal region. The precursor has a broad spectrum around $f \sim 5$ kHz [7] starting ~ 500 μs before the ELM onset. In JT-60U 200-500 μs before an ELM a mode at 5-25 kHz is observed in the reflectometer signal originating just inside the LCFS [8]. On MAST periodic excursions in the interferometer line integral density trace have been observed prior to an ELM [9]. These structures have a frequency of 20-30 kHz and are found to pre-exist the appearance of filaments in the fast camera images. These excursions have a toroidal mode number (n = 10-20), which is similar to that obtained from the camera images when the filaments become visible. There is evidence that these structures start off broad and become narrower as they push out of the plasma and become visible [9]. Further evidence for a structure inside the LCFS prior to the ELM comes from beam emission spectroscopy measurements on DIII-D. This shows evidence for a poloidally and radially localised structure rotating in the poloidal direction [10].

In summary, there is clear evidence for localised structures that exist inside and close to the LCFS that occur just before the onset of an ELM. From the mode number and size of these structures it seems reasonable to assume that these are the precursors to the filament structures that appear at the ELM onset. The characteristics of these filaments will now be discussed in the following sections.

3. Filament size

One of the original pieces of evidence for the toroidal/poloidal non-uniformity of ELMs came from the peaks observed in the ion saturation current (I_{SAT}) measured during an ELM using reciprocating Langmuir probes. A typical signal from ASDEX Upgrade is shown in figure 1. As can be seen, the I_{SAT} signal shows four or more distinct features for one single ELM as observed by the D_α signal. Each peak lasts for $\delta t\sim$ 50-80 μs and they are separated by $\Delta t\sim$ 100 μs. Similar features have been observed in the ion saturation current on other devices and are summarised in table 1. In reference [11] it was shown that as the toroidal velocity of the plasma increases the width and separation of the filaments decrease hence allowing an interpretation of the peaks as localised structures rotating past the probe. It has subsequently been confirmed by correlating visible images with the I_{SAT} data that the peaks are due to the interaction of moving filaments with the probe [12]. If the radial and/or toroidal/poloidal velocity of the filaments is known then the width and separation of the peaks can be used to estimate the size and quasi-mode number of the filaments. Table 1 shows how device dependent assumptions have been made leading to different estimates for the size of the filaments.

Table 1 Measurements of the temporal width (δt) and separation (Δt) of peaks observed in ion saturation current traces during ELMs on a variety of devices. These measurements have been combined with toroidal (V_ϕ) or radial (V_r) velocities to estimate the size (δ) and mode number of the filaments.

Device	δt (μs)	Δt (μs)	Velocity used (kms^{-1})	Size (cm)	Mode number	Ref.
AUG	50-80	100-200	$V_\phi = 10$	$\delta_\perp = 8$-13	10-20	[2]
DIII-D	50-100	150-250	$V_\phi = 13$	$\delta_\perp = 8$-16	10	[10]
MAST	15	60	$V_\phi = 10$	$\delta_\perp = 7$	10-20	[11]
NSTX	10	20-50	$V_\phi = 10$	$\delta_\perp = 7$	5-15	
JET	100-150	300	$V_\phi = 80$	$\delta_\perp = 100$	2	
			$V_\phi/(5-10)$	$\delta_\perp = 10$-20	10-20	[13]
	185		$V_r = 0.7$	$\delta_r \leq 8$		[14]
JT-60U	10-25		$V_r = 0.5$-1.5	$\delta_r = 0.5$-4		[15]
			$V_\phi = 30$	$\delta_\perp = 4$-10		

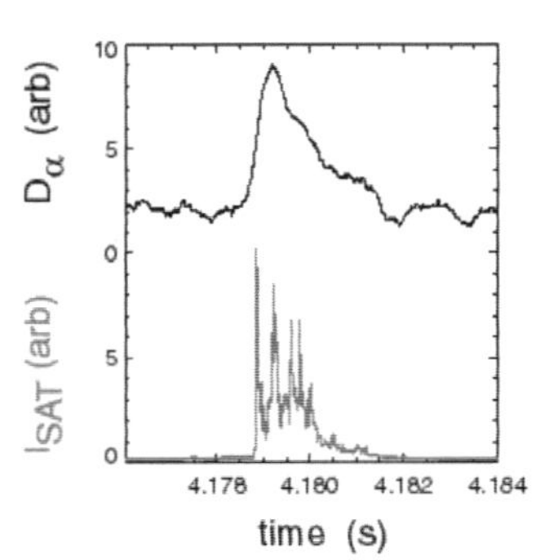

Figure 1 The target D_α and mid-plane ion saturation current during an ELM on ASDEX Upgrade.

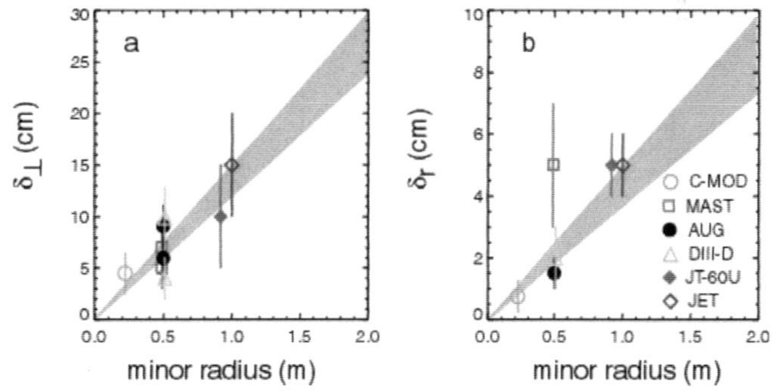

Figure 2 The a) perpendicular and b) radial size of the filaments as a function of machine minor radius. The shaded area represents the $\pm 1\sigma$ region of a fit to the data. In b) the open square (MAST data) has been omitted from the fit.

As can also be seen from figure 1, the I_{SAT} signal is composed of several peaks which are superimposed on top of a base level that begins after the first peak. This base level follows the general trend of the D_α signal suggesting it is related to the particles being released into the SOL due to the ELM. The peaks are observed for up to 1ms, much longer than the time for the pedestal losses to occur. Similar long pulse trains of peaks are observed in the I_{SAT} data from DIII-D [10] and JET [13].

However, on MAST, which has a large gap to the first wall/limiter, the length of time that peaks are observed for is much shorter, typically 100-150 μs. This raises the question of the effect of the wall on the filaments. In fact recent results from ASDEX Upgrade have shown that secondary filaments can be produced when the original filaments impact with the limiter [16].

It has been noted in JET [14] and MAST [17] that the temporal width of the I_{SAT} peaks increases with distance from the LCFS. In reference [14] it was argued that this was due to the dispersion of the filament, however, as will be shown below it is possibly due to the fact that the toroidal/poloidal motion of the filament decreases as it expands away from the LCFS.

Direct evidence for the size of the filaments has come from a variety of sources, including, visible and infrared imaging, Thomson Scattering and beam emission spectroscopy; the results are summarised in table 2. The size of the filament in the radial direction is denoted as δ_r and $\delta_\perp$ is defined as the size perpendicular to both the field line and the radial direction. In all cases the radial extent is smaller than the perpendicular extent. The mode numbers are similar to those derived from the I_{SAT} measurements, however, the observed perpendicular widths tend to be smaller.

Although there is a large spread in the data, these are the only data available in order to make extrapolations to future devices such as ITER. Figure 2a plots the measured perpendicular size of the filaments ($\delta_\perp$) as a function of minor radius. Although there is a large spread in the values the data are consistent with an increase in size with increasing machine. A linear fit has been performed to the data and the shaded area represents the $\pm 1\sigma$ region, which can be used to predict filament sizes on ITER of between 23 and 30 cm. Figure 2b is a plot of the radial filament size (δ_r) with minor radius. Obtaining a simple scaling with minor radius is not so simple here since the radial extent on MAST is larger than that on AUG and DIII-D but similar to that on JET and JT-60U. However, if the results from MAST (open square) are excluded by arguing that the magnetic geometry is different on a spherical tokamak then it is possible to extract a linear fit to the remaining data, which would indicate that the filaments would have a radial extent of between 7 and 10 cm on ITER.

Table 2 Measured radial (δ_r) and perpendicular ($\delta_\perp$) filament sizes from a variety of diagnostics and devices.

Device	Method used	δ_r (cm)	$\delta_\perp$ (cm)	Mode Number	Ref.
AUG	TS	1-2	6		[18]
	IR outer limiter		2-6		[19]
	Visible		5-8	10-20	[2]
	RP Mag	1	3		[20]
C-MOD	Visible	0.5-1	> 4.5		[21]
DIII-D	BES	> 2	2-4		[10]
	Visible			18	[22]
MAST	TS	4-6			[9]
	Visible		3-7	10-20	[23]
JET	TS	4-5			[24]
	IR outer limiter		10-20	11-16	[25]
JT-60U	Visible	3-5			[15]

4. Filament propagation

The poloidal/toroidal motion of the filaments determines the region of interaction of the filaments with the first wall/limiter, while their radial motion, energy content and loss mechanism determine the magnitude of the interaction. There is a general consensus about the loss mechanism for the filaments: the electrons cool more rapidly than the ions and the density is removed by ion parallel processes. This is based on a number of observations, including: 1) that the divertor ELM energy flux rise time is correlated with the ion transport time [26], 2) that the particle content of the filaments decreases exponentially with distance from the LCFS and 3) that the electron temperature in the filament decreases more rapidly [10] than the ion temperature [13][27].

Initial studies of the toroidal and poloidal velocity during a Type I ELM were made from measurements of the edge rotation using Charge Exchange Recombination spectroscopy on DIII-D [28] and helium Doppler spectroscopy on MAST [11]. In both cases a rapid drop in the edge rotation is observed during the rise time of the D_α light associated with the ELM. The measurements are obtained using an integration time which is long (270 μs DIII-D, 125 μs MAST) compared to the filament lifetime, so it is not possible to determine from these measurements alone if it is the filaments and/or the bulk plasma which is slowing down.

Table 3 Measurements of the radial (V_r) and toroidal velocity (V_ϕ) of the filaments from a variety of diagnostics and devices.

Device	Method	<Vr> (kms^{-1})	V_ϕ (kms^{-1})	Comments	Ref.
AUG	$t(J_{SAT} peak)$- $t(D_\alpha)$	0.4		Vr increases with Δ_r	[2]
	$t(probe_1)$- $t(probe_2)$		10-20	Decreases during ELM	[20]
	$t(probe_1)$- $t(probe_2)$	> 1		Even for large Δ_r	[20]
	TS	0.8			[18]
	Visible				
C-MOD	Visible	1-2		Later filaments slower	[21]
DIII-D	$E \times B$	0.2-0.6		Decreasing with Δ_r	[10]
	BES	8		V_r at filament ejection	[10]
	Reflectometer	0.5			[31]
	Divertor tile currents		13.5		[10]
	CER		40	Decreases with t and Δ_r	[28]
MAST	Visible	1-9		Increases with t and Δ_r	[23]
			25-0	Decreases with t and Δ_r	[23]
	CER		20-5	Decreases with t and Δ_r	[11]
	TS	8			[29]
NSTX					
JET	$t(probe_{lim}) - t(D_\alpha)$	0.7			[14]
	$t(probe_{RP}) - t(D_\alpha)$	0.75			[32]
	$E \times B$	1-2		Decreases with f_{ELM}	[33]
JT-60U	$t(J_{SAT} peak)$- t_0	0.4-2.9		Increases with Δ_r	[15]

Direct observation of the toroidal motion of the filaments has been reported on MAST using fast visible imaging [29] [30]. These show that the filaments start off at the plasma edge rotating at the velocity of the pedestal. After 50-100 μs, the filaments decelerate toroidally and accelerate radially outwards. As the filaments propagate radially they follow the direction of the local field line and remain approximately constant in width ($\delta_\perp$). Similar observations have recently been repeated on ASDEX Upgrade [16]. Filaments are observed, which start off rotating toroidally in the co-current direction and after sometime the toroidal rotation of an individual filament slows and soon afterwards it hits the limiter (12 cm from the LCFS). Although it is not possible to directly track the radial position of the filaments they are observed for up to 60 μs before hitting the limiter (12 cm distant), which gives a lower limit of the mean radial velocity of 2 kms^{-1}.

The measurements of the toroidal/poloidal and radial velocities from all devices are summarised in table 3. As can be seen in all devices, where data are available, agree that the filaments rotate initially with velocities near to the pedestal values and then decrease with time and distance into the SOL. Where there is more disagreement is in the radial propagation. For example, on MAST it is claimed from an analysis of the visible images that the filaments accelerate away from the edge but on DIII-D measurements of the electric field indicate that the filaments decelerate.

The radial propagation of the filaments has been determined by a number of techniques and it is interesting to look at the difficulties of each. For example, on MAST, initial estimates of the radial velocity were made using the time difference between the start of the rise time of the D_α light and the peak in the I_{SAT} at the mid-plane reciprocating probe as a function of distance from the LCFS [17]. A higher statistics analysis [11] showed that the data were in fact consistent with the velocity increasing with distance and hence compatible with the filaments moving radially with accelerations in the range (0.5-1.5) $\times 10^7$ ms^{-2}. Due to the large shear in the magnetic field at the outboard mid-plane on MAST, the pitch angle of the field line changes rapidly with distance from the LCFS and hence it is possible to determine the radial location of the filaments to within 1cm [29]. This radial location accuracy was cross checked by tracking filaments until they interacted with the reciprocating probe placed at various distances from the LCFS. Analysis of the image data again showed that the velocity of the filaments increased with distance from the LCFS, but in this case the accelerations were in the range (0.7-2.5)$\times 10^8$ ms^{-2} [23]. The images reveal that the reason for the discrepancy between the two results is due to the fact that the filaments do not all leave the LCFS at the same time and they do not begin to separate until 50-100 μs after the start of the rise in the D_α light. Hence the propagation time to the probe was over estimated in the I_{SAT} analysis and the subsequent velocity/acceleration underestimated.

A similar effect is seen in the ASDEX Upgrade data in that the mean radial velocity calculated using the difference between the peak in the I_{SAT} and the start in the rise of the D_α light [2] is smaller than the velocity calculated using the difference in time between two radially separated probes [20].

On DIII-D, the radial velocity of the ELM filaments has been determined from measurements of the poloidal electric field (E_θ) and using $V_r = E_\theta B / B^2$. This shows that the velocity decreases from ~ 500 ms^{-1} at the LCFS to ~ 120 ms^{-1} near the wall (4 cm from the LCFS) [10]. This is in sharp contradiction with the velocities of > 1 kms^{-1} at distances greater than 10 cm on MAST, ASDEX Upgrade and JT-60U (see table 3) The difference could be due to the difficulties of measuring the electric field in the ELM filaments or due to the fact that the radial velocity is not simply driven by $E \times B$ or because the wall/limiter may have an influence on the dynamics [16].

5. Energy content of the filaments

In order to make a quantitative prediction for the amount of power arriving outside the divertor, in addition to the information on the size and motion of the filament, a knowledge of how much energy is in the filament at the time it separates from the LCFS is needed. In order to measure the energy content of the filament the density, electron and ion temperatures and the filament volume are required. On MAST the radial density and electron temperature profiles of filaments, obtained from Thomson scattering, have been combined with the measured perpendicular size and the assumption that the filament extends between upper and lower X-points to calculate the energy content assuming $T_i = T_e$ [9] (although as discussed below T_i ~2-3T_e may be a better assumption). The maximum energy content of a single filament, observed close to the LCFS, is 2.5 % of ΔW_{ELM}. Recently filament structures have been observed in the Thomson scattering radial profiles on JET [24]. In the cases reported the maximum energy content of a filament is 2 kJ from an ELM with ΔW_{ELM} = 100 kJ. These observations would suggest that the maximum amount of ELM energy transported to the wall would be 25 % ($T_i = T_e$) or 50 % ($T_i = 3T_e$) for 10 filaments. Note that this assumes all the filaments have the maximum size observed and hence is likely an over estimate.

6. Parameters effecting the radial evolution

Rather than reporting individual filament motion the radial expansion of the filaments is often captured in terms of a radial e-folding length of particles and energy. In the simplest picture where the filaments propagate out radially with a constant velocity (V_r) and lose particles on ion parallel transport timescales ($\tau_{//} = L_{//}/c_s$, $L_{//}$ is the connection length and c_s the plasma sound speed) then the particle e-folding length (λ) can be expressed as $\lambda \sim V_r\tau_{//} \sim V_rL_{//}/c_s$. Initial evidence that the electron temperature in the ELM filaments decayed rapidly with distance from the LCFS came from reciprocating probe measurements on DIII-D [10]. These results showed that the ELM filament electron temperature is nearly that of the pedestal (400 eV) when it exits the plasma near the LCFS but falls exponentially with a decay length of 1-1.5cm. Measurement of the electron temperatures 4 cm from the LCFS on JET showed that high temperature electrons do not reach this region, the maximum temperature being measured was ~ 40 eV [33]. In contrast, measurements with a retarding field analyser probe on JET have indicated ion temperatures of 100-150 eV [13]. This large ratio of ion temperature to the electron temperature in the far SOL has also been observed on ASDEX Upgrade [27]. The heat flux to the reciprocating probe obtained from infrared measurements has been compared with the corresponding particle flux measured by Langmuir probes in order to estimate ion temperatures of ~ 60 eV.

In order to turn the I_{SAT} values into a measurement of the density of particles in the filaments a knowledge is needed of both the electron and ion temperature. Often in analyses of L-mode plasmas it is assumed that $T_i = T_e$, but as we have noted above this is not true for the ELM filaments. In DIII-D, the density in the filaments was calculated from the I_{SAT} and the measured electron temperature assuming $T_i = T_e$ [10]. The electron density, calculated in this way, was found to fall off more slowly than the electron temperature. In addition the density e-folding length was found to depend on the density of the plasma having $\lambda \sim$ 3cm at high density and $\lambda \sim$ 10 cm at low density. This was the first indication that the radial efflux could depend on a plasma parameter.

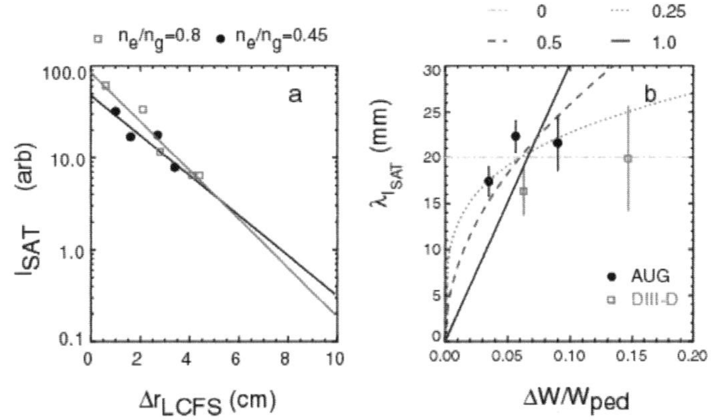

Figure 3 a) The ion saturation current as function of Δr_{LCFS} derived from published data from DIII-D for two shows with different fractional Greenwald (n_g) densities. b) λ_{ISAT} as a function of $\Delta W/W_{ped}$ derived from the data shown in a) and combined with the data from ASDEX Upgrade [16]. The superimposed curves are described in the text.

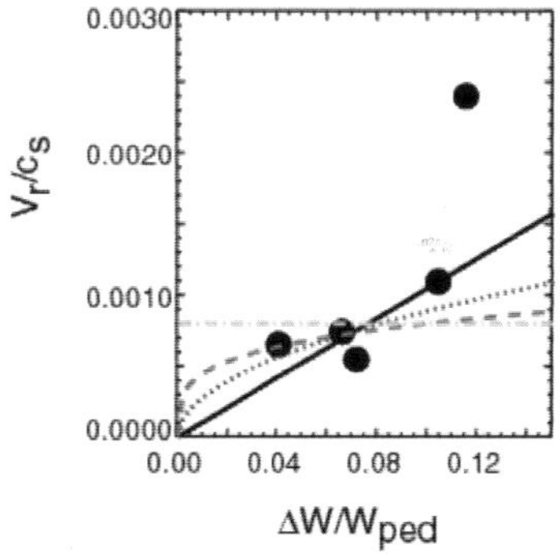

Figure 4 The filament radial velocity measured on DIII-D as a fraction of the ion sound speed as a function of $\Delta W/W_{ped}$. The superimposed curves are described in the text.

In order to determine the power load to the limiters what is needed is the power e-folding length. On ASDEX Upgrade it has been shown that the e-folding lengths in ion saturation current and power are the same [27]. Since it is the ion saturation current that is measured in most experiment, the radial profiles of I_{SAT} have been re-calculated from the published DIII-D density and temperature measurements (see figure 3a). The difference between the I_{SAT} slopes for the low ($\lambda_{ISAT} = 20 \pm 5$ mm) and high ($\lambda_{ISAT} = 16 \pm 2$ mm) density cases is much less pronounced. Using the measured dependence of the fraction of the pedestal energy released by an ELM ($\Delta W_{ELM} / W_{ped}$) on plasma density on DIII-

D [34], the dependence of the e-folding length on ELM size has been calculated. Figure 3b shows a plot of the ion saturation current e-folding length versus the ELM size expressed as a fraction of the pedestal energy. Also shown in figure 3b are recent results from ASDEX Upgrade [16]. It is interesting to note that the e-folding lengths are similar to those observed in DIII-D. Although any dependence of the e-folding length is small compared to the scatter in the underlying data, which is reflected in the error bars, the same trend as in the DIII-D data is observed i.e. a small increase in the e-folding length with $\Delta W_{ELM}/W_{ped}$. Superimposed on the figure are curves of the form: $\lambda \propto (\Delta W/W_{ped})^{\alpha}$ with $\alpha = 0$ (dashed-dot), 0.25 (dashed), 0.5 (dotted) and 1 (solid). From these data it is difficult to rule out any of these dependencies, however, fits with $\alpha = 0.5$ or 1 are strongly disfavoured.

Similar measurements of the ion saturation current e-folding length have been made as a function of $\Delta W_{ELM}/W_{ped}$ on MAST [16]. The curve that gives the poorest description of the data is $\alpha = 1$. If it is assumed that a single value of α should be valid for the dependence of λ_{ISAT} on $\Delta W_{ELM}/W_{ped}$ in DIII-D, ASDEX Upgrade and MAST then the best value of α is 0.25, although no dependence (i.e. $\alpha = 0$) can not be ruled out. In reference [16] it has been suggested that a better fit may be of the form $\lambda = \lambda_{fil} + A.(\Delta W_{ELM}/W_{ped})^{\alpha}$, where λ_{fil} represents the radial size and energy distribution with the filament. The values of λ_{fil} obtained from fits to MAST and ASDEX Upgrade data are at least consistent with the radial size of the filaments being larger in MAST [16].

As was discussed above, on DIII-D the radial expansion velocity of the filaments has been measured using reflectometry [31]. This device measures the velocity of a particular density layer and not explicitly the velocity of the filament. In reference [35] these measurements were collected together and plotted as a function of $\Delta W_{ELM}/W_{ped}$. These data are plotted in figure 4, where curves representing $\frac{V_r}{c_S} \propto \left(\frac{\Delta W}{W_{ped}}\right)^{\alpha}$ with $\alpha = 0$, 0.25, 0.5 and 1 have been superimposed. As can be seen any dependence relies, to a large extent, on the validity of the highest point. A similar scaling has been investigated in MAST using measurements of the radial velocity of the filaments from visible images. In this analysis V_r/c_s is found to be approximately constant with $\Delta W_{ELM}/W_{ped}$ [16].

The interchange instability predicts that more intense filaments travel faster and that $\frac{V_r}{c_s} \sim \sqrt{\frac{2\delta_{\perp}\Delta W}{RW_{ped}}}$ [36]. In reference [37] this parameterisation has been compared with far SOL average power widths, measured by IR, obtained during wall-outer gap scan experiments in JET. Assuming that the filament size ($\delta_{\perp}$) does not change with ELM size, the above parameterisation predicts a scaling for the power e-folding length of $\lambda^{W}_{ELM} \approx 35(\Delta W_{ELM}/0.12W_{ped})^{0.5}$mm [38]. Note that the dependence of the e-folding length on $\Delta W_{ELM}/W_{ped}$ has not been tested directly, instead it has been used to predict the fraction of ELM energy deposited on the wall, which has been compared with measurements from JET [38][39]. The experimental measurements show that the fraction of ELM energy reaching the divertor approaches unity at small $\Delta W_{ELM}/W_{ped}$ but decreases to ~ 50 % at large values of $\Delta W_{ELM}/W_{ped}$ [38]. In reference [37], a parallel loss model of the ELM filament evolution [4] was used to predict an expression for the energy arriving at the wall (i.e. not arriving at the target) of the form $W_{Wall}/W_0 = \exp[-(\Delta_{ped}/2 + \Delta_{sol})/\lambda^{W}_{ELM}]$ where W_0 is the initial energy content of the filament, W_{Wall} the energy deposited on the wall and Δ_{ped} and Δ_{sol} are the radial widths of the pedestal and SOL respectively. The model used in [38], which assumes $W_0 = \Delta W_{ELM}$, agrees with the JET data for $\Delta W_{ELM} / W_{ped} \geq 0.15$ but predicts a smaller flux to the divertor than is measured for smaller ELMs. In reference [39] it has been shown that a better fit to the data is possible if $\lambda \propto (\Delta W_{ELM} / W_{ped})^{\alpha}$ with $\alpha \geq 1$ is used. However, such large values of α are incompatible with the direct measurements presented in figure 3b.

In summary, direct measurements show that the ion saturation current e-folding lengths depend weakly on the size of the ELM ($\Delta W_{ELM} / W_{ped}$). In contrast, the fraction of energy arriving at the divertor in JET, calculated assuming a purely exponential decay of the filament energy with time (radius), requires both a strong increase in λ with $\Delta W_{ELM} / W_{ped}$ and also that the fraction of energy carried by the filaments is larger than what is measured experimentally. Understanding the reason for these discrepancies should be an urgent task for future experiments.

7. Summary

In this paper a summary of the results on the size and motion of filaments observed during type I ELMs has been presented. There is evidence that these filaments can be detected inside the LCFS prior to ELMs. The size of the filaments has been studied on a variety of devices and there is evidence that the filaments do not have a circular cross section instead they are elongated in the perpendicular (poloidal) direction. The results from all the tokamaks presented are consistent with the perpendicular size increasing linearly with the minor radius of the machine giving an estimate for their size on ITER as $\delta_{\perp}$ = 23-30 cm. Obtaining a scaling for the radial size is more difficult, but the best estimate for ITER, again assuming this scales with the size of the machine, is δ_r = 7-10 cm.

Turning now to the motion of the filaments, there is general agreement that the filaments start off rotating toroidally/poloidally with velocities close to that of the pedestal. This velocity then decreases as the filaments propagate radially. Where there is less agreement is on the radial propagation, however, weighing up all the evidence it is most likely that the filaments have at least their initial velocity when they are far out into the SOL and before they have interacted with the nearest limiter surface. Once the filament interacts with the nearest limiter then it is more difficult to predict its evolution. Whether the filaments travel with a constant velocity or accelerate away from the LCFS needs data from other devices.

There is a general consensus that the dominant loss mechanism is through parallel transport and the transport to the wall is through the radial propagation of these filaments. Measurements of the filament energy content show that each filament contains up to 2.5 % of the energy released by the ELM at the time it separates from the LCFS and it would appear that the dominant energy loss has occurred while the filaments are still attached to the LCFS [30]. Finally, the parameters that determine this radial propagation and how this effects the power e-folding length and hence the power loading to non-divertor components have been discussed. The ion saturation current e-folding lengths, which are assumed to represent the power e-folding lengths [27], show a weak, if any, dependence on $\Delta W_{ELM} / W_{ped}$. The dependence that best fits the data, if there is any dependence at all, is $\lambda^{W}_{ELM} \propto (\Delta W_{ELM} / W_{ped})^{\alpha}$ with $\alpha = 0.25$. However, measurements of the fraction of energy arriving at the divertor require larger values of α and larger energy content of the filament, particular at larger values of $\Delta W_{ELM} / W_{ped}$.

Acknowledgement

UKAEA authors were funded jointly by the United Kingdom Engineering and Physical Sciences Research Council and by the European Communities under the contract of Association between EURATOM and UKAEA. The views and opinions expressed herein do not necessarily reflect those of the European Commission.

References

[1] Loarte A *et al.* 2002 *Plasma Phys. Control. Fusion* **44** 1815
[2] Kirk A *et al.* 2005 *Plasma Phys. Control. Fusion* **47** 995
[3] Lipshultz B *et al.* 2007 *Nucl. Fusion* **47** 1189
[4] Fundamenski W and Pitts R A 2006 *Plasma Phys. Control. Fusion* **48** 109
[5] Connor J W *et al.* 1998 *Phys. Plasmas* **5** 2687
[6] Snyder P B *et al.* 2004 *Plasma Phys. Control. Fusion* **46** A131
[7] Suttrop W *et al.* 1996 *Plasma Phys. Control. Fusion* **38** 1407
[8] Oyama N *et al.* 2004 *Nucl. Fusion* **44** 582
[9] Scannel R *et al.* 2007 *Plasma Phys. Control. Fusion* **49** 1431
[10] Boedo J A et al. 2005 *Phys. Plasmas* **12** 072516
[11] Kirk A *et al.* 2005 *Plasma Phys. Control. Fusion* **47** 315
[12] Kirk A *et al.* 2006 *Plasma Phys. Control. Fusion* **48** B433
[13] Pitts R A *et al.* 2006 *Nucl. Fusion* **46** 82
[14] Fundamenski W *et al.* 2004 *Plasma Phys. Control. Fusion* **46** 259
[15] Asakura N *et al.* 2006 *Proc of 21st IAEA Fusion Energy Conf., (Chengdu)* IAEA-CN-149/EX9-2
[16] Kirk A *et al.* 2007 'Comparison of the spatial structure of type I ELMs in ASDEX Upgrade and MAST' *this conference*
[17] Kirk A *et al.* 2004 *Plasma Phys. Control. Fusion* **46** 551
[18] Kurzan B *et al.* 2005 *Phys. Rev. Lett.* **95** 145001
[19] Herrmann A *et al.* 2004 *Plasma Phys. Control. Fusion* **46** 971
[20] Schmid A *et al.* 2007 *Proc. 34th EPS conf. on Plasma Physics and Controlled Fusion (Warsaw)* P1-026
[21] Terry J L *et al.* 2007 *J. Nucl. Mater.* **363-365** 994
[22] Fenstermacher M *et al.* 2004 *Proc of 20th IAEA Fusion Energy Conf. (Villamora)* IAEA-CN-116 /EX2-5Rb
[23] Kirk A *et al.* 2004 *Plasma Phys. Control. Fusion* **46** A121
[24] Beurskens M *et al.* paper in preparation.
[25] Andrew P *et al.* 2007 *Proc. 34th EPS conf. on Plasma Physics and Controlled Fusion* I1-005.
[26] Eich T *et al.* 2005 *J. Nucl. Mater.* **337-339** 669
[27] Herrmann A *et al.* 2007 *J. Nucl. Mater.* **363-365** 528
[28] Wade M *et al.* 2005 *Phys. Rev. Lett.* **94** 225001
[29] Kirk A *et al.* 2006 *Phys. Rev. Lett.* **96** 185001
[30] Kirk A *et al.* 2007 *Plasma Phys. Control. Fusion* **49** 1259
[31] Zeng L *et al.* 2007 *Plasma Phys. Control. Fusion* **49** 1259
[32] Gonzales B *et al.* 2003 *Plasma Phys. Control. Fusion* **45** 1627
[33] Silva C *et al.* 2005 *J. Nucl. Mater.* **337-339** 722
[34] Leonard A W *et al.* 2003 *J. Nucl. Mater.* **313-316** 768
[35] Loarte A *et al.* 2006 *Proc of 21st IAEA Fusion Energy Conf, (Chengdu)* IAEA-CN-149/IT/P1-14
[36] Garcia O E *et al.* 2006 *Physics of Plasmas* **13** 082309
[37] Fundamenski W and Pitts R A 2007 *J. Nucl. Mater.* **363-365** 319
[38] Pitts R A *et al.* 2006 *Proc of 21st IAEA Fusion Energy Conf. (Chengdu)* IAEA-CN-149/EX/3-1
[39] Eich T et al. 2007 *Plasma Phys. Control. Fusion* **49** 573

Comparison of the filament behaviour observed during type I ELMs in ASDEX Upgrade and MAST

A Kirk[1], A Herrmann[2], B Ayed[1,3], T Eich[2], H W Muller[2], G F Counsell[1], S Lisgo[1], M Price[1], A Schmid[2], S Tallents[1,3] and H Wilson[4]

1 EURATOM/UKAEA Fusion Association, Culham Science Centre, Abingdon, Oxon OX14 3DB, UK
2 Max-Planck Institut für Plasmaphysik, EURATOM Association, Garching, Germany
3 Imperial College of Science, Technology and Medicine, University of London, London SW7 2BZ, UK
4 University of York, Heslington, York YO10 5DD UK

E-mail: Andrew.kirk@ukaea.org.uk

Abstract. A study of the evolution of the filaments observed during Type I ELMs on ASDEX Upgrade and MAST is presented. The filaments start off rotating toroidally/poloidally with velocities close to that of the pedestal. This velocity then decreases as the filaments propagate radially. On both devices the ion saturation current e-folding lengths of the filaments show a weak, if any, dependence on the size of the ELM ($\Delta W_{ELM}/W_{ped}$). On MAST the measured radial velocities of the filaments also show at most a weak dependence on $\Delta W_{ELM}/W_{ped}$.

1. Introduction

The energy released from the core plasma during Type I Edge Localised Modes (ELMs) is a key area of study for ITER [1], where the resultant power loadings, both inside and outside the divertor region, have an important impact on the choice of plasma facing materials. Although it is clear that filament structures exist during ELMs (see [2] and reference therein) it is still uncertain what parameters determine how much energy these filaments transport to the wall and what determines their size and how they propagate. This is mainly due to the lack of experimental data. If we can improve our understanding of at least some aspects of the ELM event, we may be able to reduce the uncertainty in the predictions for ITER.

In order to increase the data available that can be used to determine the parameters affecting the radial expansion of the filaments, IEA-ITPA co-ordinated experiments have been performed on MAST and ASDEX Upgrade. Initial studies investigated if the difference in the ion saturation current (I_{SAT}) e-folding length (λ_{ISAT}) observed on the two devices (ASDEX Upgrade: $\lambda_{ISAT} \sim$ 2-2.5 cm MAST: $\lambda_{ISAT} \sim$ 3-4 cm) could be due to the difference in the toroidal field and/or connection length in the machines. Scans were performed on both devices of B_T and I_P independently, keeping the plasma density fixed, but no measurable change in λ_{ISAT} was observed [2]. In this paper, the results of scans over other plasma parameters are presented.

The layout of this paper is as follows: in section 2 new results on filament propagation will be presented. In section 3 a study of the radial efflux and the plasma parameters that may influence it will be discussed. In section 4 a simple simulation is used to illustrate how the radial motion of the filaments combined with parallel transport gives rise to the observed radial profiles.

2. Filament propagation

The poloidal/toroidal motion of the filaments determines the region of interaction of the filaments with the first wall/limiter, while their radial motion, energy content and loss mechanism determines the magnitude of the interaction. There is a general consensus that the dominant loss mechanism, once the filaments have separated from the LCFS, is through ion parallel processes. This is based on evidence from a number of sources, including: 1) that the divertor ELM energy flux rise time is correlated with the ion transport time [3], 2) that the particle content of the filaments decreases exponentially with distance from the LCFS and 3) that the electron temperature in the filament decreases more rapidly [4] than the ion temperature [5][6].

Direct observation of the toroidal motion of the filaments has been reported on MAST using fast visible imaging [7] [8]. These show that the filaments start off at the plasma edge rotating at the velocity of the pedestal. After 50-100 μs, the filaments decelerate toroidally and accelerate radially outwards. As the filaments propagate radially they follow the direction of the local field line and remain approximately constant in poloidal width.

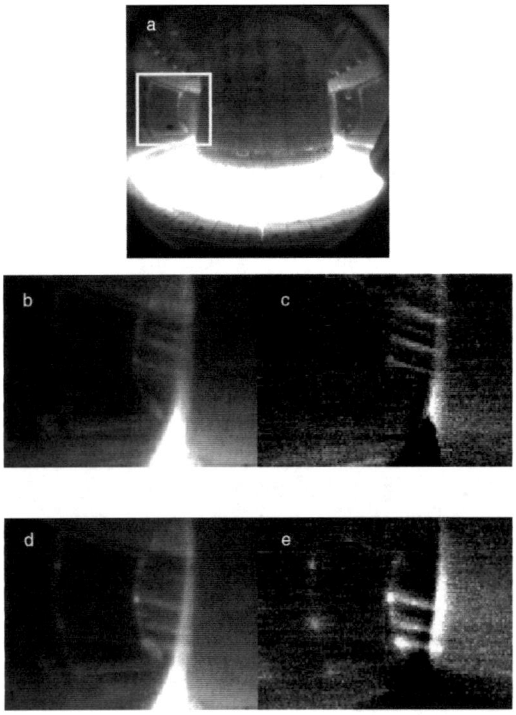

Figure 1 Visible images obtained on ASDEX Upgrade. a) full view of the plasma with the region of fast acquisition shown as a box. b) image obtained during the start of the rise time of the target D_α due to an ELM and d) an image obtained 50 μs later during the same ELM. c) and e) are background subtracted versions of b and d respectively.

Similar observations have recently been repeated on ASDEX Upgrade using the MAST Photron Ultima APX-RS camera. A wide angle view of the plasma was used as shown in figure 1a. In the analysis presented here, the reduced region (128×96 pixels) shown as the outlined box in figure 1a was recorded. This allowed images to be captured throughout the entire shot (5 s) at a frame rate of 90 kHz with an exposure time of 10 μs. In order to be able to track the filaments before they interacted with in-vessel components a large gap (> 12cm) was used between the outboard plasma edge and the nearest limiter. In addition, to make the filaments more visible, deuterium gas puffing was used from the limiter in the region of view. Figure 1b shows an image obtained just after the start of the rise of the target D_α light associated with an ELM. Clear stripes are observed, which are on the outboard (low field side) edge of the plasma. In order to make the filaments more visible and to aid in analysis, a background subtraction is performed, which produces the image shown in figure 1c. Figure 1d and e show the original and background subtracted images obtained 50 μs (5 frames) later where the filaments have rotated toroidally in the co-current direction (or poloidally downwards) and 3 of the filaments have interacted with the ICRH limiter. The interactions show up as bright spots in the image.

In order to make the analysis of the motion quantitative, magnetic field lines in the Scrape Off Layer (SOL) have been projected on to the image. The filaments are observed to be aligned with the local field line; however, the pitch angle on ASDEX Upgrade at the mid-plane does not vary much with distance from the LCFS. Hence it is not possible to determine accurately the radial location of the filaments using such a technique. Only the mean radial velocity can be estimated by using the time that a filament takes from appearing until it reaches the limiter. The toroidal location and widths of the filaments are determined by calculating the intensity along each projected field line as a function of toroidal angle. The peaks observed in the derived distribution are then analysed to determine the widths and location of the filaments.

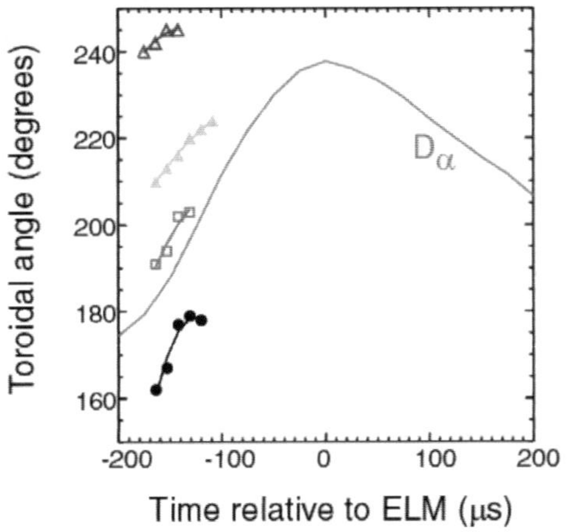

Figure 2 The toroidal angle of four filaments, tracked until they interact with the limiter, as a function of time during the ELM observed on ASDEX Upgrade. Superimposed is the divertor D_α signal.

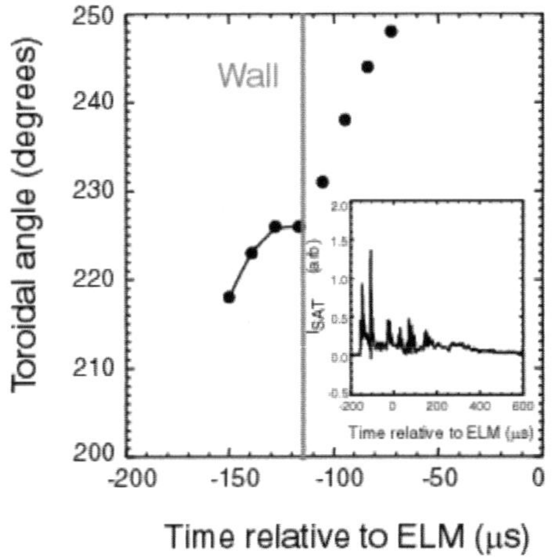

Figure 3 An example on ASDEX Upgrade of the toroidal angle of a filament as a function of time during the ELM for a filament that does not stop when it interacts with the limiter. The inset shows the ion saturation current measured by a Langmuir probe just in front of the limiter.

The width of the filaments perpendicular to the field line varies between 5 and 8 cm and as was observed on MAST [7], there is no evidence that the filament broadens as it travels out towards the limiter. The propagation of four filaments observed during a single ELM are shown in figure 2. The filaments reach the limiter ~ 120 μs before the peak in the divertor D_α light. This is consistent with the ion parallel transport timescales for this shot which is in the range 100 – 150 μs. The location of the filaments has been tracked in subsequent frames until they hit the limiter. All the filaments start off rotating toroidally in the co-current direction and after some time the toroidal rotation of an individual filament slows and soon afterwards hits the limiter. The toroidal location has been fitted as a function of time to yield the initial toroidal velocity (which is in the range 10-30 kms^{-1}) and the toroidal deceleration (which is in the range $(1\text{-}5)\times 10^{8}$ ms^{-2}). Filaments are observed for up to 60 μs before hitting the limiter (12 cm distant), which gives a lower limit of the mean radial velocity of 2 kms^{-1}. However, on MAST the filaments are found only to leave the LCFS when the toroidal rotation slows [7][8]. If it was assumed that the same happened on ASDEX Upgrade then the transit time would decrease to < 30 μs and hence increase the mean radial velocity to > 4 kms^{-1}. Although the filaments are observed to hit the limiter, once they have made contact with it there is no evidence that they continue to propagate radially into the limiter shadow. This is consistent with the very fast fall off in the heat deposition on the side of the limiter observed in ASDEX Upgrade [9]. Hence the radial velocity of the filaments must reduce quickly on contact with the limiter.

If the filaments continue to be tracked after they contact the wall sometimes they stop but in other cases the filament motion changes once it has touched the wall and in fact its toroidal/poloidal motion increases again. This is illustrated in figure 3 in which a filament is tracked for 40 μs before interacting with the wall during which time its toroidal/poloidal motion decreases. However, in subsequent frames, a filament is observed to resume its toroidal motion. Whether this is the same filament or a field line lit up with material emitted from the limiter is not certain. This effect is observed to occur for ~ 20 % of the filaments. It occurs most frequently when the interaction with the limiter is strong or when the gap between the plasma edge and the wall is small. This suggests it is related to the energy content of the filament at the time it interacts with the limiter. It is not possible to identify the radial location of the filament, however, the interaction spot on the limiter is not observed to broaden, suggesting either that the energy content of the filament has been much reduced or that there is a radial gap between the filament and the limiter. These long lived "pseudo" filaments (i.e. they only exist because the original filament interacted with the limiter) could explain why the peaks in the I_{SAT} signal recorded by Langmuir probes near the limiter continue for up to 400 μs on AUG (see the inset in figure 3) and for up to 1 ms on JET [5] and also why the $E \times B$ determined velocities appear to decrease at large distance from the LCFS or in some cases become negative [10].

3. Parameters affecting the radial evolution

Rather than reporting individual filament motion the radial expansion of the filaments is often captured in terms of a radial e-folding length of particles and energy. In the simplest picture where the filaments propagate out radially with a constant velocity (V_r) and lose particles on ion parallel transport timescales ($\tau_{//} = L_{//}/c_s$, $L_{//}$ is the connection length and c_s the ion sound speed) then the particle e-folding length (λ) can be expressed as $\lambda \sim V_r\tau_{//} \sim V_rL_{//}/c_s$. The implications of an accelerating/decelerating filament will be discussed in the next section. To date the only direct measurement of the effect of plasma parameters on the e-folding length comes from DIII-D [4], which

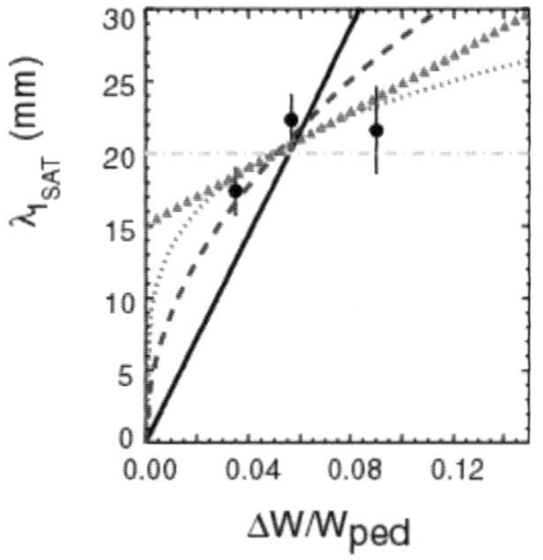

Figure 4 λ_{ISAT} as a function of $\Delta W/W_{ped}$ for data from ASDEX Upgrade. Superimposed on the figures are curves representing equation (1) with $\alpha = 0$ (dashed-dot), 0.25 (dashed), 0.5 (dotted) and 1 (solid). The triangles represent a fit to the form $\lambda = \lambda_{fil} + A\cdot(\Delta W/W_{ped})$.

Journal of Physics: Conference Series **123** (2008) 012012 doi:10.1088/1742-6596/123/1/012012

has shown that the electron density e-folding length was found to depend on the density of the plasma having λ~3cm at high density and λ~10 cm at low density. In order to determine the power loading to the limiters, the power e-folding length is required. Previously, measurements on ASDEX Upgrade have shown that the e-folding length of the ion saturation current (λ_{ISAT}) measured by Langmuir probes is similar to that of the heat flux (λ_q) measured by infrared thermography [6]. In this paper measurements of the ion saturation current as a function of distance have been used to determine λ_{ISAT}, which is assumed to represent the power e-folding length of the radial efflux. The similarity between λ_q and λ_{ISAT} implies a large ratio of ion temperature to the electron temperature in the far SOL and indicates that the reduction of the energy content in the filament is due to convective losses [6].

In this paper we report results from recent studies in which scans have been performed over plasma parameters including ELM size and pedestal temperature produced by changing the density and/or the injected beam power. Figure 4 shows a plot of λ_{ISAT} on ASDEX Upgrade as a function of the fraction of the pedestal energy released by an ELM ($\Delta W_{ELM}/W_{ped}$). Although any dependence of the e-folding length is small compared to the scatter in the underlying data, a small increase in the e-folding length with $\Delta W_{ELM}/W_{ped}$ cannot be excluded. Previous analyses [11][12][13] have suggested that the dependence of the e-folding length should have the form:

$$\lambda \propto \left(\frac{\Delta W_{ELM}}{W_{ped}} \right)^{\alpha} \qquad (1)$$

Superimposed on figure 4 are curves representing equation (1) with $\alpha = 0$, 0.25, 0.5 and 1. From this data set alone it is difficult to rule out any of these dependencies, however, $\alpha = 1$ is highly unlikely.

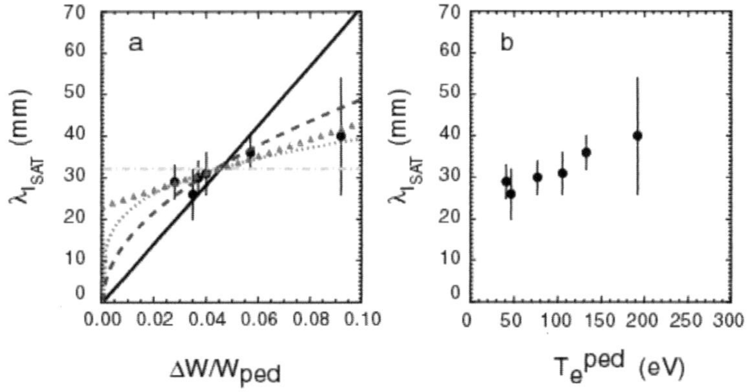

Figure 5 λ_{ISAT} as a function of a) $\Delta W/W_{ped}$ and b) T_e^{ped} for data from MAST. Superimposed on a) are curves representing equation (1) with $\alpha = 0$ (dashed-dot), 0.25 (dashed), 0.5 (dotted) and 1 (solid). The triangles represent a fit to the form $\lambda = \lambda_{fil} + A \cdot (\Delta W / W_{ped})$.

Similar measurements of λ_{ISAT} have been made as a function of plasma parameters on MAST. As can be seen from figure 5a and b there is a slight trend of increasing I_{SAT} e-folding length with increasing $\Delta W_{ELM}/W_{ped}$ and T_e^{ped} respectively. Superimposed on figure 5a are curves representing equation (1). The curve that gives the poorest description of the data is $\alpha = 1$. If it is assumed that a single value of α should be valid for the dependence of λ_{ISAT} on $\Delta W_{ELM}/W_{ped}$ in ASDEX Upgrade (figure 4) and MAST (figure 5a) then the best value of α is 0.25, although no dependence (i.e. $\alpha = 0$) cannot be ruled out. It is interesting to consider whether or not equation (1) has the correct form or whether, as may be indicated by the data, the correct form should have a constant term in addition to any dependence on ELM size i.e. $\lambda = \lambda_{fil} + f(\Delta W / W_{ped})$. Where λ_{fil} represents the initial radial dimension and distribution of energy within the filament. In this picture λ_{fil} would scale with the radial size of the filament. The triangles shown in figure 4 and 5a represent the results of a fit to the form $\lambda = \lambda_{fil} + A \cdot (\Delta W / W_{ped})$, which yields λ_{fil} = 15±3 mm (AUG) and 23 ±3 mm (MAST). In reality the size of the filaments should not be treated as a constant off-set but should be convoluted with the radial motion of the filaments. Nevertheless it is interesting to note that the trend in λ_{fil} is similar to the trend in radial size of the filaments being smaller on ASDEX Upgrade (1-2cm) than on MAST (4-6 cm).

Note the higher $\Delta W_{ELM} / W_{ped}$ discharges in figure 5a have a higher T_e^{ped}. At higher temperatures the filament density should decrease more quickly due to parallel transport. This should then reduce the e-folding length, which is not observed in figure 5b. One explanation might be that the velocity (acceleration) of the filaments increases with T_e^{ped}. In order to investigate this further the velocity of the filaments has been measured directly by tracking the evolution of the filaments observed in the visible images for shots with different $\Delta W_{ELM} / W_{ped}$ and T_e^{ped}. As was reported in [8], although not all the filaments within a given ELM have the same size, the mean and sigma of the distribution of sizes does not change with $\Delta W_{ELM} / W_{ped}$ i.e. on average there is no evidence that larger ELMs have larger filaments. The radial location of each filament during an ELM has been found as a function of time. Since individual filaments leave the LCFS at different times [7][8], t_0 has been defined as the time at which each particular filament is last in contact with the LCFS. For each value of $\Delta W_{ELM} / W_{ped}$ the filaments in several ELMs have been tracked and the radial location (Δr_{LCFS}) of each filament as a function of time has been determined. The data have been binned in 10 μs time intervals relative to the t_0 for each filament. For each time period and each value of $\Delta W_{ELM} / W_{ped}$ the mean value and the R.M.S. of the distribution has been determined. Figure 6 shows a plot of the mean value of Δr_{LCFS} as a function of time relative to t_0 for filaments with low ($\Delta W_{ELM} / W_{ped}$ = 0.028) and high ($\Delta W_{ELM} / W_{ped}$ = 0.092) ELM energy losses. As can be seen, on average the filaments in ELMs with high $\Delta W_{ELM} / W_{ped}$ expand away from the LCFS faster than the ones at low $\Delta W_{ELM} / W_{ped}$. The radial expansion has been fitted according to a radial acceleration (solid curve) or constant velocity (dashed curve).

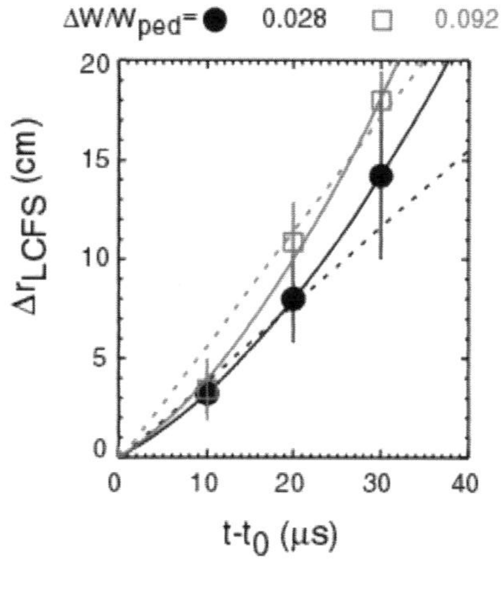

Figure 6 The radial distance of filaments from the LCFS (Δr_{LCFS}) as a function of time relative to the point at which the filament left the LCFS (t_0) for ELMs with $\Delta W/W_{ped}$ = 0.028 and 0.092 on MAST. The curves represent fits to the radial expansion as due to an acceleration (solid) or constant velocity (dashed).

Figure 7a shows a plot of the acceleration as a function of $\Delta W_{ELM} / W_{ped}$. Superimposed are curves representing $a_r \propto (\Delta W_{ELM} / W_{ped})^{\alpha}$ with $\alpha = 0$ (dashed-dot), 0.25 (dashed), 0.5 (dotted) and 1 (solid). The data are best described when $\alpha = 0.25$. Although the accelerating hypothesis best describes the underlying radial profiles, figure 7b plots the results of the fits assuming a constant velocity as a fraction of the ion sound speed versus $\Delta W_{ELM} / W_{ped}$. The general conclusion is that there is little evidence that the radial velocity as a function of the ion sound speed depends strongly on $\Delta W_{ELM} / W_{ped}$.

In reference [14] it is argued that the polarisation currents, present in the filaments, cannot be short-circuited through the targets because the parallel currents required exceed the ion saturation current at the plates. This then leads to a prediction that the filaments will accelerate away from the edge of the plasma. The acceleration is determined by the balance between the ∇B drift current inside the filament and the polarization current in the SOL plasma. The impact of the polarization current is amplified by the distortion of the magnetic flux tube, especially near the X-point. The calculations show: 1) good agreement with the measured acceleration on MAST, 2) that as is observed [2], the accelerations should be larger on MAST than on ASDEX Upgrade and 3) predict that for a given magnetic configuration the acceleration $a_r \propto \frac{(T_i + T_e)}{R}$. Figure 7c shows a plot of the radial acceleration as a function of the pedestal electron temperature on MAST. There is a slight increase in a_r with T_e^{ped} but it is not linear (solid curve) as would be predicted in reference [14]. However, the data are compatible with a curve of the form $a_r = a_0 + f(T_e)$, with $a_0 \approx 1.3x10^8$ ms^{-2}. This initial drive term (a_0) may be due to the acceleration that is predicted in the non-linear ballooning mode theory [15].

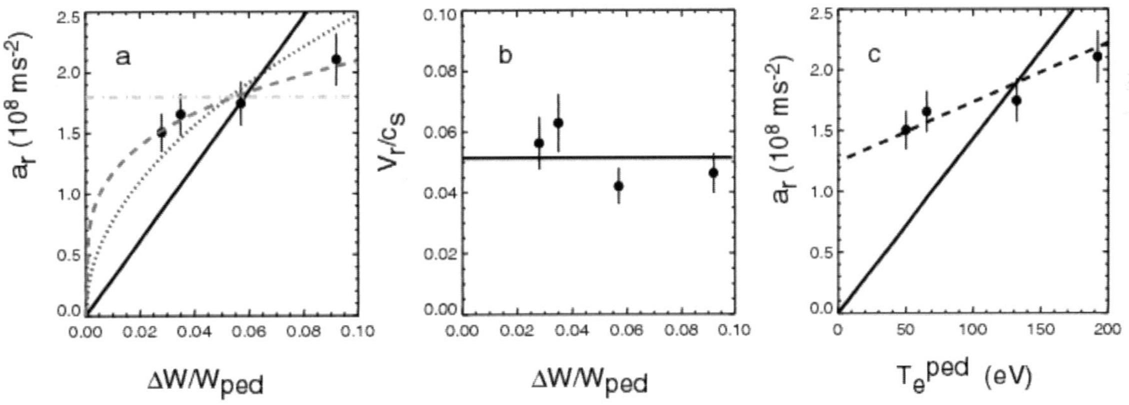

Figure 7 Analysis of the filaments observed in visible images on MAST. a) Radial acceleration as a function of $\Delta W/W_{ped}$, b) the radial velocity as a fraction of the sound speed as a function of $\Delta W/W_{ped}$ and c) radial acceleration as a function of the electron temperature pedestal.

4. Modelling the radial e-folding lengths and power loads

In order to get a better understanding of how the radial motion of, and parallel transport within, a filament gives rise to the observed radial decays a Monte Carlo simulation has been performed. Initially it is assumed that a filament has been created with length $2L$, uniformly populated with particles. Each particle has a velocity as derived from a Maxwellian velocity distribution according to an initial temperature and it is assumed that ion convective transport is dominant. Each particle is tracked until it leaves the end of the filament. As a function of time the number of particles and the mean absolute velocity of the particles within ±0.1 L of the centre of the filament are calculated to represent the density and ion temperature at the mid-plane of the filament. Figure 8a shows the number of particles at the mid-plane from two simulations for an ASDEX Upgrade like plasma with L = 10m and T_i = 500 and 1000 eV. The first thing that can be observed is that the number of particles at the mid-plane remains approximately constant for 15 (25) ms for the high (low) temperature case. This is because during this period there are effectively the same numbers of particles entering the middle section as are leaving it and it is only after this time that the loss out of the top and bottom of the filament plays a role at the mid-plane. This effect was also observed in the more refined kinetic modelling shown in figure 2 of reference [16].

If it is now assumed that the filament moves with a constant radial velocity of 1 kms^{-1} this would then lead to the radial profiles of ion saturation current (calculated by multiplying the density by the square root of the temperature i.e. ignoring the electron temperature which is assumed to be much less than the ion temperature) shown in figure 8b. The profiles show an exponential decay for values of Δr_{LCFS} > 3cm, with λ_{ISAT} similar to that observed in the data (this was made possible by choosing V_r = 1kms^{-1}). However, in addition to the exponential decay, at small Δr_{LCFS} a flat region is produced which is not observed in the data. This could indicate that either the initial assumptions of the model are incorrect (e.g. the filament is not uniformly filled along its length) or the filament is filled but does not separate from the LCFS for some time after formation or that there is some loss mechanism missing in the model or because the filaments do not travel with a constant radial velocity but accelerate away from the LCFS.

If instead of a constant velocity, the filaments are assumed to accelerate away from the LCFS with a = 3x10^7 ms^{-2} or to decelerate according to the measurements of the electric field on DIII-D [4] then the radial profiles for the ion saturation current shown in figure 8c and d respectively are produced. In the case of the acceleration (figure 8c) the filament now spends longer initially at small values of Δr_{LCFS} and so instead of producing a flat region it produces profiles that would be best fit with two exponentials, similar to what has been observed in data from JT-60U [17]. In contrast the decelerating filament still produces a flat initial region but then falls off much more quickly than the measured e-folding lengths.

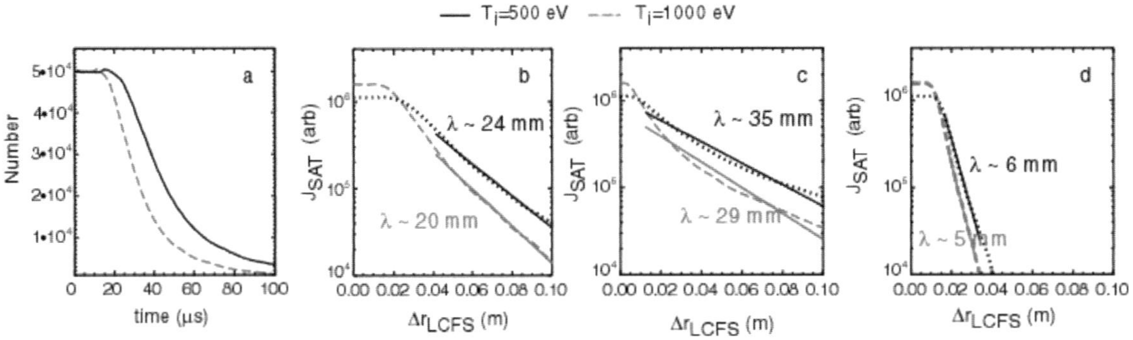

Figure 8 a) the evolution of the particles at the mid-plane as a function of time within a simulated filament with two temperatures. The radial distribution of the ion saturation current at the mid-plane for a simulated filament b) moving with a constant radial velocity of 1 kms^{-1}, c) accelerating radially with a = 3x10^7 ms^{-2} and d) decelerating from an initial velocity of 800 ms^{-1}.

The double exponential structure predicted for the accelerating filaments makes comparison with the single exponential fits applied to the data difficult. However, fits to the region of Δr_{LCFS} > 2 cm with a single exponential, give e-folding lengths that are similar but slightly larger than those observed experimentally. They could be improved by a better choice of the acceleration or the length of the filament; however, this amount of tuning would not be justified given the simplicity of the model.

In order to make a quantitative prediction for the amount of power arriving outside the divertor using a simulation of the kind described above, in addition to the information on the size and motion of the filament, a knowledge of how much energy is in the filament at the time it separates from the LCFS is needed. On MAST the radial density and electron temperature profiles of filaments, obtained from Thomson scattering, have been combined with the measured perpendicular size and the assumption that the filament extends between upper and lower X-points to calculate the energy content assuming $T_i = T_e$ [18]. The maximum energy content of a single filament, observed close to the LCFS, is 2.5 % of ΔW_{ELM}.

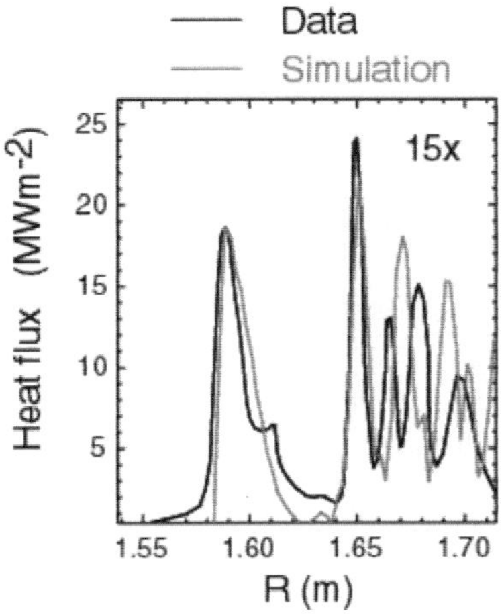

Figure 9 The simulated and observed divertor target radial power profile based on the observed evolution of the filaments.

There are large error bars on these numbers due to the uncertainties in the volume of the filament and the unknown ion temperature. However, they have been used as input to a Monte Carlo simulation of the power arriving at the divertor target in MAST [8]. On MAST, the first wall/limiter is a long way from the plasma and hence effectively all the energy released by an ELM arrives at the divertor. In order to test the validity of this model it would be better to simulate a device such as ASDEX Upgrade which has made measurements of the power due to the filaments arriving at the remote parts of the target [19] and limiters [9]. The simulation assumes T_i = 500 eV and starts off with 12 filaments each with an oval cross section with 2cm radial extent and 6cm in the perpendicular direction. For a period of 50 μs these filaments rotate with the pedestal velocity and release particles into the SOL. After this time individual filaments decelerate toroidally and accelerate away radially (with $a_r = 3\text{x}10^7$ ms^{-2}), each carrying up to 2.5 % of the total number of particles lost. As these filaments move radially, the particles are tracked until they reach the divertor target or interact with the wall which is located at a distance of 4 cm from the plasma edge. The number and velocity distribution of the particles arriving at the target and/or wall are used to calculate the effective power density profile that would be observed during the integration time of the infrared cameras. The resulting target profile at a particular toroidal location, chosen to best agree with the experimental data, is shown in figure 9. Note in the experimentally measured data the power density is artificially increased by a factor of 15 in the region $R > 1.64$ m due to a surface layer on the tile in this region [19] and therefore, for comparison, the simulation has also been increased by the same factor in this region. The size and number of the structures are similar to the observed profiles. The computed profile varies with toroidal location and the experimental one from ELM to ELM. In order to calculate the amount of energy arriving in the remote part of the target ($R > 1.64$m) the experimental data has been averaged over 20 ELMs and the model data has been averaged toroidally. The result is that 7 % is predicted from the simulation compared to the 5 % measured. In order to calculate the amount of energy arriving at the limiter a 2-D profile of the AUG limiter has been incorporated into the model. If a particle in the filament intercept with the limiter before arriving at the target its energy is assumed to have been deposited at the limiter. The energy deposited on the limiter is 18 % in the simulation compared to the 25 % measured using IR [9].

5. Summary

In this paper results from ASDEX Upgrade and MAST on the propagation of filaments observed during type I ELMs have been presented. The new images obtained on ASDEX Upgrade show how the filaments start off rotating toroidally in the co-current direction and after sometime the toroidal rotation of an individual filament slows and soon afterwards it hits the limiter (12 cm from the LCFS). Although it is not possible to track the radial position of the filaments directly they are observed for up to 60 μs before hitting the limiter, which gives a lower limit of the mean radial velocity of 2 kms^{-1}. Once the filament interacts with the nearest limiter then it is more difficult to predict its evolution. On both devices the ion saturation current e-folding lengths of the filaments show a weak, if any, dependence on the size of the ELM ($\Delta W_{ELM} / W_{ped}$). On MAST the measured radial velocities of the filaments show at most a weak dependence on $\Delta W_{ELM} / W_{ped}$. The results from a simple model based on parallel transport have been presented and shows the effect that the different radial propagation hypotheses have on the fall-off length. This model has also been used to explain the amount of power arriving at the remote parts of the target and first wall for a single shot in ASDEX Upgrade. However, it should be stressed that such models need to be refined and tested on other discharges at different $\Delta W_{ELM} / W_{ped}$ and on other devices.

Acknowledgement

UKAEA authors were funded jointly by the United Kingdom Engineering and Physical Sciences Research Council and by the European Communities under the contract of Association between EURATOM and UKAEA. The views and opinions expressed herein do not necessarily reflect those of the European Commission.

References

[1] Loarte A *et al.* 2002 *Plasma Phys. Control. Fusion* **44** 1815
[2] Kirk A *et al.* 2005 *Plasma Phys. Control. Fusion* **47** 995
[3] Eich T *et al.* 2005 *J. Nucl. Mater.* **337-339** 669
[4] Boedo J A *et al.* 2005 *Phys. Plasmas* **12** 072516
[5] Pitts R A *et al.* 2006 *Nucl. Fusion* **46** 82
[6] Herrmann A *et al.* 2007 *J. Nucl. Mater.* **363-365** 528
[7] Kirk A *et al.* 2006 *Phys. Rev. Lett.* **96** 185001
[8] Kirk A *et al.* 2007 *Plasma Phys. Control. Fusion* **49** 1259
[9] Herrmann A *et al.* 2004 *Plasma Phys. Control. Fusion* **46** 971
[10] Silva C *et al.* 2005 *J. Nucl. Mater.* **337-339** 722
[11] Loarte A *et al.* 2006 *Proc. Of 21st IAEA Fusion Energy Conf. (Chengdu)* IAEA-CN-149/IT/P1-14
[12] Fundamenski W and Pitts R A 2007 *J. Nucl. Mater.* **363-365** 319
[13] Eich T *et al.* 2007 *Plasma Phys. Control. Fusion* **49** 573
[14] Rozhansky V and Kirk A 2008 *Plasma Phys. Control. Fusion* **50** 025008
[15] Wilson H R and Cowley S C 2004 *Phys. Rev. Lett.* **92** 175006
[16] Fundamenski W *et al.* 2004 *Phys. Control. Fusion* **46** 259
[17] Asakura N *et al.* 2005 *J. Nucl. Mater.* **337-339** 712
[18] Scannel R *et al.* 2007 *Plasma Phys. Control. Fusion* **49** 1431
[19] Eich T *et al.* 2003 *Phys. Rev. Lett.* **91** 195003

A pedestal temperature model with self-consistent calculation of safety factor and magnetic shear

T Onjun[1], T Siriburanon[1] and O Onjun[2]

[1] Sirindhorn International Institute of Technology, Thammasat University, Klongluang, Pathumthani, Thailand
[2] Department of Science Service, Ministry of Science and Technology, Ratchathewi, Bangkok, Thailand

E-mail: thawatchai@siit.tu.ac.th

Abstract. A pedestal model based on theory-motivated models for the pedestal width and the pedestal pressure gradient is developed for the temperature at the top of the *H*-mode pedestal. The pedestal width model based on magnetic shear and flow shear stabilization is used in this study, where the pedestal pressure gradient is assumed to be limited by first stability of infinite *n* ballooning mode instability. This pedestal model is implemented in the 1.5D BALDUR integrated predictive modeling code, where the safety factor and magnetic shear are solved self-consistently in both core and pedestal regions. With the self-consistently approach for calculating safety factor and magnetic shear, the effect of bootstrap current can be correctly included in the pedestal model. The pedestal model is used to provide the boundary conditions in the simulations and the Multi-mode core transport model is used to describe the core transport. This new integrated modeling procedure of the BALDUR code is used to predict the temperature and density profiles of 26 *H*-mode discharges. Simulations are carried out for 13 discharges in the Joint European Torus and 13 discharges in the DIII-D tokamak. The average root-mean-square deviation between experimental data and the predicted profiles of the temperature and the density, normalized by their central values, is found to be about 14%.

1. Introduction

It is known that when a plasma discharge is heated with sufficiently high heating power, it exhibits a spontaneous transition from a regime of confinement called low-mode (*L*-mode) to a regime with markedly improved confinement called high-mode (*H*-mode) [1]. The change in the plasma first appears at the edge of the plasma, where a region of steep gradients in temperature and density is formed. This steep gradient region, called pedestal, is caused by a transport barrier that forms near the edge of the plasma. The pedestal region is important because the height of the pedestal strongly influences the confinement of the core plasma. It has been found in integrated modeling simulations that the same core transport model used in simulations of *L*-mode discharges can also be used in simulations of *H*-mode discharges, with equally good agreement with experimental data, as long as the simulation boundary conditions are given at the top of the pedestal at the edge of the *H*-mode discharges [2]. It has also been found in both experiments and simulations of *H*-mode plasmas that the height of the pedestal has a significant impact on the shape of the temperature and density profiles and, consequently, a large effect on the global confinement scaling [3-5]. However, the need to use experimental data to provide temperature and density at the boundary of the integrated modeling simulations reduces the predictive capability of these simulations. Therefore, models are needed for the pedestal boundary conditions in order to make the integrated modeling simulations more predictive. It is important to develop more completely predictive integrated modeling codes in order to understand plasma confinement in present day tokamaks and to simulate new fusion experiments and future fusion reactor designs such as ITER.

Models that can be used to predict the temperature and density at the top of the pedestal have recently been developed and calibrated against experimental data [6-9]. In reference [6], several pedestal temperature models were developed using a theoretical approach. The magnetic shear and safety factor are calculated at one pedestal width away from the separatrix, where the safety factor profile is assumed. The predictions of the best model agree with experiment data with 30% RMSE. Note that the definition of RMSE can be found in reference [6]. Later, in reference [10], one of the best pedestal temperature models in reference [6] was chosen to combine with predictive models for the core plasma to simulate existing tokamak experimental results. That pedestal temperature model is based on magnetic and flow shear stabilization width model. This pedestal model was implemented in the BALDUR integrated modeling code [11] using an automated procedure to simulate the transition from the *L*-mode to *H*-mode, once the heating power rises above the *H*-mode threshold. When the plasma is in the *H*-mode state, the predicted values for the temperature and density at the top of the pedestal are used as boundary conditions in the simulations. The average relative rms deviation between experimental data and the predicted profiles of temperature and density, normalized by central values, is found to be about 10%. Even though the procedure developed in reference [10] successfully reproduced various experimental data, the confidence in its predictions, especially when it is applied to a new plasma regime such as burning plasma experiment like ITER, is quite limited. One weak point of this procedure in reference [10] is the estimation of magnetic shear and safety factor in the pedestal temperature calculation, which is different from that used in the rest of the code. Note that the magnetic shear and safety factor play important roles in the maximum pedestal pressure gradient estimation. In addition, the pedestal temperature models in reference [6] were compared directly with the pedestal experimental data, which it is known to contain high error values, in order to determine a pedestal width constant. In this work, we examine a new approach for calculating magnetic shear and safety factor to be used in an estimation of a pedestal temperature. The pedestal model used in this experiment is one of the best pedestal temperature models in reference [6]. This pedestal temperature model is derived based on the "magnetic shear and flow shear stabilization width concept" and the "1st stability regime of infinite n ballooning mode pressure gradient limit concept". The magnetic shear and safety factor used in the pedestal model are taken directly from the BALDUR prediction, where the time-dependent equilibrium and plasma profile effects are properly included. Moreover, the pedestal width constant will be determined by minimizing the average relative rms deviation of the ion core temperature profile, in which tends to be more accurate measurements than the pedestal measurements.

This paper is organized as follows: Brief descriptions for a BALDUR integrated predictive modeling code, Multi-mode transport models, and pedestal models are given in section 2. A statistical analysis of the temperature and density profiles produced by simulation, using the MMM95 and the pedestal model together, compared with experimental data, is presented in section 3, while conclusions are given in section 4.

2. BALDUR integrated predictive modeling code

The BALDUR integrated predictive modeling code is used to compute the time evolution of plasma profiles including electron and ion temperatures, deuterium and tritium densities, helium and impurity densities, current density, neutrals, and fast ions. These time-evolving profiles are computed in the BALDUR integrated predictive modeling code by combining the effects of many physical processes self-consistently, including the effects of transport, plasma heating, particle influx, boundary conditions, the plasma equilibrium shape, and sawtooth oscillations. Fusion heating and helium ash accumulation are computed self-consistently. The BALDUR simulations have been intensively compared against various plasma experiments, which yield an over all agreement with 10% relative rms deviation [2, 12]. In the BALDUR code, fusion heating power is determined using the nuclear reaction rates and a Fokker Planck package to compute the slowing down spectrum of fast alpha particles on each flux surface in the plasma [11]. The fusion heating component of the BALDUR code also computes the rate of the production of thermal helium ions and the rate of the depletion of deuterium and tritium ions within the plasma core. In this work, an MMM95 anomalous core transport model in BALDUR will be used to carry out simulations. The descriptions of this core transport model and the pedestal temperature model are described below.

2.1. Multi-mode core transport model

The MMM95 (Multi-Mode Model 95) [13] is a combination of theory-based transport models which consists of the Weiland model for the ion temperature gradient (ITG) and trapped electron modes (TEM), the Guzdar–Drake model for drift-resistive ballooning modes, as well as a smaller contribution from kinetic ballooning modes. The Weiland model for drift modes such as ITG and TEM modes usually provides the largest contribution to the MMM95 transport model in most of the plasma core. The Weiland model is derived by linearizing the fluid equations, with magnetic drifts for each plasma species. Eigenvalues and eigenvectors computed from these fluid equations are then used to compute a quasilinear approximation for the thermal and particle transport fluxes. The Weiland model includes many different physical phenomena such as effects of trapped electrons, T_e/T_i ratio, impurities, fast ions, and finite β. The resistive ballooning model in MMM95 transport model is based on the **ExB** drift-resistive ballooning mode model by Guzdar–Drake at 1993, in which the transport is proportional to the pressure gradient and collisionality. The details of these models can be found in reference [13]. The contribution from the resistive ballooning model usually dominates the transport near the plasma edge. Finally, the kinetic ballooning model is a semi-empirical model, which usually provides a small contribution to the total diffusivity throughout the plasma, except near the magnetic axis. This model is an approximation to the first ballooning mode stability limit. All the anomalous transport contributions to the MMM95 transport model are multiplied by κ^{-4}, since the models were originally derived for circular plasmas. More information about inclusion of elongation effect can be obtained in reference [13].

2.2. Pedestal models

A complete description of the pedestal width model, based on the magnetic and flow shear stabilization, that is used in this paper to predict the temperature at the top of the pedestal at the edge of the type I ELMy *H*-mode plasmas is given in reference [14]. In the development of the pedestal temperature models described in reference [6], two ingredients are required: the pedestal width (Δ) and the pedestal pressure gradient ($\partial p/\partial r$). If the pedestal density (n_{ped}) is known, the temperature at the top of the pedestal (T_{ped}) can be estimated as

$$T_{\text{ped}} = \frac{1}{2n_{\text{ped}}k}\left|\frac{\partial p}{\partial r}\Delta\right| = \frac{\Delta}{2kn_{\text{ped}}}\frac{\alpha_c B_T^2}{2\mu_0 R q^2}, \tag{1}$$

where k is the Boltzmann constant, μ_0 is the permeability of free space, α_c is the normalized critical pressure gradient, B_T is the magnetic field, R is the major radius and q is the safety factor. The width of the pedestal is assumed to be determined by a combination of magnetic and flow shear stabilization of drift modes [14]:

$$\Delta = C_W \rho s^2, \tag{2}$$

where C_W is a constant of proportionality, ρ is the ion gyroradius, and s is the magnetic shear. Note that the magnetic shear in equation (2) is calculated at the top of the pedestal. The pedestal pressure gradient is assumed to be uniform throughout the pedestal region and it is limited by the first stability limit of infinite n ballooning mode, so that the normalized critical pressure gradient for the pedestal region is given by

$$\alpha_c = 0.4s\left(1+\kappa_{95}^2\left(1+5\delta_{95}^2\right)\right), \tag{3}$$

where κ_{95} is the elongation at the 95% flux surface, and δ_{95} is the triangularity at the 95% flux surface. Therefore, the pedestal temperature takes the following form:

$$T_{\text{ped}}(keV) = 0.323C_W^2\left(\frac{B_T}{q^2}\right)^2\left(\frac{M_i}{R^2}\right)\left(\frac{\alpha_c}{n_{\text{ped},19}}\right)^2 s^4, \tag{4}$$

where $n_{\text{ped},19}$ is the electron density at the top of the pedestal in units of 10^{19} m^{-3}, and M_i is the ion mass. The constant C_W is the constant of proportionality in equation (2). The constant C_W is chosen so as to optimize the agreement between the experimental data of ion temperature profile and the simulation results for ion temperature profile by minimizing the average relative rms. This pedestal model is implemented in the 1.5D BALDUR integrated predictive modeling code, where the safety factor and magnetic shear are solved self-consistently in both core and pedestal regions. With the self-consistently approach for calculating safety factor and magnetic shear, the effect of bootstrap current can be correctly included in the pedestal model.

3. Simulation results and discussion

BALDUR integrated modeling simulations were carried out for the 26 *H*-mode discharges (13 JET and 13 DIII-D) in various scans, using a combination of the new pedestal temperature model together with the MMM95 core transport model. These discharges are taken from the International Profile Database [15]. In this work, the electron pedestal temperature is assumed to be equal to the ion pedestal temperature. The pedestal density is taken directly from the experiment. For each of the profiles (ion temperature, electron temperature, and electron density), the normalized deviation, ε_j, between the jth experimental data point $X_j^{\exp}$ and the simulation result $X^{\text{sim}}(R_j)$ at the major radius R_j of the corresponding experimental data point is defined as

$$\varepsilon_j = \frac{X^{sim}(R_j) - X_j^{\exp}}{X_{Max}^{\exp}}. \tag{5}$$

In order to give all the deviations across each profile equal weight, each deviation is normalized by the maximum experimentally measured value for that given profile, $X_{Max}^{\exp}$. An alternative choice of normalizing each temperature deviation by the local temperature, for example, would magnify deviations near the edge of the plasma, where temperatures are lower. The relative rms deviation σ and the offset f between the profile resulting from each simulation and the corresponding experimental data is defined as

$$\sigma = \sqrt{\frac{1}{N}\sum_{j=1}^{N}\varepsilon_j^2} \tag{6}$$

and

$$f = \frac{1}{N}\sum_{j=1}^{N}\varepsilon_j, \tag{7}$$

where N is the number of experimental data points in a profile. The relative rms deviation (σ) and offset are evaluated for each of the temperature profiles for the discharges in consideration.

In figure 1, the average relative rms deviations for the ion temperature profiles of 26 *H*-mode discharges are plotted as a function of the pedestal width constant C_W. It can be seen that this function is close to a quadratic equation. The best fit with the quadratic equation is found to be:

$$\sigma(C_W) = 60.0C_W^2 - 56.1C_W - 27.4\,. \tag{8}$$

The minimum point for the function in equation (8) can be found at the pedestal width constant of 0.47 and the lowest average rms deviation is about 14.4%.

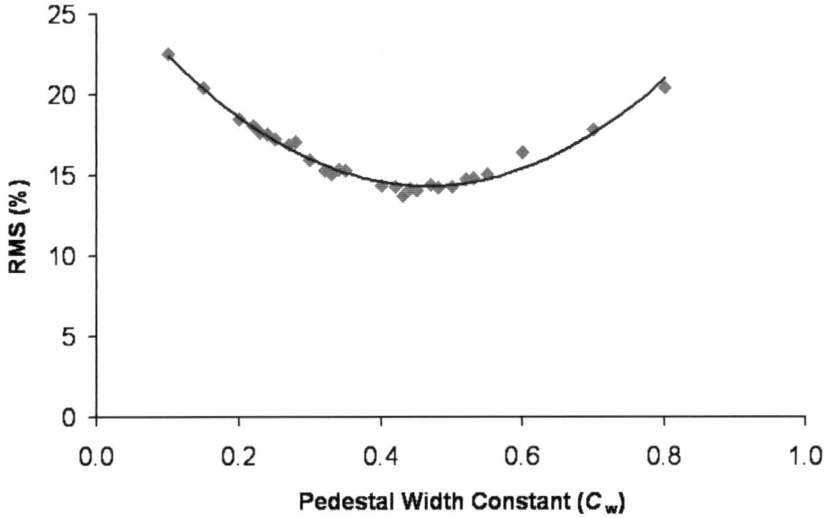

Figure 1. The average relative rms deviation for the ion temperature is plotted as a function of the pedestal width constant C_W.

The details of the statistical analysis using relative rms deviations for simulations using the pedestal model based on the magnetic and flow shear stabilization with C_W= 0.47 for ion and electron temperatures are presented in figures 2 and 3, respectively. Note that the DIII-D discharges are from discharge number 77557 to discharge number 99411 and the JET discharges are from discharge number 32745 to discharge number 38415. It can be seen in figure 2 that the relative rms deviations for ion temperature vary from one discharge to another, and from the profile to another, with a minimum of about 3.56% and a maximum of about 37.82% for ion temperatures and with a minimum of about 5.13% and a maximum of about 51.42% for electron temperatures. The average relative rms deviation σ_{avg}—averaged over 26 discharges —for the ion temperature is about 14.39% for the ion temperature profiles and about 15.51% for the electron temperature profiles. It is worth noting that in this paper, an average relative rms is chosen to represent the results. However, a median of relative rms could be another good choice. The discussion of this can be made in the future report. It can be noticed that for DIII-D discharge 90108 and JET discharges 34340, 35171, and 35174, the relative rms deviation is higher than 20%. The predicted pedestal in DIII-D discharge 90108 and JET discharge 34340 are in the range of experimental value. However, the predictions of the central ion and electron temperatures are quite low. For JET discharges 35171 and 35174, the predictions of pedestal are significantly higher than those obtained in the corresponding experiment. This is subjected to a further investigation.

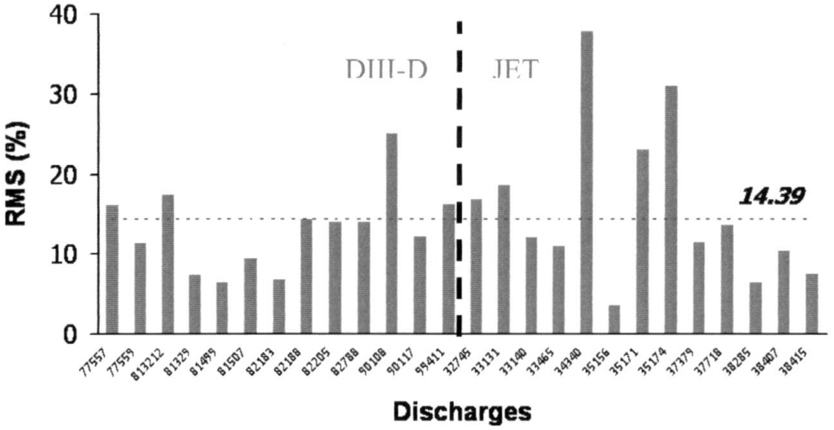

Figure 2. Relative rms deviations (%) for the ion temperature profiles produced by simulations using the MMM95 transport model combined with the pedestal model (using C_W=0.47) compared with experimental data for 26 *H*-mode discharges from DIII-D and JET experiments.

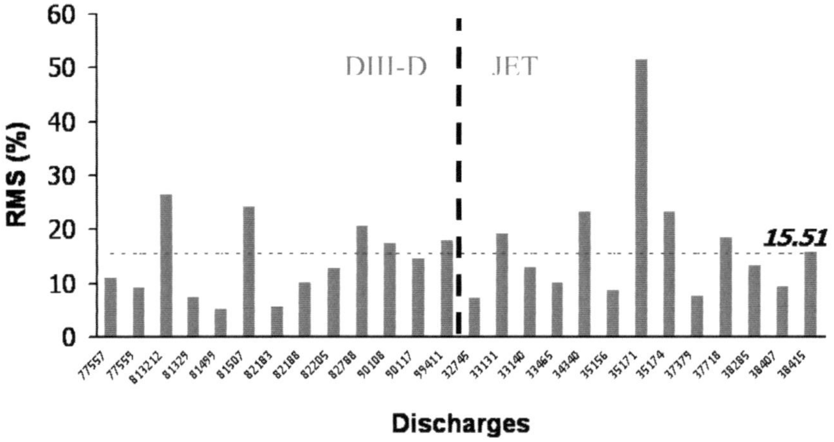

Figure 3. Relative rms deviations (%) for the electron temperature profiles produced by simulations using the MMM95 transport model combined with the pedestal model (using C_W=0.47) compared with experimental data for 26 *H*-mode discharges from DIII-D and JET experiments.

In figures 4 and 5, the relative offsets for simulations using the pedestal model with C_W= 0.47 are presented for ion and electron temperatures, respectively. It can be seen in figure 4 that the offset for ion temperature also vary from one discharge to another, and from the profile to another, with a minimum of about -1.84% and a maximum of about 0.84% for ion temperatures and with a minimum of about -2.03% and a maximum of about 1.11% for electron temperatures. The average offset over 26 discharges for the ion temperature is about -0.20% for the ion temperature profiles and about 0.01% for the electron temperature profiles.

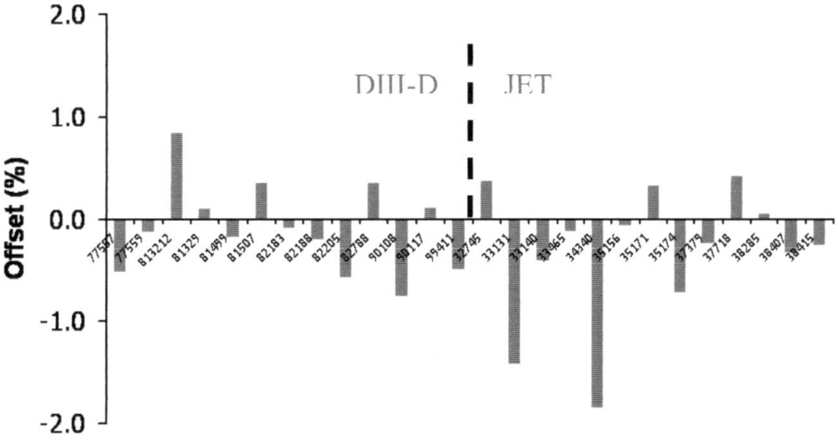

Figure 4. Relative offset (%) in the ion-temperature-profile comparison for C_W=0.47.

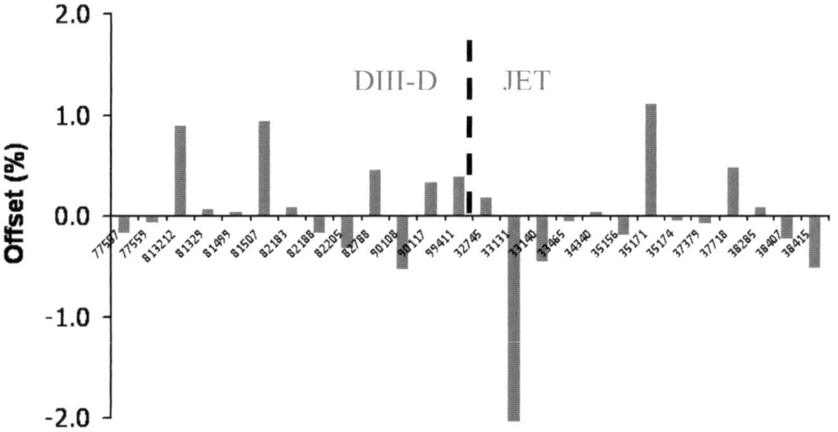

Figure 5. Relative offset (%) in the electron-temperature-profile comparison for C_W=0.47.

Ion temperature profiles are shown in figure 6 from the simulation using the new pedestal model based on magnetic and flow shear stabilization and the experimental data for a DIII-D discharge. It can be seen that the simulation profile yields good agreement with experimental data. For figure 7, ion temperature profile from the simulation using the new pedestal model is compared with experimental data for a JET discharge. It can be seen that the simulation profile also yields good agreement with experimental data. As shown in figure 7 to figure 8, the conditions at the boundary of the simulations that used the new pedestal model nearly match the corresponding experimental data. The major cause of the difference between the simulated profiles and the experimental data is to be clarified in near future; core transport models, models of sources/sinks or the new pedestal model.

The summary agreement between the predicted pedestal and experiment values is shown in figure 8. It can be seen that the predictions are in the range of the experimental values. The model tends to over predict the pedestal values in DIII-D experiments and underpredict the pedestal values in JET experiments. This might due to different nature of the pedestal in those tokamaks. The RMSE of the predicted pedestal temperature and the experimental data for this set is about 43.5%, which is higher than that found in reference [10]. The increase of RMSE might due to the error in the calculation of magnetic shear and safety factor in BALDUR code. The pedestal prediction with this model depends sensitively on a calculation of magnetic shear and safety factor since both parameters appear in pedestal width and pedestal pressure estimations. However, the discussion on the issue of the magnetic shear and safety factor calculation in BALDUR code is beyond the scope of this paper. It rather leaves this discussion for the future work.

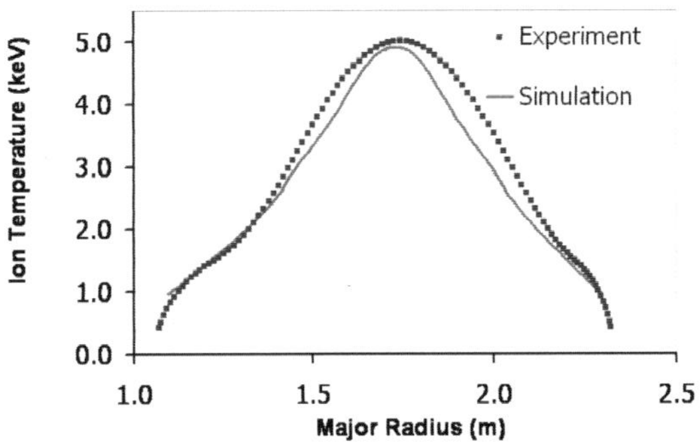

Figure 6. The ion temperature profile is shown as a function of major radius for simulations of the low κ DIII-D discharge 81499. The closed squares represent experimental data, while the solid curve is the result of simulation.

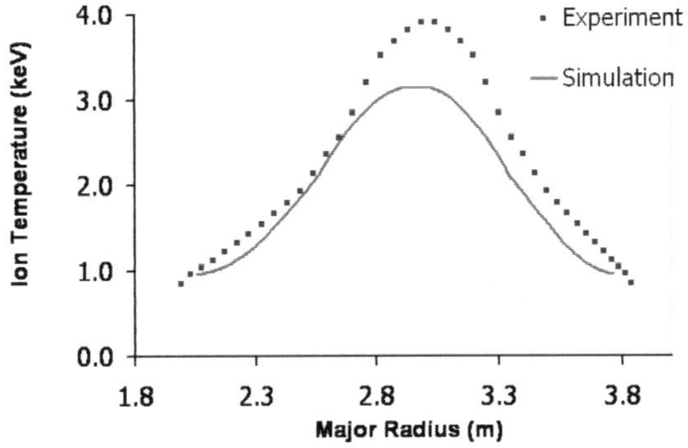

Figure 7. The ion temperature profile is shown as a function of major radius for simulations of the low β JET discharge 38407.

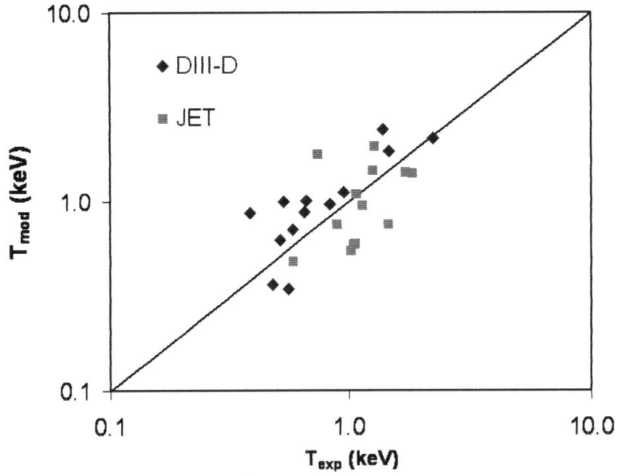

Figure 8. Comparison between the predicted ion temperature and corresponding experimental values from 26 *H*-mode discharges.

4. Conclusions

A pedestal model based on theory-motivated models for the pedestal width and the pedestal pressure gradient is developed for the temperature at the top of the *H*-mode pedestal. The pedestal width model based on magnetic shear and flow shear stabilization is used in this study, where the pedestal pressure gradient is assumed to be limited by first stability of infinite *n* ballooning mode instability. This pedestal model is implemented in the 1.5D BALDUR integrated predictive modeling code, where the safety factor and magnetic shear are solved self-consistently in both core and pedestal regions. With the self-consistently approach for calculating safety factor and magnetic shear, the effect of bootstrap current can be correctly included in the pedestal model. The pedestal model is used to provide the boundary conditions in the simulations and the Multi-mode core transport model is used to describe the core transport. This new integrated modeling procedure of the BALDUR code is used to predict the temperature and density profiles of 26 *H*-mode discharges. Simulations are carried out for 13 discharges in the Joint European Torus and 13 discharges in the DIII-D tokamak. It is found that the average root-mean-square deviation between experimental data and the predicted profiles of the temperature and the density, normalized by their central values, is about 14%.

Acknowledgements

T. Onjun is grateful to Prof. Arnold H. Kritz and Dr. Glenn Bateman at Lehigh University for their generous supports. This work is supported by the National Research Council of Thailand (NRCT), Thailand Toray Science Foundation and the Thammasat University Research Fund.

References

[1] Wagner F *et al.* 1982 *Phys. Rev. Lett.* **49** 1408
[2] Hannum D *et al.* 2001 *Phys. Plasmas* **8** 964
[3] Greenwald M *et al.* 1997 *Nucl. Fusion* **37** 793
[4] Suttrop W *et al.* 1997 *Plasma Phys. Control. Fusion* **39** 2051
[5] Kinsey J *et al.* 2003 *Nucl. Fusion* **43** 1845
[6] Onjun T *et al.* 2002 *Phys. Plasmas* **9** 5018
[7] Onjun T 2006 *Laser and Particle Beams* **24** 113
[8] Guzdar P N *et al.* 2005 *Phys. Plasmas* **12** 032502
[9] Stacey W M 2002 *Contrib. Plasma Phys.* **42** 283
[10] Bateman G *et al.* 2003 *Plasma Phys. Control. Fusion* **45** 1939
[11] Singer C E *et al.* 1988 *Comput. Phys. Commun.* **49** 275
[12] Onjun T *et al.* 2001 *Phys. Plasmas* **8** 975
[13] Bateman G *et al.* 1998 *Phys. Plasmas* **5** 1793
[14] Sugihara M *et al.* 2000 *Nucl. Fusion* **40** 1743
[15] Boucher D *et al.* 2000 *Nucl. Fusion* **40** 1955

Edge stability of stationary ELM-suppressed regimes on DIII–D

T H Osborne[1], P B Snyder[1], K H Burrell[1], T E Evans[1], M E Fenstermacher[2], A W Leonard[1], R A Moyer[3], M J Schaffer[1] and W P West[1]

[1] General Atomics, PO Box 85608, San Diego, CA 92186-5608, USA
[2] Lawrence Livermore National Laboratory, Livermore, CA 94550-9234, USA
[3] University of California-San Diego, La Jolla, CA 92186, USA

E-mail: osborne@fusion.gat.com

Abstract. We discuss the MHD stability of the H-mode pedestal region in two edge localized mode (ELM)-suppressed H-mode regimes on DIII-D, the quiescent (Q)H-mode and resonant magnetic perturbation (RMP) H-mode. The QH-mode is obtained at low density with most of the neutral beam power injected counter to the plasma current, and is characterized by a continuous, low toroidal mode number (n), edge harmonic oscillation (EHO). QH-mode is observed to transition back to ELMing H-mode as the rotational shear is decreased by increasing the co-current neutral beam power fraction or by increasing the plasma to conducting wall distance, consistent with a model for the EHO as a peeling instability with added drive from rotational shear, which saturates through self transport of momentum or damping of rotation through drag of the mode on the conducting wall. In RMP H-mode, an $n = 3$ coil provides the non-axisymmetric perturbation. Two regimes of RMP ELM suppression are observed. At low collisionality with large resonant field perturbation, the pedestal pressure gradient and current density are reduced below the peeling-ballooning mode stability limit; while at higher collisionality with large non-resonant perturbation there is little change in the pedestal parameters and type II ELMs are enhanced.

1. Introduction

The two principal large scale instabilities in the H-mode edge transport barrier (ETB) are the pressure gradient, p', driven ballooning mode and the current density, J, driven peeling mode. At ratios of p'/J common in experiments, the most unstable mode has both peeling and ballooning characteristics. The theory of the peeling-ballooning mode is well established [1] and there is strong experimental support that this instability sets the threshold for the onset of the ELM [2,3]. When ideal MHD is extended with two-fluid theory, it is found that diamagnetic effects can have a strong impact on the mode growth rate [4]. The mode growth rate remains relatively constant when diamagnetic effects are weak, $\omega_{*i} < 2\gamma_0$, where γ_0 is the growth rate computed without two fluid effects, and the ion diamagnetic frequency is given by $\omega_{*i} = (nT_i/en_i)(\partial n_i/\partial\Psi)$, where n is the toroidal mode number, T_i, is the ion temperature, n_i is the ion density, and Ψ is the MHD poloidal flux. When diamagnetic effects are strong, $\omega_{*i} > 2\gamma_0$, the growth rate of the mode is strongly reduced, so that $\gamma_0/(\omega_{*i}/2) = 1$ sets a threshold for instability (figure 1). The variation of the toroidal mode number along the $\gamma/(\omega_{*i}/2) = 1$

stability threshold plays a role in the interpretation our results. For these experiments $n \approx 10$ where peeling and ballooning mode branches merge, and n decreases as the modes become more peeling-like and J driven, while n increases as the modes become more ballooning-like and p' driven. ELMs are observed when the equilibrium point reaches either the J or p' dominated boundaries. The current density in the ETB is predominantly bootstrap current, which is proportional to p', but also decreases with increasing collisionality, ν_* [5]. Increasing ν_* in the ETB, for example by raising the density, then moves the equilibrium from the low-n, J driven boundary to the intermediate-n, p' driven boundary. Such a variation of ELM precursor mode number with density is observed experimentally [6].

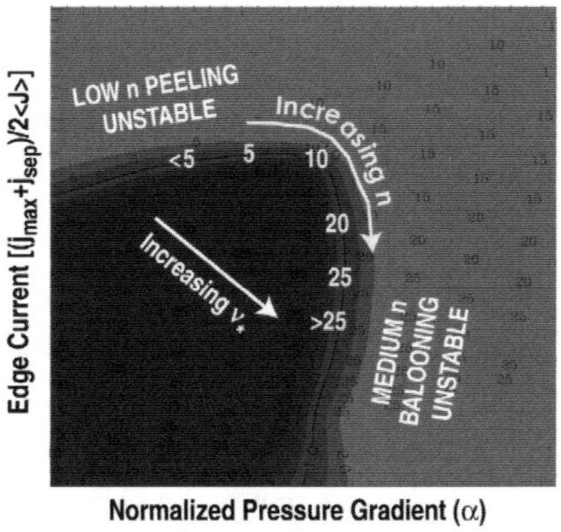

Figure 1. Peeling-ballooning mode stability diagram. Contours of growth rate over 1/2 ion diamagnetic frequency as a function of normalized pressure gradient and normalized current density. $\gamma/(\omega_{*i}/2) = 1$ separates the stable (blue) from unstable region (red). Toroidal mode number, n, of most unstable mode increases in going from current to pressure gradient driven boundaries. Increasing collisionality, ν_*, reduces j-bootstrap favoring higher n.

On DIII-D two ELM-suppressed H-mode regimes, the quiescent or QH-mode [7], and the H-mode with resonant magnetic perturbation (RMP) H-mode [8] are under study. In these regimes the undesirable bursts of enhanced transport accompanying the ELMs are replaced by a continuous loss. In the case of QH-mode, evidence is presented which suggests that an additional drive mechanism, toroidal rotational shear, coupled with a rotational damping mechanism, such as wall drag, leads to saturation of modes at low n before the current density can reach the level required to trigger an ELM. In this case the saturated peeling-ballooning mode itself provides the transport necessary to avoid the ELM. In the case of the RMP H-mode, an externally applied helical magnetic field, which is expected to break up the magnetic surfaces in the ETB region, directly provides the transport required to keep the edge p' and J below the ELM instability threshold [8].

2. Experimental method

The pressure profiles used for the stability analysis are determined experimentally as the sum of the electron pressure, the deuteron pressure, the pressure of fully stripped carbon (which is the only significant impurity in DIII-D), the fast ion pressure computed using the NFREYA module of the ONETWO transport code [9], and the lowest order contribution of the rotational energy to the pressure. The extent of the edge steep gradient region of the electron profiles on the outboard midplane is 1 to 2 cm, while it is somewhat wider, 1.5 to 3 cm, for the ion profiles. The electron pressure is determined using the DIII-D multi-pulse Thomson scattering system which measures the electron density and temperature at 37 points along a vertical chord every 12.5 ms. When mapped to the outboard midplane, the spatial resolution of the Thomson scattering system is about 0.6 cm in the ETB region. The ion temperature, C^{6+} density, and rotational speed for C^{6+} ions are measured with the DIII-D charge exchange recombination spectroscopy system (CER) that has 0.5 cm spatial resolution in the ETB region and can be run at up to 0.274 ms time resolution. The deuteron density is computed from the electron density, the fast ion density, and the C^{6+} density, assuming C^{6+} is the only impurity. The fast ion and rotational pressures are generally small in the ETB.

Profile data is averaged over a steady state phase of a discharge, or from a particular phase of the inter-ELM period for many similar ELMs, by combining profiles as a function of poloidal flux based on equilibrium, computed at the individual times of the profile measurements. By mapping to a temporally local equilibrium, small sweeps in the separatrix location can be used to improve the effective experimental radial point density in the ETB. The absolute location of the electron profiles relative to the separatrix has significant uncertainty due to the sensitivity of the equilibrium to uncertainties in the magnetics measurements. UEDGE divertor modeling simulations of a number of DIII-D discharges result in T_e profiles which are well represented by a hyperbolic tangent form with the separatrix located at one-half the width of the hyperbolic tangent outside of its inflection point [10]. This prescription was used to locate the separatrix relative to the averaged n_e and T_e profiles.

Equilibria used for stability analysis were generated with the EFIT code [11] with the pressure profile set as described above. The current density profile in the core plasma is determined from motional stark effect (MSE) measurements. The MSE system cannot be used to determine the current density profile in the ETB due to the large static electric field in this region. A lithium beam probe system, based instead on Zeeman splitting, has been used to measure the current density profile in the ETB for a limited number of discharges. These measurements indicate that the current density profile in the ETB is in good agreement with the neoclassical value [12]. Therefore, for the stability analysis that follows, the neoclassical value given by Sauter [5] for the bootstrap current is assumed.

3. ELITE stability calculations

The ELITE code [13] is used for the ETB stability calculations. ELITE is a 2D, linear, eigenvalue code based on ideal magnetohydrodynamic (MHD) utilizing a generalization of ballooning mode theory that makes use of poloidal harmonic localization to achieve high efficiency. ELITE computes the growth rates for a spectrum of modes with different toroidal mode numbers covering the range of n values of importance in the ETB. The eigenfunction of the peeling-ballooning instability at a given n number generally covers the entire ETB region, combining a range of poloidal mode numbers rather than being characteristic of a single rational surface. After computing the stability, $\gamma/(\omega_{*i}/2)$, of the experimental equilibrium, variations of J and p' in the ETB of this equilibrium are made and $\gamma/(\omega_{*i}/2)$ is recomputed to map out the stability boundary (figure 1).

4. QH-mode

QH-mode [7] is an ELM-suppressed regime of high energy confinement with constant density and radiated power. It requires operation at low density or collisionality [5], $\nu_e^* \sim 0.1$, and injection of at least 80% of the neutral beams in the toroidal direction opposite to the direction of the plasma current. QH-mode is generally accompanied by the edge harmonic oscillation (EHO), which is observed on magnetic probes as a series of low-n harmonics with typically $n = 1$ to 4 the dominant component. When the EHO is not present, as for example when the power is reduced below the power threshold, the discharge returns to standard ELM-free H-mode and the density rapidly rises, suggesting that the EHO provides the particle transport to keep QH-mode in steady state. Similar enhancements in impurity and momentum transport are also associated with the EHO [14].

4.1. QH-mode edge stability

Evaluation of the peeling-ballooning stability for discharges covering a range of cross-sectional shapes shows that the QH-mode edge is always near the low toroidal mode number, $n \leq 5$, peeling mode instability boundary (figure 2). This association with low-n instability suggests a connection between the EHO and the peeling-ballooning mode. The existence of an upper density operational limit suggests that low-n peeling mode instability may be a required feature of QH-mode. Increasing the density increases the collisionality for a given H-mode pedestal pressure, $\nu_*^{PED} \propto (n^{PED})^3/(p^{PED})^2$, which suppresses the bootstrap current and shifts the instability point toward the high-n ballooning mode branch (figure 1). At fixed collisionality $n^{PED} \propto (p^{PED})^{2/3}$, so that stronger plasma shaping, which raises the peeling-ballooning mode stability limit (figure 2) should allow higher density operation. This was verified in experiments in which the QH-mode density limit doubled as the triangularity was increased from 0.65 to 0.85. At the higher triangularity in this experiment the density in QH-mode was as high as typical discharges with ELMs, $n_e^{PED}/n_{Greenwald} = 0.5$, where $n_{Greenwald}(10^{20}\,m^{-3}) = I_P(\mathrm{MA})/\pi a(\mathrm{m})^2$ [22]. ELITE modeling confirms that it is possible to reach a higher density while maintaining access to the low-n peeling mode unstable regime at higher triangularity [16].

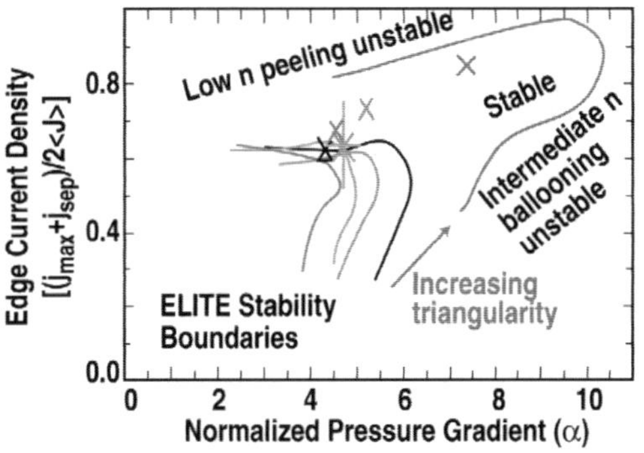

Figure 2. Peeling-ballooning mode stability boundaries for different QH-mode discharge shapes (lines) shows that the operational point for the ETB (Xs) always lies near the low-n peeling unstable boundary.

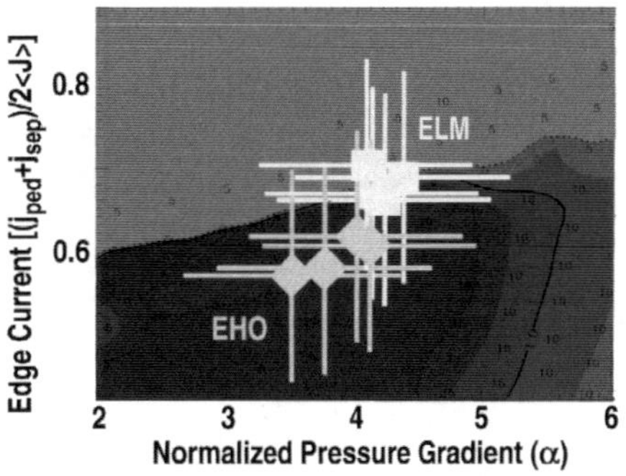

Figure 3. Discharges with small co-NBI power (diamonds) are in QH-mode with EHO and further in the stable region (dark blue) than discharges with large co-NBI (squares) where ELMs appear.

The proximity of the QH-mode edge to the low-n peeling stability boundary suggests the EHO may represent a saturated state of this instability. Observation of density fluctuations using beam emission spectroscopy showed the EHO to be localized to the region of the ETB, with no phase inversions that would suggest a non-ideal mode [14]. Further evidence that the EHO is associated with edge current density is given by current ramp experiments [15], where a downward current ramp turned off the EHO, while an upward ramp triggered ELMs; indicating not only that the EHO is current driven, but also that it is destabilized at currents below those required to trigger an ELM. Other evidence of this relationship between the EHO and ELM stability limits comes from a series of discharges in which the fraction of neutral beam power in the direction of plasma current was increased until ELMs were triggered, while keeping the total power input and other characteristics of the discharge fixed. Stability analysis of these discharges shows that the cases with ELMs lie very close to the low-n peeling mode stability limit while the QH-mode discharges with the EHO are situated somewhat below this limit (figure 3), suggesting the possibility of an additional drive mechanism in QH-mode.

4.2. Rotational shear and QH-mode

Toroidal rotational shear may provide the additional instability drive to trigger the EHO at edge currents below what would be required for the purely current driven peeling mode. ELITE modeling indicates that while rotational shear is stabilizing for modes at intermediate to higher toroidal mode number, it is destabilizing for modes at low-n [16]. Saturation of the instability might occur through self transport of momentum or drag of the mode on the conducting wall. The rotational shear is clearly reduced in the presence of the EHO. In the current ramp-down experiment described above for example, when the EHO is switched off the rotational shear rapidly increases.

Previous results indicated that the rotational shear in the ETB is significantly higher in QH-mode with counter-injection compared to ELMing H-mode discharges with pure co-injection at the same density [17]. Since the rotational shear is reduced in the presence of the EHO, this implies that high rotational shear is a property of counter-injection in general, and may explain why counter-injection is required for QH-mode. Low density operation would also increase the rotation speed by reducing the plasma inertia.

The importance of rotational shear was verified in an experiment where the rotational shear was varied by varying the fraction of neutral beam power (NBI) in the co-current direction (figure 4). In this experiment the NBI was switched from pure counter to some fraction co-current after QH-mode had been obtained, keeping the total heating power fixed. As the co-NBI fraction increased the rotational shear decreased. The initially $n = 3$ dominant EHO turned off progressively earlier as the co-NBI fraction was increased. Below co-NBI fractions of 20% the EHO returned intermittently as $n = 1$ or $n = 2$. Above 20% co-NBI, ELMs appeared, with ELMs coming sooner in discharges with higher co-NBI power fraction. Once the ELMs began, the rotational shear was reduced to near zero and the EHO did not reappear. The pressure gradient and current density also increased as the co-NBI fraction increased. This increase in the pedestal pressure came primarily through an increase in the pedestal density and expansion of the steep gradient region of the density inwards which began with the start of co-NBI. In the higher co-NBI fraction cases where ELMs returned, the primary $n = 3$ EHO turned off a few 100 ms before the first ELM. In some cases the first ELM then occurred with no EHO present, but in others the EHO returned initially as $n = 2$, then switched to $n = 1$ and subsequently grew to large amplitude over a few ms time scale as a precursor to the ELM.

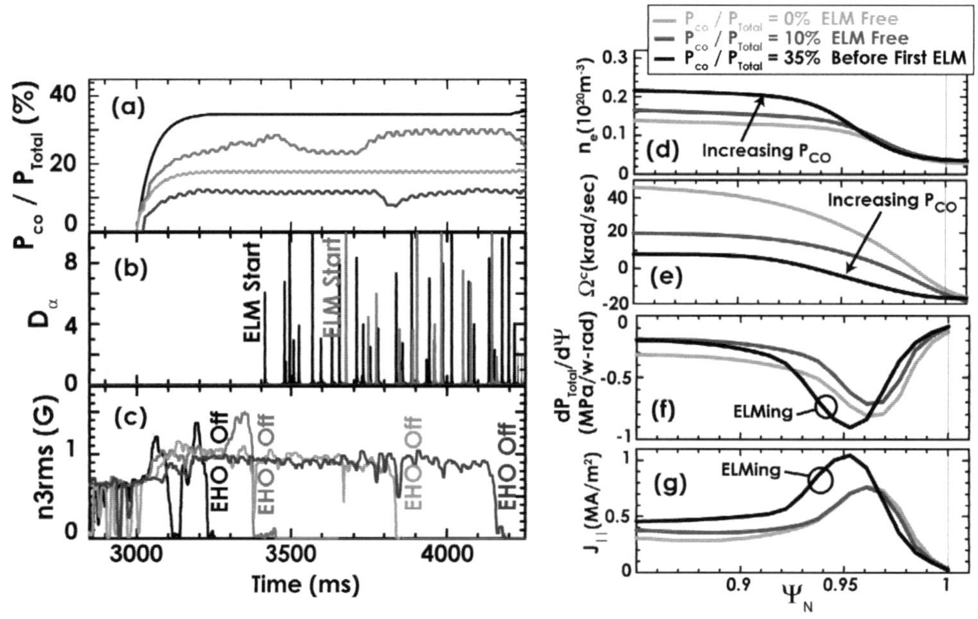

Figure 4. As co-NBI power fraction increases (a), $n = 3$ dominated EHO turns off earlier (c). ELMs are triggered at the two highest co-NBI power fraction (b). Rotational shear (e) decreases as co-NBI increase for 0% co-NBI (gray), 10% (blue) and 35% (black). Pressure gradient (f) and current density (g) rise with increasing co-NBI mainly through increase in density (d).

The importance of drag of the mode on the conducting wall as a factor in EHO saturation was consistent with results from an experiment in which the distance between the vessel wall and the plasma was varied. In this experiment, the EHO was switched off and ELMs returned as the plasma-wall separation at the outboard midplane was increased from 12 to 19 cm. The ETB pressure and current density profiles were similar at these two plasma-wall separation values, and ELITE analysis indicates both lie near the low-n peeling instability threshold. The rotational shear was larger in the discharge with small plasma-wall separation [figure 5(a)], however some change in the overall plasma shape also occurred as the wall separation was varied. ELITE analysis of the small plasma-wall separation discharge shows that rotational shear is destabilizing as expected, and that the peak growth rate occurs at $n = 5$ consistent with the dominant EHO harmonic [figure 5(b)].

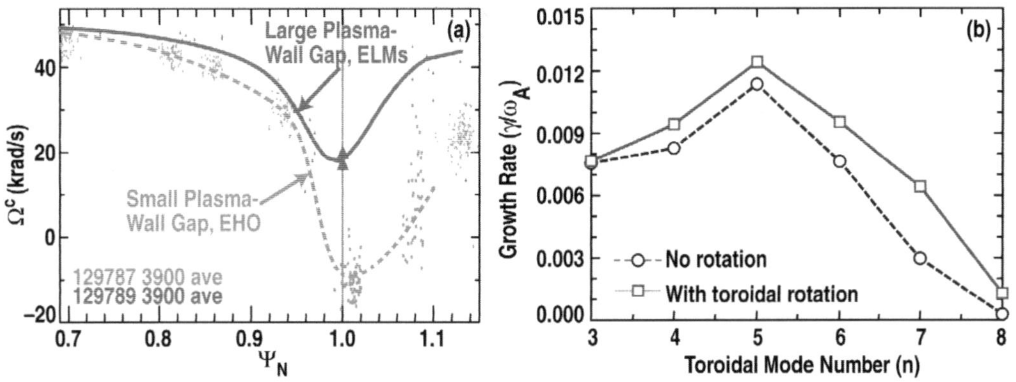

Figure 5. (a) Angular toroidal rotation velocity for carbon ions versus normalized poloidal flux. Discharge with small plasma-wall gap (green-dash) has larger rotational shear in ETB region, $0.9 < \psi_N < 1.0$, compared to larger plasma-wall gap discharge (red). (b) ELITE stability calculation of growth rate versus toroidal mode number for small plasma-wall gap discharge [green in (a)], showing rotational shear (red-solid) increases growth rate relative to no rotation case (black-dash). Growth rate peaks at $n = 5$ in agreement with observed EHO primary mode number.

5. RMP H-mode

As has been previously reported [18] the DIII-D resonant magnetic perturbation, RMP, coil is effective at suppressing ELMs. The DIII-D RMP coil is composed of two coil sets located symmetrically above and below the vessel midplane on the large major radius side. Each coil set is made up of six coils where toroidally adjacent coils are connected to carry equal and opposite currents producing an $n = 3$ perturbation. The RMP coil was effective at eliminating ELMs in low collisionality, $v_e^* \sim 0.1$, discharges, with the coil configured to produce the maximum field perturbation resonant with the equilibrium field. This configuration is expected to induce the largest magnetic islands in the ETB, giving regions of stochastic field and enhanced transport with the goal of controlling the pedestal pressure and thereby avoiding the peeling-ballooning instability. Resonance and ELM suppression is achieved with the upper and lower coil sets mirroring each other (even parity) when the discharge safety factor at the 95% flux surface, $q_{95} \approx 3.7$. In this regime both large Type I ELMs and small Type II ELMs are eliminated. An odd parity coil connection is not effective at suppressing ELMs at $q_{95} \approx 3.7$ and $v_e^* \sim 0.1$. Resonance is also achieved at $q_{95} \approx 7.5$ with the current in the upper and lower coil sets rotated 60° with respect to each other, odd parity. In this case, ELMs are strongly reduced in amplitude. In both the $q_{95} \approx 3.7$ and 7.5 cases, significant effects on the H-mode pedestal are observed [18]. RMP ELM suppression also occurred at $v_e^* \sim 1$, and $q_{95} \approx 3.7$ with the RMP coils in odd parity. Under these conditions, the resonant part of the field is reduced by roughly a factor of 7 compared to an even parity connection for the same coil current, although the total field from the coil in the ETB is comparable. In this case there is very little effect on the ETB pressure profile and, although Type I ELMs can be completely eliminated, small Type II ELMs are enhanced in amplitude.

5.1. RMP H-mode at $\nu_e^* \sim 0.1$, $q_{95} \approx 3.7$, even RMP coil parity

Previous results in low triangularity, $\delta = 0.5(\delta_{UPPER} + \delta_{LOWER}) = 0.26$, discharges have shown that at $\nu_e^* \sim 0.1$ the RMP suppresses ELMs by reducing the pressure gradient and current density in the ETB below the peeling-ballooning mode stability threshold [18]. That effect is confirmed here for discharges with ITER similar cross-sectional shapes at average triangularity $\delta = 0.53$ [figure 6(a)]. The pressure gradient and current density limits in the higher triangularity, ITER similar shape, discharges were however, unexpectedly, reduced. This reduction in the instability threshold may be driven by an unexplained expansion in the ETB width in the ITER similar shape discharges compared to the lower triangularity discharges [figure 6(b)]. ELITE modeling results for the ITER similar shape

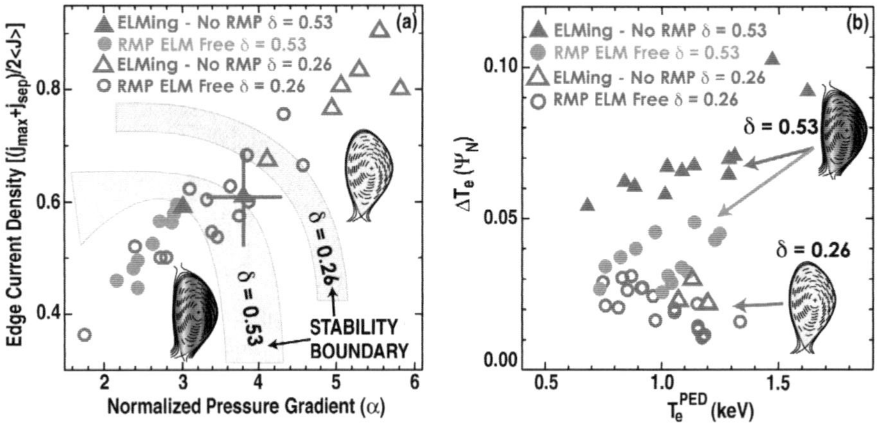

Figure 6. (a) Peeling-ballooning instability boundaries and data points for triangularities $\delta = 0.26$ (open points), and $\delta = 0.53$ (filled points) for RMP ELM free (circles) and no-RMP ELMing discharge (triangles). (b) ETB width (width of steep gradient region in T_e) is larger at $\delta = 0.53$ (filled points) compared to $\delta = 0.26$ data (open points).

indicate the pressure gradient limit is reduced when the ETB pressure width, Δ, expands as $p'_{CRIT} \propto \Delta^{-0.25}$ consistent with previous results for $\delta = 0$ [2]. This effect is a result of shifting the pressure gradient and current peaks outwards to more strongly shaped surfaces as Δ is reduced and from improved stability of the more radially extended lower n modes. The reduction in stability limit and expansion in width of the ETB in the higher triangularity, ITER similar shape discharges are offsetting so that similar pedestal pressure was obtained in these discharges compared to previous results at lower triangularity. At the minimum RMP coil current required for ELM suppression, the pedestal pressure for both the low and high triangularity cases was about 60 % of the peak pedestal pressure before an ELM without the RMP, and about 80% of the pedestal pressure when averaged over ELMs. The main effect of the RMP is to reduce the pedestal electron density for both the low and high triangularity shapes. There is also a narrowing of the steep gradient region for the electron profiles. Despite the reduction in pedestal pressure there was little change in the overall energy confinement enhancement factor in RMP ELM suppressed discharges, as a result of some peaking of the density profile and increase in the ion temperature.

As discussed in section 2, the location of the foot of the electron temperature profile relative to the separatrix has significant uncertainty. In discharges with the RMP, one might expect a region of stochastic field and hence low pressure gradient to exist just within the separatrix. To investigate the effect of such a region on stability, ELITE analysis was carried out with the electron profiles shifted inward relative to the separatrix. If the vacuum region, outside of which $j = 0$ in the ELITE calculation, is maintained at the original separatrix, equivalent to assuming that the stochastic region is filled with low pressure plasma, this has a strong stabilizing effect on the peeling modes as a result of moving the

Journal of Physics: Conference Series **123** (2008) 012014 doi:10.1088/1742-6596/123/1/012014

ETB current density away from the plasma boundary. If instead the plasma boundary is also shifted inwards, equivalent to treating the stochastic region as vacuum, both the peeling and ballooning modes become more unstable as a result of the shift of the pressure gradient and current density to less strongly shaped surfaces.

Increasing density with gas puffing in RMP ELM suppressed discharges raised the pedestal density and pressure and brought back the ELMs. At first the ELMs were small and high frequency but they became larger as the density was increased. The reappearance of ELMs is consistent with stability analysis which indicated the pressure gradient reached the ballooning limit with $n \sim 25-30$. If the RMP coil is switched off after the density rise, the ELMs become much larger and less frequent, the pedestal expands, and the most unstable mode is computed to have $n \sim 15$.

5.2. RMP H-mode at $\nu_e^* \sim 1$, $q_{95} \approx 3.7$, odd RMP coil parity

Even though the resonant magnetic field perturbation is small in the case of odd RMP coil parity at $q_{95} \approx 3.7$, complete Type I ELM suppression was achieved with this configuration in $\nu_e^* \sim 1$, higher triangularity discharges. When the ELMs were suppressed, the H-mode energy confinement enhancement factor remained unchanged. The electron density, n_e, electron temperature, T_e, and ion temperature profiles, T_i, also changed very little, both overall and in the pedestal region, in the RMP coil ELM suppressed phase compared to the ELMing case with no-RMP (figure 7). There was typically some rise in the effective charge, Z_{eff}, across the profile with the RMP coil on. The toroidal rotation speed measured for the carbon impurity, Ω_φ^C, is dramatically reduced with the RMP coil on over the entire plasma cross-section. No cases were observed for which the total pressure, P_{TOT}, in the pedestal region in the RMP coil ELM suppressed phase was significantly different from the profile before a Type I ELM with the RMP coil off under otherwise similar discharge conditions. As would be expected from the similar pedestal pressure profiles with and without the RMP coil, peeling-ballooning mode stability calculations give similar growth rates and toroidal mode numbers. In cases where the Type I ELMs were only reduced in frequency rather than being completely suppressed, the energy loss and profile change associated with the Type I ELMs, and the magnetic fluctuations at the ELM were similar with the RMP coil on and off.

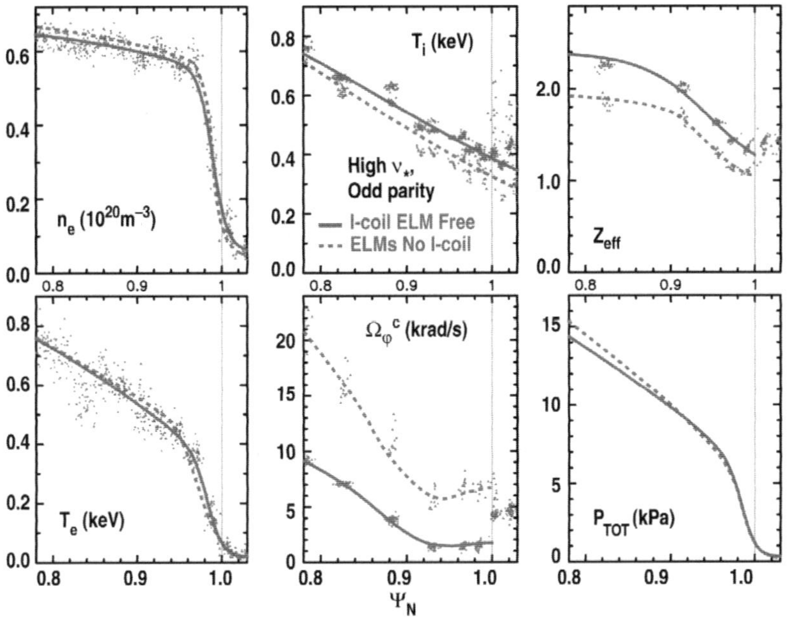

Figure 7. Profiles near the ETB as a function of normalized poloidal flux, ψ_N, for $\nu_e^* \sim 1$, odd parity RMP ELM suppressed discharge (green) and just before a Type I ELM (red-dashed) with no RMP. The separatrix location is marked.

A typical feature of H-mode discharges on DIII-D is the appearance of a series of small Type II ELMs leading up to a larger Type I ELM. The Type II ELMs become larger and fill the time interval between Type I ELMs, while the Type I ELM energy loss decreases, as the density is raised. In contrast to the results on ASDEX-Upgrade [19], the Type II ELMs on DIII-D do not completely replace the Type I ELMs in discharges without the RMP coil except at pedestal densities near the Greenwald density [22] where the pedestal pressure is strongly reduced. The Type II ELMs are more localized near the separatrix than the Type I ELMs as determined from profile reconstruction and soft x-ray observations, and have a much larger ratio of midplane to divertor magnetic fluctuations. For the Type I ELMs the midplane magnetic signals often have a coherent structure which is consistent with the peeling-ballooning mode instability [20]. During the RMP coil on phase, in contrast to the Type I ELMs which remain relatively unchanged, the Type II ELMs appear to be enhanced (figure 8), appearing larger on edge soft x-ray signals and in magnetic fluctuations.

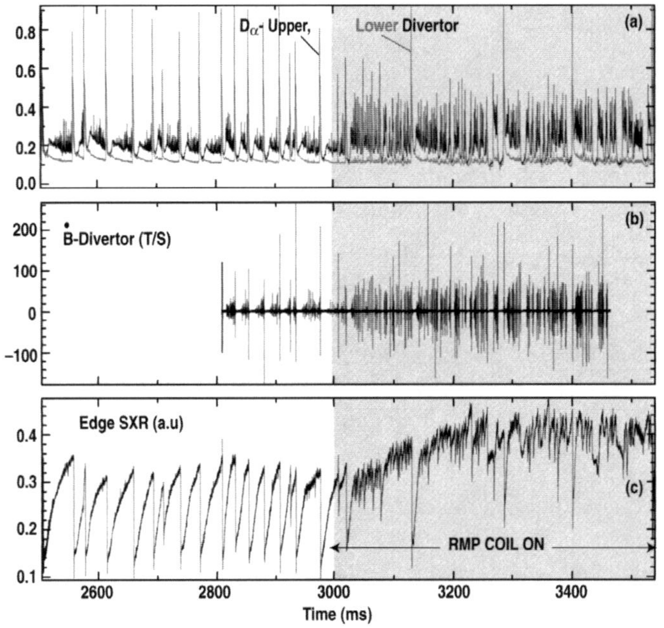

Figure 8. The size of small, Type II, ELMs increase with RMP coil on in $\nu_e^* \sim 1$, odd RMP-coil parity discharge on (a) D_α, (b) magnetic probes in divertor region, and (c) edge SXR.

5.3. RMP H-mode at $\nu_e^* \sim 0.1$, $q_{95} \approx 7.5$, *odd RMP coil parity*

With odd coil parity the resonant field is maximum at $q_{95} \approx 2.3$ and minimum at $q_{95} \approx 3.7$. There is a secondary maximum at $q_{95} \approx 7.5$ with RMP field about 3.6 times smaller than the primary maximum. By increasing the RMP coil current at $q_{95} \approx 7.5$, a resonant field amplitude about 80% of that required for ELM suppression at $q_{95} \approx 3.7$, was produced. In these $q_{95} \approx 7.5$ discharges, the pedestal pressure, J and p' were reduced nearly a factor of 2 compared to the no RMP case. In contrast to the $q_{95} \approx 3.7$ case, small ELMs with $\Delta W_{ELM} / W_{PED}$ reduced by more than a factor of 2 compared to the no RMP case persisted. These ELMs were a mixture of small and very small types suggesting the Type I ELM energy loss had decreased.

6. Conclusions

The pressure gradient and current density in the ETB of QH-mode discharges with strong counter rotation is close to the value obtained in ELMing discharges with co-rotation or balanced injection. In the ELMing case, the nonlinear phase of the peeling-ballooning instability leads to explosive growth, while for QH-mode a saturated mode is observed. This stark difference might be explained by the presence of an additional free energy source for the instability with weaker nonlinear drive and/or stronger transport effects. We present the hypothesis that velocity shear drive at low n is the additional free energy source for peeling-ballooning modes. This hypothesis is supported by several observations: 1) the mode number of the EHO, 2) the location of the QH-mode operating point near, but somewhat below the current limit for exciting an ELM, (measured from profiles and consistent with the current ramp experiment) 3) the low density requirement for QH-mode and the increase in that limit with strong plasma shaping, 4) the operational requirement of counter-injection where the velocity shear is inherently higher, and the loss of QH-mode as velocity shear is reduced with more balanced beam, 5) the damping of rotational shear when the EHO is present, 6) and the loss of QH-mode with plasma wall distance larger than a threshold value. On the basis of these results the possibility of obtaining QH-mode in ITER was investigated. Access to the low n peeling unstable regime in the ITER base case scenario would require $n_e^{PED} \leq 0.4 \times 10^{20}\ \mathrm{m}^{-3}$ [16]. This value is well below what is required to produce the desired fusion power output, $n_\mathrm{e}^{\mathrm{PED}} \approx 0.9 \times 10^{20}\ \mathrm{m}^{-3}$, however, it is possible that stronger shaping would allow access to this regime at higher density in ITER as it did in DIII-D. Neutral beams in ITER will be used for current drive as well as heating and therefore are to be oriented in the co-current direction. Our results do not rule out the possibility of QH-mode with only co-NBI given sufficient rotational shear.

The ability of the relatively simple RMP coil set to suppress Type I ELMs while maintaining good energy confinement is a positive result for future tokamaks. At ITER relevant collisionality, $\nu_e^* \sim 0.1$, the pedestal pressure was controlled and maintained below the peeling-ballooning mode stability limit while providing enough particle transport to avoid density or impurity accumulation. The control of pedestal pressure obtained in the low collisionality, even parity discharges may be a direct result of transport produced by the field stochastization. However, if this were the case it is not obvious why the strongest effect would be on the density profile since parallel transport should strongly affect the temperature.

In the case of ELM suppression at $\nu_\mathrm{e}^* \sim 0.1$, where the part of the RMP coil field resonant with the equilibrium is small, the evidence suggests an indirect effect through the ELM instability. The Type II ELMs, which appear to be enhanced with the RMP coil on, may provide enough transport to avoid reaching the stability limit for the Type I. However, from this point of view, it is difficult to understand the fact that the pedestal pressure profile is quite similar in discharges where the Type I ELMs were completely suppressed compared to the pressure profile just before a Type I ELM with the RMP coil off. No evidence was found suggesting that the peeling-ballooning mode associated with the Type I ELM was transformed into a continuous instability. No such continuous instability was observed, while very similar rapidly growing MHD activity was seen on the outboard midplane probes before a Type I ELM, and the Type I ELM was associated with about the same energy loss, with the RMP coil on and off. These results suggest that the Type II ELM has a very similar pressure gradient limit to the Type I and therefore is also likely the result of the peeling-ballooning instability but that for some reason the nonlinear phase of the instability is weaker or access to a transport channel is reduced. The Type II ELMs, or the RMP coil directly, might alter a feature other than the pressure profile which is critical for triggering large non-linear growth of the peeling-ballooning mode. It is possible that there is a difference in the pedestal current density with the RMP coil on and off, since this was not directly measured and is known to play a role in the peeling-ballooning stability. Changes in the rotation speed with the RMP coil on can be ruled out as altering the ELM instability since the Type I ELMs disappear before the rotation changes [21].

Acknowledgments
This work was supported by the U.S. Department of Energy under DE-FC02-04ER54698, W-7405-ENG-48, and DE-FG02-04ER54758.

References

[1] Wilson H R, Cowley S C, Kirk A and Snyder P B 2006 *Plasma Phys. Control. Fusion* **48** A71
[2] Snyder P B *et al.* 2004 *Plasma Phys. Control. Fusion* **46** A131
[3] Ferron J R *et al.* 2000 *Phys. Plasmas* **7** 1976
[4] Huysmans G T A 2005 *Plasma Phys. Control. Fusion* **47** B165
[5] Sauter O, Angioni C and Lin-Liu Y R 1999 *Phys. Plasmas* **6** 2834
[6] Osborne T H *et al.* 2000 *Plasma Phys. Control. Fusion* **42** A175
[7] Burrell K H *et al.* 2002 *Plasma Phys. Control. Fusion* **44** A253
[8] Evans T E *et al.* 2005 *Nucl. Fusion* **45** 595
[9] St John H E *et al.* 1994 *Proc. 15th Int. Conf. on Plasma Physics and Controlled Nuclear Fusion Research Seville (IAEA Vienna, 1995)* IAEA-CN-60/D-P22 603
[10] Porter G D, Moller J, Brown M and Lasnier C 1998 *Phys. Plasmas* **5** 1410
[11] Lao L L *et al.* 2005 *Fusion Sci. Technol.* **48** 968
[12] Thomas D M *et al.* 2004 *Phys. Rev. Lett.* **93** 065003
[13] Wilson H R, Snyder P B, Huysmans G T A and Miller R L 2002 *Phys. Plasmas* **9** 1277
[14] Burrell K H *et al.* 2005 *Phys. Plasmas* **12** 056121
[15] West W P *et al.* 2005 *Nucl. Fusion* **45** 1708
[16] Snyder P B *et al.* 2006 *Proc. 21st Intl Conf on Plasma Physics and Controlled Nuclear Fusion Research (Chengdu)* (IAEA Vienna) Paper TH/4-1Ra
[17] Burrell K H *et al.* 2004 *Plasma Phys. Control. Fusion* **46** A165
[18] Evans T E *et al.* 2006 *Nature Physics* **2** 419
[19] Stober J *et al.* 2001 *Nucl. Fusion* **41** 1123
[20] Perez C P *et al.* 2004 *Nucl. Fusion* **44** 609
[21] Evans T E *et al.* 2004 *Phys. Rev. Lett.* **92** 235003-1
[22] Greenwald M *et al.* 1988 *Nucl. Fusion* **28** 2199

Journal of Physics: Conference Series **123** (2008) 012015 doi:10.1088/1742-6596/123/1/012015

Effect of toroidal field ripple and toroidal rotation on H-mode performance and ELM characteristics in JET/JT-60U similarity experiments

N Oyama[1], G Saibene[2], Y Kamada[1], K Kamiya[1], A Loarte[2], J Lönnroth[3], V Parail[4], Y Sakamoto[1], A Salmi[3], R Sartori[2], K Shinohara[1], H Takenaga[1], H Urano[1], P de Vries[4], M Yoshida[1], the JT-60 Team[1] and JET EFDA Contributors[5]

[1] Japan Atomic Energy Agency, Naka, Ibaraki-ken, 311-0193 Japan
[2] EFDA Close Support Unit – Garching
[3] Association Euratom-Tekes, Helsinki University of Technology, Finland
[4] Euratom/UKAEA Fusion Association, Culham Science Centre, Abingdon OX14 3EA, UK
[5] Appendix of M. L. Watkins, et al., Fusion Energy 2006 (Proc. 21st Int. Conf. Chengdu) IAEA (2006)

oyama.naoyuki@jaea.go.jp

Abstract. The effect of the toroidal field (TF) ripple and edge toroidal rotation (V_T) on H-mode and pedestal performance as well as on ELM characteristics are investigated both in JET and JT-60U using matched plasma shape. In JT-60U, the amplitude of TF ripple (δ_T) was reduced from ~1.2% to ~0.5% (averaged value at B_T = 2.2 T) after the installation of ferritic steel tiles (FSTs). In addition with FSTs, δ_T can also be varied by changing the toroidal magnetic field for a given separatrix position. In JET, the TF system can be configured to feed different currents to the odd and even coil sets out of 32 TF coils. In this operation mode, δ_T can actively be varied by selecting the appropriate differential current between each coil set. In both devices, it is observed that the edge V_T, which was in the same direction as the plasma current (co-V_T), reduced with increasing δ_T. Even at the same δ_T of 0.5%, the achievable edge V_T in JT-60U was still lower than that in JET due to higher neutral beam fast ion losses in JT-60U. A series of power and density scans performed at several δ_T indicated that plasmas with smaller ripple amplitude and/or larger co-V_T are favourable to achieve higher p_{ped} and H_H factor in both devices. As for ELM characteristics, larger co-V_T seems to increase the ELM energy loss together with the reduction of the ELM frequency.

1. Introduction

Inter-machine comparison is a very powerful tool to identify the physics mechanisms determining the plasma behavior, as well as a way to validate the assumptions behind the physics based scaling used for the prediction of the plasma parameters of ITER. Both in JET and JT-60U, inter-machine experiments have been performed to compare H-mode pedestal performance and ELM characteristics using matched plasma shape. In previous similarity experiments, the matching of pedestal dimensionless parameters in JT-60U were found to be difficult to scale to those in JET, and matched dimensionless parameters could be obtained only with low heating power JET plasmas [1]. MHD stability analysis shows that the pedestal MHD stability of both tokamaks is similar and probably cannot explain the observed difference in ELMy H-mode performance [2], the toroidal field (TF) ripple and consequently different toroidal rotation profiles have been pointed out as major remaining differences between two devices, as shown in figure 1. Therefore, new similarity experiments focused on TF ripple and toroidal rotation profile have been performed in both devices.

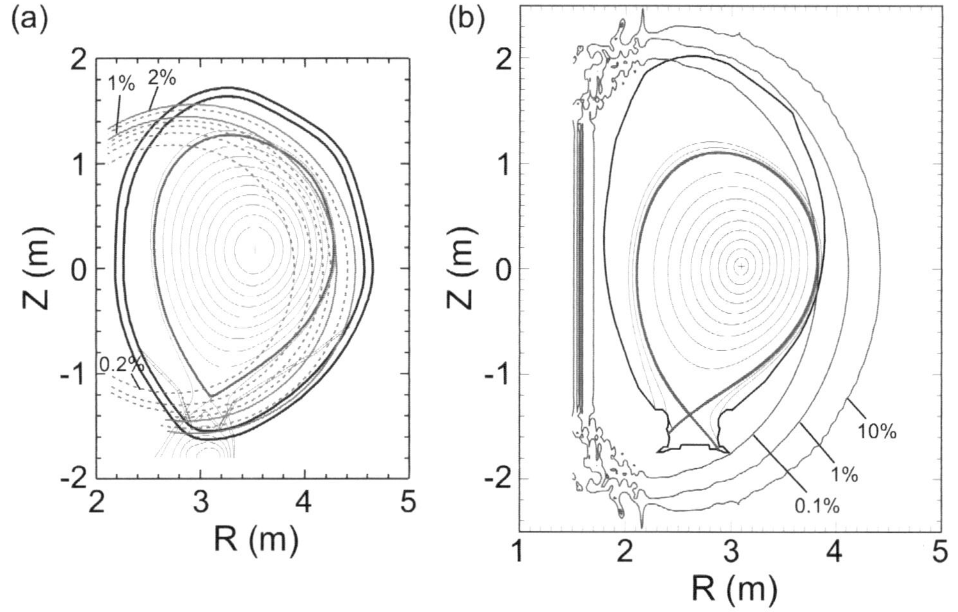

Figure 1. Plasma configuration for similarity experiments together with ripple amplitude (δ_T) in JT-60U (a) and JET (b). Here, δ_T is defined by $(B_{T,max} - B_{T,min})/(B_{T,max} + B_{T,min})$. The number of TF coils is 18 in JT-60U and 32 in JET.

The origin of the TF ripple is the finite number of TF coils. In JT-60U, 18 circular TF coils are used and therefore the contour plot of ripple amplitude (δ_T) shows a circular shape, while JET is equipped with 32 D-shape coils resulting in very small δ_T with D-shaped contours, as shown in figure 1. For comparison, the number of TF coil in ITER is 18, which is the same number as in JT-60U, resulting in a ripple δ_T at the reference separatrix position of ~1.2% [3]. Nonetheless, fast ion (particle) confinement in ITER is extremely good, and fast ion losses are predicted to be negligible, at least for H-mode profiles and in absence of core MHD instabilities. Recent experiments in JT-60U show that V_T is one of the key parameters to determine the pedestal performance and ELM characteristics [4, 5]. Thus, if V_T is essential parameter in pedestal performance and ELM characteristics, δ_T itself is of minor importance for ITER plasmas. However, the heat load to the first wall due to the loss of α-particles in reversed shear plasmas is expected to exceed the tolerable level. Therefore, to reduce the ripple loss power below 1% of α-heating, the installation of ferritic steel plate has been planned in ITER [3]. Since the remaining δ_T near the mid-plane is expected to be ~0.5% in ITER with ferritic steel plates, both effects of TF ripple and V_T on H-mode performance, and ELM characteristics are an important and belong to an urgent research area.

The paper is organized as follows. After the introduction, the experimental conditions for varying δ_T and the response of the plasma toroidal rotation at the pedestal (V_T) are described in section 2. The effects on the H-mode performance and pedestal due to changes in δ_T and the V_T are described in section 3. The ELM characteristics, in particular ELM frequency dependence and ELM amplitude, are compared in section 4. Section 5 gives a the results of detailed comparison of the pedestal and H-mode performance in terms of the effect of the injected torque in JET and the effect of local ripple amplitude near the X-point in JT-60U. Finally, a summary is given in section 6.

2. Experimental condition for ripple experiments

The method to vary δ_r differed between JET and JT-60U. The TF ripple in JT-60U was reduced by using ferritic steel tiles (FSTs), which are optimized for ripple reduction at toroidal magnetic field (B_T) of <2 T [6]. Since the magnetic field produced by FSTs is saturated at 1.78 T for $B_T \geq 0.6$ T, δ_r can be varied with B_T in JT-60U. In the previous similarity experiments, $B_T = 3.2$ T was used so as to match non-dimensional parameters between two devices at 1.08 MA and 1.8 MA. From the view point of the availability of edge diagnostics in JET and of the level of ripple reduction in JT-60U, two levels of B_T, 2.2 T for the lowest δ_r and 3.2 T for the direct comparison to the previous experiment, were selected at a fixed plasma current of 1.08 MA. Taking into account the constraints related to the matching of dimensionless plasma parameters as described in [1], plasma parameters of 1.14 MA / 2.0 T were chosen for the JET ripple experiment. In contrast, the TF system in JET can be configured to feed different currents to the odd and even set of coils out of 32 TF coils. In this operation mode, δ_r can actively be varied by selecting the appropriate differential current between each set of coils. In this experiment, four levels of δ_r, 0.1%, 0.5%, 0.75% and 1%, were used. Beside different methods to vary δ_r, both devices also have a different NBI geometry and hence a different value of NB injected torque as well as a different dynamic range of fast ions losses, which is the source of effective counter torque due to the formation of an inward electric field [7] as well as a different value of NB injected torque. The specified experimental conditions in each device including the changes in V_T are described in this section.

2.1. Ripple experiments in JT-60U

The reduction of TF ripple using FSTs is shown in figure 2(a) and 2(b). Because of breaking of the 18-fold toroidal symmetry, the usual definition of ripple amplitude, $\delta_r = (B_{T,max} - B_{T,min})/(B_{T,max} + B_{T,min})$, is not valid in JT-60U after the installation of FSTs as shown in figure 2(b). Therefore, to compare ripple level with and without FSTs, we define a 'quasi' ripple well region as whether there is a local minimum of the magnetic field within ϕ_s-$\Delta\phi < \phi < \phi_s$+$\Delta\phi$ at a given toroidal angle ϕ_s, where $\Delta\phi$ is half of the period of the TF coil installation [6]. As can be seen in figure 2(a), the 'quasi' ripple well region becomes smaller with FSTs, even if evaluated at the P3 section where there is the maximum local ripple as shown in figure 2(b). Since there are no FSTs near the X-point (X_p), it is noted that a small level of local ripple ($\delta_r \sim 0.24\%$) remains in that region for this plasma configuration.

Figure 2(c) shows the geometry of NBIs in JT-60U. There are 4 tangential NBIs (two co-injection and two ctr-injection) with a tangency radius (R_T) of ~2.7 m and 7 perpendicular (perp) NBIs with R_T ~ 0.75 m. In addition to these positive-ion based NBIs (P-NBIs) with the acceleration voltage (V_{acc}) of ~85 kV, there are two negative-ion based NBIs (N-NBIs) with V_{acc} of 350-370 kV. The reduction of fast ion losses from each P-NBI, which is evaluated using fully three-dimensional magnetic field orbit following Monte Carlo (F3D-OFMC) [8], are summarized in figure 2(d). In this figure, the loss power ratio is defined by $(P_{CX} + P_{ripple} + P_{orbit})/(P_{inj} - P_{shine})$, where P_{CX}, P_{ripple}, P_{orbit}, P_{inj} and P_{shine} are, respectively, the loss power due to charge exchange, the loss power through ripple trapped loss, the loss power due to unconfined orbits, the injected NBI power and the loss power due to shine through. It is noted that the fast ion losses for perp-NBIs with a slightly tilted injection angle to the plasma current ($co_\perp$) are smaller than other perp-NBIs with a slight counter injection ($ctr_\perp$) component. Therefore, to minimize fast ion losses, co-tangential NBIs and some $co_\perp$-NBIs are mainly used in this experiment.

In the analysis for JT-60U plasmas, we used F3D-OFMC to evaluate the power deposition profile, the loss of fast ions, the shine through loss and the net torque injection. It is noted that F3D-OFMC only calculates the transfer of the torque from fast ions to bulk plasma by collisions.

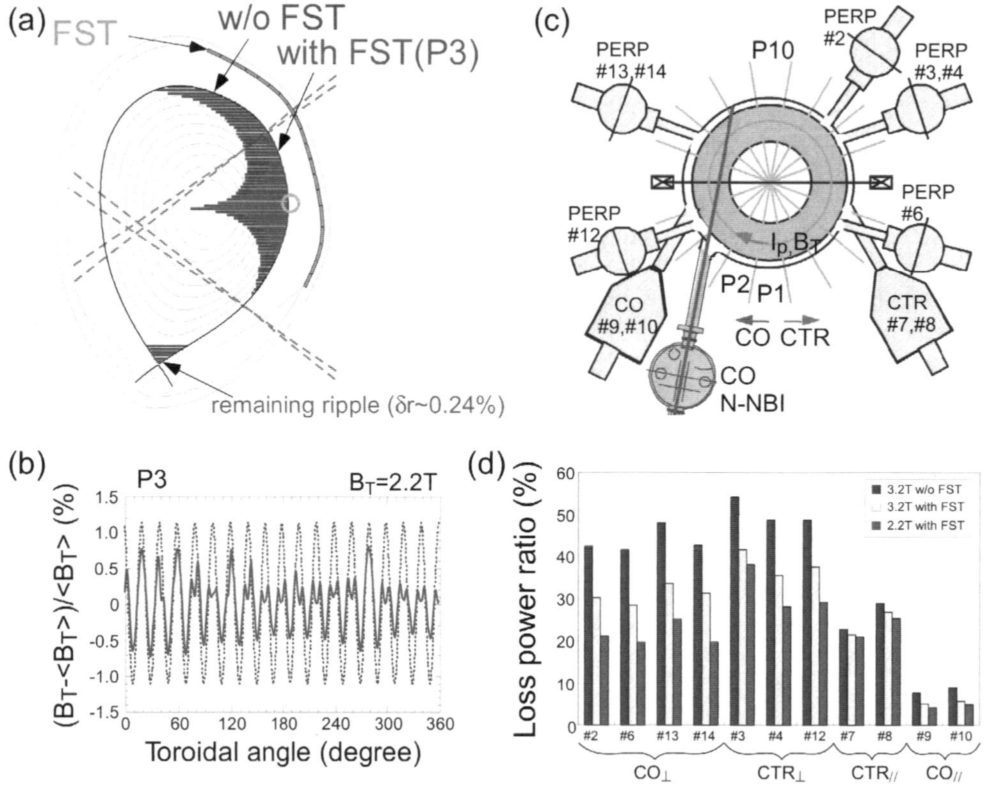

Figure 2. (a) Comparison of 'quasi' ripple well structure with and without FSTs. The location of FSTs in the poloidal cross section and the beam trajectories of two ion sources from one perp-NBI unit (dashed lines) are also shown. (b) Toroidal variation of the calculated toroidal magnetic field strength at R = 4.3 m and Z = 0.2 m marked by circle in (a). The dotted curve shows the data with TF coils alone and the solid curve shows the data with FSTs. (c) Port arrangement of NBI system (top view). The tangent radii of tangential- and perp-NBI are $R_T \sim$ 2.7 m and $R_T \sim$ 0.75 m, respectively. (d) Comparison of fast ion losses evaluated with F3D-OFMC code in three B_T cases, 3.2 T without FSTs ($\delta_r \sim$ 1.2%), 3.2 T with FSTs ($\delta_r \sim$ 0.65%), 2.2 T with FSTs ($\delta_r \sim$ 0.5%). Perp-NBIs of 2, 4, 6, 12, 14 and 3, 13 are injected from upper and lower port, respectively.

2.2. Ripple experiments in JET

Since the operation with large δ_r has the potential to cause severe damage to in-vessel components, the local heat loads to in-vessel components were assessed by F3D-OFMC before the experiment. The analysis indicated that δ_r could be increased up to 1% in plasmas with 1.14 MA / 2.0 T, remaining within tolerable power density on all critical JET in-vessel components. Figure 3(a) shows the contour plot of δ_r with the current ratio of I_{odd} / I_{even} = 0.52, which produces δ_r = 1% at the outer mid-plane. It is noted that the remaining local ripple near X_p was δ_r < 0.1% in the case of δ_r = 1% at the outer mid-plane as shown in figure 3(a), while being $\delta_r \sim$ 0.24% in JT-60U with $\delta_r \sim$ 0.5% at the outer mid-plane with FSTs as shown in figure 2(a).

Figure 3(b) shows the geometry of the JET NBI system. There are 16 independent injectors located at two toroidal sections, octant 4 and 8. The V_{acc} of these beams varies between 70 and 130 kV. Since there are two different groups of injectors with a different tangency radius; so-called 'tangential' beams with $R_T \sim$ 1.85 m and so-called 'normal' beams with a smaller $R_T \sim$ 1.31 m, the NB injected torque can be varied by changing the combination of NBIs. Nevertheless, the normal beam in JET is more

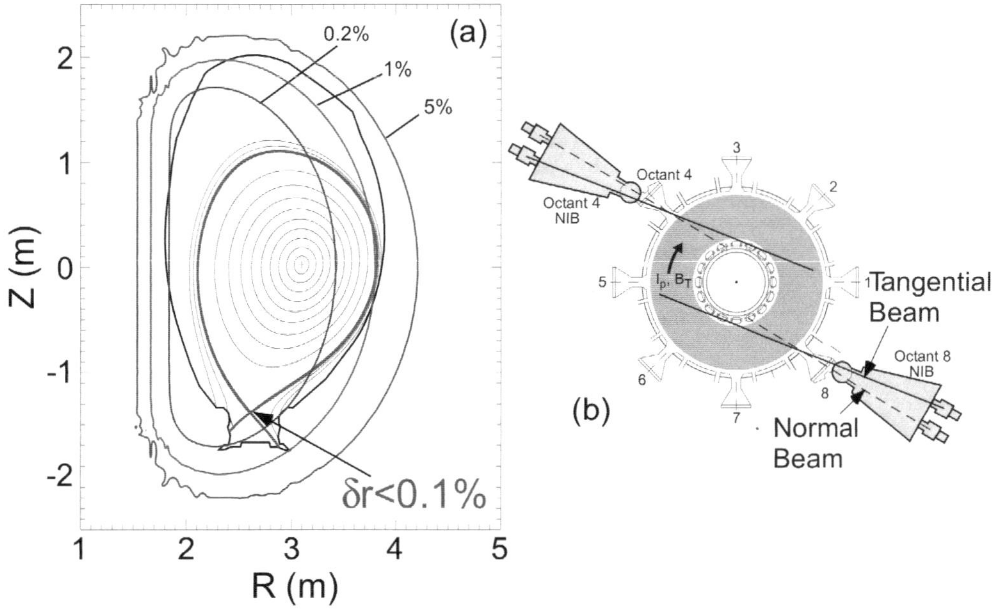

Figure 3. (a) Contour plot of ripple amplitude with the current ratio of $I_{odd} / I_{even} = 0.52$. (b) Port arrangement of NBI system in JET (top view). Tangential NBI ($R_T \sim 1.85$ m) and normal NBI ($R_T \sim 1.31$ m) are installed. Each system has 8 independent injectors.

tangential than perp-NBIs in JT-60U. Therefore, in higher heating power plasmas, the torque input from NBIs in JET was slightly higher than in JT-60U as described in next subsection.

In this paper, the power deposition profile, the loss of fast ions and the net torque injection for JET plasmas are evaluated with ASCOT [9] using the birth profile calculated by PENCIL [10, 11]. The shine through loss power is also given by PENCIL. Since a benchmark between F3D-OFMC and ASCOT indicated good agreement of the two codes, we could compare evaluated profiles with ASCOT and F3D-OFMC.

2.3. Changes in toroidal rotation by ripple amplitude and fast ion losses

Figure 4 compares the toroidal rotation frequency measured at the top of the T_i (ion temperature) pedestal between the two devices. The total torque injected by NBI was 5.5-8.5 Nm in JET except for plasmas with low heating power ($P_{NBI} < 4$ MW), and in JT-60U, ~5.5 Nm without FSTs and 6-7.5 Nm with FSTs. It is noted that the injected torque in JET was proportional to the heating power, but that in JT-60U was varied only in a narrow range, because the number of tangential NBIs was fixed and the heating power was varied by changing the number of perp-NBIs (that provide very little input torque).

As a result, the toroidal rotation frequency in JT-60U plasmas varied mainly with the loss power of the fast ions as shown in figure 4(a). When the heating power was increased by adding perp-NBIs, the loss of fast ions also increased resulting in a larger ctr-rotation. Therefore, a wide variation in the rotation frequency was observed in plasmas with similar ripple amplitude (e.g. $\delta_r = 1.2\%$) as shown in figure 4(c). After the installation of FSTs, fast ion losses were reduced as shown in figure 2(d). Plasmas with less ctr-rotation or co-rotation were obtained with similar NBI torque input, as reported in [12]. Nevertheless, higher fast ion loss power fractions than JET was observed.

In JET, the fast ions loss fraction did not vary much over the δ_r scan as shown in figure 4(a). Even in plasmas with $\delta_r = 1\%$, only 0.5 MW was lost from the injected power of 7.4 MW. Therefore, the toroidal rotation frequency in JET plasmas mainly related to the δ_r as shown in figure 4(c), which is consistent with other ripple experiments in JET [13]. For a similar level of torque input in both devices, the upper boundary of the achieved rotation frequency decreased with increasing δ_r. But, the variation of the rotation frequency at constant δ_r in JET was narrower than in JT-60U. The fact that there is a strong relation between δ_r and V_T in JET plasmas implies that it will be difficult separating the influences of δ_r and V_T. It is also noted that there was only a few data with matched rotation frequency between the two devices as seen in figure 4(c) at $\delta_r = 0.5\%$.

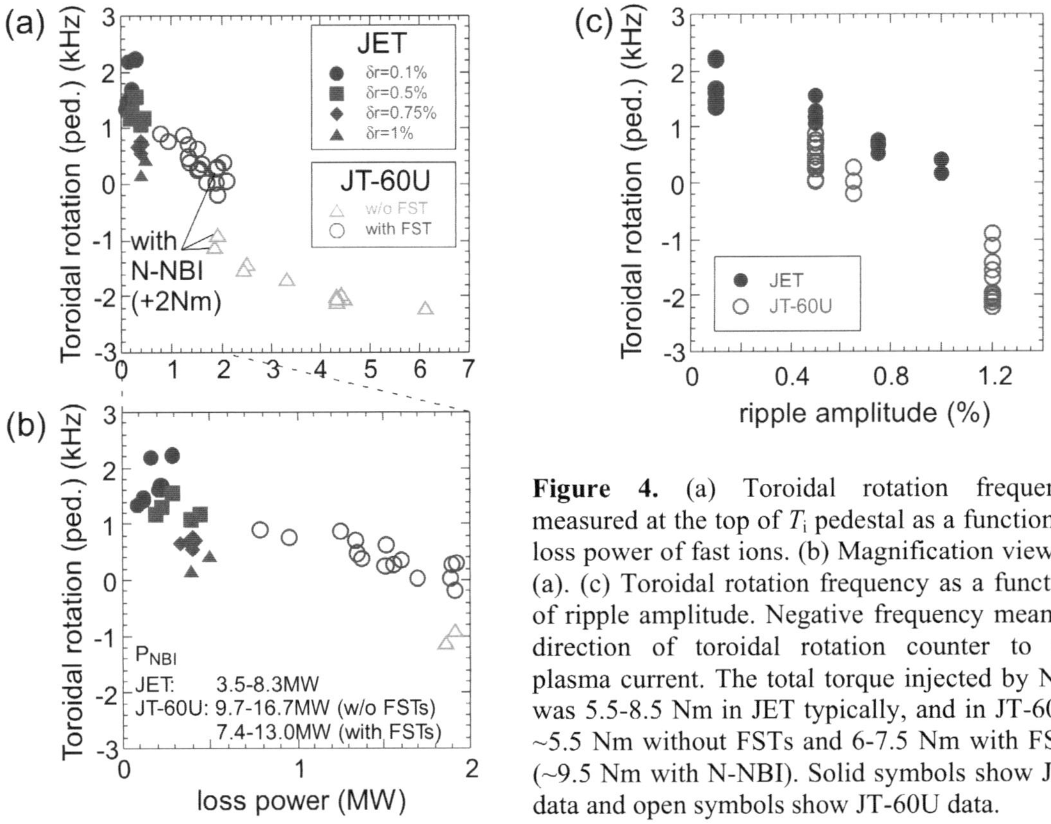

Figure 4. (a) Toroidal rotation frequency measured at the top of T_i pedestal as a function of loss power of fast ions. (b) Magnification view of (a). (c) Toroidal rotation frequency as a function of ripple amplitude. Negative frequency means a direction of toroidal rotation counter to the plasma current. The total torque injected by NBI was 5.5-8.5 Nm in JET typically, and in JT-60U, ~5.5 Nm without FSTs and 6-7.5 Nm with FSTs (~9.5 Nm with N-NBI). Solid symbols show JET data and open symbols show JT-60U data.

3. H-mode and pedestal performance

Under the conditions described in section 2, a series of heating power and density scans was performed in both devices. For JT-60U, pedestal electron density (n_e^{ped}) and electron temperature (T_e^{ped}) were measured with a Thomson scattering system, while pedestal ion temperature (T_i^{ped}) was measured with charge-exchange recombination spectroscopy (CXRS). For JET, on the other hand, n_e^{ped}, T_e^{ped} and T_i^{ped} were measured with an edge chord of an FIR interferometer, electron cyclotron emission (ECE) radiometer and CXRS, respectively.

Figure 5 compares n_e^{ped} and T_e^{ped} in the ripple scan experiments. In JT-60U, the achievable electron pressure at the pedestal (p_e^{ped}) was similar both with and without FSTs (also high & low B_T with FSTs) without gas puffing. A moderate gas puffing to increase the plasma density degraded p_e^{ped} by ~20%, and in the latter phase of the discharge ELM characteristics changed at n_e^{ped}~2.4×10^{19}m^{-3} from type I ELMs to type I+III ELMs. Therefore, it was still difficult to achieve high n_e^{ped} in JT-60U plasmas even with FSTs, where the averaged δ_r was 0.5% at $B_T = 2.2$ T. In the previous experiment without FSTs, an improvement of p_e^{ped} was observed in plasmas with N-NBIs as indicated by open triangles in figure 5(a) [1]. When the same heating scenario using N-NBIs was applied to the plasma with FSTs at $B_T =$ 3.2 T, a higher p_e^{ped} than in plasmas heated by only P-NBIs was achieved. The fact that higher p_e^{ped} can be achieved both with and without FSTs indicated that there is unknown effect of the injection of N-NBI on the pedestal performance.

In JET, the achievable p_e^{ped} was gradually degraded with increasing δ_r as shown in figure 5(b), although the reference data with no ripple (δ_r = 0.1%) are at lower p_e^{ped} than the dataset obtained in the previous experiments reported in [1]. The exact reason for this difference between the two experimental campaigns has so far not been understood. Since one of the differences between JET and JT-60U is the achievable density, the response to gas puffing was also investigated in JET at δ_r = 0.1% and δ_r = 0.75%. As indicated by dotted circles in figure 5(b), the n_e^{ped} in plasmas at δ_r = 0.75% could be increased to similar values as obtained at δ_r = 0.1% by using the same amount of gas puffing. A small degradation of p_e^{ped} was observed in JET plasmas with gas puffing. The degradation of p_e^{ped} in JT-60U plasmas was larger than in JET, even with a smaller gas puffing rate. It is noted that the achieved density with ripple was always smaller than that without ripple under the same amount of gas puffing. In addition to this, when the density feedback control was applied, the required gas puffing rate was higher with ripple than without ripple. These observations indicate a density pump-out with ripple, which is consistent with other ripple experiments reported in [14].

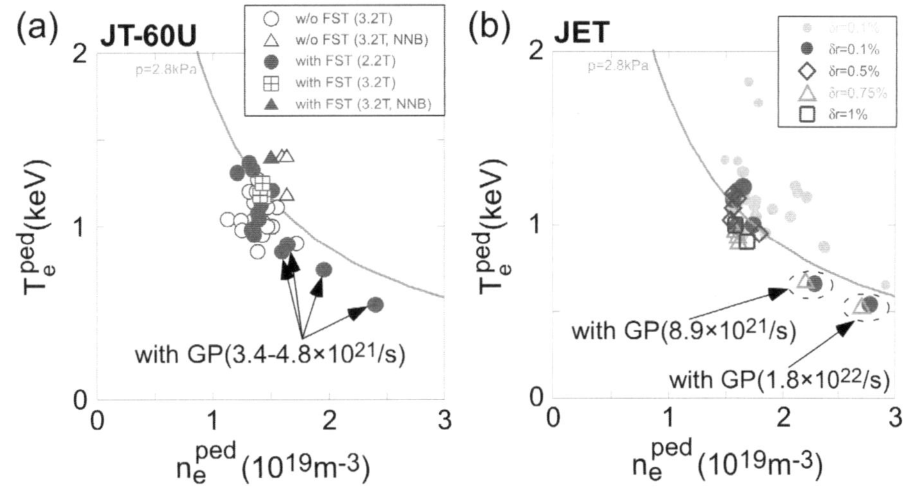

Figure 5. Comparison of pedestal density and temperature in ripple scan experiments in JT-60U (a) and in JET (b). Gray circles in (b) show dataset obtained in previous experiments reported in [1].

In order to understand the effect of δ_r and V_T, the total pedestal pressure ($p^{ped} = n_e^{ped}T_e^{ped} + n_i^{ped}T_i^{ped}$) evaluated with a spatially constant Z_{eff} and the thermal confinement enhancement factor are compared between two devices as functions of δ_r and V_T in figure 6. Although the unselected dataset for JT-60U shows no clear trend in the achievable p^{ped} between with and without FSTs (except for the plasmas heated by N-NBIs), the effect of heating power on p^{ped} should be considered. When we compare only data with similar P_{NET} = 6.3-7.3 MW marked by solid symbols in figure 6, an improvement of p^{ped} was observed in both figures 6(a) and (b). This kind of global behavior was also observed in the dedicated ripple experiment in JT-60U [4]. In JET, the achievable p^{ped} gradually degraded with increasing δ_r and/or gradually improved with increasing V_T. The data selected by heating power also shows the same trend in both figures 6(a) and (b) due to the strong relation between δ_r and V_T in JET plasmas. It is noted that the rate of change of p^{ped} as a function of V_T was similar between JET and JT-60U. The comparison at the same ripple and V_T shows that the H_H in JT-60U was smaller than in JET, while p^{ped} was similar. This experimental result implies a different core transport.

The response of the H_H factor was similar to that observed in p^{ped} as shown in figure 6(c) and (d), where the upper limit of H_H factor was gradually degraded with increasing δ_r and/or gradually improved with increasing V_T. In addition to this, the H_H factor was degraded by ~20% with gas puffing, which is usually observed in JET plasmas without large ripple. All these observations are consistent with other ripple experiment reported in [14].

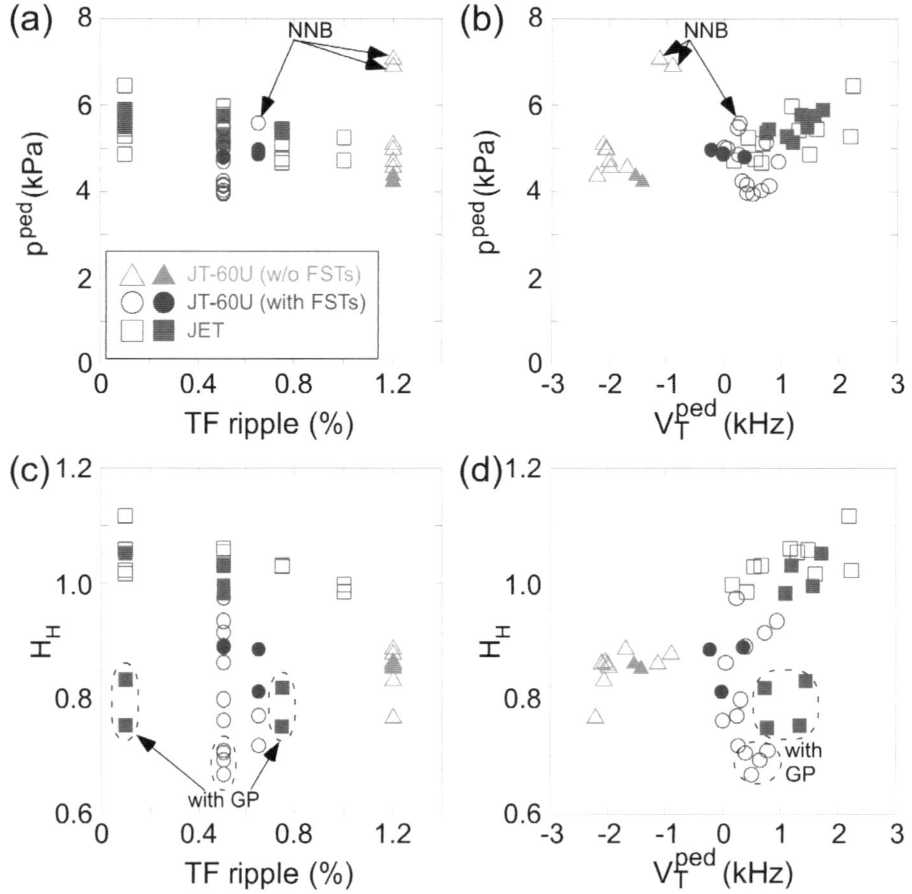

Figure 6. Comparison of the total pedestal pressure (p^{ped}) as a function of δ_r (a) and V_T (b). Comparison of the thermal confinement enhancement factor (H_H) as a function of δ_r (c) and V_T (d). Solid symbols show data selected by P_{NET} = 6.3-7.3 MW.

In dedicated ripple experiments in JT-60U using two different plasma configurations [4], the effect of TF ripple on p^{ped} has been identified between plasmas with $\delta_r \sim 2\%$ and $\delta_r \sim 1\%$ using a large volume configuration (the separatrix of the configuration was close to the wall and $V_P \sim 75$ m^3). No significant effect of TF ripple between plasmas with $\delta_r \sim 0.4\%$ and $\delta_r \sim 0.2\%$ was found with the small volume configuration (inward shifted plasma configuration with $V_P \sim 52$ m^3), while the effect of V_T was clearly observed in both plasma configurations. The difference between plasmas with $\delta_r \sim 0.1\%$ and $\delta_r \sim 0.5\%$ was small as shown in figure 6(a) and (c). While in JET, degradation in p^{ped} and H_H factor was observed at $\delta_r \sim 0.3\%$ [14]. To determine the acceptable level of δ_r, further investigation will be needed.

The effect of local ripple near X_p was also investigated in JT-60U, because ASCOT simulations that include the effect of ripple induced thermal ion losses indicates that the ripple near X_p might influence the plasma confinement [15]. This was investigated experimentally by shifting up the plasma position up vertically. The δ_r near X_p was reduced from 0.24% to 0.14%, while ripple amplitude near outer mid-plane was almost unchanged. However, there was no clear difference in plasmas profiles (n_e, T_e and T_i) both in the pedestal and the core. Therefore, the influence of local ripple near the X_p on H-mode pedestal performance seems to be small.

4. ELM characteristics

In JT-60U plasmas with FSTs, the ELM amplitude (ΔW_{ELM}) increased by 50-150%, while the ELM frequency (f_{ELM}) decreased by only ~20%. Therefore, the ELM loss power, $P_{ELM} = f_{ELM} \times \Delta W_{ELM}$, increased by 30% for a given loss power through the separatrix (P_{sep}) as shown in figure 7(a). Higher P_{ELM} / P_{sep} implies a reduction of inter-ELM transport, which is clearly observed in figure 7(b). This figure compares two discharges obtained from plasmas with similar f_{ELM} and P_{sep}, one with and one without FSTs. Although both ΔW_{ELM} and pedestal stored energy (W_{ped}) were higher with FSTs, the normalized ELM energy loss ($\Delta W_{ELM} / W_{ped}$) still increased from <5% to 5-10% because of the larger change in ΔW_{ELM}. The observed increase of a recovery rate of stored energy (dW/dt) by a factor of two can compensate approximately twice larger ΔW_{ELM}. This plasma response is consistent with the observation that the increased ΔW_{ELM} at the same f_{ELM} as found in plasmas without FSTs. The global behavior of ELM characteristics are compared in figure 7(c), where the normalized ELM frequency (f_{ELM} / P_{sep}) is plotted as a function of V_T. In contrast to pedestal and H-mode performance, a figure of f_{ELM} / P_{sep} as a function of δ_r only shows a wide scatter. Moreover, the data obtained from JET/JT-60U similarity experiments shows the same trend as that from the dedicated ripple experiment with large volume configuration marked by squares. Therefore, the V_T seems to be a leading parameter to change ELM characteristics, as reported in [4, 5]. The effect of V_T was also found in a grassy ELM regime in JT-60U [16].

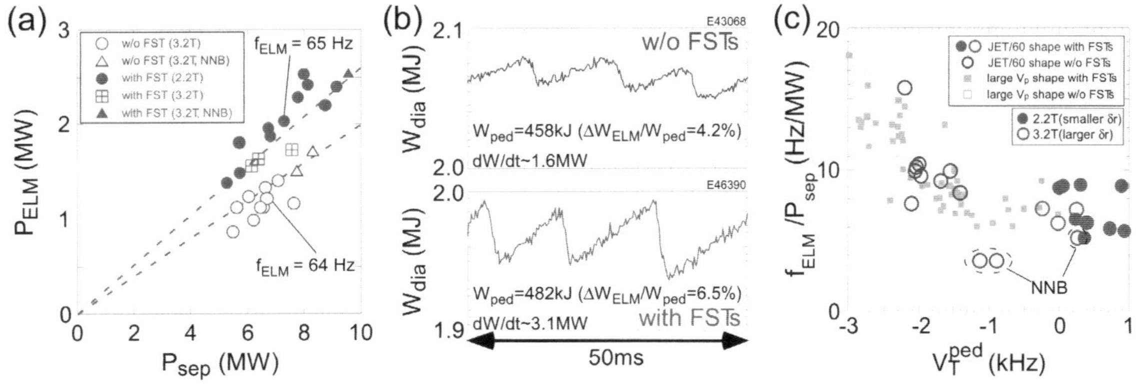

Figure 7. Comparison of ELM characteristics in JT-60U. (a) ELM loss power (P_{ELM}) as a function of loss power through the separatrix (P_{sep}). (b) Comparison of the time evolution of plasma stored energy for the points indicated in (a). (c) Normalized ELM frequency as a function of V_T. Circles in (c) show data obtained from JET/JT-60U similarity experiments and squares in (c) show data obtained from dedicated ripple experiments with a large volume configuration. In these figures, only plasmas without gas puffing are plotted.

In JET plasmas, the f_{ELM} in reference data without ripple ($\delta_r = 0.1\%$) was slightly higher than that in the previous similarity experiments reported in [1] somehow. Nevertheless, global changes in the ELM characteristics similar to those reported for JT-60U have also been observed in the ripple scans as shown in figure 8. The vertical variation in figure 8(a) at $\delta_r = 0.1\%$ seems to be correlated by the different V_T as shown in figure 8(b). Since this kind of ELM behavior was also found in a dedicated ripple experiment in JET using a typical JET plasma configuration, the effect of V_T on ELM amplitude and frequency seems to be a common feature in two devices. The fact that the torque input is proportional to P_{sep} can help to separate δ_r and V_T in the comparison of normalized ELM frequency.

Because of the higher f_{ELM} and lower ΔW_{ELM}, the diamagnetic loop measurements, while operating with different currents in each JET TF coils set, could not detect the drop of the stored energy. Therefore, the fast drop of T_e^{ped} (ΔT_e^{ped}) measured with ECE radiometer is used as a representative of ΔW_{ELM}. Figure 9 compares three discharges with different δ_r (0.1%, 0.5% and 1%), which are marked by (a), (b) and (c) respectively in figure 8(b). Lower heating power was applied to the plasma with larger δ_T, resulting in smaller P_{sep} at larger δ_T. However, the f_{ELM} increased with increasing δ_T (V_T). On the other hand, the drop in T_e^{ped} decreased with decreasing V_T (increasing δ_T) together with the reduction of ELM averaged T_e^{ped} corresponding to the reduction of p^{ped} as shown in figure 5(b). Here, the normalized reduction of T_e^{ped} ($\Delta T_e^{ped} / T_e^{ped}$) was 10.9%, 8.0% and 7.5% for figure 9(a), (b) and (c), respectively.

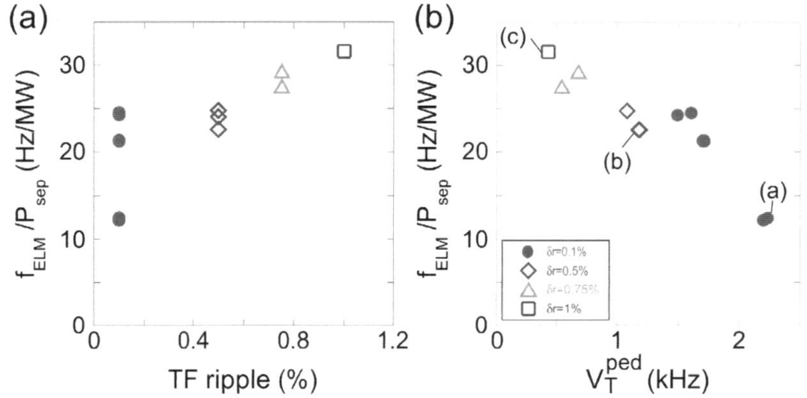

Figure 8. Normalized ELM frequency observed in JET plasmas as a function of δ_T (a) and V_T (b). Labels in (b) show discharges plotted in figure 9. Only plasmas with P_{NET} > 5 MW are plotted.

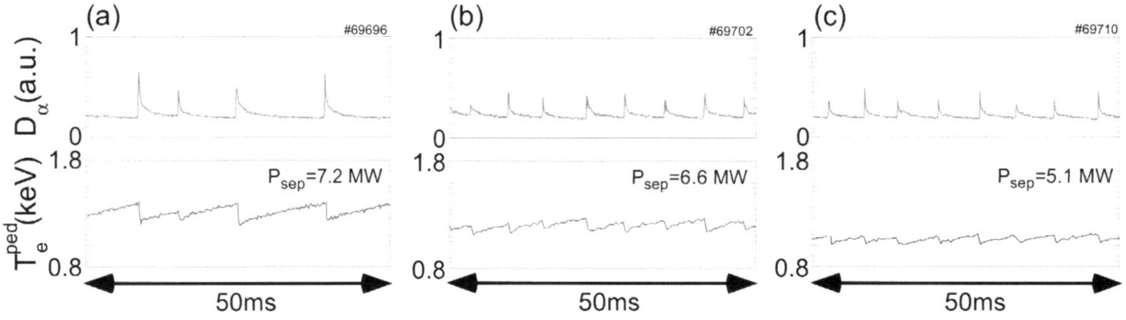

Figure 9. Comparison of the time evolution of the divertor D_α signal and pedestal temperature during the ripple scan in JET. (a), (b) and (c) show plasmas with δ_T = 0.1%, 0.5% and 1%, respectively. The corresponding V_T for each discharge is shown in figure 8.

5. Discussion

In the previous sections, the influence of the TF ripple and the toroidal rotation on the pedestal performance and ELM characteristics are described. In general, the pedestal performance affects the total plasma performance due to the well known stiffness of core profiles in the ELMy H-mode plasmas. In previous similarity experiments, on the other hand, it was found that core plasma profiles were different between JET and JT-60U even in cases where the pedestal value were very well matched between the two devices [1]. An apparent deviation from profile stiffness was also observed in JET plasmas, specifically when the toroidal rotation and torque input were varied by changing the combination of tangential and normal beams at a fixed TF ripple amplitude of δ_T= 0.5%, which is the same value as in JT-60U with FSTs.

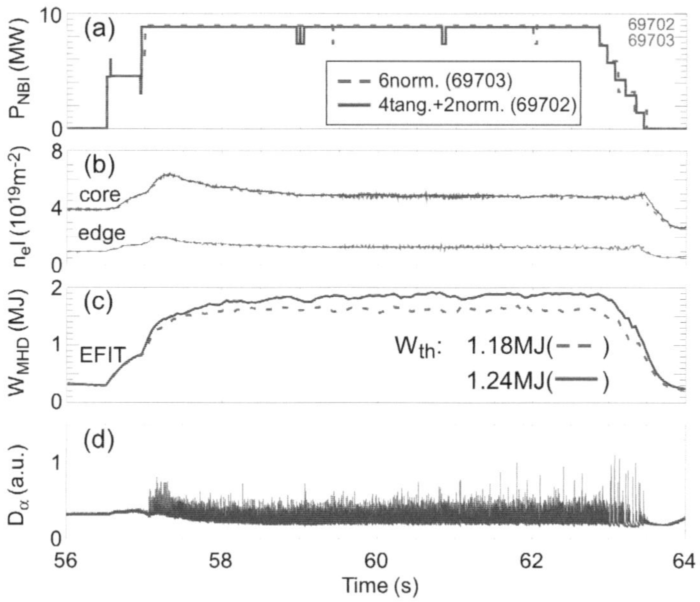

Figure 10. Waveforms of two JET plasmas for the comparison of tangential and normal injection. (a) NBI heating power. P_{NET} = 7.3 MW in 60702 and 7.2 MW in 69703. (b) line-integrated electron density in the core and edge chord. (c) MHD energy evaluated with EFIT. (d) Divertor D_α signal. Thermal stored energy evaluated by JETTO with kinetic plasma profiles was 1.18 MJ for normal injection case and 1.24 MJ for tangential+normal injection case.

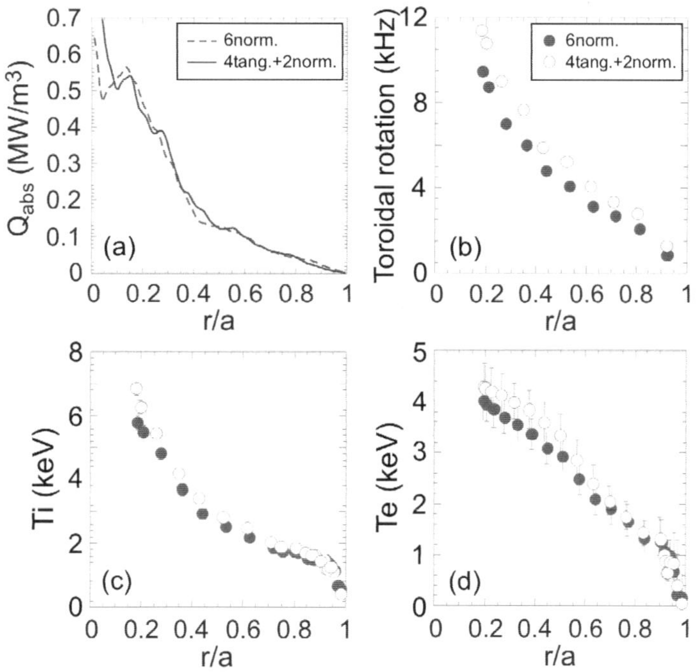

Figure 11. Comparison of plasma profiles in JET between two otherwise identical plasmas, one heated by only normal NBIs (69703, closed symbols) and by dominant tangential NBIs (69702, open symbols). (a) Power deposition profile evaluated with ASCOT. (b) Toroidal rotation frequency profile. (c) and (d) show ion and electron temperature profiles, respectively.

Figure 10 shows waveforms of two discharges, where only 6 normal beams are injected in 69703, while a combination of tangential (4 units) and normal (2 units) beams are used in 69702. A comparison of selected plasma parameters for the two discharges is shown in figure 10. As illustrated figure 10(b) and (d), averaged density and divertor D_α signal were almost identical. The fast ion loss power was higher by ~200 kW for 69703, but this was almost compensated by slightly higher P_{NBI} and P_{OH}, leading to almost identical power deposition profiles as shown in figure 11(a) (evaluated with JETTO [17] and ASCOT). Nevertheless, there is a clear difference in the plasma stored energy (both MHD and thermal) as shown in figure 10 (c).

A small, but obvious difference was observed in the temperature profiles as shown in figure 11 (c) and (d), in particular for T_e profile. It is noted that the difference in the core profiles cannot be explained by the stiffness, because the pedestal parameters (n_e^{ped}, T_e^{ped} and T_i^{ped}) are almost the same. The rotation profile was also different as shown in figure 11(b). However, the difference in the toroidal rotation at the top of pedestal was so small that no change in p^{ped} is consistent with the result in figure 6. A similar response of the core plasma profiles has been observed in JT-60U after the installation of FSTs in plasmas with an internal transport barrier (ITB). When the plasma rotation was changed from ctr-rotation to co-rotation by changing the combination of tangential NBIs, the core electron transport gradually decreased while keeping the transport in peripheral region constant (outside ITB including the pedestal region) [18]. Since there is no transport model explaining an improved core confinement by such a small change in the plasma rotation so far, further analysis and experiments are required to obtain better understanding of the core transport properties, which is also important to obtain a reliable prediction of ITER plasma.

6. Summary

Dedicated ripple experiments were performed in JET and JT-60U using a matched plasma shape. After the installation of FSTs in JT-60U, δ_r near the outer midplane was reduced to ~0.5% at $B_T = 2.2$ T. In JET, on the other hand, δ_r can be varied by selecting the appropriate differential current between odd and even set of coils out of 32 TF coils, providing in this case four levels of $\delta_r = 0.1\%$, 0.5%, 0.75% and 1%. Although the same level of ripple amplitude was successfully obtained in both devices, the V_T in JET was still higher than that in JT-60U due to a different level of fast ion losses. Since a correlation between δ_r and V_T was found in JET plasmas, it is difficult to separate the two parameters, δ_r and V_T, in JET experiments.

A series of power and density scans indicated that plasmas with lower δ_r and/or larger co-V_T are favorable to achieve higher p^{ped} and H_H factor in both devices. As for ELM characteristics, larger co-V_T seems to increase the ELM energy loss together with the reduction of the ELM frequency. Sustaining a high n_e^{ped} was still difficult in JT-60U plasmas even with FSTs ($\delta_r \sim 0.5\%$), while n_e^{ped} in JET plasmas at $\delta_r = 0.75\%$ could be increased to similar value as at $\delta_r = 0.1\%$ using the same amount of gas puffing. Therefore, there are still some differences between the two devices. In order to obtain a better prediction for ITER plasmas, further investigations into the effects of the TF ripple and the toroidal rotation are necessary, and determining the acceptable level of ripple amplitude in ITER is still difficult based on these experimental results in both devices. In addition to this, the development of a reliable model of momentum transport and of the source of effective torque at the plasma edge due to ripple losses is also important to predict ITER plasma performance and ELM characteristics.

References

[1] Saibene G, Oyama N, Lönnroth J-S, Andrew Y, de la Luna E,Giroud C, Huysmans G T A, Kamada Y, Kempenaars M A H, Loarte A *et al.* 2007 *Nucl. Fusion* **47** 969
[2] Lönnroth J-S, Parail V, Hynönen V, Johnson T, Kiviniemi T, Oyama N, Beurskens M, Howell D, Saibene G, de Vries P *et al.* 2007 *Plasma Phys. Control. Fusion* **49** 273
[3] Tobita K, Nakayama T, Konovalov S V and Sato M 2003 *Plasma Phys. Control. Fusion* **45** 133
[4] Urano H, Oyama N, Kamiya K, Koide Y, Takenaga H, Takizuka T, Yoshida M, Kamada Y and the JT-60 Team 2007 *Nucl. Fusion* **47** 706

Journal of Physics: Conference Series **123** (2008) 012015 doi:10.1088/1742-6596/123/1/012015

[5] Kamiya K, Urano H, Oyama N and Kamada Y 2007 *Plasma and Fusion Research* **2** 005
[6] Shinohara K, Suzuki Y, Sakurai S, Masaki K, Fujita T and Miura Y 2006 *Plasma Fusion Res.* **1** 007
[7] Koide Y *et al.* 1993 Proc. 14th Int. Conf. on Plasma Physics and Controlled Nuclear Fusion Research 1992 (Wuerzburg, 1992) vol 1 p 777 (Vienna: IAEA)
[8] Shinohara K, Kawashima H, Tsuzuki K, Urata K, Sato M, Ogawa H, Kamiya K, Sasao H, Kimura H, Kasai S *et al.* 2003 *Nucl. Fusion* **43** 586
[9] Heikkinen J A and Sipil S K 1995 *Phys. Plasmas* **2** 3724
[10] Challis C D, Christiansen J P, Cordey J G, Gormezano C, Gowers C W, Kramer G J, O'rourke J J, Stubberfield P M, Taroni A, Tibone F *et al.* 1992 *Nucl. Fusion* **32** 2217
[11] Stubberfield P M and Watkins M L 1987 *"Multiple Pencil Beam"* JET-DPA (06) / 87
[12] Yoshida M, Koide Y, Takenaga H, Urano H, Oyama N, Kamiya K, Sakamoto Y, Kamada Y and the JT-60 Team 2006 *Plasma Phys. Control. Fusion* **48** 1673
[13] de Vries P C, Salmi A, Parail V, Giroud C, Andrew Y, Biewer T M, Crombé K, Jenkins I, Johnson T and Kiptily V *et al.* 2008 *Nucl. Fusion* **48** 035007
[14] Saibene G, Sartori R, de Vries P, Loarte A, Lönnroth J-S, Parail V, Salmi A, Thomas P R, Andrew P, Andrew Y *et al.* 2007 Toroidal field ripple effects on H-modes in JET and implications for ITER *Proc. 34th EPS Conference on Plasma Physics* (Warsaw) O4.001
[15] Parail V, Evans T, Johnson T, Lönnroth J-S, Oyama N, Saibene G, Sartori R, Salmi A, de Vries P, Becoulet M *et al.* 2006 *Proc. 21st IAEA Fusion Energy Conf. (Chengdu, China)* TH/P8-5
[16] Oyama N, Kamada Y, Isayama A, Urano H, Koide Y, Sakamoto Y, Takechi M, Asakura N and the JT-60 Team 2007 *Plasma Phys. Control. Fusion* **49** 249
[17] Cenacchi G and Taroni A 1988 *"JETTO: A free boundary plasma transport code"* ENEA RT/TIB/88/5
[18] Oyama N, Isayama A, Suzuki T, Koide Y, Takenaga H, Ide S, Nakano T, Asakura N, Kubo H, Takechi M *et al.* 2007 *Nucl. Fusion* **47** 689

The features of the electron heat transport during high power ECRH & SMBI on HL-2A

H J Sun, X T Ding, L H Yao, J Rao, Z T Liu, Y Liu, Y Huang, C F Dong, W Li, X R Duan, Q W Yan and Yong Liu

Southwestren Institute of Physics, Sichuan, Chengdu 610041, P.R.China

E-mail: dingxt@swip.ac.cn

Abstract. With high power ECRH and supersonic molecular beam injection (SMBI), the electron heat transport has been investigated on HL-2A. This paper will show features of the electron transport during ECRH, especially new non-local phenomena triggered by SMBI, in which an electron transport barrier is formed and the confinement is improved. The diffusion coefficient both from steady-state analysis (χ^{hp}) and the heat pulse propagation (χ^{pb}) increase with the ECRH power and decrease with the electron density, which is in agreement with the L-mode scaling. The ratio χ^{hp}/χ^{pb} is about 2.8 in Ohmic discharges; it increases to 3.3 with low power ECRH and decreases to 1.7 with the high power ECRH. The non-local transport phenomena induced by SMBI have been observed firstly on HL-2A. The density limit beyond which the phenomena disappear is apparently higher in the ECRH regime than in the Ohmic regime. Both the bolometer radiation and the H_α emission decrease when the non-local effect appears. After the appearance of the non-local effect, it can be found that the region outside r/a ~ 0.5, just like usual result, is cooled by SMBI, while the core T_e increasing gives a surprise. The core T_e is increased by about 18 %. This suggests that the injection of SMBI triggers the internal transport barriers.

1. Introduction

Electron heat transport in burning plasmas is one of the great challenges to realize controlled nuclear fusion. There are some phenomena, such as profile consistency, ITB and non-local effect, which have been debated for years. Up to now, and in spite of a wealth of results, these debates have not lead to a clear conclusion. During the last few years, the electron heat transport has been extensively investigated in tokamaks [1, 2, 3], in particular supported by the electron cyclotron resonance heating (ECRH) system. The good local power deposition of ECRH in the plasma permitted a detailed study of the electron heat transport [4, 5, 6]. New experimental observation in the study helps us to understand heat transport deeply. Heat transport study with the high power ECRH in HL-2A could provide complementarities for the database of transport analysis.

From the first observation of the non-local effect at TEXT-U in 1995 [7] to later reports on many tokamaks [8, 9, 10, 11], the non-local phenomena evoke much attention of the scientists who focus on plasma transport. Then, different edge cooling experiments have been widely performed in studying the new conundrum: impurity injection by laser ablation, ice pellet injection, carbon-based molecules injection, etc. But there is no report about the effect induced by a gas puffing. In 2006, the non-local effect has been observed after the SMBI in HL-2A. This is a new observation that a cold gas pulse can also induce the non-local effect. This means the non-local effect can be investigated by the new experimental method. In this paper, we show the non-local effect induced by the SMBI both in ECRH regime and in the ohmic regime, and give preliminary analysis results about it.

In this paper, we focus on the electron heat transport with the high power ECRH and the SMBI in HL-2A. The paper is divided as follows: in section 2, the experimental arrangements are described. Section 3 gives the experimental results about electron heat transport in L-mode plasma and the non-local phenomena triggered by the SMBI, and both results are discussed in this section. The summary is presented in the last section.

2. Experimental arrangement

The HL-2A is a divertor tokomak (the major radius R = 1.65 m; the minor radius a = 0.4 m) which can be operated in a limiter or a single null-divertor configuration with the following parameters: toroidal magnetic field B_t < 2.8 T; the average electron density and plasma current are $n_e = (1\text{-}5) \times 10^{19}\,\mathrm{m}^{-3}$ and I_P = (250-430) kA, respectively [12]. More than 30 diagnostics have been developed on HL-2A in recent years. Among them, the key diagnostics for heat transport study are soft-x-ray array, Electron Cyclotron Emission (ECE) diagnostics and diamagnetic measurement. Four soft-x-ray arrays, each with twenty channels, are used. The x-ray signal results from plasma bremsstrahlung and recombination processes, both of which are strongly dependent on temperature and density. The ECE spectrum is measured by a 2 mm sweeping heterodyne radiometer. The range of the scanning frequency is between 104 and 181 GHz. Its temporal and spatial resolutions are 4 ms and 4 cm, respectively. There is also a high temporal resolution ECE system (up to 1 μs) which covers about 20 cm along the plasma midplane. In the SMBI experiment with B_t = 2.31 T, a whole second harmonic spectrum can be obtained. Under the usual discharge parameter in HL-2A, the calculation of optical thickness shows that inside $r/a \sim 0.7$ the plasma could be considered as the black body and the intensity of ECE 2nd harmonic is only proportional to the electron temperature. A one-point measurement Thomson Scattering system is developed to measure core plasma T_e of HL-2A tokamak, with the time resolution of 100 ms.

ECRH system with four 68 GHz / 500 kW / 1 s gyrotrons has been built up in HL-2A [13]. The ECRH power with fundamental O-mode up to 2 MW is injected from the low field side of the device. The deposit position of ECRH is determined by the toroidal magnetic field. When toroidal magnetic field varies from 2.43 T to 2.2 T, the resonance point can be replaced from the plasma core to the point at r = 16 cm. The wave energy deposits in a range of 3cm. The efficiency of the transmission system is more than 80 %, so at least 1.6 MW power can be injected into plasmas.

A supersonic molecular beam injection (SMBI) has been successfully developed in the HL-1M tokamak [14]. SMBI is an improvement method of conventional gas puffing (GP). The source consists of a small chamber and a Laval nozzle. The working gas can be kept at a definite pressure P_0 and temperature T_0 in the chamber. Particles of the gas are accelerated by imposed pressure through the nozzle to get into the vacuum chamber of the tokamak. It is an attempt to enhance the penetration depth and fuelling effect. The penetration depth is very important to trigger the non-local transport phenomena. H_α intensity profile measured by H_α detector array [15] and the density profiles measured by microwave reflectometry indicate that the penetration depth of the SMBI is more than r/a = 0.7 [16].

3. Experimental results

3.1. Electron heat transport during high power ECRH

High power ECRH discharges (larger than 1MW) have been investigated on HL-2A in 2006 experiment. Perfect heating efficiency has been observed and the temperature climbs to about 5 keV with the 1.6 MW ECRH. Figure 1 shows the results of a high power ECRH discharge with B_t = 2.42 T, $n_e = 1.7 \times 10^{19}\,\mathrm{m}^{-3}$, P_{ECRH} = 1.3 MW and I_P = 300 kA. The electron temperature during the ECRH is almost two times higher than that in the ohmic heating phase, both the Thomson scattering and the ECE diagnostic show the similar result.

The electron heat diffusivities both from a steady-state analysis and a heat pulse propagation analysis have been obtained on the high power ECRH discharges. All the data are selected from HL-2A experimental data with P_{ECRH} = 1 - 1.6 MW, B_t = 2.31 T, n_e = (1-3) × 10^{19} m^{-3}. An estimate for χ_e^{pb} at r/a = 2/3 can be obtained from the global power balance by using [17]

$$\chi_e^{pb} = a^2 / (4\tau_e) \quad (1)$$

where τ_e is the energy confinement time, defined by $\tau_e = W/P$; here, W is the total storage energy, P is the total absorted power, and a is the minor radius.

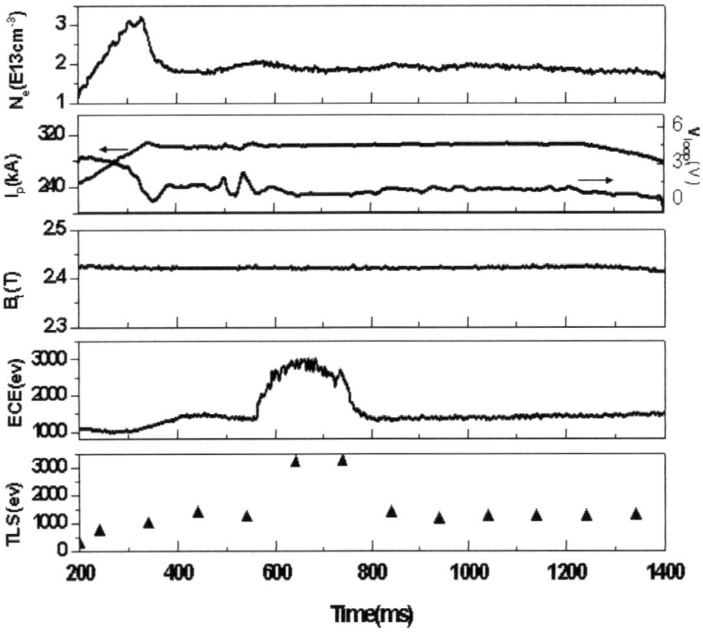

Figure 1. Shot 5830, one high power ECRH discharge. B_t = 2.42 T, n_e = 1.7 × 10^{19} /m^3, P_{ECRH} = 1.3 MW and I_P = 300 kA. From the top down: line average electron density N_e, plasma current I_p and plasma loop voltage V_{loop}, toroidal magnetic field B_t, T_e from ECE, T_e from Thomson Scattering, respectively

In perturbative transport study, the response of the steady-state plasma to a small perturbation will be investigated. Usually, the sudden change of the central part of the T_e profile as a result of the sawtooth activity is used as initial perturbation. By assuming that the relaxation process of the perturbed T_e profile is governed by a simple diffusion, the heat diffusivity can be obtained by the following a simplified relation [18]

$$\chi_e^{hp} = (r^2 - r^2_{mix}) / (8\, t_p) \quad (2)$$

where r is the radius at which the Te perturbation is measured, r_{mix} is the sawtooth mixing radius and t_p is the time at which ΔT_e reaches its maximum. In this article, χ_e^{hp} is studied at r/a = 2/3 in the confinement region, where corresponds to that evaluated by the power balance analysis.

The power balance analysis shows that the χ_e before ECRH, $\chi_e^{ohm\text{-}pb}$ is 0.7 ± 0.3 m^2/s, which is more than 2 times lower than that with the high power ECRH: $\chi_e^{ECRH\text{-}pb}$ = 1.8 ± 0.4 m^2/s. The χ_e^{hp} estimated from the heat pulse propagation in ohmic discharges is 2.0 ± 0.5 m^2/s. During ECRH, it increases to 3.5 ± 0.8 m^2/s. In figure 2(a), χ_e^{pb} is plotted versus input the ECRH power and it increases with the increasing power. As shows in figure 2(b), the heat pulse propagation studying show the similar result. It is a selection of divertor discharges with I_p = 350 kA, B_t = 2.31 T, n_e = (2-3) × 10^{19} m^{-3}. As can be seen from figure 3, both χ_e^{pb} and χ_e^{hp} decrease with the electron density.

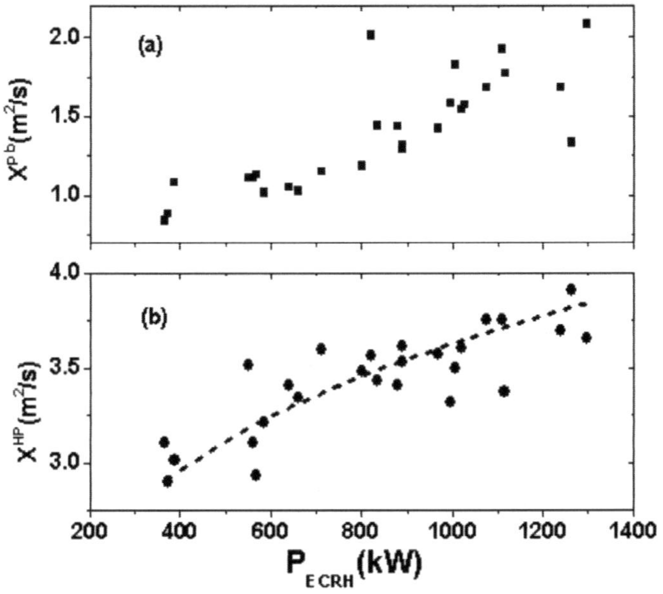

Figure 2. (a) χ_e^{pb} as a function of input ECRH power P_{ECRH}. (b) χ_e^{hp} versus ECRH power P_{ECRH}. The dashed line in (b) represents the relation between χ_e^{hp} and P_{ECRH}. A selection of divertor discharge with I_p = 350 kA, B_t = 2.3 T, n_e = (2-3) x 10^{19} m^{-3} .

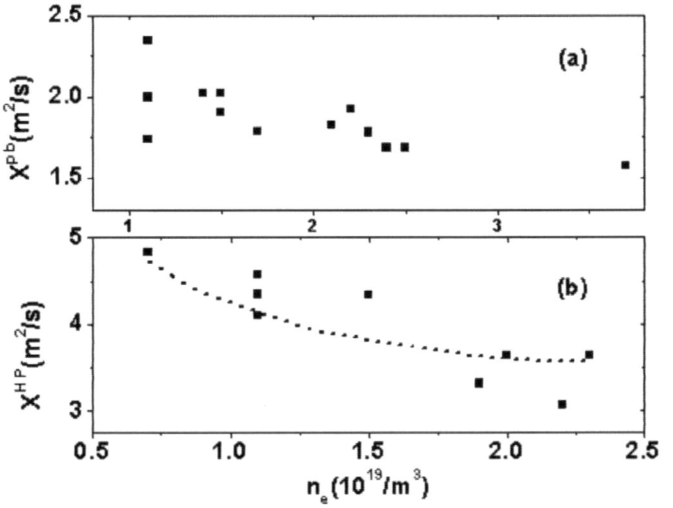

Figure 3. (a) Measured χ_e^{pb} values as a function of electron density n_e. (b) χ_e^{hp} as a function of electron density n_e, with P_{ECRH} ≈ 1.3 MW, I_p = 350 KA, B_t = 2.3 T. The dashed line in (b) represents the relation between χ_e^{hp} and n_e.

A discrepancy appears between the two methods: in Ohmic discharges the χ_e^{pb} derived from power balance analysis is lower than the χ_e^{hp} derived from heat pulse propagation analysis by a factor of 2.8. From figure 4, it can be found that the factor will increase to 3.3 with lower power ECRH (400 kW - 700 kW). However the factor deceases with the high power ECRH and it decreases to 1.7 when the power is higher than 1 MW. This discrepancy has been reported in many other tokamak and ranges from a factor of 1 to 5 [19, 20, 21], a factor of 10 even was reported in TFTR [22].

In HL-2A, a threshold exists in the normalized gradient length R/L_T ~ 12.5 and a "medium" level of profile stiffness have been observed, as shown in figure 5. The word "medium" is used to differentiate from the "strong" stiffness profile whose gradients are very close to the threshold everywhere. According to the empirical model for electron heat transport based on the existence of a threshold [23], it is not difficult to understand the heat transport increasing with the ECRH power in HL-2A. In order

Journal of Physics: Conference Series **123** (2008) 012016 doi:10.1088/1742-6596/123/1/012016

to keep temperature profile "stiff", once the auxiliary heating power increases, the heat transport increases to prevent the temperature gradient exceeding the threshold. From the results of the heat pulse propagation analysis in HL-2A, the dependence $\chi_e \propto P^{0.4}$ and $\chi_e \propto n_e^{-1}$ (see figure 2(b) and figure 3(b)) are consistent with the results of other tokamaks, which predicts $\chi_e \propto P^{0 \text{ to } 0.5}$ and $\chi_e \propto n_e^{-(0 \text{ to } 1)}$ [24].

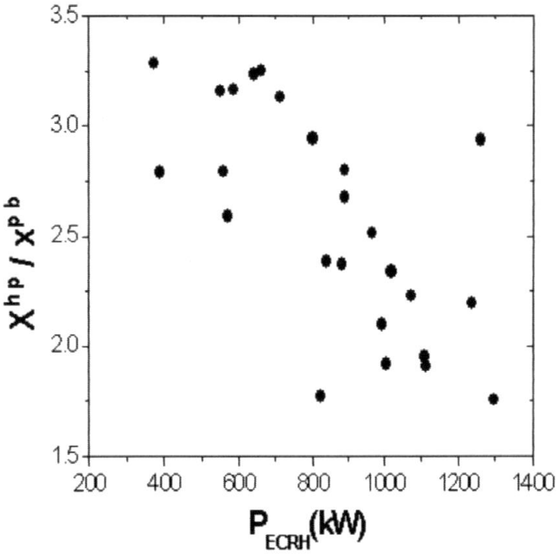

Figure 4. The ratio of χ^{pb} and χ^{hp} versus the ECRH power, with I_p = 350 KA, B_t = 2.31 T, n_e = (2-3) × 10^{19} m^{-3}.

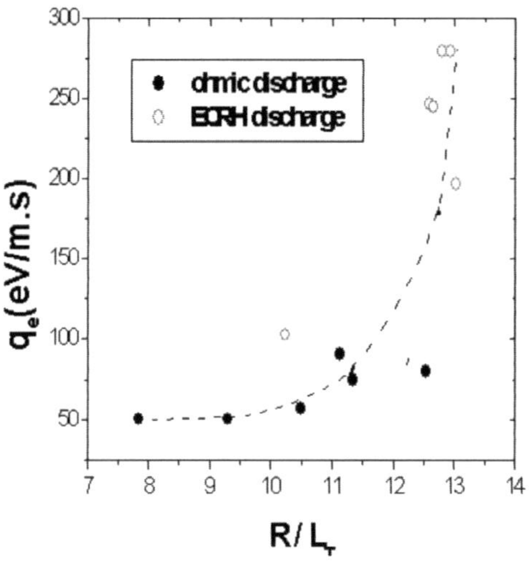

Figure 5. Electron heat flux versus the critical versus the normalized gradient length R/L_T.

3.2. Non-local phenomena triggered by SMBI during ECRH

A typical shot of SMBI experiment which induces the non-local effect is shown in figure 6, with B_t = 2.31 T, n_e = 1.0 × 10^{19} m^{-3}, I_p = 250 kA, P_{ECRH} = 800 kW, t_{ECRH} = 600 - 900 ms, t_{SMBI} = 810 ms. From the time evolution of T_e at different radii, it can be found that after the injection of SMB, the region outside r ~ 20 cm is cooled by it, while the core T_e increases. The core T_e is increased by about 18 % and the duration of the process is about 40 ms. From the temperature profile evolution, the reverse position is about 8 cm outside the q = 1 surface, as shown in figure 7. The inverse position is several centimetres outside the q = 1 surface, the calculation of EFIT shows that the position is inside the q = 2 surface.

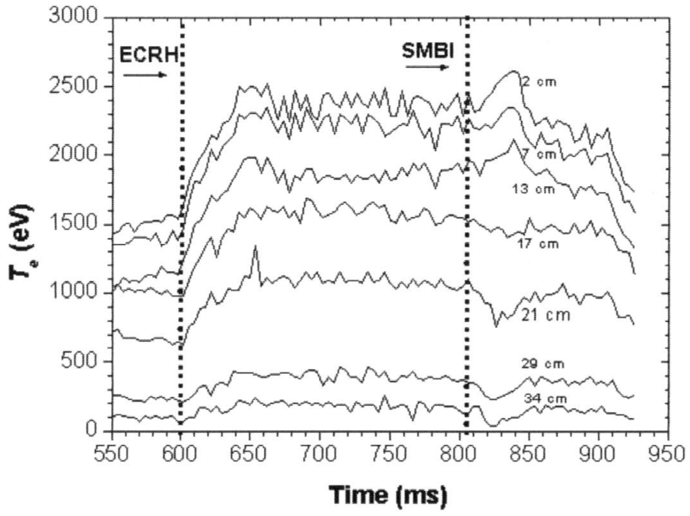

Figure 6. ECE T_e time traces for shot 6234 at several radii with SMBI during ECRH. (B_t = 2.31 T, n_e = 1.0 × 10^{19} m^{-3}, I_p = 250 kA, P_{ECRH} = 800 kW, t_{SMBI} = 810 ms.)

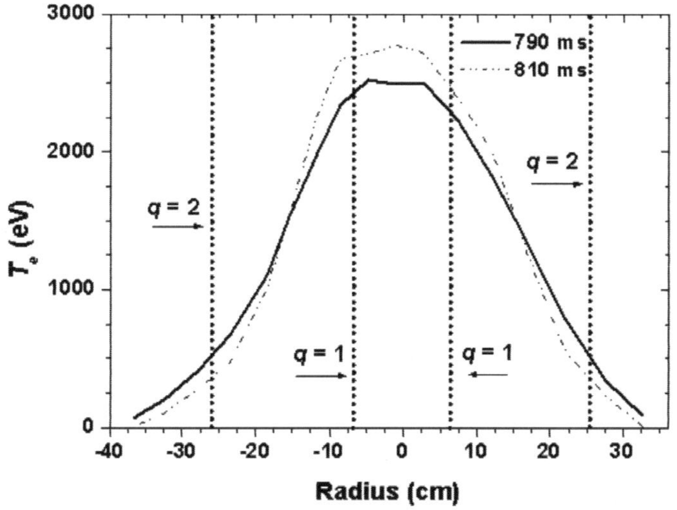

Figure 7. Evolution of electron temperature profile of shot 6234. The solid line represents the time before SMBI and the dashed is the time after SMBI.

A strong dependence of the non-local effect induced by SMBI on plasma density is observed: the effect appears only in a low density plasma, the most discharges are around 1.0 × 10^{19} m^{-3} and it disappears when the density is larger than 2.0 × 10^{19} m^{-3}. In figure 8, three SMBs are injected before ECRH, during ECRH and after ECRH, respectively. The density before the injection of the first SMB almost equals to that of the second, the density before the last injection is much lower than the former two, as shown in figure 8(b). The non-local effect appears after the injection of the latter two SMBs, while the first one doesn't induce such an effect. The effect always occurs with low density and the density limit becomes larger during the ECRH.

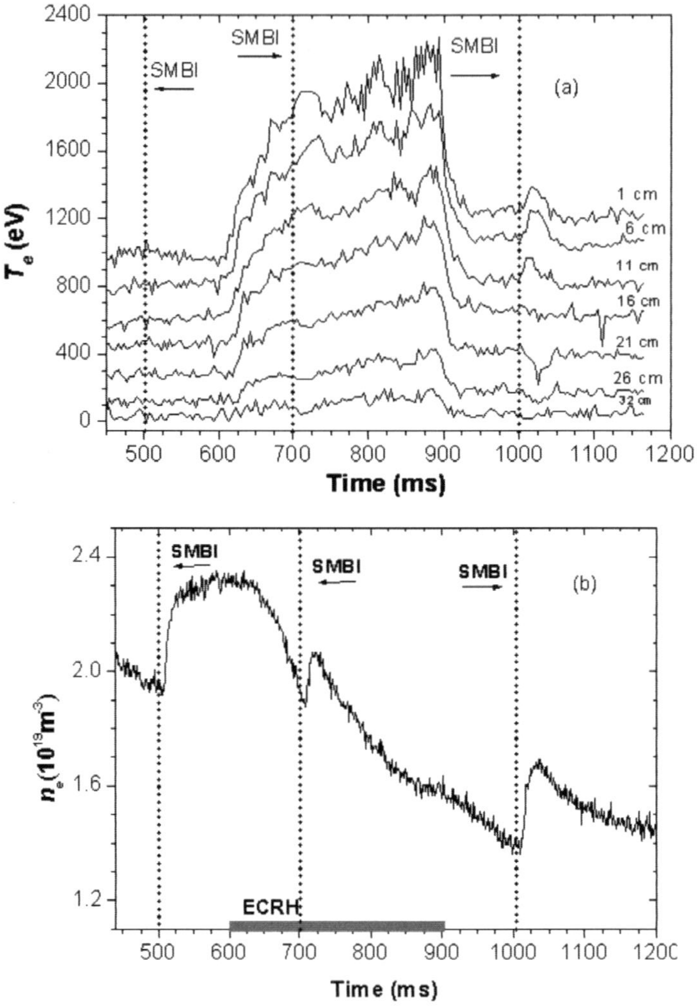

Figure 8. (a) ECE T_e time traces for shot 6351 with three SMB injections, with $B_t = 2.36$ T, $I_p = 300$ kA, $P_{ECRH} = 800$ kW. (b) the density time evolution in shot 6351.

In comparison with present-day experiments on different tokamaks, the reported results are quite different with various tokamaks. The non-local effect is enhanced by ECRH in HL-2A. The auxiliary heating (with neutral beam injection in TFTR and ECRH in TEXT) progressively weakens the non-local effect, while the effect is significantly enhanced in the presence of ECRH in RTP and of LHCD in Tore Supra. In addition, the effect induced by SMBI in HL-2A lasts much longer than that induced by pellet injection in other similar size tokomak: the duration of HL-2A is about 40 ms, while it is only 5 ms in RTP.

Improved particle confinement and decreasing of the thermal radiation have been observed during the non-local effect. In figure 9, the bolometer channels represent radiation loss in low field side and high field side, respectively. The H_α channels represent the H_α emission close to (the northern channel) and far from (the eastern channel) the injection port. It can be found that both the bolometer radiation and the H_α emission decrease when the non-local effect appears. Meanwhile, the electron density significantly increases after the injection. It is clear that the plasma confinement has been improved.

This suggests that the injection of SMB triggers internal transport barriers, which weakens the inward propagation of the cold pulse and the outward heat diffusion. Further studies will be conducted in the coming experiment in HL-2A.

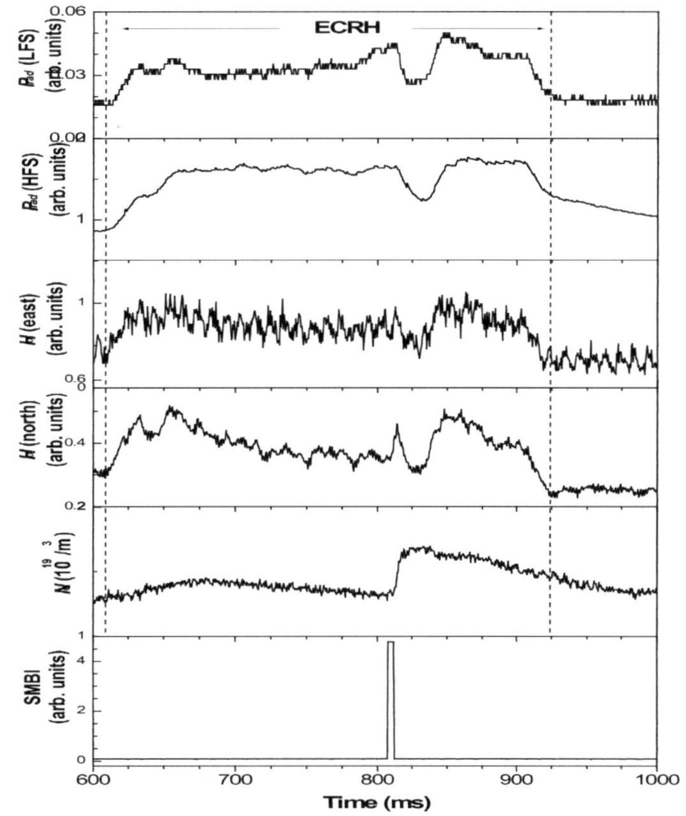

Figure 9. Parameters evolution in shot 6234 with SMBI during ECRH. From up to down: the low field side bolometer, the high field side bolometer, the east H_α, the north H_α, the electron density, the SMBI signal.

Regarding the so-called non-local effect, various models have been proposed, from empirical to theory based. There is a non-local thermal conduction model [25] which is consistent with the electron temperature profile consistency and related to the local thermal conductivity at the location $r/a \approx 0.7$. It is assumed that the cold pulse will induce the non-local effect once it deposites at that position. The SMB with directional motion and uniform speed can provide good localization and penetration into a certain region [16]. So the SMBI can induce the non-local effect, while the normal gas puffing can't.

4. Summary

From the results of the heat pulse propagation analysis in high power O-mode ECRH discharges of HL-2A, the dependence $\chi_e \propto P^{0.4}$ and $\chi_e \propto n_e^{-1}$ are consistent with L-mode scaling, which predicts $\chi_e \propto P^{0 \text{ to } 0.5}$ and $\chi_e \propto n_e^{-(0 \text{ to } 1)}$. The ratio χ^{hp}/χ^{pb} is about 2.8 in ohmic discharge, it increases to 3.3 with low power ECRH and decreases to 1.7 with high power ECRH. So the diagonal matrix element in ECRH discharges is more important than that in ohmic discharges. A 'medium' profile stiffness has been observed In HL-2A plasma, the threshold of critical gradient length is about 12.5.

With low density, the non-local effect of edge cooling has been induced by the injection of SMB in the ECRH regime and the ohmic regime. In both regimes, the existence of a threshold on plasma density has been observed. The density limit in ECRH regime is apparently higher than that in the ohmic regime. After the appearance of the non-local effect, it suggests that an electron transport barrier has been probably formed at the position just outside the $q = 1$ surface and plasma confinement has been improved.

Acknowledgments
The authors would like to thank Dr. X.L. Zhou and J.Q. Dong, for helpful advice and assistance on physical analysis, as well as other colleague for providing the necessary measurement data.

References

[1] Ryter F, Leuterer F, Pereverzev G, Fahrbach H U, Stober J, Suttrop W and ASDEX Upgrade Team 2001 *Phys. Rev. Lett.* **86** 2325
[2] Ryter F *et al.* 2006 *Plasma Phys. Control. Fusion* **48** B453
[3] Garbet B *et al.* 2004 *Plasma Phys. Control. Fusion* **46** B557
[4] Jacchia A, De Luca F, Cirant S, Sozzi C, Bracco G, Bruschi A, Buratti P, Podda S and Tudisco O 2002 *Nucl. Fusion* **42** 1116
[5] Hoang G T, Bourdelle C, Garbet X, Giruzzi G, Aniel T, Ottaviani M, Horton W, Zhu P and Budny R V 2001 *Phys. Rev. Lett.* **87** 125001
[6] Ryter F, Imbeaux F, Leuterer F, Fahrbach H U, Suttrop W and ASDEX Upgrade Team 2001 *Phys. Rev. Lett.* **86** 5498
[7] Gentle K W *et al.* 1997 *Phys. Plasmas* **4** 3599
[8] Kissick M W, Callen J D and Fredrickson E D 1998 *Nucl. Fusion* **38** 821
[9] Hogeweij G M D, Mantica P, Gorini G, Kloe J D, Lopes Cardozo N J and the RTP Team 2000 *Plasma Phys. Control. Fusion* **42** 1137
[10] Zou X L *et al.* 2000 *Plasma Phys. Control. Fusion* **42** 1067
[11] Mantica P, Galli P, Gorini G, Hogeweij G M, Kloe J D, Lopes Cardozo N J and RTP Team1999 *Phys. Rev. Lett.* **82** 5048
[12] Liu Y *et al.* 2005 *Nucl. Fusion* **45** S239
[13] Zhou Jun *et al.* 2006 *Plasma Sci. Technol.* **8** 344
[14] YAO Lianghua *et al.* 1998 *Nucl. Fusion* **38** 631
[15] Ding X T *et al.* 2006 *Review of Scientific Instruments* **77** 10F528
[16] Sun H J, Duan X R, Li W, Yao L H, Lu J, Xiao W W, Pan Y D and Ding X T 2006 *33th EPS Conf. on Controlled Fusion and Plasma Physics* **301** P-2.173
[17] Tubbing B J D *et al.* 1987 *Nucl. Fusion* **27** 1843
[18] Soler M, Callen J D 1979 *Nucl. Fusion* **19** 703
[19] Bell J D, Dunlap J L, Pare V K, Callen J D, Howe H C, Lazarus E A, Murakmi M and Thoms C E 1984 *Nucl. Fusion* **24** 997
[20] Sillen R M J, Piekaar H W, Oyevaar T, Gorbunov E P, Bagdasarov A A and Vasin N L 1986 *Nucl. Fusion* **26** 303
[21] Tubbing B J D, Barbian E, Campbell D J, Hugenholtz, C A J, Niestadt R M, Oyevaar T and Piekaar H W 1985 *12th EPS Conf. on Controlled Fusion and Plasma Physics* **9F-I** p215
[22] Fredrickson E D, Callen J D and Mcguire K 1986 *Nucl. Fusion* **26** 849
[23] Lopes Cardozo N J 1995 *Plasma Phys. Control. Fusion* **37** 799
[24] Garbet B *et al.* 2004 *Plasma Phys. Control. Fusion* **46** 1351
[25] Qu W X *et al.* 2000 *Nuclear Fusion and Plasma Physics in Chinese* **20** 129

Dependence of the H-mode power threshold on toroidal plasma rotation in the DIII-D tokamak

P Gohil[1], G R McKee[2], D Schlossberg[2], L Schmitz[3] and G Wang[3]

Presenting author's e-mail: gohil@fusion.gat.com
[1]General Atomics, P.O. Box 85608, San Diego, California 92186-5608 USA
[2]University of Wisconsin-Madison, Madison, Wisconsin USA
[3]University of California-Los Angeles, Los Angeles, California USA

E-mail: gohil@fusion.gat.com

Abstract. The power required to induce the transition from L-mode to H-mode plasmas (L-H transition) is dependent on the applied beam torque and the edge toroidal plasma rotation. For upper single null discharges in which the ion grad-B drift is away from the X-point, the L-H transition power threshold is reduced by nearly a factor of 3 by changing from predominantly co-injection (~6 MW) to predominantly counter-injection (~2.2 MW). A similar, but less prominent dependence is observed in lower single null discharges, in which the ion grad-B drift is towards the X-point, where the power threshold is reduced by nearly a factor of 2. Over a similar torque range of 2.0 Nm to –0.5 Nm, the USN discharges exhibit a reduction in threshold power of over 50% and the LSN cases show a power reduction of about 30%. The threshold power decreases with decreasing edge toroidal rotation and, at low edge toroidal rotation, large changes in the shear of the poloidal velocity of the edge turbulent eddies are observed prior to the L-H transition that may be strong enough to induce the transition. For the first time, the L-H transition has been induced at constant input power below the nominal threshold power by reducing the input torque from all co-beams to balanced beams. The variation in threshold power with applied torque is not found to be dependent on the ion orbit losses in the plasma.

1. Introduction

Understanding the influence of the plasma rotation and its shear on the physics of H-mode plasmas is important for increasing the performance of these plasmas. This is of particular relevance to future fusion devices, such as ITER, for which operational and performance predictions are being extrapolated and scaled from present day experimental results. The ability to produce high performance H-mode plasmas in these future devices will depend on knowledge of how the H-mode power threshold varies as a function of various plasma parameters and configurations [1]. Investigations in this area have been advanced at DIII-D following major modifications to the neutral beam injection (NBI) system to allow for simultaneous NBI in both the co and counter directions to the direction of the plasma current. Subsequently, it is possible to vary the input NBI power to the plasma for fixed applied beam torque and, conversely, vary the applied NBI torque at constant input power. The present experiments at DIII-D have been performed for plasma configurations in which the ion grad-B drift direction is away from the dominant X-point in the plasma and also for cases in which the ion grad-B drift direction is towards the dominant X-point. This is because studies have shown that the direction of the ion grad-B drift with respect to the X-point affects the H-mode power threshold [2]. It should be noted that the H-mode power threshold scaling formulae derived to date for the most favourable ion grad B drift case (i.e. towards the X-point) involve global plasma parameters [1,3]. However, H-mode plasmas are the result of the formation of a transport barrier at the edge of the plasma, which is susceptible to and affected by the local conditions at the plasma edge, especially with regard to quantities such as the edge radial electric field [4,5]. The experiments described in this paper have focussed on detailed measurements of plasma edge quantities in order to resolve the physics behind the H-mode power threshold behaviour. In particular, these include high spatially resolved measurements of the edge toroidal and poloidal rotation, the radial electric field and the edge plasma turbulence.

In the present experiments, the power required to induce the transition from L-mode to H-mode plasmas is found to be dependent on the applied beam torque and the edge toroidal plasma rotation. For upper single null discharges in which the ion grad-B drift is away from the X-point, the L-H transition power threshold is reduced by nearly a factor of 3 by changing from predominantly co-injection (~6 MW) to predominantly counter-injection (~2.2 MW). A similar, but less prominent dependence is observed in lower single null discharges, in which the ion grad-B drift is towards the X-point, where the power threshold is reduced by nearly a factor of 2, from ~3 MW with predominantly co-injected beams to ~1.7 MW with counter-injected beams. Further, at low plasma rotation, edge turbulence measurements indicate the presence of a significant shear in the poloidal rotation of the edge turbulent eddies in the L-mode, which may facilitate the transition to H-mode at low power values. These experiments also investigated the hypothesis that fast ion orbit differences between co and counter injected beams may account for the observed differences in power threshold. Ions born from co-current beams are better confined and localized more towards the core compared with counter-current beam ions. This can lead to changes in the edge potential and electric field, which may facilitate the H-mode transition at low power levels with counter beam injection. However, the experimental results show that prompt ion orbit losses do not affect the H-mode threshold power. Overall, the above results pose significant implications for the low toroidal rotating plasmas expected in ITER for which the assumptions and extrapolations for H-mode threshold power need to be re-examined in light of these torque and plasma rotation dependencies.

This paper describes experiments in which the H-mode threshold power has been measured for different values of the applied beam torque and for different plasma configurations. Section 2 describes details of how the H-mode threshold power varies with the applied beam torque and the edge toroidal rotation. Section 3 describes experiments in which, for the first time, the H-mode transition has been induced by performing a scan in the applied beam torque at constant input power. Section 4 contains details on how the plasma turbulence is affected as a function of the input torque and section five describes the effects of fast ion losses on the transition. The summary and conclusions are presented in Section 6.

2. Threshold power dependence on edge toroidal rotation

These experiments were performed in deuterium plasmas with a plasma current, I_p, of 1.0 MA, a toroidal magnetic field, B_T, of 2.0 T, an elongation, κ, of 1.7, an edge safety factor, q_{95} , of around 5 and a line averaged density in L-mode of 2.4x 10^{19} m^{-3}. The plasma configurations used were upper single null (USN) and lower single null (LSN) diverted plasmas in which the magnetic geometry has a null at both vertical ends, but with the plasma biased upwards or downwards by +/->5 cm, so making the upper or lower null active, respectively. In this manner, these plasmas could be described as unbalanced double-null discharges with either a dominant upper null or lower null. The ion grad-B drift remained in the same direction (i.e. downwards) for both configurations (USN and LSN). Correspondingly, the USN discharges had the ion grad-B drift away from the active null or X-point and the LSN discharges had the ion grad-B drift towards the X-point. The plasma configurations were not optimized to make use of the divertor cryopumps.

The present NBI system on DIII-D consists of seven horizontal beams, five of which are injected in the counter-clockwise direction (when viewed from above), and two beams which are directed in the opposite (i.e. clockwise) direction [6]. For the normal counter-clockwise direction of the plasma current in DIII-D, this meant that the five co-directed sources were in the same direction as the plasma current and two beams were injected counter to the plasma current. All of the beams are injected with their centerlines being near the nominal midplane of the plasma. Four of the beams (3 co and 1 counter) are injected at 47 degrees to the tangential plane of the magnetic axis of the plasma at the point of intersection and shall be referred to as tangential beams in the rest of the paper. Three of the beams (2 co and 1 counter) are injected at 63 degrees to the tangential plane and will be referred to as perpendicular beams. The beam acceleration voltages were varied from 81 kV to 60 kV during the experiments (between shots) so that the power per beam could be changed if less power was required.

The initial set of conditions investigated were with USN plasmas, which have higher L-H power thresholds compared to LSN plasmas at similar conditions [7,8]. The total beam power was increased in incremental steps of 0.25 of a beam source, i.e., between 400-700 kW/step, depending on the beam voltage settings. Note that ion orbit losses were not taken into account in the NBI power estimations and the quoted NBI powers are the values entering the vessel. The power steps were achieved by modulating the beams at various duty cycles of maximum cycle time of 40 ms, so that each 0.25 step corresponded to an additional 10 ms within the 40 ms cycle window. This modulation time scale is less than the fast-ion slowing-down time and so any instantaneous variation of the beam power is averaged out over a longer time scale by the plasma. The duration of each power step was 320 ms, which was many confinement times, so allowing the plasma to respond to the time-averaged power increases. This also allowed enough time for high-quality fluctuation measurements to be obtained. The power steps were applied whilst maintaining a given value of plasma torque. This was achieved by balancing out the torque from a co beam by injecting the same (but opposite) torque from a counter beam. Increases in tangential co beams were balanced out by the same increase in counter tangential beams and similarly for the perpendicular beams. In this manner, a power scan could be started at a given value of applied torque, which was then maintained during successive increasing steps in the beam power. This allowed for a more accurate determination of the threshold power without any variation due to a changing beam torque during the scan. The uncertainty in the torque estimations was around 10%. For the cases with all co-NBI or all counter-NBI, it was necessary to ramp the power and torque steps together since the two quantities are not independent. Power scans were started at a level well below the LH power threshold in the quasi-stationary phase of the discharge and then the power was ramped up, at constant torque, in 320 ms steps to trigger the LH transition.

The H-mode power threshold with the highest torque input was determined by injecting all co-beams into an USN plasma. Figure 1 shows the time histories of various plasma quantities as the beam power input is stepped up. The edge toroidal rotation, at a location of 1 cm inside the separatrix, slowly increases as the power is increased in L-mode until the H-mode transition occurs at 2150 ms with an input power of 5.9 MW and an injected torque of just over 4 Nm, as calculated from the beam geometry and the beam deposition calibrations. The co-injected beams in this case were applied with an accelerating voltage of mostly 75 kV. The edge toroidal rotation is determined from charge exchange recombination (CER) spectroscopic measurements performed on the emission from the C VI ion (λ=5290.5 A), which is produced as a result of charge exchange between heating beam atoms and fully stripped carbon atoms [9]. At the plasma edge in DIII-D, the spatial resolution of the toroidal rotation measurements (as well as poloidal rotation measurements from vertically viewing chords) is about 6 mm. Figure 1 shows that both the beam power and the beam torque continue to increase, since only co-injected beams are used and there are no counter beams to balance out any increases in torque due to the increasing the co-injected power. The transition, as indicated by the sudden drop in the edge recycling D_α signal, occurs at an edge toroidal rotation of about 25 km s^{-1}. Figure 2 shows the H-mode transition for the case of increased applied power at constant input torque. The power is slowly increased at near zero applied torque [figure 2(b)] from 1200 ms until the transition occurs at 1506 ms. Both sets of beams (co-injected and counter-injected) in this case were operated at 60 kV. The H-mode transition at zero torque occurs at about 3 MW. Note that the edge toroidal rotation measured about 1 cm inside the separatrix is much lower than the all co-injected case and is around 5 km s^{-1}.

Journal of Physics: Conference Series **123** (2008) 012017 doi:10.1088/1742-6596/123/1/012017

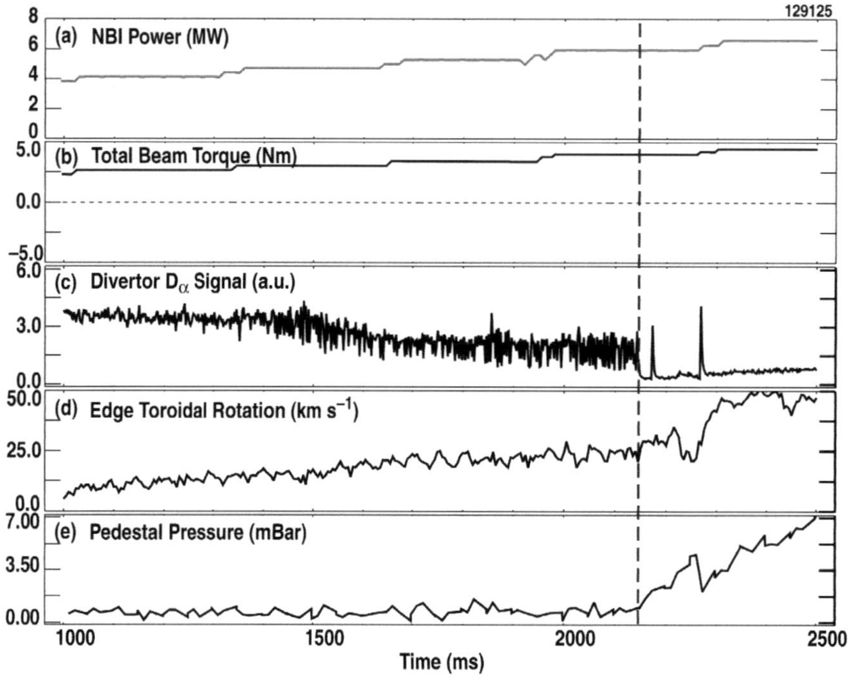

Figure 1. Time histories of various plasma quantities for a plasma discharge during the application of all co-current neutral beams. (a) Total NBI power, (b) total beam torque, (c) D_α signal from the divertor region, (d) toroidal rotation of a CER chord located 1 cm inside the separatrix, (e) electron pressure at the location of the pedestal. The time of the L-H transition is marked by the dashed vertical line.

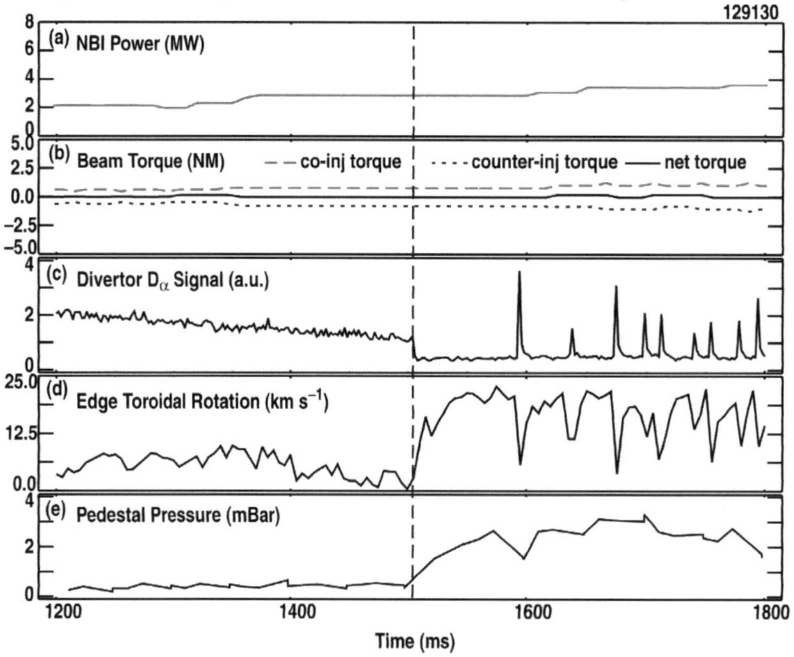

Figure 2. Time histories of various plasma quantities for a plasma discharge during the application of NBI power at net zero torque. (a) Total NBI power, (b) total beam torque together with the co- and counter-injected torque components, (c) D_α signal from the divertor region, (d) toroidal rotation of a CER chord located 1 cm inside the separatrix, (e) electron pressure at the location of the pedestal. The time of the L-H transition is marked by the dashed vertical line.

Note that he target plasmas (i.e. before the application of beams) for this cases were repeat discharges at the same operational parameters (i.e. Ip, Bt, target density, etc). The only conditions that were varied were the parameters for the NB input, such as power and torque. The Da signals for these repeat discharges were identical before the time of first beam application. Then any variation in the Da signal from discharge to discharge is the result of the combination of beams used and the affect of the applied power and torque.

The H-mode power threshold was obtained for many cases with increasing power steps at constant applied beam torque, but with different starting torques for the same set of plasma parameters and conditions. Only the plasma configuration was changed from USN to LSN plasmas. The results of these scans are shown in figure 3. The H-mode power threshold exhibits a clear and significant increase with the applied beam torque of nearly a factor of 3 from about 2.2 MW with predominantly counter-injected beams (applied torque of -0.5 Nm) to about 6 MW with all co-injected beams (applied torque of about 4.1 Nm). An increase in threshold power with applied beam torque is also observed in LSN plasmas, but is less profound showing an increase of nearly a factor of 2 from about 1.8 MW (applied torque of -1.3 Nm) to about 3.3 MW (at applied torque of 2 Nm).

The results shown in figure 3 clearly indicate a dependence of the H-mode power threshold on the applied beam torque. However, they do not, in themselves, reveal the important plasma quantity that may be influencing the threshold power behavior. In order to determine the role of any edge quantity, a detailed analysis of various quantities at the plasma edge was performed for the many (a total of 9) scans performed for figure 3. Such quantities included the evolution of the electron and ion temperatures and densities, the toroidal and poloidal rotation and the edge electron pressure in the L-mode prior to the H-mode transition. A key edge quantity that showed a clear correlation with the threshold power was the edge toroidal rotation. Figure 4 shows the dependence of the power threshold on the toroidal plasma rotation measured 1 cm inside the location of the separatrix in the L-mode phase just before the H-mode transition. The toroidal rotation is not simply determined for one CER chord for all the cases, but is instead determined from the actual profile of the toroidal rotation at the plasma edge, from which the rotation value is taken at the location of 1 cm from the separatrix. Figure 5 shows the edge toroidal rotation profiles measured for 2 USN cases at 4 Nm and 0 Nm input torque. There appears to be a near monotonic increase in the H-mode power threshold with edge rotation at low to medium rotation values (for both the USN and LSN discharges). A flattening off in the

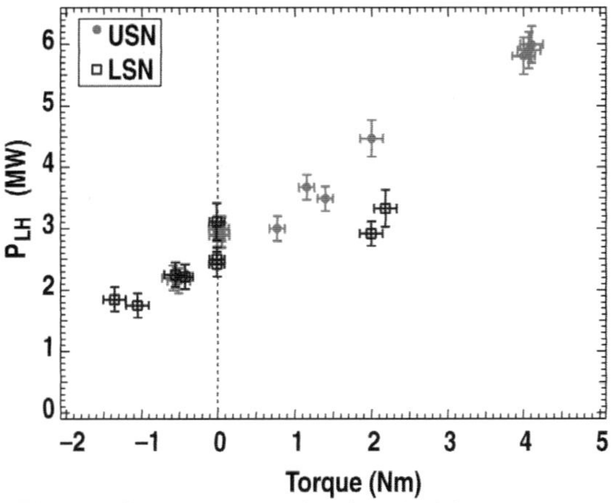

Figure 3. The H-mode threshold power as a function of the input torque from neutral beams. The dependence is shown for plasma discharges biased in the upwards direction (USN) and for plasmas biased downwards (LSN).

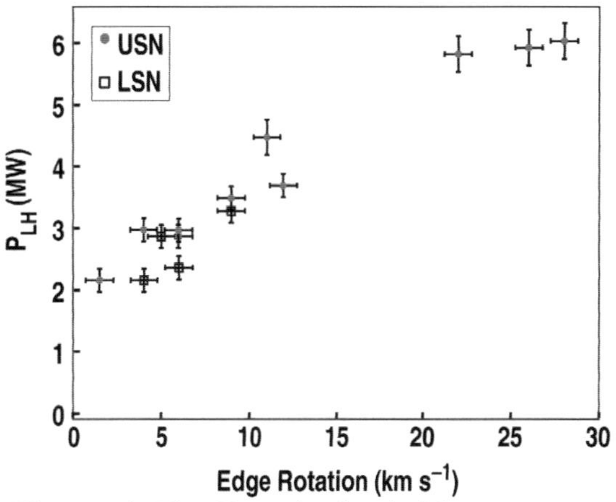

Figure 4. The H-mode threshold power as a function of edge toroidal rotation. The toroidal rotation is measured 1 cm from the separatrix, inside the plasma. The dependence is shown for both USN and LSN plasmas.

dependence is observed at high rotation values. Note that the edge rotation is still positive for the cases with negative torque input because there is some intrinsic positive rotation in the plasma even without any beam input which is larger than the rotation induced by the negative torque. It may be said that a dependence on the toroidal rotation may not be surprising given that the main driver is the applied beam torque, which directly affects the toroidal rotation. The H-mode transport barrier is an edge barrier and the edge rotation measured 1 cm from the separatrix in the L-mode phase appears to be clearly correlated with the formation of this barrier.

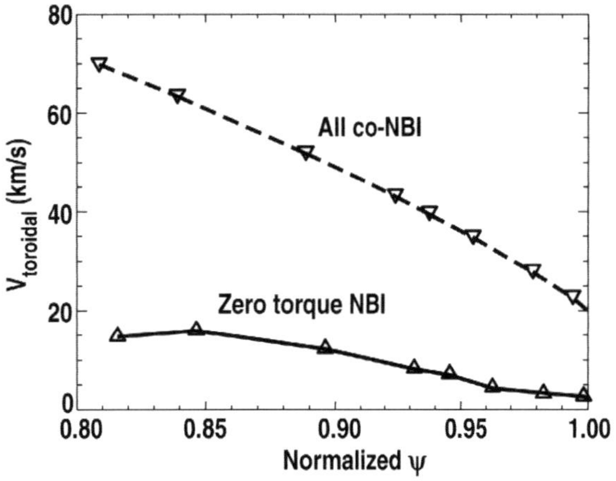

Figure 5. Edge toroidal rotation profiles just before the H-mode transition for USN discharges shown in figures 1 and 2. The location of 1 cm inside the separatrix corresponds to approximately normalized $\psi = 0.98$.

3. H-mode transition induced by torque scan at constant power

Given the strong correlation of the H-mode power threshold with the edge toroidal rotation, an attempt was made to induce the H-mode transition at power levels below the nominal threshold value by changing the applied torque during the discharge. This experiment was performed with an USN plasma, in which the power was injected into a co-rotating discharge below its power threshold i.e. in L-mode. Figure 3 shows that the H-mode threshold power at 2 Nm is about 4.5 MW. Subsequently, the total power was kept at a constant level below this value (i.e. at 3 MW) whilst the applied beam torque was changed. The initial beam configuration consisted of all co-injected beams with a total power of about 3 MW and applied torque of about 2 Nm. The applied torque was then varied during the discharge by successively applying increased counter torque (with counter-injected beams), whilst simultaneously reducing the co-injected torque (by reducing the amount of co-injected beams) and also maintaining a near constant total beam power by careful selection of the co and counter beams used. The time histories of the applied beam power and torque can be seen in figure 6. There is a very small droop in the total beam power over the time of interest as a result of a small mismatch in the net power from the various individual beams. The plasma clearly stays in L-mode during the time the net torque is in the co-direction. Then, as more counter-injected beams are applied for the given power, the edge toroidal rotation decreases and the H-mode transition occurs at about 1687 ms, at the point when the edge toroidal rotation is close to the critical rotation for the H-mode as determined from figure 4. Note that the values shown in figure 4 were determined from power scans at constant torque, whereas the rotation value at the H-mode transition shown in figure 6 is obtained through a different approach by which the rotation is varied at constant power. This result marks the first occurrence in which an H-mode transition has been produced at constant power, below the nominal threshold power, by changing the edge rotation.

A clearer understanding of the physics behind the H-mode transition described in figure 6 was obtained by more detailed analysis of the toroidal rotation in L-mode just prior to the L-H transition. Figure 7 shows an expanded time scale of figure 6, just before the H-mode transition, showing the time histories of the toroidal rotation for several CER spatial chords at various locations inside the

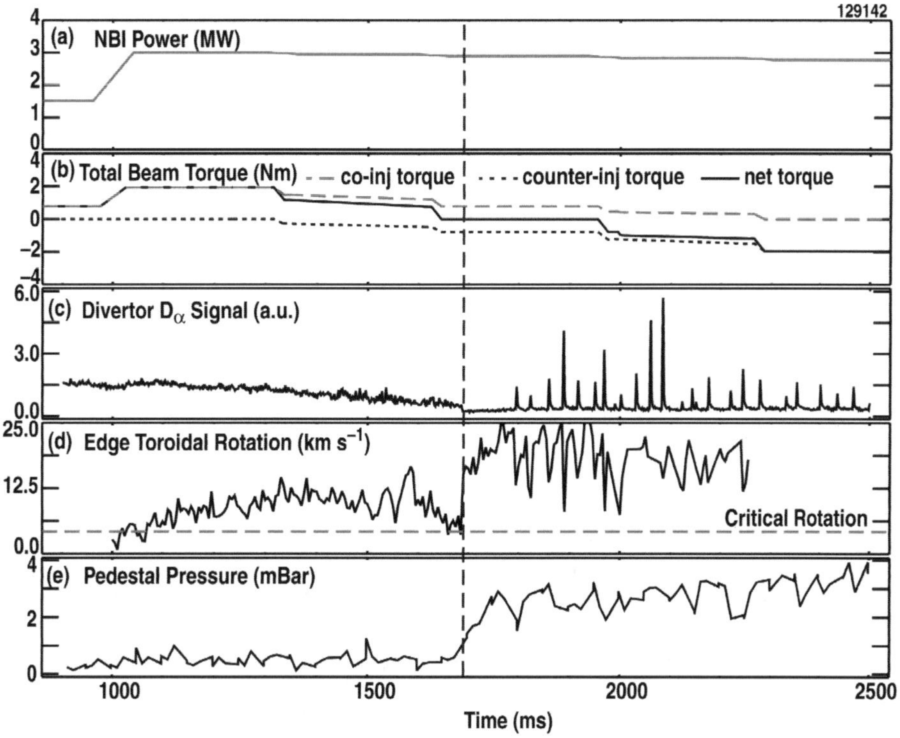

Figure 6. The time histories of various plasma quantities for a plasma discharge during the application of NBI at constant power and varying torque. (a) Total NBI power, (b) total beam torque together with co- and counter-injected torque components, (c) D_α signal from the divertor region, (d) toroidal rotation of a CER chord located 0.7 cm inside the separatrix, and (e) electron pressure at the pedestal location. The time of the L-H transition is marked by the dashed vertical line.

separatrix. At the start time of figure 7, the gradient in the edge toroidal rotation, as determined by the separation between the various toroidal chords [figure 7(c)], is small. However, the gradient in toroidal rotation increases with time, as can be seen by the increasing separation between the chords at –0.7 and –2.6 cm. This, in turn, leads to an increased shear (i.e. gradient) in the edge radial electric field, E_r, as can be seen in the increased separation of E_r between locations (c.f. at –1.0 cm and –2.3 cm) for the E_r measurements at the plasma edge [figure 7(d)]. The E_r values were determined from CER spectroscopic measurements of the edge pressure gradient and the edge toroidal and poloidal rotation velocities. The increased shear E_r in due to changes in the edge toroidal rotation may be sufficient to then cause the H-mode transition observed at 1687 ms.

A working hypothesis on the increased shear in E_r as the edge rotation is reduced is as follows: at low edge rotation, the relative influence of changes in the pressure gradient term (i.e., diamagnetic contribution) and/or poloidal flow variations are more significant to E_r and its shear than for an E_r profile dominated by a large toroidal rotation contribution. This behavior would be consistent with the decrease in the H-mode power threshold with decreasing edge rotation.

4. Plasma fluctuation measurements from Beam Emission Spectroscopy

Two-dimensional measurements of the density fluctuation spectra and spatial phase relationships were obtained near the outboard midplane in these discharges with the upgraded Beam Emission Spectroscopy diagnostic [10]. BES observes localized long-wavelength ($k_\perp \rho_I < 1$) density fluctuations by measuring the Doppler-shifted D_α beam emission intensity that is excited by collisions between beam atoms and plasma ions and electrons. Measurements were obtained with a 5 x 6 channel grid covering an approximately 4 x 7 cm region near $0.9 < r/a < 1.0$ at a spatial resolution of approximately 1 cm in the radial and 1.2 cm in the poloidal direction [11]. These time-resolved 2D measurements allow for detailed examination of turbulence and turbulence flows during the time leading up to and across the L-H transition.

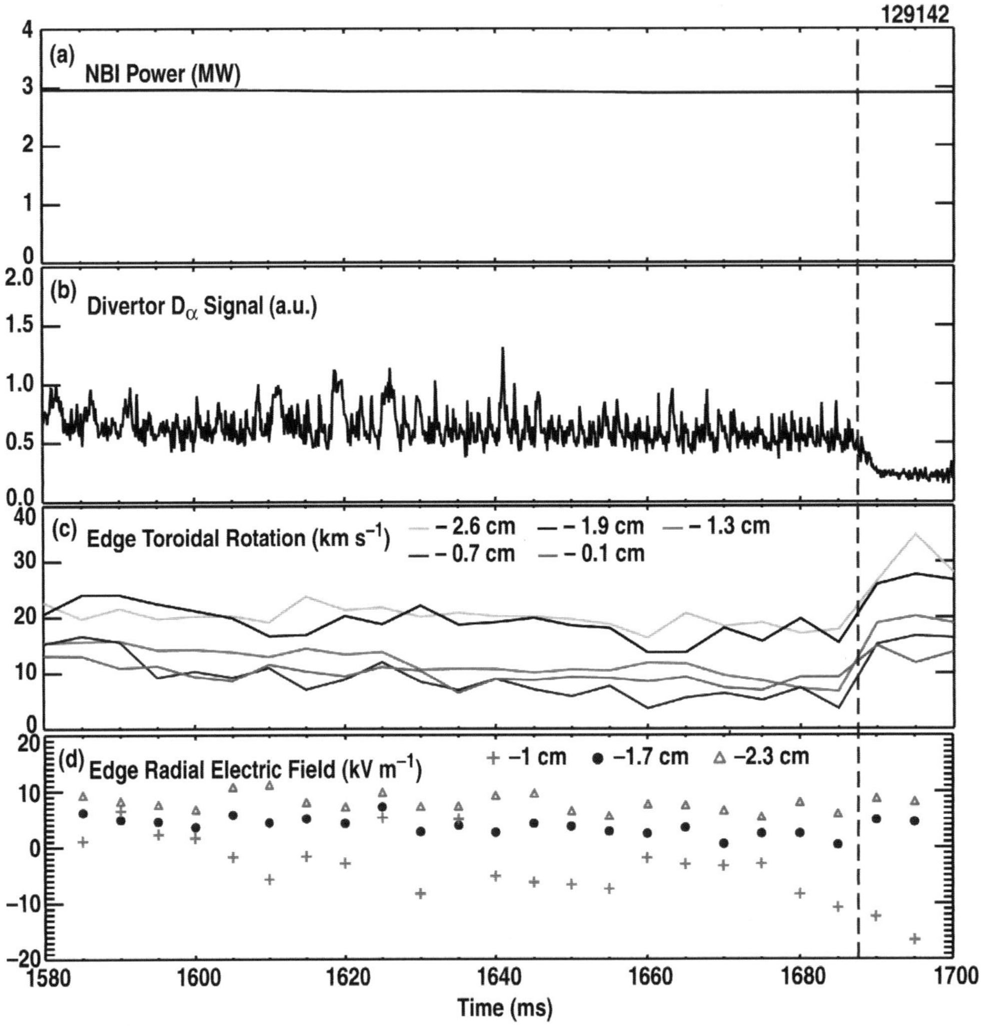

Figure 7. The time histories of various plasma quantities for the plasma discharge shown in figure 6, but with an expanded time scale. (a) The total NBI power, (b) the D_α signal from the divertor region, (c) the toroidal rotation of multiple, successive CER chords located at varying distances inside the separatrix, (d) the radial electric field as measured by successive CER chords inside the separatrix. The time of the L-H transition is marked by the dashed vertical line.

The power spectra and poloidal phase relationships are shown in figure 8. Figure 8(a) and (b) show the density fluctuation power spectra at r/a=0.9 and r/a=1.0 for a co-injected and balanced discharge, while figure 8(c) and (d) show the phase shift at r/a=0.9 and r/a=1.0, respectively, between two poloidally separated measurements (ΔZ=1.2 cm) for the same two discharges. In each case, these measurements are averaged over a 100 ms period in the L-mode phase just prior to the L-H transition. These measurements show a dramatic difference in the edge turbulence and turbulence flow patterns for these two plasma rotation conditions. In the co-rotating discharge (solid line), the spectra extend out to higher frequency, primarily due to the increased Doppler-shift for these lab-frame measurements. The frequency-integrated fluctuation amplitude is seen to be larger in the co-injection case. This is consistent with the significantly higher power level injected into this discharge in this phase just before the L-H transition.

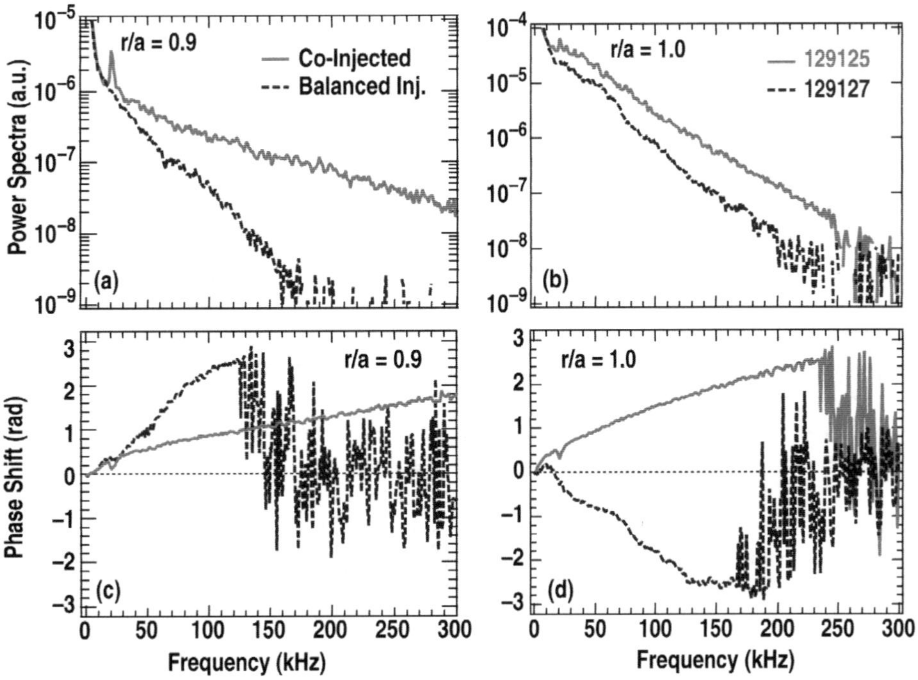

Figure 8. Density Fluctuation power spectra and phase for BES channels at two radial locations (r/a=0.9 and 1.0). (a) and (b) Density fluctuation power spectra at r/a=0.9 and 1.0, respectively, for an all co-injection USN discharge and for a balanced injection USN discharge. (c) and (d) Phase shift at r/a=0.9 and 1.0, respectively, between two poloidally separated measurements (ΔZ=1.2 cm) for the same two discharges.

The phase measurements show a dramatic difference between the two plasma rotation conditions. In the co-injected discharge, the phase shift is seen to be monotonically increasing with frequency, reflecting the approximately uniform motion of the turbulent eddy structures as a result of ExB and diamagnetic drifts. The smaller phase shift at r/a=0.9 in the co-injected discharge compared with that at r/a=1.0 results from a higher poloidal turbulence flow velocity at the interior location. A positive phase shift indicates flow in the ion diamagnetic drift direction in the lab frame that is the same as the usual ExB direction for co-injected plasmas on DIII-D. In the case of the balanced injection discharge, the phase shift changes from positive at r/a=0.9 to negative at r/a=1.0. This indicates that the turbulence undergoes a complete flow reversal near this edge region of the plasma in the balanced-injection case. This also indicates a significant poloidal flow shear between these two radial locations.

The large difference in edge turbulence poloidal flow and in particular the large flow shear observed in the balanced injection plasma at significantly lower injection power may facilitate the L-H transition and would be consistent with the developing shear in E_r in the L-mode observed by the CER system at low toroidal rotation [see figure 7(d)]. Further analysis is underway to compare the ExB shearing rates and decorrelation rates for these plasmas.

5. Effects of fast ion orbit losses

One hypothesis for the reduced H-mode power threshold with decreasing torque may be the variations in the fast ion orbit loss as a result of the different mixes of co and counter beams used. Co-current beam ions are better confined and localized more towards the plasma core compared with counter-current beam ions. For example, the fast ion confinement time can be roughly 40% greater for the co tangential injection than for the counter tangential injection from analysis using a plasma transport code [12]. Furthermore, the counter tangential beams have greater ion orbit losses than the counter perpendicular beams. Altogether, this can lead to changes in the potential and electric field and consequently the rotation profiles, independent of, or in addition to that induced by the total torque. This issue was resolved by performing experiments in which the beam power was scanned for various configurations (i.e. mixes) of co- and counter-injected beams. These experiments consisted of varying the total beam power at a fixed torque value for the following cases: (a) using the counter perpendicular beam as the baseline beam with additional power steps at net zero torque (i.e. with balanced co and counter tangential beams); (b) repeating the power scan, but now using the counter tangential beam as the baseline beam with additional power steps at zero torque (i.e. with balanced co and counter perpendicular beams); (c) performing a power scan at net zero torque throughout by using only a combination of the counter perpendicular beam with a co- perpendicular source (i.e. net zero torque) or increasing the power using only a combination of the counter tangential beam with a co tangential beam (i.e. net zero torque) will vary the injection into the loss cone. The relative losses between the perpendicular and tangential beam pairings are between 10–15%.

The results from the above scans are summarized in Table 1, which shows the power required to produce the H-mode transition as well as the time of the transition for different scans and different combinations of tangential beams (co and counter) and perpendicular beams (co and counter). If there is an influence from the fast ion losses as a result of using the tangential beams compared to the perpendicular beams, then this effect would show up in a reduced power for the tangential beams compared to the perpendicular beams and also in an earlier time for the transition given that the times of the power steps were the same for both the tangential and perpendicular beams. However, as can be clearly seen in Table 1, it appears that there is no significant difference in the H-mode power threshold or the timing of the transition (of roughly 100 ms) between the tangential and perpendicular beams. Furthermore, detailed analysis of the radial electric field at the plasma edge revealed no clear differences between the different sets of beams for the various scans performed for Table 1. Therefore, the variation in the threshold power with applied torque appears not to be dependent on the ion orbit losses in the plasma.

Table 1. The L-H transition power and time for various neutral beam configurations. The various scans consist of power scans at constant torque (for both USN and LSN discharges) and a torque scan at constant power (USN discharge). The tangential beams are a combination of co- and counter-injected tangential beams and the perpendicular beams represent a combination of co- and counter-injected perpendicular beams.

	Tangential Beams		Perpendicular Beams	
Experiment	**L-H Transition Power (MW)**	**L-H Transition Time (ms)**	**L-H Transition Power (MW)**	**L-H Transition Time (ms)**
Power scan at near zero torque (USN)	2.9	1460	3.0	1484
Power scan at near zero torque (LSN)	2.4	1616	2.4	1684
Power scan at torque –0.6 Nm (USN)	2.2	1826	2.2	1719
Power scan at torque –0.5 Nm (LSN)	2.2	1721	2.2	1747
Torque scan at power 3 MW (USN)	3.0	1694	2.9	1687

6. Summary

The power required to induce the H-mode transition is dependent on the applied beam torque and, more specifically, the edge toroidal plasma rotation. For upper single null discharges in which the ion grad-B drift is away from the X-point, the L-H transition power threshold is reduced by nearly a factor of 3 by changing from predominantly co-injection (~6 MW) to predominantly counter-injection (~2.2 MW). A similar, but less prominent dependence is observed in lower single null discharges, in which the ion grad-B drift is towards the X-point, where the power threshold is reduced by nearly a factor of 2. The threshold power decreases with decreasing edge toroidal rotation. For the first time, the L-H transition has been induced at constant input power below the nominal threshold power by reducing the input torque from all co-beams to balanced beams. At low toroidal rotation, a shear in E_r develops in the L-mode as a result of an increased gradient in the toroidal rotation just prior to the L-H transition and large changes in the poloidal velocity shear of the edge turbulent eddies are observed prior to the L-H transition that may be strong enough to induce the transition. The variation in threshold power with applied torque is not found to be dependent on the ion orbit losses in the plasma. The implications for ITER are favourable given the low rotation plasmas expected in ITER. However, the assumptions and extrapolations of the H-mode threshold power scaling laws to ITER need to be re-examined in light of these torque and plasma rotation dependencies. An important aspect of these experiments is to consider how these results compare to auxiliary heating methods that have no torque inputs such as RF heating. This will be especially relevant for ITER, which will have a significant amount of its total auxiliary heating power been applied through RF heating. Future experiments at DIII-D will be aimed at comparing these results with predominantly RF heated discharges for the formation of the H-mode barrier.

References

[1] "Progress in the ITER Physics Basis" 2007 *Nucl Fusion* **47** S1
[2] Carlstrom T N et al 1997 *Proc 24th EPS Conference on Controlled Fusion and Plasma Physics (Berchtesgaden)* Vol 21A, Pt III (EPS) p 1089
[3] Takizuka T for the ITPA H-mode Power Threshold Working Group 2004 *Plasma Phys Control Fusion* **46** A227
[4] Groebner R J et al 1990 *Phys Rev Lett* **64** 3915
[5] Gohil P et al 1998 *Nucl Fusion* **38** 93
[6] Scoville J T 2007 *Fusion Sci. Technol.* **52** 398
[7] Ryter F and H-mode Threshold Database Group 2002 *Plasma Phys Control Fusion* **44** A415
[8] Luce T et al 2006 *Proc 21st IAEA Fusion Energy Conference (Chengdu)* PD-3
[9] Gohil P et al 1992 *Proc 14th Symposium on Fusion Engineering (San Diego)* Vol 2 (IEEE) p 1199
[10] Gupta D et al 2004 *Rev Sci Instrum* **75** 3493
[11] McKee G et al 2007 *Plasma and Fusion Research* accepted for publication
[12] DeGrassie J 2007 private communication

Journal of Physics: Conference Series **123** (2008) 012018 doi:10.1088/1742-6596/123/1/012018

Change of fluctuation properties during non-local temperature rise in LHD from 2d phase contrast imaging

C A Michael†, K Tanaka, L Vyacheslavov[1], A Sanin[1], K Kawahata, S Okajima[2], N Tamura, S Inagaki[3], T Fukuda[4], M Yoshinuma and M Yokoyama

National Institute for Fusion Science, 322-6 Oroshi-cho, Toki, 509-5292 Japan
† Present address: UKAEA, Abingdon, Oxfordshire OX14 3DB, United Kingdom
[1]Budker Institute for Nuclear Physics, 630090, Novosibirsk, Russia
[2]Chubu University, Kasugai, 487-8501, Japan
[3]Kyushu University, Kasuga, Fukuoka, 816-8560, Japan
[4]Osaka University, Suita, Osaka, 565-0871, Japan

E-mail: `Clive.Michael@ukaea.org.uk`

Abstract. Fluctuation properties are analyzed in the Large Helical Device (LHD) for discharges exhibiting a non-local core temperature rise in response to edge cooling by injection of a Tracer Encapsulated Solid Pellet (TESPEL). Edge ion gyro-scale density fluctuation amplitudes, measured with a 2D CO_2 phase contrast imaging system, are found to increase as the core temperature is rising and is therefore anti-correlated with a reduction of the total heat flux. This suggests that another mechanism, such as turbulence de-correlation has a stronger role on the heat flux than change of the density fluctuation level. The core fluctuation properties, in particular the phase velocity, change immediately in response to the edge temperature reduction, and possibly change of gradient, unresponsive to core gradients, suggesting a possible non-local transfer mechanism between the edge and core. Comparing different cases, it appears that the excitation of a fluctuation branch with higher core frequency and phase velocity (~6km/s) is an essential ingredient for the core T_e rise. After pellet injection, the amplitude of higher k components increases, implying a reduction of correlation length. One possible interpretation is that there may be increased $E \times B$ shear.

1. Introduction

In both Tokamaks and Heliotron/Stellarator devices, electron heat transport can exhibit peculiar behavior in some instances, such as profile stiffness [1, 2], in steady state, and non-local transport, [3, 4, 5, 6], under perturbation. Some theoretical explanations for these phenomena assert that there may exist long range correlations between turbulence, as well as non-local turbulence triggering. The system is thought to require a global treatment rather than a simple local, diffusive-like picture [7, 8]. Most of the experimental evidence has come only from change of the profiles; little data has been presented about the changes in the turbulence properties, which may provide new information to explain these phenomena. In this work, measurements of turbulence from a 2D CO2 laser phase contrast imaging system are presented for discharges exhibiting non-local transport induced by pellet perturbation.

In the Large Helical Device (LHD), non-local transport can be observed after edge cooling by injection of a Tracer Encapsulated Solid Pellet (TESPEL). Specific details and classification of this behavior in LHD has been reported in detail in [9, 10]. Similar behavior has been observed in Tokamaks [11, 12]. In summary, for LHD, in low collisionality, high field discharges, a TESPEL pellet is injected into the edge, penetrating up to $\rho \sim 0.8$. The edge temperature decreases, while the core temperature rises, with a delay whose time-scale depends on collision frequency and magnetic configuration. The local transient thermal diffusion coefficient takes on negative values, invalidating a local diffusive transport model. The core heat flux has a much closer correlation with the temperature gradient at the edge than in the core, consistent with a non-local transport model.

Some explanations are based the idea of turbulence spreading; that is, that the local fluctuation amplitude is not simply a function of local intensity of turbulence, but rather, satisfies a diffusion-like equation [7]. An elegant model of this behavior has been proposed by [13], based on fractional calculus, demonstrating directly the non-local interaction and global behavior, based on random-walk turbulence simulations. However, in published literature, the results of this model do not predict a non-local temperature rise. Other models can predict up-gradient transport [8], but are still insufficient to produce a core temperature rise. Generally, there are still many possible theories for these phenomena and no single theory can universally explain all the observations [6].

The 2d phase contrast imaging diagnostic can measure the spatial profile of a quantity closely related to the density fluctuation amplitude with wave-numbers in the ion gyro-scale range $(1 - 9\text{cm}^{-1})$. While electron gyro-scale fluctuations are expected to play a role in electron energy transport, ion gyro-scale fluctuations are also expected to play a part. The main result from measurements presented here is that the ion gyro-radius scale fluctuation amplitude increases when the total heat flux reduces causing a temperature rise. This is opposite of the classical picture of fluctuation suppression causing confinement improvement, rather, that there is a change of the mutual phase and/or coherence between density, velocity and temperature fluctuations associated with an increase of density fluctuations. Changes of the poloidal phase velocity and radial wave-number spectrum are also associated with this phenomenon and may play a key role in producing the temperature rise.

In section (2), the phase contrast imaging measurements will be introduced, together with other diagnostic systems used in this analysis. In section (3), the non-local temperature rise phenomenon is described, and two reference discharges are introduced, where a non-local rise occurs and when a drop occurs. In section (4), PCI fluctuation measurements for these discharges are presented, and the time history of fluctuation response is compared with the flux and gradients. Finally, in section (5), the results are summarized and an interpretation is given.

2. 2D Phase contrast imaging system and other diagnostics

Fluctuation properties are measured using a 2D phase contrast imaging system. This can measure the spatial profile of the density fluctuation amplitude and also can resolve the k spectrum within the band $1 - 9\text{cm}^{-1}$. It employs a 2D imaging principle to split the line-integrated fluctuation signal into contributions from different layers along the line of sight, according to the "magnetic shear" principle [14, 15, 16]. The sightline is vertical passing at $R = 3.603$m, so penetrates from the edge to the core, depending on the magnetic axis position of the plasma (R_{ax}), so that core and edge fluctuation components (from both top and bottom) can be separated. The system cannot, however, fundamentally recover the local fluctuation amplitude because of line-integration effects unless the spectrum is isotropic, which it generally is not [17]. The line-integrated $\tilde{N}$ signal is split into contributions along the line of sight and is related to the local amplitude $\tilde{n}$ according to $\tilde{N}^2(\rho) = \tilde{n}^2(\rho) l_z l_{res}$, where l_z represents the ratio of fluctuation spectral density propagating exactly perpendicular to the probing beam to the total fluctuation power (related to longitudinal correlation length), and l_{res} is an instrumental resolution which increases with measured fluctuation wavelength. For typical measurements, the peak wavelength is such that the instrumental width is around half a radius, or less near the edge because the average wavelength is generally smaller than in the core.

Electron temperature is measured with an ECE polychromator [18]. For the plasmas studied here with configuration $R_{\rm ax} = 3.5$m, the accessible region, according to the measured frequencies, is between $\rho = 0.2$ and $\rho = 0.8$. It has been confirmed that the ECE is almost always in the blackbody regime, even in the edge [19], so that the deduced perturbed heat flux is not effected by change of opacity.

3. Non-local electron heat transport phenomenon

While electron heat transport may have non-local and non-diffusive characteristics (which may be evident by profile stiffness), a heat pulse experiment is one clear method to illustrate non-diffusive/non-local transport. In low density, high temperature LHD discharges, the plasma temperature response, measured by ECE [18], to pellet injection shows a non-diffusive response. In particular, the edge temperature decreases, as expected, while the core temperature increases. This response is characteristic of both Tracer Encapsulated Solid Pellet (TESPEL) injection (however the injection is only the polystyrene outer with no internal tracer), as well as in hydrogen ice pellet injection. Nevertheless, TESPEL injection allows small controlled perturbation to the edge. Detailed descriptions and classification of this phenomenon have been given in [9, 10, 20]; here we introduce this phenomenon with respect to the discharges studied in this paper, which were from the 10th (2006-7) experimental campaign, while further discharge types were studied before, but without PCI fluctuation measurements.

In this study two different scenarios are compared: when pellet injection induces a non-local core temperature rise, and when it produces a core temperature drop. The main extrinsic differences between these are the introduction of an extra Ar gas puff just before pellet injection. Though the Ar gas puff alone can produce a non-local temperature rise, the combination with a pellet produced a drop in this case. Since there is a clear difference of the time evolution of fluctuation phase velocity, this comparison provides clear evidence of the link between fluctuation phase velocity and the non-local temperature rise.

The time history of electron temperature and density in the non-local temperature rise case (#71160) is plotted for various ρ in figures 1 (a) and (b). Conditions for this shot are: $R_{\rm ax} = 3.5$m, $B = 2.93T$, $n_{e0} \sim 0.4 \times 10^{19}{\rm m}^{-3}$, and T_{e0}, before pellet injection, is $\sim$ 4keV, with ECR heating only (subscript $_{e0}$ denotes core value) by all three sets of gyrotrons of frequencies 168GHz (0.6MW), 84GHz (0.62MW) and 82.7Ghz (0.36MW) which deposit on axis at the fundamental and second harmonics respectively. The injected TESPEL pellet size is 700μm. The changes in the turbulence properties here are very similar that of an Ar gas puff (as reported in [20]). The edge temperature decreases, while the core temperature increases. The pellet ablation is evident from an increase of the edge density. This is cross-checked with spectroscopic measurements of the emission from the pellet, which show it ablates within 1ms and penetrates only up to $\rho \approx 0.8$. The density pulse propagates inwards, in addition to the core temperature rise. On the other hand, density and temperature evolution for shot #71169, where the core temperature decreases upon pellet ablation, is plotted in figures 2 (a) and (b). Conditions for this shot are: $R_{\rm ax} = 3.5$m, $B = 2.93T$, $n_{e0} \sim 0.6 \times 10^{19}{\rm m}^{-3}$, $T_{e0} \sim$ 4keV, with only ECR heating. An prompt additional Ar gas-puff is introduced prior to pellet injection at $t = 0.5$s as presented in [20]. Comparing the density between these two shots, figure 1 (b) with 2 (b), the density does not rise in the edge with Ar gas puff as clearly as without Ar gas puff, possibly related to a difference of the edge temperature due to cooling by the Ar gas puff.

For the shot with non-local temperature rise, a transient transport analysis is carried out in order to compare with fluctuation properties. The heat flux is evaluated from the power balance equation:

$$\frac{dW(\rho)}{dt} + \nabla \cdot q = S_{\rm ht}(\rho) - S_{\rm at}(t,\rho) \tag{1}$$

where $W = 3/2nk_BT_e$ is the stored electron energy density (k_B being the Boltzman constant), q is the electron heat flux, where the radial component only is analyzed and positive indicates outward flux. $S_{\rm ht}$ is the power source due to heating, and $S_{\rm at}$ is the lower lost due to atomic processes associated with pellet ablation. This equation is then split into equilibrium and perturbed components, i.e. $q = q_{\rm eq} + q_{\rm pert}$ such that $\nabla \cdot q_{\rm eq} = S_{\rm ht}(\rho)$, and,

$$\frac{dW(\rho)}{dt} + \nabla \cdot q_{\rm pert} = -S_{\rm at}(\rho,t) \tag{2}$$

This separation is valid if the heat source is not perturbed by the change. To a degree, the heat source may be perturbed by the significant density and temperature change in the edge, however this is likely to be small on account of the ECRH beam refraction being small at this low density. Moreover, there may be a time delay between the resonant production of energetic electrons and their relaxation with the thermal population, indicating any prompt change is not due to change of heating. The radiation measured by a bolometer shows a spike only within a few ms after pellet ablation followed by a reduction of by about 50kW from the pre-injection value, relaxing back to the initial value over the timescale of the temperature relaxation. This indicates a change of bremsstrahlung and/or line radiation after pellet ablation independent of $S_{\rm at}$. The magnitude of the change is however significantly smaller than the $dW/dt \sim$ 500kW after pellet ablation and the heating power of $\sim$ 1.5MW. Therefore, $q_{\rm pert}$ is evaluated from:

$$q_{\rm pert}(r) = \underbrace{-r^{-1}\int_0^r 1.5\frac{\partial n(r')}{\partial t}T(r')r'dr'}_{q_n} + \underbrace{(-r^{-1})\int_0^r 1.5n(r')\frac{\partial T(r')}{\partial t}r'dr'}_{q_T} + \underbrace{(-r^{-1})\int_0^r S_{\rm at}(r')r'dr'}_{q_S}, \tag{3}$$

where the terms q_n and q_T are the components evaluated from density and temperature changes, individually. Previous analysis has focused on evaluating only q_T, though towards the edge, q_n is also important because of pellet fuelling. Furthermore, it is not possible to evaluate q_S, because of uncertainties in the atomic processes, but since the pellet does not penetrate into $\rho < 0.8$, $q_S = 0$ in this region. For $\rho > 0.8$, the lack of accounting for q_S means that the true $q_{\rm pert}$ is more negative (i.e. inwards) than the measured $q_{\rm pert}$.

Because temperature measurements are not available for $\rho > 0.8$, a different approach must be taken for evaluating $q_{\rm pert}(\rho = 1)$. Given that the diamagnetic stored energy is

$$W_{\rm dia} = \int 1.5k_B(n_eT_e + n_iT_i)dV, \tag{4}$$

$q_{\rm pert}(\rho = 1)$ is evaluated from:

$$q_{\rm pert}(\rho = 1) = -0.5[(2\pi)^2aR_{maj}]^{-1}dW_{\rm dia}/dt, \tag{5}$$

assuming $T_i = T_e$ and $n_e = n_i$, where a is the average plasma minor radius $a = 0.6$m and R_{maj} is the major radius of 3.5m.

The values of $q_{\rm pert}$, including the component q_T for $\rho = 0.5, 0.8, 1.0$ are plotted in figure 1 (c). At $\rho = 0.8$, q_T is initially outwards, with a duration much longer than the pellet ablation time, then becomes inwards, though considering $q_n + q_T$, the flux is actually inwards owing to the heat pulled in by the inward particle flux. The q_T at $\rho = 0.5$ is always inwards, while

$q_{\mathrm{pert}} = q_T + q_n$ has an outwards transient just after pellet injection, followed by a phase of inwards flux. The transient outwards flux at the beginning is a result of a slight reduction in density at $\rho = 0.5$ as in figure 1 (b), but as this is only slight, it may be an artefact of the Abel inversion procedure. The calculated flux at $\rho = 1$ from the diamagnetic stored energy is smaller than that at $\rho = 0.8$, since the area through which the energy is flowing is larger at $\rho = 1$, but may possibly be slightly underestimated because $T_i < T_e$ in these EC heated discharges (here the kinetic electron stored energy component alone is 80% of the diamagnetic stored energy). At all locations, the components q_n are badly noisy because of low density. The main point is that the flux in the core, as well as at the edge, responds promptly to pellet injection.

The normalized temperature and density gradients at $\rho = 0.5, 0.8$ are plotted in figure 1 (d). The temperature and density gradient change rapidly at $\rho = 0.8$. The magnitude of the change of temperature gradient is only slight, and in fact appears to change sign, possibly because of some systematic profile error of the ECE measurement. Nevertheless the time-scale for the change is the important information here. Even though the temperature gradient may not be reliable at $\rho = 0.8$, there is distinctly a sudden decrease in dT/dt after pellet injection. In fact, it is suspected that the temperature decrease is more pronounced outside $\rho = 0.8$ because of pellet cooling, but measurements are not available here due to limited access of the ECE measurement. At $\rho = 0.5$, the density gradient hardly changes in response to pellet injection while the temperature gradient changes over a larger timescale.

Therefore, comparing flux and gradients, it is clear that there is a strong correlation between both core and edge flux and the edge temperature gradient, and there is little correlation between core gradients and core flux, suggesting a non-local response from the edge to the core. Further analysis of the gradient/flux relationship shows phases of negative and transient diffusion coefficient, defined as $\chi_{tr} = \partial(q_{\mathrm{pert}}/n)/\partial(-\nabla T)$ [10, 20]. However, the fluctuation level does not appear to change in any clear way with χ_{tr}.

4. Analysis of fluctuation properties

PCI signals are analyzed as follows. The spatially resolved density fluctuation amplitude spectrum, $S(\omega, k, \rho; t)$, where ρ is flux coordinate, is calculated from a cartesian/polar transformation of $S(\omega, k_x, k_y; t)$, the 3D Fourier transform of the time series of the image sequence $\tilde{N}(x, y, t)$, for a selected time window. The transformation is $k^2 = k_x^2 + k_y^2$, angle $\theta = \tan^{-1}(k_y/k_x)$, and ρ is determined uniquely from θ from the known magnetic field line pitch angle dispersion. More detail is given in [16].

A contour plot of $S(k, \rho)$ (integrated over all frequencies) for #71160 at $t = 0.58, 0.615$s, before and after pellet injection, is shown in figures 3 (a) and (c). Negative (positive) ρ indicates the component measured below (above) the midplane. The dashed grid characterizes the spatial-k resolution. The amplitude is seen to be strongly peaked at the edge, reducing towards the core, although with some increased instrumental broadening because the average k reduces towards the core, possibly because the ion gyro-radius is larger there.

The angular frequency ω is converted to phase velocity via $v = \omega/k$, and integration over all measurable k is performed to generate the dataset $S(v, \rho; t)$, and is plotted for times before and after pellet ablation in figures 3 (b) and (d). It is clear that there is a distribution of velocities for a single ρ. The phase velocity, being the component perpendicular to the line of sight, is observed to reverse about the mid-plane, characteristic of the poloidal propagation component, and is mostly up/down symmetric, however not completely; this point is discussed later. Before pellet ablation, in (b), the phase velocity is higher at the edge than the core, having a positive shear. On the other hand, after pellet ablation, (d), the velocity around $\rho = 0.5$ appears to have become bi-modal, with the existence of a branch at higher velocity with substantially higher phase velocity than before pellet ablation. The velocity of the peak of the distribution at $\rho = 0.5$ has increased from $\sim$ 2km/s to $\sim$ 4km/s. Since the velocity around $\rho = 1$ has reduced slightly,

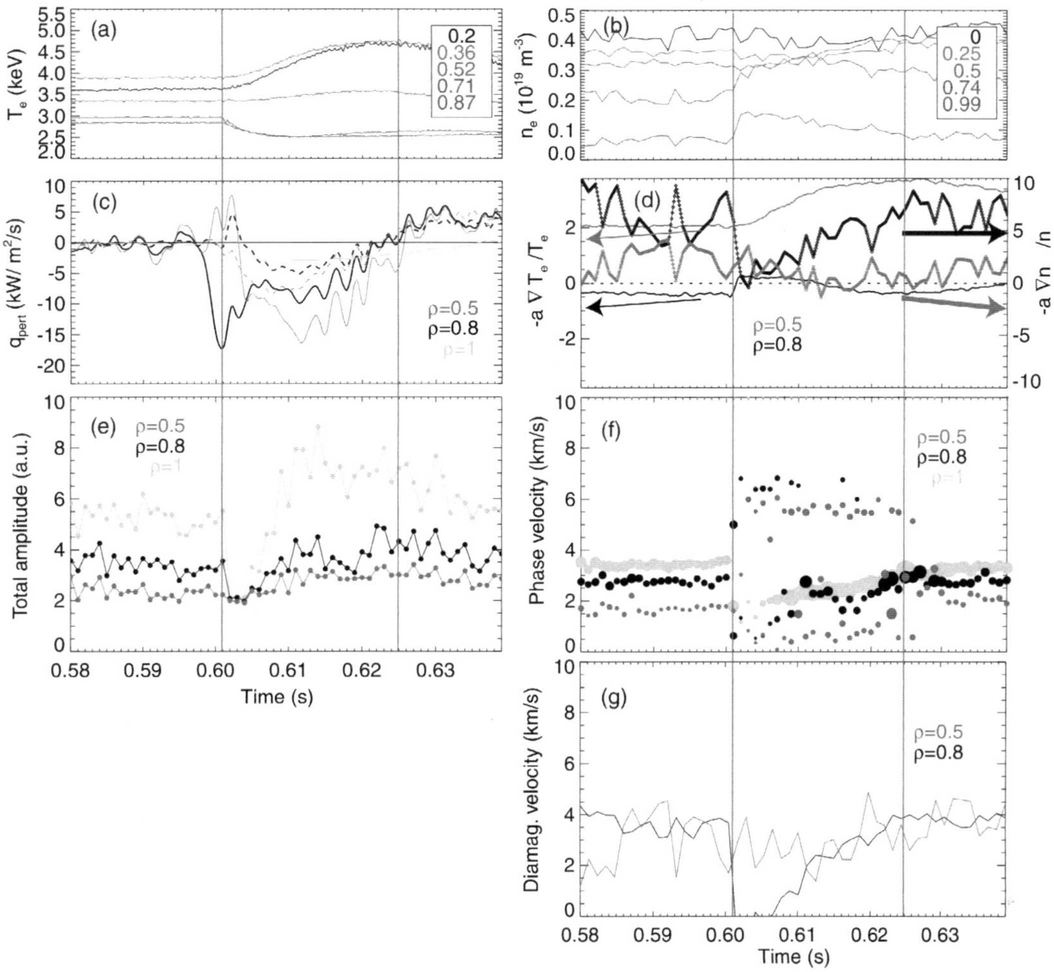

Figure 1. Time evolution of profiles, transport, gradients and fluctuations in discharge #71160 exhibiting a core temperature rise in response to pellet injection. (a) Temperature for various ρ measured with ECE diagnostic, (b) density for various ρ from abel inversion of FIR interferometer measurements; SNR ratio is somewhat marginal in this case, (c) calculated total $q_{\rm pert} = q_n + q_T$ (solid lines) and component q_T (dashed lines) at $\rho = 0.5, 0.8, 1.0$ (red,black,green) (the value at $\rho = 1.0$ is derived from diamagnetic loop voltage), (d) normalized temperature (thin lines) and density (thick lines) gradients at $\rho = 0.5, 0.8$, (e) fluctuation amplitude at $\rho = 0.5, 0.8, 1$ (red,black,green),(f) branch-separated average fluctuation phase velocity at same locations, where the size of the dots represent the relative amplitude of that branch, (g) time evolution of diamagnetic velocity computed from density/temperature profiles at $\rho = 0.5, 0.8$.

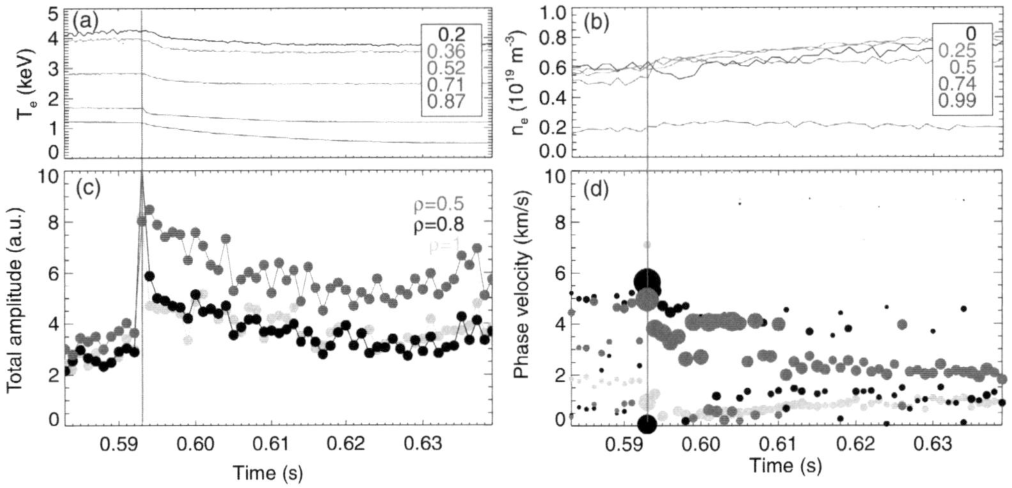

Figure 2. Time evolution of profiles, and fluctuations in the case where the core temperature decreases in response to pellet injection and Ar gas-puff injection at $t = 0.5 - 0.51s$ (#71169). (a) Temperature for various ρ, (b) density for various ρ, (c) fluctuation amplitude at $\rho = 0.5, 0.8, 1.0$ (red,black,green) and (d) branch-separated fluctuation phase velocity at same locations, where the size of the dots indicate the relative amplitude of the branch.

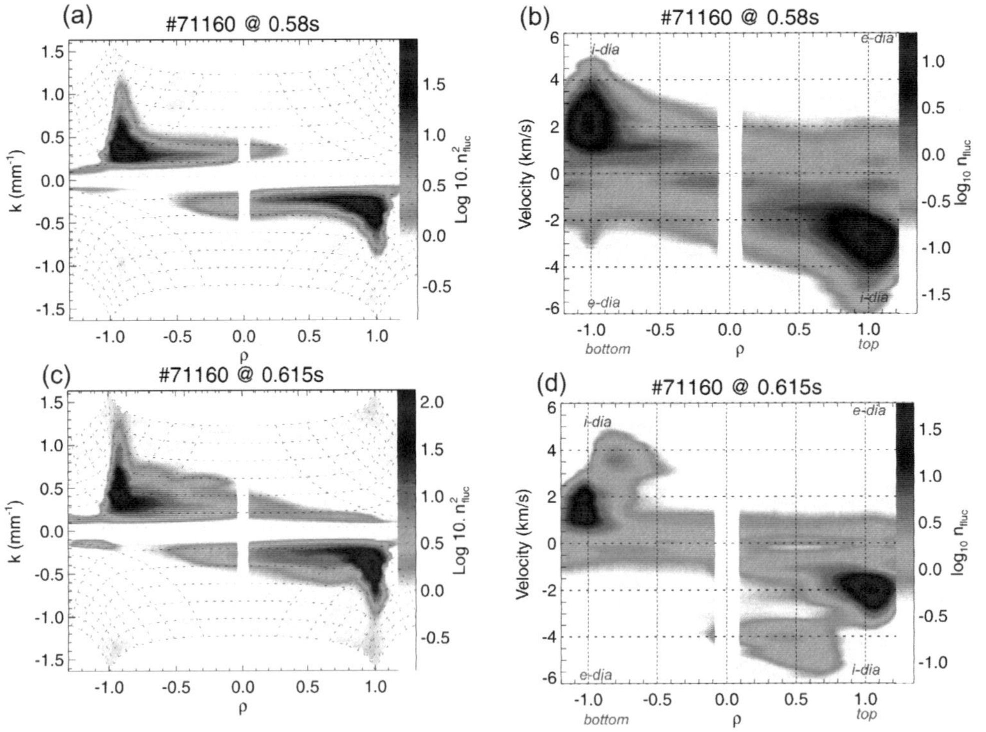

Figure 3. Analysis of fluctuation properties for single time-windows (a,b) before pellet ablation and (c,d) after pellet ablation. Fluctuation amplitude distribution (log color scale) over ρ, k in (a,c) and over ρ, v (log color scale) in (b,d).

this gives a strong negative velocity shear (as well as a small positive velocity shear for the lower velocity branch) over the outer region of the plasma.

There are distinct changes in the asymmetry properties (amplitude, phase velocity) after pellet ablation. In particular, in figure 3 (d), there is a component with $v \approx -4$km/s peaked at location $\rho = +0.5$ on the top, while on the bottom there is a component at $v \approx 4$km/s peaked at a different radial location, $\rho = -0.8$. On the other hand, the component peaked at $\rho = +1$ with $v \approx -2$km/s is more symmetric with respect to the component at $\rho = -1$ with $v \approx 2$km/s. This could be explained as follows. The diagnostic measures only components which propagate exactly horizontally, or perpendicular to the injected laser beam. Because the laser beam passes outside the magnetic axis for this magnetic configuration ($R_{\mathrm{ax}} = 3.5$m), the top measured components propagate both poloidally and (slightly) radially outwards and the bottom measured components propagate poloidally and (slightly) radially inwards. Therefore the asymmetry may be interpreted as an asymmetry of the radial wavenumber spectrum of turbulence. In figures 1,2 and 4, we plot only quantities on the upper side, however the results are similar if lower side quantities are plotted.

Having shown the amplitude and velocity measurements for time slices, attention is now turned towards the dynamical evolution of the amplitude and velocity for $\rho = 0.5, 1.0$, by analyzing 1ms time windows as a function of time. The change of phase velocity and k, ω spectrum before and after pellet ablation are subtle and can be shown more clearly by analyzing the time evolution of the spectrum as a function of velocity, ω and k in figure 4. The splitting of the velocity distribution into a bimodal distribution is clear in figure 4 (a). The width of the frequency and k spectra expands to higher values, as evident in figures 4 (c) and (d). At $t = 0.611s$, after pellet ablation, the dispersion relation $S(k, f)$ is plotted in (b). Lines showing phase velocity of 2km/s and 6km/s are indicated for reference. The higher velocity branch exists at high frequency and at lower k while the lower velocity branch has a dominant component at low frequency, low k but also a significant component at higher k. In addition, the frequency and k width of the lower velocity branch is much broader than that of the high velocity component, implying it has a shorter intrinsic decorrelation length/time [21]. Examining figure 4 (a), it appears that the amplitude of both higher and lower velocity branches is significantly oscillating in time. Such oscillation of the amplitude envelope may be an indirect indication of the action of a Zonal flow or other large scale structure in a predator-prey system [22]. More details of this analysis applied to the non-local temperature rise in LHD is given in [23].

The dynamical evolution of the total fluctuation amplitude and scalar velocity at $\rho = 0.5, 0.8, 1.0$ is plotted for the core temperature rise case in figures 1 (e) and (f), to be compared with the evolution of profiles, flux and gradients. Scalar velocities for each ρ are extracted from figure 4 (a) using a simple branch cutting technique, which cuts between the local minima in $S(v)$ and calculates the average velocity within each branch. This allows for the possibility that there may be more than one velocity for a single ρ. The relative size of each point in figure 1 (f) indicates the relative amplitude for each branch, defined by the square root of the integral between the local minima of $S(v)$.

Analysis of the total amplitude at $\rho = 0.8, 1.0$ is considered as follows. There is a reduction until $t = 0.605$s, followed by an increase beyond the initial level at around $t = 0.61$s, after which time the perturbed heat flux at both locations is still inwards, or in other words, the total heat flux is smaller than the pre-pellet-injection value. The tendency is most clear at $\rho = 1.0$, although here the heat flux can only be measured from diamagnetic voltage. Though the increase above the initial level is smaller at $\rho = 0.8$, the flux is more reliably derived from the pressure. This increase is opposite to the widely held view that confinement improvement should be associated with density fluctuation suppression. In order to explain this, since $q_{\mathrm{anom}} = \langle \tilde{n}\tilde{v} \rangle T + \langle \tilde{v}\tilde{T} \rangle n$, there must be some de-correlation between density and velocity, or velocity and temperature. This de-correlation is difficult to measure directly, but there is evidence for a reduction of correlation

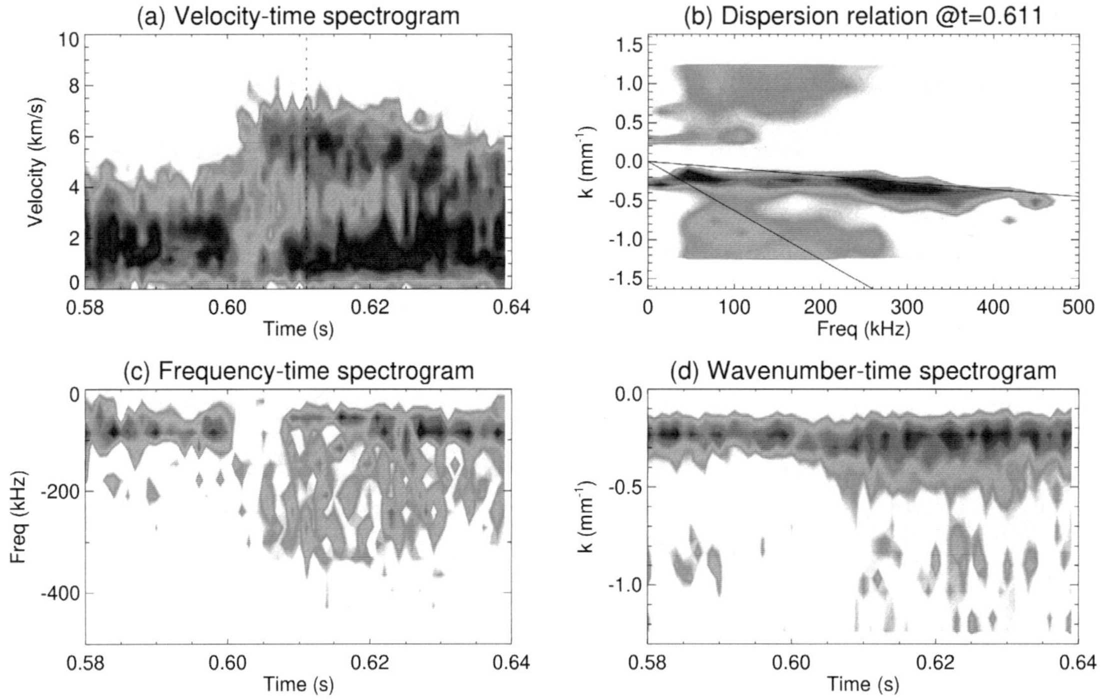

Figure 4. Time history of (a) velocity (linear color scale), (c) frequency (log color scale) and (d) wavenumber (log colour scale) resolved spectra for #71160 at $\rho = 0.5$. The dispersion relation $S(f, k)$ is indicated in (b) for $t = 0.611$s, together with two lines of phase velocity 2km/s and 6km/s.

length (equivalent to the width of the k spectrum) as shown in figure 4 (d) and a change of phase velocity shear, which might be an indication of change of $E \times B$ shear. Since the correlation length may also be influenced by the part of the k spectrum below the instrument cutoff, the change in k width may be more pronounced than measured here. The amplitude at $\rho = 0.5$ does not change significantly within the statistical time variation, even though $q_{\text{pert}}(\rho = 0.5)$ is inwards. (There is a distinct increase in higher frequency components, effecting the velocity, but their amplitude is not enough to have a significant effect the total amplitude). This is further evidence that the total fluctuation amplitude is not always the most important parameter in determining the transport.

The analysis of the scalar average velocity of each branch at $\rho = 0.5, 0.8, 1.0$ is considered as follows. After pellet injection, the velocity at $\rho = 1.0$ remains single-valued, and decreases rapidly by a factor of around 2, recovering slowly. On the other hand, the velocity at $\rho = 0.5, 0.8$ immediately splits into two branches, one with high velocity, and one with low velocity. For $\rho = 0.5$, this split is maintained until $t = 0.625$s, which correlates with the time at which q_{pert} switches from inwards to outwards. The change in velocity at $\rho = 0.5$ occurs much more quickly than the change in the local temperature and density gradients as shown in figure 1 (d). This indicates a possible non-local interaction from the edge, where the temperature and density gradients change immediately.

It has been shown that the measured phase velocity profile often conforms closely with the measured and calculated $E \times B$ velocity [24, 25], though is offset by ion and/or electron diamagnetic drifts. The change in phase velocity after pellet ablation may be due to either of these contributions, (1) $E \times B$ rotation, (2) diamagnetic drift velocity, or (3) both. Though $E \times B$ rotation is not measured here, its change may be inferred from the change in phase velocity. To clarify the effect of the change in diamagnetic velocity, this is computed from the density, temperature profiles and is plotted for $\rho = 0.5, 0.8$ in figure 1 (g). It is clear that there is no distinguishable change at $\rho = 0.5$, while there is a strong reduction of $\sim$ 4km/s at $\rho = 0.8$. At $\rho = 0.5$, the data suggests case (1): that the increase may be due to $E \times B$ rotation, and that rotation rise is clearly non-locally triggered from the edge. At $\rho = 0.8$, the story is more complicated. The fluctuation splits into 2 branches, one with higher velocity, and one with lower velocity. The change in velocity of each branch after pellet ablation is around +3km/s for the upper and −2km/s for the lower branch, which is comparable to the change in diamagnetic velocity. One possible interpretation is that there are 2 simultaneously propagating modes with respect to plasma $E \times B$ rotation, the higher speed branch in the ion diamagnetic direction (with respect to the plasma rotation), the lower speed branch in electron diamagnetic direction (with respect to the plasma rotation). However, according to this interpretation, the difference of velocity between the 2 modes (measured here to be $\approx$ 5km/s) should be equal to $2v_{dia}$, while the measured v_{dia} reduces towards zero immediately pellet ablation. This might suggest that either one or both of the modes are propagating at speeds significantly different from the diamagnetic drift velocity, or that the diamagnetic drift velocity may not be evaluated correctly at this radius due to unreliable temperature gradient from ECE (which changes sign here after pellet ablation as in figure 1 (d). Tests of the spatial resolution of PCI measurements indicate that $\rho = 0.5, 0.8$ can be distinguished (so the changes are not strongly cross-coupled) and are exemplified in the difference in the time history of the upper branch of the phase velocity.

Regardless of the interpretation, it appears from figure 1 (f) that there is a strong phase velocity shear for the higher velocity branch, located between $\rho = 0.8 - 1.0$ for the initial time $t = 0.6 - 0.61$s and then the shear layer moves inwards to $\rho = 0.5 - 0.8$ for $t = 0.61 - 0.625$s. If this were related to $E \times B$ shear, then this could explain confinement improvement and turbulence decorrelation. Such kind of non-local $E \times B$ shear mechanism has not been widely considered theoretically. One such model [26] describes the production of velocity shear a critical ingredient in the non-local temperature rise.

We now analyze the change of fluctuation amplitude and phase velocity when the core temperature decreases, as plotted in figures 2 (c) and (d). For both the $\rho = 1.0, 0.8, 0.5$, the fluctuation amplitude increases promptly (most pronounced at $\rho = 0.5$), and there is no transient decrease phase as in #71160. More importantly, an increase in velocity is not observed at any location after pellet ablation unlike for the non-local temperature rise. Analysis of several other discharges, without Ar gas-puff, show the same behavior when the core temperature decreases: no increase in velocity at $\rho = 0.5, 0.8$. This change is even evident from the raw signals in terms of an increase or not of high frequency fluctuation power. Therefore, comparison of #71160 and #71169 shows that increase of velocity at $\rho = 0.5$ is a necessary ingredient for the non-local temperature rise.

5. Summary and interpretation

The main results are summarized as follows:

- The excitation of a high velocity component in addition to a low velocity component towards the core, which has negative phase velocity shear, is necessary for confinement improvement and core temperature rise. Whether this is due to a change in bulk plasma $E \times B$ rotation, or change fluctuation velocity in the plasma frame, is unproved at this stage. There is no distinguishable change of diamagnetic drift in figure 1 (g) at $\rho = 0.5$, though there is a significant perturbation at $\rho = 0.8$ which may therefore affect the measured shear. In the future, improved CXRS measurements may be made to confirm the role of $E \times B$ velocity, however so far the SNR has been marginal in these regimes. It may turn out that the source of phase velocity shear may not be so important for confinement improvement.

- Fluctuation level can be larger when $q_{\rm pert}$ is inwards, or equivalently when the total flux is reduced, contrary to the widely held belief that fluctuation suppression is necessary for confinement improvement. It is speculated that decorrelation of temperature, density and potential resulting from increase in magnitude of $E \times B$ shear has a stronger effect than increased fluctuation level.
- A non-local trigger exists for the increase of core velocity and drive of strongly negative velocity shear. This may relate to turbulence spreading [7], although the total fluctuation level is not equivocally seen to increase, (only in the high frequency band), or possibly is consistent with SOC models [26]. Another possible simple interpretation is that profiles are modified such that a transition from the electron to ion root exists in the plasma, transiently forming an electron internal transport barrier (eITB). The discharge conditions for non-local temperature rise are close to that for eITB [10]. Some discharges which exhibit a much more prompt core temperature rise than shown here appear to have a bifurcation in the fluctuation phase velocity profile around $\rho = 0.7$, indicating they may already possess an eITB before pellet injection. In such discharges, the change of edge profiles in response to the pellet may immediately increase the $E\times B$ shear at $\rho = 0.7$, leading to a prompt increase in core temperature. In the discharge #71160 presented here, the velocity shear layer is formed initially at the edge and then moves in (as in figure 1 (f)), therefore introducing a time delay before the maximum core temperature.

A possible mechanism is described. Because temperature is not measured towards the edge, speculation is made that the temperature decreases immediately after pellet injection near the edge leading to a local flattening of the temperature profile. Thomson scattering measurements in other discharges have partially confirmed this. This may be responsible for the reduction of edge fluctuation level initially. The change in profile at the edge facilitates a non-local change of phase velocity at $\rho = 0.5$ resulting in improved confinement, allowing the temperature to increase in the core, up to a point where the higher velocity branch barrier is no longer maintained, the confinement degrades and the temperature finally relaxes to its pre-pellet-injection state.

There are still many unexplained phenomena associated with the non-local temperature rise. Further experimental and theoretical investigation is necessary to clarify the mechanism and relieve some of the speculation. The presence of a low frequency turbulence modulator has been recently demonstrated [23], and, may be another element to consider.

Acknowledgements

This work was supported by a grant-in-aid from the Japan Society for the Promotion of Science (JSPS). The authors are thankful for valuable comments from the reviewers and other people from the poster session of the workshop.

References

[1] Garbet X, Mantica P, Ryter F, Cordey G, Imbeaux F, Sozzi C, Manini A, Asp E, Parail V and Wolf R 2005 *Plasma Phys. Control. Fusion* **47** 957

[2] Ohyabu N, Yamada H, Ida K, Watanabe K, Funaba H and Morisaki T 2002 *Plasma Phys. Control. Fusion* **44** A211

[3] Ryter F, Neu R, Dux R, Fahrbach H U, Leuterer F, Pereverzev G, Schweinzer J, Stober J, Suttrop W, De Luca F *et al.* 2000 *Nucl. Fusion* **40** 1917

[4] Gentle K W, Bravenec R V, Cima G, Gasquet H, Hallock G A, Phillips P E, Ross D W, Rowan W L, Wootton A J, Crowley T P *et al.* 1995 *Phys. Plasmas* **2** 2292

[5] Kissick M W, Callen J D and Fredrickson E D 1998 *Nucl. Fusion* **38** 821

[6] Callen J D and Kissick M W 1997 *Plasma Phys. Control. Fusion* **39** B173
[7] Hahm T S, Diamond P H, Lin Z, Rewoldt G, Gurcan O and Ethier S 2005 *Phys. Plasmas* **12** 90903–1
[8] Naulin V, Nielsen A H and Rasmussen J J 2005 *Phys. Plasmas* **12** 122306–1
[9] Inagaki S, Tamura N, Tokuzawa T, Ida K, Itoh K, Neudatchin S V, Tanaka K, Nagayama Y, Kawahata K, Yakovlev M and Sudo S 2006 *Plasma Phys. Contr. F.* **48** A251–7
[10] Tamura N, Inagaki S, Tanaka K, Michael C, Tokuzawa T, Shimozuma T, Kubo S, Sakamoto R, Ida K and Itoh K 2007 *Nucl. Fusion* **47** 449
[11] Zou X L, Geraud A, Gomez P, Mattioli M, Segui J L, Clairet F, De Michelis C, Devynck P, Dudok de Wit T, Erba M *et al.* 2000 *Plasma Phys. Control. Fusion* **42** 1067
[12] Galli P, Gorini G, Mantica P, Hogeweij G M D, de Kloe J and Cardozo N J L 1999 *Nucl. Fusion* **39** 1355
[13] del Castillo-Negrete D 2006 *Phys. Plasmas* **13** 82308–1
[14] Sanin A, Tanaka K, Vyacheslavov L, Kawahata K and Akiyama T 2004 *Rev. Sci. Instrum.* **75** 3439
[15] Vyacheslavov L, Tanaka K, Sanin A, Kawahata K, Michael C and Akiyama T 2005 *IEEE Trans. Plas. Sci.* **33** 464
[16] Michael C, Tanaka K, Vyacheslavov L, Sanin A, Kawahata K and Okajima S 2006 *Rev. Sci. Instrum.* **77** 10E923
[17] Michael C *et al.* 2007 *Plasma Fusion Res* **2** S1034
[18] Kawahata K, Nagayama Y, Inagaki S and Ito Y 2003 *Rev. Sci. Instrum.* **74** 1449
[19] Inagaki S, Nagayama Y, Sasao H, Vries P d, Ito Y, Kawahata K, Narihara K, Kubo S, Komori A, Yamada H *et al.* 1999 *26th EPS Conf. on Contr. Fusion and Plasma Physics* (Maastricht) p 1341
[20] Tamura N, Inagaki S, Ida K, Tanaka K, Michael C, Tokuzawa T, Yamada I, Narihara K, Shimozuma T, Kubo S *et al.* 2007 Core electron temperature rise due to Ar gas-puff in EC-heated LHD plasmas *these proceedings*
[21] Coda S 1997 Ph.D. thesis MIT
[22] Itoh S, Itoh K, Sasaki M, Fujisawa A, Ido T and Nagashima Y 2007 *Plasma Phys. Control. Fusion* **49** L7–10
[23] Inagaki S, Tamura N, Ida K, Nagayama Y, Kawahata K, Sudo S, Itoh S, Itoh K and Komori A 2008 Correlation Structure and Transition of Heat Transport in Toroidal Plasmas with Non-Local Coupling *submitted to Phys. Rev. Lett.*
[24] Tanaka K, Michael C, Sanin A, Vyacheslavov L, Kawahata K, Murakami S and Wakasa A 2006 *Nucl. Fusion* **46** 110
[25] Michael C, Tanaka K, Vyacheslavov L, Sanin A, Yamagishi O, Yoshinuma M, Ida K, Yokoyama M, Kawahata K and Okajima S 2006 *33rd EPS conference on Plasma physics and controlled fusion* (Rome) p P4.114
[26] Carreras B A, Newman D, Lynch V E and Diamond P H 1996 *Phys. Plasmas* **3** 2903

Journal of Physics: Conference Series **123** (2008) 012019 doi:10.1088/1742-6596/123/1/012019

IBW heating experiments in HT-7 tokamak

Y Shi, J Fu, H Gao, F Wang, W Gao, H Liu, G Zhong, Y Duan, Q Xu, A Ti, G Xu, Z Wu, X Gao, B Wan, C Qin, Y Zhao and HT-7 team

Institute of Plasma Physics, Chinese Academy of Sciences, Hefei 230031, China

E-mail: yjshi@ipp.ac.cn

Abstract. The Experiments with Ion Berstein wave (IBW) heating has been carried out in HT-7 superconducting tokamak. Core electron heating and peaking profile of electron temperature can be observed in the beginning 40 ms after injection of IBW. The temporary heating efficiency of IBW increases with the toroidal magnetic field (B_T). The electron temperature drops gradually with the increasing electron density after 40 ms of IBW. The electron density increases by a factor up to 2.4 and the profile of electron density becomes wide. Improved particle confinement induced by IBW is one reason for the density increase. Impurity release during IBW heating is also a contribution for the density increase. Suppression of MHD by IBW is also observed. The fluctuation and transport in the SOL of HT-7 tokamak have been investigated using a Langmuir probe array. The polodial shear flow induced by IBW maybe the reason for improved particle confinement.

1. Introduction

Ion Bernstein wave (IBW) is a hot plasma wave with a frequency close to an ion cyclotron harmonic frequency, which is coupled directly to the plasma thermal ions and electrons by the ion cyclotron damping or electron Landau damping (ELD) that occurs near the resonant layer [1]. Theory and experiments have showed the advantage of IBW heating in different tokomaks. Both ion and electron heating via linear and nonlinear processes were observed in IBW experiments on quite a few devices [2-10]. IBW heating was also investigated in HT-7 deuterium plasma [11-16]. Enhanced particle confinement and suppression of MHD by IBW are usual phenomena in HT-7.

The HT-7 device is a medium-sized superconducting tokamak with a major radius of R = 1.22 m and is of a limiter configuration with two full poloidal limiters and one inner toroidal belt limiter, all of which are water-cooled and made of special doped graphite coated by SiC film with a minor radius of 27cm. In the HT-7, the ICRF heating system has a 1.5 MW continuous wave (CW) output power capability and the operation frequency range is from 15 MHz to 110 MHz（30 MHz for IBW）. The ICRF system includes RF generator, transmission lines, two stub tuners and two antennas (ICRF antenna and IBW antenna). IBW antenna is an asymmetric quadruple T-antenna with a central feeding and shorted ends. It was installed in the horizontal mid-plane on the low field side and oriented in the toroidal direction. The radii of central conductor and Faraday shielding are 32 cm and 28.5 cm respectively. The antenna center conductor is coated by TiN, and incorporated with graphite-doped protectors and Faraday shielding to reduce impurities.

2. Experimental results of IBW heating

The waveform of a typical discharge with 1.94 toroidal field (B_T) and 27 MHz IBW heating is shown in figure 1. The power of IBW was about 320 kW at flat top. The electron temperature was measured with a sixteen-channel electron cyclotron emission (ECE) grating polychromator (GPC) system [17]. The abrupt drop of D_α after the injection of IBW denoted the improved particle confinement. The electron temperature at r = 3 cm increased in the initial 40 ms of IBW heating. At the same time, the electron temperature at r = -12.7 cm was almost unchanged. The heat exchange time between electron and deuterium ion was about 100 ms. The electron should be directly heated by IBW. It can be seen in figure 2a the core plasma electron was heated by IBW and the profile of electron temperature become more peaked compared to that at OH heating. But the electron temperature of whole plasma dropped very much when electron density reached the flat top. Here, a simple temporary heating efficiency of IBW is defined as following,

$$\eta_{IBW} = \Delta T_e \bar{n}_{e0} / P_{IBW}$$

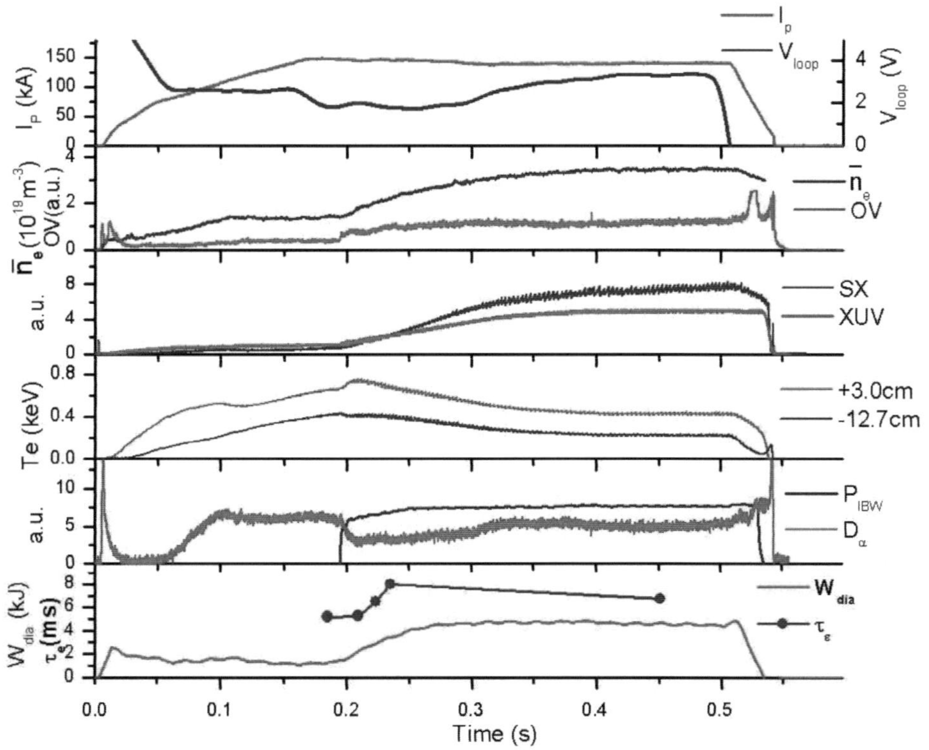

Figure 1. Waveform for an_IBW discharge. From top to bottom: I_P, the plasma current, V_{loop}, loop voltage, n_e, the center line average electron density, OV, the intensity of OV lines, SX, the intensity of soft x-ray at center, XUV, the radiation power measured with XUV detector, T_e, the electron temperature at r = 3 cm and r = -12.7 cm, P_{IBW}, the power of IBW, D_α, the radiation of D_α, W_{dia}, the store energy measured by diamagnetism, τ_e, the energy confinement time.

Journal of Physics: Conference Series **123** (2008) 012019 doi:10.1088/1742-6596/123/1/012019

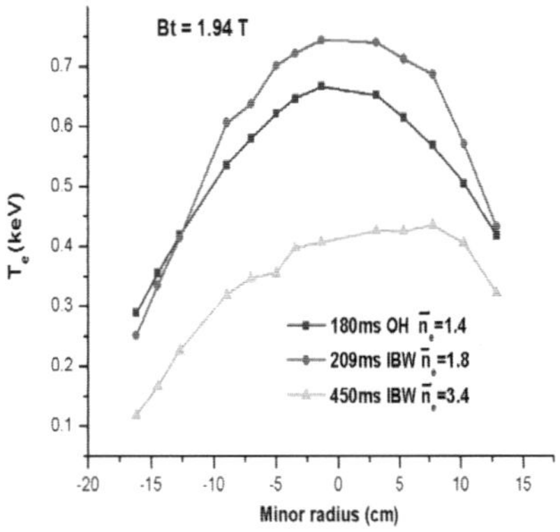

Figure 2a. The profiles of the electron temperature of the same shot in figure 1. IBW was injected at 190 ms.

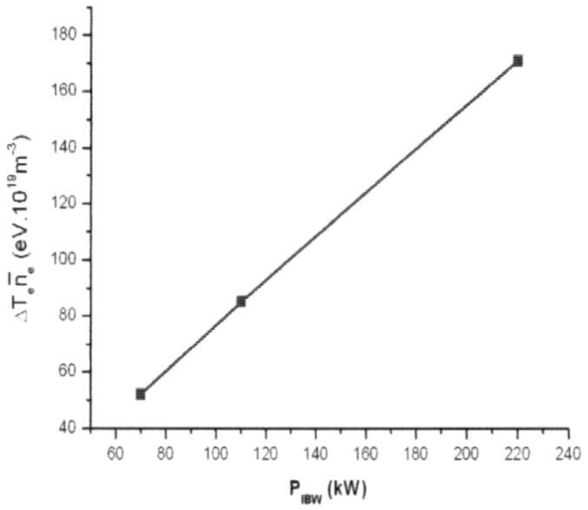

Figure 2b. The heating effect of IBW with different power at B_T = 1.94 T.

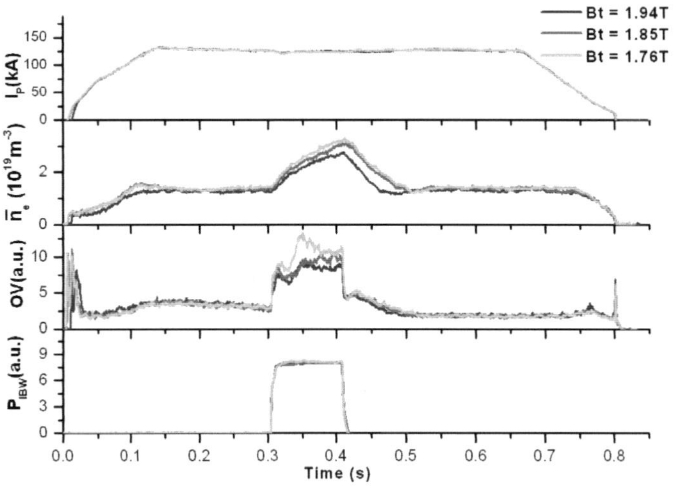

Figure 3. Waveforms for IBW shots with different B_T.

where ΔT_e is maximum value in IBW heating phase, $\bar{n}_{e0}$ is the center line average density at the time of ΔT_e, P_{IBW} is the power of IBW at that time. The heating effect of different power of IBW with the same target plasma parameters ($\bar{n}_{e0}$ (OH) = $1.4\times10^{19}\text{m}^{-3}$, B_T = 1.94 T) are show in figure 2b. The heating efficiency of IBW is consistent and not changed with the power of IBW. The heating effect of IBW with different B_T was also investigated. Figure 3 shows the waveform of three discharges with IBW heating. All of parameters are the same except B_T. The power of IBW was about 300 kW at flat top in these shots. The profiles of T_e at B_T = 1.85 T and 1.76 T are shown in figure 4a. It can be seen the core plasma was still heated in lower toroidal filed. It seems like the power deposition profile on electron of 27 MHz IBW at this target plasma was unchanged with toroidal magnetic filed. On the other hand, the temporary heating efficiency of IBW increased with B_T (as shown in figure 4b). At the same time, both the growth rate and the maximum value of electron density during IBW heating were higher in lower B_T plasma (as shown in figure 4). The impurity (OV) had the similar behavior as electron density.

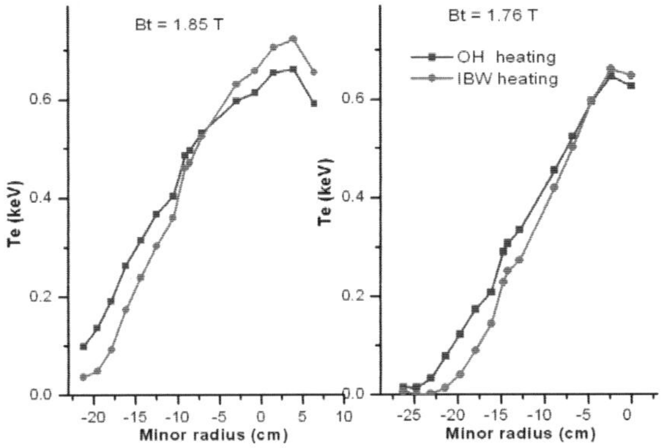

Figure 4a. The profiles of the electron temperature of the same shots in figure 3.

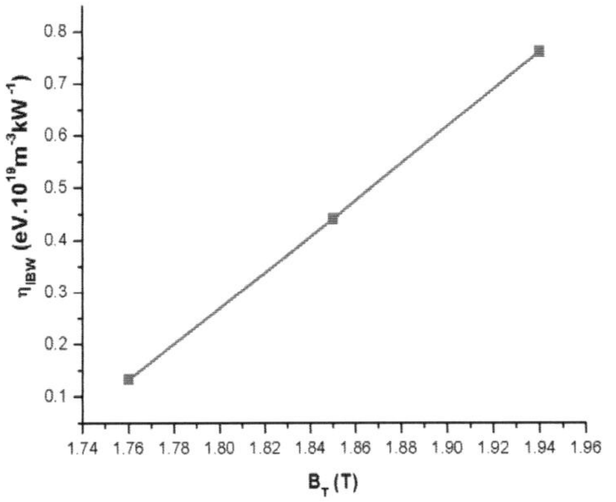

Figure 4b. The heating efficiency of IBW with different power B_T.

The electron density in HT-7 was measured with a five-channel far-infrared (FIR) hydrogen cyanide (HCN) laser interferometer [18]. The exact density profile is quite difficult to be deduced from this diagnostics. The ratio of n_{e0}/n_{e20cm} (n_{e20cm} is the line average density at r = 20 cm) can denote the peaked degree of the electron density profile. Figure 5a shown the time evolution (the same discharge as in figure 1) of n_{e0}/n_{e20cm}, n_{e0}, and T_e. The electron temperature dropped gradually with the increasing electron density after 40 ms of IBW. The n_{e0} increased from 1.4 to 3.4 $\times 10^{19}$m^{-3} while the center T_e dropped to half as that in OH heating. The electron density profile became wider and wider in the initial 80 ms of IBW heating and kept the width in the later time, however, the radiation profiles (as shown in figure 5b) measured with a XUV bolometer array became more peaked with time. There are two possible reasons for the peaked XUV profiles. One maybe the impurity concentration in center plasma. The other maybe the edge electron temperature during IBW heating became too low and radiation from this part cannot be effectively detected by XUV array. The impurity spectra measured with a PI SP-300i spectrometer are also shown in figure 6. The light impurity line can be observed from the spectra. Boronization is main wall cleaning method in HT-7. The boron film in these discharges with IBW heating is neither fresh nor old. The distinct increase of carbon and boron lines from the peripheral plasma (r = 20 cm) indicated that quite a number of impurities entered to the plasma during IBW heating. Compared to the peripheral plasma, only OV and CVI charge-exchange line from the center plasma increased during IBW heating. The reason is only the highly stripped impurity ions can penetrate the center plasma due to the higher electron temperature in center plasma. So the contribution from the impurity is also a considerable mechanism for the density increase during IBW heating. The maximum heating efficiency of 27 MHz IBW was no more than 0.75 (as shown in figure 4b). The limited heating effect of IBW was counteract and greatly exceeded by the cooling effect due to the increasing electron density and impurity radiation, which might be the main reason for the gradual decrease of the electron temperature during the later phase of IBW heating.

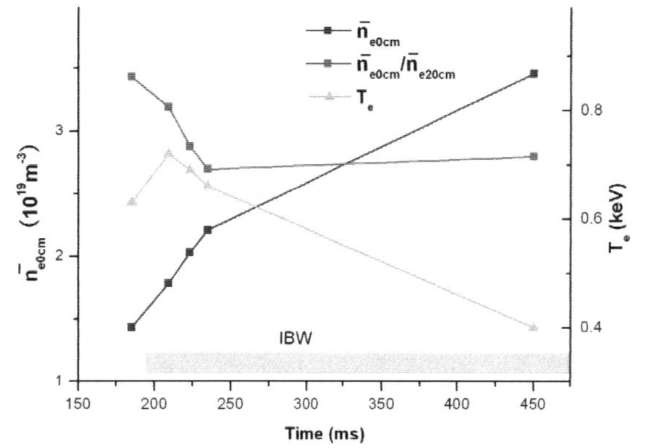

Figure 5a. Time evolution of of $\bar{n}_{e0}/\bar{n}_{e20cm}$, $\bar{n}_{e0}$, and T_e

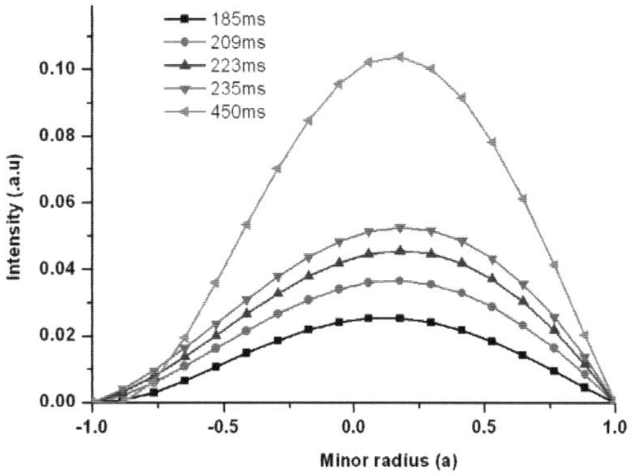

Figure 5b. The profiles of radiation power measured with XUV array.

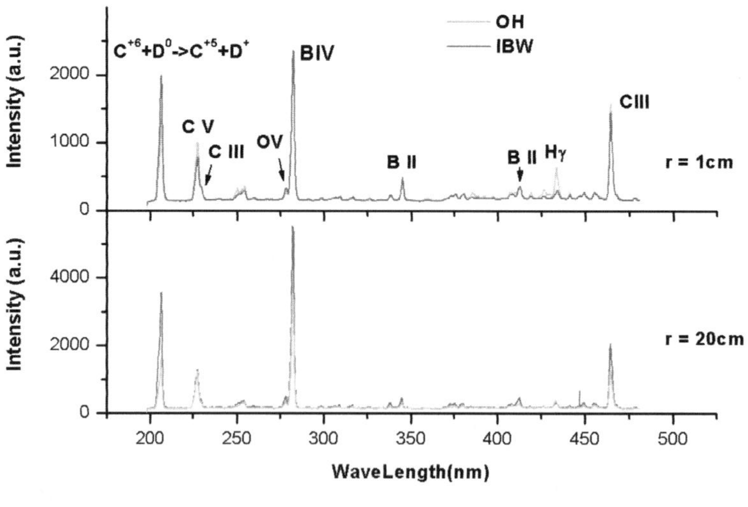

Figure 6. The spectra of IBW shot. The intensity had been normalized by electron density.

Improved particle confinement and suppression of MHD are ordinary phenomena in IBW heating plasma in HT-7. It can be seen in figure 7 the drop of D_α and suppression of MHD simultaneously appeared during IBW heating. The fluctuation and transport in SOL had been investigated using a Langmuir probe in previous 30 MHz IBW heating experiments in HT-7 [13, 16, 19]. Both heat flux and particle flux in SOL decreased during IBW injection (as shown in figure 8a). B_T was 1.92 T in the 30MHz IBW heating experiment with Langmuir probe, which made the $5/2\Omega_D$ layer located in the plasma edge of $r \approx 26$ cm on the low field side, where an IBW is expected to induce sheared flows. Such sheared flows were indeed observed with Langmuir probe. The results of Langmuir probe showed the averaged poloidal phase velocity of turbulence turned from the ion diamagnetic direction to the electron diamagnetic direction in the SOL after IBW switched on. An additional inward (negative) E_r was generated in the plasma edge during IBW heating (as shown in figure 8b). The structure of clear E_r wells and the obvious increase of E_r shear can be seen in this figure. The shearing rate, $\omega_{E\times B}$, of a turbulent structure in IBW-heating phase exceeded the ambient turbulence decorrelation rate, $\Delta\omega_D$, for the drift-wave-like turbulence. This strong shear decorrelation effect produced a distinct weak turbulence regime in the boundary plasma and might account for the improved particle confinement.

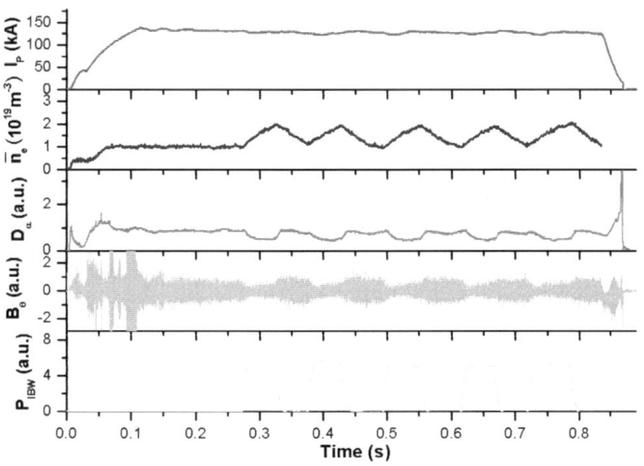

Figure 7. The discharge with IBW modulation.

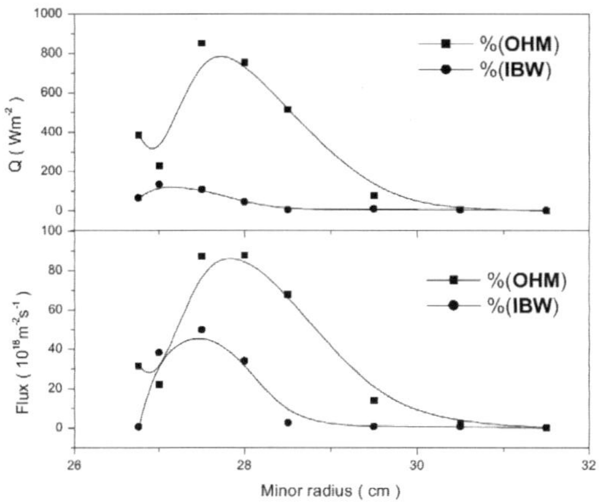

Figure 8a. Profiles of heat flux (up part) and particle flux (low part) in SOL.

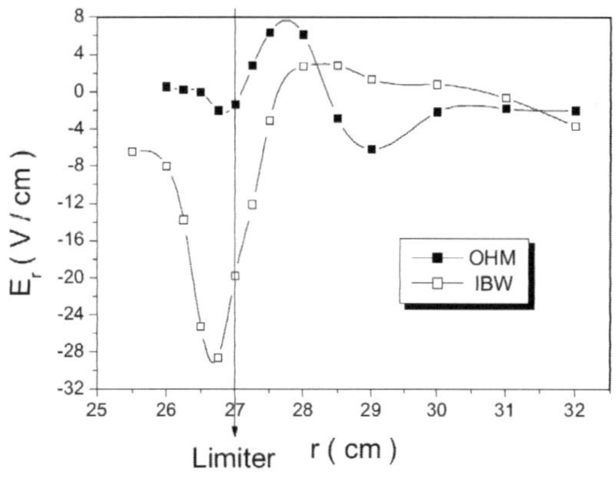

Figure 8b. Profiles of the radial electric filed in SOL.

Journal of Physics: Conference Series **123** (2008) 012019 doi:10.1088/1742-6596/123/1/012019

3. Summary

The electron heating effect by 27 MHz IBW was investigated in the deuterium plasma on HT-7. The core plasma electron heating could be obviously observed in the initial 40 ms during IBW injection. It seems like the IBW power deposition profile on electron at this target plasma was unchanged with B_T. But the temporary heating efficiency of IBW increased with B_T. Except for the improved particle confinement, the contribution from the impurity is also a considerable source for the density increase during IBW heating. The limited heating effect of IBW was counteract and greatly exceeded by the cooling effect due to the increasing electron density and impurity radiation, which might be the main reason for the gradual decrease of the electron temperature during the later phase of IBW heating. With the measurement of Langmuir probe, an additional inward (negative) E_r was observed in the plasma edge during IBW injection. This strong E_r shear decorrelation effect produced a distinct weak turbulence regime in the boundary plasma and might account for the improved particle confinement.

References

[1] Ono M 1993 *Phys. Fluids* B**5** 241
[2] Ono M *et al.* 1985 *Phys. Rev. Lett.* **54** 2339
[3] Ogawa Y *et al.* 1987 *Nucl. Fusion* **27** 1379
[4] Seki T *et al.* 1992 *Nucl. Fusion* **32** 2189
[5] Ono M *et al.* 1988 *Phys. Rev. Lett.* **60** 294
[6] Jiang T W and Ono M 1990 *Bull. Am. Phys. Soc.* **35** 1954
[7] Takase Y, Moody J D, Fiore C L, McDermott F S, Porkolab M and Squire J 1987 *Phys. Rev. Lett.* **59** 1201
[8] LeBlanc B *et al.* 1995 *Phys. Plasmas* **2** 741
[9] Berry L A, Jaeger E F and Batchelor D B 1999 *Phys. Rev. Lett.* **82** 1871
[10] LeBlanc B *et al.* 1999 *Phys. Rev. Lett.* **82** 331
[11] Zhao Y P, Li J G, Xie J K, Meng Y D, Wu X C, He Y X, Xue D Y, Deng X and Shao Y G 1997 *Chin. Phys. Lett.* **14** 916
[12] Li J G *et al.* 1999 *Nucl. Fusion* **39** 973
[13] Mao J S, Xu G S, Phillips P, Song M, Zhao Y P, Gao X, Asif M and Luo J R 2003 *Plasma Science Technology and Engineering* **3** 445
[14] Wan B N *et al.* 2004 *Nucl. Fusion* **44** 400
[15] Qin C M, Zhao Y P, Shen W C, Liu H Q, Xu G S, Lv B, Xu J and HT-7 team. 2005 *Plasma Sci. Tech.* **7** 3002
[16] Xu G S, Wan B N, Song M, Ling B L, Li C F, Li J G and HT-7 Team 2002 *Fusion Energy: Proc. 19th Int. Conf. (Lyon)* (Vienna:IAEA) CD-ROM file EX/P3-08
[17] Ling B L, Xu Y C and Wan B N 2005 *Journal of Microwaves* **21** 66
[18] Jie Y X, Gao X, Cheng Y F, Yang K and Tong X D 2000 *Plasma Sci. Tech.* **2** 105
[19] Bao Y 2002 *Ph.D. thesis* (Institute of Plasma Physics, Chinese Academy of Sciences)

Journal of Physics: Conference Series **123** (2008) 012020 doi:10.1088/1742-6596/123/1/012020

Temporal evolution of particle transport of Super Dense Core plasma in LHD

K Tanaka[1], C Michael[1], L N Vyacheslavov[2], K Ida[1], R Sakamoto[1], H Yamada[1], J Miyazawa[1], S Masuzaki[1], T Morisaki[1], Y Suzuki[1], M Kobayashi[1], I Yamada[1], T Tokuzawa[1], K Kawahata[1], K Narihara[1], N Oyabu[1], A Komori[1], K Muraoka[3] and the LHD Experimental Group[1]

[1] National Institute for Fusion Science, Toki 509-5292, Japan
[2] Budker Institute of Nuclear Physics, 630090, Novosibirsk, Russia
[3] Chubu University, 1200 Matsumoto, Kasugai, Aichi 487-8501, Japan

E-mail: ktanaka@nifs.ac.jp

Abstract. A highly peaked density profile was obtained in pellet-injected discharges in LHD. The peaking factor, which is the ratio of the central to volume-averaged densities, increased from around 0.8 in the gas puff fuelled phase, up to more than 2.0 after multiple pellet injection. The core density reached to several times $10^{20}m^{-3}$. Temporal changes of density profiles were measured using a CO_2 laser imaging interferometer. The change of the particle flux was estimated from temporal variations of measured density profiles. The diffusion coefficient (*D*) and the convection velocity (*V*) were then obtained from temporal traces of the relationship between the normalized density gradient and the normalized particle flux. Thus-obtained diffusion coefficients increased monotonically toward the plasma edge, while the convection velocity was inwardly directed. These observations are clearly different from those in the low collision regime, where *D* was spatially almost constant and *V* was outwardly directed. Three different regimes of particle transport were identified after the pellet injection. The first phase is characterized by a constant density peaking just after the pellet injection. In this phase, negative slopes between the normalized particle flux and the normalized density gradient were found, implying that *D* and *V* could not be determined. The second phase is at an additional density peaking, where the normalized particle flux increased with an increase in the normalized density gradient. In this phase, the core diffusion coefficient ($\rho < 0.6$) was reduced compared with that before the pellet injection, and an inwardly directed pinch was observed. The third phase is with the density broadening, where the normalized particle flux was reduced with a decrease in the normalized density gradient. Here, the core diffusion coefficients were further reduced, and the core pinch velocity ($\rho < 0.8$) became close to zero. The transition from the first (the constant peaking phase) to the second phase (the additional peaking phase) occurred almost simultaneously in the entire region of the plasma, while the transition from the second to third phase (the broadening phase) occurred from the outer region to the inner region of the plasma. The collisionality dependence of *D* and *V* at $\rho = 0.4\sim0.7$ was studied in the wide range of collisionality ($\nu_b^* = 0.3\sim30$) from density modulation experiments and the SDC analysis. A minimum value of *D* was likely to exist in the plateau regime ($\nu_b^* = 1\sim5$) together with almost zero convection velocity. This may be favourable for future high density reactor operations of the heliotron reactor.

Journal of Physics: Conference Series **123** (2008) 012020 doi:10.1088/1742-6596/123/1/012020

1. Introduction

Large Helical Device (LHD) is a superconducting heliotron. The aim of the LHD project is to demonstrate scientific feasibility of helical reactors and also to provide comprehensive understanding of toroidal plasmas from comparison between heliotron and tokamak plasma behavior. One of the big advantages of LHD is good capability of steady state operation and of performance at high density. There is no operational limit in density like the Greenwald density limit in tokamaks, as the density limit in LHD is determined by the balance of input power and radiation loss. Recently, an extremely high density (~several times of $10^{20}m^{-3}$) was achieved by multiple pellet-injection with local island divertor [1]. This extreme high density shot is called "Super Dense Core (SDC)" plasma. The magnetic configuration and discharge condition were optimized to achieve as high density as possible [2,3]. SDC plasma suggests a possibility of a high density (~$5\times10^{20}m^{-3}$) and moderate temperature (~5 keV) fusion reactor scenario. However, details of particle transport characteristics have not yet been understood. In this article, particle transport of SDC was quantitatively studied from the temporal evolution of density profiles for the first time.

2. Experimental results

2.1 Temporal behavior of SDC plasma

Figure 1 shows temporal behavior of electron temperature and density. Here, the plasma was produced by 82.7, 84 and 168 GHz ECH at t = 0.15~0.67 sec then additionally heated by 3.6 MW co-injected and 6.7 MW counter-injected NBI at t = 0.3~2.17 sec. The magnetic configuration was the so called standard configuration, where the magnetic axis position (R_{ax}) was at 3.75 m and the toroidal magnetic field at the magnetic axis was 2.64 T.

Electron temperature profiles were measured using YAG Thomson scattering [4] and electron density profiles were measured using a 65-channel CO_2 laser imaging interferometer [5,6]. The CO_2 laser imaging interferometer covered from the plasma center to the plasma boundary. The spacing between channel chords were 7.5 mm [6]. Both profiles are a function of magnetic flux surfaces.

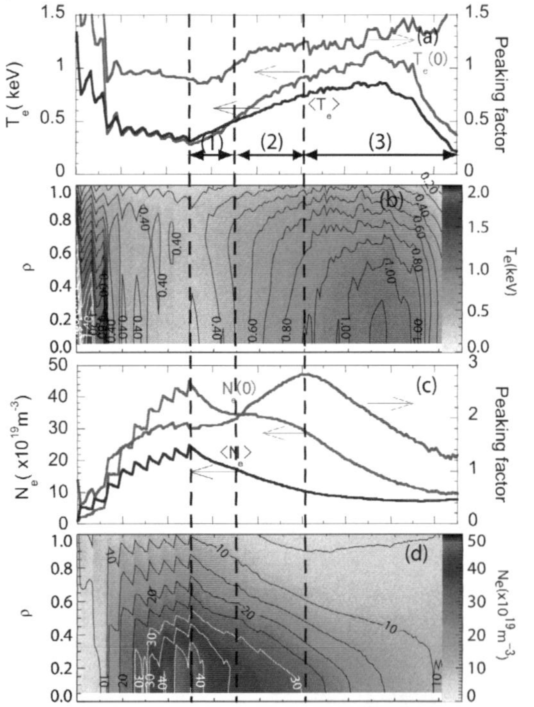

Figure 1. Temporal histories of (a) central electron temperature ($T_e(0)$), volume-averaged electron temperature $<T_e>$ and the peaking factor, (b) contours of equi-electron temperatures in a flux surface coordinate ($T_e(\rho)$), (c) central electron density ($N_e(0)$) and volume-averaged electron density $<N_e>$, and (d) contours of equi-electron densities in a flux surface coordinate ($N_e(\rho)$).

The magnetic flux was determined from the data base of equilibrium configurations calculated using the VMEC code [7]. In the magnetic flux coordinate used in the VMEC equilibrium calculation, the position of a flux surface was indicated by ρ. The value of ρ was defined as the square root of the ratio of a surface inside a particular toroidal flux against that of the last closed flux surface (LCFS). The value of ρ was approximately equal to the ratio of averaged radius of each flux to that of LCFS. The appropriate flux surface was selected from the Abel inversion of measured density profiles along laser chords, using least square fitting with regularization [6]. The procedure was carried out so as to minimize the discrepancy between measured line densities and line densities calculated by integration of reconstructed radial density profiles. The radial density was calculated every 10 msec. Thomson scattering data were obtained using four different lasers having repetition rates of 20 msec and 100 msec. T_e profiles obtained using Thomson scattering were converted into flux surface coordinates obtained as was described above. Then, T_e profiles were fitted with the 8th order polynomial function and temporarily interpolated at interferometer measurement times in order to compare T_e (from Thomson scattering) and N_e (from interferometer) profiles. The volume-averaged density was calculated inside LCFS. The peaking factors of T_e and N_e were defined as the ratio of the central value at $\rho = 0$ against the volume-averaged value.

The time history of density profiles is characterized by three different phases (phase (1), (2) and (3)) after pellet-injection which are separated by dashed lines in figure 1, where the phases were defined when the peaking factors changed. The phase (1) (t = 1.3~1.5 sec) is the constant peaking phase after the final pellet-injection. The central and volume-averaged densities both decreased resulting in keeping almost the same ratio, while electron temperature profiles started peaking. The phase (2) (t = 1.5~1.8 sec) is the additional peaking phase, where the density peaking factor increased again without pellet-injection. The phase (3) (t = 1.8~2.17 sec) is the broadening phase, and the density peaking factor started to decrease.

Electron density and temperature profiles before pellet-injection and profiles at the transitions of three phases are shown in figure 2. A hollow density profiles was observed for the gas puff fueling phase [9,10]. This hollow density profile is an observation particularly in outwardly shifted configurations, where neoclassical transport becomes the larger. Experimentally estimated outward convection velocities were comparable with neoclassical values [10]. Peaked density profiles were observed with pellet-injected discharges as shown in figure 2 (c). Discrete changes of density gradients

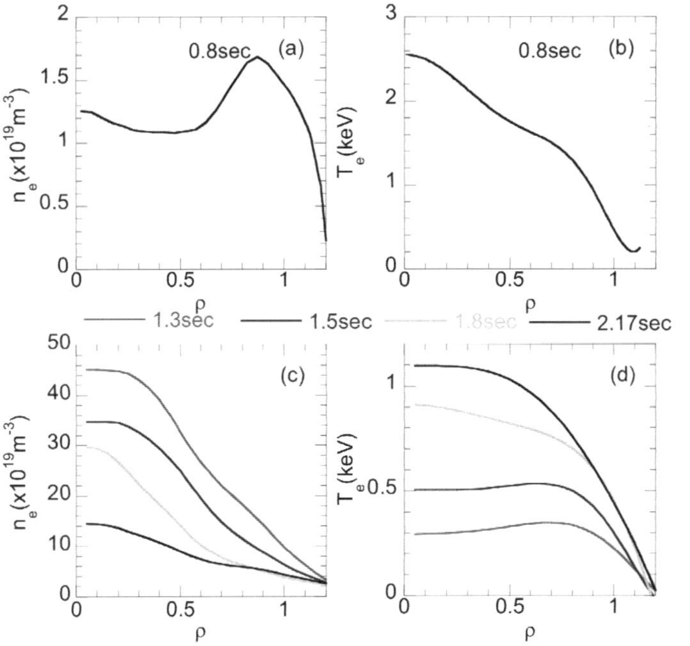

Figure 2. (a) N_e and (b) T_e profiles just before pellet-injection, and (c) N_e and (d) T_e profiles at three different phases. (a) was obtained from a 13 channel FIR laser interferometer [8], while (c) was obtained from a 64 channel CO_2 laser imaging interferometer [5,6].

were seen at $\rho = 0.6$ at $t = 1.8$ and 2.17 sec as shown in figure 2 (c). Changes of the peaking factor in the additional peaking and broadening phases are due to the change of density profiles at $\rho < 0.6$.

The observed dramatic variation of electron density profiles after the final pellet-injection indicates that particle transport changed temporally.

2.2 Analysis of particle transport

The equations of particle balance are

$$\frac{\partial N_e}{\partial t} = -\nabla \cdot \Gamma + S = -\frac{1}{r}\frac{\partial}{\partial r} r\Gamma + S \qquad (1)$$

$$\Gamma = -D\nabla N_e + N_e V \qquad (2)$$

Here, N_e is the local density, Γ is the particle flux, S is the particle source rate, and r is the spatial coordinate and is an averaged-radius of each flux surface. For simplicity of analysis, a cylindrical coordinate was employed. In equation (2), D is the diffusion coefficient and V is the convection velocity. Equation (1) is rewritten as

$$\Gamma(r) = \frac{1}{r}\int_0^r r\left(S - \frac{\partial N_e}{\partial t}\right)dr \qquad (3)$$

Particle fueling is from external gas puffing, wall recycling, beam fueling, and pellets injection. Pellets ablate within several milliseconds, thus pellets do not contribute to further particle fueling after several milliseconds from the final pellet-injection. The neutral beam injection in the discharges analyzed in this article was based on the negative ion tangential injection with a beam energy of 180 kV. Even in the low density regime (having a line-averaged density of $1\sim2\times10^{19}$ m^{-3}), the particle source from 11 MW NBI, where deposition profile was centralized and was around 0.6×10^{20} /m^{3}s, does not contribute to the peaking of density profiles [9,10,12]. In SDC discharge, the effect of the beam fueling is much smaller because density is much higher. Although external gas fueling and recycling are difficult to estimate, both are localized close to the plasma boundary. According to the three dimensional Monte Carlo simulation in LHD [10,11,12], the particle source is localized around $\rho = 1.1$ and decreases by one order of magnitude at $\rho = 1.0$ when an electron density is near to 1.5×10^{19} m^{-3}, and electron temperature is close to 250 eV at the same magnetic configuration. Therefore, it is reasonable to neglect particle source from gas fueling and wall recycling inside of LCFS ($\rho = 1.0$). Thus, the particle source rate S in equation (3) was neglected in the analysis, and the particle flux was calculated from temporal evolution of density profiles.

Equation (2) can be re-written as follows.

$$\frac{\Gamma}{N_e} = -D\frac{\nabla N_e}{N_e} + V \qquad (4)$$

The change of density profiles provides a data set of the relationship between Γ/N_e and $\nabla N_e/N_e$. If the straight line is fitted to an experimentally obtained trace of equation (4), the diffusion coefficient can be obtained from the slope of the trace, and the convection velocity can be obtained from an offset of the fitted line.

The density profiles derived from the Abel inversion of CO_2 laser interferometer data and their temporal evolution were not smooth. For stable analysis, the temporal profiles of radial densities were fitted by two-dimensional spline curves using the least square approximation. The comparison of reconstructed density profiles and fitted profiles is shown in figure 3.

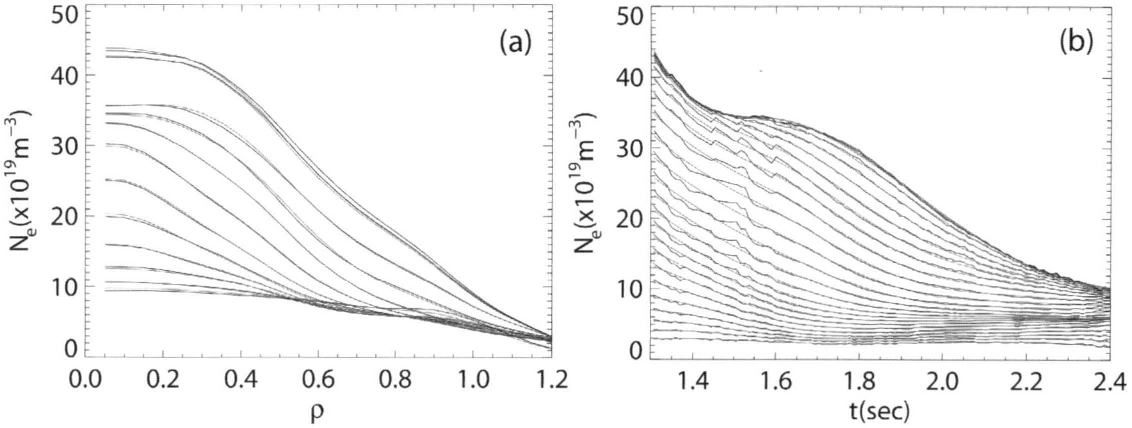

Figure 3. (a) Reconstructed radial density profiles (in black) and spline fitted profiles (in red), and (b) temporal traces of local densities from reconstructed profiles (in black) and from spline fitted data (in red). The data in (b) are of ρ = 0.025~1.025 (ρ = 0.05 step).

Figure 4 shows traces of Γ/N_e versus $-\nabla N_e/N_e$. (see figure 1 (c) for the three different phases). Data were obtained every 10 msec from t = 1.3 sec, namely just after the final pellet-injection, to t = 2.17 sec, namely the time of the NBI switch off. Shapes of traces vary depending on the radial location as is shown in figure 4. Temporally, the normalized particle flux went from high values, reaching minima to further maxima. The gradients and offset values of the traces between Γ/N_e versus $\nabla N_e/N_e$ changed before and after maxima and minima of normalized fluxes. These times are when particle transport changed, and we call these times as "transition times" of particle transport. The minima and maxima of Γ/N_e correspond to the first and second transitions, respectively.

As stated previously regarding to figure 1, three different phases (from t = 1.3 sec until the first transition, from the first to the second transition and from the second transition until t = 2.17 sec) were observed. The important point is that the transitions depend on radial locations. Figure 5 shows the spatio-temporal changes of the transitions. The first transition began almost simultaneously in the whole region, while the second transition started from the edge and moved to the core.

The data before the first transition (the constant peaking phase) is marked in green in figure 4. During this time period, the traces show negative derivatives and diffusion coefficients cannot be defined.

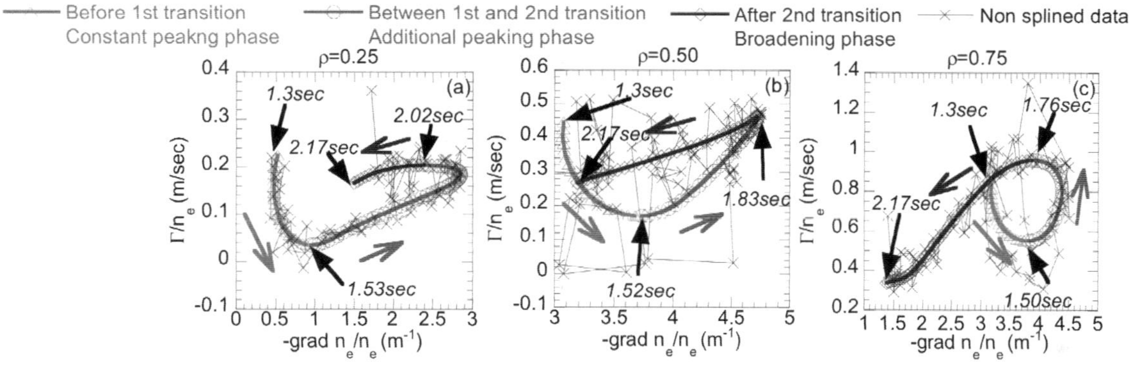

Figure 4. Experimentally obtained traces between Γ/N_e and $-\nabla N_e/N_e$, at (a) ρ = 0.25, (b) ρ = 0.5, and (c) ρ = 0.75. Colored symbols are from spline fitted profiles, while black crosses are from reconstructed profiles without spline fitting.

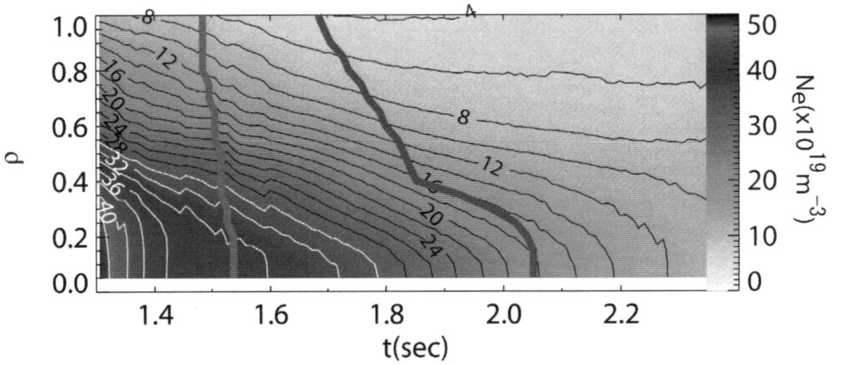

Figure 5. Spatio-temporal changes of transition times. The red and blue lines indicate the first and the second transitions, respectively.

The data between the first and the second transitions (the additional peaking phase) are marked in red in figure 4. The most of the data points showed positive slope at ρ = 0.25 and 0.50 as shown in figures 4 (a) and (b). At ρ = 0.5, the slope deviates from a straight line and follows a curbed line. This indicates that the diffusion coefficient and the convection velocity changed in time. On the other hand, it is difficult to determine any positive gradient at ρ = 0.75, thus the diffusion coefficient and the convection velocity can not be defined during this period.

The data after the second transition (the broadening phase) is marked in blue in figure 4, and almost a straight line can be fitted. This indicates that the diffusion coefficient and the convection velocity were temporally constant.

The diffusion coefficient and the convection velocity were obtained between the first and second transitions (the additional peaking phase) and after the second transition (the broadening phase). The profiles of *D* and *V* thus obtained are shown in figure 6. Values of *D* and *V* were derived from fitted lines when correlation coefficients were larger than 0.5 and positive values of *D* were obtainable.

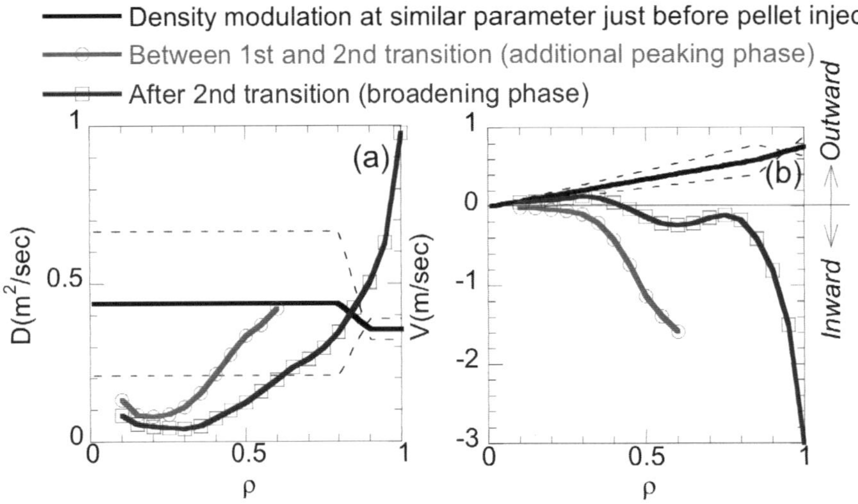

Figure 6. Profiles of (a) the diffusion coefficient (*D*) and (b) the convection velocity (*V*), obtained from density modulation experiments (black lines) and from the analysis of the SDC discharge after pellet-injection at time periods between the first and the second transition (red line) and after the second transition (blue line). Negative values of *V* indicates an inwardly directed pinch. Dotted lines indicate ranges of uncertainty in the determination of *D* and *V* from density modulation experiments

Although traces between Γ/N_e versus $\nabla N_e/N_e$ were not constant at some locations, a linear line was fitted to determine representative values of D and V.

The profiles of D and V from density modulation experiments are also plotted in black lines in figure 6. Electron temperature and density profiles at the density modulation experiments were almost the same as the ones just before the first pellet-injection (figures 2 (a) and (b)). The diffusion coefficient and the convection velocity were determined from the propagation of the modulated density and the density profile in an equilibrium state. The core and edge values of D and V were determined from the analysis. The analysis method was described in reference [12]. Because the profiles of D and V from the density modulation experiments represent D and V before the first pellet injection as stated above, we can argue the differences of D and V before and after pellet-injection.

As shown in figure 6, D and V profiles changed dramatically after the first transition (during the additional peaking phase) compared with D and V before pellet-injection. The diffusion coefficient increased monotonically toward the edge and the convection direction was reversed from outward to inward. Small values of the diffusion coefficient in the central region, together with an inwardly directed pinch, induced the additional peaking. After the second transition (the broadening phase), the convection velocity was nearly zero at $\rho < \sim 0.8$, resulting in an only diffusive process in this region. This caused a reduction of the density gradient in this region, and reduced the peaking factor.

In a low collisionality regime, the collisionality dependence of D and V had systematically been studied [10]. Figure 7 shows comparison of D and V from the density modulation experiments in the low collisionality regime and values from the SDC discharge in the high collisionality regime. Here, the electron ion-collision frequency was normalized by the trapped electron bounce frequency (ν^*_b), expressed as $\nu^*_b = \nu_{ei}/(\varepsilon_t^{3/2}\nu_T/qR)$. Here, ε_t is an inverse aspect ratio, ν_T is an electron thermal velocity, q is a safety factor and R is a major radius. The values of D and V were averaged between $\rho = 0.4\sim0.7$. The values of D and V in the additional peaking phase of the SDC discharge were averaged at $\rho = 0.4\sim0.6$, because D and V cannot be obtained from the analysis at $\rho > 0.6$ of additional peaking phase.

On the other hands, in the high collisionality regime, the diffusion coefficient from the SDC discharge shows a positive collisionality dependence. This is the same dependence with the neoclassical diffusion coefficient in Pfirsch-Schluter. regime. The convection velocity in high collisionality showed a negative collisionality dependence. The negative convection (inwardly directed convection) increases with collisionality.

Figure 7 suggests the existence of a minimum value of the diffusion coefficient at around $\nu_b^* = 1\sim5$, and the convection velocity may be at minimum of around zero when the diffusion coefficient takes

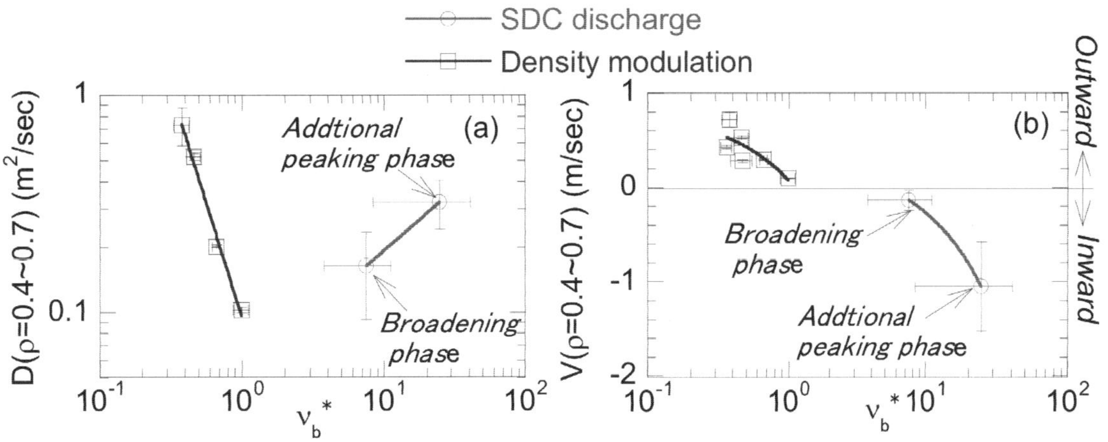

Figure 7. Collisionality dependences of (a) D and (b) V. Values from density modulation experiments (in blue) and from the SDC discharge (in red) are shown.

the minimum value. This diffusion minimum with zero convection point will be favorable for the future reactor operation of high density and moderate temperature ($n_e(0)$ ~ 5×10^{20}m^{-3}, $Ti(0)$ ~ 5 keV) scenario, because good particle confinement without impurity accumulation may be possible.

3. Discussion and summary

The behavior of particle transport in the pellet-injected super dense core (SDC) plasma was investigated from the temporal change of density profiles. As the result, three different phases were identified which occurred after multiple pellet-injection; namely, the 1st phase was characterized by a decrease of the normalized flux together with an increase of negative normalized density gradient, occurring after the final pellet-injection, the 2nd phase occurred by a transition simultaneously in the whole region of the plasma where the normalized flux increased together with an increase in the negative normalized density gradient, and finally the 3rd phase occurred by the second transition when both the diffusion coefficient and the inward pinch reduced. This second transition appeared initially at the edge then penetrated to the core region.

Peaked density profiles were achieved by pellet injection at first. Then, additional peaking occurred by a small value of the diffusion coefficient in the core and an enhanced diffusion coefficient in the edge and an inwardly directed pinch. After the second transition, both the diffusion coefficient and the inwardly directed pinch were reduced.

The collisionality dependence of D and V suggest a plasma operation having a minimum value of D together with zero convection at ν_b^* = 1~5. An operation at around this collisionality may be a candidate of future heliotron reactor operation.

The analysis used spatio-temporal derivatives of radial densities. The accuracy of spline fitting affects the results. In order to confirm results obtained here, optimization of the fitting is necessary. A coupling of particle transport with heat transport is also important and will be studied in future.

Acknowledgements

This work was supported by the National Institute for Fusion Science budget NIFS05ULHH511

References

[1] Oyabu N *et al.* 2006 *Phys. Rev. Lett.* **97** 055002-1
[2] Yamada H *et al.* to be published in *Plasma Phys. Control. Fusion*
[3] Sakamoto R *et al.* to be published in *Journal of Plasma and Fusion Research*
[4] Narihara K 2001 *Rev. Sci. Instrum.* **72** 1122
[5] Tanaka K *et al.* 2004 *Rev. Sci. Instrum.* **75** 3429
[6] Tanaka K *et al.* 2007 *Plasma and Fusion Research* **2** S1033 (http://www.jspf.or.jp/PFR/PFR_articles/pfr2007S1.html)
[7] Hirshman S P and Whitson J C 1983 *Phys. Fluids* **26** 3553
[8] Kawahata K, Tanaka K, Ito Y, Ejiri A and Okajima S 1999 *Rev. Sci. Instrum.* **70** 707
[9] Takenaga H, Tanaka K, Muraoka K *et al. Nucl. Fiusion* **48** (2008) 075004
[10] Tanaka K *et al.* 2007 *Fusion Sci. Tech.* **51** 97
[11] Shoji M *et al.* 2005 *J. Nucl. Mater.* **337–339** 186
[12] Tanaka K *et al.* 2006 *Nucl. Fusion* **46** 110
[13] Murakami S *et al.* 2002 *Nucl. Fusion* **42** L19

Journal of Physics: Conference Series **123** (2008) 012021 doi:10.1088/1742-6596/123/1/012021

Observation of m/n=2/1 magnetic island on the foot point of electron internal transport barrier using soft x-ray CCD camera in the Large Helical Device

T Kobuchi[1], Y Liang[2], K Ida[3], M Yoshinuma[3], K Y Watanabe[3] and LHD group[3]

[1]Department of Quantum Science and Energy Engineering, Tohoku University, Sendai, Miyagi, 980-8679, Japan
[2]Institut fuer Plasmaphysik, Forschungszentrum Juelich, D-52425, Germany
[3]National Institute for Fusion Science, Toki, Gifu, 509-5292, Japan

E-mail: `takashi.kobuchi@qse.tohoku.ac.jp`

Abstract. The existence of m/n=2/1 magnetic islands in the plasma with an electron internal transport barrier (ITB) is identified in tangential soft x-ray emission images measured with a soft x-ray CCD camera using the Fourier-Bessel expansion reconstruction technique in the Large Helical Device. A clear m/n=2/1 magnetic island is observed in the discharge with ECRH and NBI in the direction antiparallel to the equivalent plasma current (counter-NBI), where the magnetic shear is expected to become small enough to cause the formation of a magnetic island. On the other hand, no magnetic island is observed in the ECRH and NBI in the direction parallel to the equivalent plasma current (co-NBI), where the magnetic shear is expected to be sufficiently high.

1. Introduction

The internal transport barriers (ITBs) characterized by the large temperature gradient interior of the plasma have been observed in many tokamaks [1-6]. The ITBs in the electron thermal energy have been observed with dominant electron cyclotron heating in plasmas with negative magnetic field shear in many tokamaks and stellarator/heliotorn plasmas. The understanding of thermal transport barriers physics in plasmas is important to obtain high performance plasma with high temperature and improved energy confinement. In this paper, the role of magnetic shear in formation of the electron internal transport barrier is discussed. The electron ITB has been observed associated with the transition from ion root to the electron root of the radial electric field (E_r) in low-density plasmas heated by neutral beam injection (NBI) and electron cyclotron resonance heating (ECRH) in many stellarators [7-9] In the Large Helical Device (LHD), the ITB with respect to electron thermal transport were observed when the ECRH power was highly localized on the center of the plasma sustained by the NBI [10, 11]. Although the electron ITB is observed regardless of the direction of NBI, both in the direction parallel (co-injection) and antiparallel (counter-injection) to the equivalent plasma current, the radial profile of electron temperature observed shows differences between the co-NBI and counter-NBI plasma because of the existence of magnetic islands. In the counter-NBI discharge, as the foot point of ITB is located just on the inner side of the magnetic island, a clear foot point is typically observed. This is in contrast to the unclear foot point of ITB in co-NBI discharge, where there is no magnetic island. The existence of a magnetic island also affects the hardness of the transition from L-mode to ITB mode. A sharper transition is observed in the discharge with counter-NBI than the discharge with co-NBI. Transport analysis shows that both a standard electron thermal diffusivity and an incremental electron thermal diffusivity are reduced significantly in the ITB [12]. The improvement of electron transport in the ITB is resulted from the relation of these two diffusivities. In the LHD, the size of 2/1 magnetic island can be controlled by the cancellation magnetic field using external field coils and it has been demonstrated that the 2/1 magnetic island contributes to the formation of the ITB. Studies of the magnetic island physics and active control of magnetic islands have been reported [13-15]. The shear of the rotational transform is strongly dependent on the radial profile of the net toroidal plasma current [16]. The weak shear of the rotational transform causes the growth of the magnetic island. The measurement system with the multichannel detector sensitive to electron cyclotron emission (ECE) or soft x-rays has been applied to observed the rotating island using tomography techniques [17, 18]. A high-speed x-ray camera with a micro-channel plate (MCP) detector coupled with a scintillator and a fast intensified charge-coupled device (ICCD) are employed to measure a tangential x-ray image for the study of magnetohydrodynamics (MHD) instabilities [19]. In the LHD, a tangential x-ray CCD camera has been applied for measurement of the magnetic axis [20, 21] and the magnetic island.

In this paper, measurement of the magnetic island with a soft x-ray CCD camera for ITB plasmas in the LHD is described. The relationship between the magnetic island and the magnetic shear of the rotational transform in the ITB plasma sustained by co-NBI and counter-NBI with additional ECRH is discussed.

2. Experimental setup and analysis

The LHD is a heliotron type device, which has superconducting coils with poloidal and toroidal period numbers of $L/M = 2/10$, plasma major radius $R_{\rm ax}^{\rm v}$ of 3.5-4.1 m, and averaged minor radius of the last closed magnetic surface a of 0.6 m [22]. The target plasma has been sustained using three NBI (two counter-NBI and one co-NBI) with a beam energy of around 140-180 keV and absorbed power of 1-3 MW. The plasma with the electron ITB is formed by injection of ECRH [82.7 GHz/250 kW, 84 GHz/420 kW and 168 GHz/400 kW] highly localized on the magnetic axis as shown in figure 1 (a).

The radial profile of the electron density ($n_{\rm e}$) is measured with a 13-channel far-infrared (FIR) laser interferometer system at vertically elongated cross-section, while the radial profile of the electron temperature ($T_{\rm e}$) is measured with a 200-channel YAG-laser Thomson scattering system at the horizontally elongated cross-section in the LHD [23, 24].

The net toroidal plasma current is measured with a Rogowski coil. The co-NBI and counter-NBI drive the net toroidal current in the direction parallel and antiparallel to the equivalent plasma, respectively (see figure 1 (b)). (Here the positive current represents the current parallel to the equivalent plasma current.) The net toroidal plasma currents at the time slice of the soft x-ray image measurements are +3 kA/T and -15 kA/T with co-NBI and counter-NBI, respectively. The current radial profiles used in this calculation for the discharge with positive and negative current are $(1-\rho^2)$ and $(1-\rho^8)^2 - 0.9(1-\rho^2)^3$, respectively, as shown in figure 2. The radial profiles of toroidal current used here are consistent with the measurements of rotational transform by Motional Stark effect spectroscopy, which suggest the peaked current profile for co-NBI phase and hollow profile in counter-NBI [16]. The difference in the radial profile of the current profile is due to the difference in deposition profiles of NBI. The deposition of counter-NBI is peaked off-axis, while the beam deposition of co-NBI is peaked on-axis. The current profile in the discharge with counter-NBI tends to be hollow because the inductive

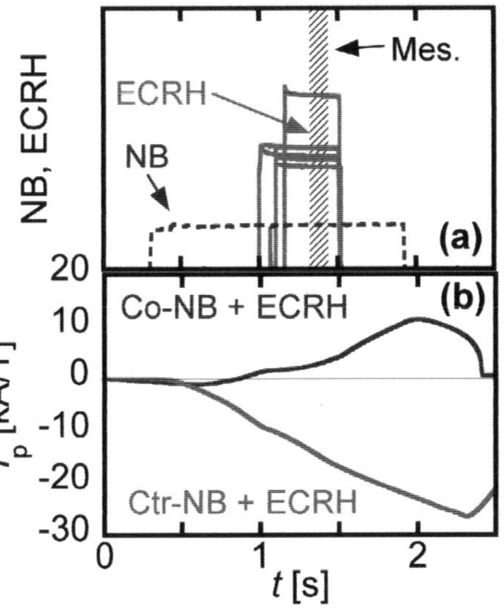

Figure 1. (a) Waveform of ECRH and NBI in this experiment. (b) Time evolution of the net toroidal current in the plasma sustained by co-NBI and counter-NBI, respectively

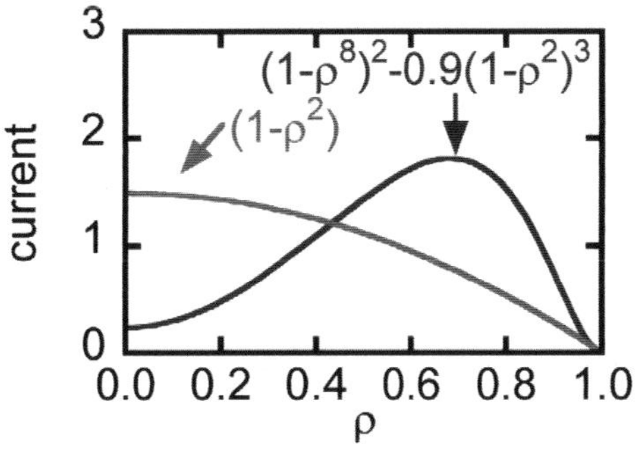

Figure 2. Radial profiles of net toroidal current used in this calculation of the rotational transform profile with the VMEC code. The current radial profiles are assumed (1-ρ^2) and $(1-\rho^8)^2-0.9(1-\rho^2)^3$ to the positive current and negative current, respectively

current (return current) flows in the direction to compensate the current driven by NBI near the plasma center, where the current diffusion time scale is large due to high temperature. The soft x-ray CCD camera system, which consists of pinholes, Be filters, a shutter and a CCD detector sensitive to soft x-rays, has been installed on a tangential view port on the LHD to measure the shape of the magnetic flux surface from a tangential soft x-ray image [25]. By choosing the appropriate combinations of pinhole size and thickness of Be filters, the soft x-ray image can be measured for the plasmas over a wide range of electron temperatures and densities. The soft x-ray images were measured with an exposure time of 0.1 s and with an interval of 0.55 s. The interval is limited by the maximum readout speed of the CCD camera. It should be noted that this soft x-ray CCD camera is applied to measure non-rotation magnetic islands, and not to measure the rotating magnetic island often observed in tokamaks.

The magnetic island structure can be expressed by introducing the x-ray emissivity $g(\rho,\theta,\phi)$ with the Fourier-Bessel expansion [26] as

$$g(\rho,\theta,\phi)=\sum_{m=0}^{M}\sum_{n=0}^{N}\sum_{l=0}^{L}[a_{m,n}^{l}\cos(m\theta-n\phi)+b_{m,n}^{l}\sin(m\theta-n\phi)]J_m(\lambda_m^{l+1}\rho), \tag{1}$$

where ρ, θ and ϕ are the normalized minor radius and the poloidal and toroidal angels, respectively. λ_m^l is the l-th zero of m-th order Bessel function $J_m(z)$. Here, the number of expansion terms, L, is taken to be 6, which is sufficiently large to reconstruct the soft x-ray emission profile measured [20]. The soft x-ray intensity, I_{ij}^c, can be calculated by integrating the soft x-ray emission, $g(\rho,\theta,\phi)$, numerically along the line-of-sight for each zone from the wall to the pinhole using the database of the MHD equilibrium calculated with three-dimensional free boundary code VMEC [27], as

$$I_{ij}^c=\int g(\rho,\theta,\phi)ds. \tag{2}$$

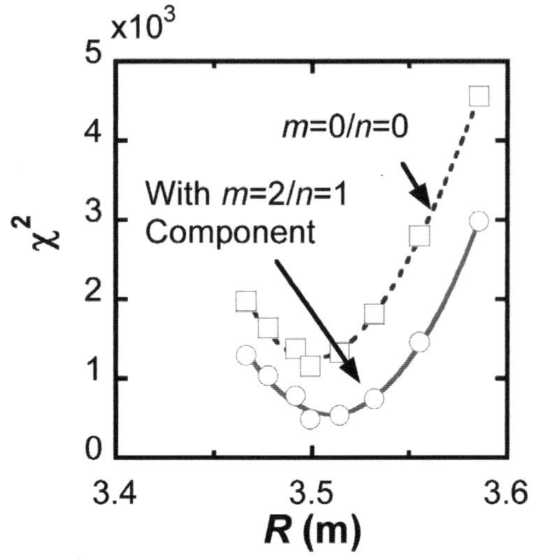

Figure 3. (color online) Distribution of the differences between the calculation and measurement of the soft x-ray intensity, χ^2, as a function of the magnetic axis $R_{\rm ax}$. Comparison of χ^2 distribution for m/n=0/0 mode with m/n=2/1 mode.

The expansion coefficients of $a^l_{m,n}$ and $b^l_{m,n}$ are determined by the best fit of calculated intensity I^c_{ij} to the intensity I^m_{ij} measured with the soft x-ray CCD camera. Differences between these two intensities measured and calculated using best fit $a^l_{m,n}$ and $b^l_{m,n}$ are given as

$$\chi^2 = \frac{\sum_{i=1}^{N_x}\sum_{j=1}^{N_y}(I^m_{ij} - I^c_{ij})^2}{N_x N_y}. \tag{3}$$

Here, N_x and N_y are the number of spatial zones in horizontal and vertical directions, respectively. Figure 3 shows the square of the difference χ^2 between two-dimensional profiles of the calculated emission and that measured with the soft x-ray CCD camera as a function of the magnetic axis of the MHD equilibrium calculated with the VMEC code, which gives the minimum difference. Then, the x-ray intensity measured is fitted with the intensity integrated along the line-of-sight assuming the x-ray emission is constant on the same magnetic flux surface.

The database of the MHD equilibrium consists of a few hundred magnetic configurations with several different pressure profiles. Many MHD equilibria are calculated to cover the range of pressure profiles from a peaked to a broad profile, which is realized in the LHD.

3. Observation of the m/n=2/1 magnetic island using the soft x-ray emission image measured the soft x-ray CCD camera

The tangential x-ray images are measured with the soft x-ray CCD camera with a pinhole 0.4 mm in diameter and a Be filter 190 μm in thickness from a low-density ($n_{\rm e} = 0.5 \times 10^{19}{\rm m}^{-3}$) ITB plasma. The vacuum magnetic axis, $R^{\rm v}_{\rm ax}$, is 3.5 m and the magnetic field, $B_{\rm t}$, is 2.854 T. The plasma net current, $I_{\rm p}$, measured with the Rogowski coil is +3 kA/T and -15 kA/T at the time slice of soft x-ray image measurements in the discharge with co-NBI + ECRH and counter-NBI +ECRH, respectively. Figures 4 (a) and (b) show the tangential images of the soft x-ray intensity measured with the soft x-ray CCD camera for the ITB plasma sustained by co-NBI and counter-NBI with additional ECRH, respectively. Figures 4 (c) and (d) show contour plots of the two-dimensional tangential x-ray intensity profiles measured with the soft x-ray CCD camera for the plasma with the ITB sustained co-NBI and counter-NBI with the additional ECRH, respectively. It is difficult to identify the structure of magnetic islands clearly in the raw data, as shown in figure 4, because of the integration effect of the viewing line of the CCD camera.

To be clarify the magnetic island structure in the ITB plasma, three poloidal/toroidal mode structures of m/n=1/1, 2/1, 3/1 are included in the x-ray emissivity $g(\rho, \theta, \phi)$ expressed in equation (1), then the analysis that the best fitted MHD equilibrium is taken from database, is curried out. The squares and circles indicate the results of analysis evaluated from the Bessel function including the poloidal/toroidal mode structure with m/n=0/0 mode and m/n=0/0 mode with m/n=2/1 mode, respectively. The minimum of χ^2 for the mode structure including m/n=2/1 is smaller than that with the mode number m/n=0/0. This result suggests that the magnetic island exists in the ITB plasma sustained by counter-NBI.

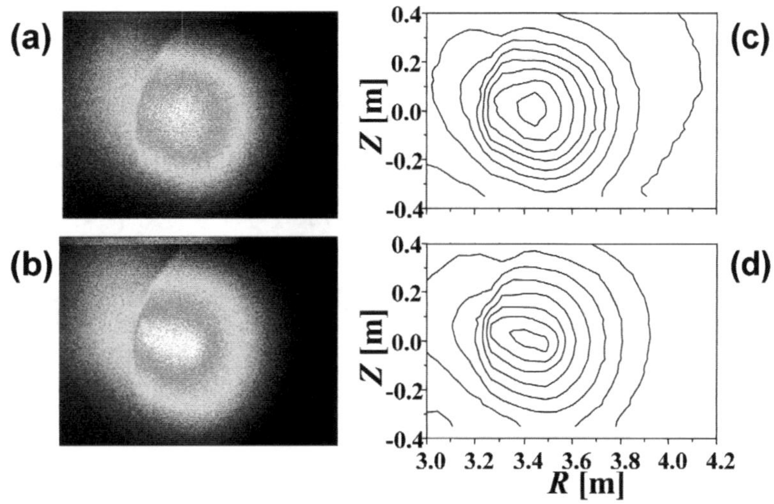

Figure 4. (color online) Tangential soft x-ray emission images measured with the soft x-ray CCD camera for the ITB plasmas sustained by (a) co-NBI and (b) counter-NBI with the additional ECRH. (c), (d) Contour plots of the tangential soft x-ray emission images measured with the soft x-ray CCD camera.

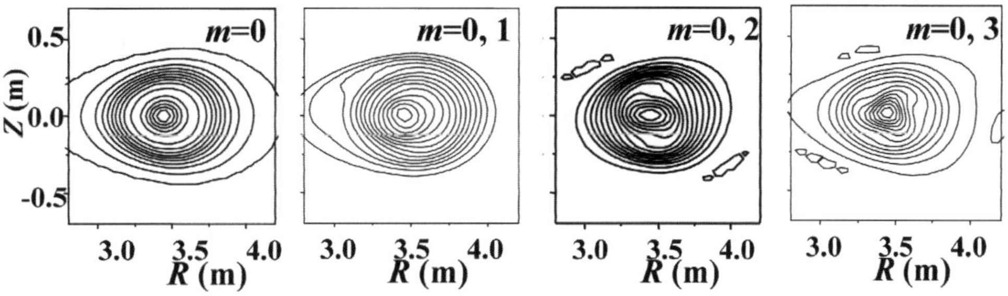

Figure 5. Contour plots of the soft x-ray emission reconstructed from the tangential soft x-ray emission images at a horizontal elongated cross-section for each mode structure including m/n=1/1, m/n=2/1 and m/n=3/1.

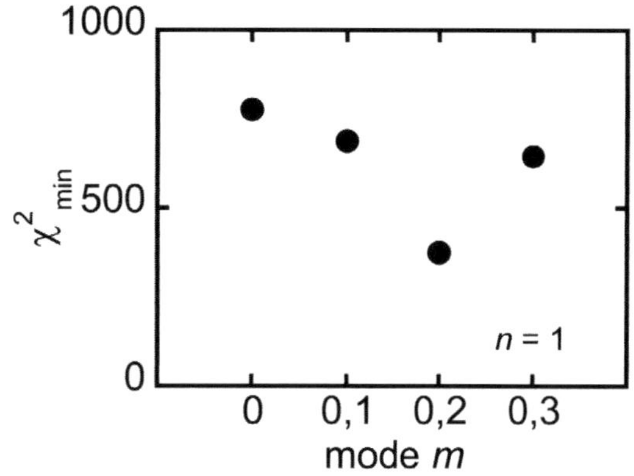

Figure 6. Comparison with the minimum χ^2 ($\chi^2_{\rm min}$) for analysis with different poloidal mode structure, m/n=0/0, and including m/n=1/1, m/n=2/1, and m/n=3/1 mode components.

Two-dimensional soft x-ray emission profiles are reconstructed from the soft x-ray images measured with the soft x-ray CCD camera using best fitted coefficients of the Bessel function including each mode. The contour plots of two-dimensional x-ray emission profile at a horizontal elongated cross-section for each mode structure including m/n=1/1, m/n=2/1 and m/n=3/1 are shown in figure 5. The magnetic island for each mode is reconstructed in the contour plots of the two-dimensional soft x-ray emission.

Figure 6 shows a comparison of the minimum χ^2, $\chi^2_{\rm min}$, evaluated for the mode number m/n=0/0 with that considered including three mode numbers, m/n=1/1, m/n=2/1 and m/n=3/1, respectively. There is only a slight improvement in the χ^2 for the cases with m=1 and m=3 modes compared with the case with m=0 mode. A large decrease in the minimum χ^2 is seen by taking m/n=2/1 mode. This result clearly shows the existence of the m/n=2/1 magnetic island in the ITB plasma.

4. Properties of the rotational transform in the plasma with ITB sustained by co-NBI and counter-NBI with additional ECRH

In the LHD, the magnetic islands with m/n=2/1 mode have been observed from the tangential soft x-ray emission images measured with a soft x-ray CCD camera in the plasma with the electron ITB. The radial profile of the normalized soft x-ray emission on the mid-plane for the ITB plasma with counter-NBI shows a flattening region in the magnetic island, while no such flattening region is observed in the discharge with co-NBI, as shown in figures 7 (a) and (b).

It should be noted that the large gradient of emission (indicating large gradient of electron temperature) near the core region compensate for the flattening in the region of the magnetic island in the discharge with counter-NBI. Therefore, the central electron temperature in the discharge with counter-NBI is as high as that in the discharge with co-NBI, despite the existence of a flat region at the magnetic island. The radial profile of soft x-ray emission is consistent with the electron temperature measured by YAG Thomson scattering.

Figure 8(a) shows the radial profiles of the rotational transform calculated with VMEC code for the vacuum magnetic field and the magnetic field with a negative current and a positive current. The current profile used in the calculation of the rotational transform is $(1-\rho^2)$ for the positive current of +3 kA/T and $(1-\rho^8)^2-(1-\rho^2)^3$ for the negative current of -15 kA/T, as shown in figure 2.

Although the rotational transforms in the outer region $\rho > 0.4$ change as expected from the sign of total current (increase in ι for positive current and decrease in ι for negative current), the rotational transform near the magnetic axis always increases. This is because the change in rotational transform due to the shift in the magnetic axis is associated with the change in net

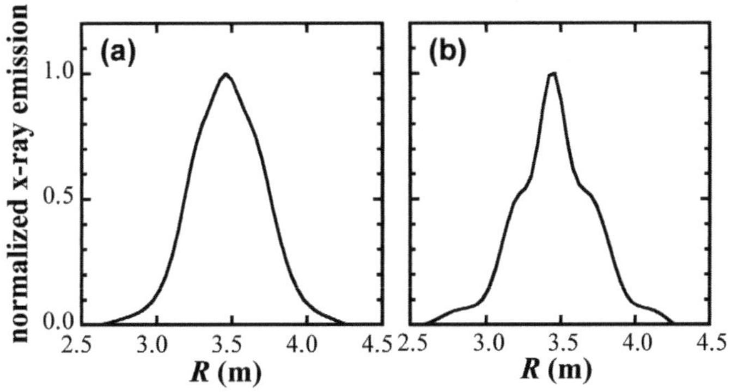

Figure 7. Radial profiles of the normalized soft x-ray emission reconstructed from the tangential soft x-ray emission images measured with the soft x-ray CCD camera for the ITB plasmas sustained with (a) co-NBI + ECRH and (b) counter-NBI + ECRH.

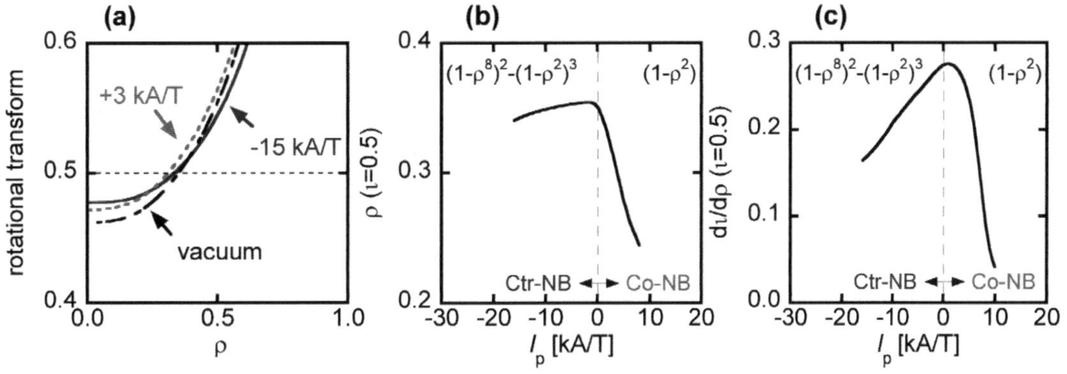

Figure 8. (color online) (a) The radial profile of rotational transform for vacuum magnetic field and magnetic field with net toroidal plasma current of +3kA/T with current radial profile of $(1-\rho^2)$ and -15 kA/T with $(1-\rho^8)^2-(1-\rho^2)^3$. (b) Normalized minor radius(ρ) at the rotational transform, ι, of 0.5. (c) Magnetic shear ($\mathrm{d}\iota/\mathrm{d}\rho$) at the location of $\iota = 0.5$.

current [28]. In this magnetic field configuration in LHD, when the net current is positive, the magnetic axis shifts outward, and rotational transform tends to decrease. However, the increase in rotational transform due to the peaked positive current is typically larger than the decrease due to magnetic axis shift. In contrast, when the net current is negative, the magnetic axis shifts inward and rotational transform increases and overcomes the decrease of rotational transform due to the inward shift of the magnetic axis.

Figures 8(b) and (c) show the location (ρ) of the rotational transform of 0.5 for the net toroidal plasma current and the shear of the rotational transform ($\mathrm{d}\iota/\mathrm{d}\rho$). The $\iota = 0.5$ location moves gradually inside and the magnetic shear becomes weaker as the magnitude of the negative current is increased. This decrease in magnetic shear enhances the growth of the magnetic island. Therefore, the magnetic island is expected to grow in the discharge with counter-NBI. However, as the magnetic shear becomes weeks, the rational surface moves to the plasma center, where the detection of magnetic island id difficult because of the small gradient of temperature and x-ray emission. That the magnetic island is not observed in the discharge with co-NBI because of the decrease in magnetic shear or disappearance of the $\iota = 0.5$ surface is expected.

5. Conclusion

The Fourier-Bessel expansion reconstruction technique has been applied to soft x-ray measurement with a soft x-ray CCD camera to identify the non-rotating m/n=2/1 magnetic island in the plasma with electron ITB in the LHD. The magnetic island of m/n=2/1 mode in the ITB plasma has been observed in the discharge with counter-NBI, where the magnetic shear decreases due to the hollow NBI-driven current. In contrast, the magnetic island is not observed in the discharge with co-NBI, where the magnetic shear is sufficiently high. The two-dimensional soft x-ray image measured with the soft x-ray CCD camera confirms that the flattening observed near the foot point of ITB is due to the existence of the magnetic island.

Acknowledgments

The authors would like to acknowledge the LHD technical group for their excellent operation of LHD. This work is partly supported by a Grant-in-aid for Scientific research (15206106) of MEXT Japan. This work is also partly supported by NIFS06LUBB510.

References

[1] Fujita T, Ide S, Shirai H, Kikuchi M, Naito O, Koide Y, Takeji S, Kubo H and Ishida S 1997 *Phys. Rev. Lett.* **78** 2377
[2] De Baar M R, Hogeweij G M D, Cardozo N J L, Oomens A A M and Schuller F C 1997 *Phys. Rev. Lett.* **78** 4573
[3] Buratti P *et al.* 1999 *Phys. Rev. Lett.* **82** 560
[4] Gunter S *et al.* 2000 *Phys. Rev. Lett.* **84** 3097
[5] Pietrzyk Z A, Angioni C, Behn R, Coda S, Goodman T P, Henderson M A, Hofmann F and Sauter O 2001 *Phys. Rev. Lett.* **86** 1530
[6] Challis C D *et al.* 2001 *Plasma Phys. Control. Fusion* **43** 861
[7] Fujisawa A *et al.* 1999 *Phys. Rev. Lett.* **82** 2669
[8] Stroth U, Itoh K, Itoh S-L, Hartfuss H and Laqua H 2001 *Phys. Rev. Lett.* **86** 5910
[9] Ida K *et al.* 2001 *Phys. Rev. Lett.* **86** 3040
[10] Shimozuma T *et al.* 2003 *Plasma Phys. Control. Fusion* **45** 1183
[11] Ida K *et al.* 2003 *Phys. Rev. Lett.* **91** 085003
[12] Ida K *et al.* 2004 *Phy. Plasmas* **43** 861
[13] Ohyabu N *et al.* 2002 *Phys. Rev. Lett.* **88** 055005
[14] Narihara K *et al.* 2001 *Phys. Rev. Lett.* **87** 135002
[15] Navratil G A, Cates C, Mauel M E, Maurer D, Nadle D, Taylor E, Xiao Q, Reass W and A, Wurden G A 1998 *Phy. Plasmas* **5** 1855
[16] Ida K, Yoshinuma M, Watanabe K Y, Kobuchi T and Nagaoka K 2005 *Rev. Sci. Instrum.* **76** 053505
[17] Nagayama Y, Taylor G, Fredrickson E D, Budny R V, Janos A C, Mansfield D K, McGuire K M and Yamada M 1996 *Phys. Plasmas* **3** 2631
[18] Sauthoff N R, Von Goeler S and Stodiek W 1978 *Nucl. Fusion* **18** 1445
[19] Von Goeler S, Kaita R, Bitter M, Fuchs G, Poier M, Bertschinger G, Koslowski H R, Toi K, Ohdachi S and Donne A 1999 *Rev. Sci. Instrum.* **70** 599
[20] Liang Y *et al.* 2002 *Plasma Phys. Control. Fusion* **44** 1383
[21] Kobuchi T, Ida K, Yamada H, Yokoyama M, Watanabe KY, Sakakibara S and Yoshinuma M 2006 *Plasma Phys. Control. Fusion* **48** 789
[22] Iiyoshi A, Fujiwara M, Motojima O, Ohyabu N and Yamazaki K 1990 *Fusion Technol.* **17** 169
[23] Narihara K, Yamada I, Hayashi H and Yamauchi K 2001*Rev. Sci. Instrum.* **72** 1122
[24] Kawahata K, Tanaka K, Ito Y, Ejrii A and Wylde R J 1999 *Rev. Sci. Instrum.* **70** 695
[25] Liang Y, Ida K, Kado S, Minami T, Okamura S, Nomura I, Watanabe K Y and Yamada H 2001 *Rev. Sci. Instrum.* **72** 717
[26] Nagayama Y 1987 *J. Appl. Phys.* **62** 2702
[27] Hirshman S P and Whitson J C 1983 *Phys. Fluids* **26** 3553
[28] Kobuchi T, Ida K, Yoshinuma M, Watanabe K Y, Sakakibara S, Yamada H, Tanaka K and Kawahata K 2004 *Proc. 31st EPS Conference on Plasma Phys. London* 62

Clear transition to high-T_e state with an electron internal transport barrier creation in EC Heated LHD plasmas

T Shimozuma[1], S Kubo[1], H Igami[1], Y Yoshimura[1], T Notake[2], S Inagaki[3], N Tamura[1], K Ida[1], M Yokoyama[1], I Yamada[1], K Narihara[1], T Mutoh[1], A Komori[1] and LHD experimental group

[1] National Institute for Fusion Science, Toki, Gifu, 509-5292, Japan
[2] Fukui University, Fukui 910-8507, Japan
[3] Kyushu University, Kasuga, Fukuoka, 816-8580, Japan

E-mail: shimozuma.takashi@LHD.nifs.ac.jp

Abstract. A transition to a high electron temperature state in the core accompanied by an electron internal transport barrier (eITB) was clearly observed in neutral beam (NB) sustained plasmas with strongly focused on-axis electron cyclotron heating (ECH) in Large Helical Device (LHD). During stepwise injection of ECH power, the core electron temperature started to build up spontaneously and became high temperature in the core (high-T_e state), while the temperature outside of $\rho \simeq 0.3$ drastically decreased. The transition dynamics of the core temperature is analyzed by the time evolution of electron temperature profiles and heat flux changes based on the data from multi-channel electron cyclotron emission. Effect of an initial island size (m/n=2/1) on the high-T_e transition was experimentally investigated by controlling the width by an external coil field. The results show that the island size possibly affects a threshold value of ECH power and collisionality on the high-T_e transition.

1. Introduction

In LHD, transition phenomena to a high electron temperature state (high-T_e state) in the core region has been observed by centrally focused ECH injected into a plasma sustained by NBI. The transition to the high-T_e state is sometimes accompanied by a formation of an electron internal transport barrier (eITB). This is characterized by strongly peaked electron temperature profiles and large positive radial electric field, E_r, in the core region, which is the common feature observed in many helical devices [1, 2, 3, 4]. These phenomena are interpreted as a transition to the "electron root" solution of the ambipolarity condition for E_r under the neoclassical transport. Recently this improved confinement has been called "core electron-root confinement (CERC)" [5, 6].

Main features of the high-T_e transition in LHD are summarized as follows [7, 8]. It is characterized by a high central electron temperature up to 10 keV and steep temperature gradient around ρ (normalized minor radius) $\simeq 0.3$. The appearance of those states depends on the injection direction of NBI. Heat transport analyses by using ECH power modulation clarified that the structure of the ITB existed only in the counter-injected NBI sustained plasmas. There is a threshold ECH power for the transition to the high-T_e state. Sudden transitions to the high-T_e state are triggered by external perturbations, such as additional ECH power, decrease in density, a small pellet injection, and so on. The presence of $m/n = 2/1$ island or $\iota/2\pi = 0.5$ ($q = 2$) rational surface in the counter-NBI plasma configuration possibly facilitates the transition to the high-T_e state and the formation of the ITB with a steep temperature gradient.

In this paper, a clear transition of a high-T_e state are reported, in which the time evolution of the electron temperature was spatially resolved by using a multi-channel ECE diagnostic. Such transition with an eITB has been clearly observed in neutral beam (NB) sustained plasmas with strongly focused on-axis ECH. During stepwise injection of ECH power, the core electron temperature started to build up spontaneously and became high temperature in the core, on the other hand the temperature outside of $\rho \simeq 0.3$ drastically decreased. This phenomenon helps to understand the temporal and spatial behavior of a formation process of eITB in the helical device. The transition phenomena to the high-T_e state with an ITB formation were mainly observed in plasmas sustained with counter-injected NBI [8]. It is speculated that this process includes the first expansion of a magnetic island ($m/n = 2/1$) by low power ECH injection, subsequent disappearing or healing of the island by additional ECH power and final transition to high-T_e in the core and eITB formation by realization of so called" electron-root " . We tried to contract and expand the island by the external coil field to investigate the effect of the seed island size on the transition. The results show that the island size possibly affects the threshold values of ECH power and collisionality on the high-T_e transition. During ECH power injection, some crashes in the core temperature and flattening of temperature profile frequently occurred and they inhibit attainment of higher central electron temperature. Clarification of the eITB formation process contributes an achievement of higher electron temperature and high performance plasmas by ECH.

This paper organized as follows. In Sec. 2, experimental devices and configuration including magnetic and heating condition are briefly summarized. In Sec. 3, experimental observations will be given in detail, together with the analyses of heat flux change. Experimental results of an effect of the initial island size on high-T_e transition are also described. Finally, a summary will be presented in Sec. 4.

2. Experimental setup and configuration

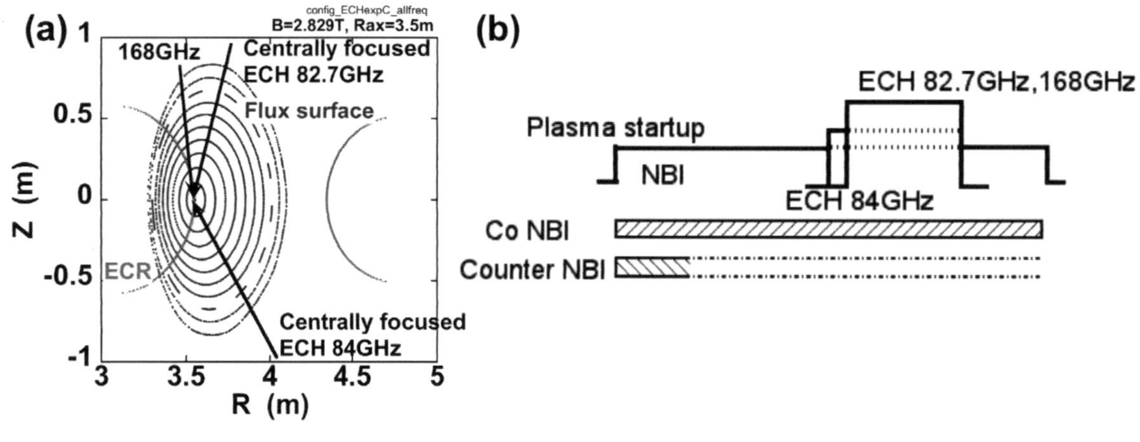

Figure 1. (a) Configuration of ECH injection. Magnetic flux surfaces, ECR layer and injection direction of millimeter-waves from each antenna are illustrated. (b)Time sequence of NBI and ECH pulses. Direction of beam-driven currents could be changed by switching NB injectors in co- and counter-direction.

LHD is a heliotron-type device ($\ell = 2, m = 10$) with superconducting helical coils and poloidal coils. In this experiment the major and the averaged minor radii of the produced plasmas are typically 3.5 m and 0.6 m, respectively. The ECH system consists of two 82.7 GHz and two 84 GHz gyrotrons for the fundamental resonance heating, and four 168 GHz gyrotrons for the second harmonic heating at 3 T [9].

In the present experiments, 82.7 GHz and 168 GHz power were injected from the antenna installed in the upper port (U-antenna), and 84 GHz power was launched from the antenna in the lower port (L-antenna), as shown in figure 1 (a). The electron cyclotron resonance (ECR) layers for each frequency are located near the magnetic axis on the equatorial plane, which are denoted by ECR in this figure. The ECR layers are almost the same for each frequency. That is the fundamental resonance for 84 GHz and 82.7 GHz and is the second harmonic resonance for 168 GHz. The U- and L- antennas were installed in the toroidal section with a vertically elongated cross section. The beam waist radii at the focal point were 15 mm in the radial direction for the U-antennas, and 30 mm for the L-antennas.

The experiments were performed at the magnetic configuration with magnetic field B_0 = 2.829T and the magnetic axis of R_{ax} = 3.5m. A schematic of the time sequence in NBI and ECH is illustrated in figure 1 (b). The target plasmas were produced and sustained by NBI with the port-through power of about 2.4MW, and an additional ECH was turned on at the flat top of the discharge in figure 1 (b). Direction of beam-driven currents could be changed by switching NB injectors in co- and counter-direction. The ECH power was additionally injected into a plasma in the steady state with the electron temperature of about 2 keV and the electron density of 0.6×10^{19} m^{-3}. The ECH pulse formed a stair-like pattern such that 84GHz power (0.4MW) was turned on from t = 1.3 s and 82.7GHz and 168GHz powers (0.72MW) were turned on from t = 1.5 s, and both pulses were terminated at t = 2 s.

The electron temperature profiles were measured by using a Thomson scattering system at about 130 positions in the equatorial plane of the horizontally elongated toroidal section with a spatial resolution of 20 – 35 mm [10]. For electron temperature measurements with a higher time resolution, a multi-channel radiometer system was used to detect electron cyclotron emission (ECE) radiated from the low magnetic field side (30 channels, spatial resolution: 10 – 70 mm) [11]. The absolute error of T_e determined by the absolute calibration of ECE is about 20 % and the relative error of the T_e determined by the noise levels of ECE is only 1 %.

3. Experimental results

3.1. Spontaneous T_e transition and crash

During additional ECH pulse, a spontaneous transition to the high-T_e state was clearly observed in a counter-NBI sustained plasma with an ITB formation. Figure 2 shows a time evolution of some plasma parameters. A target plasma was initiated by co- and counter-NBI at first, then co-NBI pulse was switched off at t = 0.5 s while counter-NBI pulse was extended to t = 2.1 s. ECH power was injected from t = 1.3 s and from t = 1.5 s stepwisely. A plasma current in the counter direction continued to increase up to the end of the NBI pulse. As shown in T_e measured by ECE, a transition to high-T_e in the core region is obviously noticed at t = 1.67 s, when a rapid increase of core temperature and a little step-down of temperature in outer region (ρ > 0.34) occurred almost simultaneously. The inversion radius of the temperature change is about $\rho \simeq 0.3$. Precise time evolution of the electron temperature profile by ECE is shown in figure 3. The plasma sustained by counter-NBI has a rather flat profile in the core region (t = 1.6 s). The second peak of the temperature appeared around $\rho \sim 0.4$ just before the transition (t = 1.65s). After the transition, the central temperature was growing up, while the second peak was shrinking. This behavior of the temperature profile looks to be inverted at $\rho \simeq 0.3$. Position of a transport barrier seems to be the position of the inversion radius $\rho \simeq 0.3$ inside which temperature increases and outside which it decreases.

In a co-NBI sustained plasma, a big crash of electron temperature in the core region was observed. Figure 4 shows a time evolution of some plasma parameters. A target plasma was initiated and sustained by co-NBI only. ECH power was injected in the same manner as counter-NBI case. Plasma current in the co-direction continued to increase up to the end of the NBI pulse. The final attained absolute value of the plasma current is almost the same as that in counter case, though the direction of current was inverted. As shown in T_e measured by ECE, a big T_e crash in the core region suddenly occurred at $t \sim 1.79$ s, when a drastic decrease of core temperature and a little step-up of temperature in outer region occurred

Journal of Physics: Conference Series **123** (2008) 012022 doi:10.1088/1742-6596/123/1/012022

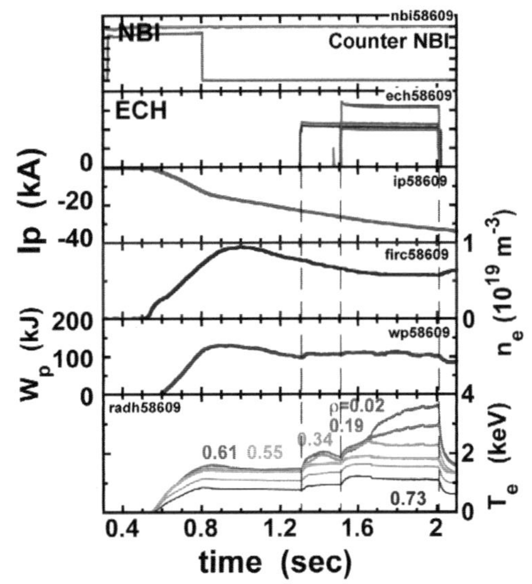

Figure 2. Time evolution of some plasma parameters for counter-NBI plasma with ECH. Spontaneous transition to the high-T_e state occurred at $t = 1.67$ s. Figures show NBI pulse, ECH pulse forms, plasma current I_p, line-averaged electron density n_e, plasma stored energy W_p and electron temperature by ECE (normalized radial position is indicated in the figure) from top to bottom.

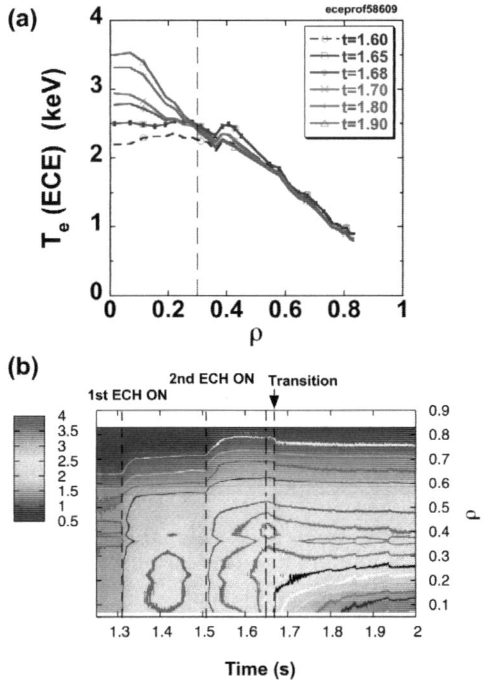

Figure 3. (a)Time evolution of electron temperature profile measured by ECE during the transition(from $t = 1.6$ s to 1.9 s). The relative error determined by noise level are typically plotted for the data points of $t = 1.65$ s. (b) Contour plot of electron temperature profile.

almost simultaneously. Precise time evolution of the electron temperature profile is shown in figure 5 from $t = 1.6$ s to 1.9 s. The plasma sustained by co-NBI has a rather peaked profile in the core region with higher temperature than in counter-NBI case at $t = 1.6$ s from the beginning. The inversion radius of the temperature behavior is about $\rho \simeq 0.23$ as shown in the figure. It is obviously smaller than that in the counter-NBI case. The difference of the inversion radius is supposed to arise from the position of a lower order rational surface, $\iota/2\pi = 0.5$, in special, induced by plasma current.

3.2. Temporal and spatial analyses of electron heat flux

During ECH pulse, ECH power is assumed to be constant. Ray tracing calculation shows that it is absorbed within $\rho \lesssim 0.2$. Then variation of an electron heat flux through a given normalized radius, ρ, and at time, t, will be given by a following equation outside of the ECH deposition layer.

$$\delta Q_e(\rho, t) \propto -\frac{1}{\rho}\int_0^{\rho}\left(\frac{3}{2}\right)\frac{\partial}{\partial t}\Big\{n_e(\rho', t)T_e(\rho', t)\Big\}\rho' d\rho' \tag{1}$$

This corresponds to a time derivative of the total kinetic energy stored within a given radius ρ. Using temperature data measured by ECE and density data by FIR interferometer, temporal and spatial variation

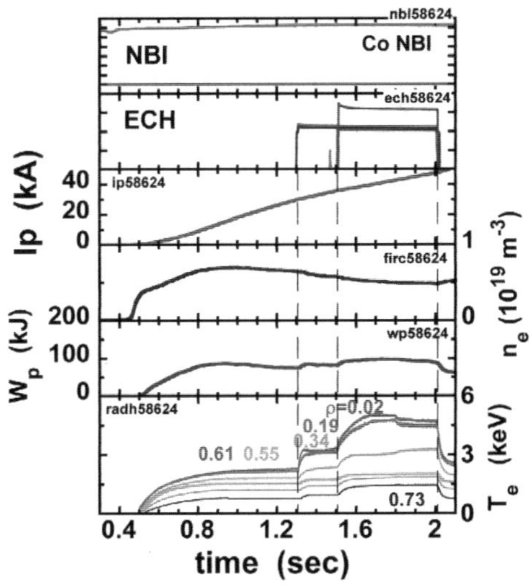

Figure 4. Time evolution of some plasma parameters for co-NBI plasma with ECH. A sudden crash of core T_e occurred at $t \sim 1.79s$. Figures show NBI pulse, ECH pulse forms, plasma current I_p, line-averaged electron density n_e, plasma stored energy W_p and electron temperature T_e by ECE (normalized radial position is indicated in the figure) from top to bottom.

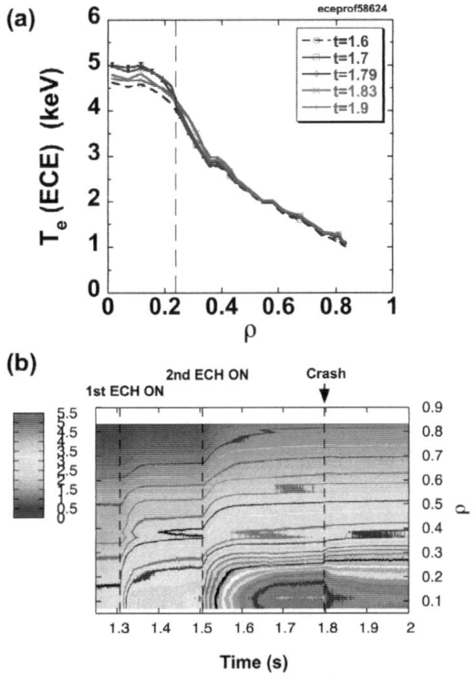

Figure 5. (a) Time evolution of electron temperature profile measured by ECE during the crash (from t = 1.6 s to 1.9 s). The relative error determined by noise level are typically plotted for the data points of t = 1.7 s. (b) Contour plot of electron temperature profile.

of δQ_e can be calculated. Difference of in- and out-flux through a thin layer $2\Delta\rho$ between $\rho - \Delta\rho$ and $\rho + \Delta\rho$ was calculated and defined as $\Delta\delta Q_e(\rho, t)$.

The values, $\Delta\delta Q_e(\rho, t)$, are calculated for the high-T_e transition and temperature crash shots described in the previous subsection. In these analyses $\Delta\rho$ was selected to be 0.01. Difference between in- and out-heat fluxes at ρ is plotted as a function of time for the shot of the high-T_e transition in figure 6. Electron temperature is gradually increasing after on-timing of the second ECH pulse (t = 1.5 s), keeping a flat profile in the core region. From about 20 ms before a jump of core electron temperature at $t \sim 1.67s$, the value $\Delta\delta Q_e$ is increasing gradually between ρ = 0.25 and 0.45 (from t = 1.65 s to 1.67 s). More precisely, the slow decrease of temperature between ρ = 0.3 and 0.55 is observed as shown in figure 3 (b). This fact implies that a recovering of confinement has been already begining around the central part of the plasma from t = 1.65 s gradually. Then abrupt dip of $\Delta\delta Q_e$ up to $\rho \sim 0.3$ and its increase at $\rho > 0.45$ can be noticed. It is suggested that a kind of transport barrier of electron heat transport was established at $\rho \sim 0.3$ at this moment and that the confinement of the electron energy was improved within $\rho \sim 0.3$. The first small crash of the core temperature happened at t = 1.8 s, when the reverse process occurred, that is, increase of $\Delta\delta Q_e$ in the core ($\rho < 0.31$) and a rapid dip of flux deference in $\rho > 0.45$. The inversion radius of these phenomena such as the transition and crash is the same.

A opposite behavior of heat flux difference $\Delta\delta Q_e$ was noticed in the core temperature crash of the co-NBI sustained plasma case. A rapid build-up and slow relaxation of the heat flux occurred within

Journal of Physics: Conference Series **123** (2008) 012022 doi:10.1088/1742-6596/123/1/012022

$\rho < 0.23$ at $t \sim 1.79$ s. On the contrary, a rapid decrease and its relaxation were observed in the outside region ($\rho > 0.25$) as shown in figure 7. The minimum of the flux difference has a time delay of several millisecond in the outer region. This implies that a heat flux flowing out from the core propagates outward.

These temporal and spatial behavior of the heat flux helps to understand a dynamics that occurs in plasmas with extremely localized ECH.

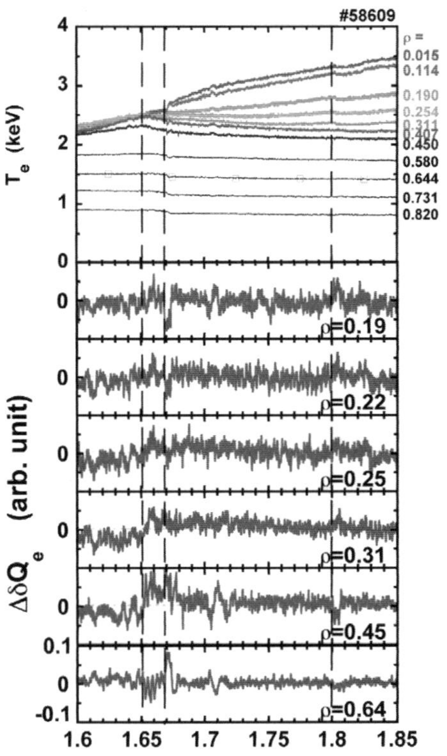

Figure 6. Time evolution of T_e and difference of in- and out-flux of heat $\Delta\delta Q_e$ during high-T_e transition (shot#58609) for expanded time region from $t = 1.6$ s to $t = 1.85$ s. Normalized radial position ρ for flux difference is indicated in the figure, $\rho = 0.19, 0.22, 0.25, 0.31, 0.45, 0.64$, from top to bottom.

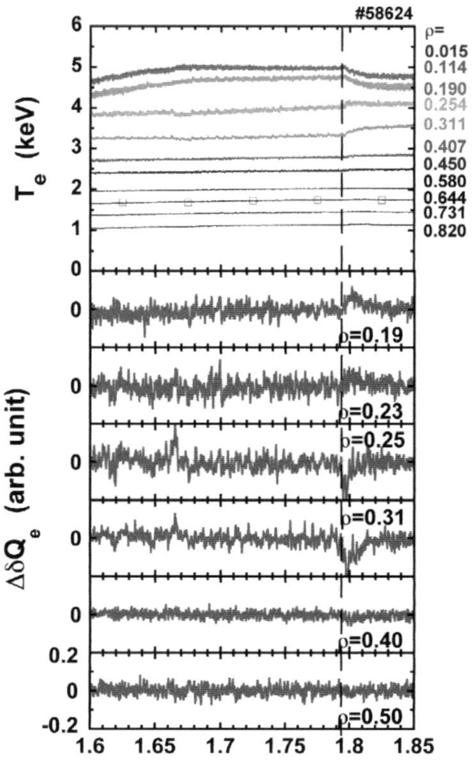

Figure 7. Time evolution of T_e and difference of in- and out-flux of heat $\Delta\delta Q_e$ during core temperature crash (shot#58624) for expanded time region from $t = 1.6$ s to $t = 1.85$ s. Normalized radial position ρ for flux difference is indicated in the figure, $\rho = 0.19, 0.23, 0.25, 0.31, 0.40, 0.50$, from top to bottom.

3.3. Effect of seed island size for high T_e transition

It is noticed that the presence of $m/n = 2/1$ island or $\iota/2\pi = 0.5$ ($q = 2$) rational surface in the counter-NBI plasma configuration may facilitate the transition to the high-T_e state and the formation of the ITB with a steep temperature gradient [8, 9].

There are some natural islands produced by existing error fields in LHD. Using an electron gun as the electron beam source and the multiple probe array, their position, width, and phase at a given toroidal section can be successfully identified in a high magnetic field state (B=2.75 T). Calculations based on the magnetic field line tracing can reproduce these islands if the given error field sources are assumed.

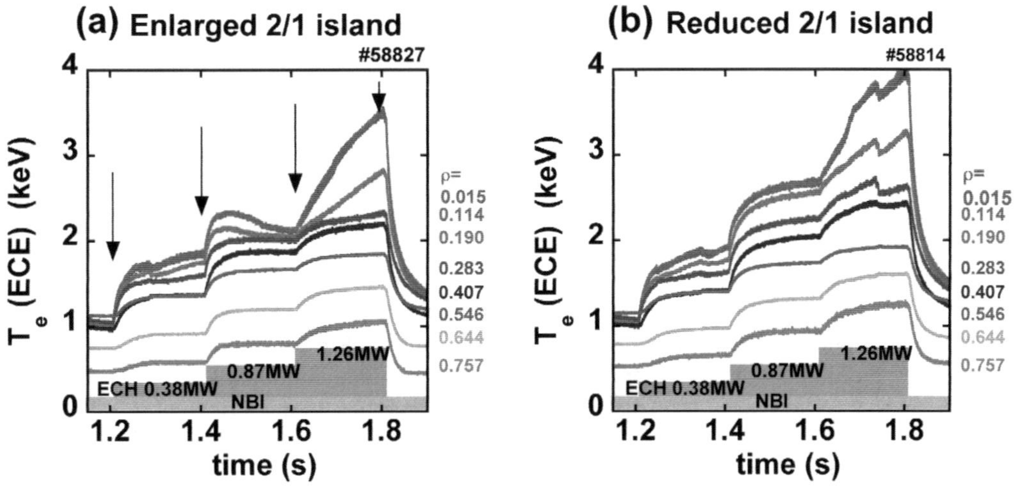

Figure 8. Time evolution of electron temperature during stair-like ECH power injection for (a) with enlarged 2/1 island and (b) with reduced island. The measured positions of the temperature correspond to $\rho = 0.015, 0.114, 0.190, 0.283, 0.407, 0.546, 0.644, 0.757$ from top to bottom.

For local island divertor experiments, LHD has ten pairs of perturbation coils located above and below the LHD vacuum vessel. Arranging the perturbation coil currents from the calculated expectations can change the size of these natural islands. With an arrangement of the perturbation coil currents, the width of these islands can be changed by a factor of two. The maximum width of the 2/1 island is controlled from 58 mm that is the natural island size to 29 mm (reduced island) and to 83 mm (enlarged island) by the application of the external magnetic field produced by the perturbation coils. Target plasmas were produced and sustained by counter-NBI power and ECH power was additionally injected stairlikely as shown in figure 8. Figures show time evolutions of electron temperature measured by ECE for the EC heated plasma with enlarged 2/1 islands in figure 8 (a) and with reduced islands in (b). ECH power was turned on by each 0.2s from t = 1.2 s and the power level increased from 0.38 MW (t = 1.2 – 1.4 s), 0.87 MW (t = 1.4 – 1.6 s) and finally to 1.26 MW (t = 1.6 – 1.8 s). Although the time period of 0.2 s was insufficient so that the temperature might attain a steady value, temperature values just before the next power steps were adopted as data for power dependence. Those timings are indicated by arrows in figure 8 (a).

ECH power normalized by line-averaged electron density $P_{\mathrm{ECH}}/\bar{n}_e$ is plotted as a function of electron temperature gradient at ρ = 0.25 as a representative of ITB foot position in figure 9. Because $P_{\mathrm{ECH}}/\bar{n}_e$ represents a product of heat diffusivity χ_e and temperature gradient in steady state, a slope of a straight line drawn from the origin to data point roughly expresses the heat diffusivity χ_e.

All curves show almost the linear increase of T_e gradient as the ECH power increases at low power level. However, there are specific power ranges where the temperature gradient stays almost constant or even decreases (which means an increase of heat diffusivity). The larger is the island size, the wider the power range becomes. These phenomena stem from the extreme flattening of the core temperature profile as observed at t=1.6s in figure 8(a). The spontaneous expansion of the m/n=2/1 island towards the

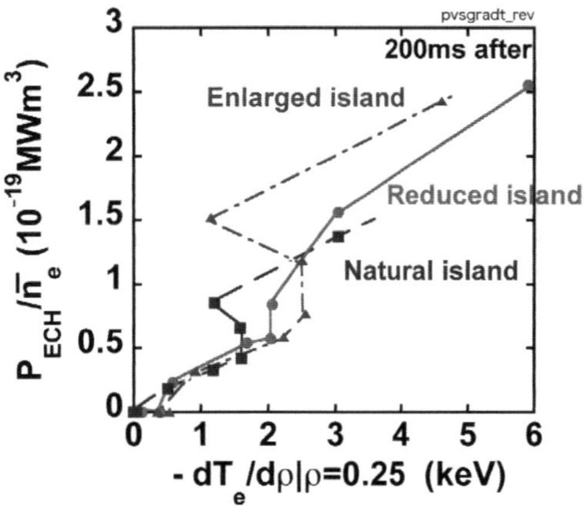

Figure 9. ECH power normalized by line-averaged electron density is plotted as a function of electron temperature gradient $-dT_e/d\rho$ at $\rho = 0.25$. The data are plotted for each case of plasmas with natural island, reduced island and enlarged island

plasma center is speculated as one of causes. Then a jump in the temperature gradient occurs at higher density-normalized ECH power, which implies a recovery of the heat diffusivity or a change towards another confinement state. There are threshold power at which temperature gradient suddenly increases. It is not clear in the case of reduced island. The natural island case looks to have the minimum threshold power among three cases. The optimum size of the island possibly exists to facilitate such transition.

4. Summary

During stepwise power injection of centrally localized ECH in the plasma sustained counter-NBI, the core electron temperature within $\rho \simeq 0.3$ started to build up spontaneously and became high temperature (high-T_e state) in the core, on the other hand the temperature outside of $\rho \simeq 0.3$ drastically decreased. This phenomenon directly shows temporal and spatial behavior of the formation process of an eITB (or CERC) in the helical device. The transition phenomena to the high-T_e state with an ITB formation were mainly observed in plasmas sustained with counter-injected NBI. The second peak of the temperature appeared around $\rho \sim 0.4$ just before the transition. After the transition, central temperature was growing up, while the second peak was shrinking. It is suggested that this process is affected by the change of heat flow around a magnetic island ($m/n = 2/1$) induced by ECH injection.

There is a noticeable difference in the inversion radius of the electron temperature change between for the high T_e transition in the counter-NBI plasma and for the temperature crash in the co-NBI plasma. This fact is considered to result from a change of the rotational transform profile caused by an NBI-driven current. The exact positions of the inversion radius, however, differ from those of calculated positions based on the assumed current profiles. Determination of current profiles will be required experimentally.

Effect of $m/n = 2/1$ island size on the high-T_e transition was investigated by controlling the island size. The width of the island was reduced and enlarged by external coil field by about factor of two. The results suggested that there is an optimum size of island to facilitate a transition phenomenon in core T_e. When increasing ECH power, confinement of electron energy was once degraded and then recovered or changed to another confinement state. This confinement degradation phase relevant to ECH power seems to be wider with the size of the island. Detailed study of a role of islands on the high-T_e transition mechanisms is one of urgent issues.

Acknowledgments
The authors gratefully acknowledge the LHD technical group for their sophisticated operation of LHD, and the heating and diagnostic devices. They also thank Professors S. Sudo and O. Motojima for their continuous guidance and encouragement. This work has been supported by NIFS under NIFS06ULRR501,502,503.

References

[1] Fujisawa A 2003 *Plasma Phys. Control. Fusion* **45** R1

[2] Ida K, Shimozuma T, Funaba H, Narihara K, Kubo S, Murakami S, Wakasa A, Yokoyama M, Takeiri Y, Watanabe K Y, Tanaka K, Yoshinuma M, Liang Y, Ohyabu N and LHD experimental group 2003 *Phys. Rev. Lett.* **91** 085003

[3] Estrada T, Krupnik L, Dreval N, Melnikov A, Khrebtov S M, Hidalgo C, van Milligen B, Castejón F, Ascasíbar E, Eliseev L, Chmyga A A, Komarov A D, Kozachok A S and Tereshin V 2004 *Plasma Phys. Control. Fusion* **46** 277

[4] Maaßberg H, Beidler C D, Gasparino U, Romé M, Dyabilin K S, Marushchenko N B, Murakami S 2000 *Phys. Plasmas* **7** 295

[5] Yokoyama M, Maaßberg H, Beidler C D, Tribaldos V, Ida K, Castejón F, Estrada T, Fujisawa A, Minami T, Shimozuma T, Takeiri Y, Herranz J, Murakami S and Yamada H 2006 *Fusion Sci. and Tech.* **50** 327

[6] Yokoyama M, Maaßberg H, Beidler C D, Tribaldos V, Ida K, Estrada T, Castejón F, Fujisawa A, Minami T, Shimozuma T, Takeiri Y, Dinklage A, Murakami S and Yamada H 2007 *Nuclear Fusion* **47** 1213

[7] Shimozuma T, Kubo S, Idei H, Yoshimura Y, Notake T, Ida K, Ohyabu N, Yamada I, Narihara K, Inagaki S, *et al.* 2003 *Plasma Phys. Control. Fusion* **45** 1183

[8] Shimozuma T, Kubo S, Idei H, Inagaki S, Tamura N, Tokuzawa T, Morisaki T, Watanabe K Y, Ida K, Yamada I, Narihara K, Muto S, Yokoyama M, *et al.* 2005 *Nuclear Fusion* **45** 1396

[9] Shimozuma T, Kubo S, Yoshimura Y, Igami H, Nagasaki K, Notake T, Inagaki S, Ito S, Kobayashi S, Mizuno Y, Takita Y, Saito K, Seki T, Kumazawa R, Watari T, Mutoh T and LHD experimental group 2006 *Fusion Sci. and Tech.* **50** 403

[10] Narihara K, Yamada I, Hayashi H and Yamauchi K 2001 *Rev. Sci. Instrum.* **72** 1122

[11] Nagayama Y, Kawahata K, England A, Ito Y, Bretz N, McCarthy M, Taylor G, Doane J, Ikezi H, Edlington T and Tomas J 1999 *Rev. Sci. Instrum.* **70** 1021

Journal of Physics: Conference Series **123** (2008) 012023 doi:10.1088/1742-6596/123/1/012023

Core electron temperature rise due to Ar gas-puff in EC-heated LHD plasmas

N Tamura[1], S Inagaki[2], K Ida[1,3], K Tanaka[1], C Michael[1], T Tokuzawa[1,3], I Yamada[1], K Narihara[1], T Shimozuma[1], S Kubo[1], H Igami[1], T Fukuda[4], K Itoh[1], Y Nagayama[1,3], K Kawahata[1,3], S Sudo[1,3], A Komori[1,3] and LHD team[1]

[1]National Institute for Fusion Science, 322-6, Oroshi-cho, Toki, Gifu 509-5292, Japan
[2]Research Institute for Applied Mechanics, Kyushyu University, 6-1, Kasuga-Koen, Kasuga, Fukuoka 816-8580, Japan
[3]The Graduate University for Advanced Studies, Department of Fusion Science, 322-6, Oroshi-cho, Toki, Gifu 509-5292, Japan
[4]Center for Atomic and Molecular Technologies, Osaka Univeristy, 2-1, Yamadaoka, Suita, Osaka, Japan

E-mail: ntamura@LHD.nifs.ac.jp

Abstract. A rise of the core electron temperature T_e in response to a controlled Ar gas-puff is observed in electron cyclotron heated plasmas of the Large Helical Device. The increase in core T_e is about 1 keV after the Ar gas-puff. Since density peaking does not take place after the Ar gas-puff, this phenomenon is different from radiatively improved mode. Transient response analysis shows that a close similarity in the relationship between the electron heat flux and the T_e gradient appears between the core T_e rise after the Ar gas-puff and that due to the nonlocal effect induced by a pellet injection. Thus non-locality in the electron heat transport is essential for the core T_e rise after the Ar gas-puff.

1. Introduction

Both spontaneous and externally-triggered changes of transport structure have been observed in magnetically confined toroidal plasmas. Examples of externally-triggered change as well as spontaneous change in the transport include the formation of an internal transport barrier and the transition from the low confinement mode (L mode) to the high confinement mode (H mode). An abrupt increment in electron temperature T_e in the core region just after edge cooling, which is so-called "nonlocal transport phenomenon", can also be considered as one example of externally-triggered change in the transport, since the core T_e rise due to the edge cooling is, even transiently, accompanied by an improvement in transport [1]. The nonlocal T_e rise has been observed in many tokamaks so far [2, 3, 4, 5, 6] and more recently in a helical device [7, 8]. Its physical mechanism of that, however, still remains an open question. Since local transport is governed by local turbulence according to the paradigm of anomalous transport in the magnetically confined toroidal plasmas, a suppression of turbulence in the core region is a trigger for the core T_e rise in response to the edge cooling, but it is not clear how the edge perturbation acts on the core turbulence in the timescale shorter than the diffusive one. A multi-scale coupling [9] is a promising candidate to explain this edge-core interaction. In order

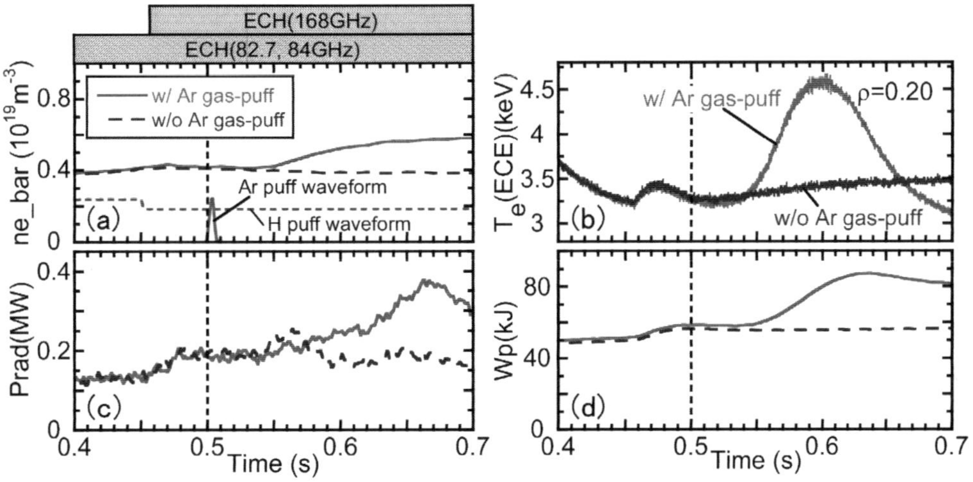

Figure 1. Temporal evolutions of (a) the line-averaged electron density and (b) the electron temperature at $\rho = 0.20$, (c) the radiated power $P_{\rm rad}$ and (d) the diamagnetically measured plasma stored energy $W_{\rm p}$ in the case with Argon (Ar) gas-puff (solid lines) and without that (dashed lines). In (a), waveforms of hydrogen (H) gas-puff and Ar gas-puff are also plotted.

to elucidate the mechanism of the nonlocal $T_{\rm e}$ rise, a further experimental study on the nonlocal $T_{\rm e}$ rise with various perturbation techniques is important. In this paper, a core $T_{\rm e}$ rise after the controlled Ar gas-puff is presented and discussed.

2. Experimental results

The experiments are carried out in the Large Helical Device (LHD) [10], which has a heliotron type magnetic configuration, with a magnetic axis position of $R_{\rm ax} = 3.5$ m, an average minor radius of $a = 0.6$ m and a magnetic field at the axis of 2.93 T. Only the electron cyclotron heating (ECH) with the total injected power of 1.6 MW is applied for the plasma startup and its sustainment. The focus of all ECH beams (the applied frequencies: 82.7, 84 and 168 GHz) is adjusted to be near the $R_{\rm ax}$. The $T_{\rm e}$ is higher than the ion temperature $T_{\rm i}$, and the electron loss channel dominates the ion loss channel. The plasma current attributed to the bootstrap current is very small. As a working gas, hydrogen (H) is continuously injected into the LHD vacuum vessel. A 32-channel heterodyne radiometer is used to measure temporal behaviors of $T_{\rm e}$ with high time resolution ($\Delta t \sim 1$ μs). Typical spatial resolution of the heterodyne radiometer is about 1 cm at the position of the half plasma minor radius [11]. The $T_{\rm e}$ measured by the heterodyne radiometer is in good agreement with that by YAG Thomson scattering (TS) system [12] in this experiment. Spatial resolution of the YAG TS system is about 2 - 4 cm and temporal resoultion of that can be down to 5 ms in a high-repetition mode. A 13-channel far-infrared (FIR) interferometer [13] with a chord separation of 90 mm and a laser beam width of about 50 mm is used to measure temporal behaviors of electron density $n_{\rm e}$ and the profile of $n_{\rm e}$ is reconstructed by using the Abel inversion technique based on the data measured. The temporal behavior of a total radiated power is measured by a wide angle metal foil resistive bolometer [14] with temporal resolution of 10 ms. The Ar gas puff is done for the measurement with an X-ray crystal spectrometer [15]. The amount of Ar gas is determined such as not to have an effect on the background level of Ar in the core, which is confirmed by almost no change in the temporal behavior of Ar XVIII intensity measured with the X-ray crystal spectrometer.

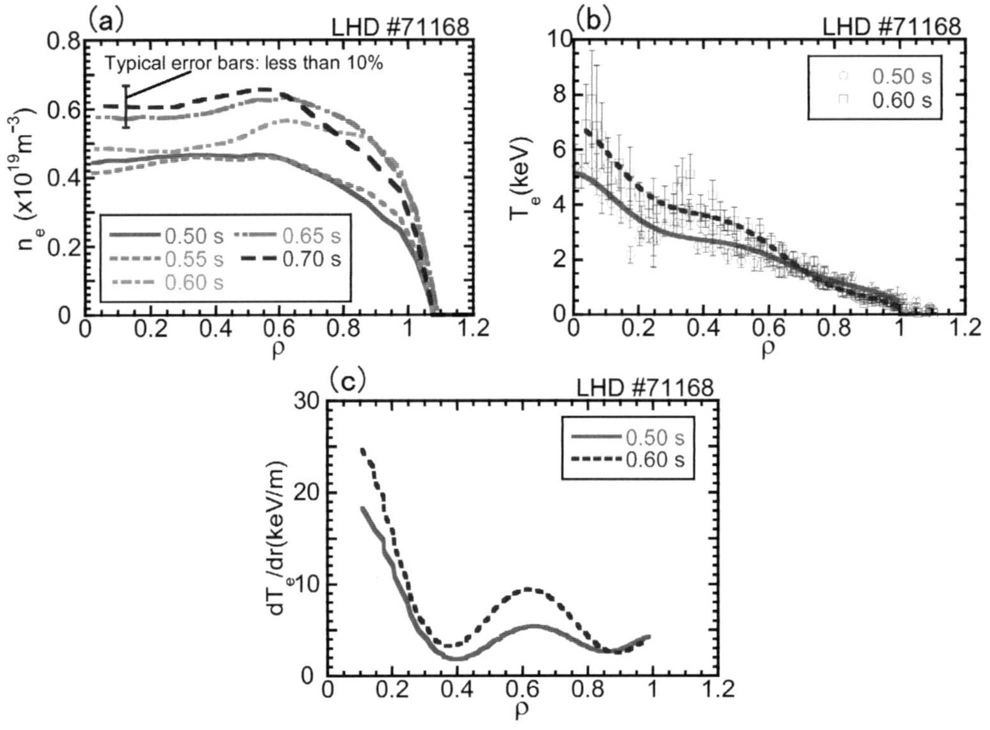

Figure 2. Radial profiles of (a) the electron density measured with the FIR intereromete, (b) the electron temperature T_e measured with the YAG Thomson scattering system and (c) T_e gradient at $t = 0.5$ s and at $t = 0.6$ s.

Figure 1 shows a typical result of the controlled Ar gas-puff experiment with the ECH plasma of LHD. As a reference, a typical experimental result without Ar gas-puff under the same experimental conditions is also shown in figure 1. In this experiment, the value of line-averaged electron density $\bar{n_e}$ before the Ar gas-puff is set around 0.4×10^{19} m^{-3}. The Ar gas-puff is started from $t \sim 0.5$ s and its duration is about 7 ms. After the controlled Ar gas-puff, the $\bar{n_e}$ starts to increases from $t \sim 0.55$ s and the increment of $\bar{n_e}$ reaches about 40 %. Unexpectedly, the core T_e also increases with some delay but drastically and the increment of the electron temperature δT_e at $\rho = 0.20$ reaches about 1.3 keV at $t \sim 0.6$ s. Accordingly, a plasma stored energy W_p, which is diamagnetically measured ($\Delta t = 0.1$ ms), increases by up to about 50 %. At the time of peak T_e ($t \sim 0.6$ s), the total radiated power is increased by about 30 %. There is no significant change in the T_i evaluated from the data obtained by the X-ray crystal spectrometer. Moreover, in the rising phase of the core T_e, neither significant change in low-m magnetohydrodynamics (MHD) modes nor shift of the magnetic axis is observed. The enhancement of T_e, however, is not sustained and goes down below the level before the Ar gas-puff. As can be seen in figure 2, at $t = 0.6$ s, T_e profile measured with the YAG TS system becomes more peaked compared to the base T_e profile at $t = 0.5$ s, even taking the error bars into account. The change in the electron temperature gradient ∇T_e appears prominently around $\rho \sim 0.6$. On the other hand, almost no change in the n_e profile is observed within the accuracy of the Abel inversion (the assessed error bars are typically less than 10 %) when the core T_e starts to increase, at $t \sim 0.55$ s. Then the n_e profile becomes a hollow transiently and it finally turns back to the original shape.

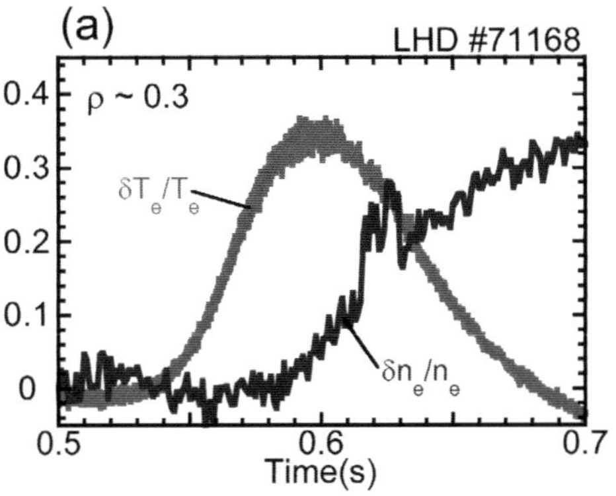

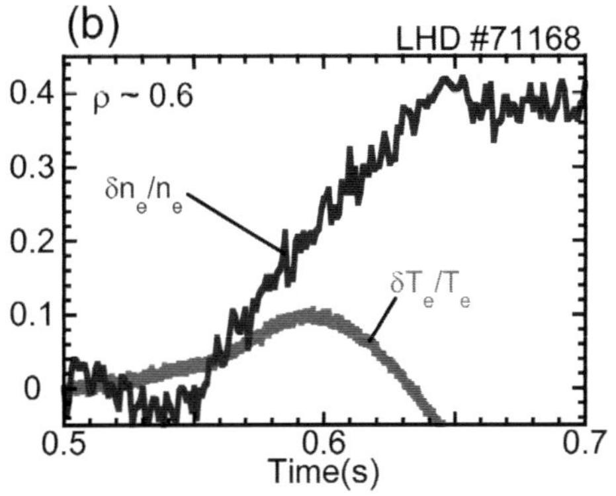

Figure 3. Temporal evolutions of the increment of the electron density (the electron temperature) normalized by the base electron density (electron temperature) (at $t = 0.50$ s) after the controlled Ar gas-puff at (a) $\rho \sim 0.3$ and (b) $\rho \sim 0.6$.

3. Transient transport analysis and discussion

Although the core T_e rise, in other words the improvement in core plasma energy confinement, is invoked by the impurity injection into the plasma edge, this phenomenon cannot be categorized as the radiatively improved (RI) mode [16, 17]. Because a peaked density profile, which is one of the features in the RI mode plasma, is not established after the controlled Ar gas-puff. Moreover, an ion temperature gradient (ITG) driven turbulence, which is considered to be surppressed in the transition to the RI mode from the L mode [18] , can be originally weak in the present ECH plasma.

In LHD, the core T_e rise in response to the edge cooling has already been observed in pellet (tracer-encapsulated solid pellet(TESPEL) or tiny solid H_2 pellet) injection experiments [1, 8]. When the pellet (TESPEL or tiny solid H_2 pellet) is injected into the LHD plasma with low density and high temperature, the T_e in the core region increased almost simultaneously. In order to make a detailed comparison between the core T_e rise after the Ar gas-puff and that induced by the TESPEL injection, a transient transport analysis is performed for the both cases. The perturbed equation derived from the equations for the conservation of energy and particles can be written as [19]

$$\frac{3}{2} n_e \frac{\partial \delta T_e}{\partial t} = -\nabla \cdot \delta q_e - T_e \nabla \cdot \delta \Gamma_e. \tag{1}$$

Here δq_e denotes the perturbed electron heat flux and $\delta \Gamma_e$ does the perturbed electron particle flux, respectively. In regard to the perturbation for heat source, the change in the ECH absorbed power is negligibly small after the edge perturbation. In regard to the perturbation for particle source, it can be igonored so far as the core region is considered. A large change in the n_e profile, especially in the core region, is obtaiend by the Ar gas-puff, in contrast to the almost no change by the TESPEL injection. In order to estimate the effect of the perturbed particle flux on the analysis in the case of the Ar gas-puff, the first term and second one of the right-hand side of equation (1) is compared;

$$\frac{T_e \nabla \cdot \delta \Gamma_e}{\nabla \cdot \delta q_e} \approx \frac{\delta n_e / n_e}{\delta T_e / T_e} \left(\frac{L_{\delta T_e}}{L_{\delta n_e}}\right)^2 \frac{D}{\chi}, \tag{2}$$

where δn_e is the increment of the electron density, $L_{\delta T_e}$ ($L_{\delta n_e}$) is the perturbed electron temperature (electron density) scale lengh, D and χ are the diffusivity for particles and heat. Figure 3 shows the temporal evolutions of $\delta n_e / n_e$ and $\delta T_e / T_e$ after the controlled Ar gas-puff at

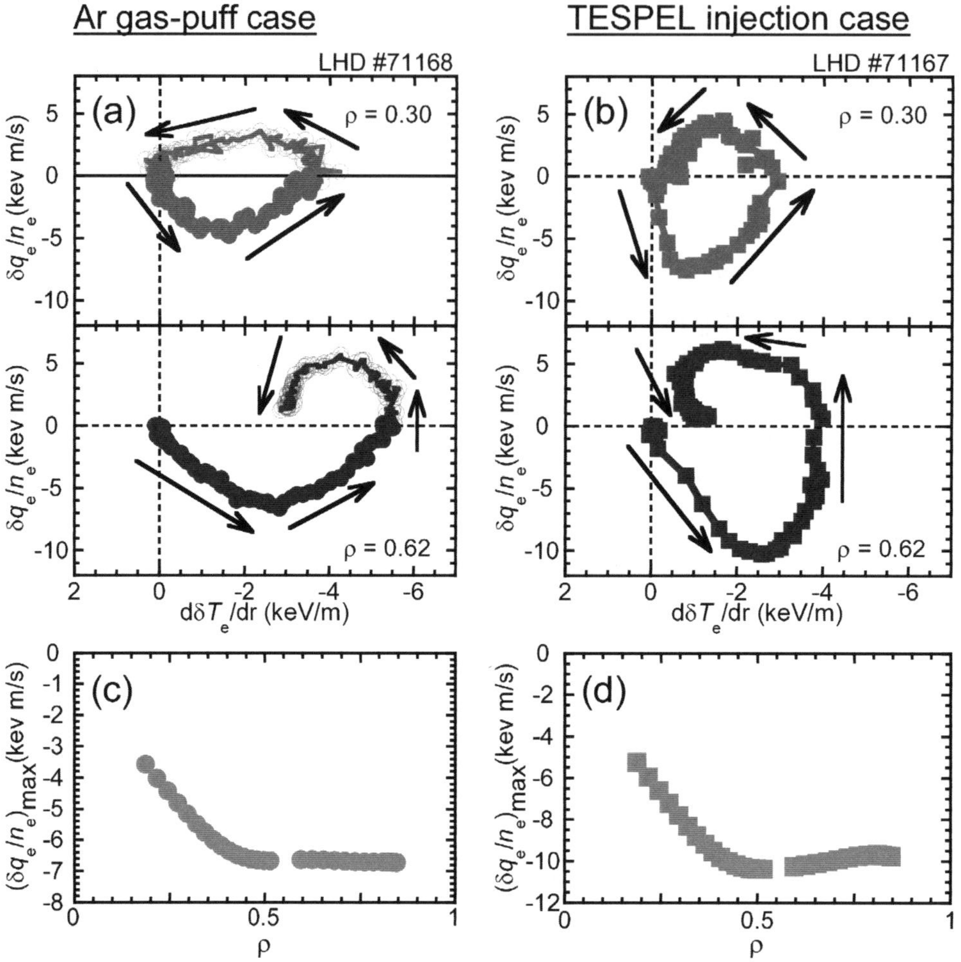

Figure 4. (*a*, *b*) Relation of the electron heat flux perturbation normalized by the base electron density to the gradient of the electron temperature perturbation after the edge perturbation at $\rho = 0.30$ and 0.62. Each data point is obtained at 1 millisecond intervals. In (*a*), the data points after $t = 0.6$ s are indicated by open symbols. (*c*, *d*) Radial profiles of the minimum of the electron heat flux perturbation normalized by the base electron density. The Ar gas-puff case is shown in (*a*, *c*) and the TESPEL injection case shown in (*b*, *d*).

$\rho \sim 0.3$ and ~ 0.6. At $\rho \sim 0.3$, the $\delta T_e/T_e$ is predominantly larger than the $\delta n_e/n_e$ until around $t = 0.6$ s. Meanwhile, at $\rho \sim 0.6$, the $\delta n_e/n_e$ is comparable to the $\delta T_e/T_e$ until $t = 0.57$ s and then starts to dominate the $\delta T_e/T_e$. As far as the anomalous transport dominates, $D/\chi < 1$ can be valid. Therefore the perturbed particle flux effect on the analysis can be ignored, in so far as the analysis is limited to the core region until the core T_e reaches the maximum, $t \sim 0.6$ s, even if the $L_{\delta T_e}$ is comparable to the $L_{\delta n_e}$. Consequently, the δq_e can be expressed as

$$\delta q_e(r,t) = -\frac{1}{r}\int_0^r \frac{3}{2} n_e \frac{\partial \delta T_e(r',t)}{\partial t} r' dr'. \quad (3)$$

Figure 4(*a*) shows the relations of the perturbed electron heat flux normalized by the base electron density $\delta q_e/n_e$ to the gradient of the electron temperature perturbation $d\delta T_e/dr$ at $\rho = 0.30$ and 0.62 for the case with the Ar gas-puff, which is shown in figure 1. In figure 4(a), the data points obtained after $t = 0.6$ s, which are likely to be influenced by the perturbed particle flux, are indicated by open symbols. The relations for the case with the TESPEL injection, which is obtained under the same experimental conditions as the Ar gas-puff case, are shown in figure 4(b). In both figures, each data point is obtained at intervals of 1 milliseconds. Regardless of cooling method, similar hysteresis loops appear in the relation of the $\delta q_e/n_e$ to the $d\delta T_e/dr$ at $\rho = 0.30$. At the very initial phase of the reduction in the $\delta q_e/n_e$, there is little change in the local electron temperature gradient, which is very difficult to understand with a simple diffusive nature (Accoring to the diffusive nature, the data points in figure 4(a, b) should lie on a straight line from the bottom left to the top right.). Thus the transport in this phase is characterized by nonlocal. Also, at $\rho = 0.62$, similar dynamic responses in the flux-gradient space appear, even though the final value of the $\delta\nabla T_e$ in the flux-gradient space with Ar gas-puff is different from that with the TESPEL injection (In the both cases, it does not come back to the level before the edge perturbation.). As shown in figures 4(c, d), the reduction of the $\delta q_e/n_e$ is obtained over a wide region of the plasma for the both cases and the minimum of $\delta q_e/n_e$ appears at almost the same location ($\rho \sim 0.5$) far from the initially perturbed region ($\rho > 0.7$). The duration time between the initial edge perturbation and relaxation to a final state is almost twice as long for the Ar gas-puff case ($\sim$ 180 ms) compared with that for the TESPEL injection case ($\sim$ 80 ms). The attainable value of the electron temperature gradient in the Ar gas-puff case is higher than that in the TESPEL injection case, even though the reduction of the $\delta q_e/n_e$ in the Ar gas-puff case is lower than that in the TESPEL injection case. This experimental result suggests that the deviation of the q_e/n_e and the ∇T_e from the steady state is independent of the time scale of the variation.

The anomalous transport is considered to be mainly attributed to microturbulences, whose characteristic lengths are on a scale of the electron/ion gyroradius. Thus, in order to describe reasonably the nonlocal response of the electron heat transport, it is important to consider the multi-scale (from the micro- to the meso- or the macro-scale) interaction [9]. The existence of zonal flows has been verified experimentally in the toroidal helical plasmas, in particular the H-1 Heliac plasma [20], and those have been shown to be nonlinearly driven by drift wave turbulence [21]. Recently, a numerical simulation demonstrated the multi-scale interaction among the micro-, the meso- and the macro-scale fluctuation [22]. In this simulation, the macro MHD mode is nonlineary triggered as a result of the interaction between the turbulence and zonal flow. Nevertheless, further experimental observations and theoretical investigations are required to elucidate the relationship and interaction between the nonlocal T_e rise and the turbulences.

4. Summary

In LHD, the core T_e rise after the controlled Ar gas-puff is observed in the low density and high temperature plasmas heated only by ECH. The transient transport analysis reveals that the dynamic response of the electron heat transport during the core T_e rise after the Ar gas-puff is beyond a simple diffusive nature. A close similarity in the relation of the $\delta q_e/n_e$ to the $\delta\nabla T_e$ is found between the core T_e rise after the Ar gas-puff and the nonlocal T_e rise in response to the pellet injection. Thus the nonlocality in the electron heat transport is essential for the core T_e rise in the low density and high temperautre plamsas of LHD after the edge cooling regardless of method. The Ar gas-puff induced core T_e rise suggests that the deviation of the q_e/n_e and the ∇T_e from the steady state is independent of the timescale of the variation.

Acknowledgments

The authors acknowledge all of the technical staff of NIFS for their excellent support. They also would like to thank Prof. O. Motojima (Director of NIFS) for his continuous encouragement.

This work is partly supported by a Grant-in-Aid for Scientific Research (B) (No. 19340179) from Japan Society for the Promotion of Science, a Grant-in-Aid for Young Scientists (B) (No. 19740349) from MEXT Japan and a budgetary Grant-in-Aid No.NIFS07ULHH510 of the National Institute for Fusion Science.

References

[1] Tamura N *et al* 2005 *Phys. Plasma* **12** 110705
[2] Gentle K W *et al* 1995 *Phys. Rev. Lett.* **74** 3620
[3] Kissick M W, Callen J D and Fredrickson E D 1998 *Nucl. Fusion* **38** 821
[4] Galli P, Gorini G, Mantica P, Hogeweij G M D, Kloe J de, Lopes Cardozo N J, RTP team 1999 *Nucl. Fusion* **39** 1355
[5] Zou X L *et al* 2000 *Plasma Phys. Control. Fusion* **42** 1067
[6] Ryter F *et al* 2000 *Nucl. Fusion* **40** 1917
[7] Inagaki S *et al* 2006 *Plasma Phys. Control. Fusion* **48** A251
[8] Tamura N *et al* 2007 *Nucl. Fusion* **47** 449
[9] Itoh K, Itoh S-I, Spineanu F, Vlad M O and Kawasaki M 2003 *Plasma Phys. Control. Fuison* **45** 911
[10] Motojima O *et al* 1999 *Phys. Plasma* **65** 1843
[11] Kawahata K, Nagayama Y, Inagaki S, Ito Y and the LHD Experimental Group 2003 *Rev. Sci. Instrum.* **74** 1449
[12] Narihara K, Yamada I, Hayashi H and Yamauchi K 2001 *Rev. Sci. Instrum.* **72** 1122
[13] Kawahata K, Tanaka K, Ito Y, Ejiri A and Okajima S 1999 *Rev. Sci. Instrum.* **70** 707
[14] Peterson B J, Kostrioukov A Yu, Ashikawa N, Liu Y, Xu Y, Osakabe M, Watanabe K Y, Shimozuma T, Sudo S and the LHD Experimental Group 2003 *Plasma Phys. Control. Fusion* **45** 1167
[15] Morita S and Goto M 2003 *Rev. Sci. Instrum.* **74** 2375
[16] Messiaen A M *et al* 1996 *Phys. Rev. Lett.* **77** 2487
[17] Jackson G L *et al* 1999 *J. Nucl. Mater.* **266-269** 380
[18] Tokar M Z, Jaspers R, Weynants R R, Koslowski H R, Krämer-Flecken A, Messiaen A M, Ongena J and Unterberg B 1999 *Plasma Phys. Control. Fusion* **41** L9
[19] Cardozo N J L 1995 *Plasma Phys. Control. Fusion* **37** 799
[20] Fujisawa A *et al* 2004 *Phys. Rev. Lett.* **93** 165002
[21] Shats M G and Solomon W M 2002 *Phys. Rev. Lett.* **88** 045001
[22] Ishizawa A and Nakajima N 2007 *Phys. Plasmas* **14** 040702

Development of hydrogen storage electrode for plasma biasing in the Tohoku University Heliac

H Utoh[1], K Nishimura[2], H Takahashi[2], Y Tanaka[1], H Aoyama[1], K Iwazaki[1], H Umetsu[1], K Ishii[1], S Inagaki[3], M Yokoyama[2], Y Suzuki[2], A Okamoto[1], T Kobuchi[1], S Kitajima[1] and M Sasao[1]

[1] Department of Quantum Science and Energy Engineering, Tohoku University, Sendai 980-8579, Japan
[2] National Institute for Fusion Science, Toki 509-5292, Japan
[3] Research Institute for Applied Mechanics, Kyushu University, Fukuoka 816-8580, Japan

E-mail: hiroyasu.u@ppl2.qse.tohoku.ac.jp

Abstract. In the Tohoku University Heliac (TU- Heliac), a new type of hydrogen storage electrode made of a palladium (Pd) was developed. This new gold (Au)-coated palladium electrode can produce high-density plasma up to ~40 shots per hydrogen gas charge. The Pd-Au electrode is the most useful tool for production of high-density plasma, compared with previous electrodes (Ti, V, and Pd). A reduced-size Au-coated palladium electrode was also developed. The electrode current was controllable in the lower voltage region and the controllability of electrode current in the small electrode current range was confirmed in only one discharge. Compared with the hot cathode experiments, the reduced-size Pd-Au electrode had the ability to produce a biased plasma, which was similar to the characteristics of the biased plasma by the hot cathode made of LaB_6, even though the surface area of the electrode for the plasma was much smaller than that the hot cathode.

1. Introduction

The H mode transition has been observed in many devices [1–5]. The radial electric field is considered important to understand the nature of LH transition. Electrode biasing experiments are effective tools for the study of improved confinement mode [6, 7]. It has been reported that the H mode can be triggered by applying sufficient voltage to an electrode located inside the plasma [8, 9], to control the driving force for plasma rotation. It is necessary to control an electrode current externally using a current control power supply. A hot cathode made of LaB_6 is a useful electron emission source for electrode biasing experiments. In our previous experiments on the Tohoku University Heliac (TU-Heliac), the role of the radial electric field on the transition to improved modes has been investigated by electrode biasing experiments. In positive biasing experiments with a stainless steel (SUS) electrode (electron collection), the transition to improved confinement mode was observed clearly when the radial electric field was formed [10]. Furthermore, by negative biasing with a hot cathode electrode (electron emission), the radial electric fields can be actively controlled by changing the electrode current, I_E [11–13]. The local maxima in viscosity were found at poloidal Mach number around $-M_p \sim 1$–3, as predicated by neoclassical theory [14–17]. The poloidal Mach number $M_p = -(E_r/B_0) / \Theta v_{th}$, $\Theta = \varepsilon / q$, where E_r is the radial electric field, B_0 is the magnetic field on the axis,

$\varepsilon = \langle r \rangle / R_0$ is the toroidal ripple, q is the safety factor, and v_{th} is the ion thermal velocity. At $-M_p \sim 4$, where the torque in the poloidal direction exceeded the local maxima in viscosity, the following characteristics were observed: increases in electron density and energy confinement time, formation of strong electric field, and suppression of fluctuation level. These phenomena were also observed in Compact Helical System (CHS) at low magnetic field [18]. However, a hot cathode made of LaB_6 (especially of the solenoid type) is fragile to electromagnetic force, and thus in the strong magnetic field biasing experiments on medium or large devices, we cannot adopt a hot cathode made of LaB_6 as the electrode with controllability of electrode current.

Consequently, a new type of electrode made of hydrogen storage metal has been developed for particle injection (electron, ion, and neutral particle). Using an electrode made of a hydrogen storage metal, such as titanium (Ti) or vanadium (V), the following can be expected: (1) the electrons injected from the negative-biased electrode allow biasing experiments for studying the LH transition by control of the electrode current, (2) the electrons/neutral particles injected from the negative-biased electrode provide production of high-beta plasma and high-density plasma, and (3) ions accelerated from the positive-biased electrode allow simulation of the orbit loss of high-energy particles if hydrogen can be stored successfully in and released from the electrode. There is no hard limit of the high-density operation in stellarators, unlike the Greenwald density limit in tokamaks, which is a great advantage for future reactor operation. Therefore, it is important to study high-density plasma production and to explore the high-density limit in stellarators. In TU-Heliac, the highest density is about 1×10^{13} cm^{-3}, which can be achieved by pre-filling with working gas. In additional gas puffing, broadening of neutral gas requires some time (a few milliseconds), and the plasma density increase slowly (a few milliseconds). Therefore, the discharge time (10 ms) is insufficient to increase plasma density by additional gas puffing. In contrast, biasing electrodes made of hydrogen storage metals can increase the density at the plasma core rapidly (~0.5 ms) [19, 20]. In addition, the TU-Heliac has a non-planar magnetic axis system, which is expected to stabilize high-beta plasma due to the self-stabilization effect [21–23]. Negative biasing experiments using this hydrogen storage electrode have the potential to realize high-beta experiments in some small-sized devices (*e.g.*, TU-Heliac).

In the TU-Heliac, using a titanium (Ti) or vanadium (V) electrode after hydrogen gas charging in negative electrode biasing, the electron density and the line intensity of H_α increased by about 10-fold as compared with those before biasing, the electrode current I_E increased by about 100-fold as compared with the ion saturation current [19, 20]. Specifically, with vanadium electrode biasing, the radial distribution of the electron density sloped steeply at the electrode position, and a strong negative radial electric field was formed outside the electrode. The mechanism of high-density plasma production suggests the injection of electrons/neutral hydrogen from the Ti electrode into the plasma, caused by influx of high-energy particles into the electrode. When the V electrode was negatively biased in an Ar plasma, under the low magnetic field, the β value increased up to about 0.5%, allowing the realization of a new field of high-β experiments. One of the most important performance-related factors for hydrogen storage electrodes is long life time, which means that we can produce a number of high-density shots per hydrogen gas charge. The Ti electrode can produce only 3 high-density plasma shots per hydrogen gas charge, while the V electrode can produce up to 24 shots of high-density plasma per hydrogen gas charge. However, the V electrode has a disadvantage (hydrogen embrittlement), which causes reduction of hydrogen gas evolution from the electrode surface.

Therefore, we developed a palladium (Pd) electrode, which is free from the problem of hydrogen embrittlement. If the palladium electrode could produce release of hydrogen gas efficiently, it would be expected to be useful not only for high-density plasma production but also for biasing experiments to study LH transition by controlling the electrode current. Here, we propose a new type of electrode, the results of biasing experiments with this new electrode, and comparison of similar experiments with other electrodes (Ti, V, and LaB_6).

2. Experimental setup

2.1. TU-Heliac

The TU-Heliac is a small standard heliac device (major radius R = 0.48 m, average plasma radius a = 0.07 m), with a toroidal magnetic period n = 4 [24, 25]. The heliac configurations are produced by three sets of coils: toroidal field coils, a center conductor coil, and vertical field coils. Various magnetic configurations can be formed easily by selecting the ratios of coil currents, and Fourier components can be varied over a wide range. The target plasma was produced by Joule heating with an 18.9 kHz alternating current in additional poloidal coils, with effective input power of about 3 kW. The vacuum vessel was filled with a working gas (H_2, He, Ar) before discharges. The magnetic field on axis, B_0, was 0.3 T in standard operation. Plasma parameters were measured with a Langmuir probe (triple probe: electron density, n_e; electron temperature, T_e; floating potential, V_f; plasma space potential, V_s) at toroidal angle $\phi = 0°$ and a 6-mm microwave interferometer (electron line-density, $n_e l$) at $\phi = 90°$. The line intensity of H_α,, $I_{H\alpha}$, was measured with a 25-cm monochromator coupled with a photomultiplier tube at $\phi = 159°$.

2.2. Hydrogen storage electrode

The experimental setup of the hydrogen storage electrode is shown in Figure 1 (a) along with the cross-section of the magnetic flux surfaces in a vacuum at toroidal angle $\phi = 0°$. The hydrogen storage electrode was inserted vertically into the plasma from the top at $\phi = 0°$. It was biased positively or negatively against the vacuum vessel through a field-effect transistor (FET). The hydrogen storage electrode head (10 mm in diameter and 7 mm in length) was mounted at the end of the copper shaft housed in a glass tube as an insulating sheath (Figure 1 (b)). The hydrogen storage electrode was conditioned in the upper chamber (conditioning chamber) separated by a gate-valve. The temperature of the hydrogen storage metal head could be controlled by heating a tungsten (W) wire (0.15 mm in diameter) and was measured with a thermocouple gauge. The W wire and the thermocouple gauge were installed inside the electrode head.

The titanium stores and releases hydrogen at about 700°C, and hydrogen storage and release processes do not occur at room temperature. In contrast, vanadium and palladium store and release hydrogen at room temperature (20~250°C). The surface barrier (oxide layer) interferes with the storage and release of hydrogen gas. The surface barrier of palladium is much lower than those of other hydrogen storage metals, such as titanium or vanadium.

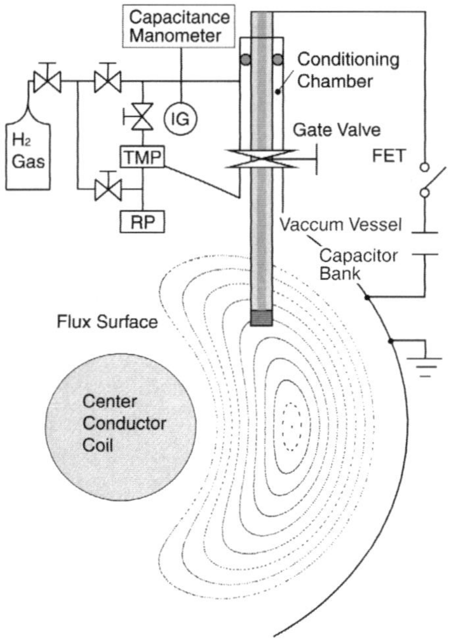

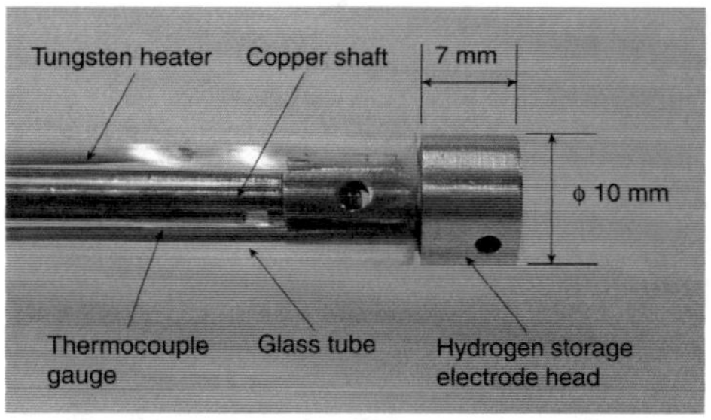

Figure 1. (a) The position of the hydrogen storage electrode and the computed cross-section of the magnetic flux surfaces in vacuum at $\phi = 0°$. The hydrogen storage electrode was conditioned in the upper chamber (conditioning chamber) separated by a gate-valve, and was inserted vertically into the plasma from the low magnetic field side.
(b) The hydrogen storage electrode head (10 mm in diameter and 7 mm in length) was mounted at the end of the copper shaft housed in a glass tube as an insulating sheath.

Journal of Physics: Conference Series **123** (2008) 012024 doi:10.1088/1742-6596/123/1/012024

3. Experimental results

3.1. Gold-coated Palladium electrode

In treatment of the palladium electrode, the electrode was first heated to about 300°C for 24 h. After baking, the electrode was conditioned in hydrogen gas (p_{in} = 247 kPa) at 30°C in the conditioning chamber shown in Figure 1 (a). These conditions of temperature and gas pressure for the Pd electrode were almost the same as those for the Ti and V electrodes, but the decrease in hydrogen gas pressure p_{in} in the Pd electrode ($\Delta p_{in} \approx 48$ kPa) was about double and about 5-fold higher than those in the V electrode ($\Delta p_{in} \approx 25$ kPa) and the Ti electrode ($\Delta p_{in} \approx 10$ kPa), respectively. The number of Pd atoms in the electrode, N_{Pd}, was 2.9×10^{22}. The total number of hydrogen atoms absorbed in the Pd electrodes was estimated from the decrease in hydrogen gas pressure after treatment, and was equivalent to about 2.0×10^{22} hydrogen atoms (N_H), and the hydrogen concentration in the Pd electrode ($x = N_H / N_{Pd}$) was about 0.7. On the other hand, hydrogen concentrations ($x = N_H / N_{Ti,\ V}$) were about 0.2 and 0.3 in the case of Ti and V electrodes, respectively [20]. When the Pd electrode was biased negatively after treatment, the high-density plasma was produced, similar to the Ti or the V electrode experiments. However, the production of high-density plasma was observed only twice with the Pd electrode after one treatment. When the Pd electrode was inserted into the vacuum vessel, the Pd electrode constantly released hydrogen gas, and therefore a controlled working gas pressure could not be achieved.

To avoid the above uncontrollable release of hydrogen gas from the electrode, we developed a new type electrode made of gold (Au)-coated palladium (Pd-Au). The Pd was coated with Au about 100 nm thick using an ion coater, which was variable by changing the discharge time. The electrode surface in contact with the plasma directly was coated with Au. In addition, when the electrode was conditioned, the hydrogen gas pressure (p_{in} = 19.8 kPa) was reduced to p = 14.3 kPa at 36°C ($\Delta p_{in} \approx$ 5.5 kPa). The release of hydrogen gas from the gold-coated palladium electrode was less than half that of the uncovered Pd electrode. Furthermore, the new type of electrode can produce up to ~40 shots of high-density plasma per hydrogen gas charge. The typical time evolution of plasma parameters in Ar discharge is shown in Figure 2. When the Pd-Au electrode was biased negatively from 4 ms to 10 ms, the electrode current, I_E, increased up to about 100 A, and the electron density, n_e, at $\rho = 0.54$ increased by about 5-fold as compared with those before biasing. The differences among the plasma space potential signals measured at three difference radial potions showed that the strong negative radial electric field was formed by the Pd-Au electrode. The line intensity of H_α, $I_{H\alpha}$, at the central chord increased rapidly by about 10-fold as compared with that before biasing. In this experiment, the negative bias voltage was supplied against the vacuum vessel by a constant-voltage power supply. The threshold in the electrode bias voltage for production of high-density plasma was about 100 V.

Figure 3 shows the relation between the electrode bias voltage for production of high-density plasma and the electron density of target plasma before biasing in Ar discharge. The electron density was measured with the triple probe around $\rho = 0.28 \sim 0.54$. The closed circles, open circles, open triangles, and open squares indicate the density of the gold-coated palladium electrode, the uncovered Pd electrode, the V electrode, and the Ti electrode, respectively. In the case of Pd-Au electrode biasing, the bias voltage and target plasma density necessary for production of high-density plasma were lower than for other electrodes. These results show that the Pd-Au electrode can easily produce high-density plasma. Figure 4 shows the relation between the electrode bias voltage for production of high-density plasma and the ratio of the electron density with biasing $n_{e,\ bias}$ to the electron density before biasing $n_{e,\ no\ bias}$ in Ar discharge. In the case of Pd-Au electrode biasing, we can see that the rate of increase in electron density was similar to those for the other electrodes (Ti, V, and Pd) in the lower bias voltage range than other electrodes.

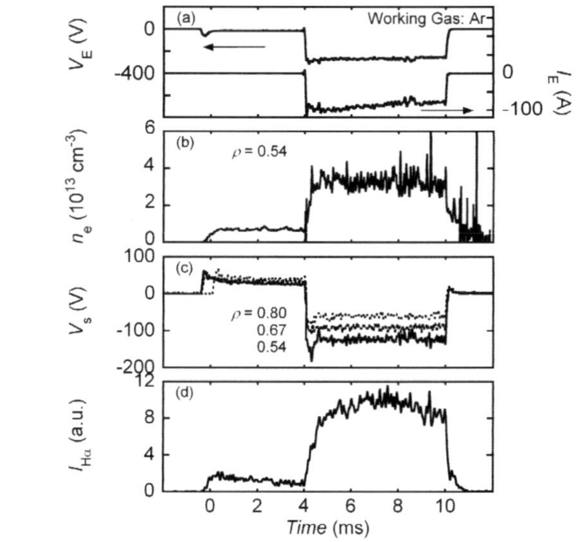

Figure 2. Time evolutions of plasma parameters: (a) electrode voltage, V_E and electrode current, I_E; (b) electron density, n_e, at $\rho = 0.54$; (c) Differences among the plasma space potential signals, V_s, at $\rho = 0.54$, 0.67, and 0.80; and (d) line intensity of H_α, $I_{H\alpha}$, at the central chord.

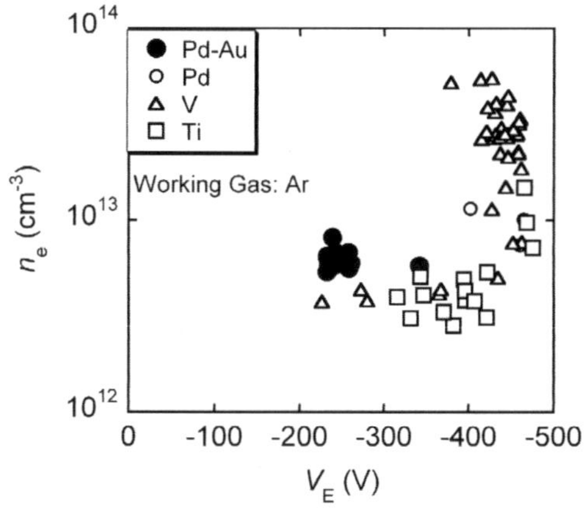

Figure 3. Relations of the electrode bias voltage for production of high-density plasma and the electron density of target plasma before biasing in Ar plasma. The electron density was measured with a triple probe around $\rho = 0.28 \sim 0.54$.

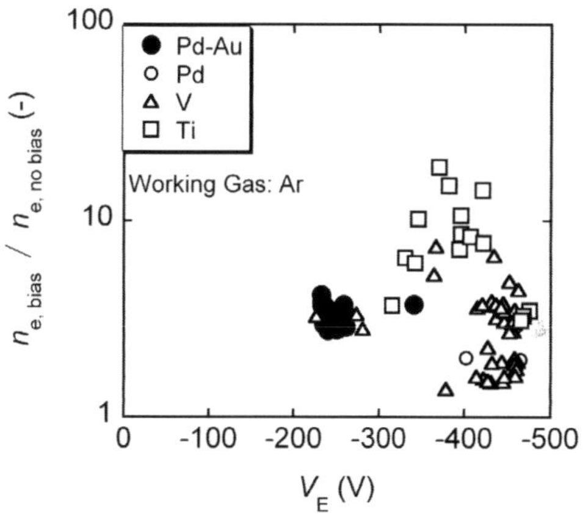

Figure 4. The relation of the electrode bias voltage for production of high-density plasma and the ratio of the electron density with biasing $n_{e,\,bias}$ to the electron density before biasing $n_{e,\,no\,bias}$ in Ar discharge. The close circles, open circles, open triangles and open squares denote the density of the Pd-Au electrode, the density of the Pd electrode, the density of the V electrode and the density of the Ti electrode.

Table 1 shows the performance of each of the electrodes made of hydrogen storage metal, titanium, vanadium, palladium, and gold-coated palladium [19, 20]. "Number of shots" indicates the number of rounds of production of high-density plasma per hydrogen gas charge, "Threshold of electrode biasing voltage" indicates a threshold in the electrode bias voltage for production of high-density plasma, "Minimum electrode current" indicates the minimum electrode current in high-density plasma, and "Residual gas pressure" indicates release of hydrogen gas from the electrode. Compared with other electrodes (Ti, V, Pd), the characteristics of the Pd-Au electrode were as follows: (1) maximum number of shots per hydrogen gas charge, (2) the comparatively small amount of hydrogen gas release from the electrode, (3) lowest threshold in the electrode bias voltage for production of high-density plasma, and (4) lowest electrode current in high-density plasma. These results indicate that the Pd-Au electrode is useful for production of high-density plasma.

Table 1. The performance of each hydrogen storage metal electrode.

Electrode material	Number of shots	Residual gas pressure (Pa)	Threshold of electrode voltage – V_E (V)	Minimum electrode current – I_E (A)
Ti	3	$\sim 5\times10^{-6}$	~300	~100
V	21	$\sim 2\times10^{-4}$	~200	~100
Pd	2	$\sim 8\times10^{-4}$	~150	~10
Pd-Au	~40	$\sim 6\times10^{-5}$	~100	<10

3.2. Size reduction of a gold-coated Palladium electrode

The previous electrode (Pd-Au electrode; Type 1) is a good tool for the production of high-density plasma, but the electrode current was too large to allow its control in the study of LH transition at poloidal Mach numbers around $-M_p \sim 1$–3. It was necessary to reduce the electrode current to less than 10 A. Therefore, we reduced the size of the gold-coated palladium electrode. The reduced-sized Pd-Au electrode (Type 2) head was 10 mm in diameter and 2 mm in length, and the surface area in contact with plasma S was about 1.4 cm^2, which was half that of the previous electrodes (Type 1; $S \sim 3.0$ cm^2). This small-sized electrode could have less influence on the target plasma.

A high electrode current mode was observed in biasing experiments using the reduced-size Pd-Au electrode (Type 2). The electrode current of type 2 electrode was half that of the type 1 electrode. The controllability of the electrode current was examined by sweeping the electrode voltage with an external control signal. Figure 5 shows the typical time evolution of (a) the electrode voltage V_E, (b) the electrode current I_E, (c) the plasma space potential V_s and (d) electron line density $n_e l$. In this experiment, the electrode voltage ramped up from 0 to –150 V at *Time* = 4 to 10 ms. When the electrode voltage was about –100 V (*Time* ~ 8.5 ms), the electrode current increased suddenly and electron line density also increased. Furthermore, the electrode current increased in proportion to the electrode voltage at V_E = –100 ~ –150 V. A clear threshold was observed in the electrode bias voltage for high electrode current mode.

Figure 6 shows the dependence of the electrode current on the electrode voltage in negative biasing experiments with both the type 1 (#69907 ~ #69910, closed circles) and type 2 (#71450, closed triangles and #71452, open squares) electrodes in the He plasma. The electrode current was proportional to the surface area of the electrode around V_E = –100 ~ –200 V. In this figure, we can see that the electrode current increased in proportion to the electrode voltage. In addition, in type 2 electrode biasing experiments, we confirmed the controllability of electrode current in the small electrode current range (I_E = –3 ~ –12 A) in only one discharge (#71450 or #71452).

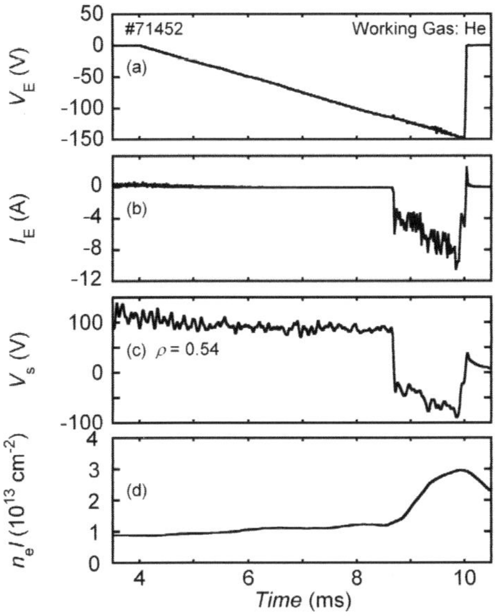

Figure 5. The typical time evolution of (a) the electrode voltage V_E, (b) the electrode current I_E, (c) the plasma space potential V_s and (d) electron line density $n_e l$. In this experiment, the electrode voltage ramped up from 0 to −150 V at *Time* = 4 to 10 ms.

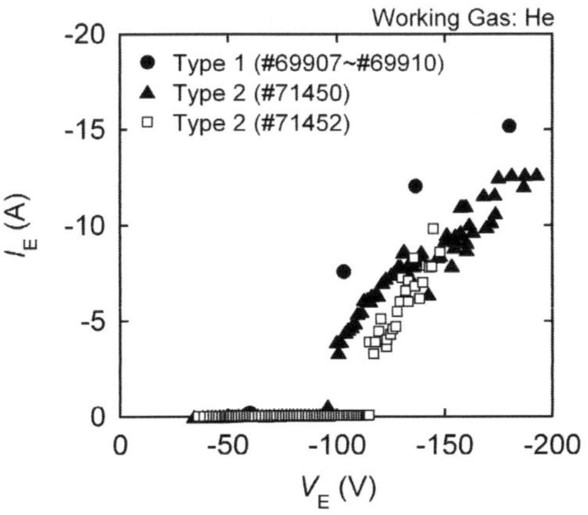

Figure 6. The dependence of the electrode current on the electrode voltae in negative biasing experiments with both type 1 (#69907 ~ #69910, closed circles) and type 2 (#71450, closed triangles and #71452, open squares) electrodes.

3.3. Comparison with theLaB_6 hot cathode experiments

To examine the plasma parameters produced by type 2 electrode biasing, we compared the results with those of hot cathode biasing experiments. In the TU-Heliac, to measure the plasma profile, it is necessary to use the triple probe. The high-precision measurgement using the triple probe requires a similar discharge many times. The gold-coated palladium electrode had a longer life time, and it was available to measure the plasma profile and for comparison with the hot cathode biasing experiments. The LaB_6 hot cathode (10 mm in diameter and 17 mm in length) was inserted horizontally into the plasma from the low field side at toroidal angle $\phi = 270°$. Figure 7 shows the typical time evolution of (a) the electrode voltage V_E and the electrode current I_E in the hot cathode and (b) the reduced-size Pd-Au electrode (Type 2) biasing experiments. At the same electrode voltage, even though the surface area of type 2 electrode S was one third that of the hot cathode, the electrode current of type 2 was 4-fold that of the hot cathode. Figure 8 shows the radial profiles of (a) the electron density n_e, (b) the electron pressure $n_e T_e$, and (c) the plasma space potential V_s measured with the triple probe, in hot cathode (circle) and type 2 electrode (square) biasing experiments, both before (open) and during biasing at $V_E = -150$ V (closed) in the He plasma. The electron density in the type 2 electrode biasing experiment was higher, while the electron pressure was almost the same as that in the hot cathode biasing experiment. In the type 2 electrode biasing experiment, the electron pressure was higher around the electrode position, and it was lower in the plasma core. This was because the electron temperature was lower at the plasma core. The decrease in temperature was considered to be due to low-temperature electron emission from the electrode and to measurement errors by triple probe measurement. In both the hot cathode and type 2 electrode biasing experiments, a strong negative radial electric field was formed around the electrode. Even though the surface area of the electrode for the plasma was much smaller than the hot cathode (S = 4.2 cm^2), the reduced-size Pd-Au electrode (Type 2) had the ability to produce a biased plasma with similar characteristics of the biased plasma by the hot cathode made of LaB_6.

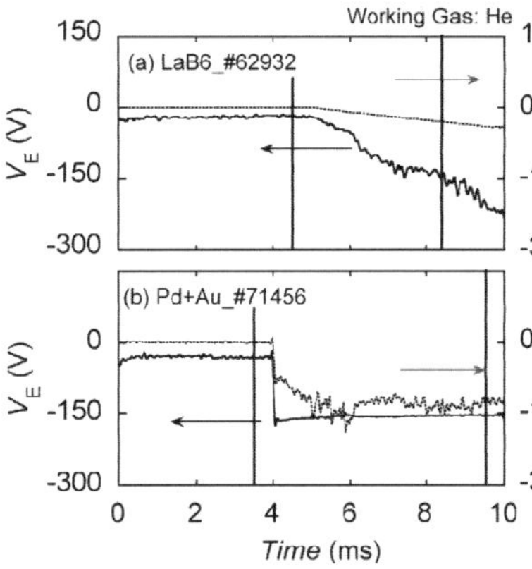

Figure 7. The typical time evolution of (a) the electrode voltage V_E and the electrode current I_E in the hot cathode and (b) the reduced-size Pd-Au electrode (Type 2) biasing experiments.

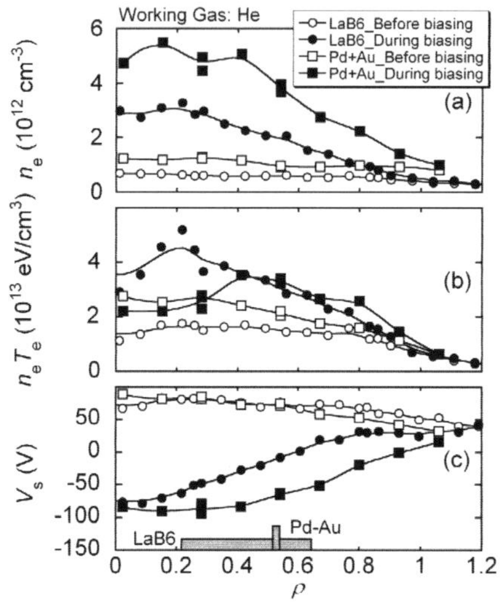

Figure 8. The radial profiles of (a) the electron density n_e, (b) the electron pressure n_eT_e, and (c) the plasma space potential V_s measured with the triple probe in hot cathode (circles) and type 2 electrodes (square) biasing experiments, both before (open) and during biasing at $V_E = -150$ V (closed) in the He plasma.

4. Summary

In the Tohoku University Heliac (TU- Heliac), we developed a new type of a hydrogen storage electrode made of a palladium (Pd). The release of hydrogen gas from the Au-coated palladium electrode was less than half that of the uncovered Pd electrode, and the new Au-coated palladium electrode could produce up to ~40 shots of high-density plasma per hydrogen gas charge (longer life time). In the case of Pd-Au electrode biasing, the bias voltage and target plasma density necessary for production of high-density plasma were lower, while the rate of increase in electron density was similar to those for the other electrodes (Ti, V, and Pd) in a lower bias voltage range than for the other electrodes. These results indicate that the Pd-Au electrode is the most useful tool for production of high-density plasma, as compared with the previous electrodes (Ti, V, and Pd). We developed a reduced-size Au-coated palladium electrode, the electrode current of which was controllable in the lower voltage region ($V_E = -100 \sim -200$ V), and we confirmed the controllability of electrode current in the small electrode current range ($I_E = -3 \sim -12$ A) in only one discharge. In comparison with the hot cathode experiments, although the surface area of the electrode for the plasma was much smaller than for the hot cathode, the reduced-size Pd-Au electrode (Type 2) had the ability to produce a biased plasma with similar characteristics to that produced by the hot cathode made of LaB_6.

Acknowledgments
This work was supported in part by the LHD Joint Planning Research program at the National Institute for Fusion Science and by MEXT under the Scientific Research of Priority Areas, "Advanced Diagnostics for Burning Plasmas," a grant-in-aid from the Ministry of Education, Science, and Culture of Japan (No. 17360439), and a Research Fellowship from the Japan Society for the Promotion of Science (JSPS) for Young Scientists (No. 19-7419).

References

[1] Wagner F, Becker G, Behringer K, Campbell D, Eberhagen A, Engelhardt W, Fussmann G, Gehre O, Gernhardt J, Gierke G V *et al.* 1982 *Phys. Rev. Lett.* **49** 1408
[2] Burrell K H, Ejima S, Schissel D P, Brooks N H, Callis R W, Carlstrom T N, Colleraine A P, DeBoo J C, Fukumoto H, Groebner R J *et al.* 1987 *Phys. Rev. Lett.* **59** 1432
[3] Kotschenreuther M, Dorland W, Beer M A and Hammett G W 1995 *Phys. Plasmas* **2** 2381
[4] Erckmann V, Wagner F, Baldzuhn J, Brakel R, Burhenn R, Gasparino U, Grigull P, Hartfuss H J, Hofmann J V, Jaenicke R *et al.* 1993 *Phys. Rev. Lett* **70** 2086
[5] Toi K, Ohdachi S, Morisaki T, Sakakibara S, Morita S, Minami T, Tanaka K, Yamada H, Nishimura K, Matsuoka K *et al.* 1996 *Plasma Phys. Control. Fusion* **38** 1289
[6] Askinazi L G, Golant V E, Lebedev S V, Rozhanskij V A and Tendler M 1992 *Nucl. Fusion* **32** 271
[7] Wang E Y, Wang Xin, Diebold D A, Doczy M, Edgell D, Hershkowitz N, Intrator T, Kant G, Oliva S, Regan S *et al.* 1995 *Nucl. Fusion* **35** 467
[8] Taylor R J, Brown M L, Fried B D, Grote H, Liberati J R, Morales G J, Pribyl P, Darrow D and Ono M 1989 *Phys. Rev. Lett.* **63** 2365
[9] Weynants R R and Taylor R J 1990 *Nucl. Fusion* **30** 945
[10] Inagaki S, Kitajima S, Takayama M, Nakamura E, Yoshida T and Watanabe H 1997 *Japan. J. Appl. Phys.* **36** 3697
[11] Kitajima S, Nakanishi S, Harada K, Sugawara K and Hashizume H 2001 *J. Plasma Fusion Res. Ser.* **4** 391
[12] Kitajima S, Nakanishi S and Hashizume H 2002 *Int. J. Appl. Electromagn. Mech.* **13** 381
[13] Kitajima S, Takahashi H, Tanaka Y, Utoh H, Sasao M, Takayama M, Nishimura K, Inagaki S and Yokoyama M 2006 *Nucl. Fusion* **46** 200
[14] Itoh S-I and Itoh K 1988 *Phys. Rev. Lett.* **60** 2276
[15] Shaing K C and Crume E C Jr 1989 *Phys. Rev. Lett.* **63** 2369
[16] Shaing K C 1993 *Phys. Fluids* B **5** 3841
[17] Shaing K C 1996 *Phys. Rev. Lett.* **76** 4364
[18] Takahashi H, Utoh H, Kitajima S, Isobe M, Suzuki C, Takeuchi M, Ikeda R, Tanaka Y, Yokoyama M, Toi K *et al.* 2007 *Fusion Science and Technology* **51** 54
[19] Utoh H, Nishimura K, Inagaki S, Takahashi H, Tanaka Y, Takenaga M, Ogawa M, Shinde J, Shinto K, Kitajima S *et al.* 2005 *Proceedings of the 32nd EPS Conference on Plasma Physics* P2_066
[20] Utoh H, Nishimura K, Inagaki S, Takahashi H, Tanaka Y, Takenaga M, Ogawa M, Shinde J, Iwazaki K, Okamoto A *et al.* 2006 *Fusion Science and Technology* **50** 434
[21] Bazhanova A E, Glagolev V M and Shafranov V D 1967 *Soviet-Phys. Tech. Phys.* **11** 1177
[22] Nagao S, Sugita K, Watanabe H, Goto Y and Kondo Y 1969 *J. Phys. Soc. Jpn.* **27** 1082
[23] Mikhailovskii A B and Shafranov V D 1974 *Soviet. Phys. JEPT.* **39** 88
[24] Kitajima S, Takayama M, Zama T, Takaya K, Takeuchi N and Watanabe H 1991 *Japan. J. Appl. Phys.* **30** 2606
[25] Zama T, Kitajima S, Takayama M, Takeuchi N and Watanabe H 1993 *Japan. J. Appl. Phys.* **32** 349

Journal of Physics: Conference Series **123** (2008) 012025 doi:10.1088/1742-6596/123/1/012025

Simulation study of density dynamics effect on the ELM behavior with TOPICS-IB

N Hayashi, T Takizuka, N Aiba, T Ozeki and N Oyama

Japan Atomic Energy Agency, Naka Fusion Institute, Naka 311-0193, Japan

E-mail: hayashi.nobuhiko@jaea.go.jp

Abstract. Density dynamics effect on the edge-localized-mode (ELM) behavior has been studied by using an integrated simulation code TOPICS-IB based on a 1.5 dimensional transport code with a stability code of peeling-ballooning modes and a transport model of scrape-off-layer (SOL) and divertor plasmas. Neutral models of core and SOL-divertor regions have been integrated with the TOPICS-IB. The TOPICS-IB successfully simulates the behavior of the whole plasma with density dynamics. It is confirmed by varying the density and the temperature instead of artificially enhancing the collisionality in the previous study that the experimentally observed collisionality dependence of the ELM energy loss is caused by the bootstrap current and the SOL transport. Additionally, ion convective and charge-exchange losses are found to enhance the collisionality dependence by the equipartition effect. On the other hand, the ELM particle loss is found to be almost independent of the collisionality, as shown in experiments, due to the density profile before the ELM and the increase of the SOL density during the ELM in addition to the bootstrap current effect.

1. Introduction

Edge localized modes (ELMs) induce sometimes very large heat load to divertor plates and cause the reduction of plate lifetime. Analyses from multi-machine experiments showed that the ELM energy loss increases with decreasing the collisionality [1-3]. Experimental analyses also suggested that the collisionality dependence of the ELM energy loss is mainly caused by the collapse of the electron temperature profile during an ELM crash (electron conductive energy loss) and the collapse of ion temperature and density profiles (ion conductive and convective energy losses) has smaller effect [1-3]. Contrary to the ELM energy loss, the ELM particle loss is almost independent of the collisionality in experiments [1,3]. Although effects of the bootstrap current and the scrape-off-layer (SOL) transport on the ELM energy loss have been discussed [1,2], the physical understanding and quantitative evaluation are not fully accomplished so far.

The integrated simulation code is one of effective methods to study the ELM mechanism [4-9]. For example, an integrated code COCONUT, which integrates 1.5 dimensional (1.5D) core transport code JETTO and 2D divertor code EDGE2D, has been developed and is being used for the study of ELM dynamics [8]. On the other hand, we have developed an integrated code TOPICS-IB [9] with a dynamic five-point model for SOL-divertor plasmas [10] and a stability code for peeling-ballooning modes, MARG2D [11]. The TOPICS-IB is based on a 1.5D core transport code TOPICS extended to the integrated simulation for burning plasmas. The TOPICS-IB successfully simulated a series of transient behaviors of an H-mode plasma; the pedestal growth, an ELM crash and the recovery of the pedestal. At the ELM crash, the increase of SOL temperature mitigates the radial edge gradient and lowers the ELM energy loss. The collisionality dependence of the ELM energy loss was found to be caused by both the edge bootstrap current and the SOL parallel transport. The collisionality dependence caused by the parallel transport was also pointed out by the analytical model [12]. The steep pressure gradient inside the pedestal top broadens the region of the ELM enhanced transport through the broadening of eigenfunctions and enhances the ELM energy loss [13]. Figure 1 shows TOPICS-IB results of (a) the ELM energy loss, ΔW_{ELM}, normalized by pedestal energy, W_{ped}, as a function of the normalized collisionality ν^* for two cases A and B with different pressure gradient inside the pedestal top at the normalized minor radius $\rho = 0.925$ at the ELM onset in figure 1(b). In figure 1(a), the TOPICS-IB reproduced the magnitude of the ELM energy loss and its collisionality dependence in the data of multi-machine and JT-60U experiments [1,14]. In this simulation, however, the density profile is fixed for simplicity. The density collapse enhances the values of $\Delta W_{ELM}/W_{ped}$ by about 50 % through the convective energy loss under the assumption of the similar collapse to the temperature one.

In this paper, we study the effect of the density dynamics on the ELM behavior and the resultant particle and energy losses. The dynamics of the pedestal density strongly connects with the neutral recycling. We integrate neutral models in both the TOPICS and the five-point model. The collisionality dependence of the ELM particle and energy losses is investigated by varying the density and the temperature instead of artificially enhancing the collisionality in the previous study. In the next section, we explain the TOPICS-IB and the integration with neutral models. An integrated simulation result of the whole plasma with the density dynamics is shown in section 3. The pedestal density and temperature are varied by the gas puffing in section 4. The collisionality dependence of ELM particle and energy losses is clarified in sections 5 and 6.

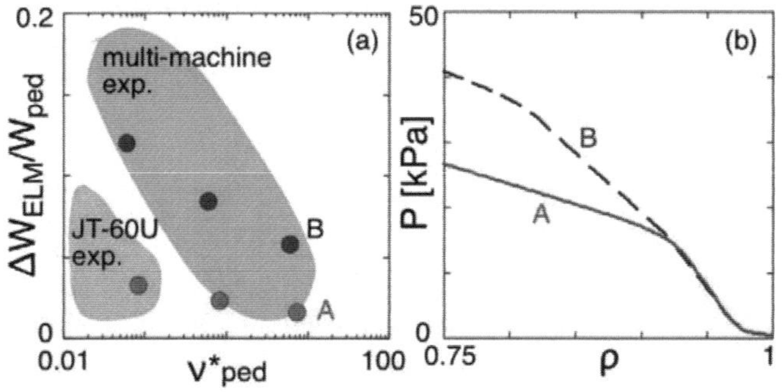

Figure 1. (a) $\Delta W_{ELM}/W_{ped}$ as a function of ν^* for two cases A and B with pressure profiles in (b) where shaded regions denote experimental data [1,14]. (b) Pressure profiles at the ELM onset with different pressure gradients inside pedestal top at $\rho = 0.925$.

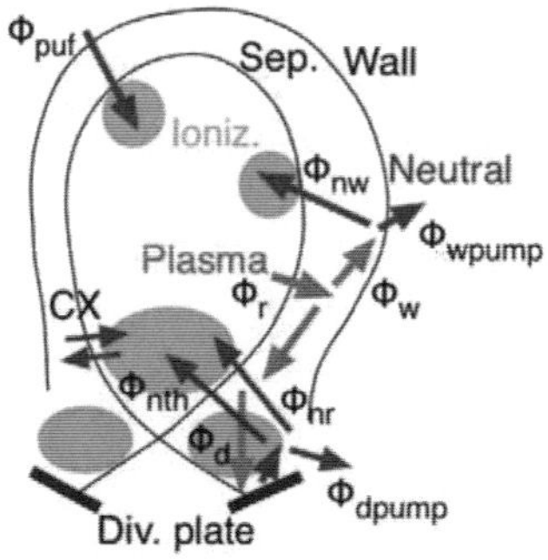

Figure 2. Particle flow in whole region.

2. Integrated code TOPICS-IB for the ELM dynamics

The ELM dynamics is investigated by the TOPICS-IB, in which the TOPICS is coupled with the ELM model [4] and the SOL-divertor model [10]. Details of the TOPICS-IB are shown in [9]. Some essential features, neutral models and those integration are explained as follows.

2.1. TOPICS

TOPICS self-consistently solves the 1D transport and current diffusion equations and the Grad-Shafranov equation of the MHD equilibrium on the 2D plane. The transport equations are the continuity equation for the deuterium ion density, n_i, the power balance equations for the electron temperature, T_e, and the ion temperature, T_i, on the coordinate of the normalized minor radius, ρ, defined by the square root of the toroidal flux. In this paper, particle and thermal diffusivities are assumed as $D = D_{neo,i} + D_{ano}$ and $\chi_{e,i} = \chi_{neo,i} + \chi_{ano,e,i}$, where $D_{neo,i}$ and $\chi_{neo,i}$ denote neoclassical ion diffusivities. The anomalous diffusivities D_{ano} and $\chi_{ano,e,i}$ are simply given as an empirical formula based on JT-60U experiments, $D_{ano} = \chi_{ano,e} = 0.18\ (1+2\rho^3)\ (1+P_{NB}^{0.5})\ \mathrm{m^2/s}$ and $\chi_{ano,i} = 2\ \chi_{ano,e}$, where P_{NB} is the neutral beam power in the unit of MW. In order to produce the H-mode pedestal structure, the transport near the edge is reduced to the neoclassical level ($D_{ano} = \chi_{ano,e,i} = 0$). The region of the reduced transport is prescribed for simplicity so that the pedestal width, Δ_{ped}, is fixed constant.

Neutrals are treated by the 2D Monte-Carlo method, in which the ionization due to electron and ion impacts, and the charge-exchange (CX) are taken into account. The neutrals are assumed to establish its steady-state on each step of the time-evolution of plasmas for simplicity. This assumption is based on that the transit time of neutrals is much shorter than the typical ELM duration of a few 100 μs. For example, the transit time of neutrals with the energy of 3 eV across a length of 0.2 m is 17 μs. The particle flux of neutrals across the separatrix derived from the particle balance model in subsection 2.4 is used as an input for this Monte-Carlo code.

2.2. ELM model

The ELM model [4] has been developed by coupling the TOPICS with a linear MHD stability code MARG2D [11]. In the present simulation, stabilities of n = 1-50 modes are examined by the MARG2D at given time-intervals along the pedestal growth, where n is the toroidal mode number. When some modes become unstable, an ELM is assumed to occur. The ELM enhanced diffusivities, D_{ELM} and χ_{ELM}, are added on the basis of radial profiles of eigenfunctions of unstable modes, where $D_{ELM} = \chi_{ELM} = \chi_{ELM}^{max} \times (\Sigma_n\ \xi_{r,n}^{\ 2})/N$, where χ_{ELM}^{max} is the maximum value, $\xi_{r,n}$ denote the radial displacement of the plasma by the unstable mode with specific n, and N is the total number of the unstable modes with various n. The profile of the radial displacement $\xi_{r,n}$ is assumed to be the sum of poloidal mode components of the eigenfunction, $\xi_{r,n,m}$, i.e., $\xi_{r,n} \propto \Sigma_m\ \xi_{r,n,m}$, where $\xi_{r,n}$ is normalized by its maximum value. The ELM enhanced transport is maintained for a time interval τ_{ELM} given as a parameter.

2.3. SOL-divertor model

The five-point model [10] for SOL-divertor plasmas is based on the integral of time dependent fluid equations, i.e., particle, momentum, electron and ion temperatures, generalized Ohm's law and current equations. The model geometry is an open magnetic flux-tube nearest to the separatrix. The flux tube is divided into four regions: the SOL region divided into two regions at the stagnation point where the parallel particle flux is zero and two divertor regions on either side of the SOL region. The integral fluid equations in each region are reduced to a set of nonlinear equations with physical variables at five positions (stagnation point, upstream throats of divertor regions and sheath entrances). Exponential radial profiles are assumed as $f(\rho) = f(0)\ \exp(-\rho a/\lambda_f)$ where f denotes each of n, T_e, T_i and a is the minor radius. Radial scale lengths, λ_n, λ_{Te}, λ_{Ti}, are solved in the model. A kinetic-effect model (heat flux limit) based on results from the particle code is introduced. The five-point model can deal with the in-out/up-down asymmetry of divertor plasmas, but the symmetry is assumed in this paper for simplicity. Radial electron particle flux across the separatrix, Γ_{re}, and radial heat fluxes of electrons and ions across the separatrix, Q_{re} and Q_{ri}, obtained in the TOPICS are used as inputs for the five-point model. Heat and particle diffusivities in the five-point model, $D_\perp$, $\chi_{e\perp}$ and $\chi_{i\perp}$, are set equal to those values at the separatrix in the TOPICS. The five-point model calculates the SOL electron density at the separatrix, n_{eSOL}, and temperatures, T_{eSOL} and T_{iSOL}, which are used as boundary conditions in the TOPICS. During the ELM crash, the radial scale lengths vary along the time evolution and become long according to the ELM enhanced diffusivity at the separatrix. The collisionality dependence in the five-point model appears through the parallel heat conduction and the equipartition energy flow between electrons and ions.

The particle source density due to recycling neutrals in the divertor region, S_d, is given by $S_d = \eta'_r\Gamma_d/L_d$ where Γ_d and L_d denote a particle flux to the divertor plate and a length of the divertor region along the magnetic field line, respectively. The divertor recycling coefficient η'_r, which was assumed to be constant in [10], is modeled by $\eta'_r = (1-f_{dpump}-f_{dr})(1-\exp(-\theta L_d/\lambda_{iz}))$ where θ is the pitch of the magnetic field, $\lambda_{iz} = \upsilon_n/(n_{ediv}<\sigma\upsilon>_{iz})$ is the ionization mean-free-path of neutrals with the velocity $\upsilon_n = (T_n/m)^{0.5}$. The fraction f_{dpump} is pumped out from the divertor and the fraction f_{dr} radially escapes the divertor region and goes back to the core region. The ionization rate coefficient $<\sigma\upsilon>_{iz}$ is a strong function of the divertor electron temperature, T_{ediv}. This model is similar to that used in [15], which is validated by the 2D divertor transport code B2-EIRENE. In the present model, the particle source in the SOL regions is not taken into account.

2.4. *Integration of neutral models*
We consider the particle balance in core and SOL-divertor regions and integrate neutral models in the TOPICS and the five-point model. Figure 2 illustrates a simple model of the particle balance. Plasma particle flow across the separatrix, Φ_r, goes to the divertor plates and the first wall. The plasma flow to the divertor plates, Φ_d, is calculated by $\Phi_d = 4\pi R\theta\int_0^{Lr}\Gamma_d(r)dr$ from results of the five-point model, where R and L_r denote the major radius and a distance between the separatrix and the first wall, respectively. The radial profile of Γ_d is defined as $\Gamma_d(r) = n_{idiv}\exp(-r/\lambda_n)((T_{ediv}\exp(-r/\lambda_{Te})+3T_{idiv}\exp(-r/\lambda_{Ti}))/m)^{0.5}$. The plasma flow to the first wall, Φ_w, is calculated by $\Phi_w = (D_\perp n_{iSOL}/\lambda_n)\exp(-L_r/\lambda_n)S_{sep}$ where S_{sep} is the area of the separatrix surface. A part of neutrals produced at the divertor plates, $\eta_r\Phi_d$, are reionized in the divertor region. The global divertor recycling coefficient η_r is recalculated from the flux-tube value in the five-point model as $\eta_r = 1-\Gamma_d(0)\lambda_n(1-\exp(-L_r/\lambda_n))(1-\eta'_r)/(\int_0^{Lr}\Gamma_d(r)dr)$, which is derived by integrating radially particle equations in the five-point model. The remained part of neutrals produced at the divertor plates is pumped out from the divertor ($\Phi_{dpump} = f_{dpump}\Phi_d$) or escapes from the divertor region to the radial direction ($\Phi_{nr} = f_{dr}\Phi_d$) or to the divertor throat ($\Phi_{nth} = (1-f_{dpump}-f_{dr}-\eta_r)\Phi_d$). Then, these fractions Φ_{nr} and Φ_{nth} go to the core region. Neutrals produced at the first wall is pumped out from the wall ($\Phi_{wpump} = f_{wpump}\Phi_w$) or escapes from the SOL to the core region ($\Phi_{nw} = (1-f_{wpump})\Phi_d$). In addition to neutrals from the divertor and the first wall, neutrals are added by gas puffing in the core region (Φ_{puf}). Neutrals escaping from the core region due to the CX are assumed to go back to the core region.

3. Integrated simulation results
The ELM activity is simulated for JT-60U like parameters: $R = 3.4$ m, $a = 0.9$ m, $\kappa = 1.5$, $\delta = 0.26$, $I_p = 1.5$ MA, $B_t = 3.5$ T, $q_{95} = 4.3$-4.6 and $\beta_N \sim 0.7$-1.1. The ELM parameters are chosen as $\tau_{ELM} = 200$ μs, $\chi_{ELM}^{max} = 100$ m^2/s and $\Delta_{ped} = 0.05$ on ρ. The values of τ_{ELM} and Δ_{ped} are typical values in JT-60U and other experiments [1,14]. On the other hand, the value of χ_{ELM}^{max} is chosen to obtain ELM energy and particle losses comparable with JT-60U experiments [14]. As shown in [9,13], these parameters vary the ELM energy loss but the collisionality dependence does not change under the assumption of constant parameters. The pumping and escaping fractions of neutrals in SOL-divertor regions are chosen as $f_{dpump} = 0.01$, $f_{wpump} = 0$ and $f_{dr} = 0.02$ in the present study. The distance between the separatrix and the first wall L_r=0.4 m and the constant neutral energy $T_n = 3$ eV are assumed.

Figure 3 shows the time evolution of profiles of the electron density and temperature (a)(b) before, (c)(d) during and (e)(f) after an ELM crash. The pedestal top and foot are located at $\rho \sim 0.925$ and 0.975, respectively (thus $\Delta_{ped} = 0.05$). The ion temperature at the pedestal top, $T_{iped} \sim 2$ keV, is higher than the electron one $T_{eped} \sim 1.3$ keV at the ELM onset. Figure 4 shows the time evolution of (a) the volume averaged electron density, $<n_e>$, (b) the electron part of the stored energy, W_{se}, the ion part, W_{si}, the total stored energy, W_s, (c) Γ_{re}, (d) Q_r, (e) n_{eSOL}, the divertor plasma density, n_{ediv}, (c) T_{SOL} and the divertor plasma temperature, T_{div}. The simulation starts at nearly stationary state and the reduction of

Journal of Physics: Conference Series **123** (2008) 012025 doi:10.1088/1742-6596/123/1/012025

diffusivities for the pedestal begins at t = 0.03 s. Along the pedestal growth, the density and temperature increase in the pedestal region in figures 3(a) and (b) and the averaged density and the stored energies increase in figures 4(a) and (b). The reduction of diffusivities transiently decreases the radial fluxes and the SOL-divertor plasma densities and temperatures in figures 4(c)(d)(e) and (f). The ion SOL temperature is higher than the electron one, while the divertor plasma temperatures is lower than the SOL ones due to the high recycling divertor plasma.

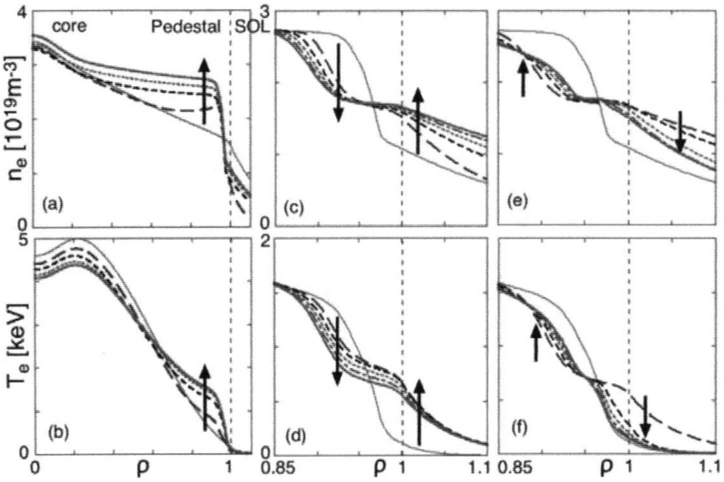

Figure 3. Time evolution of n_e and T_e profiles (a)(b) before, (c)(d) during and (e)(f) after an ELM crash. In (a)(b), lines denote 50 ms interval from t = 0 s to 0.2 s in figure 4. In (c)(d), 40 μs interval from t = 0.2 s to 0.2002 s. In (e)(f), 0.2 ms interval from t = 0.2 s to 0.201 s.

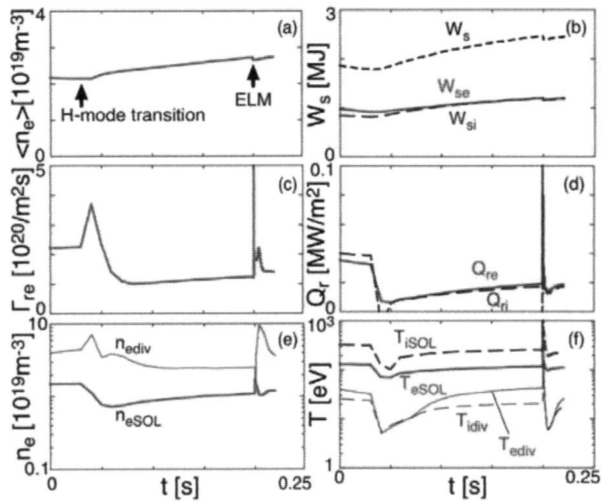

Figure 4. Time evolution of (a) $<n_e>$, (b) W_{se}, W_{si}, W_s, (c) Γ_{re}, (d) Q_{re}, Q_{ri}, (e) n_{eSOL}, n_{ediv}, (f) T_{eSOL}, T_{ediv}, T_{iSOL} and T_{idiv}. Profiles are shown in figure 3.

In the progress of the pedestal growth, the high-n modes (n = 10,11) become unstable and an ELM occurs at $t = t_{ELM}$ = 0.2 s. The profile of the ELM enhanced diffusivity is shown in figure 7(c) (case B). The collapse of the density and temperature profiles occurs in the pedestal region in figure 3(c) and (d). The drop of n_e at the pedestal top is about 30 % of n_{eped} and T_e drop is about 40 % of T_{eped}, which are comparable with those in experiments [2-4]. Pedestal particle and energy flow into the SOL and the SOL-divertor densities and temperatures rapidly increase as shown in figures 5(a)(b)(c) and (d). The increase of the SOL density and temperature mitigates radial edge gradients and lowers the ELM particle and energy losses. Figures 5(e) and (f) show the time evolution of (e) the ELM particle loss, δN, normalized by the pedestal particle content, N_{ped}~1.4×10^{21}, (f) the ELM energy loss, δW_s, normalized by the pedestal energy, W_{ped} ~ 1.2 MJ. The ELM particle loss occurs only in the ELM duration of τ_{ELM} = 200 μs. The resultant ELM particle loss is less than 10 % of the pedestal particle and is comparable to those obtained in experiments [1,3,14]. On the other hand, the ELM energy loss continues about 1 ms after the ELM duration. The resultant ELM energy loss is less than 10 % of the pedestal energy and is comparable to those in JT-60U [14]. Because of the flat temperature (pressure) gradient inside the pedestal top at the ELM onset in figure 3(b), like the case A in figure 1, the ELM energy loss in this simulation is small compared with multi-machine experiments [1].

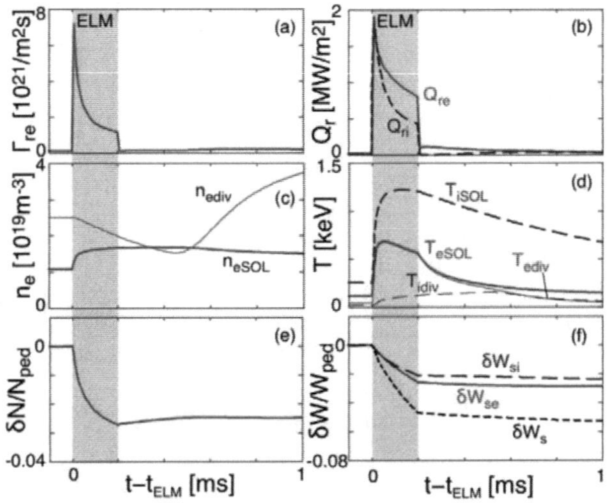

Figure 5. Time evolution of (a) Γ_{re}, (b) Q_{re}, Q_{ri}, (c) n_{eSOL}, n_{ediv}, (d) T_{eSOL}, T_{ediv}, T_{iSOL}, T_{idiv}, (e) δN, (f) δW_{se}, δW_{si} and δW_{s} during and after an ELM crash on $t - t_{ELM} = 0$-200 μs (shaded region) for a simulation of figure 4.

4. Pedestal density and temperature variations by gas puffing

In order to study the collisionality dependence of ELM particle and energy losses, the density and the temperature are varied by the gas puffing. Figure 6 shows profiles of (a) the electron density, (b) the electron temperature and (c) the total pressure, P, at the ELM onset for various amounts of gas-puffing flow (A) $\Phi_{puf} = 0$, (B) 0.3×10^{22}, (C) 0.9×10^{22} and (D) 2×10^{22} s^{-1}. The case B is the same as shown in figures 3-5. The gas puffing increases the pedestal density and decreases the pedestal temperature. Thus, the normalized electron collisionality of the pedestal plasma, $\nu^*_{ped} = \pi R q_{95}/\lambda_{ee}$ where λ_{ee} is the electron mean free path, increases from 0.11 to 2.1 in figure 6. The gas puffing also reduces the density pedestal width by increasing the particle source near the separatrix in figure 6(a), while pedestal widths of the temperature and the total pressure are almost constant in figures 6(b) and (c). At the ELM onset, the pedestal pressure decreases with increasing the density except for the case A.

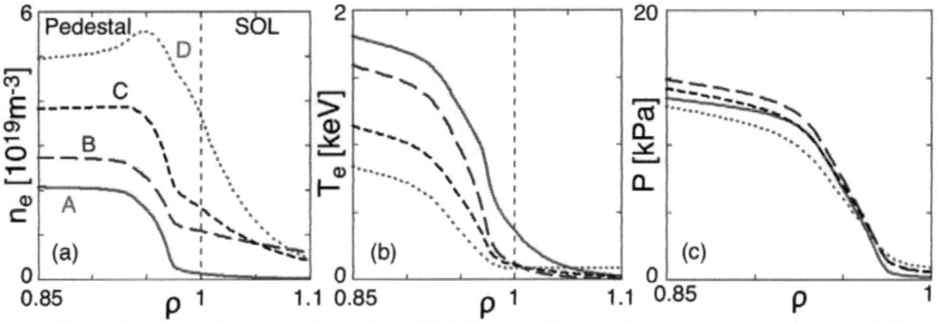

Figure 6. Profiles of (a) n_e, (b) T_e and (c) P at ELM onset for various amounts of gas-puffing flow (A) $\Phi_{puf} = 0$, (B) 0.3×10^{22}, (C) 0.9×10^{22} and (D) 2×10^{22} s^{-1}.

Figure 7 shows profiles of (a) the bootstrap current, j_{BS}, (b) the magnetic shear, s, at the ELM onset and (c) $D_{ELM}(\chi_{ELM})$ during the ELM. For higher collisionality, the bootstrap current decreases and the magnetic shear increases except for the case A in figures 7(a) and (b). The increase of the magnetic shear reduces the width of eigenfunctions of unstable modes. Unstable mode numbers vary from medium to high values; (A) n = 10,11,14, (B) 10,11, (C) 26-38 and (D) 48-50. Only in the case D, the infinite ballooning mode is already unstable. Even if the transport are assumed to be also enhanced by the infinite ballooning mode, the simulation result in the case D does not change very much because the pressure profile at the onset of the infinite ballooning mode is a little lower than that in figure 6(c) and the width of the unstable region is similar to that of the eigenfunctions in the result of figure 7(c). As a result of higher collisionality, the bootstrap current suppression reduces the area of the ELM enhanced transport in figure 7(c). This behavior is the same as found in the previous study [9], in which the collisionality is artificially enhanced in the bootstrap current model.

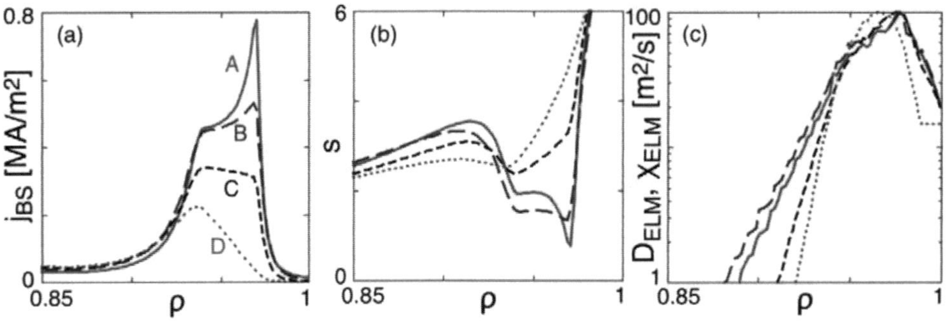

Figure 7. Profiles of (a) j_{BS}, (b) s at the ELM onset and (c) $D_{ELM}(\chi_{ELM})$ during ELM for the same cases as in figure 6.

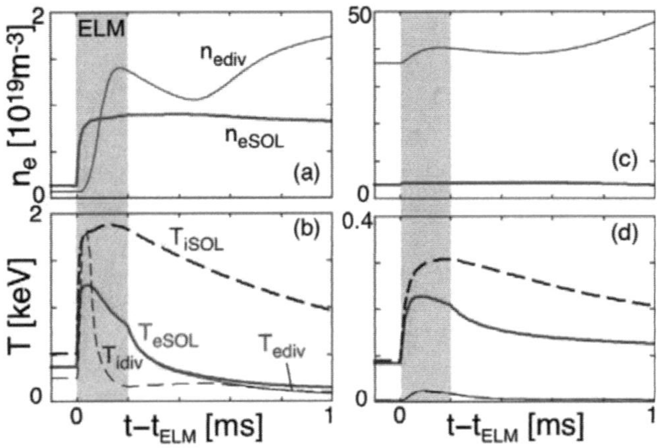

Figure 8. Time evolution of (a) n_{eSOL}, n_{ediv}, (b) T_{eSOL}, T_{ediv}, T_{iSOL} and T_{idiv} during and after an ELM crash for case A in figure 6. (c)(d) Time evolution for case D.

The results in the case A shown above are different from the other cases B-D. This difference results from the recycling condition of divertor plasmas. Figure 8 shows the time evolution of densities and temperatures of SOL-divertor plasmas during and after the ELM crash in the cases A and D. The time evolution in the case B is already shown in figures 5(c) and (d). Without the gas-puffing in the case A, the divertor plasmas are under the low recycling condition with low density and high temperature before the ELM. The divertor plasma density is lower than the SOL density and the electron temperature is almost constant along the field line, $T_{eSOL} \sim T_{ediv}$. The gas puffing changes the condition before the ELM from the low recycling in the case A to the high recycling in the cases B-D where the divertor density is higher than the SOL density and the divertor temperature is lower than the SOL temperature. The ELM transiently changes the above conditions of the SOL-divertor plasmas. For lower amount of the gas puffing and hence lower collisionality, the ion SOL temperature is higher than the electron one due to the ineffectiveness of equipartition energy flow from ions to electrons. The low recycling condition before the ELM in the case A results in the different ELM behavior from the other cases, as shown in figures 6, 7, 9 and 11-12. Because the low recycling divertor is not desirable from the view point of the damage of divertor plates, the divertor seems to be under the high recycling condition in the majority of experiments. Thus, hereafter, the results in the cases B-D will be mainly discussed.

5. Collisionality dependence of ELM particle loss

The collisionality dependence of the ELM particle loss is next investigated by using the results in the previous section. Figure 9(a) shows the collisionality dependence of the ELM particle loss, ΔN_{ELM}, where the outflow loss across the separatrix, ΔN_Γ, and the particle source, ΔN_s, are also shown. Here, the ELM particle loss ΔN_{ELM} is defined by the magnitude of δN at which δN becomes the lowest value, ex. $t - t_{ELM} = 0.2$ ms in figure 5(e). The ELM particle loss is determined by the balance between the outflow loss and the source as shown in figure 9(a). Components of the particle source from the divertor, ΔN_{sdiv}, from the first wall, ΔN_{swal}, and from the gas puffing, ΔN_{spuf} are shown in figure 9(b). The major part of neutrals, which becomes the source, comes from the divertor, especially for the high collisionality. Except for the case A, the outflow loss increases a little with the collisionality while the source increases in figure 9(a). As a result, the ELM particle loss decreases only a little with the collisionality. Thus, the ELM particle loss is almost independent of the collisionality.

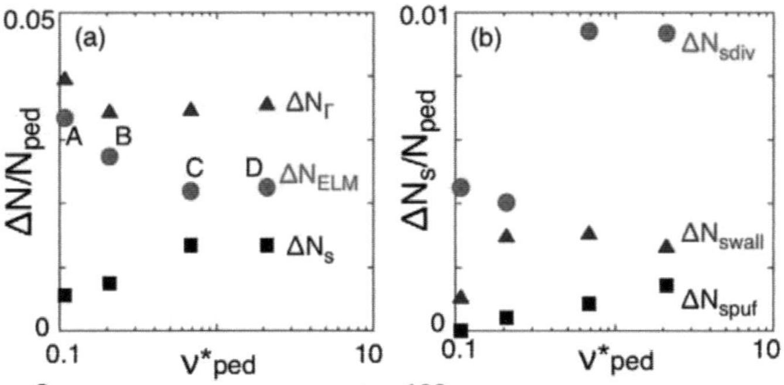

Figure 9. Collisionality dependence of (a) ΔN_{ELM}, ΔN_Γ, ΔN_s, (b) ΔN_{sdiv}, ΔN_{swal} and ΔN_{spuf} for the same cases as in figure 6. All values are normalized by N_{ped}.

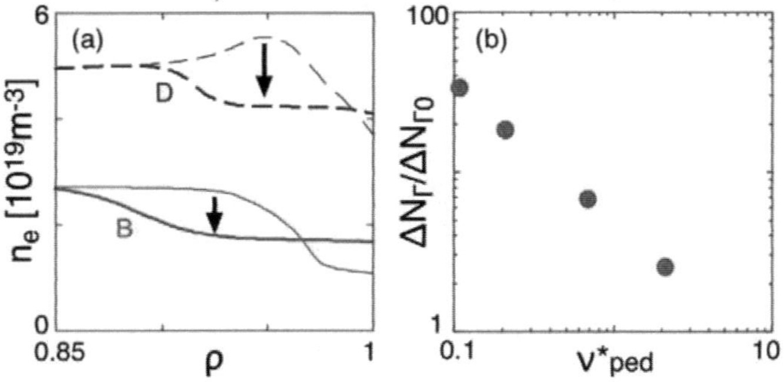

Figure 10. (a) Profiles of n_e before and after an ELM crash for cases B and D in figure 6. (b) Collisionality dependence of $\Delta N_\Gamma/\Delta N_{\Gamma 0}$.

This collisionality independence is due to the density profile before the ELM and the increase of the SOL density during the ELM in addition to the bootstrap current effect as shown in figure 10(a). As mentioned in the previous section, the density pedestal width becomes narrow due to the particle source near the separatrix for the high collisionality. Additionally, the increase of the SOL density at $\rho = 1$ becomes small for the high collisionality as shown in the comparison between the case B and D in figure 10(a). The increase of the SOL density depends on a ratio of ΔN_Γ to the outflow loss without the ELM, $\Delta N_{\Gamma 0}$. For higher collisionality, the outflow loss becomes small compared with that without the ELM as shown in figure 10(b), which results in the small increase of the SOL density. The SOL-divertor plasmas are less influenced by the ELM particle loss for the high collisionality. Due to the bootstrap current effect on the ELM enhanced transport discussed in the previous section, the collapse

Journal of Physics: Conference Series **123** (2008) 012025 doi:10.1088/1742-6596/123/1/012025

region becomes narrow in $\rho < 0.95$ for the high collisionality in figure 10(a). As a result, from the case B to D, the magnitude of the particle loss becomes about twice. But the pedestal particle N_{ped} also increases about twice. Thus, the value of $\Delta N_{ELM}/N_{ped}$ does not change much.

6. Collisionality dependence of ELM energy loss

The collisionality dependence of the ELM energy loss is examined by using the results in section 4. Figure 11(a) shows the collisionality dependence of the ELM energy loss ΔW_{ELM}, its electron component, ΔW_e, and ion one, ΔW_i, where a shaded line denote the ELM energy loss obtained by artificially varying the collisionality in models of the bootstrap current and the SOL-divertor plasmas for a plasma with $\nu^* \sim 0.7$. Here, the ELM energy loss ΔW_{ELM} is defined by the magnitude of δW_s at which δW_s becomes the lowest value, ex. $t - t_{ELM} = 1$ ms in figure 5(f). When the collisionality increases by one order of magnitude, the electron and ion energy losses become about half except for the case A. As a result, the total energy loss becomes about half. The electron component of the ELM energy loss and its collisionality dependence are mainly determined by the conductive energy outflow loss across the separatrix, ΔW_{ecd}, as shown in figure 11(b). The electron convective loss, ΔW_{ecv}, is small and its collisionality dependence is weak compared with the conductive one. The weak dependence of the electron convective loss on the collisionality is due to the ELM particle loss shown in figure 9(a). The behavior of the electron conduction is the same as found in the previous study [9], in which the collisionality is artificially enhanced in the SOL-divertor model as well as the bootstrap current one. Thus, the shaded line in figure 11(a) is explained only by the electron energy loss with the assumption of the constant ion loss. In figure 11(a), however, the reduction of the total energy loss is larger that caused by the bootstrap current and the SOL transport. This large reduction is due to the ion energy loss.

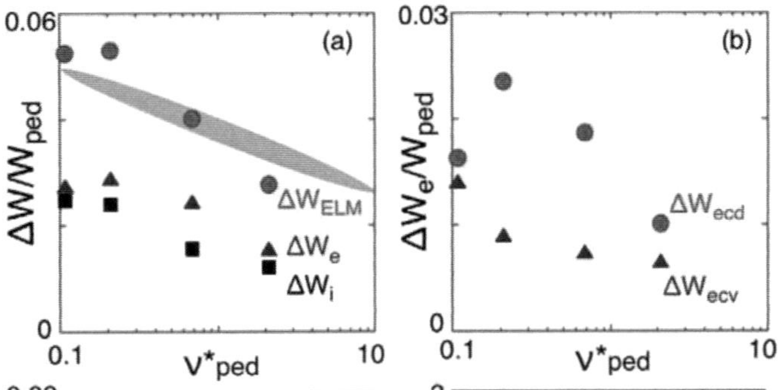

Figure 11. Collisionality dependence of (a) ΔW_{ELM}, ΔW_e, ΔW_i, (b) ΔW_{ecd} and ΔW_{ecv} for the same cases as in figure 6, where a shaded line denote ΔW_{ELM} obtained by artificially varying collisionality in models of bootstrap current and SOL-divertor plasmas. All values are normalized by W_{ped}.

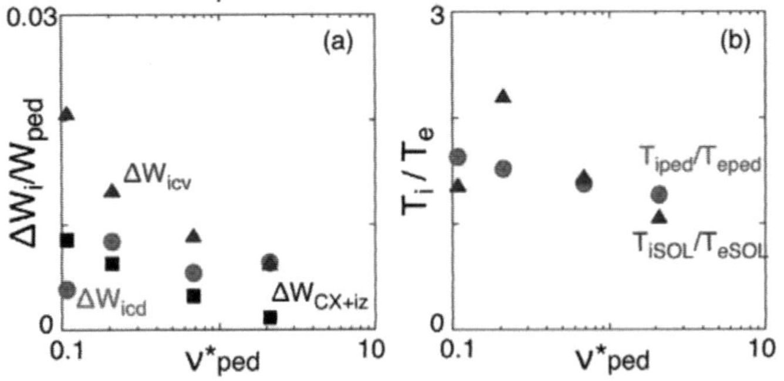

Figure 12. Collisionality dependence of (a) ΔW_{icd}, ΔW_{icv}, ΔW_{CX+iz}, (b) T_{iped}/T_{eped} and T_{iSOL}/T_{eSOL} at the ELM onset for the same cases as in figure 6. In (a), all values are normalized by W_{ped}.

Figure 12 (a) shows the collisionality dependence of the ion conductive energy outflow loss, ΔW_{icd}, the convective loss, ΔW_{icv}, and the energy loss due to the CX corrected by the energy source due to the ionization, ΔW_{CX+iz}. In the ion energy loss, the convective loss is larger than the conductive and CX ones. The ion convective and CX losses decrease with increasing the collisionality, while the conductive loss does not change much. The collisionality dependence of both the convective and CX losses is caused by the equipartition effect as shown in figure 12(b). For lower collisionality, the pedestal and SOL ion temperatures become higher than the electron ones. If the ion temperature is equal to the electron one, the collisionality dependence of the ion convective and CX losses become weak like the electron convective loss in figure 11(b). As a result, the ion convective and CX losses enhance the collsionality dependence of the ELM energy loss.

7. Conclusion

Density dynamics effect on ELM behavior has been studied by using the TOPICS-IB. Neutral models of core and SOL-divertor regions have been integrated with the TOPICS-IB. The TOPICS-IB successfully simulates the behavior of the whole plasma with density dynamics. The ELM particle loss is found to be almost independent of the collisionality, as shown in experiments, due to the density profile before the ELM and the increase of the SOL density during the ELM in addition to the bootstrap current effect. It is confirmed by varying the density and the temperature instead of artificially enhancing the collisionality in the previous study that the collisionality dependence of the ELM energy loss is caused by the bootstrap current and the SOL transport. Additionally, the ion convective and CX losses are found to enhance the collisionality dependence by the equipartition effect.

In order to validate our models, simulations for various tokamaks and comparisons with experiments and nonlinear simulations are our future works. The model improvement, such as the transport model which determines the pressure gradient inside the pedestal top as well as the pedestal width, and so on, is necessary. We will consider effects of the localization of ELMs at the low-field side midplane and the filamentary structure [16,17], which are not taken into account in the present model as the same in other integrated codes [5-8]. The ELM cycle following the first ELM shown in this paper will be studied in future.

Acknowledgments

We would like to thank Dr. R. Hiwatari for fruitful discussion and comments and Mr. I. Kamata for developing and running the integrated code. We are grateful to JT-60 team members for collaboration. This work was partly supported by JSPS, Grant-in-Aid for Scientific Research. This work was also carried out under the collaborating research program at National Institute for Fusion Science. We would like to thank Drs A. Takayama and Y. Tomita for taking care of the NIFS program.

References

[1] Loarte A *et al.* 2003 *Plasma Phys. Control. Fusion* **45** 1549
[2] Leonard A W *et al.* 2003 *J. Nucl. Mater.* **313-316** 768
[3] Wade M R, Burrel K H, Hogan J T, Leonard A W, Osborne T H, Snyder P B and Coster D 2005 *Phys. Plasmas* **12** 056120
[4] Ozeki T, Aiba N, Hayashi N and Takizuka T 2006 *Fusion Sci. Tech.* **50** 68
[5] Onjun T, Kritz A H, Bateman G, Parail V, Wilson H R and Dnestrovskij A 2005 *Phys. Plasmas* **12** 012506
[6] Lonnroth J S *et al.* 2006 *Contrib. Plasma Phys.* **46** 726
[7] Pankin A Y *et al.* 2006 *Nucl. Fusion* **46** 403
[8] Wiesen S, Parail V, Corrigan G, Fundamenski W, Loarte A, Lonnroth J, Sibene G, Sartori R and JET EFDA contributers 2008 *Contrib. Plasma Phys.* in press
[9] Hayashi N, Takizuka T, Ozeki T, Aiba N and Oyama N 2007 *Nucl. Fusion* **47** 682
[10] Hayashi N, Takizuka T and Hosokawa M 2007 *J. Nucl. Mater.* **363-365** 1044
[11] Aiba N, Tokuda S, Takizuka T, Kurita G and Ozeki T 2007 *Nucl. Fusion* **47** 297
[12] Tokar M Z, Gupta A, Kalupin D and Singh R 2007 *Plasma Phys. Control. Fusion* **49** 395
[13] Hayashi N, Takizuka T, Aiba N, Ozeki T and Oyama N 2008 *Contrib. Plasma Phys.* in press
[14] Oyama N *et al.* 2004 *Nucl. Fusion* **44** 582
[15] Hiwatari R, Kuzuyama Y, Hatayama A, Okano K. Asaoka Y, Zhu S and Tomita Y 2005 *J. Nucl. Mater.* **337-339** 386
[16] Kirk A *et al.* 2004 *Phys. Rev. Lett.* **92** 245002
[17] Snyder P B, Wilson H R and Xu X Q 2005 Phys. Plasmas 12 056115

Selection rule for turbulent structural formation: Study of magnetized cylindrical plasmas

N Kasuya[1], M Yagi[2], K Itoh[1] and S-I Itoh[2]

[1] National Institute for Fusion Science, Toki 509-5292, Japan
[2] Research Institute for Applied Mechanics, Kyushu University, Kasuga 816-8580, Japan

E-mail: kasuya@nifs.ac.jp

Abstract. The three-field reduced fluid model was extended to describe the resistive drift wave turbulence in magnetized cylindrical plasmas. There include multiple energy exchange paths, and two kinds of nonlinear steady states were obtained in simulations with a fixed particle source: one with a zonal flow generated by modulational coupling of various unstable modes, which suppress transport, and the other with a streamer, which is a azimuthally-localized vortex structure and induces convective transport, formed by parametric coupling. These structures are formed selectively, depending on the ion-neutral collision frequency, which is a damping force of a zonal flow and has a stabilizing effect of linear instability in this model. This is a minimal model for analyzing the turbulent structural formation mechanism by mode coupling in cylindrical plasmas, and competitive nature of structural formation was revealed.

1. Introduction

Turbulent plasmas form a variety of meso-scale structures such as a zonal flow and a streamer [1,2]. The turbulent structures regulate micro-scale fluctuations and affect the level of anomalous transport in fusion plasmas. Therefore, the formation and self-regulated mechanism of the turbulent structures should be taken into consideration to understand the transport processes in H-mode transport barriers. The preferential formation of zonal flows and streamers has been studied, considering the spectral anisotropy of turbulent fluctuations [3,4]. Drift wave instability is one of the candidates that induce anomalous transport in magnetically-confined inhomogeneous plasmas, and its characteristics have been studied to understand the global transport phenomena [5].

Plasma experiments in a simple linear configuration have been revisited recently for quantitative understandings of the structural formation mechanism by turbulence [6-14]. Drift wave turbulence is driven by the density gradient, which is universal in confined plasmas, and has been observed in linear configurations [7,8,15]. Two-dimensional measurements have revealed the feature of turbulent structures [16,17] and their formation mechanisms by mode coupling [15,18]. Nonlinear simulations of drift wave turbulence in linear devices have been carried out to understand the fundamental mechanism of structural formation by comparison with experiments [19-22]. Measurements of the drift wave turbulence are also carried out in a toroidal configuration and the wave pattern has been identified [23].

We use a minimal model for analyzing the turbulent structural formation mechanism by mode coupling. The three-field reduced fluid model has been extended to describe the resistive drift wave turbulence in cylindrical plasmas. A zonal flow and streamer are formed in nonlinear states in numerical simulations, and their formation mechanisms have been studied. One of the structures is selected to be formed in the turbulence, and their selection rule is obtained using a parameter, which controls a damping force of the zonal flow.

The paper is organized as follows. In section 2, the plasma configuration and the set of model equations for the analyses are described. The nonlinear calculation gives formation of the zonal flow and streamer, and their formation mechanism is discussed in section 3. The bifurcation of the turbulent structures using the ion-neutral collision frequency as a control parameter is presented in section 4, and the effect of structural formation on particle transport is discussed in section 5. Finally, we summarize our results in section 6.

2. Model

We have been developing a three-dimensional numerical simulation code called 'Numerical Linear Device' (NLD), which describes the resistive drift wave turbulence in a linear device [19]. The plasma configuration and the model equations for simulating the linear device, such as the Large Mirror Device (LMD) in Kyushu University [6], are described in this section.

2.1. Geometry

The plasma has a simple cylindrical shape with device length λ, and the magnetic field has only the component in the axial direction with uniform intensity B (figure 1). In each end of the experimental device, functions of plasma production or plasma control by voltage biasing are installed, which are not taken into account in the present version of NLD. According to experiments, high density ($n_e > 1\times10^{19}$[m^{-3}]) and low temperature ($T_e < 5$ [eV]) plasmas in an argon discharge are analyzed. The density of neutral particles is high even in the plasma core region [24], so the effect of neutral particles is taken into consideration.

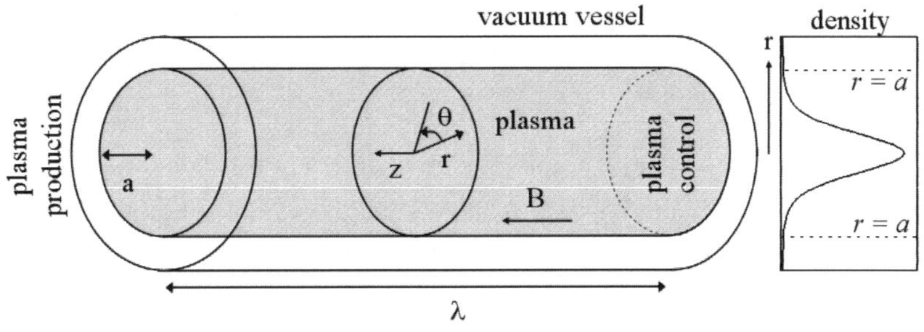

Figure 1. Plasma configuration in NLD, which has a cylindrical shape. A peaked plasma density profile is assumed.

2.2. Equations

The three-field (density, potential and parallel velocity of electrons) reduced fluid model is adopted. The continuity equation, the vorticity equation and Ohm's law can be used to obtain the fluctuating density, potential and parallel velocity of electrons [25]:

$$\frac{dN}{dt} = -\nabla_{//}V - V\nabla_{//}N + \mu_N \nabla_\perp^2 N + S, \tag{1}$$

$$\frac{d\nabla_\perp^2\phi}{dt} = \nabla N \cdot \left(-\nu_{in}\nabla_\perp\phi - \frac{d\nabla_\perp\phi}{dt} \right) - \nu_{in}\nabla_\perp^2\phi - \nabla_{//}V - V\nabla_{//}N + \mu_W \nabla_\perp^4\phi, \tag{2}$$

$$\frac{dV}{dt} = \frac{M}{m_e}\left(\nabla_{//}\phi - \nabla_{//}N\right) - \nu_e V + \mu_V \nabla_\perp^2 V, \tag{3}$$

where $N = \ln(n / n_0)$, $V = v_{//} / c_s$, $\phi = e\varphi / T_e$, n is the density, n_0 is the density at $r = 0$, $v_{//}$ is the electron velocity parallel to the magnetic field, c_s is the ion sound velocity, φ is the electrostatic potential, T_e is the electron temperature, $d / dt = \partial / \partial t + [\phi,]$ is the convective derivative, S is a particle source term, M / m_e is mass ratio of ion and electron, ν_{in} is ion-neutral collision frequency, $\nu_e = \nu_{ei} +$

ν_{en} is the sum of ion-electron and electron-neutral collision frequency, and μ_N, μ_V, μ_W are artificial viscosities. The ion cyclotron frequency Ω_{ci} and Larmor radius measured by the electron temperature ρ_s are used for the normalizations of the time and distance, respectively. This set of equations gives an extension of the Hasegawa-Wakatani equations [26] in two ways. (i) It includes an inertial term of electron and a resistive term caused by neutrals in equation (2), and (ii) three-dimensional dynamics are considered. Since the dissipation associated with the parallel motion of electrons affects the drift wave instability, the inertial term is retained. Equations (1) – (3) compose the set of model equations for analyzing the resistive drift wave in NLD.

2.3. Boundary conditions

The equations are solved in the cylindrical coordinate with spectral expansion in the azimuthal and axial directions assuming periodic boundary condition

$$f = \sum_{m,n} \tilde{f}_{mn}(r) \exp\left[i\left(-\omega_{mn} t + m\theta + \frac{2\pi n}{\lambda} z \right)\right], \quad (4)$$

where m and n are the azimuthal and axial mode number, respectively, and f implies $\{N, \phi, V\}$. The boundary condition in the radial direction are set to $f = 0$ at $r = 0$, a when $m \neq 0$, and $\partial f / \partial r = 0$ at $r = 0$, $f = 0$ at $r = a$ when $m = 0$, and $r = a$ gives an outer boundary of the plasma column. We only solve the plasma region $0 \le r \le a$, and do not solve the vacuum region $r > a$ in this model.

3. Structural Formation

Using the set of model equations described in the previous section, nonlinear simulations have been performed to examine the saturation mechanism of the resistive drift wave turbulence. Nonlinear steady states have been obtained with a fixed particle source. Two kinds of turbulent structures are formed, depending on the value of ν_{in}, and their formation mechanisms are explained in this section.

3.1. Simulation parameters

For simulations of the drift wave turbulence in a linear device, the following parameters are used: $B = 0.1$ [T], $T_e = 2$ [eV], $a = 10$ [cm], $\lambda = 1.7$ [m], $\mu_N = 1 \times 10^{-2}$, $\mu_V = \mu_W = 1 \times 10^{-4}$. Larger μ_N is used to accelerate formation of the background density. Values of μ_V, μ_W are little not to effect on the drift wave instability [19]. Using these parameters, ν_e is estimated to be $\nu_e = 310$ [19]. There is ambiguity of the absolute value of collision frequency ν_{in}, which depends on the neutral density. Therefore, ν_{in} is used as a parameter for controlling the instability in our simulations. Simulations are performed with modes $(m, n) = (0, 0)$ and $m = \pm 1$ - ± 16, $n = \pm 1$ - ± 16, i.e., 1025 modes in total. In simulations discussed in this paper, modes with $(m, n) = (3 - 5, 1)$ have the largest amplitudes, so this number of modes must be taken in calculations at least. The calculation with a fixed particle source has been carried out, where the time independent source profile is given by

$$S(r) = \frac{4 S_0 \mu_N}{L_N^{\ 2}} \left[1 - \left(\frac{r}{L_N} \right)^2 \right] \exp\left[-\left(\frac{r}{L_N} \right)^2 \right], \quad (5)$$

with $S_0 = 5.0$, $L_N = 5$ [cm]. A density profile peaked at $r = 0$ is formed by this particle source, and destabilizes the resistive drift wave.

A resistive drift wave can be excited with a small ion-neutral collision frequency [19]. Linear analysis gives linear growth rates and eigenfrequencies. Modes with low m and only $n = 1$ are unstable with these parameters. The destabilized modes increase their amplitudes, and their nonlinear coupling generates turbulence. The density flattening occurs in the core region ($0.1 < r / a < 0.7$), which comes from an enhanced transport owing to the fluctuations [19]. The dispersion relation of the linear eigenmodes with $n = 1$ shows weak dispersion in low m ($m < 4$), and $\partial \omega / \partial k_\theta \sim 0$ with $m = 4 - 6$. These ranges of modes couples to each other to form zonal flows and streamers, so mechanisms of nonlinear excitation of turbulent structures can be studied by analyzing this turbulence, as explained in the following subsections.

3.2. Zonal flow formation

The case with small ion-neutral collision frequency ν_{in} is described in this subsection. Nonlinear coupling of modes with weak dispersion (modulational coupling) generates the (0, 0) mode, which affects the stability of modes. If ν_{in} is small, in nonlinear steady states, modulational coupling of unstable modes with lower mode numbers plays an important role. Figure 2 shows the time evolution of the fluctuation energy of the potential, decomposed into Fourier components, with ν_{in} = 0.02. Modes with (m, n) = (0, 0), (2 – 4, 1) are shown, which have the same level of amplitude. The mode amplitudes are fluctuating, and the mode with a maximum amplitude exchanges from time to time.

Figure 2 shows a bursty oscillation of the fluctuation energies. The energy of the (0, 0) mode repeat increasing and decreasing, and those of the other modes also varies, accordingly. The mechanism of the oscillation is understood by observing the relationship between $E_\phi(0, 0)$ and E_ϕ of the dominant mode. The trajectory is plotted in $E_\phi(0, 0)$ - $E_\phi(2, 1)$ phase space, when the (2, 1) mode is dominant (from 2500 to 3000), in figure 3. Firstly, unstable modes, such as that with (m, n) = (2, 1), increase (growing phase (i)). Then, nonlinear coupling causes the growth of the (0, 0) mode (potential generation (ii)). The generated mean potential stabilizes the unstable modes, which turn to decrease (stabilizing phase (iii)). Without the nonlinear source, the (0, 0) mode can not be sustained (mean damping phase (iv)), and once the mean potential becomes small enough, the $m \neq 0$ modes begin to grow again (phase (i)). In this way, the limit cycle oscillation is generated (as was reviewed in [1]), and the steady state is sustained. Figure 4 shows the time evolution of the contour of the fluctuating potential. The dominant mode is replaced from (4, 1) to (2, 1) at $t \sim 2520$, and the vortex structures are tilted and broken by the (0, 0) shear flow at $t \sim 2720$.

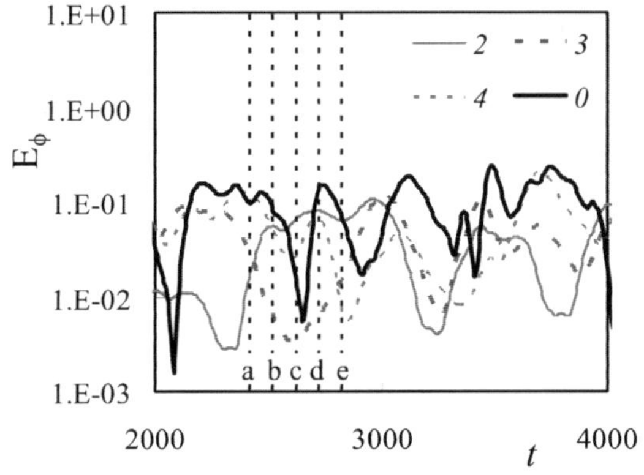

Figure 2. Time evolutions of Fourier modes of the fluctuation energy in the nonlinear steady state. The fluctuation energies of the electrostatic potential of Modes (m, n) = (0, 0), (2 – 4, 1) are shown when ν_{in} = 0.02.

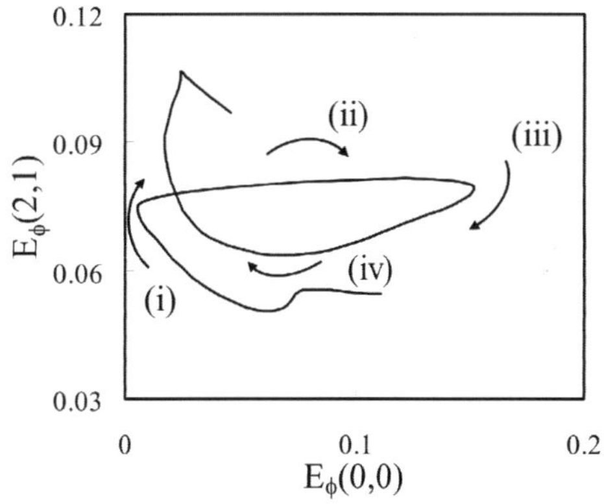

Figure 3. Relationship between $E_\phi(0,0)$ and $E_\phi(2,1)$, which forms a limit cycle. This is the plot from t = 2550 to 3300.

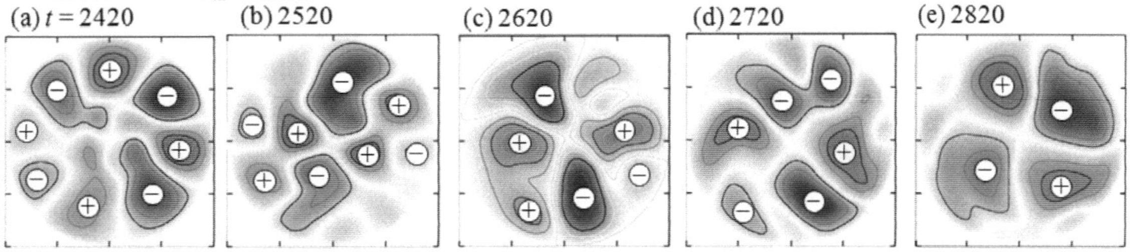

Figure 4. Contour plots of the potential at the times indicated by the dashed lines in figure 2. These contours are draw with excluding ϕ_{00} component.

3.3. Streamer formation

The case with large ion-neutral collision frequency ν_{in} is described next. If the collision frequency is large, the zonal flow remains stable, owing to a strong collisional damping. Three-wave coupling including modes, which has the wave number and frequency close to each other (parametric coupling), can resonantly takes place in this case. Figure 5 shows the time evolutions of the fluctuation energy of the potential, decomposed into Fourier components, with $\nu_{in} = 0.1$. Modes with $(m, n) = (4, 1)$ and (5, 1) are dominant, and their parametric coupling plays an important role in this case. The mode with a maximum amplitude exchanges from time to time, and evolutions of the mode amplitudes are rather periodic compared with those in the small ν_{in} case in the previous subsection.

Figure 6 shows the time evolution of the contour of the fluctuating potential. There formed a streamer, which is a strong vortex localized in the θ direction. The vortex structure is sustained for a much longer duration than the drift wave oscillation period. Figure 7 shows a time evolution of the θ profile of the radial E × B convective velocity. The localized structure, streamer, induces localized convective transport in the radial direction as shown in figure 7. There are two kinds of time scales involved: one is the characteristic time scale for the micro-scale structural (drift wave) propagation, and the other is that for the meso-scale structural (streamer) propagation. There are points where the maximum mode exchanges from one to the other. These points appear as irregular nodes such as at $t \sim 6500$ and $\theta \sim 3 / 2 \pi$ in figure 7. A continuous flow pattern of the micro-scale structure is bent or broken at these points.

Three-dimensional mode coupling is essential for preservation of a streamer structure. Figure 8 (a) shows the dependency of the linear growth rate of $n = 1$ mode on the azimuthal mode number. These growth rates are calculated with the density profile at $t = 8000$ and $\phi_0 = 0$. Modes with $m = 3 - 8$ are unstable. Figure 8 (b) shows the potential energy spectrum on m with $n = 1 - 3$ at $t = 8000$. The linearly unstable modes with $m = 3 - 8$ have largest amplitudes, and couplings of these largest modes

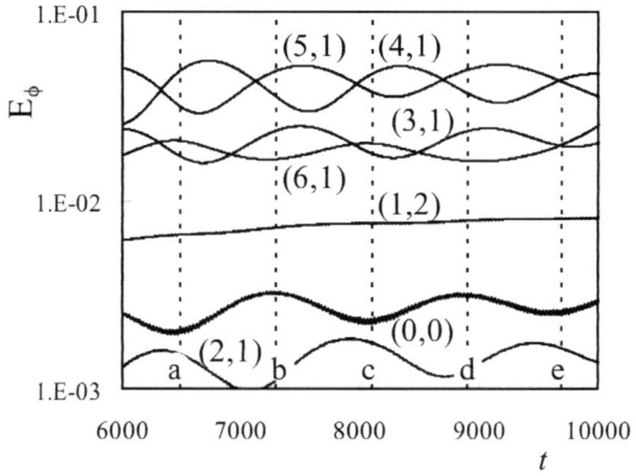

Figure 5. Time evolution of Fourier modes of the fluctuation energy in the nonlinear steady state. The fluctuation energies of the electrostatic potential of modes $(m, n) = (0, 0)$, $(2 - 7, 1)$, $(1, 2)$ are shown when $\nu_{in} = 0.1$.

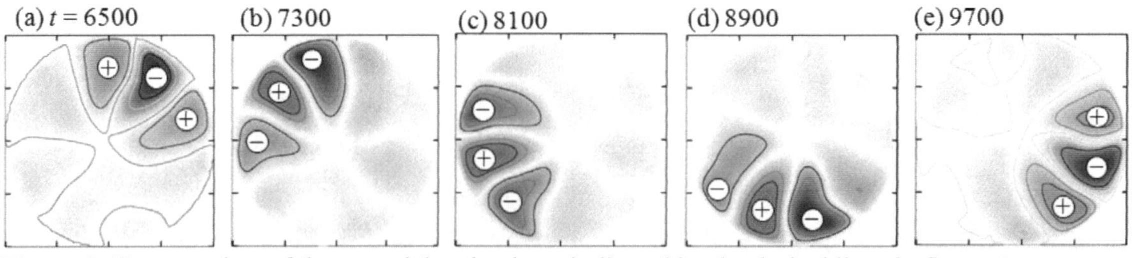

Figure 6. Contour plots of the potential at the times indicated by the dashed lines in figure 5.

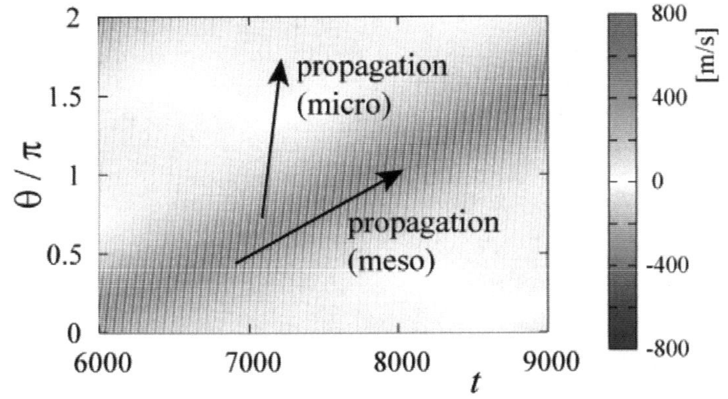

Figure 7. Time evolution of the θ profile of the convective velocity at $r = 0.5a$ and $z = 0.375\lambda$. The case with $\nu_{in} = 0.1$ is shown.

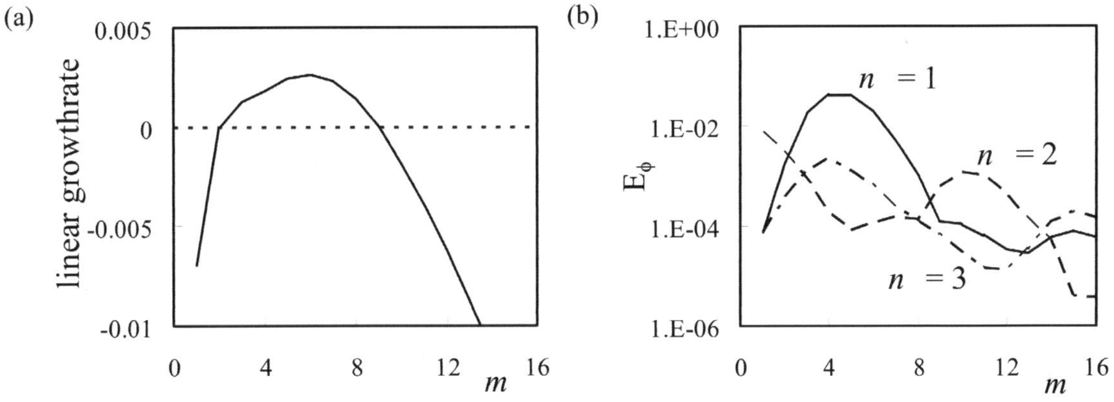

Figure 8. (a) Dependency of the instantaneous linear growth rate of mode $n = 1$ on the azimuthal mode number m with $\nu_{in} = 0.1$. These growth rates are calculated with the density profile at $t = 8000$ and $\phi_0 = 0$. (b) Potential energy spectrum on m with the axial mode numbers $n = 1 - 3$ at $t = 8000$.

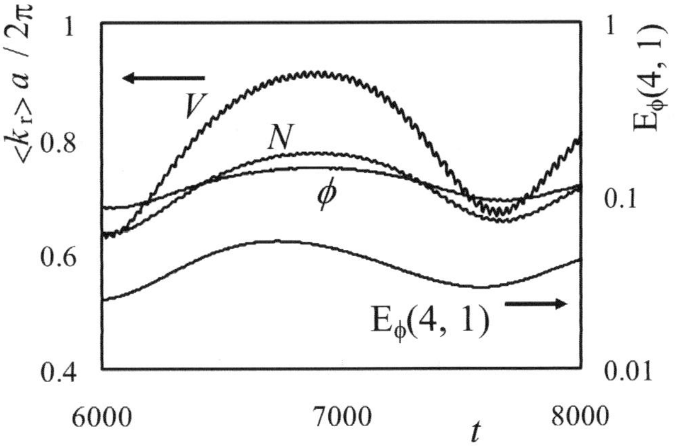

Figure 9. Time evolutions of the effective radial wave number $<k_r>$ of N, ϕ, V. The time evolution of $E_\phi(4,1)$ is also plotted.

induce the other peak at $(m, n) = (1, 2)$ and $(10, 2)$ in the spectrum. The formation of the streamer structure is related with this coupling of dominant modes. Modes with $(m, n) = (4, 1)$ and $(5, 1)$ are coupled with each other by mediator mode (1, 2) to form the streamer. This is the coupling in k_θ and k_z space, which gives matching of (4, 1) and (5, 1) to rotate with the same velocity, though the phase velocities of the linear eigenmodes are different from each other. The other coupling such as between modes with (4, 1) and (6, 1), affects to give an additional perturbation to the dominant modes, and the oscillation of the amplitudes, shown in figure 5, is induced. Modes with $n = 0$, which are not taken into the calculations, also can act as mediator modes. The case with other paths of energy exchange will be discussed in future works. In k_r space, the effective radial wave number, defined to be

$$\langle k_r \rangle = \sqrt{\int \left| \frac{\partial f}{\partial r} \right|^2 r dr \Big/ \int |f|^2 r dr}\ , \tag{6}$$

gives the index of the radial profile, where f implies $\{N, \phi, V\}$. Figure 9 shows the time evolutions of $\langle k_r \rangle$ of the (4, 1) mode. When the amplitude of the (4, 1) mode increases, $\langle k_r \rangle$ increases, accordingly. The increase of $\langle k_r \rangle$ corresponds to the induction of higher k_r components, which comes from nonlinear couplings with other modes. The evolution of $\langle k_r \rangle$ follows the evolution of the mode amplitude (figure 9). The larger k_r tends to stabilize the mode, so the relationship between $\langle k_r \rangle$ and the mode amplitude suggests that the profile modification affects the mode saturation. In this way, the streamer structure is sustained with three-dimensional mode coupling.

The streamers are characterized to be i) their typical scale k_θ: finite, $k_r \sim k_z \sim 0$, $\omega \ll \omega$(micro turbulence), ii) formed by nonlinear processes, and iii) localized in the direction of wave propagation (θ direction) [1]. An azimuthally-localized structure needs several Fourier modes to be constructed. The streamer in our simulation consists of rather small number of dominant contributors, (4, 1) and (5, 1), which are coupled with each other by means of the mediator mode (1, 2), and form quasi-mode like structure. The axial mode number of the turbulent structure discussed here is $n = 1$, which is small enough to satisfy the definition of the streamer. On the other hand, the typical scale length of the drift wave ρ_i is not so small compared with the plasma radius a ($\rho_i / a \sim 0.09$). Therefore, the turbulent structure has a radial scale length close to the plasma radius, not the meso scale. However, the rotation frequency is $f_{st} = 3\times10^{-4}$, which is much smaller than that of the drift wave $f_d = 0.02$ ($f_d / f_{st} \sim 70$). In this way, the characteristics of streamers are satisfied without the typical space length, and we call the structure 'streamer'.

4. Selective formation

Two kinds of turbulent structures are formed in the nonlinear steady states, as shown in the previous section. The dispersion relation depends on the plasma parameters, and our choice of parameters, described in section 3.1, allows both kinds of structural formation to be taken place by modes with rather small m. Therefore, selective formation of the turbulent structures can be discussed, using the ion-neutral collision frequency as a control parameter. A collision frequency scan is carried out from $\nu_{in} = 0.01$ to 0.12. Collision frequency ν_{in} brings two effects. That is, decreasing ν_{in} not only strengthens the driving force of the drift wave instability [19], but also weakens a damping force of the zonal flow represented by the third term in the right hand side of equation (2). These two effects together dictate the formation of the turbulent structure and the level of fluctuations in nonlinear steady states.

Figure 10 shows the dependencies of fluctuation energies of the potential on ν_{in} in nonlinear steady states. The energy components of (0, 0) mode, which is the zonal flow, and the sum of (4, 1) and (5, 1) modes, which form the streamer, are shown. The amplitudes are oscillating with period $T_d = O(10)$ and, the density profile, affected by the fluctuations, evolves with the time constant $T_q = O(1000)$. Therefore, the values shown in figure 10 are averaged over the duration longer than T_d and shorter than T_q. The zonal flow amplitude is small and the streamer modes are excited when ν_{in} is large. As ν_{in} is decreased, the zonal flow begins to be excited. The inflection point is given to be $\nu_{inc} \sim 0.052$

in this case. Figure 10 shows that the streamer can be formed without the zonal flow formation. When the zonal flow is formed, the E × B shearing of the zonal flow breaks the phase locking of the modes, so the streamer is not formed, even though amplitudes of these modes are larger than that of the zonal flow. When ν_{in} is much smaller ($\nu_{in} < 0.01$), the distribution of the fluctuation energy to each mode is different. This is because of the modification of the density profile to change the structures in the micro, meso and macro scales, simultaneously. This global bifurcation will be studied in future works.

There are two dominant energy exchange paths for structural formation by mode coupling of $m \neq 0$ mode. One is that to (0, 0) mode to form the zonal flow, and the other is that to the mediator mode ((1, 2) in this case) to form the streamer. These two kinds of structural formation mechanisms are involved, but only one of the structures can appear in steady states from their competitive nature. A simple analytical model of the energy balance [27] can give the bifurcation. We use a model including one unstable mode (drive) and two paths of mode coupling to a stable mode (sink). The energy balance in three modes, source of instability, mediator to form a streamer and zonal flow, is solved. The critical value of the zonal flow damping rate Γ_c is obtained, which depends on the damping rate of the mediator mode and coupling constants of the energy exchange from the drive to the mediator and zonal flow. When the zonal flow damping rate is larger than Γ_c, only the streamer mode is excited. When the zonal flow damping rate is smaller than Γ_c, only the zonal flow mode is excited, and the energy increases linearly, as the zonal flow damping rate becomes small. The value of ν_{in} represents the strength of the zonal flow damping rate, on which the selection of the turbulent structures depends. The detailed of the analytical model will be reported in the separated article. In this way, the bifurcation of the structures in turbulent plasmas has been represented using the dictating parameter (ν_{in} in this case).

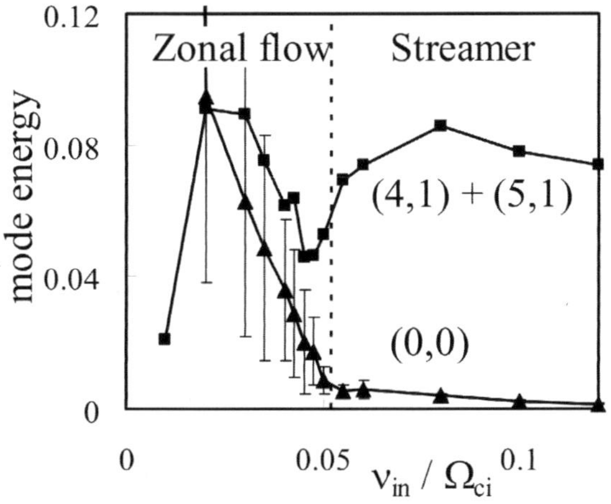

Figure 10. Dependencies of the fluctuation energies of the electrostatic potential on ν_{in} in nonlinear steady states.

5. Effects on particle transport

The turbulent structure affects the level of particle transport. It is well known that zonal flows suppress turbulent transport, and streamers induce convective transport. The effect of the structural formation in our cases is discussed in this section.

Figures 11 (a) and (b) show the dependency of the instantaneous linear growth rate and fluctuation energy of the potential on ν_{in}, respectively. This linear growth rate is evaluated with the density profile in the nonlinear steady state, and constant potential profile in space, i.e., only the destabilized effect of the density profile and no stabilizing effect of the potential profile is involved. The fluctuation energy is temporal average of the sum of the each potential energy of the Fourier modes except (0, 0) in the nonlinear steady state. The effects of nonlinear couplings and the stabilization by formation of the potential profile are involved to give this magnitude of the energy. The error bars in figure 11 correspond to the intermittent variation, and are larger in the zonal flow case. When E_ϕ increases, the turbulent flux increases, correspondingly, and *vice versa*. In the streamer case, the linear growth rate, and therefore, the fluctuation energy are small and constant. On the other hand, the linear growth rate and the fluctuation energy increase as ν_{in} decreases in the zonal flow case. These tendencies imply that the profile change in the zonal flow case greatly destabilize the system and could induce a large turbulent particle flux. However, the particle source is the same through the ν_{in} scan, so the turbulent fluxes have the same level, not depending on ν_{in}, as shown in figure 11 (c), which describes the dependency of the turbulent particle flux $\left\langle \tilde{N}\tilde{V}_r \right\rangle$ on ν_{in}. The level of particle transport does not change, in spite of the strong instability, because of turbulent suppression by the zonal flow.

When ν_{in} is large, fluctuation amplitudes are rather small, so the turbulent conductive transport is small, but convective transport is generated by the streamer. We also carried out a simulation, in which the mediate mode (1, 2) is artificially removed after the saturation with $\nu_{in} = 0.1$, when the streamer is formed with the (1, 2) mode. The streamer is not sustained and only a single mode becomes dominant without the (1, 2) mode. In this no streamer case, the fluctuation energy is 20 % larger than those in the streamer case, though the turbulent flux and density profiles are almost the same (the maximum of the difference of the density gradient is 5 %). That means the particle flux per the same fluctuation energy is larger in the streamer case. The discrepancy shows the drive of larger convective transport by the streamer. Energy condensation in the azimuthal direction by streamer formation enhances the radial transport in this case. In this way, the particle transport is affected by the selective formation of the turbulent structures, the zonal flow and the streamer, in the case when the magnitude of the particle flux is fixed (flux driven system).

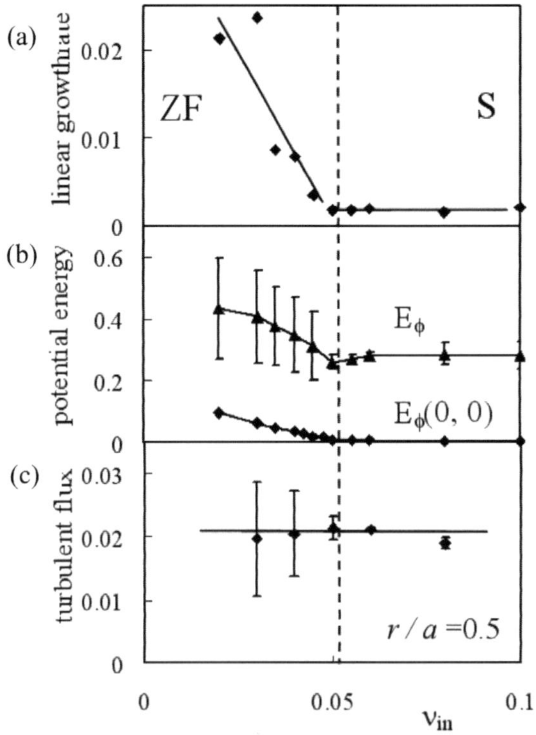

Figure 11. Dependency of (a) the instantaneous linear growth rate of (4, 1) mode, evaluated with the density profile in the nonlinear steady state, (b) fluctuation energy and energy of (0, 0) mode of the potential, and (c) turbulent flux at $r / a = 0.5$ on ν_{in}.

6. Summary

We have carried out the nonlinear simulation of the resistive drift wave in cylindrical plasmas, using a three-dimensional turbulence simulation code NLD. Turbulence with a zonal flow or streamer was obtained in the nonlinear steady states. In the case with smaller ν_{in}, the zonal flow is formed by nonlinear mode coupling. In the steady state, dynamical energy exchange between (0, 0) and the other unstable modes plays an important role for mode saturation. In the case with larger ν_{in}, the streamer is formed by coupling between a pair of unstable modes. A self-bunching of the perturbations takes place and the azimuthally-localized structure is sustained. Simulations showed the selective formation of the zonal flow and streamer using ν_{in} as a control parameter. The value of ν_{in} represents the strength of the damping force of the turbulent structures, and the bifurcation of the formed structures in turbulent plasmas has been obtained.

Acknowledgments

Authors acknowledge discussions with Prof. A. Fukuyama, Prof. Y. Kawai, Dr. S. Shinohara, Dr. A. Fujisawa, Dr. Y. Nagashima, Dr. T. Yamada, Dr. S. Inagaki, Prof. P. H. Diamond, and Prof. G. R. Tynan. This work is partly supported by the Grant-in-Aid for Specially-Promoted Research (16002005) and for Scientific Research (19360418) of MEXT, by Research Fellowships of the Japan Society for the Promotion of Science for Young Scientists, and by the collaboration programs of NIFS (NIFS06KDAD005, NIFS07KOAP017, NIFS06KNXN079) and of RIAM of Kyushu University.

References

[1] See reviews, e.g. Diamond P H, Itoh S -I, Itoh K and Hahm T S 2005 *Plasma Phys. Control. Fusion* **47** R35
[2] Yoshizawa A, Itoh S -I and Itoh K 2003 *Plasma and fluid turbulence* (Bristol: IOP Publishing).
[3] Champeaux S, Diamond P H 2001 *Phys. Lett.* A **288** 214
[4] Li J Q, Kishimoto Y, Miyato N, Matsumoto T and Dong J Q 2005 *Nucl. Fusion* **45** 1293
[5] Horton W 1999 *Rev. Mod. Phys.* **71** 735
[6] Saitou Y, Yonesu A, Shinohara S, Ignatenko M V, Kasuya N, Kawaguchi M, Terasaka K, Nishijima T, Nagashima Y, Kawai Y, Yagi M, Itoh S -I, Azumi M and Itoh K 2007 *Phys. Plasmas* **14** 072301
[7] Tynan G R, Holland C, Yu J H, James A, Nishijima D, Shimada M and Taheri N 2006 *Plasma Phys. Control. Fusion* **48** S51
[8] Brochard F, Bonhomme G, Gravier E, Oldenbürger S and Philipp M 2006 *Phys. Plasmas* **13** 052509
[9] Schröder C, Grulke O, Klinger T and Naulin V 2005 *Phys. Plasmas* **12** 042103
[10] Sokolov V and Sen A K 2004 *Phys. Rev. Lett.* **92** 165002
[11] Kaneko T, Tsunoyama H and Hatakeyama R 2003 *Phys. Rev. Lett.* **90** 125001
[12] Koga M and Kawai Y 2003 *Phys. Plasmas* **10** 650
[13] Inutake M 2001 *J. Plasma Fusion Res. SERIES* **4** 75
[14] Kono M and Tanaka M 2000 *Phys. Rev. Lett.* **84** 4369
[15] Terasaka K, Shinohara S, Nagashima Y, Yamada T, Kawaguchi M, Maruta T, Inagaki S, Kawai Y, Kasuya N, Yagi M, Fujisawa A, Itoh K and Itoh S -I 2007, *Plasma Fusion Res.: Rapid Comm.* **2** 031
[16] Antar G Y, Yu J H and Tynan G 2007 *Phys. Plasmas* **14** 022301
[17] Windisch T, Grulke O and Klinger T 2006 *Phys. Plasmas* **13** 122303
[18] Brochard F, Windisch T, Grulke O and Klinger T 2006 *Phys. Plasmas* **13** 122305
[19] Kasuya N, Yagi M, Azumi M, Itoh S -I and Itoh K 2007 *J. Phys. Soc. Jpn.* **76** 044501
[20] Kasuya N, Yagi M and Itoh K 2006 *J. Plasma Phys.* **72** 957
[21] Kervalishvili G N, Kleiber R, Schneider R, Scott B D, Grulke O and Windisch T 2006 *Contrib. Plasma Phys.* **46** 739
[22] Naulin V, Garcia O E, Nielsen A H and Rasmussen J J 2004 *Phys. Lett.* A **321** 355
[23] Lechte C, Niedner S and Stroth U 2002 *New. J. Phys.* **4** 34
[24] Ignatenko M, Azumi M, Yagi M, Shinohara S, Itoh S -I and Itoh K 2007 *Jpn. J. Appl. Phys.* **46** 1680
[25] Schröder C 2003 Ph.D. thesis, Ernst-Moritz-Arndt-Univ. Greifswald
[26] Hasegawa A and Wakatani M 1987 *Phys. Rev. Lett.* **59** 1581
[27] Das A, Sen A, Mahajan S and Kaw P 2001 *Phys. Plasmas* **8** 5104

Journal of Physics: Conference Series **123** (2008) 012027 doi:10.1088/1742-6596/123/1/012027

Gyrofluid simulation on the nonlinear excitation and radial structure of geodesic acoustic modes in ITG turbulence

J Q Li[1], Y Kishimoto[1], N Miyato[2], K Miki[1], J Anderson[1] and B R Shi[3]

[1] Graduate School of Energy Science, Kyoto University, Uji, Kyoto 611-0011, Japan
[2] Japan Atomic Energy Agency, Naka Fusion Institute, Naka 311-0193, Japan
[3] Southwestern Institute of Physics, Chengdu, Sichuan 610041, China

E-mail: lijq@energy.kyoto-u.ac.jp

Abstract. The nonlinear excitation and saturation mechanism of geodesic acoustic mode (GAM), as well as its radial structure, in tokamak plasmas are investigated by applying a newly well-benchmarked gyrofluid model. At first, an *empirical* closure relation for the conventional three-field gyrofluid modeling is presented for ion temperature gradient (ITG) fluctuations and the GAMs. The zonal flow damping is precisely examined by comparing with theoretical predictions and other kinetic calculations. Then, a local code and the global version are advanced to simulate the nonlinear excitation of the GAMs by ITG fluctuations through the Reynolds stress. It is found that the GAM instability can be nonlinearly excited under the competition between the nonlinear driving and the collisionless damping. The pump amplitude threshold of the GAM instability is higher than that of the zonal flow instability. Meanwhile, the unstable GAMs are mainly saturated by the intrinsic Landau damping, which is different from the stationary zonal flow counterpart. It is testified that the sound waves are damped fluctuations in ITG turbulence. Furthermore, the radial structure of the GAMs is shown as $k_r\rho_i \le 1.0$, which is shorter than that of the pure zonal flows.

1. Introduction

Geodesic acoustic mode (GAM) is a class of toroidal eigenmode with finite low frequency [1]. It has attracted increasing attention recently due to the combination with the stationary zonal flows (zero frequency) through the geodesic curvature effects [2-4]. The GAMs are characterized in spatial structure by poloidally and toroidally symmetric potential and poloidally asymmetric density or pressure fluctuations. The latter gives rise to a time-dependent zonal flow in toroidal plasmas. GAM dynamics is important in turbulent transport studies since their low-frequency radial structure can modulate or scatter drift-wave fluctuations, which are primarily responsible for collisionless transport. On the other hand, the GAM is a damped oscillator with finite frequency $\omega_{GAM} = \sqrt{2(\Gamma+\tau)T_{eq}}\, a/R$ coupling with axisymmetrical static potential (i.e., the stationary zonal flows). Here $\Gamma = 5/3$, $\tau = T_e/T_i$, T_{eq} is the equilibrium ion temperature normalized by the central value T_{ic}, a and R are the minor and major radius of tokamak plasma, respectively. The frequency is normalized by v_{ti}/R with $v_{ti} = \sqrt{T_{ic}/m_i}$. The level of zonal flows in toroidal ITG turbulence is strongly influenced by the collisionless damping of the GAMs [5]. Hence, the GAM dynamics have been intensively studied in toroidal plasma experiments and large-scale parallel simulations in light of theoretical analyses [6-20].

The GAM is a damped eigenmode so that it is believed to be mainly excited nonlinearly similar to the stationary zonal flows in turbulence[2-4, 21], except for the case linearly driven by energetic particle[22]. It may also be degenerate with the Beta-induced Alfven eigenmode(BAE) in the long wavelength limit[23]. In this work, we numerically study the nonlinear excitation, saturation mechanism and the radial structure of the GAMs in tokamak plasmas by using newly well-benchmarked gyrofluid model. At first, an *empirical* closure relation of gyrofluid modeling is presented for the ion temperature gradient (ITG) fluctuations and the GAM components. The zonal flow damping is precisely testified by comparing with theoretical predictions and other kinetic calculations. Then, the model is applied to simulate the nonlinear excitation of the GAMs by ITG fluctuations. Further, the saturation mechanism and the radial structure of the GAMs are investigated.

2. Gyrofluid ITG-zonal flow model with new closure relation

It is well-known that fluid and gyrofluid models possess advantages of clear/simplified physics picture in theoretical analyses and of fast calculation in large-scale simulations. The gyrofluid modelling has been well developed in plasma turbulence simulation, especially in the studies of zonal flow dynamics and turbulent transport [24-26]. However, a precise closure relation for lower order moment equations is still under improvement. Especially, the residual level of the zonal flows in ITG turbulence due to the collisionless damping is sensitively depended on the closure relation of the higher order moments in gyrofluid theory [27]. A new zonal flow closure relation has been derived systematically by writing the parallel heat flux as a summation of long and short wavelength parts. Here, we present an *empirical* closure relation for the GAM components by considering the Landau damping and its parallel wavenumber, which may *implicitly* involve the effect of finite orbit width. The new gyrofluid model and closure relation are testified by comparing the linear eigen values of ITG mode with other kinetic calculations under the standard Cyclone parameters [28] and the zonal flow damping in ITG turbulence with the theoretical prediction [29].

2.1. Gyrofluid ITG model with new closure relation

Following the standard procedure, a set of three-field fluid equations of the normalized potential ϕ; parallel ion velocity $\upsilon_{//}$ and ion pressure p_i can be derived as follows under the assumption of adiabatic electron response $n_e = (n_{eq}/T_{eq})(\phi - <\phi>) = (n_{eq}/T_{eq})(1-\delta)\phi$ [30]

$$\begin{aligned}(1-\delta-\nabla_\perp^2)\partial_t\phi = &-(a/L_n)\nabla_\theta\phi - (1+\eta_i)(a/L_n)\nabla_\theta\nabla_\perp^2\phi - 2\Gamma\omega_D(\phi+p_i) \\ &-\tfrac{1}{2}\omega_D\nabla_\perp^2\phi - \partial_r <\phi> \nabla_\theta\phi - \nabla_{//}\upsilon_{//} + [\phi,\nabla_\perp^2\phi] - \mu_\perp\nabla_\perp^4\phi\end{aligned} , \quad (1)$$

$$\partial_t\upsilon_{//} = -\nabla_{//}(\phi+p_i) - [\phi,\upsilon_{//}] + \eta_\perp\nabla_\perp^2\upsilon_{//} , \quad (2)$$

$$\begin{aligned}\partial_t p_i = &-(1+\eta_i)(a/L_n)\nabla_\theta\phi - (\Gamma-1/3)\Gamma(1+\eta_i)(a/L_n)\nabla_\theta\nabla_\perp^2\phi + 4\Gamma\omega_D p_i \\ &+(\Gamma-1/2)\omega_D\nabla_\perp^2\phi - \Gamma\nabla_{//}\upsilon_{//} - \gamma_{LD}\sqrt{8T_{eq}/\pi}|k_{//}|(p_i-\phi) - [\phi,p_i] + \chi_\perp\nabla_\perp^2 p_i\end{aligned} . \quad (3)$$

The normalized perturbed quantities are conventionally defined as [15]

$$(r;\nabla_\perp;\nabla_{//};t) \to (r/\rho_i;\rho_i\nabla_\perp;a\nabla_{//};tv_{ti}/a);$$
$$(n;\phi;\upsilon_{//};p_i) \to (a/\rho_i)(n/n_c;e\phi/T_{ic};\upsilon_{//}/\upsilon_{ti};p_i/n_cT_{ic}).$$

Here the magnetic drift term $\omega_D f = 2\varepsilon(\cos\theta\nabla_\theta - \sin\theta\nabla_r)f$ for any perturbed quantity f with $\varepsilon = a/R$, Heaviside step function $\delta = 0(1)$ for ITG fluctuations (the zonal flow component $<\phi>$, $<\cdot>$ denotes the flux surface average), which represents appropriately the adiabatic electron response to ITG fluctuations and the zonal flow; $\eta_i = L_{Ti}/L_n$ with $L_n = (d\ln n_{eq}/dr)^{-1}$ and $L_{Ti} = (d\ln T_{eq}/dr)^{-1}$, i.e., the characteristic length of equilibrium density (and ion temperature); ion Larmor radius $\rho_i = v_{ti}/\omega_{ci}$ with ion cyclotron frequency $\omega_{ci} = eB_0/m_i$; $\mu_\perp$, $\eta_\perp$ and $\chi_\perp$ are the numerical normalized cross-field viscosities and thermal conductivity, which absorb the energy cascaded to short wavelength region. The dominant nonlinear terms come from the $\vec{E}\times\vec{B}$ convective nonlinearity, which are expressed by the Poisson bracket $[f,g] = (\partial_r f \partial_\theta g - \partial_\theta f \partial_r g)/r$ in circular tokamak geometry (r,θ,φ) with the radius of the magnetic surface r, the poloidal and toroidal angles θ and φ, respectively. In this work, $\tau = 1$ is used.

The kinetic Landau damping physics is also represented by Hammett-Perkins closure model for the parallel heat flux moment $q_{//} = -i\gamma_{LD}\sqrt{8T_{eq}/\pi k_{//}}\, T_i/|k_{//}|$ [24]. However, the coefficients γ_{LD} for the ITG fluctuations and for the GAM components are different. It is *empirically* determined as

$$\gamma_{LD} = \begin{cases} \Gamma - 1 & \text{for ITG} \\ 3\Gamma & \text{for GAMs} \end{cases} . \tag{4}$$

Further, the parallel wavenumber for the GAMs is chosen with the parametric dependence of the safety factor q and the inverse aspect ratio ε, which may *implicitly* involve the finite orbit width effects.

$$\nabla_{//} = ik_{//} = \begin{cases} \varepsilon(\partial_\theta/q - \partial_\varphi) & \text{for ITG} \\ (3+\Gamma)(q/1.6)^{1/4}\varepsilon^{1/2}(\varepsilon\,\partial_\theta/q) & \text{for GAMs} \end{cases} . \tag{5}$$

It is noticed that the coefficients in equations (4) and (5) for the GAMs are still adjustable, which depend on the nonlinear modification of pressure profile.

2.2. Benchmark of toroidal ITG mode with Cyclone base case parameters

The standard Cyclone base case parameters have been extensively applied to benchmark newly developed kinetic codes. However, it is difficult to precisely benchmark a three-field fluid and/or gyrofluid model due to the low-order moment closure relation for finite Larmor radius effects. In equations (1)-(3), several additional terms, which represent more precise finite Larmor radius effects, have been incorporated to have a correct spectral structure of linear ITG growth rate. A local spectral code with fixed and/or periodic boundary conditions in radial direction and corresponding global version have been advanced to calculate the nonlinear evolution equations (1)-(3). Figure 1 displays the eigen frequency and growth rate of toroidal ITG mode under the standard Cyclone base case

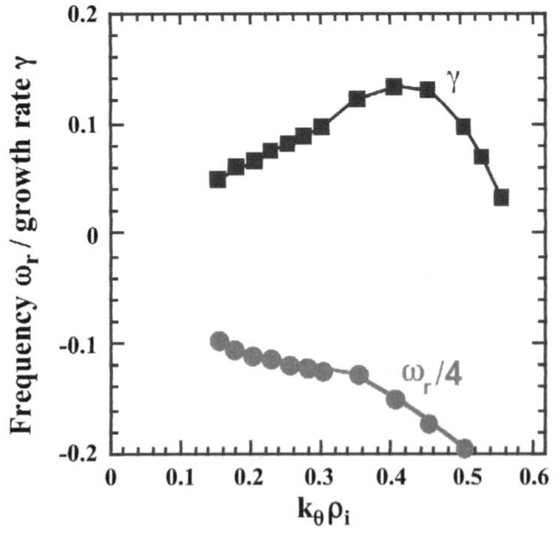

Figure 1. Growth rate and real frequency of toroidal ITG modes under the standard Cyclone base case parameters: $L_{Ti}/R = 6.9$, $L_n/R = 2.2$, $a/R = 0.18$, $q = 1.4$, $\hat{s} = 0.78$. $\mu_\perp = \eta_\perp = \chi_\perp = 0.8$.

parameters: $L_{Ti}/R = 6.9$, $L_n/R = 2.2$, $a/R = 0.18$, $q = 1.4$, $\hat{s} = rq'/q = 0.78$. Here the characteristic lengths of density and temperature and the magnetic shear $\hat{s}$ are taken at local region with maximum pressure gradient. Note that the maximum growth rate and corresponding frequency are much close to the kinetic values near the spectral peak $k_\theta \rho_i \approx 0.4$. The global mode structure near the surface of maximum pressure gradient exhibits strong ballooning characteristics, as shown in figure 2. The profiles of the safety factor q, the density and ion temperature are assumed as [31]

$$q(r) = 0.854 = 2.184(r/a)^2 \;;$$
$$n(r) = n_0 \exp\{-0.667\varepsilon \tanh[(r/a - 0.5)/0.3]\} \;;$$
$$T_i(r) = T_{i0} \exp\{-2.076\varepsilon \tanh[(r/a - 0.5)/0.3]\} \;.$$

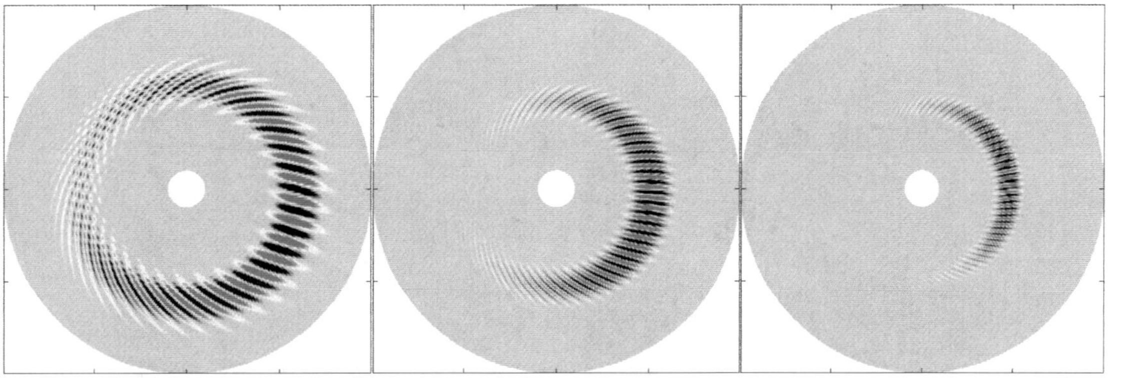

Figure 2. Eigenfunctions of toroidal ITG mode for different toroidal mode number $n = 20$ (left); $n = 40$ (center) and $n = 60$ (right). The corresponding wavenumbers are $k_\theta \rho_i = 0.175$; 0.35 and 0.525. The standard Cyclone base case parameters are used: $L_{Ti}/R = 6.9, L_n/R = 2.2$, $a/R = 0.18$, $q = 1.4, \hat{s} = 0.78$. $\mu_\perp = \eta_\perp = \chi_\perp = 0.8$, $a/\rho_i = 320$.

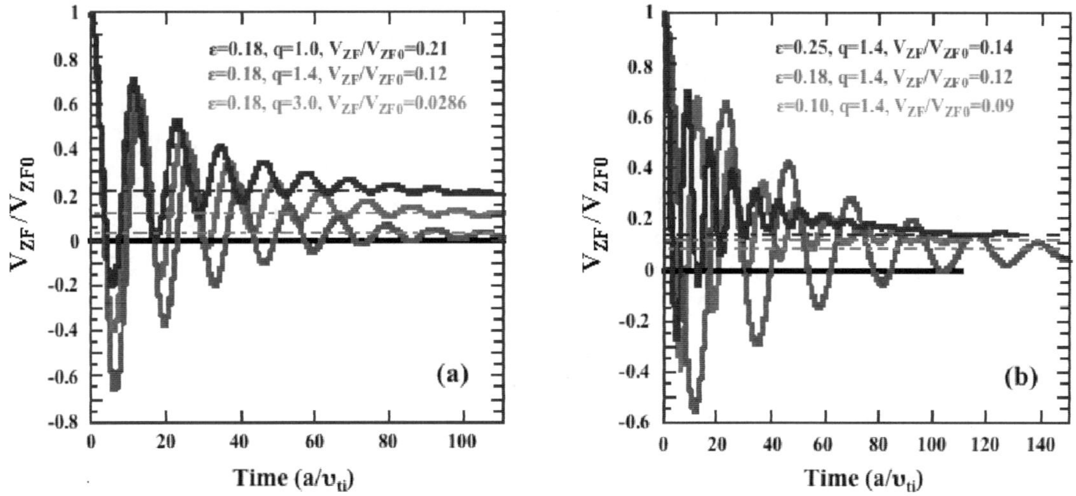

Figure 3. Time evolution of an initial static zonal flow $V_{ZF0}(t = 0) = \sin(0.19x)$ in the zonal flow damping tests for different q values with $\varepsilon = 0.18$ (a) and for different ε values with q=1.4 (b).

2.3. Benchmark of the zonal flow damping

The level of the zonal flows in turbulent fluctuations determines the regulation of turbulence and the anomalous transport. It is sensitively dependent on the collisionless Landau damping due to the toroidal coupling with the GAMs. To have a suitable coefficient of the closure relation for the GAMs, an initial value test for the zonal flow damping is performed by comparing the residual zonal flows with the analytical theory under the Cyclone base case parameters. Figure 3 illustrates the residual level of the zonal flows for different parameters q and ε. It can be seen that the new closure relation shows the zonal flow damping and a correct residual level of the zonal flows, which are well in agreement with the theoretical prediction $V_{ZF} = \left(1.+1.6q^2/\varepsilon^{1/2}\right)^{-1} V_{ZF0}$ by Rosenbluth and Hinton [29]. The wavelet energy analysis as shown in figure 4 also clearly exhibits the zonal flow damping process and the parametric dependence of the GAM frequency on $\varepsilon = a/R$.

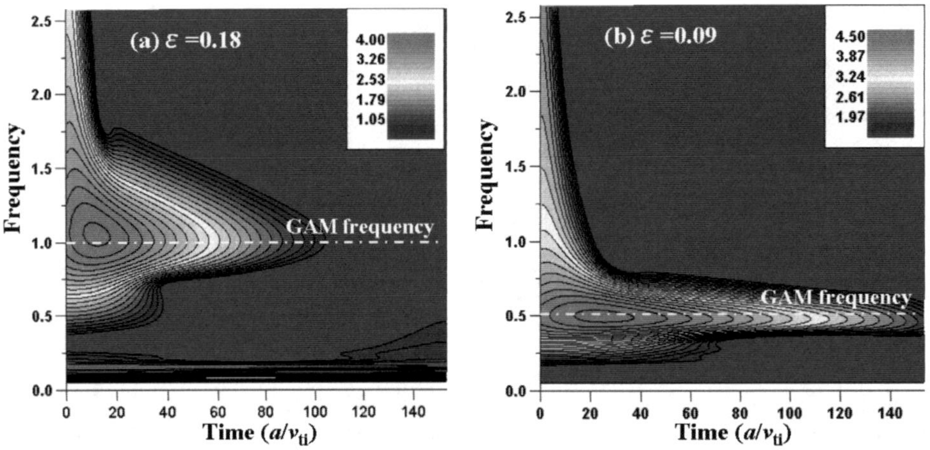

Figure 4. Wavelet energy analysis for the zonal flow damping. It produces clearly the parametric dependence of the GAM frequency on the inverse aspect ratio, i.e., $\omega_{GAM} \propto \varepsilon\ (= a/R)$. The initial static zonal flow is assumed as $V_{ZF0}(t=0) = \sin(0.19x)$, q=1.4. The contours indicate the spatially averaged $\ln(<\phi_{ZF}^2/2>)$ in the frequency-time plane.

3. GAM dynamics in ITG turbulence

The GAM is a damped eigenmode in toroidal plasmas, which is coupled with axiasymmetrical static potential through the geodesic curvature. Since the GAM is important in determining the regulation role of the stationary zonal flows and the transport level in ITG turbulence, the generation mechanism and/or the excitation process of the GAMs become an interesting topic. Generally, the GAM oscillation can be kicked through a static potential like an initial stationary zonal flow or a stationary pressure ripple. It is quickly damped through the Landau damping and the coupling with ion sound wave(SW). The decay rate is roughly estimated as $\propto q^5 \exp(-\alpha q^2)$ with constant α for small drift orbits [20, 23]. It is believed that the GAMs can be excited nonlinearly through the Reynolds stress like the stationary zonal flows. Based on equations (1)-(3), the GAM oscillator and the nonlinear source terms are expressed as follows

$$-\nabla_\perp^2 \frac{\partial \phi_{(0,0)}}{\partial t} = -2\Gamma \omega_D (\phi + p_i)_{(\pm 1,0)} - \frac{1}{2}\omega_D \nabla_\perp^2 \phi_{(\pm 1,0)} + \left[\phi, \nabla_\perp^2 \phi\right]_{(0,0)} , \qquad (6)$$

$$\frac{\partial p_{i(\pm1,0)}}{\partial t} = 4\Gamma\omega_D p_{i(0,0)} + \left(\Gamma - \frac{1}{2}\right)\omega_D \nabla_\perp^2 \phi_{(0,0)} - \Gamma\nabla_{//} \upsilon_{//(\pm1,0)} - \gamma_{LD}\sqrt{\frac{8}{\pi}}|k_{//}|(p_i - \phi)_{(\pm1,0)} - [\phi, p_i]_{\pm1,0)} \quad . \quad (7)$$

In addition, the equations of $\phi_{(\pm1,0)}$ and $\upsilon_{//(\pm1,0)}$ should be also incorporated to have a coupled system of the GAMs, ion SWs and the zonal flows. When the nonlinear terms are ignored, the damped GAM oscillator accompanied by the ion SW is formed with the frequency ω_{GAM}. Here we will simulate the nonlinear excitation of the GAMs due to the ITG fluctuations.

The GAMs are the combination of the perturbation of $(m,n) = (0,0)$ and $k_r \neq 0$ in potential field and the perturbation of $(m,n) = (\pm1,0)$ and $k_r \neq 0$ in the density and/or pressure fields. The former is mixed by the stationary zonal flows and the components with finite low frequency ω_{GAM}. The radial structure of the GAMs may also be different from the counterpart of the zonal flows. Hence, the zonal flow component $\phi_{(0,0)}$ is composed in a general sense by two parts with different frequencies, i.e.,

$$\phi_{(0,0)} = \lambda_{ZF}\phi_{(0,0)}(\omega = 0) + \lambda_{GAM}\phi_{(0,0)}(\omega = \omega_{GAM}) \quad . \quad (8)$$

The ratio of the two parts is determined by the balance between the driving force and damping sink of the GAMs. The stationary part of the zonal flows is commonly generated nonlinearly through a modulational instability. Meanwhile, it also suffers from the collisionless damping due to the coupling with the GAMs. For the part with GAM frequency, it may be driven nonlinearly through a parametric instability, namely, 3-wave interaction, due to the finite frequency. The parametric instability may involve the nonlinear terms $[\phi, p_i]_{\pm1,0)}$, $[\phi, \upsilon_{//}]_{\pm1,0)}$ and $[\phi, \nabla_\perp^2\phi]_{\pm1,0)}$ as well as the coupling with the Reynolds stress $[\phi, \nabla_\perp^2\phi]_{0,0)}$. The parametric instability results from the balance between the nonlinear driving and the Landau damping as well as the coupling with the ion SW. To have a direct simulation of the nonlinear excitation of the GAMs in ITG turbulence, several simulations have been designed to investigate the detailed physics processes.

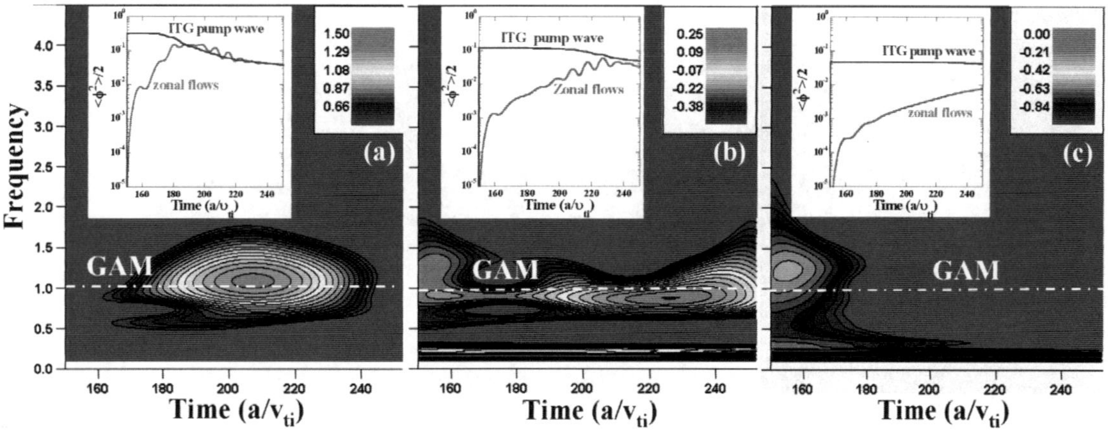

Figure 5. Wavelet energy analyses for the nonlinear excitation of the GAMs in high q plasmas by the ITG fluctuation with different pump amplitude level $<\phi_{ITG}^2/2> \approx 0.3$ (a); 0.12(b); 0.047(c). The contours exhibit the spatially averaged $\ln(<\phi_{ZF}^2/2>)$ in the frequency-time plane. Note that the nonlinear excitation of GAM instability depends on the ITG pump amplitude. The inset graphs are the corresponding time evolution of total zonal flow energy and the pump wave. The standard Cyclone base case parameters are used here except for $q = 2.6$.

3.1. Nonlinear excitation of the GAMs

The GAMs are robust in toroidal plasmas with higher q [15, 16]. To observe the parametric instability of the GAMs, ITG fluctuations with different amplitude are provided as the pump wave. In these simulations, linear ITG modes are initially excited to some level and then artificially controlled to keep a constant amplitude. In such a quasi-steady ITG fluctuation, the nonlinear excitation of the zonal flows with the GAMs can be analyzed by choosing different ways for the comparison. Figure 5 shows the dependence of the nonlinear excitation of the GAMs on the ITG pump amplitude. The inset graphs illustrate the time evolution of total zonal flow energy and the pump waves. It can be seen that the GAM instability becomes weaker as the ITG pump amplitude decreases. Further, the GAM fluctuation may be stabilized or damped even if the zonal flows still grow up for lower ITG pump amplitudes, indicating that the pump amplitude threshold for the GAM instability is higher than that of the zonal flow instability. This is understandable at least since the GAMs suffer from strong Landau damping. As the ITG pump amplitude becomes lower, weak GAMs are initially produced only through the beat wave of the pump fluctuations and strongly damped as shown in figure 5(c). In addition, the zonal flow energy decreases after the saturation due to the collisionless damping of the GAMs. Hence, the GAMs are difficult to survive in turbulent fluctuations if there exists no enough strong nonlinear driving force.

On the other hand, for the plasmas with lower q values, weak GAM instability can be observed only in higher ITG pump amplitude case, and it is quickly damped after the zonal flow saturation, as shown in figure 6(a). For the ITG pump wave with low amplitude, the GAM fluctuation is almost damped. A reference simulation without the GAM components is performed for the comparison with figure 6(b). It shows that the damped GAMs can efficiently reduce the zonal flow instability and the suppression role in ITG fluctuations even if they are very weak.

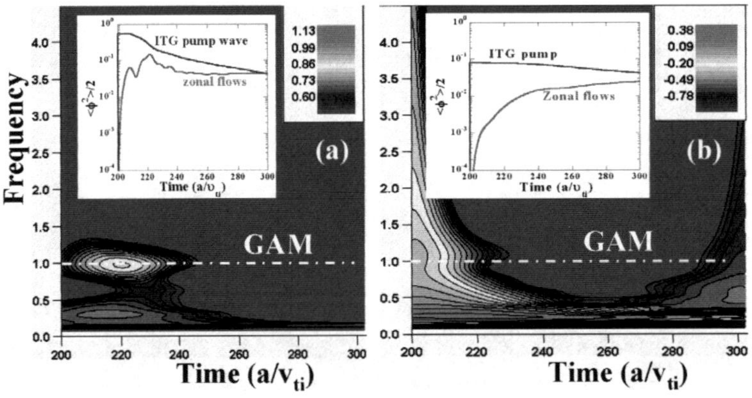

Figure 6. Wavelet energy analyses for the nonlinear excitation of the GAMs in low q plasmas by the ITG fluctuation with different pump amplitude level $<\phi_{ITG}^2/2> \approx 0.58$ (a); 0.082 (b). The standard Cyclone base case parameters are used here except for $q = 1.0$.

3.2. Saturation of the GAM instability

Generally, the zonal flow can be saturated through different mechanisms such as the tertiary instability (Kelvin-Helmholtz mode), spectral modulation of ambient turbulence and wave-packet scattering or trapping [4]. However, the GAMs themselves suffer from intrinsic Landau damping so that the GAM instability becomes weakened. While the GAM instability is determined by the competition between the nonlinear driving and the Landau damping as well as the coupling with sound wave, the saturation mechanism may be also attributed to the balance of the driving and damping. Careful comparison of the ITG pump amplitudes at the GAM saturation shows that the ITG fluctuation levels are the same at the saturation of $t \approx 210$ in figure 5(a) and $t \approx 230$ in figure 5(b). Note that the zonal flows are saturated earlier than the GAMs in figure 5(a), indicating their saturation mechanisms are different. Figure 5(a) and (b) also show that the zonal flow saturation is strongly coupled with the GAMs if the ITG pumps are weaker. Whereas the zonal flows are saturated even if the GAMs are still unstable. Since the GAM damping depends on q with the scaling $\propto q^5 \exp(-\alpha q^2)$, the ITG pump amplitude at the GAM saturation should be higher in low q plasma than that in the high q case. This point is confirmed by the observation in figure 6(a). Hence, the GAM instability is mainly saturated by the intrinsic Landau damping.

3.3. GAMs and ion sound wave in quasi-steady ITG turbulence

Although the GAMs are highly damped fluctuations, they can be sustained in a quasi-steady ITG turbulence through the balance among the nonlinear driving, collisionless damping and the coupling with the ion SW. Several simulations have been performed for different q values. In the case without the nonlinear modification of pressure profile, i.e., forbidding the quasi-linear flatting effects, the GAM fluctuations are continuously kept with the frequency ω_{GAM} after the ITG turbulence saturation. It is interestingly observed that the SWs with the frequency $\omega_{SW} \approx \omega_{GAM}/q$ are excited earlier than the GAMs and are damped after the ITG saturation even if the GAM fluctuations are sustained, as shown in figure 7. Note that the frequencies of the GAM fluctuations in the perturbed potential $\phi_{(\pm 1,0)}$, pressure $p_{i(\pm 1,0)}$ and parallel velocity $v_{//(\pm 1,0)}$ are downshift slightly from the GAM frequency in the zonal flow potential for all q values. Meanwhile, the ion SWs are damped in the quasi-steady ITG turbulence. It is also observed that some fluctuations with much lower frequency about half sound wave frequency are excited. These SW-like fluctuations may correspond to the low frequency spectra observed in some theoretical analyses and experiments [9,19,20]. When the quasi-flatting effects of the pressure profile are allowed in the simulations, the frequencies of GAM fluctuations in all perturbed components with $(m,n) = (\pm 1, 0)$ are downshift after the ITG turbulence saturation. This may show that the zonal flows have stronger coupling with the pressure profile through the GAM fluctuations.

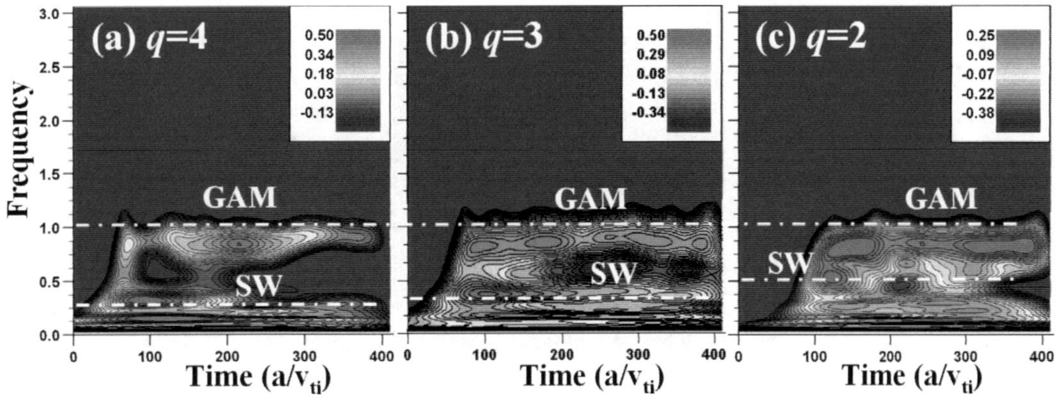

Figure 7. Wavelet energy analyses for the GAM fluctuations with the ion SW in the quasi-steady ITG turbulence. The parameters are the same as in figure 1. The contours exhibit the spatially averaged $\ln(< v^2_{//(\pm 1,0)}/2 >)$ in the frequency-time plane. The standard Cyclone base case parameters are used here except for different q values (a) q=4, (b) q=3 and (c) q=2.

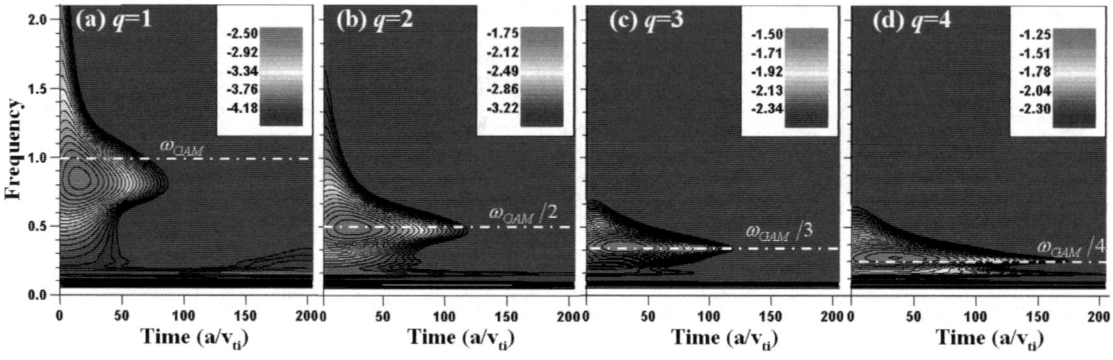

Figure 8. Wavelet energy analyses in an artificial simulation similar to the zonal flow damping test in figures 3 and 4 except that the static zonal flow is set to be constant in time. The contours exhibit the spatially averaged $\ln(< \upsilon^2_{//(\pm1,0)}/2 >)$ in the frequency-time plane. Note that the SW is excited and behaves as a damping fluctuation.

The observation that the ion SW is a damped fluctuation in the quasi-steady ITG turbulence can be further tested and analyzed through an artificial simulation. Similar to the test simulation for the zonal flow damping in figures 3 and 4, the static initial flow is set to be constant in time. It is found from figure 8 that the GAM fluctuations are artificially suppressed, but the ion SWs are excited, which well match the SW frequency for different q values. Note that these SWs are damped although the same static flow is always provided. This test result may be helpful to understand the observed damped sound wave in the ITG turbulence.

3.4. Radial spectrum of the GAMs

The GAM fluctuations are characterized by finite low frequency in the zonal flow potential and the perturbed components with $(m,n) = (\pm1,0)$. They also possess a radial structure. Two local simulations are performed for the cases without and with the GAMs under the Cyclone base case parameters for the most unstable ITG mode $k_\theta \rho_i = 0.35$ to compare the difference of the zonal flows and the GAMs. It is observed that the radial structure of the GAMs is shorter than that of the pure zonal flow fluctuation as shown in figure 9. This result approaches the observation $k_r \rho_i \sim 1.0$ in the global ITG simulation [16]. Note that some scaling of the radial structure of the GAMs has been also estimated as $k_r \rho_i \propto L_T^{1/3}$ or $L_n^{1/3}$ [21,32]. Some simulations with different L_T/R have been done and the results show almost the same spectra of the GAMs.

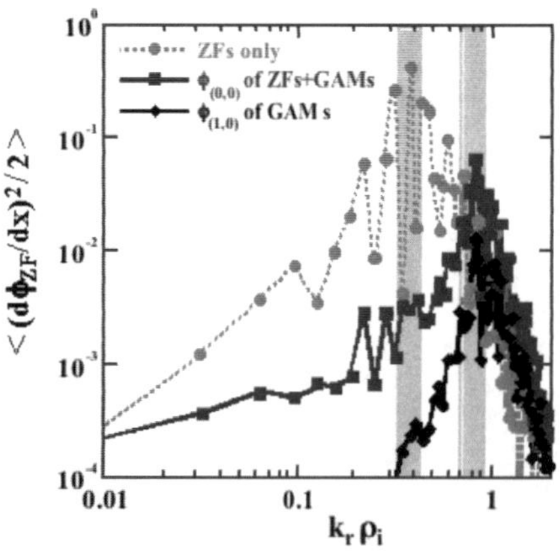

Figure 9. Radial spectra of the zonal flows in the simulations without (closed circles) and with (closed squares) the GAMs. The curve with diamonds corresponds to that of $d\phi_{(\pm1,0)}/dx$ component of the GAMs. The standard Cyclone base case parameters are used in both simulations except for $q = 2.0$ and $k_\theta \rho_i = 0.35$.

4. Summary

The precise gyrofluid model is helpful in large-scale turbulence simulations and physics analyses. In this work, a new *empirical* closure relation for the conventional three-field gyrofluid model is presented with emphasizing the zonal flow residual level due to the collisionless damping. It is shown that the model can reproduce the linear spectra of the toroidal ITG mode and the ballooning structures under the standard Cyclone base case parameters. The simulation tests for the zonal flow damping show that the key parametric dependence of the residual level of the static zonal flows on both the safety factor and the inverse aspect ratio is in agreement with the analytical prediction of the gyrokinetic theory.

As an application of the newly developed gyrofluid modelling, the nonlinear excitation of the GAMs is simulated in toroidal ITG turbulence. The spatio-temporal spectra of the zonal flows with the GAM components are analyzed by using time-depended wavelet energy analysis. It is examined that the damped GAMs can be nonlinearly driven to excite an instability by an ITG pump wave with larger amplitude, which is higher than the pump amplitude threshold of the zonal flow instability. Meanwhile, the GAM instability is mainly saturated by the intrinsic Landau damping. It is also observed that the ion SW can be excited with the GAMs in the quasi-steady ITG turbulence, but it is characterized by a damped fluctuation even if the GAMs are sustained continuously. Meanwhile, some SW-like fluctuations with much lower frequency are observed in the simulations, which may correspond to the theoretically predicted low frequency spectra by the kinetic theory and the observation in some experiments. Furthermore, it is found that the radial structure of the GAMs is scaled as $k_r \rho_i \leq 1.0$, which is shorter than that of the pure zonal flows.

Acknowledgments

J. L. would like to thank Liu Chen for the suggestion and discussion. This work was partly supported by the Grant-in-Aid from Japan Society for the Promotion of Science (No.18340186 and 19560828) and the National Science Foundation of China Grant No. 10575032.

References

[1] Winsor N, Johnson J L and Dawson J M 1969 *Phys. Fluids* **11** 2448
[2] Hallatschek K and Biskamp D 2001 *Phys. Rev. Lett.* **86** 1223
[3] Itoh K, Hallatschek K and Itoh S-I 2005 *Plasma Phys. Control. Fusion* **7** 635
[4] Diamond P H, Itoh S-I, Itoh K and Hahm T S 2005 *Plasma Phys. Control. Fusion* **47** R35
[5] Miki K, Kishimoto Y, Miyato N and Li J Q 2007 *Phys. Rev. Lett.* **99** 145003
[6] Shats M G and Solomon W M 2002 *Phys. Rev. Lett.* **88** 045001
[7] McKee G R *et al.* 2003 *Phys. Plasmas* **10** 1712
[8] Fujisawa A *et al.* 2004 *Phys. Rev. Lett.* **93** 165002
[9] Conway G D *et al.* 2005 *Plasma Phys. Control. Fusion* **47** 1165
[10] Zhao K J *et al.* 2006 *Phys. Rev. Lett.* **96** 255004
[11] Kramer-Flecken A *et al.* 2006 *Phys. Rev. Lett.* **97** 045006
[12] Ido T *et al.* 2006 *Nucl. Fusion* **46** 512
[13] Hamada Y *et al.* 2007 *Phys. Rev. Lett.* **99** 065005
[14] Scott B 2003 *Phys. Lett.* A **320** 53
[15] Miyato N, Kishimoto Y and Li J Q 2004 *Phys. Plasmas* **11** 5557
[16] Miyato N, Kishimoto Y and Li J Q 2006 *Plasma Phys. Control. Fusion* **48** A335
[17] Naulin V *et al* 2005 *Phys. Plasmas* **12** 052515
[18] Kendl A and Scott B D 2005 *Phys. Plasmas* **12** 064506
[19] Watari T, Hamada Y, Notake T, Takeuchi N and Itoh K 2006 *Phys. Plasmas* **13** 062504
[20] Gao Z, Itoh K, Sanuki H and Dong J Q 2005 *Phys. Plasmas* **12** 064506
[21] Chakrabarti N, Singh R, Kaw P K and Guzdar P N 2007 *Phys. Plasmas* **14** 052308
[22] Berk H L *et al.* 2006 *Nucl. Fusion* **46** S888
[23] Chen L and Zonca F 2007 "Radial structures and nonlinear excitation of Geodesic Acoustic Modes" *Sherwood 2007*

Journal of Physics: Conference Series **123** (2008) 012027 doi:10.1088/1742-6596/123/1/012027

[24] Hammett G W and Perkins F W 1990 *Phys. Rev. Lett.* **64** 3019
[25] Beer M A and Hammett G W 1996 *Phys. Plasmas* **3** 4046
[26] Sugama H, Watanabe T H and Horton W 2001 *Phys. Plasmas* **8** 2618
[27] Sugama H, Watanabe T H and Horton W 2007 *Phys. Plasmas* **14** 022502
[28] Dimits A M, Bateman G, Beer M A *et al.* 2000 *Phys. Plasmas* **7** 969
[29] Rosenbluth M N and Hinton F L 1998 *Phys. Rev. Lett.* **80** 724
[30] Horton W 1990 *Phys. Reports* **192** 1
[31] Idomura Y, Tokuda S and Kishimoto Y 2003 *Nucl. Fusion* **43** 234
[32] Itoh S-I, Itoh K, Sasaki M, Fujisawa A, Ido T and Nagashima Y 2007 *Plasma Phys. Control. Fusion* **49** L7

A model of GAM intermittency near critical gradient in toroidal plasmas

K Miki[1†], Y Kishimoto[1], N Miyato[2] and J Q Li[1]

[1] Graduate School of Energy Science, Kyoto University, Uji, Kyoto 611-0011, Japan
[†]Present address: Center for Astrophysics and Space Science, UCSD, La Jolla, CA 92093, USA.
[2] Fusion Research and Development Directorate, Japan Atomic Energy Agency, Naka, Ibaraki 311-0193, Japan

E-mail: kamiki@ucsd.edu

Abstract. We have constructed a four-field minimal model that describes the *growing intermittency* of turbulence associated with the geodesic acoustic mode (GAM) observed in our toroidal Landau-fluid simulations [K Miki *et al.* 2007 *Phys. Rev. Lett.* **99**, 145003]. The intermittent dynamics are well reproduced by the model for the reference parameters used in the simulation. The model can also reproduce characteristics of turbulent transport associated with the GAM, such as a single burst leading to a full quench of turbulence and also a steady state turbulence mixed with steady zonal flows and GAMs. Investigating the behaviour of the solution trajectories around the fixed points in four-dimensional phase space, we study the comprehensive properties of the model and identify the bifurcation property between Dimits shift and steady state turbulence regimes, which correspond to different eigen-states.

1. Introduction

The transport near the critical gradient is of great interest in understanding various transition dynamics such as L-H transition and formation of internal transport barriers in magnetic fusion plasmas [1]. It is known that near the critical gradient, multiple states dominated by different physical processes coexist, such as turbulent plasma with high transport and quiescent flowing plasma with high confinement. A typical phenomenon around the critical gradient is so-called *critical slowing down*, representing a concept that the time scale of phenomena is prolonged in order for the system to determine a state in such multiple states [2]. Therefore, small perturbation and/or dissipation can significantly influence the transition dynamics due to the marginal nature of the system. A nonlinear up-shift of the critical gradient, referred to as the *Dimits shift*, due to the quench of turbulent fluctuation by self-generated zonal flows is one of typical examples [3]. An intermittent dynamics has been observed near the critical gradient due to a collisional damping of zonal flows in ion temperature gradient (ITG) driven turbulence [4]. A theoretical model which primarily describes the intermittency, has been extensively studied in terms of predator-prey approach [5]. A dynamical systems approach which predicts the Dimits shift is also proposed in slab geometry [6].

Recently, geodesic acoustic mode (GAM) [7], which is an oscillatory counterpart of zonal flow in toroidal plasmas coupled with anisotropic pressure perturbation of $(m, n) = (1, 0)$, has been extensively studied, since it significantly changes the characteristics of zonal flows and then the transport [8,9], where m and n denote poloidal and toroidal wave numbers. In our previous studies [10], we found that

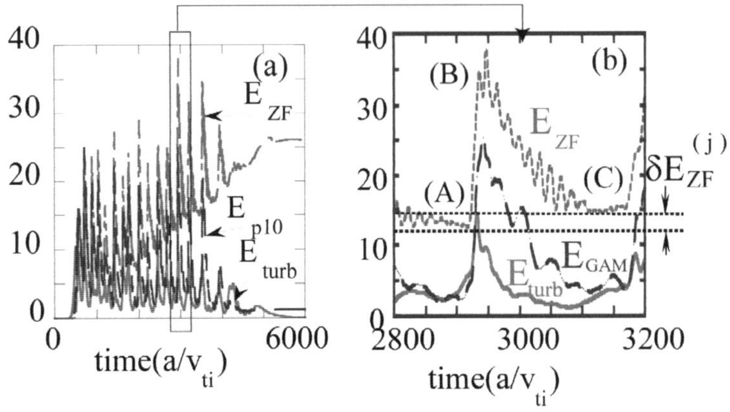

Figure 1. Temporal evolution of turbulence E_{turb}, zonal flow E_{ZF}, and p_{10} pressure perturbation energy E_{p10} for the time interval (a) (0 < t < 6000) and (b) (2800 < t < 3200). The level of zonal flow energy is found to be slightly up-shifted by $\delta E_{\mathrm{ZF}}^{(j)}$ in the j-th burst.

the GAM dynamics depends sensitively on the magnetic safety factor q-value, so that GAMs are preferentially excited in high-q edge region, whereas stationary zonal flow dominates in low-q core region. Such a dependency on q-value is consistent with recent experiments in which GAMs are observed near the edge of plasmas [11]. It is noted that those studies are performed only far from the critical gradient region.

An alternative concern is the role of such GAMs on turbulent transport near the critical gradient regime, which is closely related to the trigger of plasma instability and transition dynamics. As a typical phenomenon, the nonlinear upshift (Dimits shift) of linear critical gradient was investigated in plasmas dominated by low frequency zonal flow due to low q-value near the magnetic axis ($q \approx 1$). However the effect of high frequency zonal flow, which is equivalent to the GAMs, on the critical gradient phenomena has not been addressed so far, even in our previous studies [10].

Recently we have studied this problem based on the global toroidal Landau-fluid simulations [10] by choosing the temperature gradient, i.e. R/L_{Ti}, slightly larger than the linear instability threshold [12]. We have found a prominent intermittency associated with GAM dynamics near the critical gradient region, which is qualitatively different from that originated from the collisional damping [4]. In particular, we have found that during the intermittent bursts, stationary zonal flow increases with a slow time scale due to the accumulation of undamped residues. We refer to this phenomenon as *growing intermittency*. Similar intermittent transport phenomena due to GAMs are also observed in experiments [13,14,15].

We have proposed a four-field model consisting of four-differential equations which govern the temporal evolution of turbulence and zonal flow energies, and also anisotropic pressure perturbation and ion parallel sound velocity in order to understand the complex features as seen in figure 1 [12]. The numerical simulation based on the model successfully reproduced the growing GAM intermittency. However, contrary to the conventional two field model only dealing with the turbulence and zonal flow energy, the characteristics of the four-field model have not been investigated in detail.

In this paper, we investigate the details of the four field model as a dynamical system and explore the intrinsic potential of wider physical processes which involve the dynamics of the single burst leading to a full quench of turbulence and also a steady state turbulence mixed with zonal flows/GAMs observed in the simulation. Here, we introduce a fixed point stability analysis which is one of the basic approaches to investigate nonlinear systems [16]. Since it is not simple to derive the analytical solution directly from the present four-field model, we investigate the solution trajectories around the fixed point incorporated with an analysis on eigen-value and corresponding eigen-vector to get some insights for the complex GAM dynamics near the critical gradient. It is expected that the fixed point analysis may predict the bifurcation of the states such as in Dimits shift regime and in the steady state regime above the critical gradient due to the change of governing parameters.

The remainder of this paper is organized as follows. In section 2, we propose the extended predator-prey model. Some numerical results and analyses are shown in section 3. In section 4, the fixed point analysis is introduced to investigate detailed properties around specific fixed points in the extended predator-prey model. The transition of eigen-states due to the change of parameters, especially the backflow from ion sound wave is also discussed. In section 5, we make some concluding remarks.

2. An extended predator-prey model

Here, we present a theoretical model to reproduce the various phenomena observed in the simulation and to understand the underlying physical processes. In the framework of a fluid model [10], keeping the total energy conservation in mind, we have obtained the equation system of GAM dynamics with respect to zonal flow velocity $\langle v_{E_r} \rangle$, anisotropic pressure perturbation $\langle p \sin\theta \rangle \equiv p_{10}$, which corresponds to the GAM, and anisotropic ion sound velocity $\langle v_{\parallel} \cos\theta \rangle \equiv v_{10}$, as follows:

$$\frac{\partial \langle v_{E_r} \rangle}{\partial t} = -\langle \tilde{v}_{E_r} \tilde{\Omega} \rangle - \frac{2a}{n_{eq} R} \langle p \sin\theta \rangle, \tag{1}$$

$$\frac{\partial \langle p \sin\theta \rangle}{\partial t} = -\langle [\tilde{\phi}, \tilde{p}] \sin\theta \rangle + (\Gamma + \tau) p_{eq} \frac{a}{R} \langle v_{E_r} \rangle + (\Gamma + \tau) p_{eq} \frac{a}{qR} \langle v_{\parallel} \cos\theta \rangle - (\Gamma - 1) n_{eq} \sqrt{\frac{8T_{eq}}{\pi}} \frac{a}{qR} \langle T_i \sin\theta \rangle \tag{2}$$

$$\frac{\partial \langle v_{\parallel} \cos\theta \rangle}{\partial t} = -\langle [\tilde{\phi}, \tilde{v}_{\parallel}] \cos\theta \rangle - \frac{a}{n_{eq} qR} \langle p \sin\theta \rangle, \tag{3}$$

where $\langle\ \rangle$ denotes the flux average, and where a and R denote the minor and major radii of the tokamak plasma, respectively. Further, n_{eq}, p_{eq}, and T_{eq} represent the density, pressure and temperature profiles, which are fixed in the present work. Note that we neglected high m GAM components in equations. (2)-(3). Here, we used the following normalization and definition of the variables:

$$(t,\ U,\ G,\ V) \rightarrow \left(t/\sqrt{T_{eq}},\ \langle v_{E_r} \rangle,\ \langle psin\theta \rangle / (\sqrt{T_{eq}}\, p_{eq}),\ \langle v_{\parallel} \cos\theta \rangle \right).$$

Equations (1) –(3) are then written as follows

$$\frac{\partial U}{\partial t} = -\beta G + \hat{N}_{[\phi,\phi]}, \tag{4}$$

$$\frac{\partial G}{\partial t} = \beta' U - \gamma_{LD} G + \beta_2 V + \hat{N}_{[\phi,p]}, \tag{5}$$

$$\frac{\partial V}{\partial t} = -\beta_2' G + \hat{N}_{[\phi,v_{\parallel}]}, \tag{6}$$

where $\beta = 2a/R$, $\beta' = (\Gamma + \tau)a/R$, $\beta_2 = (\Gamma + \tau)a/qR$, $\beta_2' = a/qR$, and the Landau damping rate $\gamma_{LD} = T_{eq}\sqrt{8/\pi}(a/qR)$. The nonlinear terms are given by $\hat{N}_{[\phi,\phi]} = -\langle \tilde{v}_{E_r} \tilde{\Omega} \rangle$, $\hat{N}_{[\phi,p]} = -\langle [\tilde{\phi}, \tilde{p}] \sin\theta \rangle / p_{eq}$, and $\hat{N}_{[\phi,v_{\parallel}]} = -\langle [\tilde{\phi}, \tilde{v}_{\parallel}] \cos\theta \rangle$. $T_{eq} = 1$ is assumed for simplicity. At first, we analyse the basic feature of the growing intermittency based on equations (4)-(6). Equations (4)-(6) can be expressed as the following second order differential equation with respect to G:

$$\frac{\partial^2 G}{\partial t} + \gamma_{LD} \frac{\partial G}{\partial t} + \omega_G^2 = 0, \tag{7}$$

where $\omega_G = \sqrt{\omega_{GAM}^2 + \omega_{sound}^2} = \sqrt{\beta\beta' + \beta_2\beta_2'}$. It is noted that the nonlinear terms work as the sources to trigger the GAM dynamics. Meanwhile, instead of considering those nonlinear terms, here we simply introduce the effect of the sources as initial conditions for (U, G, V), e.g., $U(t=0) = U_0$, $G(t=0) = G_0$, and $V(t=0) = V_0$. Then equation (7) is solved as follows;

$$U = A_U \exp(-\gamma_{LD} t/2)\sin(\omega_r t + \delta_1) + C_U, \tag{8}$$

$$G = A_G \exp(-\gamma_{LD} t/2)\sin(\omega_r t + \delta_2), \tag{9}$$

$$V = A_V \exp(-\gamma_{LD} t/2)\sin(\omega_r t + \delta_1) + C_V, \tag{10}$$

where $\omega_r = \sqrt{\omega_G^2 - (\gamma_{LD}/2)^2}$, and (A_U, A_G, A_V) and (C_U, C_V) are constants determined by the initial conditions. Here, C_U and C_V are expressed $C_U = (U_0 - 2qV_0)/(1 + 2q^2)$ and $C_V = -qC_U$. δ_1 and δ_2 are the phase factors also determined by initial conditions and $\Delta\delta \equiv \delta_1 - \delta_2 \approx \pi/2$ is estimated. As seen in equations (8)-(10), each variable exhibits a damping oscillation ascribed from the GAM oscillation and decay due to the Landau damping. It is noticed that the undamped residual part can be survived in U and V as a stationary zonal flow and ion parallel velocity after the complete damping of G, which balance in accordance with $\beta' C_U + \beta_2 C_V = 0$. Therefore, it is expected that the residual part of the zonal flow accumulates during the repetition of the burst and gradually increases as seen in figure 1.

In order to reproduce the intermittent dynamics in a self-consistent manner, an analysis including the nonlinear terms incorporated with the evolution equation of turbulence field is necessary. By defining the turbulent energy as $N = \sum_{m,n} |\tilde{\phi}_{(m,n)}|^2$, we construct the following four-field equation system as a predator-prey model to represent the possible GAM dynamics :

$$\frac{\partial N}{\partial t} = \gamma_L N - \gamma_{NL} N^2 - \alpha_1 U^2 N - \alpha_2 \frac{\beta}{\beta'} G^2 N - \alpha_3 \frac{\beta\beta_2}{\beta'\beta_2'} V^2 N, \tag{11}$$

$$\frac{\partial U}{\partial t} = \frac{1}{2}\alpha_1 U N - \beta G - \frac{1}{2}\gamma_{damp} U, \tag{12}$$

$$\frac{\partial G}{\partial t} = \frac{1}{2}\alpha_2 G N + \beta' U + \beta_2 V - \gamma_{LD} G, \tag{13}$$

$$\frac{\partial V}{\partial t} = \frac{1}{2}\alpha_3 V N - \beta_2' G, \tag{14}$$

where γ_L and γ_{NL} are the growth rate of the turbulent fluctuation and the nonlinear damping rate, respectively. γ_{damp} is the collisional damping rate of the zonal flow. Note here that contrary to the conventional predator-prey model where the zonal flows are expressed in the unit of energy, e.g. U^2, equations (12)-(14) are not described in terms of U^2, G^2, and V^2, but in terms of U, G, and V, in order to represent the oscillatory characteristics of the GAM and ion parallel sound wave. $(\alpha_1, \alpha_2, \alpha_3)$ represent the strength of nonlinear coupling including the cross-coupling [17] among the zonal flows, GAMs and sound waves as well as the directions of energy flow from the turbulence N to the zonal flow U, pressure perturbation G, and ion parallel sound velocity V, respectively. In this minimal model, it is assumed that the nonlinear cross-coupling, i.e., the off-diagonal terms in equations (12)-(14) have been effectively absorbed in the diagonal terms for simplicity (see Appendix A). The collisional damping effect on the zonal flow is neglected to manifest the effect of GAM damping. Note that this model is constructed so as to satisfy the total energy conservation in the absence of linear driving source and damping terms. Namely, the energy equation is written as follows:

$$\frac{\partial}{\partial t}(N + E_U + E_G + E_V) = \gamma_L N - \gamma_{NL} N^2 - \gamma_{damp} E_U - 2\gamma_{LD} E_G , \tag{15}$$

where $E_U \equiv U^2$, $E_G \equiv (\beta / \beta')G^2$, and $E_V \equiv (\beta\beta_2 / \beta'\beta_2')V^2$.

3. Numerical analysis of four-field equation system

In the analyses of the predator-prey model, we employ the parameters as the following, $\gamma_L = 0.05$, $\gamma_{NL} = 0.003$, $\gamma_{damp} = 0$, and $q = 2.10$, which roughly follow those used in the simulation [12]. The dynamics is found to show different features depending on the relative value of α_1, α_2 and α_3, which characterizes the nonlinear energy flow from turbulence (N) to zonal flow (U), pressure perturbation (G), and ion parallel sound velocity (V). In the following, we show three cases.

3.1 Case 1 ($\alpha_1 < \alpha_2$): Here, we choose $\alpha_1 = 0.05$, $\alpha_2 = 0.5$ and $\alpha_3 = 10^{-4}$, respectively. This physically corresponds to the case in which the Reynolds stress is less dominant than the nonlinear coupling term between turbulence and p_{10} pressure perturbation. The temporal evolutions of (N, U, G) are shown in figure 2 and basic features of the GAM intermittency observed in figure1, such as quasi-period bursts, gradual increase of zonal flow and quenching of turbulence, are found to be qualitatively reproduced. Namely, the accumulation of undamped residue discussed in section 2 is confirmed. Figure 3 shows the details in a specific time interval of a burst in figure 2. As turbulence energy N reaches a critical value that satisfies the relation $\gamma_L - \gamma_{NL}N \cong E_\alpha$ ($\equiv \alpha_1 E_U + \alpha_2 E_G + \alpha_3 E_V$), the abrupt growth of E_U, E_G, (and also E_V) and successive damping accompanied with typical GAM oscillation are observed as seen in figure 3(b). The turbulence also suffers damping once the relation $\gamma_L - \gamma_{NL}N < E_\alpha$ is satisfied at $t \cong 1190$ as seen in figure 3(a). Followed by the turbulence, GAM damping starts after the relation $\gamma_{LD} > \alpha_2 N/2$ is satisfied as seen in figure 3(c).

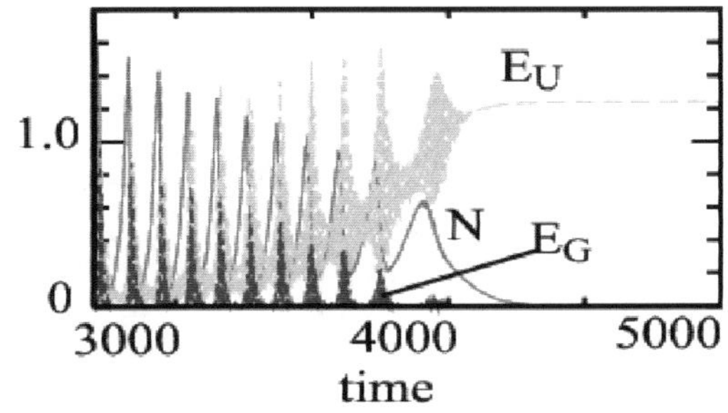

Figure 2. Time evolution of N, E_U, and E_G in the numerical calculation with the extended predator-prey model during the time (2700 < t < 5000) in the case 1.

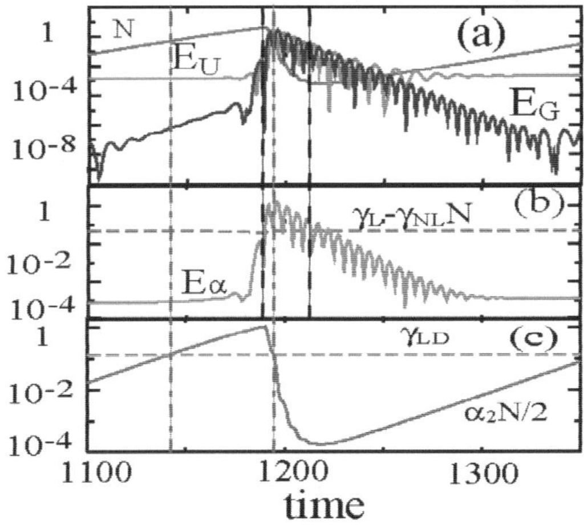

Figure 3. (a): Temporal evolution of N, E_U, and E_G for the case of the extended minimum model in the time interval of one burst (1100 < t < 1250). (b): Temporal evolution of (γ_L - $\gamma_{NL}N$) and E_α. (c): Temporal evolution of $\alpha_2 N/2$ and γ_{LD}.

The characteristics of the abrupt growth of G are investigated by solving equation (13) under the condition $\beta' U + \beta_2 V = 0$. Here it is assumed that U and V are almost balanced, and $N = N_0 \exp(\gamma_L t)$ is satisfied where N_0 is a reference amplitude when $\partial G/\partial t = 0$, i.e. $\gamma_{LD} = \alpha_2 N_0/2$ is fulfilled. The solution is approximately given by

$$G = G_0 \exp\left(\frac{\gamma_{LD}}{\gamma_L} e^{\gamma_L t} - \gamma_{LD} t\right) \cong G_0 \exp(\gamma_L \gamma_{LD} t^2 / 2), \tag{16}$$

showing faster growth than that of linear instability. The time scale of the abrupt growth is evaluated as $\tau_b \sim \sqrt{2/\gamma_L \gamma_{LD}}$. For the present parameters, $\tau_b \sim 17.5$ is measured, which is consistent with that observed in figure 3. It is noted that the typical damping rate of N is characterized by E_α after the condition $\gamma_L - \gamma_{NL} N < E_\alpha$ is satisfied. The damping of N is ceased when E_α crosses again $\gamma_L - \gamma_{NL} N$ as seen in figure 3(b) at $t \sim 1212$, and then N once again starts to grow linearly. During the phase of Landau damping, E_G is quenched while certain equilibrium levels of E_U and E_V are sustained, which are described as C_U and C_V, respectively. This process leads to the accumulation of zonal flows as shown in figure 1.

3.2 Case 2 ($\alpha_1 > \alpha_2$) : Contrary to the case1, here, we choose $\alpha_1 = 0.5$, $\alpha_2 = 0.05$, and $\alpha_3 = 10^{-4}$, respectively. The result is shown in figure 4. In this case, the Reynolds stress is dominant compared with the nonlinear coupling between turbulence and p_{10} pressure perturbation. As seen in figure 4, after the single burst of the turbulence occurs, the residual zonal flow immediately quenches both turbulence and p_{10} pressure perturbation and causes the decoupling of zonal flow and GAM. This may correspond to the single burst nature observed in the fluid simulation [12], in which the turbulence is completely suppressed by the generated zonal flow.

As seen in above cases 1 and 2, when $\alpha_2 > \alpha_1$ is satisfied, multiple bursts are exhibited successively, whereas the dynamics is dominated by a single burst when $\alpha_1 > \alpha_2$. The change of the dynamics between case 1 and case 2 is found to take place at $\alpha_1 \approx \alpha_2$. This suggests that a certain amount of energy flow directly from turbulence to p_{10} pressure perturbation may be necessary to have the growing intermittency as seen in figure 2. The underlying physical mechanism that the relative relation between α_1 and α_2 provides different dynamical features is considered as follows. As discussed in section 3, the saturation of turbulence takes place when the suppression term $-E_\alpha \cong -(\alpha_1 E_U + \alpha_2 E_G + \alpha_3 E_V)$ overwhelms the linear driving term γ_L in equation (11) in both case 1 and 2. Namely, both α_1 and α_2 (and/or $\alpha_1 E_U$ and $\alpha_2 E_G$) decide the saturation level of turbulence N, i.e. maximum level of turbulence, as a whole. On the other hand, the contribution of GAM in E_α, i.e.

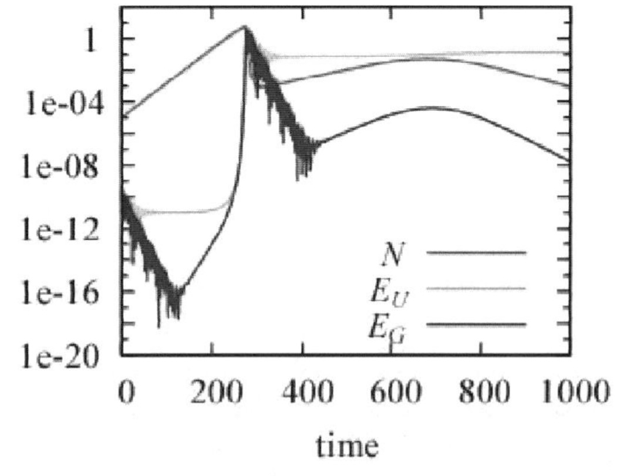

Figure 4. Time evolution of N, E_U and E_G in case of α_1 is dominant (case 2).

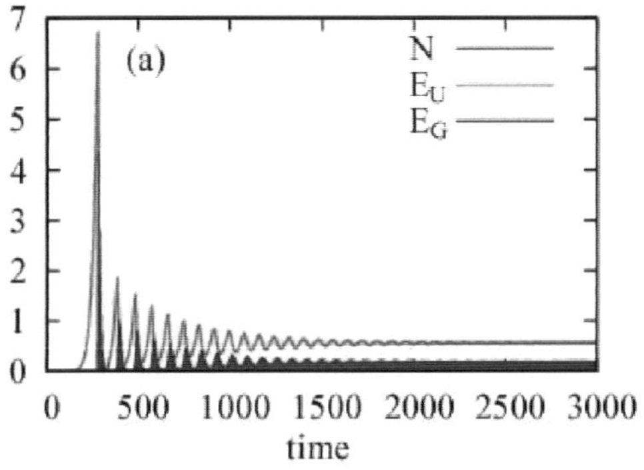

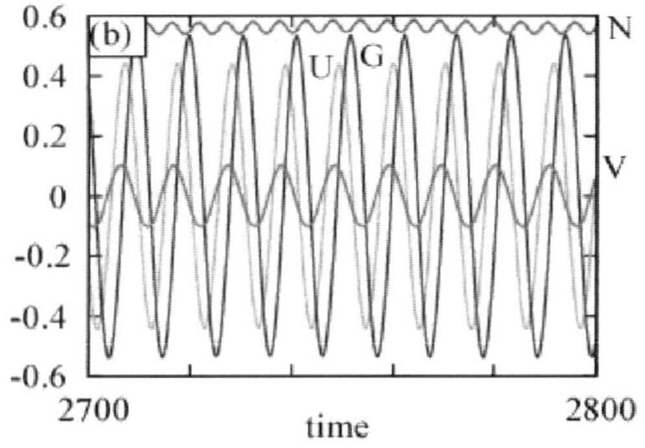

Figure 5. (a) Time evolution of N, E_U and E_G during the time $0 < t < 3000$ in case 3, (b) Time evolution of N, U, G, and V during the time $2700 < t < 2800$.

$\alpha_2 E_G$, inevitably suffer the collisionless damping even after the saturation. Due to this reason, the turbulence grows again after the quench in Case 1($\alpha_1 < \alpha_2$) leading to a multiple-burst intermittency since $\alpha_2 E_G$ is damped. On the other hand, in Case 2 ($\alpha_1 > \alpha_2$), the turbulence is suppressed mainly by the undamped zonal flow, i.e. $\alpha_1 E_U$ which does not suffer damping. Namely, in Case 2, once the quench of turbulence takes place, next burst does not happen. This feature is consistent to the analysis performed in section 2 (see equations (8)-(10)) that the residual components C_U and C_V do not depend on the initial condition G_0 which is equivalent to the energy inflow from the turbulence to p_{10} pressure perturbation through nonlinear coupling. This supports the observation that α_2 does not contribute the accumulation of undamped residues. Thus, it is found that the relative relation between α_1 and α_2 determine, "the quality of saturation" and subsequent dynamics.

3.3 Case 3: This case corresponds to the Case 1 for α_1 and α_2, but a negative value α_3 = -0.05 is chosen. In this case, the turbulence is not completely quenched, but approaches to a steady state level gradually as shown in figure 5(a). The other components such as U, G, and V show an oscillatory feature which corresponds to GAM oscillation. It is found that the phase shift between U and V is almost the same and that between U and G show the differsence nearly by $\pi/2$, which is consistent with the analytical result obtained in equations (8)-(10). Enough energy backflow from ion parallel sound velocity to turbulence can sustain steady state turbulent level against the suppression effect of zonal flow. This state will be investigated in more detail with a help of fixed point analysis in the next section.

4. Fixed point analysis

In this section, detailed properties of the extended predator-prey model are investigated by using fixed point analysis. A survey of the eigenvalues and corresponding eigenvectors around the fixed points gives comprehensive properties of solution trajectories in phase space.

Before starting the analysis, we introduce the following normalized variables and parameters:

$$(\alpha_1,\alpha_2,\alpha_3,\gamma_{damp},\gamma_{LD},\gamma_{NL}) \to (\alpha_1/\gamma_L,\alpha_2/\gamma_L,\alpha_3/\gamma_L,\gamma_{damp}/\gamma_L,\gamma_{LD}/\gamma_L,\gamma_{NL}/\gamma_L),$$
$$(\gamma_L t, N, U, \sqrt{\frac{\beta}{\beta'}}G, \sqrt{\frac{\beta\beta_2}{\beta'\beta_2'}}V) \to (t,N,U,G,V),$$
$$\omega_{GAM} = \sqrt{\beta\beta'} \to \omega_{GAM}/\gamma_L, \omega_{sound} = \sqrt{\beta_2\beta_2'} \to \omega_{sound}/\gamma_L. \quad (17)$$

Then, equations (11)-(14) are rewritten as

$$\frac{\partial N}{\partial t} = N - \gamma_{NL} N^2 - \alpha_1 U^2 N - \alpha_2 G^2 N - \alpha_3 V^2 N \,, \tag{18}$$

$$\frac{\partial U}{\partial t} = \frac{1}{2}\alpha_1 UN - \omega_{GAM} G - \frac{1}{2}\gamma_{damp} U \,, \tag{19}$$

$$\frac{\partial G}{\partial t} = \frac{1}{2}\alpha_2 GN + \omega_{GAM} U + \omega_{sound} V - \gamma_{LD} G \,, \tag{20}$$

$$\frac{\partial V}{\partial t} = \frac{1}{2}\alpha_3 VN - \omega_{sound} G \,, \tag{21}$$

Here we set $\gamma_{damp} = 0$ first for simplicity. In this model there are four types of fixed points distributed in the four-dimensional phase space given by (N, U, G, V). Those are (i) (0, 0, 0, 0), (ii) $(1/\gamma_{NL}, 0, 0, 0)$, (iii) $(0, \omega_{sound}\delta, 0, -\omega_{GAM}\delta)$, (iv) (n, u, g, v), where δ is arbitrary constant value. In (iv) n, u, g, v are nonzero constants and expressed as

$$n = \frac{\gamma_{LD}}{\alpha_2}\left[1 \pm \sqrt{1 - \frac{4\alpha_2}{\gamma_{LD}^2}\left(\frac{\omega_{GAM}^2}{\alpha_1} + \frac{\omega_{sound}^2}{\alpha_3}\right)}\right],$$

$$u^2 = \frac{2\omega_{GAM}^2 (1 - \gamma_{NL} n)}{\alpha_1^2 \gamma_{LD} n}, \quad g = \frac{\alpha_1 un}{2\omega_{GAM}}, \text{ and } \quad v = \frac{\alpha_1 \omega_{sound}}{\alpha_3 \omega_{GAM}} u \,. \tag{22}$$

Note that the fixed points in the case of (iv) only exist under conditions that n is a positive and also real value.

Next, we investigate the stability around each fixed-point. Small perturbations $\delta\vec{f} \equiv (\delta n, \delta u, \delta g, \delta v)^T$ are assumed around a fixed-point (n_0, u_0, g_0, v_0). Namely, substituting $(N, U, G, V) = (n_0+\delta n, u_0+\delta u, g_0+\delta g, v_0+\delta v)$ into equations (18)-(21) and linearizing around the fixed-point, we obtain the following equations:

$$\frac{\partial}{\partial t}\delta\vec{f} = \mathbf{M}\delta\vec{f} \,, \tag{23}$$

where $\mathbf{M}$ is a matrix defined by

$$\mathbf{M} = \begin{pmatrix} \Delta & -2\alpha_1 n_0 u_0 & -2\alpha_2 n_0 g_0 & -2\alpha_3 n_0 v_0 \\ \alpha_1 u_0/2 & (\alpha_1 n_0 - \gamma_{damp})/2 & -\omega_{GAM} & 0 \\ \alpha_2 g_0/2 & \omega_{GAM} & (\alpha_2 n_0/2) - \gamma_{LD} & \omega_{sound} \\ \alpha_3 v_0/2 & 0 & -\omega_{sound} & \alpha_3 n_0/2 \end{pmatrix}, \tag{24}$$

$$\text{with} \quad \Delta = 1 - 2\gamma_{NL} n_0 - \alpha_1 u_0^2 - \alpha_2 g_0^2 - \alpha_3 v_0^2 \,. \tag{25}$$

Assuming that the perturbations are expressed as $\delta\vec{f} \approx \delta\vec{f} e^{\lambda t}$, then, equation (23) can be written as an eigen-equation given by $(\lambda\mathbf{I} - \mathbf{M})\delta\vec{f} = 0$. Here $\mathbf{I}$ is the identity matrix. In order for all perturbations to have nontrivial solution, $|\lambda\mathbf{I} - \mathbf{M}| = 0$ must be satisfied. Calculating the eigenvalues and corresponding eigenvectors of the matrix $\mathbf{M}$ described by equation (24), we can investigate stability around the fixed point. If all eigenvalues have negative values in their real part, solution trajectories around the fixed-point direct inward, so that they are stable for any perturbation around the fixed point. On the other hand, if some eigenvalues are positive and others are negative or zero, solution trajectories direct outwards along the eigenvectors corresponding to the positive eigenvalues due to "saddle node" nature of the fixed point. In this case, solutions are stable only when the solution trajectories are along with stable eigenvectors. However, once unstable eigenvectors are included, the solution trajectories become unstable. In the following, we investigate the dynamics around each fixed point.

4.1. *Case with the fixed point (0, 0, 0, 0)*

Substituting $(n_0, u_0, g_0, v_0) = (0, 0, 0, 0)$ into equation (24), the eigenvalue λ of the matrix $\mathbf{M}$ can be calculated by solving the following eigen-equation;

$$|\lambda \mathbf{I} - \mathbf{M}| = \lambda(\lambda - 1)\left[\lambda^2 + \gamma_{LD}\lambda + (\omega_{GAM}^2 + \omega_{soumd}^2)\right] = 0. \tag{26}$$

Then, four eigenvalues are obtained as $\lambda = (0, 1, \omega_\pm)$, where $\omega_\pm$ are complex values associated with GAM frequency, which are expressed as $\omega_\pm = -\gamma_{LD}/2 \pm i\sqrt{(\omega_{GAM}^2 + \omega_{sound}^2) - (\gamma_{LD}/2)^2}$. It is assumed that $(\omega_{GAM}^2 + \omega_{sound}^2) - (\gamma_{LD}/2)^2 > 0$ is generally fulfilled. Corresponding eigenvectors ($\mathbf{u}_0$, $\mathbf{u}_1$, $\mathbf{u}_{\omega_\pm}$) are obtained as

$$\mathbf{u}_0 = \left(0, 1, 0, -\frac{\omega_{GAM}}{\omega_{sound}}\right), \qquad \mathbf{u}_1 = (1, 0, 0, 0),$$
$$\mathbf{u}_{\omega_\pm} = \left(0, 1, -\frac{-\gamma_{LD} \pm i\sqrt{4(\omega_{GAM}^2 + \omega_{sound}^2) - \gamma_{LD}^2}}{2\omega_{GAM}}, \frac{\omega_{sound}}{\omega_{GAM}}\right). \tag{27}$$

It is found that perturbation along with $\mathbf{u}_1$ is unstable and that along with $\mathbf{u}_{\omega\pm}$ is stable, so that a perturbation starting from (U, G) plane exhibits a spiral trajectory. This kind of fixed point is categorized in saddle focus. The solution trajectories are focused to the original point due to the GAM oscillation with collisionless damping, whereas the perturbation for turbulent fluctuation is advanced. As a result, the solution trajectories are driven in the direction of $\mathbf{u}_1$.

4.2. *Case with the fixed point ($1/\gamma_{NL}$, 0, 0, 0)*

For the fixed point $(n_0, u_0, g_0, v_0) = (1/\gamma_{NL}, 0, 0, 0)$, the following eigen-equation can be derived as

$$(\lambda + 1)\left[\left(\lambda - \frac{\alpha_1}{2\gamma_{NL}}\right)\left(\lambda - \frac{\alpha_2}{2\gamma_{NL}} + \gamma_{LD}\right)\left(\lambda - \frac{\alpha_3}{2\gamma_{NL}}\right) + \omega_{GAM}^2\left(\lambda - \frac{\alpha_3}{2\gamma_{NL}}\right) + \omega_{sound}^2\left(\lambda - \frac{\alpha_1}{2\gamma_{NL}}\right)\right] = 0. \tag{28}$$

Then the eigenvalues are obtained as $\lambda = (-1, \lambda_1, \lambda_2, \lambda_3)$, where λ_1, λ_2, and λ_3 are solutions of the cubic equation obtained from equation (28). The comprehensive property of the solution trajectories around the fixed point is understood as a saddle node nature because at least one root is found to be positive (unstable) as long as the condition $\gamma_{LD} < (\alpha_2/2)/\gamma_{NL} + 2\gamma_{NL}(\omega^2_{GAM}/\alpha_1 + \omega^2_{sound}/\alpha_3)$ is satisfied, which means that the stationary level of turbulence energy dominated by the collisionless damping, i.e. $N_0 = (2\gamma_{LD}/\alpha_2)$, is assumed to be lower than that dominated by the nonlinear damping of turbulence, i.e. $N_{NL} = (1/\gamma_{NL})$. Note that N_0 is obtained from equation (30). Therefore it is found in this case that a small perturbation for (U, G, V) around the fixed point exponentially increases, leading the solution trajectories to diverging. Note that this fixed point is not essential as long as the level of turbulence. $1/\gamma_{NL}$, is higher than that around other fixed points.

4.3. *Case with the fixed points (0, $\omega_{sound}\delta$, 0, -$\omega_{GAM}\delta$)*

Here we set the level of stationary component of zonal flow and ion sound parallel velocity by using the parameter δ. Note that any δ satisfies the condition of fixed points, hence the group of the fixed point forms a line in the four-dimensional phase space labeled by δ. This suggests that any states parameterized by δ where stationary zonal flow balances with ion sound parallel flow without p_{10} pressure perturbation G can exist. Substituting (n_0, u_0, g_0, v_0) = (0, $\omega_{\rm sound}\delta$, 0, -$\omega_{\rm GAM}\delta$) into equation (24), we obtain the eigenvalues $\lambda = (0, \Delta, \omega_\pm)$, where the definitions of Δ and $\omega_\pm$ are the same as those described in the subsection 4.1. The corresponding eigenvectors are solved as follows;

$$\mathbf{u}_0 = \left(0,1,0,-\frac{\omega_{GAM}}{\omega_{sound}}\right), \qquad \mathbf{u}_{\omega_\pm} = \left(0,1,-\frac{-\gamma_{LD} \pm i\sqrt{4(\omega_{GAM}^2+\omega_{sound}^2)-\gamma_{LD}^2}}{2\omega_{GAM}}, \frac{\omega_{sound}}{\omega_{GAM}}\right)$$

$$\mathbf{u}_\Delta = \left(1, \frac{\delta\omega_{sound}(\alpha_1\omega_{sound}^2+\alpha_1\Delta\gamma_{LD}+\alpha_3\omega_{GAM}^2+\alpha_1\Delta^2)}{2\Delta[\Delta^2+\gamma_{LD}\Delta+(\omega_{GAM}^2+\omega_{sound}^2)]}, -\frac{(\alpha_3-\alpha_1)\delta\omega_{sound}\omega_{GAM}}{2[\Delta^2+\gamma_{LD}\Delta+(\omega_{GAM}^2+\omega_{sound}^2)]}, -\frac{\delta\omega_{GAM}(\alpha_3\omega_{GAM}^2+\alpha_3\Delta\gamma_{LD}+\alpha_1\omega_{sound}^2+\alpha_3\Delta^2)}{2\Delta[\Delta^2+\gamma_{LD}\Delta+(\omega_{GAM}^2+\omega_{sound}^2)]}\right) \tag{29}$$

Now, we analyse the characteristics of eigen-values and eigen-vectors. One of four eigen-values, $\lambda = \Delta$ (eigen-state (I)), is unstable when $\Delta > 0$ is satisfied. Note from equation (25) that Δ is estimated as $\Delta \sim \gamma_L$ corresponding to the linear growth of turbulence as long as the stationary components of (U, V, G) are small enough. Two eigen-values given by $\lambda = \omega_\pm$ (eigen-state (II)) manifest stable GAM damping oscillation with the damping rate given by γ_{LD}. This is also obvious from the corresponding eigen-vectors $\mathbf{u}_{\omega_\pm}$ where zonal flow, anisotropic pressure and parallel ion sound perturbation are incorporated. The other one, i.e. $\lambda = 0$ (eigen-state (III)), is marginal. As is found from the eigen-vector $\mathbf{u}_0$, this state is identical to the fixed point with residual stationary zonal flow characterized by the parameter δ. Note that the time scale of eigen-state (I) is shorter than that of the eigen-state (II) by γ_{LD}/γ_L, whereas the time scale of the eigen-state (III) is much longer than that of (I) and (II) due to its marginal nature.

From these discussions, it is found that each eigen-state corresponds to the each dynamics observed during growing intermittency in figure 1, i.e. trigger of burst (→ state (I)), associated GAM damping (→ state (II)), and accumulation of residual zonal flow (→ state (III)). Note that the dynamics is moreprecisely represented by linear combination of those eigen-states, $\sum_{j=1}^4 c_j \mathbf{u}_j \exp(\lambda_j)$ with a proper choice of coefficient c_j. In order to capture the transition dynamics between them, it is useful to investigate the solution trajectories in four-dimensional phase space formed by eigen-vectors as schematically shown in figure 6. Note here that a two-dimensional plane, which is formed by $\mathbf{u}_{\omega_\pm}$ is orthogonal to $\mathbf{u}_0$ due to $\mathbf{u}_{\omega_\pm}\cdot\mathbf{u}_0 = 0$ in equation (29), whereas $\mathbf{u}_\Delta$ is generally orthogonal neither to $\mathbf{u}_0$ nor to $\mathbf{u}_{\omega_\pm}$, i.e. $\mathbf{u}_0\cdot\mathbf{u}_\Delta \neq 0$ and $\mathbf{u}_{\omega\pm}\cdot\mathbf{u}_\Delta \neq 0$. Here, we consider a fixed point (A) labeled by the parameter δ_1 or corresponding plane orthogonal to the vector $\mathbf{u}_0$. A trigger dynamics, i.e. the eigen-state (I), initiated from the point (A) is shown in figure 6 as a movement of the point along with the direction of $\mathbf{u}_\Delta$ from (A) to the point (B), which places on the plane labeled by δ_2. Note that the point (B) is away from the fixed point (C) on the axis along with the vector $\mathbf{u}_0$, so that it is subjected to the oscillation damping toward the point (C). The difference of δ during the movement from (A) to (C), i.e. $\Delta\delta = \delta_2 - \delta_1$, corresponds to the residual increase of stationary zonal flow balanced with ion parallel velocity. These dynamics also correspond to the behavior of energies in the Landau-fluid simulations as labeled by (A), (B) and (C) in figure 1(b), and the difference of δ in figure 6 can be comparable to the accumulation of residual zonal flow energy $\delta E_{ZF}^{(j)}$ in figure 1(b). Note that the increase of δ during the bursting process results from the non-orthogonal property $\mathbf{u}_0\cdot\mathbf{u}_\Delta \neq 0$. The recursion of the burst and associated accumulation of the residual zonal flow results from the repetition of this process. We can also predict the termination of growing intermittency and associated maximum amplitude of zonal flow. As the burst is repeated, the value of δ increases and once δ surpasses $\delta_{crit} = \sqrt{1/(\alpha_1\omega_{sound}^2 + \alpha_3\omega_{GAM}^2)}$, which is estimated from $\Delta = 0$ in equation (25), Δ becomes negative. Then, the eigen-state (I) becomes stable so that the bursting nature is diminished.

Note that such a bursting process occurs only when the condition $\mathbf{u}_0 \cdot \mathbf{u}_\Delta > 0$ is fulfilled. Namely, in case of $\mathbf{u}_0 \cdot \mathbf{u}_\Delta < 0$, the bursting event decreases δ, so that the accumulation does not happen. Then, from the condition $\mathbf{u}_0 \cdot \mathbf{u}_\Delta > 0$, the growing intermittency associate with GAM is realized for $0 < \alpha_1\omega_{sound}^2 + \alpha_3\omega_{GAM}^2 < 1$. Note that the condition $\alpha_1\omega_{sound}^2 + \alpha_3\omega_{GAM}^2 > 1$ leads to $\Delta < 0$. Furthermore, an interesting feature is observed in the case of $\alpha_1\omega_{sound}^2 + \alpha_3\omega_{GAM}^2 < 0$. That is, if $\alpha_3 < \alpha_{3,crit}$ $[\equiv -\alpha_1(\omega_{sound}^2/\omega_{GAM}^2)]$ is satisfied, the suppression role of zonal flow in turbulence is always cancelled by the backflow due to the ion parallel flow for any value of δ as found from equation (18). In this case, instead of the accumulation of residual zonal flow, a steady state where turbulence coexists with zonal flow and GAM is realized as seen in figure 5.

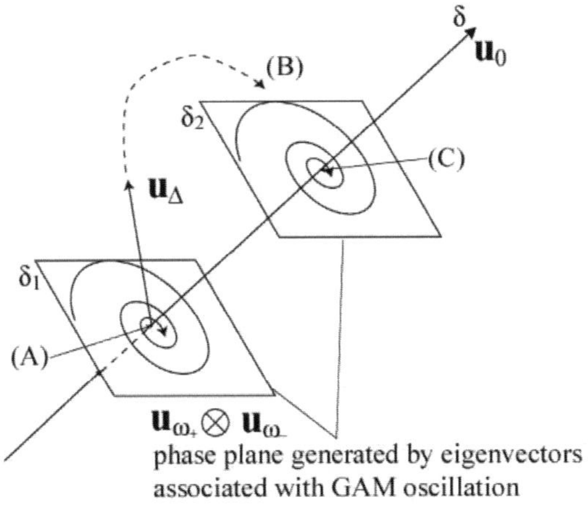

Figure 6. Illustration of the solution trajectory around $(n, u, g, v) = (0, \omega_{sound}\delta, 0, -\omega_{GAM}\delta)$. The GAM intermittency is identified by suction by GAM damping, blowing off of turbulent fluctuation, and subsequent suction by another GAM phase plane around new fixed point.

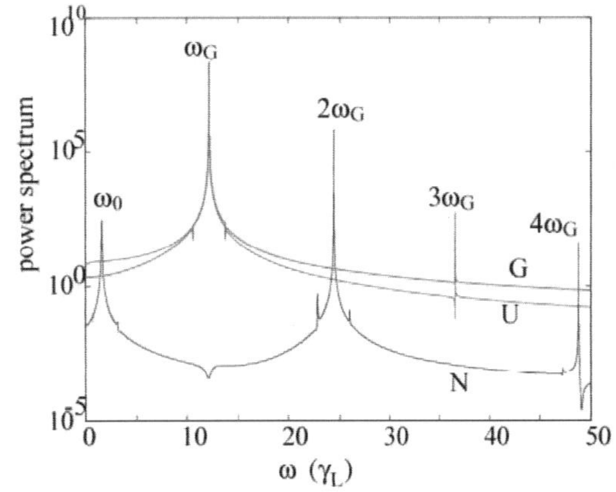

Figure 7. Power spectra of *N*, *U*, and *G* in the steady-state phase in the case of $\alpha_3 < \alpha_{3crit}$.

4.4. Case with the nonzero fixed point described in equation (22)

The eigenvalues around the fixed point (*n, u, g, v*) described by equation (22) are numerically investigated. As a reference case, the normalized parameters such as $\alpha_1 = 1.0$, $\alpha_2 = 10$, $\alpha_3 = -0.1$ are used in accordance with equation (17), so that $\omega_{GAM} = 11.55$, and $\omega_{sound} = 3.89$. In this case the fixed point is numerically estimated as $(n, u, g, v) = (2.94, 5.44, 0.691, -18.3)$ and the corresponding eigenvalues are solved as $\lambda = 4.82 \pm 18.5i$, 3.67 and -7.14×10^{-3}. Three unstable eigenvalues and one weakly stable eigen-value are obtained here. It indicates that the solution trajectories are unstable for the fixed point as a saddle node. Even in the further search of eigenvalues for variable α_3, the property of solution trajectory around this type of fixed point is found to have at least two unstable roots in four solutions. This type of fixed points is distributed far from realistic region, comparing with the other cases. Then, dynamics around the nonzero fixed point may have little effect on the GAM intermittent dynamics.

4.5. *Steady state observed when $\alpha_3 < 0$*

It is noted that the steady state solution observed in the case of $\alpha_3 < \alpha_{3crit}$ in subsection 4.3 or in figure 5 is different from those around the fix point discussed in subsection 4.4. This behavior is not categorized as the previous case such as the motion around some fixed points. In this case the turbulence N is sustained in cooperation with U, G, and V. Figure 7 shows the power spectra of N, U, and G in the case of α_1 = 1.0, α_2 = 10, α_3 = -0.2, where a steady state of turbulence and GAM is established as figure 5(b). Several peaks are observed at the characteristic GAM with frequency $\omega_G \equiv \sqrt{\omega_{sound}^2 + \omega_{GAM}^2}$. The power spectra of U, G, and V are characterized by odd harmonics of the GAM frequency, whereas that of N is characterized by the even harmonics. In the spectrum of N another peak characterized by the intermittency is also observed at $\omega_0 \approx 1.57\gamma_L$, together with the side bands corresponding to $2\omega_G \pm \omega_0$.

We can estimate the level of the steady state. In terms of the following expansions,

$$U = \sum_{n=0}^{\infty} U_n e^{i(2n+1)\omega_G t}, \quad G = \sum_{n=0}^{\infty} G_n e^{i(2n+1)\omega_G t}, \quad V = \sum_{n=0}^{\infty} V_n e^{i(2n+1)\omega_G t}, \quad N = \sum_{n=0}^{\infty} N_n e^{i2n\omega_G t},$$

and orderings $f_n \sim O(\varepsilon^n)$ for $f = (N,U,G,V)$, $\alpha_1,\alpha_3 \sim O(\varepsilon)$, $\alpha_2 \sim O(1)$, $\omega_G,\omega_{GAM},\omega_{sound},\gamma_{LD} \sim O(1)$, we obtain the following results by taking account the lowest order of equations (18)-(21),

$$N_0 = 2\gamma_{LD}/\alpha_2, \quad U_0 = i(\omega_{GAM}/\omega_G)G_0, \quad G_0 = \sqrt{2/\alpha_2}, \quad V_0 = i(\omega_{sound}/\omega_G)G_0. \tag{30}$$

The magnitudes of N_0, U_0, G_0, and V_0 are estimated from equation (30) as $N_0 \sim 0.5$, $G_0 \sim 0.45$, $U_0 \sim 0.42$, and $V_0 \sim 0.14$, respectively. These are about 10% difference from those in figure 5(b). The differences are within the ordering parameter, $\varepsilon \sim \alpha_1/\alpha_2 \sim 0.1$. The phase lag between U and V is about zero, whereas those between U and G, and G and V are about $\pi/2$, which are consistent with the results of linear response of the GAM oscillator, i.e. equations (8)-(10). This is why we have concluded that as the lowest order approximation these results are consistent with those obtained in the numerical calculation.

5. Conclusion.

We have proposed a four-field minimum model that describes the growing intermittency of turbulence associated with the GAM observed in our toroidal Landau-fluid simulations. The model can reproduce the different nature of turbulent transport by controlling the coefficients of nonlinear coupling terms. Energy inflow due to the nonlinear coupling between turbulence and zonal flows, i.e. Reynolds stress, can excite GAMs and simultaneously generate residual zonal flows balancing with ion sound parallel velocity. On the other hand, the flow due to nonlinear coupling between turbulence and p_{10} pressure perturbation excites GAMs but generates no residual zonal flows. Multiple/single-burst nature in the GAM intermittency originates from qualitatively difference between the two nonlinear coupling terms. Energy backflow due to the nonlinear coupling between turbulence and ion sound parallel velocity cancels the effect of Reynolds stress. When the backflow always overwhelms the inflow due to the Reynolds stress, steady state of turbulence appears.

We have also investigated the comprehensive properties of solution trajectories around the fixed points incorporated with eigen-value and corresponding eigen-vector. It is found that the coefficient of the nonlinear coupling term between turbulence and ion sound velocity can make a bifurcation between turbulence quench due to the accumulation of zonal flow and steady state of turbulence. This result may also predict the bifurcation of states such as Dimits shift regime and steady state regime above the critical gradient due to the change of governing parameters.

A predator-prey model we presented reproduces the Dimits shift region where turbulence is quenched through the intermittent bursts or single one. The model also reproduces a quasi-linear state turbulent transport incorporated with zonal flows. Those two different states are obtained by changing the ratio and/or the sign of (α_1, α_2, α_3) in equations (11)-(14). However, it is not straightforward to explicitly relate those parameters to that of the instability strength such as η_i and/or R/L_T, which is out of scope from our present study. In order to determine the shift quantity or to determine the critical gradient, we need to further improve the model, e.g. by coupling with transport equation [18]. Intermittent transport phenomena which are related to GAM dynamics are observed in experiments [14,15]. Comparison between the growing intermittency observed in our simulations and those experiments is our future work.

Acknowledgements

This work is supported by the Grant-in Aid from Japan Society for the Promotion of Science (No. 18340186 and No. 19560828) and National Natural Science Foundation of China (NSFC) Grant No. 10575032. This work is also supported by NIFS/NINS under the project of Formation of International Network for Scientific Collaborations.

Appendix A: Approximate treatment of the off-diagonal terms in four-field minimal model

In this appendix, we describe that as a minimal analytical modelling how we can effectively involve the impacts of nonlinear cross-coupling (i.e., off-diagonal terms) in the diagonal parts of Equations (12)-(14) for simplicity. The nonlinear terms in equations (12)-(14) originate from the convective nonlinearities such as the terms $\hat{N}_{[\phi,p]}$, $\hat{N}_{[\phi,\upsilon]}$ and $\hat{N}_{[\phi,\phi]}$. They can be expressed as diagonal and off-diagonal parts based on a quasi-linear description of modulational coupling loop. Taking the GAM equation (13) as an example, the nonlinear coupling term $\hat{N}_{[\phi,p]}$ can be decomposed into four pairs of couplings

$$\hat{N}_{[\phi,p]} \sim \left\langle\left[\tilde{\phi},\tilde{p}\right]sin\theta\right\rangle = \left\langle\left[(\tilde{\phi}^m_{SB})^*,\tilde{p}^{m\pm1}_p\right]sin\theta\right\rangle + \left\langle\left[(\tilde{\phi}^m_p)^*,\tilde{p}^{m\pm1}_{SB}\right]sin\theta\right\rangle$$
$$+\left\langle\left[(\tilde{\phi}^{m\pm1}_{SB})^*,\tilde{p}^m_p\right]sin\theta\right\rangle + \left\langle\left[(\tilde{\phi}^{m\pm1}_p)^*,\tilde{p}^m_{SB}\right]sin\theta\right\rangle \quad \text{(A1)}$$

Here the subscripts "p" and "SB" denote the pumps and sidebands in k_r space. The four terms at RHS correspond to $\Gamma_1 \propto (\tilde{\phi}^m_{SB})^* \tilde{p}^{m\pm1}_p$, $\tilde{A}_2 \propto (\tilde{\phi}^m_p)^* \tilde{p}^{m\pm1}_{SB}$, $\Gamma_3 \propto (\tilde{\phi}^{m\pm1}_{SB})^* \tilde{p}^m_p$ and $\Gamma_4 \propto (\tilde{\phi}^{m\pm1}_p)^* \tilde{p}^m_{SB}$, respectively. Note that the pumps $\tilde{\phi}_p$ and $\tilde{p}_p$ are primary toroidal ITG modes, which satisfy the linear ITG dispersion relation. Based on the nonlinear coupling, the sidebands can be expressed as

$$\tilde{\phi}^m_{SB} \propto \tilde{\phi}^m_p \tilde{\phi}_q \,, \quad \text{(A2)}$$

$$\tilde{\phi}^{m\pm1}_{SB} \propto \tilde{\phi}^{m\pm1}_p \tilde{\phi}_q \,, \quad \text{(A3)}$$

$$\tilde{p}^m_{SB} = \left[\tilde{\phi}_q{}^*,\tilde{p}^m_p\right] + \left[(\tilde{\phi}^{m\mp1}_p)^*,\tilde{p}_{\pm1}\right] \propto c^m_1 \tilde{\phi}^*_q \tilde{p}^m_p + c^m_2 (\tilde{\phi}^{m\mp1}_p)^* \tilde{p}_{\pm1} \,, \quad \text{(A4)}$$

$$\tilde{p}^{m\pm1}_{SB} = \left[\tilde{\phi}_q{}^*,\tilde{p}^{m\pm1}_p\right] + \left[(\tilde{\phi}^m_p)^*,\tilde{p}_{\pm1}\right] \propto c^{m\pm1}_1 \tilde{\phi}^*_q \tilde{p}^{m\pm1}_p + c^{m\pm1}_2 (\tilde{\phi}^m_p)^* \tilde{p}_{\pm1} \,. \quad \text{(A5)}$$

So substituting these relations into equation (A1), we have

$$\left\langle\left[\tilde{\phi},\tilde{p}\right]sin\theta\right\rangle = \Gamma_1 + \Gamma_2 + \Gamma_3 + \Gamma_4 \propto (\tilde{\phi}^{m\pm1}_p)^2 \tilde{p}_{\pm1} + \left[(\tilde{\phi}^m_p)^* \tilde{p}^{m\pm1}_p + (\tilde{\phi}^{m\pm1}_p)^* \tilde{p}^m_p \right] \tilde{\phi}_q \,. \quad \text{(A6)}$$

These two terms correspond to the diagonal (~*NG*) and off-diagonal (~*NU*) parts. Considering that the GAM is an oscillatory eigenmode, the corresponding electrostatic potential, i.e., the zonal flow $\tilde{\phi}_q$, includes two components with an oscillatory and a zero-frequency. In the case without GAMs, the nonlinear driving of the zonal flows due to the gradient of Reynolds stress is balanced by the collisional damping in the quasi-steady. When it couples with the GAM, it becomes to oscillate approximately with GAM frequency. Hence, in equation (12) for the zonal flow, although the nonlinear driving may have some influence, the coupling with GAMs is dominated by the linear GAM oscillator with spectral and phase's matching. Based on this idea, we represent the relation between the zonal flows and the GAM dynamics as $\tilde{\phi}_q \propto \tilde{p}_{\pm 1}$ within an approximation of the lowest order in an analytical minimal model. Under this assumption, we effectively absorb the off-diagonal term into the diagonal part with a modified coefficient α_i.

Interestingly, some efforts have been also made in a recent work [19] to estimate the contribution of the off-diagonal terms to the GAM dynamics based on the ballooning character of ITG modes in order to resolve the complexity due to the off-diagonal effects in a simplified modeling. Further improvement of the approximation in an advanced modelling is in progress.

References

[1] Diamond P H, Itoh S-I, Itoh K and Hahm T S 2005 *Plasma Phys. Control. Fusion* **47** R35
[2] Gebhardt W and Krey U 1980 *Phase Transitions and Critical Phenomena* (Yoshikawa Shoten, 1992), translated by S Komura; German edition (Friedr, 1980)
[3] Dimits A M *et al.* 2000 *Phys. Plasmas* **7** 969
[4] Lin Z, Hahm T S, Lee W W, Tang W M and Diamond P H 1999 *Phys. Rev. Lett.* **83** 3645
[5] Malkov M A, Diamond P H and Rosenbluth M N 2001 *Phys. Plasmas* **8** 5073
[6] Kolesnikov R A and Krommes J A 2005 *Phys. Rev. Lett.* **94** 235002
[7] Winsor N, Johnson J L and Dawson J M 1968 *Phys. Fluids* **11** 2448
[8] Hallatschek K and Biskamp D 2001 *Phys. Rev. Lett.* **86** 1223
[9] Scott B 2003 *Phys. Lett. A* **320** 53
[10] Miyato N, Kishimoto Y and Li J Q 2004 *Phys. Plasmas* **11** 5557
[11] McKee G R *et al.* 2003 *Phys. Plasmas* **10** 1712
[12] Miki K, Kishimoto Y, Miyato N and Li J Q 2007 *Phys. Rev. Lett.* **99** 145003
[13] Conway G D *et al.* 2005 *Plasma Phys. Control. Fusion* **47** 1165
[14] Melnikov A V *et al.* 2006 *Plasma Phys. Control. Fusion* **48** S87
[15] Ido T *et al.* 2006 *Nucl. Fusion* **46** 512
[16] Hirsch M W and Smale S 1974 *Differential equations, dynamical systems, and linear algebra* (New York: Academic Press).
[17] Itoh K, Hallatschek K and Itoh S-I 2005 *Plasma Phys. Control Fusion* **47** 451
[18] Kishimoto Y, Tajima T, Horton W, LeBrun M J and Kim J Y 1996 *Phys. Plasmas* **3** 1289
[19] Chakrabarti N, Singh R, Kaw P K and Guzdar P N 2007 *Phys. Plasmas* **14** 052308

Journal of Physics: Conference Series **123** (2008) 012029 doi:10.1088/1742-6596/123/1/012029

Three-dimensional neutral transport simulation in the JT-60U H-mode plasmas

Y Nakashima[1], Y Higashizono[1], H Kawano[1], H Takenaga[2], N Asakura[2], N Oyama[2], Y Kamada[2] and K Yatsu[1]

[1] Plasma Research Center, University of Tsukuba, Tsukuba 305-8577, Japan
[2] Japan Atomic Energy Agency, Naka Fusion Institute, Naka 311-0193, Japan

E-mail: nakashma@prc.tsukuba.ac.jp

Abstract. The first results of a fully three-dimensional (3-D) Monte-Carlo simulation and analysis of neutral behavior in H-mode plasmas of the JT-60U tokamak are described. In the DEGAS Monte-Carlo code, a precise mesh structure including isolated dome, baffle and divertor plates is built-up and the simulation space is extended from SOL/divertor to core plasma region. Benchmarking with 2-dimensional simulation with the DEGAS2 code successfully showed the validity of the present simulation. Result of 3-D simulation in the JT-60U plasma with gas puffing indicates the significant localization of neutrals around the gas puffer and the detail of neutral diffusion in toroidal and poloidal directions are clarified. The dependence of plasma density on the neutral penetration and ionization area are also discussed.

1. Introduction

Simulation analysis of neutral particle transport is an important issue to investigate hydrogen/deuterium recycling and transport phenomena in not only the plasma edge regions but also the core plasma region of magnetically confined plasmas. Neutral particles in periphery plasmas also play a crucial role on pedestal formation during transport barrier generation in high temperature plasmas. Neutral particle transport simulations based on Monte-Carlo methods have been widely used as a standard way to investigate neutral behavior in fusion devices. A number of neutral transport codes have been extensively utilized for modelling in divertor plasmas [1-3]. On the other hand, three-dimensional (3-D) Monte-Carlo analysis of neutral transport is an important subject apart from the modelling assuming the toroidal symmetry, such as divertor simulation, in order to evaluate the effect of the localized particle source and complicated structure of vacuum chamber wall and related components. In fusion devices with 3-D geometrical configuration, such as helical systems and non-axisymmetric mirror systems with minimum B configuration, a fully 3-D simulation is indispensable. Even in tokamak devices containing toroidal symmetry, the 3-D simulation is a powerful tool for evaluating the effect of inner-components such as antennas and the local particle source from gas puff and neutral beam injection. The DEGAS ver.63 Monte-Carlo code has been modified for fully three-dimensional simulation in the University of Tsukuba [4, 5] and has been applied to a tandem mirror device including non-axisymmetric cell and helical devices with non-axisymmetric magnetic configurations [6, 7]. Recently a simulation study on fully 3-D neutral transport in the JT-60U tokamak was started using DEGAS Monte-Carlo code in order to investigate detailed behavior of high performance H-mode plasmas in JT-60U.

Journal of Physics: Conference Series **123** (2008) 012029 doi:10.1088/1742-6596/123/1/012029

It is also an important issue to investigate the role of neutrals on the edge pedestal scaling in H-mode plasmas. So far a number of analytical and simulation studies based on the experimental data base has been extensively performed in many kinds of fusion devices and characterization of edge transport barrier in H-mode phase was investigated [8-12]. In this paper, the results of a fully 3-D Monte-Carlo simulation using DEGAS code applied to the JT-60U tokamak for the first time and analysis of neutral behavior in H-mode plasmas are described. Detailed 3-D simulation is carried out under the condition of local gas puffing and the behavior of the particle diffusion in both poloidal and toroidal directions are precisely investigated. The characteristics of neutral penetration and ionization profile are also evaluated in ELMy H-mode discharges. In Section 2, a mesh structure used in the 3-D simulation is explained and a result of the benchmark test with the DEGAS2 code is presented in Section 3. Results of 3-D simulation are shown and the localization of neutrals in poloidal and toroidal directions are discussed in Section 4. Comparison of neutral behavior between two H-mode plasmas is described from the viewpoint of neutral penetration and ionization region in Section 5 and a summary is presented in Section 6.

2. Mesh model of JT-60U for DEGAS simulation

Figure 1 shows a comparison between the poloidal cross-section views of the usual mesh model and the newly designed one for JT-60U. In the left side, the usual mesh model prepared for the DEGAS2 code [3] is presented and its consistency with the mesh model of the UEDGE plasma code [13] has been ensured. The present mesh model shown in the right side is developed based on the previous one and is extended to the plasma core, the vacuum chamber wall and the divertor regions. The mesh is divided into 32 segments in the minor-radius direction and also divided into 28 in the azimuthal direction on the poloidal cross-section. 50 segments divided in the toroidal direction enable a detailed prediction of neutral density and Dα line-emission intensity and modelling of localized particle source, such as gas puff and neutral beam injection.

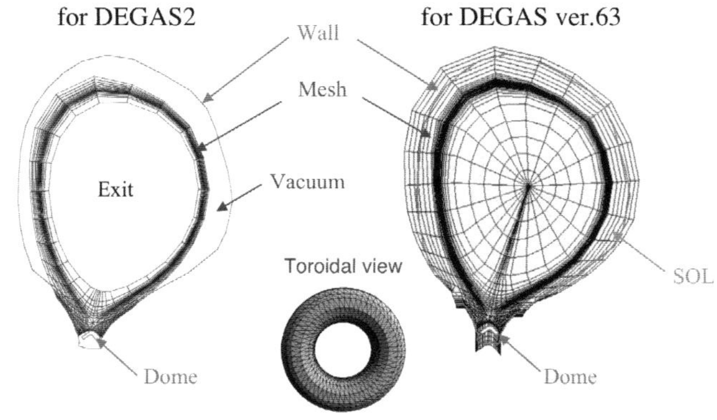

Figure 1. Mesh model of JT-60U for the DEGAS simulation. Left: usual model used in DEGAS2, Right: present model used in DEGAS ver.63.

3. Benchmark with a 2-dimensional simulation using DEGAS2

Firstly a benchmark test with 2-dimensional simulation code was carried out after building up the 3-D mesh model. Figure 2 shows the sightlines of arrays of Dα line-emission detectors installed in JT-60U together with the mesh structure on the poloidal cross-section used for the present simulation. In JT-60U, as shown in the figure, Dα line-emission detector system consists of three detector arrays. All the detector arrays are installed at the upper view port of the vacuum chamber

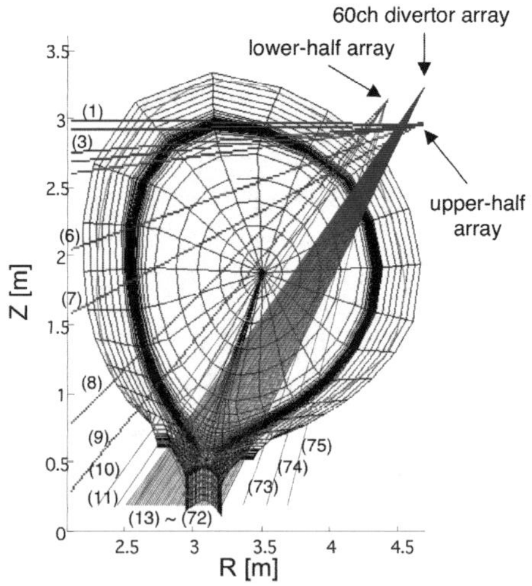

Figure 2. JT-60U mesh model and sight line of the Dα detector array.

and are located more than 2m away from the gas puffer. In this simulation background plasma parameters and the intensity of particle flux on the divertor plates are determined from the UEDGE code.

In figure 3, the profile of Dα line intensity is compared with the simulation results. All the data are plotted as a function of channel number of detectors. Triangles in the figure represent the results of 2-D simulation by use of DEGAS2 which is reproducing the experimental data (cross) by artificially giving the recycling particle flux on the first wall [14]. In the DEGAS2 simulation, a particle source from the divertor plates is dominant (9.8×10^{22} s^{-1}) compared with the others (wall: 7.8×10^{21} s^{-1}, volume source: 2.8×10^{22} s^{-1}). Circles show the result of the present 3-D simulation under the same boundary condition of the particle source flux as those in DEGAS2. The 3-D simulation result is averaged over toroidal direction and compared with the DEGAS2 result. As shown in the figure, a fairly good agreement is confirmed with that of DEGAS2, which proves the validity of this 3-D simulation. However, a noticeable discrepancy is found near the outer strike point and the private region. This seems to be caused by the small difference of mesh structures between the present version and DEGAS2.

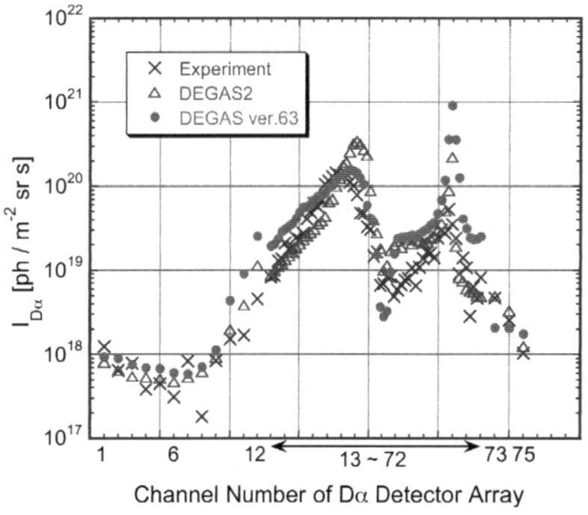

Figure 3. Comparison of the profile of Dα line intensity on the JT-60U poloidal cross-section. cross: experiment, triangle: simulation with DEGAS2, circle: present simulation.

4. Result of 3-dimensional simulation

4.1. Result of gas puff simulation

The first trial in fully 3-D simulation was performed in the gas puff simulation. In figure 4, neutral particle density profiles on the several poloidal cross-sections are shown in the case of D_2 gas puff

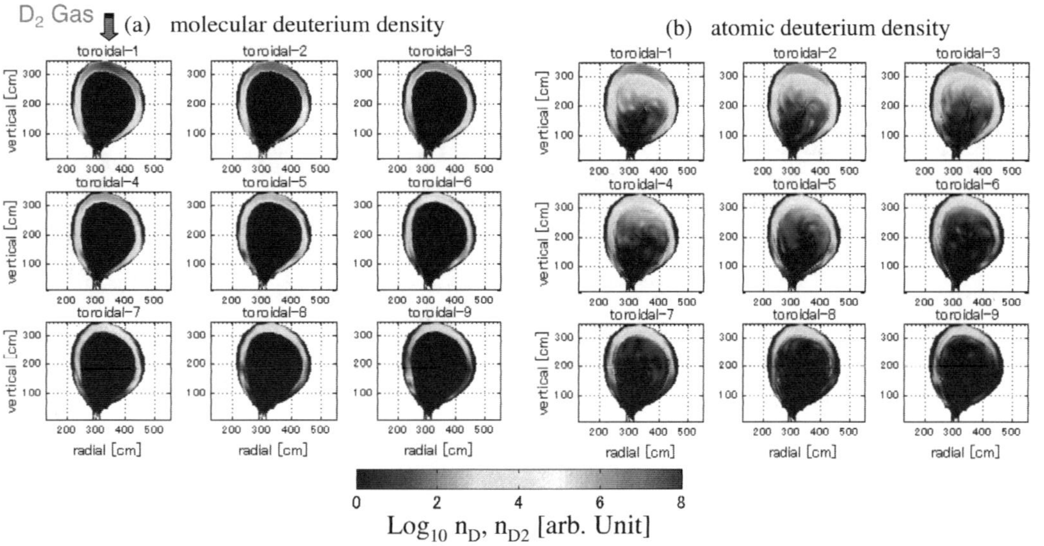

Figure 4. Results of the 3D-DEGAS simulation in the case that the gas puffing from the top of the vacuum chamber. (a) molecular deuterium density, (b) atomic deuterium density.

from the top of the JT-60U vacuum chamber. In this simulation, neither first-wall recycling source nor divertor source is given in order to highlight the effect of particle diffusion from the local gas puff. As shown in figure 4(a), deuterium molecules do not penetrate into the plasma core and is observed to poloidally diffuse along the SOL to the X-point at the bottom of the chamber attenuated by more than four order of magnitude. In the toroidal direction, the decay length of molecular density is much longer than that in the poloidal one and the reduction of the density by only three order of magnitude is estimated at the location about 210 cm (toroidal-6) apart from the gas puffer. On the other hand, deeper penetration of atomic deuterium is observed than the case of molecules. (at most two order of magnitude) However, there is no significant difference recognized between atoms and molecules in the poloidal direction.

Figure 5 shows the bird-eye view showing the Dα emission profiles in the poloidal plane, which are toroidally aligned, based on the atomic and molecular deuterium particle density. From this figure, it is recognized that the Dα emission area due to D_2 gas puff is limited within one fifth of full torus. From the above results, the influence of gas puff is estimated to be small at the Dα detector arrays. Therefore the first result of full 3-D neutral transport simulation indicates the significant localization of neutrals in the vicinity of the gas puff exit.

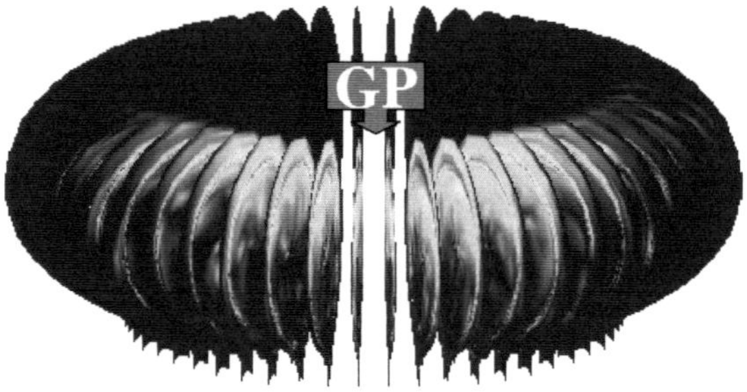

Figure 5. Bird-eye-view of the Dα emissivity profiles on the poloidal cross-section which are aligned along the toroidal direction.

4.2. Effect of SOL plasma

In order to evaluate the effect of SOL plasma on the diffusion process of neutrals in 3-D geometry, a simulation assuming a fairly high density in SOL region was carried out. In the simulation shown in figure 4, the density of the outer SOL region (from the end position of UEDGE calculation to the first

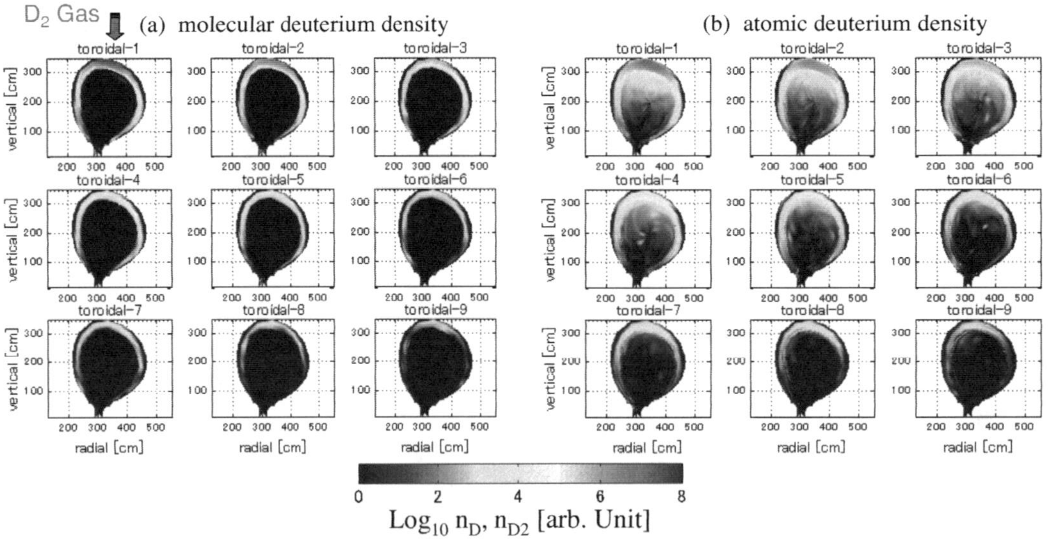

Figure 6. Results of the 3D-DEGAS simulation in the case that the gas puffing from the top of the vacuum chamber. The electron density in the outer SOL region is assumed to be 1×10^{12} cm^{-3}. (a) molecular deuterium density, (b) atomic deuterium density.

wall; $r/a \geq 1.05$ as shown in figure 8(a)) is assumed to be 1×10^{11} cm^{-3}. Figure 6 shows the simulation results of poloidal neutral density profiles executed under the outer SOL density of 1×10^{12} cm^{-3}. In this case the density is kept constant in this value from the wall to the location of $r/a \approx 1.02$. The same input values as the previous simulation are given in the other boundary conditions. From figure 6(a), it is found that the increase of SOL density prevents the diffusion of molecules in both toroidal and poloidal directions comparing with figure 4(a). However, no significant change was observed in atomic density profiles as shown in the comparison between the results of figures 4(b) and 6(b). These results can be explained by the following reason: Deuterium molecules are mainly transported to the toroidal and poloidal direction through the outer SOL region. Deuterium atoms, on the other hand, which mainly consists of Franck-Condon neutrals are produced in the vicinity of LCFS area where the plasma density is higher than that in the outer SOL region, so that the atoms are much more affected in the periphery of the core plasma than the SOL plasma.

5. Comparison of neutral behavior in H-mode plasmas

In order to investigate the plasma parameter dependence on the pedestal structure, neutral transport simulation using DEGAS was performed in two different ELMy H-mode plasmas, and the results were compared from the viewpoint of neutral penetration and ionization region. Figure 7 shows the time behavior of plasma parameters sampled from two H-mode plasma discharges. In the case with gas puffing in H-mode phase (E037854), relatively higher electron density is achieved than the case without gas puffing (E037851). The ELM frequency is observed to be higher in the case with gas puffing.

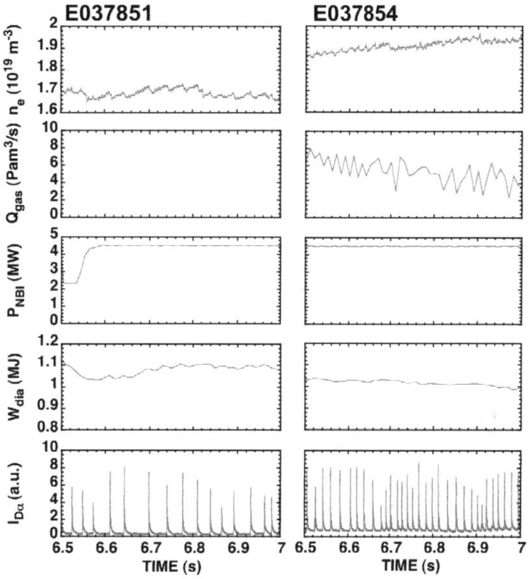

Figure 7. Time behavior of plasma parameters in two different ELMy H-mode discharges. The left figure (E037851) is the case without gas puff in the H-mode phase and the right one represents the case with gas puff.

The spatial profile of the electron density around edge transport barrier region measured in the experiments is shown in figure 8. These data are measured with Thomson scattering (circles) and reciprocating probe (triangles) [15, 16]. In the probe data, some data, which are recognized to be affected by ELM, are eliminated from the figure. Solid lines in the figure represent the density profile estimated using the UEDGE plasma code based on the obtained experimental data. The determined pedestal width (distance between LCFS and the knee position with hyperbolic tangent fitting) in the higher density plasma shot (#37854) is 3.06 cm and is estimated to be smaller than that in the lower density case (3.37 cm: #37851) by nearly 10 %.

Figure 9 shows an example of the input parameters used in the DEGAS simulation based on the above-mentioned data analyses. In this simulation, toroidal uniformity of the particle source (recycling) is given in the 3-D geometrical structure and the absolute value of the simulation results are determined by normalizing the obtained results to the measured Dα line emission intensity. On both cases, electron density and temperature in the outer SOL region is assumed to be 1×10^{11} cm^{-3} and 10 eV, respectively.

Based on the above input data, the simulation with DEGAS was carried out and the penetration of neutrals is evaluated in the two H-mode plasmas. In figure 10 atomic neutral density profiles on the midplane of the poloidal cross-section are plotted. As shown in the figure, a noticeable increase in the decay length $\Delta X_{1/e}$ of atomic density is recognized in the lower-density case (#37851).

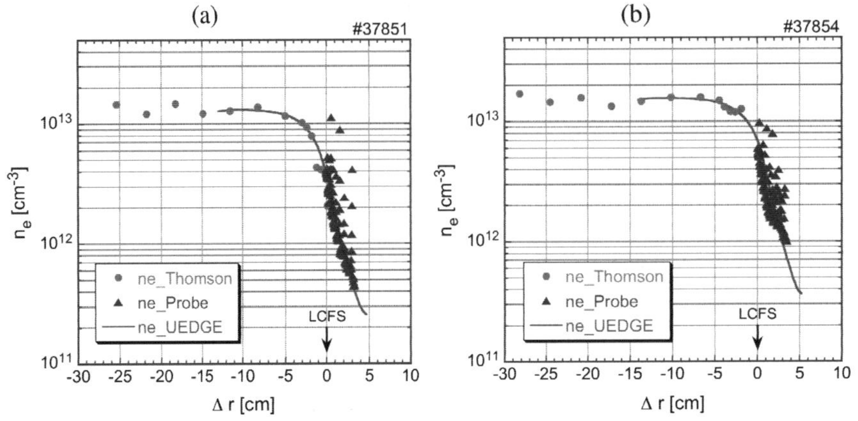

Figure 8. Edge plasma density profiles obtained from measurements and simulation. Circles: Thomson scattering, triangles: reciprocating probe and lines: UEDGE code.

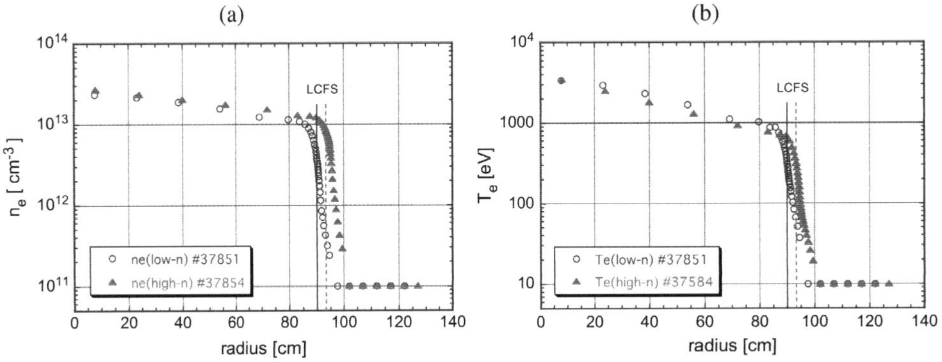

Figure 9. Input parameters for DEGAS simulation. (a) electron density, (b) electron temperature.

By using the data from UEDGE and DEGAS, the ionization profile near the edge transport barrier region is evaluated as shown in figure 11. In both cases, the ionization area is localized near the LCFS and it is found that the determined width ($\Delta r_{1/2}$; a full width half maximum value) in the lower-density case (#37851) is wider than that in the higher-density case (#37854) by 24 %.

From the above results, the density pedestal tends to be wider for longer penetration length of neutral and longer ionization length in edge region, which is similar to the results obtained in DIII-D [8]. However, ionization profiles are almost the same for both cases in the main plasma except for the region just inside the LCFS. Further systematic study is necessary to understand parameter correlation.

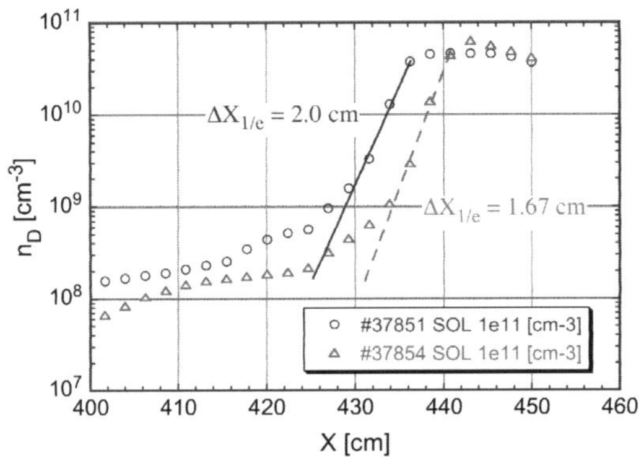

Figure 10. Penetration of neutrals near the edge region calculated by DEGAS. The SOL density is assumed to be 1×10^{11} cm^{-3}.

6. Summary

Fully 3-dimensional neutral transport simulation has been started using DEGAS Monte-Carlo code in order to investigate detailed behavior of high performance H-mode plasmas in the JT-60U tokamak. A precise mesh structure including isolated dome, baffle and divertor plates was built-up and the simulation space was extended from SOL to core plasma region. A benchmark test with 2-dimensional DEGAS2 code was successfully performed and the validity of the simulation was confirmed. The first result of 3-D simulation in JT-60U performed with gas puffing indicated the significant localization of neutrals around the gas puffer and the effect of SOL plasma on neutral diffusion in toroidal and poloidal directions are clarified. Comparison of neutral density profiles between the high and low density H-mode plasmas was carried out and the noticeable difference in neutral penetration and ionization width was recognized. The above-obtained results provide the useful aspect for pedestal physics and this analysis method also gives us important information for the three-dimensional plasma modelling in near future.

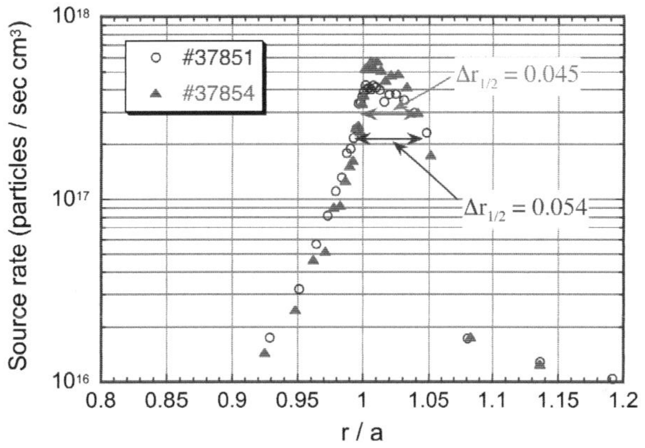

Figure 11. Ionization profile near the edge region determined from the result shown in figure 10.

Acknowledgments

The authors would like to acknowledge the all those who contributed to the JT-60U project. This work was mainly performed based on the research collaboration between Japan Atomic Energy Agency and University of Tsukuba.

References

[1] Heifetz D, Post D, Petravic M Weisheit J and Bateman G 1982 *J. Comput. Phys.* **46** 309
[2] Reiter D *et al.* 1991 *Plasma Phys. Control. Fusion* **33** 1579
[3] Stotler D P *et al.* 1994 *Contrib. Plasma Phys.* **34** 392
[4] Nakashima Y *et al.* 2005 *J. Nucl. Mater.* **337-339** 466
[5] Nakashima Y, Higashizono Y, Shoji M, Kobayashi S, Kubota Y, Yoshikawa M, Kobayashi T, Islam M K and Cho T 2006 *J. Plasma Phys.* **72** 1123
[6] Shoji M *et al.* 2004 *J. Plasma Fusion Res. SERIES* **6** 512
[7] Kobayashi S *et al.* 2004 *Proc 31th EPS conf. Plasma Phys.* (London, 2004) **28G**, P-5.097
[8] Groebner R J, Mahdavi M A, Leonard A W, Osborne T H, Porter G D, Colchin R J and Owen L W 2002 *Phys. Plasmas* **9** 2134
[9] Kallenbach A *et al.* 2004 *Plasma Phys. Control. Fusion* **46** 431
[10] Kirk A, Counsell G F, Arends E, Meyer H, Taylor D, Valovic M, Walsh M, Wilson H and the MAST Team 2004 *Plasma Phys. Control. Fusion* **46** A187.
[11] Horton L D *et al.* 2005 *Nucl. Fusion* **45** 856
[12] Hughes J W, LaBombard B, Terry J, Hubbard A and Lipshultz B 2007 *Nucl. Fusion* **47** 1057
[13] Rognlien T D *et al.* 1994 *Contrib. Plasma Phys.* **34** 362
[14] Takenaga H *et al.* 2006 *Nucl. Fusion* **46** S39
[15] Asakura N *et al.* 2002 *Plasma Phys. Control. Fusion* **44** A313
[16] Kallenbach A, Asakura N, Kirk A, Korotkov A, Mahdavi M A, Mossessian D and Porter G D 2005 *J. Nucl. Mater.* **337-339** 381

Journal of Physics: Conference Series **123** (2008) 012030 doi:10.1088/1742-6596/123/1/012030

Non-local turbulent effect on internal transport barrier collapse in reversed shear configuration

S Tokunaga[1], **M Yagi**[2], **S-I Itoh**[2] **and K Itoh**[3]

[1]Interdisciplinary Graduate School of Engineering Science, Kyushu University, Kasuga, Fukuoka, 816-8580, Japan

[2]Research Institute for Applied Mechanics, Kyushu University, Kasuga, Fukuoka, 816-8580, Japan

[3]National Institute for Fusion Science, National Institute of Natural Sciences, Toki, Gifu, 509-5292, Japan

E-mail: `toku@riam.kyushu-u.ac.jp`

Abstract. The mechanism of transport barrier collapse in reversed shear configuration is investigated using a global ion temperature gradient driven drift wave (ITG) turbulence simulation code. The transport barrier collapse event is explained in the context of non-local meso-scale interaction. It is found that the turbulence evolution and the sustainment of the transport barrier is sensitive to the safety factor profile. However, some common meso-scale modes are excited via various non-linear interaction channels whether or not a meso-scale mode in the vicinity of the q profile minimum has resonant surfaces. The non-linear interaction among these non-local modes makes long range correlation possible and carries energy from the core region to the q minimum region. It was shown that such linkage can lead to the collapse of the transport barrier and the change of the global temperature profile.

1. Introduction

Recently, the internal transport barrier (ITB) has been observed in many tokamaks with a reversed-magnetic-shear (RS) configuration. The ITB is beneficial for the steady-state operation of International Thermonuclear Experimental Reactor (ITER) and advanced tokamak reactors. The improved performance with ITB was surveyed in multi-machine comparisons [1]. The analysis of experimental ITB database has made remarkable progress in understanding the ITB formation [2]. It is reported that there exist different kinds of ITB, such as the box-type or parabolic-type, depending on the magnetic field configuration [3]. The key physics of ITB and its characteristics have been studied by the database analysis and 1.5D transport simulation [4, 5, 6]. Thus far, many works on ITB formations have been done theoretically [7] and numerically [8, 9, 10, 11]. The velocity shear and the negative magnetic shear as well as the zonal flow play an important role in the ITB formation [12]. In addition, the suppression of turbulence is affected by the coupling with the zonal flow and the geodesic acoustic modes (GAMs) [13]. Recently, turbulence spreading is also investigated in terms of the barrier propagation [14, 15, 16]. The turbulent spreading can allow turbulence to penetrate the regions where local analysis predicts a low level of turbulent transport [17]. Moreover, turbulence spreading can dynamically couple different regimes of plasma such as the edge and core, so that it gives rise to non-local effects. The important issue remaining to be clarified is the mechanism of ITB collapse. Possible causes reviewed in [23] include the barrier localized mode [18, 19], momentum transfer events [20, 21], and X-event [22].

In this paper, we report results from global simulations on ITB formation and collapse in reversed shear plasmas. Numerical calculations using a simple gyro-fluid model of the ITG turbulence are performed. We observed the formation of ITB similar to parabolic-type ITB and its expansion to the outside of $q_{\rm min}$. The edge of the ITB outside of $q_{\rm min}$ is referred to as the ITB foot in this paper. It is found that the ITB collapse occurs due to the global mode which spreads out near the $q_{\rm min}$ surface. This mode is excited for both resonant and off-resonant cases by various energy channels, however, once it is excited the sequence of the ITB collapse is similar. Non-local interaction takes place between the core region and the ITB region, which excites the global mode near $q_{\rm min}$ region. During the collapse of the region inside the ITB, the ITB foot still remains for a short time due to the velocity shear stabilization, however, when ITG turbulence overcomes it, then the ITB foot collapse is triggered.

2. Model

2.1. Model equations

We employ the simplified version of the "3+1" gyro-fluid model [24] proposed by Ottaviani *et al.* [25] and modified by Yagi [16] to investigate the ITG turbulence transport by global simulation. The set of equations is described by the ion density, the ion parallel momentum and the ion temperature evolution equations as follows:

$$\frac{dW}{dt} + \kappa_{\rm n}\frac{1}{r}\frac{\partial\Phi}{\partial\vartheta} + A\nabla_{\parallel}V = \varepsilon\hat{\omega}_{\rm d}F + \rho_*\frac{q}{\varepsilon}\mu^{\rm NC}\frac{\partial U_{\rm p}}{\partial r} - \rho_*^2\mu\nabla_{\perp}^4 F \quad (1)$$

$$\frac{dV}{dt} = -A\nabla_{\parallel}F + 4\mu\nabla_{\perp}^2 V - \mu^{\rm NC}U_{\rm p} - A\sqrt{\frac{1}{\tau}\frac{2}{5}}\sqrt{\pi}\left|k_{\parallel}\right|V + \frac{2}{5}A\frac{1}{\tau}\nabla_{\parallel}T \quad (2)$$

$$\frac{3}{2}\left(\frac{dT}{dt} + \kappa_T\frac{1}{r}\frac{\partial\Phi}{\partial\vartheta}\right) - \left(\frac{dn}{dt} + \kappa_n\frac{1}{r}\frac{\partial\Phi}{\partial\vartheta}\right) = -\frac{9}{5\sqrt{\pi}}\sqrt{\frac{1}{\tau}}A\left|\nabla_{\parallel}\right|T + \frac{2}{5}A\nabla_{\parallel}V + \chi_{\perp}\nabla_{\perp}^2 T \quad (3)$$

where, $W = n-\nabla_{\perp}^2 F$ is the generalized vorticity, $n = \Phi-\langle\Phi\rangle$ the fluctuating density, $F = \Phi+p/\tau$ the generalized potential, $\tau \equiv T_{\rm e0}/T_{\rm i0}$, $p = n+T$ the fluctuating ion pressure and $\langle\Phi\rangle$ the flux surface averaged potential, which is equivalent to $\Phi_{m=0,n=0}$ in Fourier space. The approximation $p_1 = T_{i0}(r_s)n_1 + n_0(r_s)T_{i1}$ is made where $r_{\rm s} = 0.6$ indicates the location of the $q_{\rm min}$ surface. Here $U_{\rm p} = V + \rho_*\frac{q}{\varepsilon}\frac{\partial F}{\partial r}$ represents the poloidal velocity, $\kappa_T = -d\ln T_0/dr$ and $\kappa_n = -d\ln n_0/dr$ indicate the inverse of the gradient scale length of the ion temperature and density, respectively, $\hat{\omega}_{\rm d} = 2\cos\vartheta\frac{1}{r}\frac{\partial}{\partial\vartheta} + 2\sin\vartheta\frac{\partial}{\partial r}$ is the curvature operator, $A = \varepsilon_a/\rho_*$ is the ratio of the normalized ion gyro-radius $\rho_* = \rho_s/a$ and the inverse aspect ratio $\varepsilon_a = a/R_0$. We also introduced $\varepsilon = r/R_0$ the local inverse aspect ratio. The transport coe cients are given by the ion thermal diffusivity χ, the neoclassical ion viscosity $\mu^{\rm NC}$ and the ion viscosity μ. Ion Landau damping is also taken into account as a collisionless dissipation [26].

Normalization : $t/t_{\rm B} \to t$, $r/a \to r$, $z/R_0 \to z$, $e\Phi/T_{e0} \to \Phi$, $V/c_{\rm s} \to V$, $T/T_{i0} \to T$, $\chi/(\rho_{\rm s}c_{\rm s}) \to \chi$, $\mu/(m_i n_0\rho_{\rm s}c_{\rm s}) \to \mu$, $\mu^{\rm NC}a^2/(\rho_{\rm s}c_{\rm s}) \to \mu^{\rm NC}$ is used, where $t_{\rm B} = a^2/\chi_{\rm B}$ represents the Bohm time determined by the Bohm diffusion $\chi_{\rm B} = cT_{e0}/(eB)$.

2.2. Initial conditions

We introduce a heat source term into the model. It is given by $S(r) = -8.0\times10^{-3}\,(2r^2-1)/(1-r_{\rm s}^2)^2$ with $r_{\rm s} = 0.6$ and is added to the ion temperature evolution equation (3). The equilibrium density profile is given by $n_0(r) = (1-r^2)/(1-r_{\rm s}^2)$ which is not changed during the simulation due to the relation : $n = \Phi - \langle\Phi\rangle$. The ion temperature is not separated into equilibrium and perturbative components $T = T_0 + \tilde{T}$, but rather solved as a whole.

Journal of Physics: Conference Series **123** (2008) 012030 doi:10.1088/1742-6596/123/1/012030

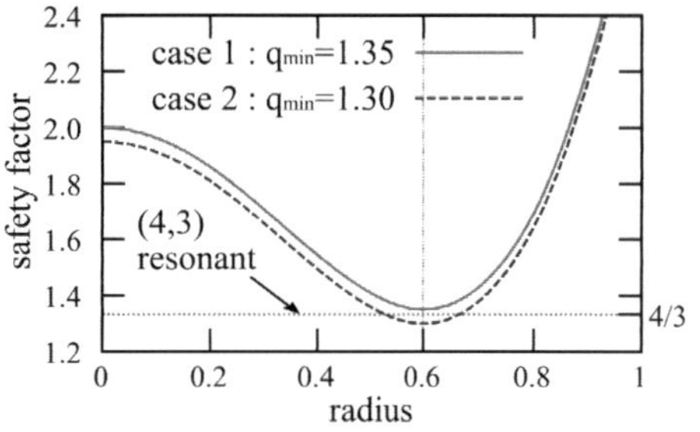

Figure 1. Safety factor profile for cases with $q_{\mathrm{min}} = 1.35$ and $q_{\mathrm{min}} = 1.30$. Magnetic shear has a null point at $r = 0.6$ in both cases. In the case with $q_{\mathrm{min}} = 1.30$, the (4,3) mode has resonant surfaces at $r = 0.528$ and 0.666.

For the safety factor q profile, we employ a reversed magnetic shear profile, proposed by Garbet *et al.* [10], which is given by $q(r) = q_{\mathrm{min}} + C_2(r^2 - r_{\mathrm{s}})^2 + C_3(r^2 - r_{\mathrm{s}}^2)^3$, with $r_{\mathrm{s}} = 0.6$, $C_2 = 4.66$, and $C_3 = -0.987$. Here, we investigate two cases with $q_{\mathrm{min}} = 1.35$, and $q_{\mathrm{min}} = 1.30$. Note that in the case with $q_{\mathrm{min}} = 1.35$, the (m,n)=(4,3) mode is off-resonant. On the other hand, in the case with $q_{\mathrm{min}} = 1.30$, the (4,3) mode has resonant surfaces around $r = 0.6$ as shown in figure 1.

3. Simulation results

3.1. Off-resonant case

At first, we investigate the case with $q_{\mathrm{min}} = 1.35$. The system starts from $T(r) = 0$, and gradually evolves according to the transport equation. In this simulation, the heat source directly supplies the energy to the $T_{m=0,n=0}$ component where m indicates the poloidal mode number and n the toroidal mode number. In the early stages of evolution, the temperature gradient is small enough that the system is stable for the ITG modes and fluctuations are suppressed to a negligible level. When the temperature gradient exceeds the threshold value of the ITG instability ($t \sim 60$), the energy of the most unstable modes including (29,20), (30,20) and so on starts to grow at $r \sim 0.4$. Then, the energies of the zonal flow (0,0) and GAM (1,0) begin to increase. These modes saturate at finite amplitude by the non-linear interaction.

3.1.1. Time evolution of fluctuating energy The top of figure 2 shows the temporal evolution of the fluctuating internal energies of some important modes during the saturation phase of the ITG modes. It should be noted that $\langle T \rangle$ cannot be separated into its equilibrium and fluctuating parts, so that we do not plot $T_{0,0}$ at the top of figure 2.

The (4,3), (7,4) and (11,7) modes are found to be non-linearly excited by the three wave

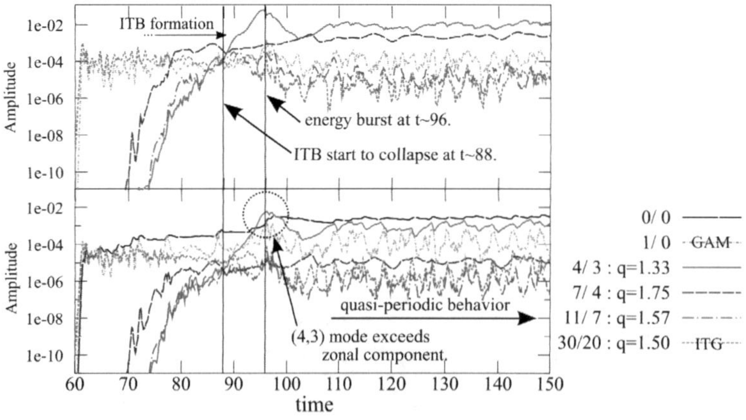

Figure 2. Top : Time evolution of fluctuating internal energy. The behavior of (4,3), (7,4) and (11,7) modes are related to the ITB collapse. Bottom : Time evolution of fluctuating parallel kinetic energy. At $t \sim 96$, the (4,3) mode exceeds the level of the zonal component.

Journal of Physics: Conference Series **123** (2008) 012030 doi:10.1088/1742-6596/123/1/012030

coupling. These modes are linearly stable, which is confirmed by solving the linearized equations and keeping the toroidal coupling. The non-linear eigenfunctions of the (7,4) and (11,7) modes are localized at each rational surface. Therefore, these are not low-n ballooning modes. Firstly, the (7,4) mode starts to grow at $t \sim 70$, then (11,7) and (4,3) modes start to grow at $t \sim 75$. These modes reach the same amplitude at $t \sim 88$ and the (4,3) mode still increases over the level of other modes. It saturates at $t \sim 96.1$. The behavior of the (4,3), (7,4) and (11,7) modes are related to the ITB collapse events as shown in the next section.

The bottom of figure 2 shows the temporal evolution of the fluctuating parallel ion kinetic energy. The zonal component (0,0) has a large amplitude as does GAM (1,0) in the saturation phase ($60 < t < 88$). At $t \sim 96$, the (4,3) mode exceeds the level of the (0,0) mode and dominates the system.

After the quench of the (4,3) mode at $t \sim 96$, the global structure of the ion temperature is changed, and all modes show quasi-periodic behavior. In this phase, intermittent bursting heat transport occurs, accompanied with a temperature profile relaxation concurrent with the avalanches [16].

3.1.2. ITB collapse The time slices of the radial profile of the turbulent transport coe cient are shown in figure 3. At $t \sim 75$, the transport barrier is formed around $r = 0.6$. The barrier starts to collapse from inside at $t \simeq 88$. However, the ITB foot ($r \simeq 0.66$) still remains till $t \sim 95$. Then it collapses abruptly and energy flows away at $t \sim 96$. As is shown in figure 2, the ITB collapse is due to the growth of the off-resonance (4,3) mode whose energy is fed by the non-linear coupling of the (7,4) and (11,7) modes. The trigger of the ITB foot collapse is related to the destabilization of the ITG modes near $r \sim 0.66$ at $t \sim 96$.

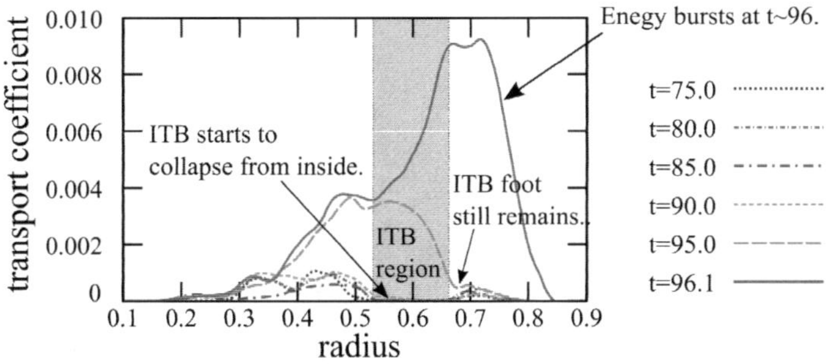

Figure 3. Turbulent transport coe cient radial profiles with $q_{\min} = 1.35$ from $t = 75$ to $t = 96.1$. Transport barrier is formed around $r = 0.6$ until $t \sim 89$.

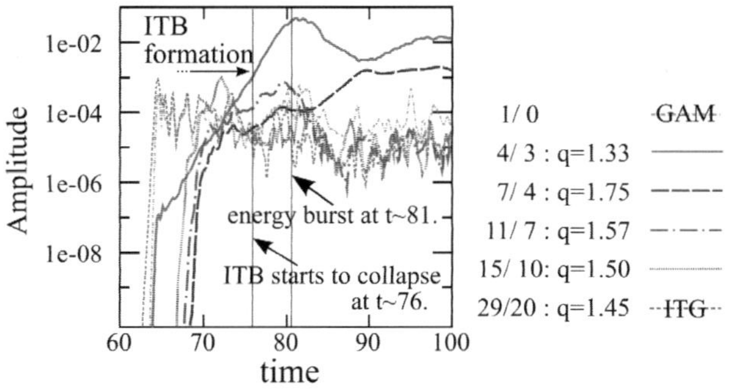

Figure 4. Time evolution of fluctuating internal energy with $q_{\min} = 1.30$. The ITB starts to collapse at $t \sim 71$. The ITB foot collapses at $t \sim 81$.

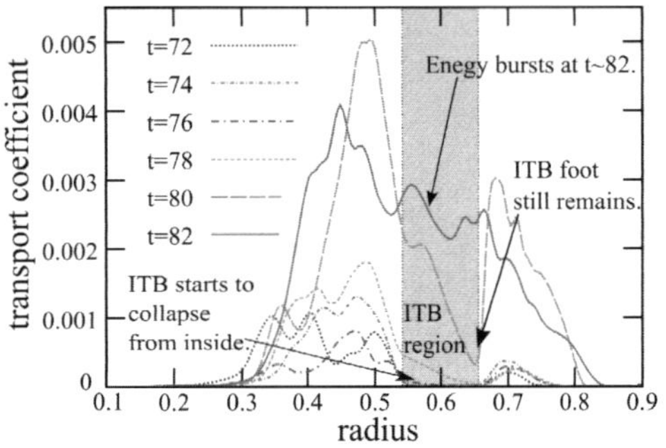

Figure 5. Turbulent transport coe cient radial profiles with $q_{\min} = 1.30$. Transport barrier is formed around $r = 0.6$ until $t \sim 76$.

3.2. Resonant case

Next, we investigate the case with (4,3) resonant surfaces near $q_{\min}$. Figure 4 and 5 show temporal evolution of the fluctuating internal energy and time slices of the radial profiles of the turbulent transport coe cient in the case with $q_{\min} = 1.30$.

In this case, the (4,3) mode starts to grow earlier than the (7,4) and (11,7) modes do in the off-resonant case. Until $t \sim 64$, it is driven by the non-linear coupling of the ITG modes such as the (30,20) and (34,23) modes; then its growth becomes slower at $t \sim 64$. Similar behavior is observed in the Neoclassical Tearing Mode(NTM) simulation [27] and is identified due to the beat interaction of drift waves [28]. Since the energy transfer channels are different from the previous case, they indicate different behavior in the growing phase ($70 < t < 76$). For example, the (7,4) and (11,7) modes are fed by (4,3) mode instead. Although the collapse of the ITB occurs at an earlier time, the sequence after $t \sim 76$ is similar to the previous case.

4. Discussion

4.1. Non-local turbulent transport effect on ITB collapse

Figure 6 shows time slices of the absolute value electrostatic potential profiles $|\Phi_{m,n}(r)| = \sqrt{\Phi^*_{m,n}\Phi_{m,n}}$ for the (4,3) ,(7,4) and (11,7) modes in the case with $q_{\min} = 1.35$. At $t \sim 74.5$, (figure 6a), the (7,4) mode is excited at its resonant surface. It develops and spreads out (figure 6b). Once it reaches the resonant surface of the (11,7) mode, the (11,7) mode starts to grow (figure 6c). At the same time, the (4,3) mode is excited by non-linear coupling of these modes which is shown in figure 2 . After that, the (4,3) mode develops around $r = 0.6$ (figure 6d). Then, the ITB starts to collapse from inside (figure 6e), however, the ITB foot still remains until $t = 95.9$. At $t \sim 96.0$, the stored energy inside the ITB abruptly bursts out (figure 6f). The (4,3) mode loses energy, however the (7,4) and (11,7) modes survive at rational surfaces (figure 6g). The connection between the (4,3) and (7,4) modes remains and interaction between them is sustained after collapse ($115 < t$ in figure 2).

Figure 7 shows the time slices of the absolute value radial profiles of these modes and the (15,10) mode in the case with $q_{\min} = 1.30$. In the earlier phase ($t \sim 62$), the (4,3) mode is non-linearly driven by the (30,20) and (34,23) modes (figure 4). Then, the (15,10) mode grows at its resonant surface at $t \sim 68$ (figure 7a) and (11,7) mode is non-linearly excited by non-linear coupling between the (15,10) and (4,3) modes (figure 7b). Subsequently, the (7,4) mode is excited by the (11,7) and (4,3) modes (figure 7c). After that, similar sequence as the previous case is observed (figure 7d-g).

Although the (4,3) mode is off-resonant in one case and is resonant in the other case, the global eigenfunction of the (4,3) mode plays an essential role for the ITB collapse. In both cases, three modes with relatively small k_r, (7,4) located in the core region, (4,3) near the $q_{\min}$ and (11,7) in between them, interact with each other. Once the connection between these modes has

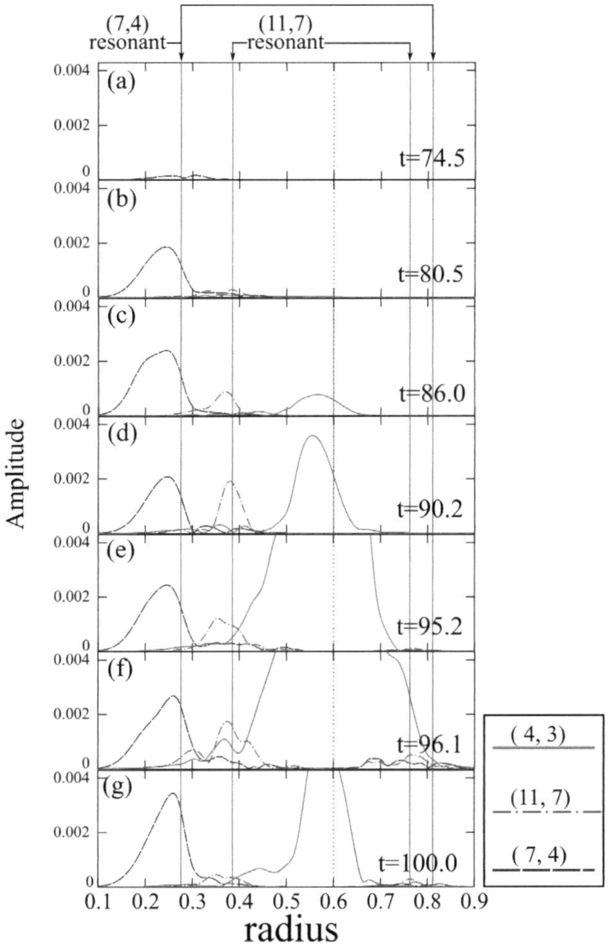

Figure 6. Time slices of absolute value radial profiles of (4,3), (7,4) and (11,7) modes with $q_{\min} = 1.35$.

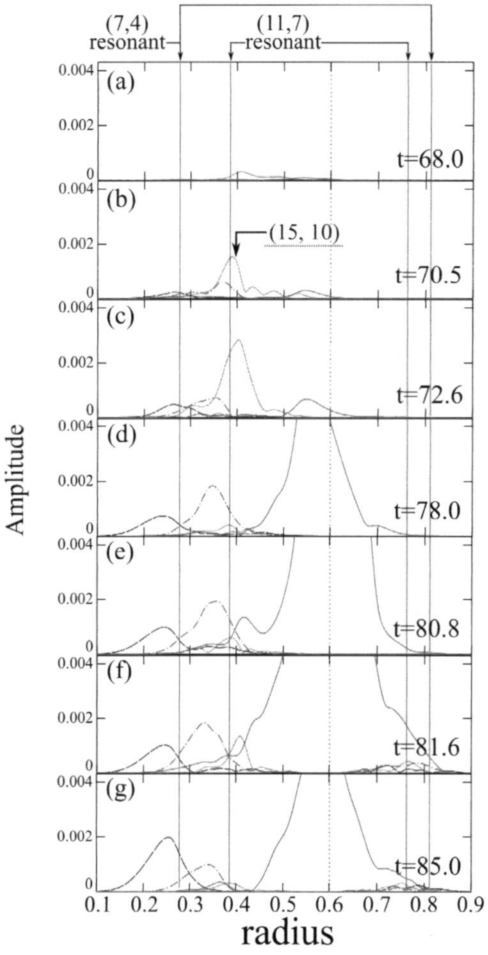

Figure 7. Time slices of absolute value radial profiles of (4,3), (7,4), (11,7) and (15,10) modes with $q_{\min} = 1.30$.

been established, the internal energy stored in the core region is transferred to the ITB region.

We perform the simulation in which all non-linear interactions are suppressed in the (4,3) mode evolution equations except for the toroidal coupling. Figure 8 shows the time evolution of the fluctuating internal energy in the case when all non-linear interactions are suppressed in the (4,3) mode evolution equations. Figure 9 shows the transport coe cient radial profiles. In this case, the (4,3) mode does not grow so much, and the ITB collapse does not occur. It indicates that the (4,3) mode which is excited by three wave coupling plays an essential role in the global collapse of the ITB.

We also perform another simulation in which the all non-linear interactions are suppressed in the (7,4) and (11,7) modes evolution equations except for toroidal coupling. In this case, other mediator modes such as the (10,6) and (14,9) modes are excited instead of (7,4) and (11,7). Then, the (4,3) mode is non-linearly excited by three wave coupling with mediator modes, and causes the collapse of the ITB.

The long range interaction among the meso-scale turbulent rolls can destroy the internal

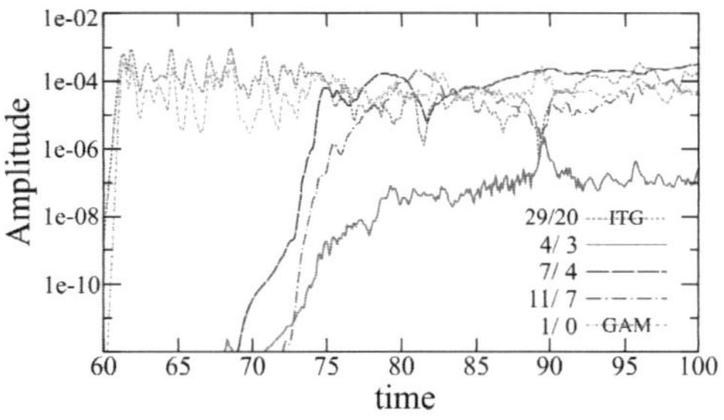

Figure 8. Time evolution of fluctuating internal energy in case when all non-linear interactions are suppressed in (4,3) mode evolution equations.

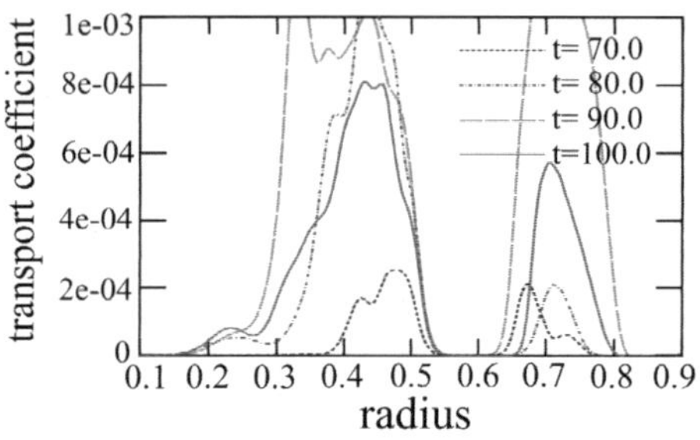

Figure 9. Radial profiles of transport coe cient. ITB remains until t=100, without the growth of (4,3) mode.

transport barrier non-linearly. The (4,3) mode is excited as the driver which leads to the energy minimum state, i.e., ITB collapse state. Our simulation shows that the channel of mediator mode generation is flexible. However, the non-local mode excited near $q_{\min}$ is intrinsically chosen by the system.

4.2. Trigger of ITB foot collapse

Figure 10(a) indicates the radial profiles of the zonal component of the parallel velocity at $t = 92.0$ and $t = 95.7$. The velocity shear exists at $r \sim 0.68$ till $t \sim 93$. The ITB foot can be sustained by the shear flow stabilization of the ITG modes before the collapse.

Recent theoretical studies showed that only $E \times B$ flow shear is the dominant contributor to the shearing rate for the flute-like fluctuations. However, both $E \times B$ flow shear and toroidal flow shear contribute to the shearing rate for the ballooning-like fluctuations in toroidal plasmas [29]. Coupling of the (0,0) component of the $v_{\|}$ and ITG modes occurs via parallel dynamics. Actually, we observe the toroidal ITG mode near the ITB foot where the finite magnetic shear (finite $k_{\|}$) exists. In the vicinity of the $q_{\min}$ region, the magnetic shear is weak and slab-like ITG might be excited, in this case, the zonal flow may play an important role for ITG suppression. We also perform a non-linear simulation suppressing the generation of the (0,0) component of $v_{\|}$ and confirm that the collapse of the ITB foot occurs at an earlier time ($t \sim 86$) due to the weak stabilization effect.

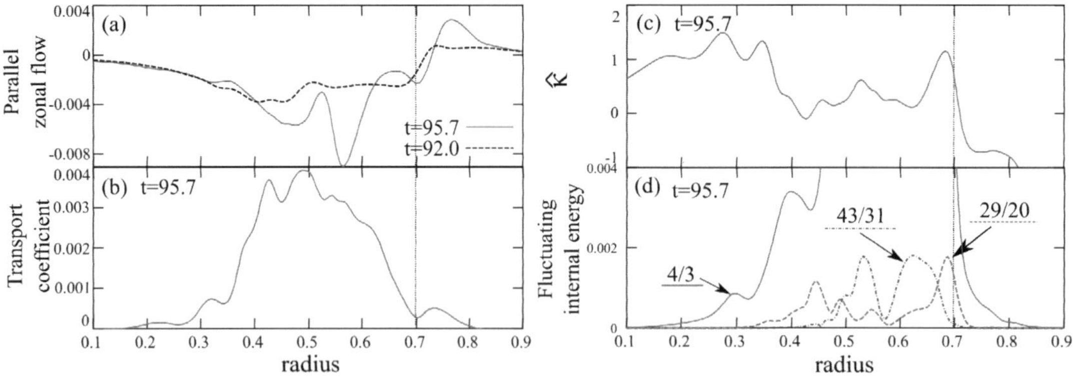

Figure 10. Radial profiles of (a) zonal component of parallel velocity, (b) transport coe cient, (c) $\hat{\kappa} = 2/3\kappa_T - \kappa_N$ the driving term of ITG mode, (d) fluctuating internal energy.

Figure 10(b) shows the profile of the transport coe cient at $t = 95.7$. The contribution from the (4,3) mode to the transport is not large outside the ITB foot ($0.7 < r < 0.8$), although the (4,3) mode spreads out in this region shown in figure 10(d).

Figure 10(c) indicates the radial profile of driving term of the ITG mode $\hat{\kappa} \equiv \frac{2}{3}\kappa_T - \kappa_n$. $\hat{\kappa}$ has a peak value at $r \sim 0.7$, i.e., the position of steep temperature gradient, which indicates the destabilization of the ITG modes near the ITB foot.

Figure 10(d) shows the absolute value of the fluctuating internal energy of the ITG modes destabilized near the ITB foot before collapse and the (4,3) mode in the $q_{\rm min} = 1.35$ case at $t = 95.7$. For $t \leq 93$, the ITG modes such as (29,20) are excited at $r \sim 0.45$, i.e., the resonant surface inside the $q_{\rm min}$ surface, but the ITG modes such as (43,31) and (29,20) have small amplitude outside of $q_{\rm min}$. In the simulation in which all non-linear interactions are suppressed in the (4,3) mode evolution equations, we observe the weak excitation of the ITG modes in the vicinity of the ITB foot. It indicates that the global collapse of the ITB driven by the (4,3) mode gives rise to the quasi-linear profile modification which destabilizes ITG modes at the ITB foot. From $t \sim 93.0$, the zonal flow profile starts to change according to the growth of the (43,31) and (29,20) modes. At $t \sim 96.0$, these modes cannot be suppressed any more by the velocity flow shear, and the collapse of the ITB foot occurs.

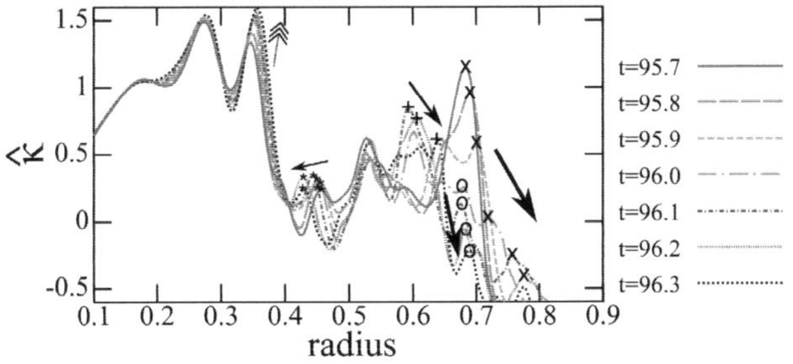

Figure 11. Time slices of radial profile of $\hat{\kappa}$. The peak at $r \sim 0.7$ indicated by (x) spreads out in a short period. Other small peaks indicated by (o,+) also spread in and out.

Accompanied with the burst, the peak of $\hat{\kappa}$ propagates outward and decays as shown in figure 11, which indicates avalanches. The temperature profile is modified by this large-scale relaxation. After the collapse of the ITB foot, the zonal flow is regenerated by excited micro turbulence. However, transport suppression inside the ITB is not as complete as before. As a result, the system indicates quasi-periodic oscillation with intermittent energy transport accompanied with avalanche as is shown in figure 12.

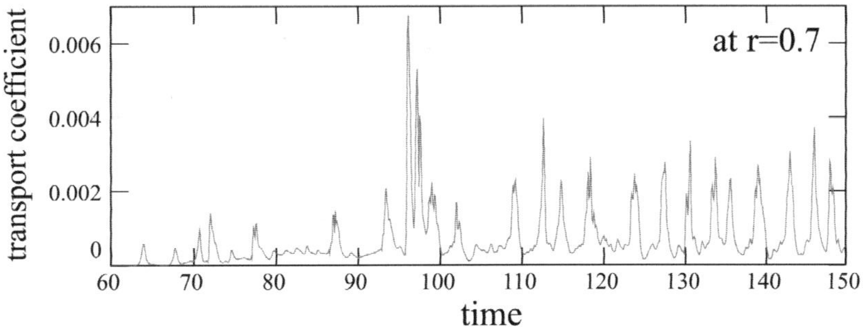

Figure 12. Intermittent heat transport measured at $r = 0.7$ in the $q_{\rm min} = 1.35$ case. The bursts are synchronized with the decay of the (4,3) mode in figure 2.

5. Conclusion

We investigate the multi-scale interaction between transport and turbulence in the reversed shear configuration, using the global ITG simulation code. It is found that the (4,3) mode plays an important role in the ITB collapse for cases with or without the (4,3) rational surface. However, the development of the (4,3) mode is controlled by the different energy channels causing the timing of the collapse to vary. For the off-resonant case, the (4,3) is excited by the non-linear coupling between the (7,4) and (11,7) modes and the ITB collapse occurs later than in the resonant case. For the resonant case, the (4,3) mode is excited by the (30,20) and (34,23) modes in the early non-linear phase, but in the later non-linear phase, similar behavior to the off-resonant case is observed. The ITB foot remains for a while during the ITB collapse due to velocity shear stabilization of the ITG modes after which it abruptly bursts out. It is concluded that the ITG modes excited near the ITB foot trigger the collapse of ITB foot. Afterwards, the avalanche occurs and the temperature profile is modified by the large-scale relaxation. Then, the system takes on a quasi-periodic oscillation with intermittent energy burst. Thus, the analysis in this article provides the basis to understand the quasi-periodic onsets of barrier collapse in plasmas with internal transport barrier [23].

The phenomena is attributed to non-local interactions with different spatial and temporal scale. We have shown one example of multi-scale interaction between transport and turbulence.

We have also shown that the timing of collapse is sensitive to the safety factor profile since the meso-scale mode is excited by the different energy channels. Therefore, it is necessary to solve Ohm's law to investigate the effect of the safety factor profile on the ITB collapse more accurately by taking into account the q profile evolution, which requires electromagnetic dynamics be included in the simulation. This is left for a future work.

Acknowledgement

This work is partly supported by the Grant-in-Aid for Specially-Promoted Research of MEXT(16002005), by the Grant-in-Aid for Scientific Research (B) of JSPS(19360415) and by the collaboration programs of NIFS (NIFS07KOAP017) and of the Research Institute for Applied Mechanics of Kyushu University.

References

[1] Litaudon X, Barbato E, Bécoulet A *et al.* 2004 *Plasma Phys. Control. Fusion* **46** A19
[2] Connor J W, Fukuda T, Garbet X *et al.* 2004 *Nucl. Fusion* **44** R1
[3] Shirai H, Kikuchi M, Takizuka T *et al.* 1999 *Nucl. Fusion* **39** 1713
[4] Sakamoto Y, Kamada Y, Ide S *et al.* 2001 *Nucl. Fusion* **41** 865
[5] Takizuka T, Sakamoto Y, Fukuda T, Fujita T, Kamada Y, Suzuki T, Ide S and Shirai H 2002 *Plasma Phys. Control. Fusion* **44** A423
[6] Hayashi N, Takizuka T, Sakamoto Y, Fujita T, Kamada Y, Ide S and Koide Y 2006 *Plasma Phys. Control. Fusion* **48** A55
[7] Itoh K and Itoh S-I 2002 *Plasma Phys. Control. Fusion* **44** A367
[8] Waltz R E, Kerbel G R, Milovich J and Hammett G W 1995 *Phys. Plasmas* **2** 2408
[9] Kishimoto Y, Kim J-Y, Horton W, Tajima T, Lebrun M J and Shirai S 1998 *Plasma Phys. Control. Fusion* **40** A663
[10] Garbet X, Bourdelle C, Hoand G T *et al.* 2001 *Phys. Plasmas* **8** 2793
[11] Candy J, Waltz R E and Rosenbluth M N 2004 *Phys. Plasmas* **11** 1879
[12] Diamond P H, Itoh S-I, Itoh K and Hahm T S 2005 *Plasma Phys. Control. Fusion* **47** R35
[13] Miyato N, Kishimoto Y and Li J 2004 *Phys. Plasmas* **11** 5557
[14] Garbet X, Laurent L, Samain A and Chinardet J 1994 *Nucl. Fusion* **34** 963
[15] Hahm T S, Diamond P H, Lin Z, Itoh K and Itoh S-I 2004 *Plasma Phys. Control. Fusion* **46** A323
[16] Yagi M, Ueda T, Itoh S-I, Azumi M, Itoh K, Diamond P H and Hahm T S 2006 *Plasma Phys. Control. Fusion* **48** A409
[17] Dimits A M, Cohen B I, Mattor N, Nevins W M, Shumaker D E, Parker S E and Kim C 2000 *Nucl. Fusion* **40** 661

[18] Koide Y *et al.* 1994 *Plasma Phys. Control. Fusion* **36** A195
[19] Takeji S *et al.* 1997 *Phys. Plasmas* **4** 4283
[20] Obsorne T H *et al.* 1995 *Nucl. Fusion* **35** 23
[21] Burrell K H *et al.* 1993 *Proc. 20th EPS Conf. on Controlled Fusion and Plasma Physics, Lisbon, 1993* (European Physical Society, Lisbon ,1993), Vol. 17C, Part I, p. 27
[22] JET Team 1993 *Proc. 14th Int. Conf. Plasma Physics and Controlled Nuclear Fusion Research, Würzburg, 1992* (IAEA, Vienna, 1993), Vol. 1, p. 329
[23] Itoh S-I, Itoh K, Zushi H and Fukuyama A 1998 *Plasma Phys. Control. Fusion* **40** 879
[24] Beer M A and Hammett G W 1996 *Phys. Plasmas* **3** 4046
[25] Ottaviani M and Manfredi G 1999 *Phys. Plasmas* **6** 3267
[26] Chang Z and Callen J D 1992 *Phys. Fluids B* **4** 1167
[27] Yagi M, Yoshida S, Itoh S-I, Naitou H, Nagahara H, Leboeuf J-N, Itoh K, Matsumoto T, Tokuda S and Azumi M 2005 *Nucl. Fusion* **45** 900
[28] Yagi M, Itoh S-I, Itoh K, Azumi M, Diamond P H, Fukuyama A and Hayashi T 2007 *Plasma and Fusion Res.* **2** 025
[29] Hahm T S and Burrell K H 1995 *Phys. Plasmas* **2** 1648

Global nature of zonal flow due to the finite band width

K Uzawa†, J Q Li and Y Kishimoto

Department of Fundamental Energy Science, Graduate School of Energy Science, Kyoto University, Gokasho, Uji 611-0011, Japan
†Present address: National Institute for Fusion Science, Toki 509-5292, Japan

E-mail: uzawa.ken@nifs.ac.jp

Abstract. The spectral effects of both micro-scale electron temperature gradient driven turbulence and zonal flow on the zonal flow generation are investigated theoretically and computationally based on the Hasegawa-Mima turbulence equation. As the minimum model describing the most primary spectral feature, we have developed an eight-wave interaction model which includes two sets of the conventional four-wave coupling systems with adjacent radial spectral difference of dk_x. It is found that the zonal flow is characterized by a global nature with an enhancement of growth rate γ_q, namely, *global zonal flow eigen-mode*, which is attributed to the appearance of new cross mode couplings due to the finite band width of zonal flow. This is qualitatively different from the monochromatic character of zonal flow described in the conventional four-wave coupling model. Direct numerical simulations of the Hasegawa-Mima equation with a zonal flow spectral structure have clearly proven the analytical results by employing a rigorous spectral code. The calculations further show that the global zonal mode is not only governed by the spectral structure of pump waves, but also their phase relation.

1. Introduction

The importance of secondary large scale structures nonlinearly generated by the ambient turbulence is now being widely recognized[1], such as zonal/mean flows, streamers, geodesic acoustic modes (GAMs), and secondary low-frequency long wavelength fluctuations. Among them, toroidally and poloidally symmetric low frequency zonal flow with $k_\theta = k_\phi = 0$, but $k_r \neq 0$ is found to play an important role leading to a suppression of turbulent transport[2, 3, 4]. The generation mechanisms of zonal flow have been intensively studied. The modulational instability has been discussed as one of the plausible candidates of the zonal flow generation. In studying the modulation process, two approaches have been developed based on the Hasegawa-Mima (HM) turbulence model[5] which describes the most primary nonlinear interaction process of inviscid drift wave turbulence. One is the method using the coherent mode coupling (CMC)[6, 7, 8, 9, 10] and the other is that using the wave-kinetic (WK) equation[11, 12, 13]. In the CMC model, the four-wave model which samples the turbulent fluctuation as a monochromatic pump wave with the excitation of two sideband components through nonlinear coupling with a zonal seed has been proposed as the minimum model. In the WK model, the action invariant of the turbulence and associated evolution are represented in terms of the eikonal equation of the drift wave turbulence based on the assumption of scale separation between ambient turbulence and zonal flow[14]. It is noted that the dispersion relation obtained from the WK model is described in terms of the spectral density of turbulence.

Recently, these two different methods have been compared for the generation of zonal flow from the ion temperature gradient (ITG) mode for monochromatic pump wave and shown that two models give qualitatively the same results even though the assumptions such as scale separation between turbulence and zonal flow are different[13]. Meanwhile, the effects of the turbulent spectra on the zonal flow generation were investigated employing the advantage of the WK model. A turbulent spectrum with a Gaussian shape has been applied in the analysis of the WK model[12]. We have also investigated the zonal flow generation by representing the turbulent spectrum with two monochromatic pump waves by keeping the total pump energy constant. From these analyses, however, no essential difference for the zonal flow instability was obtained. It is noticed that these studies still assume monochromatic wave with respect to the zonal flow. Namely, no theoretical works that take into account the effect of zonal flow spectra have been attempted.

In the present paper, we have extended the previous modulational instability analysis of the zonal flow generation originated from electron temperature gradient (ETG) driven turbulence[7] by including the effect of the zonal flow spectra in the CMC model. We propose eight waves which involves two sets of the four-wave coupling system with adjacent radial spectral difference of dk_x. Namely, we represent the effect of the zonal flow spectra by means of two monochromatic waves with the radial wave number given by k_q and $k_q + dk_x$, which couple with two monochromatic pump waves and also four sidebands. We then derive the dispersion relation and found from the analyses that the zonal flow generation is qualitatively different in two cases with and without the spectral structure of the zonal flow. Namely, due to a new cross coupling between two sets of the modulational loop, two zonal flow components are no longer independent, but couple with each other, leading to a *global zonal flow eigen-mode*. The growth rate of the global mode is found to be enhanced compared with that estimated from the four-wave model.

In order to justify the validity of the model, we have performed direct numerical simulations of the HM equation. The formation of the global zonal flow eigen-mode and associated enhancement of the instability are clearly proved. We have also performed simulations with a quasi-continuous Gaussian-type spectrum for both zonal flow and pump wave. It is noticed that the radial structure of the pump waves can be different depending on the phase relation of each Fourier harmonics even if the shape of the power spectrum is identical. Due to this fact, it is found that the radial structure of the zonal flow and associated enhancement nature of the instability are influenced not only by the spectrum shape but also by their phases. Namely, the zonal flow is preferentially excited in the region where the intensity of the pump wave is higher. From these analyses, we found that the zonal flow spectra and resultant global nature of zonal flow play an important role for zonal flow generation whereas previous works only deal with the spectrum of turbulence.

The paper is organized as follows. In section 2, we present our theoretical model of zonal flow generation including the effect of spectral of both zonal flow and turbulence. The ETG mode is considered as an example. Based on the theoretical model, we derived a dispersion relation and present the numerical results. In section 3, we present numerical simulations of the HM equation to justify our theoretical model. We also examine the effect of initial phase of turbulence on the zonal flow instability. Finally, we summarize the results and give concluding remarks in section 4.

2. Theoretical model and numerical results

2.1. Theoretical model

As a model to investigate secondary instability in the zonal flow-drift wave turbulence, the well-known HM equation is conveniently employed to analyze the modulation process due to nonlinear coupling between the zonal flow and the pump wave through the production of the sidebands. The final normalized evolution equation reads as

$$(1-\nabla_{\perp}^{2})\frac{\partial\phi}{\partial t}=\frac{\partial\phi}{\partial y}+[\phi,\ \nabla_{\perp}^{2}\phi], \tag{1}$$

where ϕ and $\nabla_{\perp}^{2}$ denote the electrostatic potentials of the drift-wave turbulence and two-dimensional Laplacian perpendicular to the magnetic field, respectively. Here, we assume the turbulence due to the ETG mode, so that the adiabatic ion response is employed. The adiabatic ion response is a direct consequence of short wavelength limit of ion scale perturbation obtained from the gyrokinetic formalism. In equation (1), space coordinate, time, and potential are normalized as follows

$$(x,\ y,\ t,\ \phi)\rightarrow\left(\frac{x}{\rho_e},\ \frac{y}{\rho_e},\ \frac{v_{te}}{L_{ne}}t,\ \frac{L_{ne}}{\rho_e}\frac{e\phi}{T_e}\right), \tag{2}$$

where $\rho_e=v_{te}/\omega_{ce}$, $v_{te}=\sqrt{T_e/m_e}$, $L_{ne}=(\mathrm{d}\ln n_e/\mathrm{d}x)^{-1}$. This equation is the same in form as the reduced version of the three-field nonlinear evolution equation of ETG driven turbulence under the lowest order and electrostatic limits in a slab geometry. The Poisson bracket presents the dominant nonlinear term which is originated from the convective nonlinearity.

As explained in section 1, two approaches, i.e., the CMC method and the WK one are employed. In these methods, a same assumption has been taken in which monochromatic wave of zonal flow is sampled. However, the generation of the zonal flow is purely a nonlinear modulation process, which essentially involves the complex coupling along whole spectral distribution under the wave number and frequency match conditions. In the description of the zonal flow generation, we should not only consider the spectral structure of the turbulent pump fluctuations, but also need to incorporate the spectral characteristics of the zonal flow. In this case, each spectral component of zonal flow may be linked successively due to the nonlinear couplings with pumps through the modulation excitation of sidebands. As a result, the zonal flow instability is then characterized as a global mode rather than a monochromatic wave.

In this section, we propose a minimal model to describe the primary nonlinear interaction involved among the ambient turbulence and zonal flow in order to capture the essential feature of zonal flow spectral effects arising in the zonal flow generation processes, i.e., the effect of finite band width. Hence, as a minimal model to characterize the spectral effects of both zonal flow and the ambient pump turbulence, we represent those spectrum by means of two monochromatic waves with a same spectral difference dk_x. The corresponding four sidebands can be produced through the nonlinear interaction similar to those appeared in the conventional four-wave coupling model, as illustrated in figure 1. This is the proposed our minimal eight-wave coupling model. It looks to be equivalent to a system involving two sets of four-wave coupling loops with a spectral shift dk_x. However, this system is not a simple superposition of two sets, a new cross coupling loop appears between these two sets through the sideband, which links two modulation processes as explained in the following. This may change the characteristics of zonal flow instability. Based on this idea, we will derive a dispersion relation of zonal flow instability in the frame of coherent mode coupling method as follows.

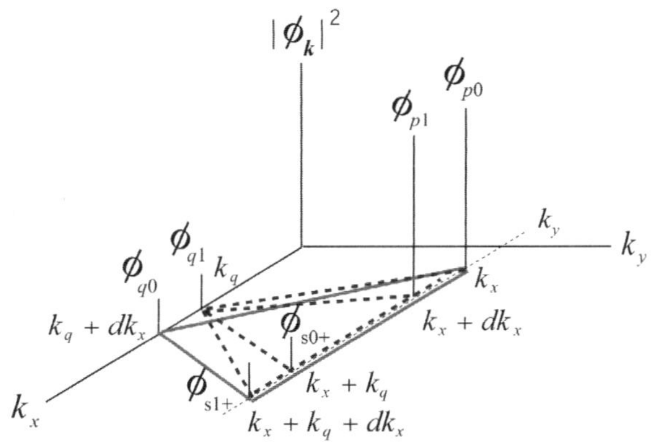

Figure 1. Schematic view of eight-wave modeling in (k_x, k_y) space. The primary spectral effects of both ambient turbulence and zonal flow are taken into account. The dotted loop1 consists of two mode coupling pairs $[\phi_{k_x,k_y},\ \phi_{k_q,0},\ \phi_{k_x+k_q,k_y}]$ and $[\phi_{k_x+dk_x,k_y},\ \phi_{k_q,0},\ \phi_{k_x+k_q+dk_x,k_y}]$. The coupling between $[\phi_{k_x,k_y},\ \phi_{k_q+dk_x,0},\ \phi_{k_x+k_q+dk_x,k_y}]$ forms another modulational loop drawn by a solid line. The two modulational loops are connected through the side band $\phi_{k_x+k_q+dk_x,k_y}$.

The pump wave packets are represented as

$$\tilde{\phi}_p(\boldsymbol{r}, t) = \sum_{j=0}^{1} \phi_{k_{xj},k_y} \exp i(k_{xj}x + k_y y - \omega_{0j}t) + c.c., \tag{3}$$

where $(k_{xj},\ k_y) = (k_x + jdk_x,\ k_y)$ are the radial and poloidal wave numbers, and the frequencies ω_{0j} satisfy the dispersion relation of the drift wave under equation (1), i.e., $\omega_{0j} = -k_y/(1 + k_{xj}^2 + k_y^2)$. Arbitrarily small perturbation is chosen as a seed of the zonal flow with two wave components

$$\tilde{\phi}_q(\boldsymbol{r}, t) = \sum_{j=0}^{1} \phi_{k_{qj},0} \exp i(k_{qj}x - \omega_{qj}t) + c.c., \tag{4}$$

where $k_{qj} = k_q + jdk_x$ are the radial wave numbers. Corresponding side bands can be produced through the nonlinear coupling between the zonal components and pump waves as expressed as follows

$$\tilde{\phi}_{s\pm}(\boldsymbol{r}, t) = \sum_{j=0}^{1} \phi_{k_{xj\pm},k_y} \exp i(k_{xj\pm}x + k_y y - \omega_{\pm j}t) + c.c. \tag{5}$$

with denotations $k_{xj\pm} = k_{xj} \pm k_q$ and $\omega_{\pm j} = \omega_{0j} \pm \omega_{qj}$. The frequencies of all sidebands are also subject to the dispersion relation of drift wave. Considering all possible nonlinear coupling among these eight waves involved in the model under equation (1), the perturbed equations of the zonal flow components and corresponding sidebands can be given as follows

$$\begin{aligned}
&- \mathrm{i}\omega_q(1 + k_q^2)\phi_{k_q,0} \\
&= k_q k_y(2k_x k_q + k_q^2)\phi_{k_x+k_q,k_y}\phi_{-k_x,-k_y} + k_q k_y(2k_x k_q - k_q^2)\phi_{-k_x+k_q,-k_y}\phi_{k_x,k_y} \\
&\quad + k_q k_y[(k_x + k_q + dk_x)^2 - (k_x + dk_x)^2]\phi_{-k_x-dk_x,-k_y}\phi_{k_x+k_q+dk_x,k_y} \\
&\quad + k_q k_y[-(-k_x + k_q - dk_x)^2 + (k_x + dk_x)^2]\phi_{k_x+dk_x,k_y}\phi_{-k_x+k_q-dk_x,-k_y}
\end{aligned} \tag{6}$$

$$-\mathrm{i}[(\omega_0+\omega_q)(1+k_0^2+k_q^2+2k_xk_q)+k_y]\phi_{k_x+k_q,k_y} = k_qk_y(k_0^2-k_q^2)\phi_{k_x,k_y}\phi_{k_q,0} \quad (7)$$

$$-\mathrm{i}[(-\omega_0+\omega_q)(1+k_0^2+k_q^2-2k_xk_q)-k_y]\phi_{-k_x+k_q,-k_y} = k_qk_y(k_0^2-k_q^2)\phi_{-k_x,-k_y}\phi_{k_q,0} + (k_q+dk_x)k_y[k_0^2-k_q^2+2(k_x-k_q)dk_x]\phi_{-k_x-dk_x,-k_y}\phi_{k_q+dk_x,0} \quad (8)$$

$$-\mathrm{i}[(\omega_0+\omega_q)\{1+(k_x+k_q+dk_x)^2+k_y^2\}+k_y]\phi_{k_x+k_q+dk_x,k_y} = k_qk_y(k_0^2-k_q^2+2k_xdk_x+dk_x^2)\phi_{k_x+dk_x,k_y}\phi_{k_q,0} + (k_q+dk_x)k_y(k_0^2-k_q^2+2k_xdk_x-dk_x^2)\phi_{k_x,k_y}\phi_{k_q+dk_x,0} \quad (9)$$

$$-\mathrm{i}[(-\omega_0+\omega_q)\{1+(-k_x+k_q-dk_x)^2+k_y^2\}-k_y]\phi_{-k_x+k_q-dk_x,-k_y} = -k_qk_y(k_0^2-k_q^2+2k_xdk_x+dk_x^2)\phi_{-k_x-dk_x,-k_y}\phi_{k_q,0} \quad (10)$$

$$-\mathrm{i}\omega_q[1+(k_q+dk_x)^2]\phi_{k_q+dk_x,0} = (k_q+dk_x)k_y[(k_x+k_q+dk_x)^2+k_y^2-k_0^2]\phi_{-k_x,-k_y}\phi_{k_x+k_q+dk_x,k_y} + (k_q+dk_x)k_y[(k_x+dkx)^2-(-k_x+k_q)^2]\phi_{k_x+dk_x,k_y}\phi_{-k_x+k_q,-k_y}. \quad (11)$$

Note that we have assumed that two complex frequencies of zonal flow components are the same, i.e., $\omega_{q0}=\omega_{q1}=\omega_q$. From equation (6) to equation (11), we can derive a dispersion relation of the zonal flow;

$$\omega_q=\omega_q(k_q;k_{xj},k_y,dk_x,|\phi_{k_{xj},k_y}|^2), \quad (12)$$

which is expressed by the eleventh order algebraic equation with respect to the complex zonal flow frequency $\omega_q=\Omega_q+\mathrm{i}\gamma_q$. The maximum growth rate is chosen from the eleven complex roots. Note that in the zonal flow equations, two components are linked through the cross coupling overlapping on the side band components with wave numbers $(k_{x1\pm},k_y)$, which depend on the finite band width of the zonal flow. This new cross coupling may enhance or reduce the zonal flow generation due to the complex modulation processes. In the following we will calculate the dispersion relation to investigate the effect of spectral band width.

2.2. Numerical results of dispersion relation

First, the growth rate of the zonal flow in this newly developed eight-wave coupling model is compared with that in the conventional four-wave model. Note that the generation of secondary structures is related to the spectral isotropy of ambient turbulence, namely, a radially longer wavelength pump wave ($k_x \ll k_y$) tends to excite zonal flow[15]. Here, we choose radial wave number k_x from 0.1 to 0.5 and poloidal wave number $k_y=0.98$ for pump fluctuation in order to excite zonal flow efficiently. The spectral difference between two turbulent wave packets or two zonal flow components is chosen to be $dk_x=0.0125$. Note that the generation of the zonal flow depends on the pump amplitude or pump energy, it is assumed that the pump energy is the same in both models to directly compare the growth rates. Hence the pump amplitude of the two components in the eight-wave coupling model is reduced by a factor $1/\sqrt{2}$. Figure 2 plots the growth rate of the zonal flow γ_q with respect to the radial wave number of pump modes k_x in both models. It shows that the growth rate of zonal flow in the eight-wave model is larger than that in the four-wave model. It is understood that this enhancement of zonal flow instability originates from a new modulation loop due to the spectral band width of zonal flow as explained

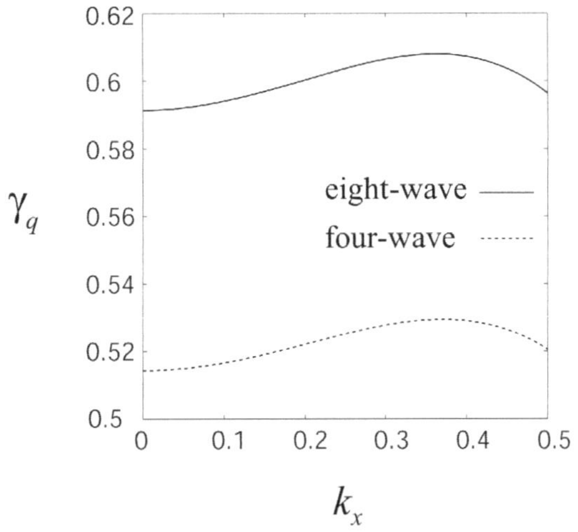

Figure 2. The growth rate γ_q of zonal flow as a function of the wave number of turbulence k_x. A solid line shows the growth rate from the eight-wave model and a dotted line is that from the four-wave model. Other parameters are $(k_y,\ |\phi_0|^2) = (0.98,\ 4.0)$ for pump wave. We choose wave numbers of the zonal flow $k_q = 0.75$ which provides the maximum growth rate. The distance between two radial wave numbers of turbulence dk_x is chosen to 0.0125.

above. It has been examined that if the modulation contribution of the new cross coupling between the pump (k_{x1}, k_y) and the zonal component $(k_{q0}, 0)$ to the sideband component $(k_{x1} + k_{q0}, k_y)$, which is the same as $(k_{x0} + k_{q1}, k_y)$, is artificially removed, the growth rate of each zonal flow component in the eight-wave model become the same as that in the four-wave model as $dk_x \to 0$. On the other hand, if monochromatic zonal flow wave is considered in the coupling system with spectral band width of the pumps but keeping the same potential energy, the result is completely the same as that in four-wave coupling model. These analyses indicate that the new cross coupling due to the spectral band width of zonal flow plays an essential role in the enhancement of zonal flow instability. This new cross coupling path can not only link the two sets of four-wave coupling loops, but also enhance the modulation process through the contribution to the sidebands.

Next, we examine the dependence of the zonal flow growth rate on the spectral difference dk_x. Figure 3 plots the growth rate of the zonal flow γ_q with respect to dk_x. The spectral difference is chosen from $dk_x = 0$ to 0.1. It is shown that the growth rate is approximately 0.594 for $dk_x < 0.1$ where two turbulent components are adjacent to each other, and the dk_x dependence of the zonal flow growth rate is found to be weak in this region. At the limit of $dk_x = 0$ which corresponds to the four-wave model, the growth rate is approximately 0.516. Note that the growth rate is discontinuous at $dk_x = 0$. The result shows that there is a qualitative difference between the models with and without zonal flow spectrum, suggesting that this *global nature of the zonal flow* should be taken into account in theoretical analyses of the zonal flow generation.

3. Numerical simulation

As shown in theoretical analysis in the minimal modeling, the zonal flow instability is enhanced due to the spectral effects of the zonal flow. In order to justify the validity of the analysis and to evaluate the role of the zonal flow spectrum in realistic turbulence, a spectral code is advanced to solve 2D HM equation. Replacing differential operators in

Journal of Physics: Conference Series **123** (2008) 012031 doi:10.1088/1742-6596/123/1/012031

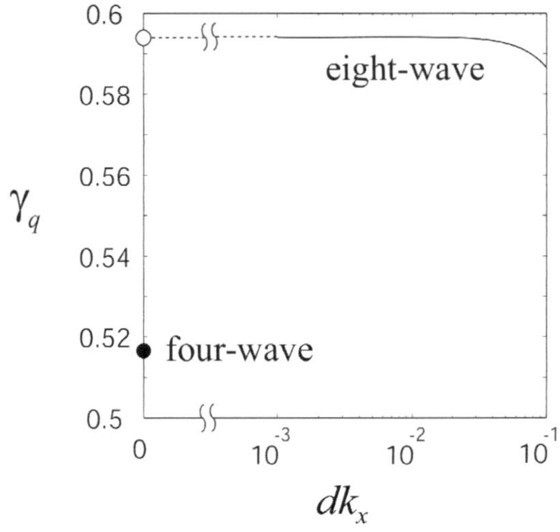

Figure 3. The growth rate of the zonal flow γ_q with respect to the difference between two radial wave numbers of turbulence dk_x. Other parameters are $(k_x,\ k_y) = (0.16,\ 0.98)$ and $|\phi_0|^2 = 4.0$ for pump wave. We choose wave numbers of the zonal flow $k_q = 0.75$ which provides the maximum growth rate.

space coordinates by radial and poloidal wave numbers, we solve the ordinary differential equations by using the 6-th order Runge-Kutta method with respect to the complex amplitude of each component $\phi(\mathbf{r}, t) = \phi_{k_x,\ k_y}(t) \exp i(k_x x + k_y y + \theta_{k_x}) + c.c.$, where θ_{k_x} indicates initial phase for different k_x in the Fourier space. The grid size δk_x and simulation box size k_{xmax} in the radial wave number space are chosen to be $\delta k_x = 0.0125$, $k_{xmax} = 4.0$, respectively. Here, single component $k_y = 1.0$ is chosen along y-direction. Along the x-direction, we assume the Gaussian-type spectral distribution in k_x space for the turbulence as $E_{k_x} = E^0_{k_x} \exp\left[-(k_x - k_{xc})^2/2(\Delta k_x)^2\right]/\Delta k_x$, where k_{xc} and Δk_x represent the center value of the spectrum and spectral width, respectively. In performing numerical calculation, we choose 27 discrete pump waves by keeping total pump energy constant, i.e, $E = \sum_{k_x,\ k_y}(1 + k_0^2)|\phi_{k_x,\ k_y}|^2/2 \simeq 4.0$, which is the same as that in the four-wave model. Here, $dk_x = 0.0125$, $k_{xc} = 0.1$ and $\Delta k_x = 0.5$, respectively.

Many zonal flow components satisfying $(k_x \neq 0,\ k_y = 0)$ can be generated through the nonlinear interaction due to the pump spectrum. Firstly, we examine the effect of the zonal flow spectrum on the zonal flow instability by performing two simulations with and without a zonal flow spectrum. The former is a multi-mode case, where we allow all zonal flow components in the numerical calculation. In the latter case, only one zonal flow component is chosen with wave number $k_q = 0.75$ which corresponds to the maximum growth rate in the conventional four-wave model. Figure 4 plots the temporal evolution of total zonal flow energy for all cases. Here, we assume a coherent phase for all initial turbulent components i.e., $\theta_{k_x} = 0$ for simplicity. The growth rate in the case with single zonal flow component is observed to be identical to that in the four-wave model. It can be also seen that the growth rate in the multi-mode case is approximately 1.7, which is more than three times larger than that in the single mode case. Figure 5 shows the growth rate of zonal flow as a function of number of zonal flow n. The zonal flow component with wave numbers $(k_q \pm (n-1)dk_x, 0)$, $n = 1,2,3,\cdots$ is included in each numerical calculation one by one. Note that $n = 1$ corresponds to the single mode case as shown in figure 4, i.e., γ_q=0.52, and the growth rate finally converges to $\gamma_q \simeq 1.7$, which corresponds to the

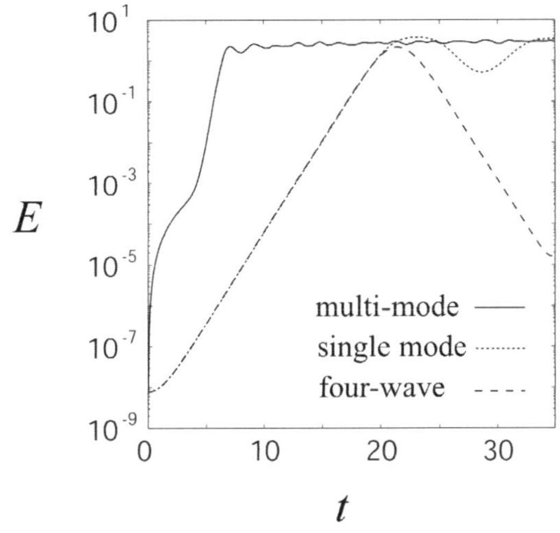

Figure 4. The temporal evolution of total zonal flow energy for all cases. We choose $(k_x,\ k_y) = (0.1,\ 1.0)$ and $E(0) = 4.0$ for pump wave. Wave number of the zonal flow seed is chosen as $k_q = 0.75$ which provides the maximum growth rate. dk_x is chosen to 0.0125. The grid size δk_x and simulation box size k_{xmax} for radial wave number space are chosen to $\delta k_x = 0.0125$, $k_{xmax} = 4.0$, respectively.

multi-mode case. It is found that the growth rate is enhanced with the number of zonal flow, and two zonal flow components ($n = 2$) are sufficient to enhance the growth rate. These results support the theoretical analysis above, showing that the eight-wave model can be reasonably proposed as a minimal model including the spectral effect of the zonal flow.

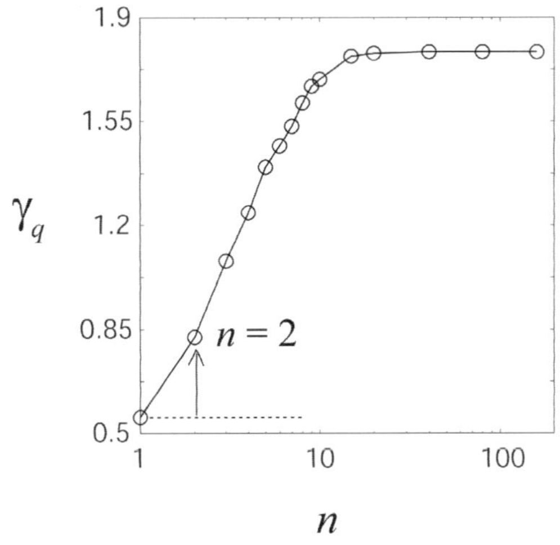

Figure 5. The growth rate of zonal flow as a function of number of zonal flow components n. Parameters are same as shown in figure 4.

Often observed in most experimental and numerical results, the realistic turbulence shows a strongly anisotropic structure. In order to represent this feature, we set a finite but non-zero random phase θ_{k_x} for the initial turbulent components in the numerical calculation. In figure 6, the temporal evolution of zonal flow energy for two different cases of θ_{k_x} is illustrated. When $\theta_{k_x} \neq 0$, growth rate is approximately 0.85, showing that the

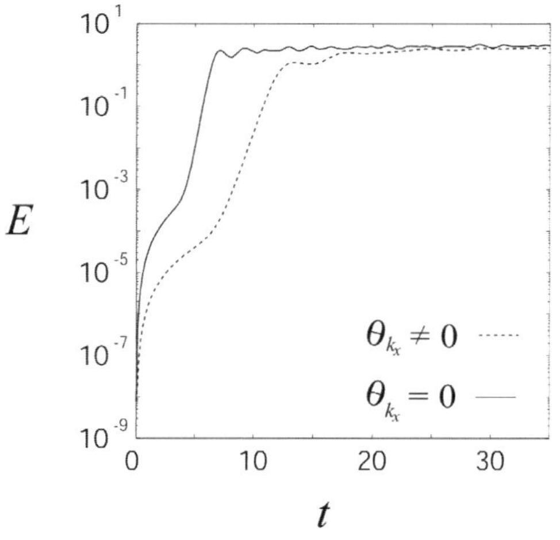

Figure 6. The temporal evolution of zonal flow energy for different initial phase of turbulence θ_{k_x}. Other parameters are same as shown in figure 4. We set a non-zero random phase for each turbulent component in $\theta_{k_x} \neq 0$ case. One typical result is plotted as a dotted line. In $\theta_{k_x} = 0$ case, all components are artificially chosen to zero (solid line).

enhancement of growth rate is weakened compared to the $\theta_{k_x} = 0$ case. This result reflects the relation of the spatial structure between turbulence and the zonal flow. Figure 7 shows potential structure of both turbulence and zonal flow in x-direction at the saturation state $t = 20$ in the cases of (a) $\theta_{k_x} = 0$ and (b) $\theta_{k_x} \neq 0$. The blue and red curves represent

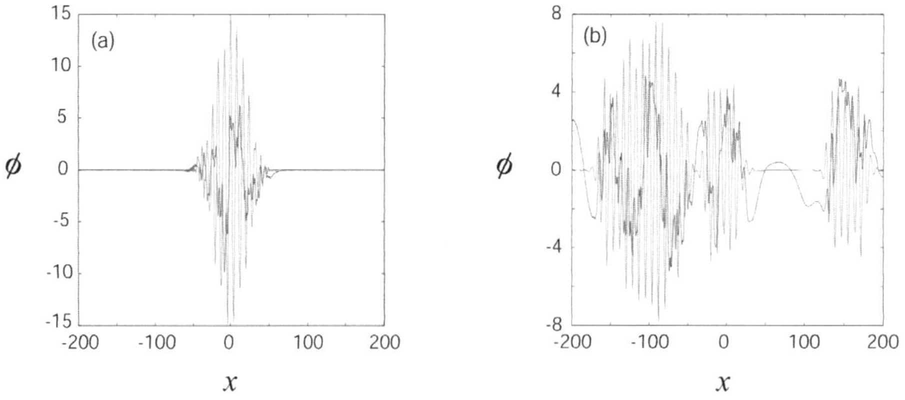

Figure 7. A potential structure of both turbulence (blue) and zonal flow (red) in x-direction at the saturation state $t = 20$ in the cases of (a) $\theta_{k_x} = 0$ and (b) $\theta_{k_x} \neq 0$.

potential of turbulence and zonal flow, respectively. When $\theta_{k_x} = 0$, the turbulence and the zonal flow are spatially localized around $-50 < x < 50$. On the other hand, they are delocalized and spreading to the whole region when $\theta_{k_x} \neq 0$. If the initial spatial structure of turbulence is localized, the zonal flow nonlinearly generated via the Reynolds stress of the ambient turbulence is also localized. As a result, the locally large amplitude turbulence may enhance the growth rate of zonal flow. It is also noted that in the case with $\theta_{k_x} \neq 0$, the zonal flow evolves gradually before it reaches the final saturation level.

4. Summary and discussion

In this work, we have theoretically and numerically investigated the spectral effects of both the turbulence and the zonal flow on zonal flow instability in micro-scale ETG turbulence based on the HM equation. Here, a coherent mode coupling method is employed for the modulational analysis of zonal flow generation. To capture the essential feature of the spectral effects of zonal flow, we have extended the conventional four-wave coupling model to a minimal eight-wave model, which includes two sets of the conventional four-wave coupling systems with a spectral shift dk_x. Assuming turbulence as a sum of two sinusoidal waves, a dispersion relation of the zonal flow instability which is represented by the eleventh order algebraic equation with respect to a complex frequency $\omega_q = \Omega_q + i\gamma_q$ is derived. Numerically solving the dispersion relation, we have shown that the spectral effect of zonal flow can increase its growth rate. The result shows that there is a qualitative difference between models with or without zonal flow spectrum. It is found that a new cross coupling loop through the overlapping of sidebands can enhance the modulation instability of zonal flow generation. The zonal components are coupled each other so that the zonal flow is characterized by a *global nature of the zonal flow* should be taken into account in some theoretical analyses on the zonal flow generation.

In order to justify the validity of the eight-wave model and to investigate the dependence of the initial phase of the turbulence on the zonal flow instability, a spectral code to solve 2D HM equation is advanced. As an example, we choose a sum of 27 discrete modes for the pump wave to approximate the Gaussian-type turbulent structure in the Fourier space. Numerical simulation shows that the more initial phase becomes coherent, the more the zonal flow instability increases. This result is understandable because the turbulence and the zonal flow are spatially localized with coherent phase relation, the interaction becomes strong locally. In the case with single zonal flow, the growth rate is the same as that from the conventional four-wave model. We conclude that the zonal flow spectrum is a more essential characteristic of the zonal flow instability whereas only the turbulence is characterized by a spectrum in some previous works. Here, we have discussed the zonal flow structure based on the micro-scale ETG driven turbulence-zonal flow system, but the present result can be applied to that of ITG driven turbulence-zonal flow system.

Acknowledgments

The authors are indebted to Dr. P. H. Diamond and Dr. M. Yagi for fruitful discussions and comments.

References

[1] Diamond P H, Itoh S-I, Itoh K and Hahm T S 2005 *Plasma Phys. Control. Fusion* **47** R35
[2] Lin Z, Hahm T S, Lee W W, Tang W M and White R B 1998 *Science* **281** 1835
[3] Hahm T S and Burrell K H 1995 *Phys. Plasmas* **2** 1648
[4] Kishimoto Y, Kim J Y, Horton W, Tajima T, LeBrun M J and Shirai H 1999 *Plasma Phys. Control. Fusion* **41** A663
[5] Hasegawa A and Mima K 1978 *Phys. Fluids* **21** 87
[6] Chen L, Lin Z and White R 2000 *Phys. Plasmas* **7** 3129
[7] Li J Q and Kishimoto Y 2002 *Phys. Plasmas* **9** 1241
[8] Lashmore-Davies C N, McCarthy D R and Thyagaraja A 2001 *Phys. Plasmas* **8** 5121
[9] Manfredi G, Roach C M and Dendy R O 2001 *Plasma Phys. Control. Fusion* **43** 825
[10] Uzawa K, Kishimoto Y and Li J Q 2006 *Plasma Fusion Res.* **1** 024
[11] Smolyakov A I, Diamond P H and Shevchenko V I 2000 *Phys. Plasmas* **7** 1349
[12] Malkov M A, Diamond P H and Smolyakov A I 2001 *Phys. Plasmas* **8** 1553
[13] Anderson J and Kishimoto Y 2007 *Phys. Plasmas* **14** 012308
[14] Vedenov A A, Gordeev A V and Rudakov L I 1967 *Plasma Phys.* **9** 719
[15] Li J Q and Kishimoto Y 2005 *Phys. Plasmas* **12** 054505

Journal of Physics: Conference Series **123** (2008) 012032 doi:10.1088/1742-6596/123/1/012032

Internal transport barrier simulation with pellet injection in tokamak and helical reactor plasmas

Y Higashiyama[1], K Yamazaki[1], J Garcia[2], H Arimoto[1] and T Shoji[1]

[1]Nagoya University, Chikusa-ku, Nagoya 464-8603, Japan,
[2] Association EURATOM-CEA, CEA/DSM/DRFC, CEA-Cadarache, F-13108 St. Paul lez Durance, France

E-mail: yamazaki@ees.nagoya-u.ac.jp

Abstract. In the future fusion reactor, the plasma density peaking is important for the increase in the fusion power gain. The density control and the internal transport barrier (ITB) formation due to the pellet injection have been simulated in tokamak and helical reactors using the toroidal transport linkage code TOTAL. Firstly, the pellet injection simulation is carried out including the neutral gas shielding model and the mass relocation model in the TOTAL code, and the effectiveness of the high field side (HFS) pellet injection is clarified. Secondly, the ITB simulation with the pellet injection is carried out with the confinement improvement model based on the **E**×**B** shear effects, and it is found that the deep pellet penetration is helpful for the ITB formation as well as the plasma core fuelling in the reversed shear tokamak reactor, but the deep pellet penetration is not effective in the helical reactor.

1. Introduction

The total fusion power strongly depends on the radial profile of plasma density, and the density control is important for the fusion reactor operation. To control the plasma density and pressure profiles, the pellet injection is considered as a prospective technique. The pellet injection is used for the plasma core fuelling and for making the density profile peaked. In JET and other tokamak experiments, it was shown that the density profile modifications disagree with pellet ablation theory [1] that assumes the pellet particles remain on the magnetic field lines where they are ionized [2]. The pellet penetration depth measured by the pellet light emission agreed well with the pellet ablation theory. This suggested that a fast outward major-radius drift may occur during the pellet ablation and toroidal symmetrization processes. To test this hypothesis the experiment of the high-field side (HFS) pellet injection has done in ASDEX-Upgrade, and it was shown that the fuelling efficiency and the penetration depth of pellets are improved [3]. The similar results are observed in DIII-D [4] and other tokamak experiments, and the HFS pellet injection is expected to be an effective technique of plasma core fuelling in future tokamak reactors.

When the ITB is formed in the plasma, it brings good confinement and rather peaked pressure profile so that the fusion power gain is increased. The ITB is observed in both tokamak and helical systems and it is classified into several high performance modes, such as high β_p mode [5], reversed shear mode [6], and pellet enhanced performance mode (PEP) [7, 8]. Transport simulation studies have been carried out focusing on the ITB formation in tokamak and helical plasmas. In helical systems the ITB model based on Bohm and GyroBohm-like transport with **E**×**B** shear flow effects has already been compared with the LHD experimental ITB [9] and this model is inspired from the JET mixed-model [10]. This model is introduced into the toroidal transport linkage TOTAL code [11, 12], and is applied to the 1-dimensional (1-D) ITB formation simulation of both 3-D equilibrium helical and 2-D equilibrium tokamak plasmas.

Both the pellet injection and the ITB formation have big influence on the density profile and the fusion power output, so that we investigate the relationship between the ITB formation and the pellet injection density control in tokamak and helical reactor plasmas by using the TOTAL simulation code. Section 2 and 3 will describe the details of the transport models and the HFS pellet injection model included in the TOTAL code, and simulation results will be shown in section 4. The conclusion and discussion will be given in section 5.

2. Transport model description

The Bohm and GyroBohm mixed transport model with the **E**×**B** shear flow effect has already been compared with experimental results of tokamak and helical plasmas with ITBs [9, 10]. The most widely accepted explanation for the ITB formation relies on the suppression of ion temperature gradient (ITG) turbulence due to **E**×**B** shear flow. The suppression of the turbulence might occur when the **E**×**B** flow shearing rate $\omega_{E\times B}$ exceeds the ITG linear growth rate γ_{ITG} . The shearing rate $\omega_{E\times B}$ is defined as [13, 14]

$$\omega_{E\times B} \cong \left| \frac{RB_\theta}{B_\phi} \frac{\partial}{\partial r} \left(\frac{E_r}{RB_\theta} \right) \right|, \tag{1}$$

where E_r , B_θ and B_ϕ are the radial electric field, the polidal and toroidal magnetic field, respectively. In tokamaks, the radial electric field E_r and poloidal rotation are not easily measured directly, and E_r is calculated from the plasma radial force balance equation under the assumption that the poloidal velocity can be given according to the neoclassical theory [15, 16]. In this paper, we employ more simple assumption for the E_r gradient as

$$\frac{dE_r}{dr} \cong -\frac{1}{en_i^2} \frac{dn_i}{dr} \frac{dp_i}{dr} , \tag{2}$$

in the H-mode condition [17], where n_i and p_i are ion density and ion pressure, respectively. In the helical reactor, the radial electric filed is determined from the ambipolarity condition of helical ripple-induced neoclassical flux [11].

The ITG growth rate γ_{ITG} is defined as [18]

$$\gamma_{ITG} = \frac{(\eta_i - 2/3)^{1/2} |s| c_i}{qR}, \tag{3}$$

where $\eta_i = L_n/L_T$ (L_n = dln_i/dr and L_T = $dlnT_i/dr$), $c_i = \sqrt{T_i / m_i}$, R is major radius, q is the safety factor, and s is the magnetic shear defined as

$$s = \frac{r}{q}\left(\frac{dq}{dr} \right). \tag{4}$$

Most of transport simulations adopt a thermal diffusion coefficient χ in the form

$$\chi = \chi_{neoclassical} + \chi_{anomalous} \times F\left(\omega_{E\times B} / \gamma_{ITG} \right), \tag{5}$$

where the coefficient $\chi_{neoclassical}$ is the neoclassical part of thermal diffusion coefficient, and $\chi_{anomalous}$ is the anomalous part. The effect of the **E**×**B** shear stabilization on the anomalous transport is introduced by an improvement factor F. In the present study, we employ the Bohm and GyroBohm mixed transport model [9] of $\chi_{anomalous}$;

$$\chi_{anomalous} = 4.0\chi_{Bohm} + 0.5\chi_{Gyrobohm}, \tag{6}$$

where
$$\chi_{Bohm} = 4\times 10^{-5} R \left| \frac{\nabla(n_e T_e)}{n_e B_\phi} \right| q^2, \tag{7}$$

and
$$\chi_{GyroBohm} = 5\times 10^{-6} \sqrt{T_e} \left| \frac{\nabla T_e}{B_\phi^2} \right|, \tag{8}$$

with a simple and widely-used form of F

$$F\left(\omega_{E\times B} / \gamma_{ITG} \right) = \frac{1}{1 + \tau \times \left(\omega_{E\times B} / \gamma_{ITG} \right)^{\gamma}}. \tag{9}$$

where n_e and T_e are electron density and electron temperature respectively. In this paper, fitting parameters τ and γ are used with $\tau = 2$, $\gamma = 4$ in tokamak case and $\tau = 15$, $\gamma = 2$ in helical case which were decided from the comparison between the JT-60U and LHD experimental data and the simulation results by using TOTAL code.

3. HFS pellet injection

The high-field-side (HFS) pellet injection is described as two processes which are the pellet ablation and the mass relocation. We simulate the HFS injection with the pellet penetration model combined with the ablation model and the mass relocation model. There are a few models which satisfactorily describe the pellet ablation and relevant experimental light emissions [19]. So, we use here the most popular one; the neutral gas shielding (NGS) model [1] given by

$$\frac{dN}{dt} = 5.2\times 10^{16} n_e^{0.333} T_e^{1.64} r_p^{1.333} M_i^{-0.333}, \tag{10}$$

where N, n_e (m^{-3}), T_e (eV), r_p (m), and M_i (u) are the number of particles in a pellet, the electron density, the electron temperature, the pellet size, and the pellet mass, respectively. The number of atoms in the pellet, N, is expressed as

Journal of Physics: Conference Series **123** (2008) 012032 doi:10.1088/1742-6596/123/1/012032

$$N = 2n_s\left(\frac{4\pi r_p^3}{3}\right), \tag{11}$$

where n_s is the molecular density of the solid hydrogen ($n_s = 3.12\times10^{28}$ m^{-3}). The ablation rate along the pellet path l is obtained as

$$\frac{dN}{dl} = 4.38\times10^3 N^{0.444} n_e^{0.333} T_e^{1.64} M_i^{-0.333} / V_p, \tag{12}$$

where V_p is the pellet injection velocity.

The mass relocation δx ($x = r/a$ is the normalized minor radius) with the plasmoid drift in the major-radius direction is described as [20]

$$\delta x = \delta r/a \sim \delta\psi/\Delta\psi, \tag{13}$$

where ψ is poloidal flux. The poloidal flux perturbation $\delta\psi$ scaling is

$$\delta\psi = q\beta B_\phi(1 + qL_c/a)^{-1} a^{-1}\delta n r_0^2 (n + \langle\delta n\rangle)^{-1}, \tag{14}$$

and total poloidal flux within the plasma boundary is

$$\Delta\psi = \psi(a) - \psi(0) \sim \int r dr / q, \tag{15}$$

where β, r_0 and L_c are toroidal beta value, cylindrical cloud with cross-field radius and height, respectively. Figure 1 shows an example of a simulation, where the ablation profile is calculated with the standard NGS model, and that is sifted in the direction to the center by δx calculated with the mass relocation model [20].

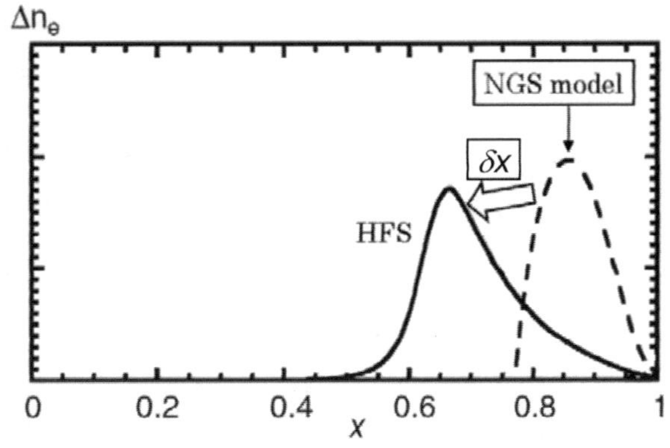

Figure 1. Schematic diagram of the HFS pellet injection calculation. The ablation profile is calculated with the standard NGS model, and that is sifted in the direction to the center by Δx calculated with the mass relocation model.

Journal of Physics: Conference Series **123** (2008) 012032 doi:10.1088/1742-6596/123/1/012032

4. Simulation results

4.1. Demonstration of HFS pellet injection

The machines parameters of the target tokamak and helical reactors are shown in table 1, which were typical designs optimized using the reactor design system PEC (Physics-Engineering-Cost) code [12]. These parameters refer to the 1GW electric power fusion reactor design. Typical results of the HFS pellet injection simulation in tokamak reactor are shown in figure 2. The size of a deuterium pellet is chosen as r_p = 2.5mm. The pellet ablation densities are shown for the different pellet injection velocity V_p. Here, the radial parabolic temperature and flat density profiles are assumed as $T(x) = T_0(1-x^2)^3$, $n(x) = n_0(1-x^2)^{0.5}$ with $T_0 = 30$keV and $\langle n_e \rangle = 10^{20}\text{m}^{-3}$. This figure shows that the HFS injection could provide rather deep fuelling in the reactor-grade tokamak plasma, and the further increase in the injection velocity improves the central fuelling. For assumed temperature and density profiles, the HFS injection with the pellet velocity of 1 km/s could provide the density increase at the normalized radius r/a ~ 0.1. Moreover, it shows that the reversed shear mode improves the central fuelling for the HFS injection based on the mass relocation model of equation (13).

Table 1. Reactor machine parameters used in this paper

	Tokamak reactor (ITER-like)	Helical reactor (LHD-like)
R_p [m]	5.2	12.5
a_p [m]	1.2	2.2
κ	2.0	-
δ	0.5	-
B_t [T]	7.1	4.6

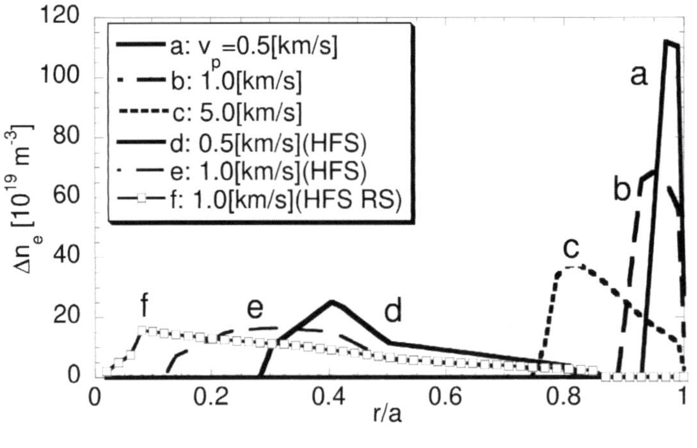

Figure 2. Model prediction for the HFS pellet injection in tokamak reactor. Ablation profiles are shown by a, b and c, and HFS injection simulation results are given by d, e, and f depending on the pellet injection velocities. The profiles d and e are in the normal shear case, and f is in the reversed shear case.

4.2. ITB simulation with pellet injection

In the previous subsection, we show that the pellet injection could provide deep fuelling in tokamak reactor by the HFS injection. So that we compare with deep (by HFS injection) and shallow (by medium field side injection) penetration depth cases in the reversed shear ITB plasmas. The results are shown in figure 3. In the present study, the high pedestal temperature is artificially chosen so that the required fusion power for the 1GW electric power can be obtained. We can see that there are clear differences between both cases. In the deep penetration case shown in the left-hand figure, the large gradient of density and pressure gradients, and the strong reduction of $F(\omega_{E\times B}/\gamma_{ITG})$ defined at equation (9), appear at $r/a \sim 0.5$-0.6, so the ITB is formed. However in the shallow case in the right-hand figure the ITB is not formed.

This reason is shown in the bottom two figures in figure 3. In this simulation, the ITB formation is determined by two parameters, $\omega_{E\times B}$ and γ_{ITG}. In both cases, γ_{ITG} is reduced at $r/a \sim 0.6$ where the magnetic shear $s \sim 0$. The suppression occurs when $\omega_{E\times B}$ exceeds γ_{ITG}, so the ITB tends to be formed at low γ_{ITG} position. However, in the shallow penetration case the rate $\omega_{E\times B}$ is small at $r/a \sim 0.6$, because the gradient of radial electric field dE_r/dr is small depending on the term dn_i/dr in equation (2). The transient density profile and the relevant clear ITB formation depend on the pellet penetration depth. The deeper pellet penetration brings the larger gradient of E_r and the larger shearing rate $\omega_{E\times B}$ at the position of small γ_{ITG}, so that the ITB is formed there. We can say that the deep pellet penetration plays an important role in the ITB formation in the model adopted here, and the HFS pellet injection is a quite effective technique for the confinement improvement as well as the core density fuelling in the reversed shear tokamak reactor.

Figure 4 shows the results of the ITB simulation in the helical reactor. It is compared with deep and shallow pellet penetration depth cases (same in tokamak). In this simulation, different from tokamak case, the deep penetration case corresponds to the high speed pellet injection and not to the HFS injection, because the effectiveness of HFS injection has not been observed so far in helical system [21]. This might be because the field connection length between the high filed side and the low filed side is quite short in comparison with tokamak case. In the helical reactor, the radial electric filed is determined from the ambipolarity condition of helical ripple-induced neoclassical flux, and the large gradient of radial electric filed and strong reduction of $F(\omega_{E\times B}/\gamma_{ITG})$ appear at $r/a < 0.4$ in both cases. In this area flat q profile make small γ_{ITG} and the inside gradient of electric field brings large **E**×**B** shear flow, so that the strong reduction in the anomalous transport is occurred. This inside negative electric field is formed autonomously in this reactor grade plasma condition and it seems to be unrelated to the pellet penetration depth. These results show that the deeper penetration is not needed for the ITB formation in the helical reactor.

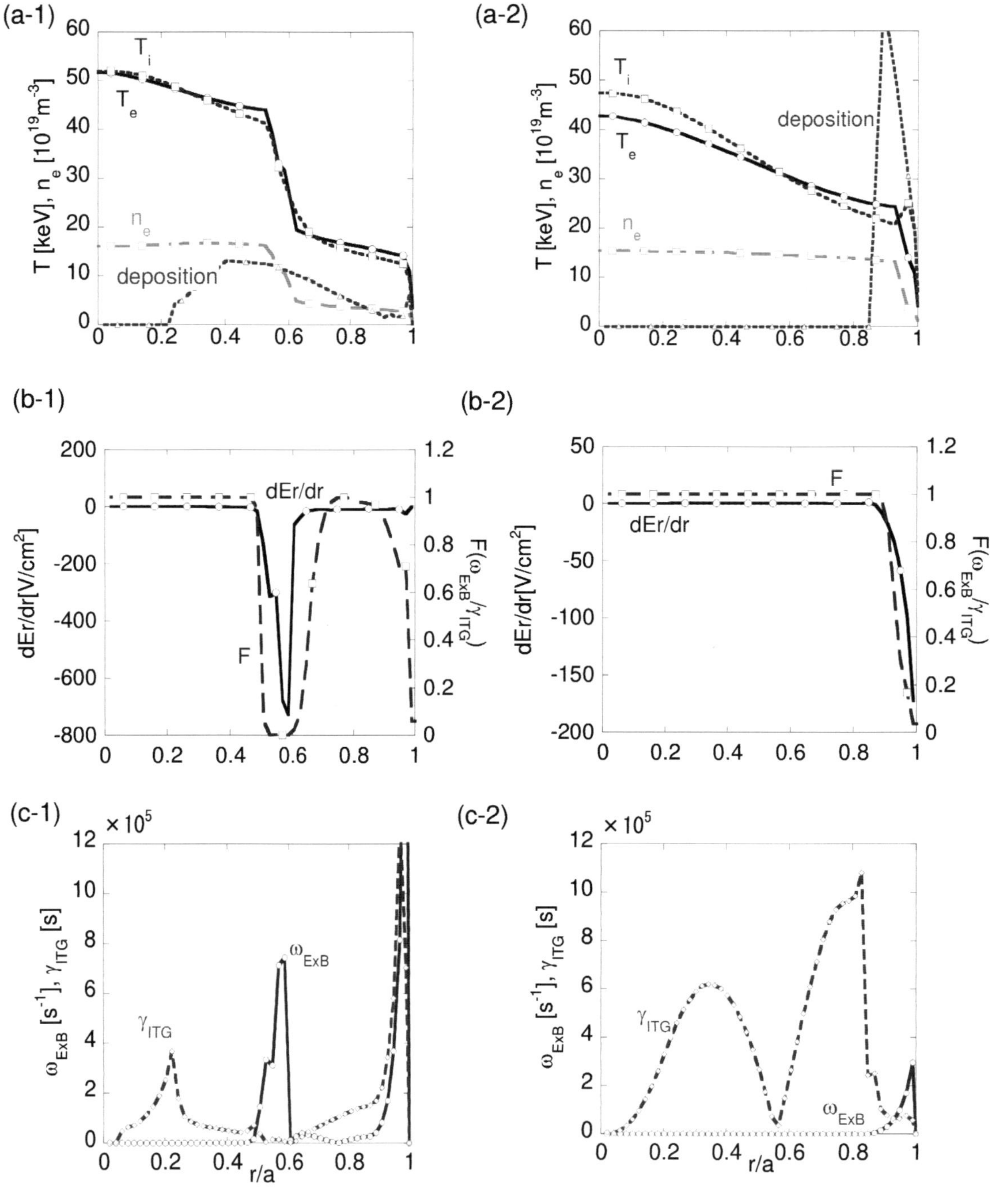

Figure 3. ITB simulation results with pellet injection in the reversed shear tokamak reactor. Left figures denote deep pellet penetration case (by HFS), and right figures are shallow penetration case (by the medium field side injection). The upper figures (a-1,2) show ion and electron temperatures, electron density, and pellet deposition profiles. The middle figures (b-1,2) show radial electric field derivative and $F(\omega_{ExB}/\gamma_{ITG})$ defined at equation (9), and the lower figures (c-1,2) show $\omega_{E\times B}$ and γ_{ITG} profiles.

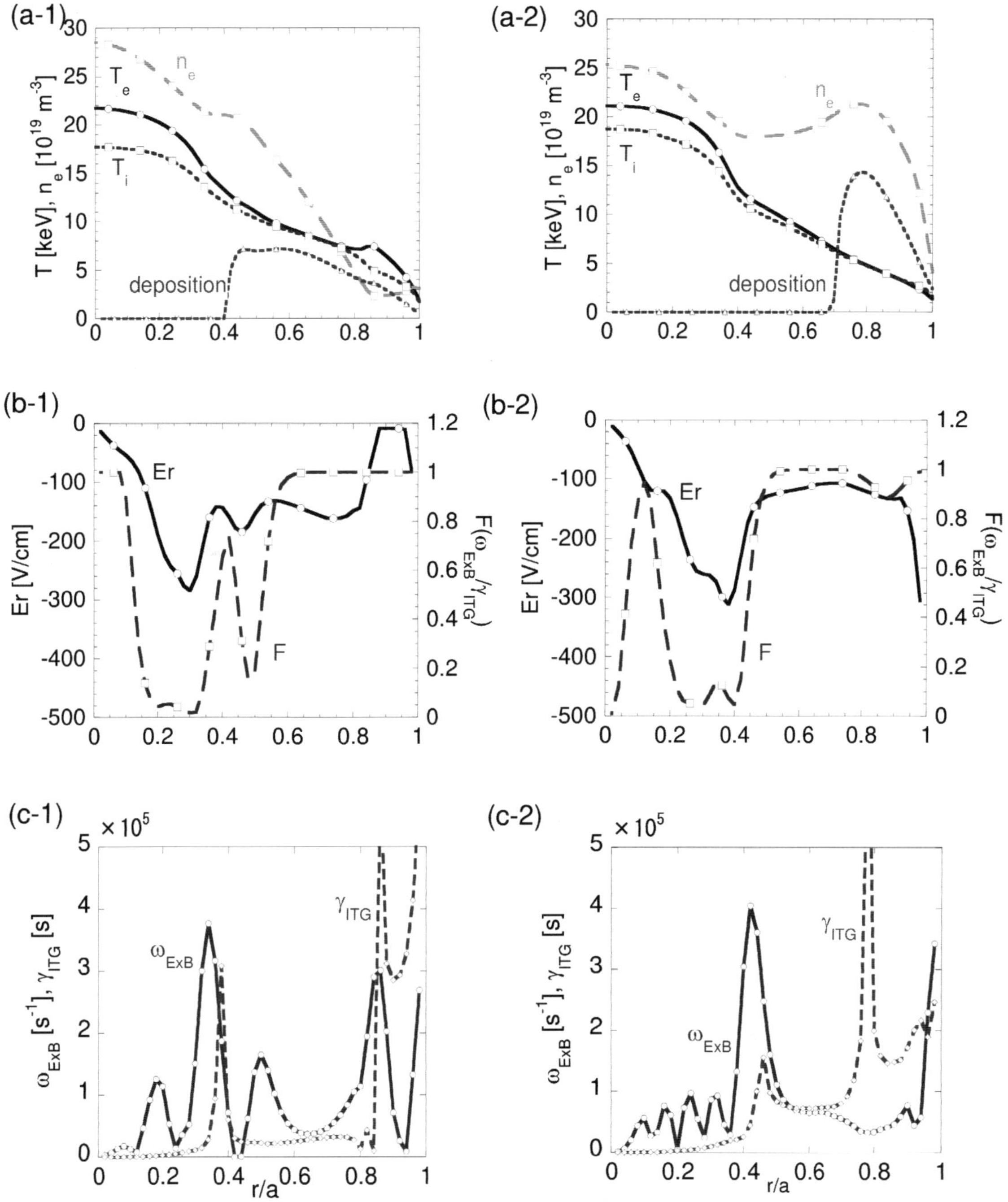

Figure 4. ITB simulation results with pellet injection in helical reactor. The right-hand figure corresponds to the shallow penetration case (pellet size = 2.5mm, injection velocity = 20 km/s), and the left-hand figure is the deep penetration case (pellet size = 2.5mm, injection velocity = 500 km/s). The upper figures (a-1,2) show ion and electron temperatures, electron density, and pellet deposition profiles. The middle figures (b-1,2) show radial electric field and $F(\omega_{ExB}/\gamma_{ITG})$ which defined at equation (9), and the lower figures (c-1,2) show $\omega_{E\times B}$ and γ_{ITG} profiles.

5. Summary and discussion

We investigated the relationship between the ITB formation and the pellet injection in tokamak and helical reactors, by means of transport simulation with the TOTAL code. An improvement factor describing the **E**×**B** shear stabilization effect was incorporated to the Bohm and GyroBohm mixed transport model. The high field side (HFS) pellet injection in the tokamak reactor was analyzed and its effectiveness was clarified. In the tokamak reactor with reversed shear profile, it was shown that the pellet penetration depth plays an important role in the ITB formation in the model adopted here, and the HFS injection would be an effective technique for the confinement improvement as well as the plasma core fuelling. In the helical reactor, we showed that the deep pellet penetration is not needed for the ITB formation because the inside structure of radial electric field which makes large **E**×**B** shear flow and strong reduction in anomalous transport is unrelated to the pellet penetration depth. However very high speed pellet injection velocity is required in order to realize the central fuelling. So that, it may be need the other fuelling technique like HFS pellet injection in tokamak about the central fuelling in helical reactor.

These results strongly depend on the transport model and the radial electric field model, and in the present simulation we used semi-empirical transport model with **E**×**B** shear confinement improvement. It should be noted that the large gradient of radial electric field E_r is a key parameter in the present confinement improvement model, and the more accurate E_r modeling different from the present simple model should be adopted in the future.

References

[1] Parks P B and Turnbull R J 1978 *Phys. Fluids* **21** 1735
[2] Bayler L R *et al.* 1992 *Nucl. Fusion* **32** 2177
[3] Lang P T *et al.* 1997 *Phys. Rev. Lett.* **79** 1478
[4] Bayler L R *et al.* 1998 *Fusion Technol.* **34** 425
[5] Koide Y *et al.* 1994 *Plasma Phys. Control. Fusion* **36** 195
[6] Levinton F M *et al.* 1995 *Phys. Rev. Lett.* **75** 4417
[7] JET Team (presented by Schmidt G L) 1989 *Proc. 12th Int. Conf. Plasma Physics and Controlled Nuclear Fusion Research (Nice, 1988)* vol 1 (Vienna: IAEA) p 215
[8] Tubbing B *et al.* 1991 *Nucl. Fusion* **31** 839
[9] Tala T *et al.* 2001 *Plasma Phys. Control. Fusion* **43** 507
[10] Garcia J, Yamazaki K, Dies J and Izquierdo J 2006 *Phys. Rev. Lett.* **96** 105007
[11] Yamazaki K and Amano T 1992 *Nucl. Fusion* **32** 633.
[12] Yamazaki K *et al.* 2006 *Fusion Engineering and Design* **81** 2743
[13] Hahm T S and Burrell K H 1995 *Phys. Plasmas* **2** 1648
[14] Zhou P, Horton W and Sugama H 1999 *Phys. Plasmas* **6** 2503
[15] Ernst D R *et al.* 2000 *Phys. Plasmas* **7** 615
[16] Kim Y B, Diamond P H and Groebner R J 1991 *Phys. Fluids* **B 3** 2050
[17] Hinton F L and Staebler G M 1993 *Phys. Fluids* **B 5** 1281
[18] Esposito B *et al.* 2003 *Plasma Phys. Control. Fusion* **45** 933
[19] Baylor L R *et al.* 1997 *Nucl. Fusion* **37** 445
[20] Polevoi A R and Shimada M 2001 *Plasma Phys. Control. Fusion* **43** 1525
[21] Sakamoto R *et al.* 2002 *Plasma Phys. Control. Fusion* **26B** P-1.074

Journal of Physics: Conference Series **123** (2008) 012033 doi:10.1088/1742-6596/123/1/012033

Power requirement for accessing the H-mode in ITER

Y R Martin[1], T Takizuka[2] and ITPA CDBM H-mode Threshold Database Working Group[3]

[1] Centre de Recherches en Physique des Plasmas, Association Euratom-Confédération Suisse, Ecole Polytechnique Fédérale de Lausanne (EPFL), CH-1015 Lausanne, Switzerland
[2] Japan Atomic Energy Agency, Naka Fusion Institute, Naka, Ibaraki-ken, 311-0193, Japan
[3] From Alcator C-Mod: J A Snipes, M Greenwald
From ASDEX/ASDEX Upgrade: F Ryter, O J W F Kardaun, J Stober
From CHS: T Akiyama
From COMPASS-D/MAST: M Valovic
From DIII-D: G Wang, J C DeBoo
From JET/EFDA: Y Andrew, J G Cordey, R Sartori, K Thomsen
From JFT-2M/JT-60U: T Takizuka, Y Miura, T Fukuda, Y Kamada, K Shinohara
From PBX-M/NSTX: S M Kaye, C Bush, R Maingi
From TCV: Y R Martin
From TUMAN-3M: S Lebedev

E-mail: yves.martin@epfl.ch

Abstract. The input power requirements for accessing H-mode at low density and maintaining it during the density ramp in ITER is addressed by statistical means applied to the international H-mode threshold power database. Following the recent addition of new data, the improvement of existing data and the improvement of selection criteria, a revised scaling law that describes the threshold power required to obtain an L-mode to H-mode transition is presented. Predictions for ITER give a threshold power of ~52MW in a deuterium plasma at a line average density $n_e = 0.5\times10^{20}\text{m}^{-3}$. At the nominal ITER H-mode density, $n_e = 1.0\times10^{20}\text{m}^{-3}$, the threshold power required is ~86MW. Detailed analysis of data from individual devices suggests that the density dependence of the threshold power might increase with the plasma size and the magnetic field. On the other hand, the density at which the threshold power is minimal is found to decrease with the plasma size and increase with magnetic field. The influence of these effects on the accessibility of the H-mode regime in ITER plasmas is discussed. Analyses of the confinement database show that, in present day devices, H-modes are generally maintained with powers exceeding the threshold power by a factor larger than 1.5, and that, on the other hand, good confinement can be obtained close to the threshold power although rarely demonstrated.

1. Introduction

Transitions from a standard confinement mode (L-mode) to an improved confinement regime (H-mode), called L-H transitions, are observed in most tokamaks. Although progress is being made in understanding the physics underlying L-H transitions, the lack of model based predictions for ITER leads to the requirement of extrapolations from present-day devices.

The international H-mode threshold database (DB) collects data from the widest possible range of tokamaks to provide a statistical approach to the L-H transition studies. The current version of the DB is populated by 7700 time slices from 14 tokamaks. Several time slices, recording values taken just before the L-H transition and labelled accordingly, are used in the statistical analysis of the threshold power such as the evaluation of an expression in the form of a power law scaling. The scaling law is then used to estimate the threshold power in ITER. The DB also contains data taken during the L-mode or H-mode phases, or at the transition from H-mode back to L-mode (H-L transitions). These data can be used, for instance, to characterise the power required to stay in the H-mode.

Although known to depend on many parameters, it is widely accepted that the L-H transition threshold power strongly depends on the plasma density, toroidal magnetic field and plasma size. A series of power law scalings, estimating the threshold power on the basis of these parameters, were published following the evolution of the DB content [1, 2, 3, 4, 5, 6]. More recently, new data from spherical tokamaks (MAST and NSTX) have widened the parameter ranges allowing for the analysis of the effect of the aspect ratio [7, 8]. On the other hand, other dependences were searched in the goal of reducing the uncertainty in the power law coefficients [9], but with limited success. Therefore, an alternative approach has been considered consisting of the extraction of power law coefficients from a reduced dataset including only ITER-like conditions [5]. For the scaling analysis of the H-mode energy confinement time, a similar approach using the ITER-like subset of the database was accomplished [8, 10].

The estimated power law scalings of the L-H transition threshold power are used to predict the power requirement, for a new device such as ITER, to enter the H-mode. The predicted power should allow the access to the H-mode but, if the power remains close to the threshold power value, the discharge would generally remain in a regime with type III Edge Localised Modes (ELMs) according to the widely accepted picture of the different ELM type characteristics [11]. This ELMy regime is characterised by only a moderate increase in the energy confinement time compared to the L-mode regime. Good confinement values, those used in the energy confinement time scalings for instance, are generally obtained when the power is significantly larger than the threshold to reach a type I ELM regime. Therefore, the power required to achieve a good confinement in ITER might then be larger than that estimated by the scaling law. The overall description of the behaviour is complicated by the power reduction due to the hysteresis effect. Indeed, once in the H-mode the required power to keep the same plasma conditions is reduced since the energy confinement time increases. This has been clearly seen for the Ohmic H-mode transition, where the loop voltage and Ohmic heating power P_{OHM} drop sharply [12]. For the additional heating cases, the net power through the plasma surface, $P_L = P_{OHM} + P_{abs} - dW/dt$, is reduced after the L-H transition because the growth of the stored energy W becomes more rapid than that before the transition.

The paper addresses these issues within the following structure: in section 2, the scaling obtained with a reduced data set is presented followed by the effect of the most recent contributing device, CHS, in section 3. Individual density dependences of the threshold power for several devices are considered in section 4. In section 5 the way in which the critical density for minimum threshold power is related to device size and magnetic field is explored. Then section 6 discusses the issue of the power required in stationary phase in H-mode regimes with good confinement properties in all devices. All results are discussed and summarised in terms of predictions for ITER in section 7.

2. Threshold power estimation

Since the last paper published at the IAEA Fusion Energy Conference in 2004 [8], the main modifications to the International Global Threshold DataBase (IGDBTH) have been the addition of new time slices (ts), corrections of existing ts and improvement of the selection criteria. New ts have been provided by NSTX, JFT-2M and CHS teams. The DB now contains 7700 ts taken during the different phases of the discharges evolution: L-mode (2357 ts), L-H transition (2667), H-mode (2268), H-L transition (404). It includes contributions from: Alcator C-Mod (1227 ts / with 334 ts at L-H transition), ASDEX (600/122), ASDEX Upgrade (636/237), CHS (6/6), Compass-D (46/30), DIII-D

(752/178), JET (3111/1291), JFT-2M (1019/241), JT-60U (109/81), MAST (20/20), NSTX (13/11), PBXM (5/5), TCV (131/94), TUMAN-3M (15/15). Corrections were applied to ASDEX-Upgrade data (injected power, absorbed power) and Alcator C-Mod data (absorbed power, plasma density, ion grad B drift direction).

A new selection criteria (SELEC2007) has been defined to select ITER like plasmas only from the DB. This was done with the aim of reducing the uncertainty in the threshold power scaling coefficient estimation. It is based on the selection criteria SELDB2 [1], made on a series of 10 binary conditions, which selects single null configurations (except PBX-M) with ion grad B drift towards the X point, deuterium plasmas, and rejects ts with too low plasma density (Alcator C-Mod: n_{e20}(unit of 10^{20}m^{-3}) < 0.8, ASDEX Upgrade: $n_{e20} < 0.2$, DIII-D: $n_{e20} < 0.2$, JET: $n_{e20} < 0.1$, JFT-2M: $n_{e20} < 0.2$, JT-60U: $n_{e20} < 0.1$), too low safety factor at the 95% flux surface ($q_{95} < 2.5$), too close to the beginning of heat pulse, too large counter-NBI fraction ($P_{ctr}/P_{NB} > 0.8$), too small gaps between plasma surface and wall (d < 5 cm), too high radiation losses ($P_{rad}/P_L > 0.5$). SELEC2007 furthermore rejects transitions obtained in Ohmic conditions since they are not relevant for ITER and also rejects Electron Cyclotron only heated discharges since this heating scheme, mainly used in small devices, regularly leads to high threshold power values (as far as the present database displays). Note that the prediction of the threshold power for ECRH in ITER is allowable with the present scaling because the equalization between T_e and T_i can be established in large devices. Configurations different from single null and plasma elongation lower than 1.2 have also been rejected.

Applying this selection criteria, one obtains the following distribution: Alcator C-Mod (115 ts), ASDEX Upgrade (175 ts), DIII-D (56 ts), JET (562 ts), JFT-2M (58 ts), JT-60U (58 ts). The large fraction of JET data sometimes plays like a fulcrum for the extrapolation to ITER.

The H-mode threshold power is known to depend on the plasma density, magnetic field and plasma size. Two descriptions of the plasma size have been used so far, the couple of major and minor plasma radii and the plasma surface area. The latter comes from threshold studies in ASDEX [13] and DIII-D [14] while the former is more appropriate to describe the effect of the aspect ratio. This description however is not useful in the frame of this study since the aspect ratio covers a rather limited range in the selected ITER like plasma configurations. A power law scaling, including the line average plasma density, the magnetic toroidal field and the plasma surface area, is then used for fitting the experimental threshold power more precisely defined as the loss power P_L which equals the sum of the ohmic power P_{OHM} and absorbed power P_{abs} minus the time variation of the total plasma energy dW/dt, minus the power loss by fast ions due to unconfined orbits and charge-exchange processes:

$$P_L = P_{OHM} + P_{abs} - \mathrm{d}W/\mathrm{d}t - P_{Floss}\,. \tag{1}$$

Fitting the power law expressions via their corresponding logarithmic expressions in the least square sense over the 1024 ts data set leads to the following expression:

$$P_{Thresh} = 0.0488\ e^{\pm 0.057}\ n_{e20}^{\ 0.717\pm 0.035}\ B_T^{\ 0.803\pm 0.032}\ S^{0.941\pm 0.019}\,, \tag{2}$$

where P_{Thresh} is the threshold power expressed in MW, n_{e20} the line average electron density in 10^{20}m^{-3}, B_T the magnetic field in T and S the plasma surface area in m^2. The uncertainties in the exponents correspond to the standard errors and the RMS value of the fit is 30.8%. This expression was estimated without the Kadomtsev constraint, but the resulting exponents almost verify it: if the density and magnetic field exponents were considered as correct ($\alpha_n = 0.717$ and $\alpha_B = 0.803$), the surface area exponent α_S would be 0.844 to satisfy the dimensionality of the expression; $8\alpha_n + 5\alpha_B - 8\alpha_S = 3$.

Another fitting expression with the use of the minor radius a (m) and the major radius R (m) instead of the surface area S is also obtained as follows:

$$P_{Thresh} = 2.15\ e^{\pm 0.107}\ n_{e20}^{\ 0.782\pm 0.037}\ B_T^{\ 0.772\pm 0.031}\ a^{0.975\pm 0.08}\ R^{0.999\pm 0.101}\,. \tag{3}$$

The RMS value of this fit is 29.5%. The reduction of the RMS value can be brought by the increase of the fitting parameters [4]. Because this reduction is only a little, we use hereafter the former expression with minimum fitting parameters (n_e, B_T and S) as a basic scaling.

The predictions for ITER are expressed in table 1, for 2 values of the plasma density while the magnetic field and surface area values are 5.3T and 678m^2, respectively. At the beginning of its operation ITER will have 73MW of additional power. This table indicates that entering the H-mode at low density of deuterium discharges should certainly be possible, but H-mode access at higher density might become marginal.

Table 1. Threshold power predicted by the scaling (2) and 95% confidence interval for ITER at two different densities of deuterium discharges.

Density [10^{20}m^{-3}]	Predicted threshold power [MW]	95% confidence interval [MW]
0.5	52	28 - 96
1	86	46 - 160

The above scaling expressions are obtained from the data of deuterium discharges and of deuterium plasmas heated by hydrogen beam. The latter plasmas are considered to consist mainly of deuterium at the L-H transition. It is widely known that the L-H transition occurs easily at a lower heating power in deuterium discharges than in hydrogen discharges. It is also found in JET that the threshold power in tritium discharges becomes further lower. The dependence of the threshold power on the ion mass number M was roughly given by $P_{\mathrm{Thresh}} \propto 1/M$ [15]. When this mass dependence is applied to the deuterium-tritium discharges for ITER, the above predicted values of P_{Thresh} can be reduced by ~ 20%.

3. Effect of CHS data

CHS is a compact, low aspect ratio, helical system with a major radius, R ~ 1m [16]. In neutral beam heated discharges, H-modes have been obtained after the formation of an edge transport barrier (ETB). Many similarities with H-modes in tokamaks were found in this helical device such as the presence of a threshold power which depends on the plasma density, magnetic field and magnetic configuration [17]. Six time slices from CHS have been included in the DB. These discharges were performed in hydrogen. The experimental threshold power value, P_L, is compared to the estimation done with the scaling (2), P_{Thresh}, as shown in figure 1. CHS data, represented by crosses, lie significantly above the scaling line. Even when their values are reduced by half in accordance with the possible mass dependence, they are still larger than the scaling. It is worth noting that these points lie close to those from MAST and NSTX, the spherical tokamaks of the DB, while the tokamak data, used for evaluating the power law scaling, are obviously found closer to the scaling line. It was suggested that the lower aspect ratio or the lower untrapped-particle fraction could increase the threshold power [7]. The data of CHS and the spherical devices, though they are not taken into account to establish the present scaling, might help the understanding of the global behaviour of the H-mode accessibility. This remains for the future work.

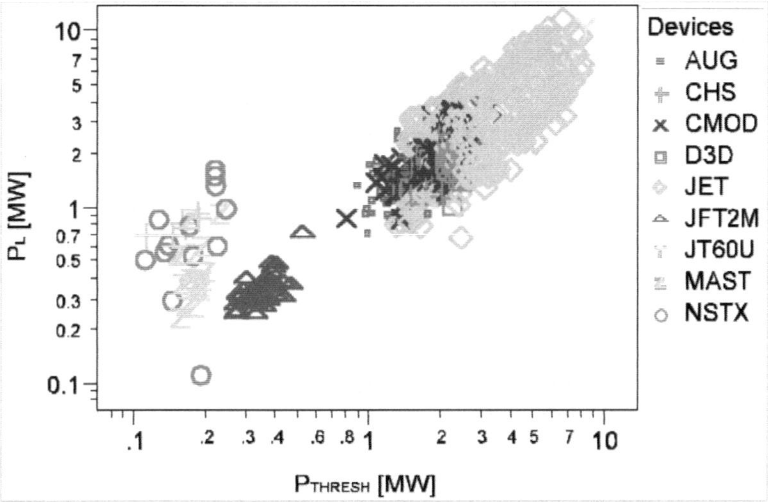

Figure 1. Actual versus fitted threshold power for tokamak time slices used in the fit (Alcator CMod, ASDEX Upgrade, DIII-D, JET, JFT-2M, JT-60U). In addition, spherical tokamaks and helical system threshold power are superimposed to show their distance to the fit (MAST, NSTX, CHS).

4. Individual density dependence

The density dependence of the threshold power plays an important role in the predictions for ITER since its planned density range will be rather large. It is currently envisaged that an H-mode in ITER will be accessed via an L-H transition at low density, followed by a ramp to the vicinity of the Greenwald density limit, while relying on possible hysteresis in power to maintain the H-mode since the power available at the beginning of ITER's operation will remain below the estimated threshold power at high density. With a mild density dependence of the threshold power the hysteresis effect might be sufficient to maintain the plasma in H-mode while the density is increased. However, if the threshold power strongly depends on the density, the hysteresis effect might not be sufficient to maintain the H-mode or even less to access good confinement. The auxiliary input power could fall below the H-L transition level as P_{Thresh} increases with density causing an H-L transition. Detailed studies of fitting residuals and recent results from individual devices revealed that the density dependence in individual devices might differ from the average dependence obtained in the global fit [18, 19]. Therefore, the density dependence is now analysed for each device, from the data contained in the threshold DB.

The density dependence was evaluated by fitting the threshold power with the plasma density and magnetic field for each individual datasets, with the same selection criteria as before, assuming that the plasma surface area S does not vary significantly for one device. The obtained density dependences show some increase with the plasma size. Alcator C-Mod data with high B_T demonstrate a strong density dependence despite its relative small size indicating an additional magnetic field influence. These density exponents α_n are shown in figure 2 as functions of (a) the plasma minor radius a and (b) the product of a and B_T, where the different symbols correspond to different values of the magnetic field for different devices as indicated in the legend. The error bars indicate the uncertainty in the exponents obtained from the fit. One can see that the gentle increase in density exponent with plasma size and magnetic field emerges over the uncertainty range.

Journal of Physics: Conference Series **123** (2008) 012033 doi:10.1088/1742-6596/123/1/012033

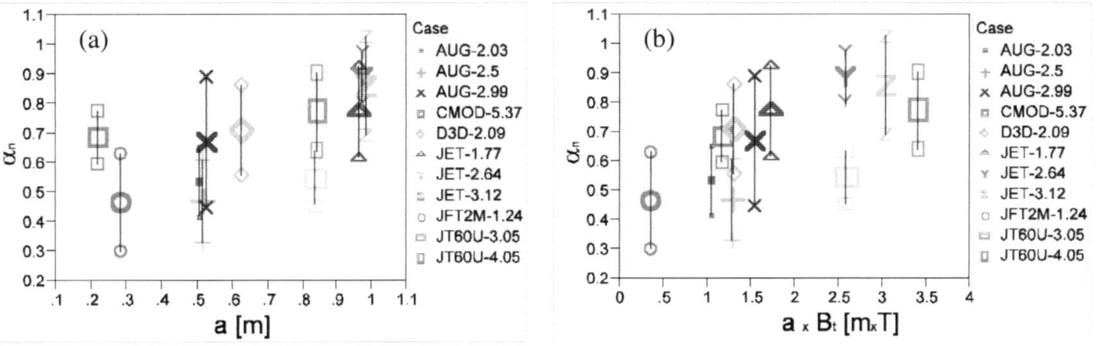

Figure 2. Dependence of density exponent α_n in P_{Thresh} on (a) the plasma minor radius a and (b) the product of a and B_T, for different devices and magnetic field values as indicated in the legend under the form 'device'-'toroidal magnetic field in T'.

Since ITER has a large radius and will operate at a high magnetic field, a strong density dependence might be expected. Extrapolations from the dataset of figure 2 lead to a density exponent of 1.3 for ITER at its nominal toroidal field value (B_T = 5.3T), with an RMS value of 10%. This estimation has to be compared to the global fit value of ~0.75. Although this new estimation of α_n = 1.3 might contain a large uncertainty, the threshold power is predicted to reach 135MW at n_e = $1.0\times10^{20}m^{-3}$, instead of the ~90MW prediction based on the global scaling. The pivot for the calculation of the high density case is fixed at the low density value ($0.5\times10^{20}m^{-3}$) because this point approximately corresponds to the overall average density value of full selected dataset.

One of the reasons of the increase of density exponent for large devices was supposed that the edge density n_e^{edge} increases rapidly with the increase of the averaged density n_e, such as $n_e^{edge} \propto n_e^{1.4}$, and as a result the threshold power governed by n_e^{edge} could increase rapidly with n_e; $P_{Thresh} \propto (n_e^{edge})^{0.7} \propto n_e^{1.0}$ [18]. On the other hand, it has been reported that the density becomes peaked with the decrease of collisionality [20]. In the ITER plasma, the collisionality will be much lower than the present plasmas and n_e^{edge} cannot rapidly increase compared with n_e. This means that the density exponent is expected to remain ~0.7 without increasing up to ~1.3. From the above considerations, uncertainty in the prediction, P_{Thresh} = 46 - 160MW at n_e = $1.0\times10^{20}m^{-3}$, is really to be taken into account for the planning of ITER operation.

5. Density at the minimum threshold power

Most devices have observed the presence of a minimum in the threshold power as a function of the plasma density [21 and references therein]. This reference indicates that the minimum threshold is found at very low edge density (~$0.1\times10^{20}m^{-3}$) in JET. In contrast, Alcator C-Mod results show that, at their high magnetic field values, the density of the minimum threshold power rises up to 0.8-1.0 $\times10^{20}m^{-3}$ [22]. Since ITER will operate at high field as well, the access to the H-mode at low density might be an issue. The threshold power DB is analysed to search for a global picture of the minimum threshold density in all devices.

For each device having data used for the global scaling (Alcator C-Mod, ASDEX Upgrade, DIII-D, JET, JFT-2M and JT-60U), one takes all L-H transition data since the absolute minimum is the interesting value. Then, for each device, a series of datasets were selected, all presenting a concentration of time slices around a particular value of magnetic field. In all these data sets, one takes the average density (and its standard deviation) of the 10 time slices with the lowest threshold values. An example for JET is given in figure 3.

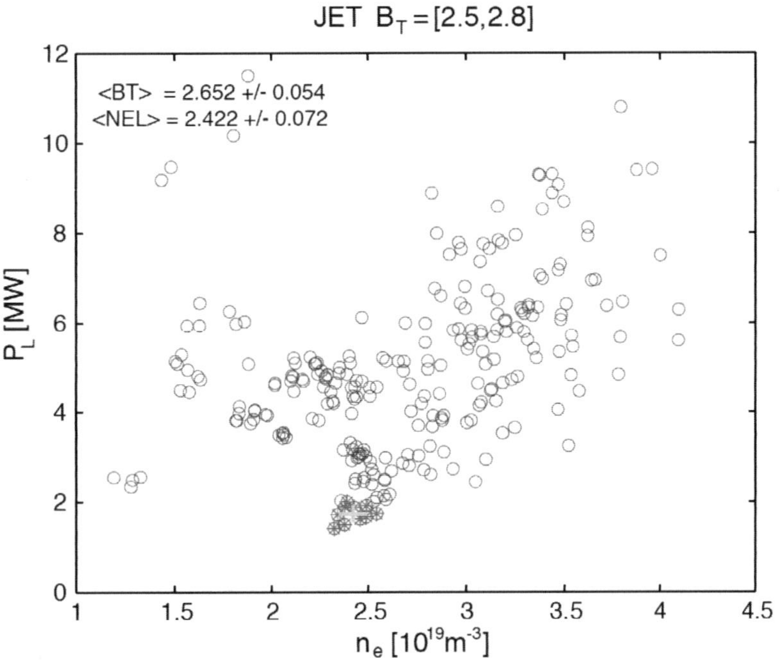

Figure 3. Threshold power as a function of plasma density n_e for JET data with toroidal field B_T in the range 2.5 - 2.8T. Red stars indicate the tens points with lowest threshold power, whose average value of B_T and n_e are shown at the left top corner.

Hence, for each device, a series of density values indicates where the threshold power is minimum for a given magnetic field value, as shown in figure 4 (a). The length of the segments represents one standard deviation each side of the average value. This figure then shows, for instance, that the minimum threshold power in Alcator C-Mod is obtained when the density is slightly greater than $1.0\times10^{20}m^{-3}$ for magnetic field value of 5T. On the contrary, the minimum threshold power is obtained at a rather low density in JET over the full range of available magnetic field values (1-3T) even if it gently increases with the magnetic field. JT-60U shows similar values at an even higher magnetic field. Smaller devices, ASDEX Upgrade and DIII-D also show a gradual increase of the density at minimum threshold with the magnetic field but the density remains much smaller than in the Alcator C-Mod case. The data for B_T = 2 - 3T appear to indicate that the larger the machine (JET and JT-60U) is the smaller the density for the minimum threshold power is, but ongoing joint ITPA experiments are still trying to assess this point. Next, we examine these density values by normalizing the Greenwald density limit n_{GW} $(10^{20}m^{-3}) = I_p/\pi a^2$ (MA/m^2). Figure 4 (b) shows that they lie in the range, $0.2 < n_e/n_{GW} < 0.4$. One might then expect that entering the H-mode at low density in ITER (n_e = $0.5\times10^{20}m^{-3} = 0.42n_{GW}$) is possible even at the full toroidal magnetic field B_T = 5.3T and plasma current I_p = 15MA.

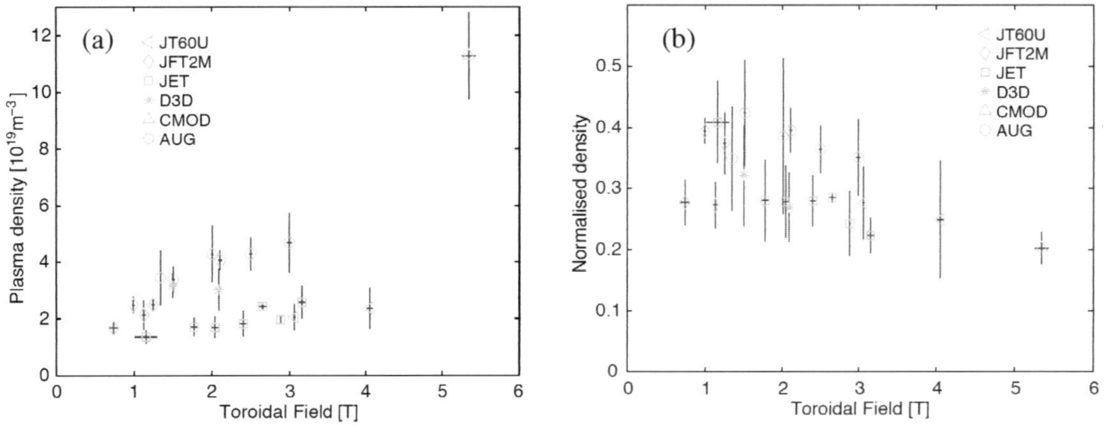

Figure 4. (a) Averaged density n_e and (b) Greenwald density ratio n_e/n_{GW} at which threshold power is minimum as a function of the toroidal magnetic field B_T for a series of tokamaks. The error bars indicate one standard deviation around averaged values along both axis.

6. Access to an H-mode with good confinement properties

The threshold power values predicted for ITER correspond to the power required to enter the H-mode but not necessarily to the power required to obtain a good H-mode in the sense of H-mode having a significantly improved confinement or, equivalently, the H-factor, $H_{IPB98(y2)}$ based on the IPB98(y2) [2] scaling, larger or equal to unity. From the operational point of view, H-modes with $H_{IPB98(y2)} \geq 1$ are often obtained when the power is still increased from the threshold value to leave the type III ELM regime to reach the ELM free regime and then the type I ELM regime [11, 23], for instance, or a grassy ELM regime [24]. Although the threshold database contains data in the H-mode phase, the time slice labelling does not allow us to select time slices at the transition between different H-mode regimes and then analyses similar to the threshold power fitting are excluded. However, the ratio between the actual power and the threshold power, estimated at the same time, can give some indication of the power used to maintain this regime. It then describes the operational range instead of the threshold and, accordingly, can be as high as desired by the operator although limited by the available power. The confinement DB, which contains data from most operational regimes obtained in the contributing devices, is available to study this issue. A result of brief analysis has suggested that good confinement can be obtained in H-mode even for the heating power lower than the threshold power [6]. In the present paper, we expand the analysis by using the most recent version IGDBH4v4 [25].

The distribution of the power ratio for 4 devices considered in previous sections and for the time slices which have been selected for the energy confinement time studies (HMWS05 = 1) in the confinement DB are shown in figure 5. These histograms include time slices with either small or large ELMs (PHASE = HSELM or HGELM). Table 2 indicates the number of time slices for each of these devices, together with the average power ratio.

Table 2. Power ratio (actual input power divided by threshold power estimated at the same time) and number of time slices for different devices.

Tokamak	Number of time slices	Average power ratio
Alcator C-Mod	31	0.88
ASDEX Upgrade	509	1.93
DIII-D	264	2.61
JET	1487	1.68

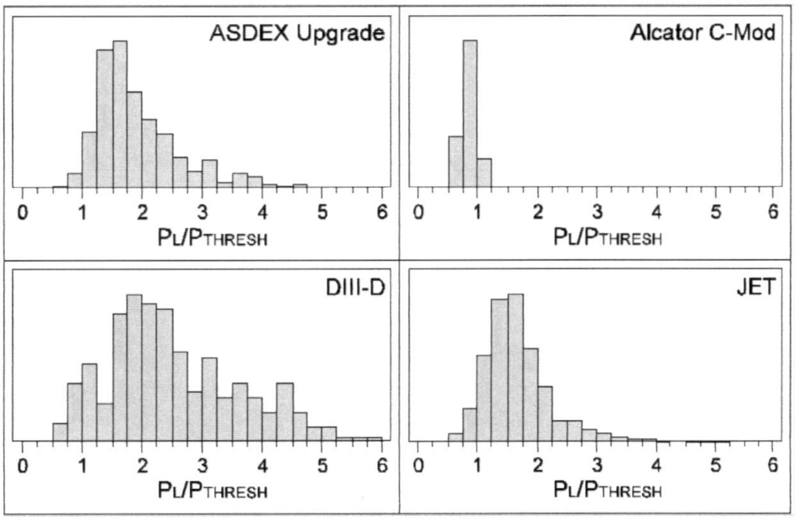

Figure 5. Distribution of power ratio P_L/P_{Thresh} for 4 devices.

Alcator C-Mod clearly benefits from the hysteresis effect, since the average value of the power ratio is smaller than 1. However, the largest population resides between 0.75 and 1.0 and is somewhat larger than the ratio of one half as presented in [4]. In the other devices the average ratio is significantly larger than 1, closer to 1.5 or even 2.5 for DIII-D. Detailed analysis of ASDEX Upgrade data can be found in [26]. In these figures, all ELM types are taken together. As already stated, Type I ELMs are expected to be found at higher power ratio. Unfortunately, ELM types are defined as 'small' or 'large' in the DB. It is not sure but probable there exists correlation between 'small' and Type III on one hand and between 'large' and Type I on the other hand. Anyway, the average power ratio for both ELM classes taken individually does not show significant differences in all devices. This result is confirmed by the relationship between the H-factor, based on the IPB98(y2) [2] scaling, and the power ratio, as shown in figure 6 where green dots correspond to 'large' ELMs and red dots to 'small' ELMs. Indeed, this figure shows that good confinement properties might be obtained at a power ratio close to unity, for both ELM types and on the other hand that increasing the power much over the threshold does not simply guarantee a good confinement. However, most devices currently operate with much more power, compared to the H-mode threshold power than what ITER might be doing. It is therefore mandatory to explore regimes with power closer to the threshold power. On the other hand deeper analysis of the confinement DB might give some results. Such analysis will be presented in a future paper.

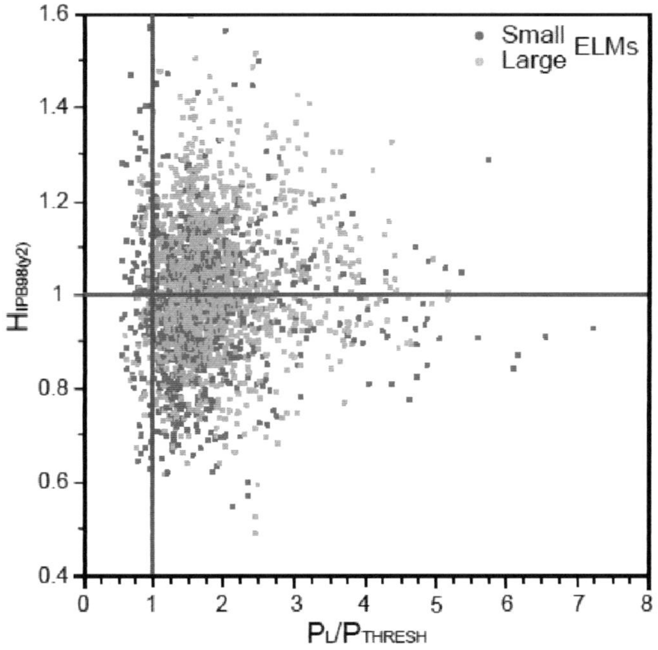

Figure 6. Confinement improvement factor ($H_{IPB98(y2)}$) vs the power ratio P_L/P_{Thresh}. Symbols in the upper left quadrant represent data with good confinement properties obtained at power values below the threshold power of the L-H transition.

7. Conclusion and discussion

New data have been provided to the threshold power database, as well as some modifications to the Alcator C-Mod and ASDEX Upgrade existing data sets. Improvements in the selection criteria have also been made. These changes lead to a new power law scaling expression in which the density dependence is approximately equal to 0.7. With this scaling, the threshold power in low density (n_e = $0.5\times10^{20}m^{-3}$) deuterium plasmas for ITER is estimated at 52MW while at the nominal density of n_e = $1.0\times10^{20}m^{-3}$, it rises to 90MW.

The selection criteria used in these analyses rejected most of small devices in which the threshold power is quite high compared with existing scaling laws [3, 4, 5, 6, 7, 8, 9]. Removing small devices with high threshold power leads to higher predicted values for ITER, through the leverage effect, and, also, to higher uncertainty since the relative extent in the extrapolation is larger. Moreover, the large variety of JET data, issued from different experimental campaigns with different divertor configurations showing significantly different threshold power dependences, causes the uncertainty in determining α_n and α_B. As a result, the 95% confidence interval ranges are rather wide; 30 - 100MW at n_e = $0.5\times10^{20}m^{-3}$ and 50 - 170MW at n_e = $1.0\times10^{20}m^{-3}$. The large threshold power value at high density is increased further when one considers a possible increase in the density exponent α_n with the plasma size and magnetic field. Extrapolating the variation of α_n up to ITER parameters, one finds it might reach 1.3, implying P_{thresh} as high as 135MW at n_e = $1.0\times10^{20}m^{-3}$, though this extrapolation has large error bars. According to this estimation, the access to the H-mode at high density might be very marginal in ITER deuterium plasmas since only 73MW will be available during the deuterium phase. On the other hand, access to H-mode at low density should not present any problem. Indeed, a favourable size effect is found when comparing the density at which the threshold power is minimum.

If one now compares the planned available power to the predicted threshold power during stationary H-mode phases one notices that the power ratio is smaller than unity while, in most present day devices, the ratio is ranging between 1.5 and 2.5. However, analysis of the confinement DB shows that the confinement improvement in H-mode is not directly related to the power ratio P_L/P_{Thrsh}. This is confirmed also by the several experimental reports, such as reference [27] where the good confinement was obtained in JET with a power ratio of 1.2.

Acknowledgments
This work could not have been done without the strong commitment of the teams operating the different devices. The authors would then like to thank them all. This work was partly supported by the Swiss National Science Foundation.

References

[1] Ryter F and H-mode Database Working Group 1996 *Nucl. Fusion* **36** 1217
[2] ITER Physics Basis 1999 *Nucl. Fusion* **39** 2175
[3] The ITER H-mode Threshold Database Working Group presented by Righi E 1998 *Plasma Phys. Control. Fusion* **40** 857
[4] Snipes J A and the International H-mode Threshold Database Working Group 2000 *Plasma Phys. Control. Fusion* **42** A299
[5] Ryter F and the H-mode Threshold Database Group 2002 *Plasma Phys. Control. Fusion* **44** A415
[6] ITPA Confinement and H-mode Threshold Database Working Group presented by Snipes J A 2002 *Proc. 19th IAEA Fusion Energy Conference (Lyon, 2002)* (Vienna: IAEA) CD-ROM file CT/P-04
[7] ITPA H-mode Power Threshold Database Working Group presented by Takizuka T 2004 *Plasma Phys. Control. Fusion* **46** A227
[8] Progress in the ITER Physics Basis 2007 *Nucl. Fusion* **47** S18
[9] ITPA H-mode Threshold Database Working Group presented by Martin Y R 2004 *Proc. 20th IAEA Fusion Energy Conference (Vilamoura, 2004)* (Vienna: IAEA) CD-ROM file IT/P3-35
[10] Cordey J G *et al.* 2005 *Nucl. Fusion* **45** 1078
[11] Zohm H 1996 *Plasma Phys. Control. Fusion* **38** 105
[12] Martin Y R and TCV Team 2002 *Plasma Phys. Control. Fusion* **44** A143
[13] ASDEX Team 1989 *Nucl. Fusion* **29** 1959
[14] Burrel K H *et al.* 1989 *Plasma Phys. Control. Fusion* **31** 1649
[15] Righi E *et al.* 1999 *Nucl. Fusion* **39** 309
[16] Matsuoka K *et al.* 1989 *Proc. 12th Int. Conf. on Plasma Physics and Controlled Nuclear Fusion Research 1988 (Nice, 1988)* vol 2 (Vienna: IAEA) p 411
[17] Akiyama T *et al.* 2006 *Plasma Phys. Control. Fusion* **48** 1683
[18] Fukuda T *et al.* 2000 *Plasma Phys. Control. Fusion* **42** A289
[19] Andrew Y *et al. submitted to Plasma Phys. Control. Fusion*
[20] Weisen H *et al.* 2006 *Plasma Phys. Control. Fusion* **48** A457
[21] Andrew Y *et al.* 2006 *Plasma Phys. Control. Fusion* **48** 479
[22] Snipes J A *et al.* 1996 *Plasma Phys. Control. Fusion* **38** 1127
[23] Sartori R *et al.* 2004 *Plasma Phys. Control. Fusion* **46** 723
[24] Kamada Y *et al.* 2002 *Plasma Phys. Control. Fusion* **44** A279
[25] McDonald D C *et al.* 2007 *Nucl. Fusion* **47** 147
[26] Ryter F *et al.* 2007 "H-Mode confinement properties close to the power threshold in ASDEX Upgrade" *this conference*
[27] Saibene G *et al.* 2002 *Plasma Phys. Control. Fusion* **44** 1769

Journal of Physics: Conference Series **123** (2008) 012034 doi:10.1088/1742-6596/123/1/012034

Projected performance of ITER based on different theoretical based pedestal temperature models

T Onjun[1], K Tharasrisuthi[1], A Y Pankin[2] and O Onjun[3]

[1] Sirindhorn International Institute of Technology, Thammasat University, Klongluang, Pathumthani, Thailand
[2] Department of Physics, Lehigh University, Bethlehem, Pennsylvania, 18015, USA
[3] Department of Science Service, Ministry of Science and Technology, Ratchathewi, Bangkok, Thailand

E-mail: thawatchai@siit.tu.ac.th

Abstract. Self-consistent modeling of the ITER has been carried out using the 1.5D BALDUR integrated predictive modeling code. In these simulations, the boundary is taken to be at the top of the pedestal, where the pedestal values are described using the theory-based pedestal scalings. These pedestal temperature scalings are based on three different pedestal width models: magnetic and flow shear stabilization, flow shear stabilization, and normalized poloidal pressure. The pedestal width scalings are combined with a pedestal pressure gradient scalings based on ballooning mode limit to predict the pedestal temperature. The developed pedestal temperature scalings are used together with a core transport model, which is a combination of an anomalous transport and a neoclassical transport. An anomalous transport is calculated either using the Mixed Bohm/gyro-Bohm (Mixed B/gB) core transport model or the Multimode (MMM95) core transport model, while a neoclassical transport is computed using the NCLASS model. At the reference design point (with 40 MW auxiliary heating: 33 MW NBI and 7 MW RF), it is found that the pedestal temperatures with different pedestal width scaling ranges from 2.4 keV to 2.8 keV. As a result, the performances with the same anomalous core transport model are almost similar. It is also found that the simulations using MMM95 yield better performance than those using Mixed B/gB. In addition, when the MMM95 is used, it appears that the ion temperature gradient (ITG) and trapped electron modes (TEM) are the most dominant modes. When the Mixed B/gB is used, it appears that the Bohm contribution is the most dominated term.

1. Introduction

The ITER project is an international collaborative effort with an aim to demonstrate the scientific and technological feasibility of fusion energy using magnetic confinement fusion concept [1]. Due to the fact that high confinement mode (*H*-mode) discharges in tokamaks generally provide excellent energy confinement and have acceptable particle transport rates for impurity control, burning fusion experiments such as ITER tokamak are designed to operate in the *H*-mode regime. The improved performance of *H*-mode mainly results from the formation of the edge transport barrier, called pedestal. Therefore, the performance of ITER depends sensitively on the pedestal values.

In the previous ITER performance study by G. Bateman and his colleagues [2], the BALDUR integrated predictive modeling code [3] with the Multi-mode (MMM95) anomalous transport model

[4] together with neoclassical transport, calculated using the NCLASS neoclassical model [5], was used to predict the plasma core profiles of ITER and, consequently, the performance of ITER. In that work, the boundary conditions — which were taken to be at the top of the pedestal — were obtained from a predictive pedestal model based on magnetic and flow shear stabilization width model and first stability regime of infinite-n ballooning modes pressure gradient model [6]. It is assumed that 40 MW of RF heating power is used as the auxiliary heating power, where 24 MW of RF heating power goes to thermal ions, and 16 MW of RF heating power goes to thermal electrons. Fast ions resulting from auxiliary heating are not considered. The heating produced by fusion reactions and the resulting fast alpha particles are added to the ohmic and auxiliary heating. The performance of ITER was evaluated in term of fusion by fusion reactions and the resulting fast alpha particles are added to the ohmic and auxiliary heating. The performance of ITER was evaluated in term of fusion Q. Note that fusion Q is the ratio of a fusion power with an applied heating power. An optimistic performance of ITER was obtained in that simulation with fusion Q of 10.6 with both ion and electron pedestal temperature of 2.7 keV. In the later ITER performance study by T. Onjun and his colleagues [7], ITER simulations were carried out using the JETTO integrated predictive modeling code with the Mixed Bogm/gyro-Bohm (Mixed B/gB) anomalous transport model [8] with NCLASS neoclassical transport. In addition, the combination of 33 MW of NBI heating power and 7 MW RF heating power (similar to ITER reference design) was used. When a fix pedestal width of 6 cm (the ion and electron pedestal temperature of 4.9 and 4.4 keV, respectively) is used, an optimistic performance of ITER with fusion Q of 16.6 was found. It was also found that the JETTO code predicts the strong edge pressure gradient, which would occur in the second stability regime of ballooning modes. Therefore, the values at the top of the pedestal used in the JETTO simulations are higher than those used in the BALDUR simulations. It is worth noting that the dynamic of pedestal width is not included in the JETTO simulations; while it is included in the BALDUR simulations.

In this work, the BALDUR integrated predictive modeling code is used to carry out simulations of the ITER plasma with the standard *H*-mode scenario. Due to the design of BALDUR code, the BALDUR simulations are started from the top of the pedestal to the center of the plasma. As a result, several important edge physics may not be properly included and might be ignored such as the effects of ELM crashes. However, it is enough to properly describe the plasma core behaviors [9, 10]. Two issues have been investigated for ITER. The first issue is the sensitivity of the ITER performance on the model for predicting pedestal temperature. Three different theoretical based pedestal temperature models are used in this investigation. These pedestal temperature models were developed using three different pedestal width concepts: magnetic and flow shear stabilization width scaling [11], flow shear stabilization width scaling [6], and normalized poloidal pressure width scaling [12]. These pedestal width scalings are combined with a pedestal pressure gradient model based on first stability regime of infinite n ballooning mode. It was found in reference [6] that these pedestal temperature models could yield equally well agreement with the pedestal data obtained from the ITPA pedestal database [13]. This exploration will provide more confidence in the BALDUR predictions. Due to the fact that ITER is designed to operate with high triangularity, low collisionality and high bootstrap current, the plasma is likely to obtain an access to second stability regime of ballooning modes. However, since the pedestal temperature models used in this work do not include effects of an access to second stability regime of ballooning modes and peeling mode instability, the results can be treated only as a low bound of the ITER performance projection. Secondly, the sensitivity of core transport prediction is explored. The BALDUR simulations are extended to be carried out with two different anomalous core transport models: the Multimode (MMM95) core transport model or the Mixed Bohm/gyro-Bohm (Mixed B/gB) core transport model. Both core transport models were derived from different approaches: MMM95 is a theoretical based anomalous transport model; while Mixed B/gB is an empirical based anomalous transport model. The brief details of both models can be found in section 2. It is worth mentioning that the simulations using both core transport models yield about 10% RMS deviation with experimental data for both *L*-mode [9] and *H*-mode plasmas [10]. This exploration will also provide a confidence on the predicted performance of ITER tokamak. In addition with those issues, there are several minor modifications on these simulations comparing with those simulations carried out in reference [2] — such as the auxiliary heating power (using 33 MW of NBI heating power and 7 MW RF heating power) and the sawtooth model (using Porcelli sawtooth triggering model [14] and modified Kadomtsev magnetic reconnection model [15]).

This paper is organized as follows: brief descriptions for a BALDUR integrated predictive modeling code, anomalous transport models, and pedestal models are given in section 2. The ITER prediction using a BALDUR integrated predictive modeling code is described in section 3, while conclusions are given in section 4.

2. BALDUR integrated predictive modeling code

The BALDUR integrated predictive modeling code is used to compute the time evolution of plasma profiles including electron and ion temperatures, deuterium and tritium densities, helium and impurity densities, magnetic q, neutrals, and fast ions. These time-evolving profiles are computed in the BALDUR integrated predictive modeling code by combining the effects of many physical processes self-consistently, including the effects of transport, plasma heating, particle influx, boundary conditions, the plasma equilibrium shape, and sawtooth oscillations. Fusion heating and helium ash accumulation are also computed self-consistently. The BALDUR simulations have been intensively compared against various plasma experiments, which yield an over all agreement with 10% relative RMS deviation [9, 10]. In BALDUR code, fusion heating power is determined by the nuclear reaction rates and a Fokker Planck package to compute the slowing down spectrum of fast alpha particles on each flux surface in the plasma [3]. The fusion heating component of the BALDUR code also computes the rate of the production of thermal helium ions and the rate of the depletion of deuterium and tritium ions within the plasma core. In this work, two core transport models in BALDUR will be used to carry out simulations of ITER. The brief details of these transport models are described below.

2.1. Mixed B/gB core transport model

The Mixed B/gB core transport model [8] is an empirical transport model. It was originally a local transport model with Bohm scaling. A transport model is said to be "local" when the transport fluxes (such as heat and particle fluxes) depend entirely on local plasma properties (such as temperatures, densities, and their gradients). A transport model is said to have "Bohm" scaling when the transport diffusivities are proportional to the gyro-radius times thermal velocity over a plasma linear dimension such as major radius. Transport diffusivities in models with Bohm scaling are also functions of the profile shapes (characterized by normalized gradients) and other plasma parameters such as magnetic q, which are all assumed to be held fixed in systematic scans in which only the gyro-radius is changed relative to plasma dimensions. The original JET model was subsequently extended to describe ion transport, and a gyro-Bohm term was added in order for simulations to be able to match data from smaller tokamaks as well as data from larger machines. A transport model is said to have "gyro-Bohm" scaling when the transport diffusivities are proportional to the square of the gyroradius times thermal velocity over the square of the plasma linear dimension. The Bohm contribution to the JET model usually dominates over most of the plasma. The gyro-Bohm contribution usually makes its largest contribution in the deep core of the plasma and plays a significant role only in smaller tokamaks with relatively low power and low magnetic field.

The version of the MixedB/gB model that is used in this paper can be briefly described below. Both the electron and ion thermal diffusivities consist of two terms. One term has Bohm scaling

$$\chi^B \equiv \rho_s c_s q^2 \frac{a(dp_e/dr)}{p_e} \Delta T_e, \tag{1}$$

while the other term has gyro-Bohm scaling

$$\chi^{gB} \equiv \frac{P_s^2 c_s (dT_e / dr)}{T_e}, \tag{2}$$

where r_s is the ion gyroradius, c_s is the sound speed, q is the safety factor, p_e is the electron pressure, a is the minor radius and T_e is the electron temperature. In the Bohm diffusivity expression, DT_e is a finite difference approximation to the normalized temperature electron temperature difference at the plasma edge

$$\Delta_{Te} = \frac{T_e(r/a = 0.8) - T_e(r/a = 1)}{T_e(r/a = 1)}. \tag{3}$$

The resulting anomalous ion and electron thermal diffusivities are constructed from the sum of these Bohm and gyro-Bohm terms, with empirically determined coefficients

$$\chi_i = 1.6 \times 10^{-4} \chi^B + 1.75 \times 10^{-2} \chi^{gB}, \tag{4}$$

$$\chi_e = 8.0 \times 10^{-5} \chi^B + 3.5 \times 10^{-2} \chi^{gB}, \tag{5}$$

and the hydrogenic and impurity charged particle diffusivity is given by

$$D \propto \frac{\chi_i \chi_e}{\chi_i + \chi_e}. \tag{6}$$

2.2. Multimode core transport model

The MMM95 model [4] is a linear combination of theory-based transport models which consists of the Weiland model for the ion temperature gradient (ITG) and trapped electron modes (TEM), the Guzdar–Drake model for drift-resistive ballooning modes, as well as a smaller contribution from kinetic ballooning modes. The Weiland model for drift modes such as ITG and TEM modes usually provides the largest contribution to the MMM95 transport model in most of the plasma core. The Weiland model is derived by linearizing the fluid equations, with magnetic drifts for each plasma species. Eigenvalues and eigenvectors computed from these fluid equations are then used to compute a quasilinear approximation for the thermal and particle transport fluxes. The Weiland model includes many different physical phenomena such as effects of trapped electrons, $T_i \neq T_e$, impurities, fast ions, and finite β. The resistive ballooning model in MMM95 transport model is based on the 1993 ExB drift-resistive ballooning mode model by Guzdar–Drake, in which the transport is proportional to the pressure gradient and collisionality. The contribution from the resistive ballooning model usually dominates the transport near the plasma edge. Finally, the kinetic ballooning model is a semi-empirical model, which usually provides a small contribution to the total diffusivity throughout the plasma, except near the magnetic axis. However, for the ITER case in this work, it is surprisingly found that the contribution from the kinetic ballooning mode plays quite a significant role in the region near the plasma core up to a radius of 1.0 m. This will be discussed in section 3. This model is an approximation to the first ballooning mode stability limit. All the anomalous transport contributions to the MMM95 transport model are multiplied by κ^{-4}, since the models were originally derived for circular plasmas.

2.3. Pedestal model

In the development of the pedestal temperature models described in reference [6], two ingredients are required: the pedestal width (D) and the pressure gradient ($\partial p/\partial r$). If the pedestal density (n_{ped}) is known, the temperature at the top of the pedestal (T_{ped}) can be estimated as

$$T_{ped} = \frac{1}{2n_{ped}k}\left|\frac{\partial p}{\partial r}\Delta\right| = \frac{\Delta}{2kn_{ped}}\frac{\alpha_c B_T^2}{2\mu_0 R q^2}. \quad (7)$$

where k is the Boltzmann constant, μ_0 is the permeability of free space, α_c is the normalized critical pressure gradient, B_T is the magnetic field, R is the major radius and q is the safety factor. In this work, three pedestal temperature models in reference [6] are selected. These pedestal temperature models yield equally well agreement with the pedestal data from the ITPA Pedestal Database. These pedestal temperature models are based on either the magnetic and flow shear stabilization width scaling ($\Delta \propto \rho s^2$) [9], the flow shear stabilization width scaling ($\Delta \propto \sqrt{\rho R q}$) [6], or the normalized poloidal pressure width scaling ($\Delta \propto R\sqrt{\beta_{\theta,ped}}$) [10], where ρ is the ion gyro radius, s is the magnetic shear, and $\beta_{\theta,ped}$ is the normalized pedestal pressure. The pedestal pressure gradient calculation is normally complicated and required a lot of details. For simplicity, the pedestal gradient is assumed to be uniform throughout the pedestal region and the pedestal gradient is limited by the first stability limit of infinite n ballooning mode, so that the normalized critical pressure gradient for the pedestal region is estimated by

$$\alpha_c = 0.4s\left(1+\kappa_{95}^2(1+5\delta_{95}^2)\right), \quad (8)$$

where κ_{95} is the elongation at the 95% flux surface, and δ_{95} is the triangularity at the 95% flux surface. The further details of these pedestal temperature models can be obtained from reference [6]. It is worth noting that these pedestal temperature models were derived from different pedestal width scalings. The pedestal width constant in each models was chosen to minimize the RMS deviation with 533 experimental data points from 4 large tokamaks obtained from the ITPA pedestal database. So, in this work the pedestal models with the chosen width constant in reference [6] are used. These pedestal temperature models include the effect of edge bootstrap current, which has an impact on magnetic shear and safety factor. This inclusion results in a non-linear behavior in the pedestal temperature model. The scheme to deal with the calculation of magnetic shear and safety factor was addressed in reference [6]. Note that the descriptions for magnetic shear and safety factor in the pedestal model are different from the rest of the code.

The pedestal density, n_{ped}, is described by a simple pedestal density model. Since the pedestal density is usually a large fraction of line average density, n_l, the pedestal density can be calculated as:

$$n_{ped} = 0.71 n_l, \quad (9)$$

This pedestal density model agrees with the pedestal data obtained from the ITPA pedestal database with 12% RMSE.

3. Simulation results and discussion

The BALDUR integrated predictive transport modeling code is used to carry out the simulations of ITER with the designed parameters (R = 6.2 m, a = 2.0 m, I_p = 15 MA, B_T = 5.3 T, κ_{95} = 1.85, δ_{95} = 0.33 and $n_l = 1.0\times10^{20}$ m^{-3}). In this work, the plasma parameters are ramped up to the target values within 15 sec, which is a faster ramp-up than that used in the ITER simulations in reference [2]. It is found that the plasma reaches the H-mode phase at the time of 4 sec. It is worth noting that even though the plasma current reaches its flat top with in 15 sec, complex interactions within the plasma itself— such as the self plasma heating by the alpha particle and redistribution of heating power after sawtooth crash — still occurs and lead to interesting observation. Note that the sawtooth oscillation is considered during the time of 15 sec to 297 sec. For each simulation, an anomalous transport is calculated either using the Mixed B/gB transport model or using the MMM95 transport model, while the neoclassical transport is computed using the NCLASS module. The boundary conditions are provided at the top of the pedestal by the pedestal model described above. It is assumed that the electron and ion pedestal temperatures are of the same values. In most simulations, the auxiliary heating power of 40 MW, which is a combination of 33 MW NBI heating power with 7 MW of RF heating power, is used. The effect of sawtooth oscillation is also included, where a Porcelli sawtooth model [14] is used to determine a sawtooth crash and a modified Kadomtsev magnetic reconnection model [15] is used to describe the effects of sawtooth crash.

Figure 1 shows the profiles for ion and electron temperatures and electron density as a function of minor radius at a time of 300 sec using Mixed B/gB and MMM95 core transport models. In these figures, these simulations are carried out using a pedestal temperature model based on the magnetic and flow shear stabilization width model. It can be seen that the ion and electron temperature profiles are peak profiles. For the density profiles, the simulation with the MMM95 transport model is quite an unusual profile. There is a small humps in the region about 60% of the plasma and contain a smaller peak at the region close to the center of the plasma. However, this type of density profile is often observed in plasma simulations with the MMM transport model [9, 10]. When different pedestal temperature model is used, it is found that the pedestal temperatures and central temperatures in the simulation using the pedestal temperature based on normalized poloidal pressure width model is the highest, while those in the simulation using the pedestal temperature based on flow shear stabilization is the lowest. The results are summarized in table 1. It is worth noting that the central temperatures obtained in the ITER simulation using the BALDUR code in reference [2] are higher than the results obtained in this work. This might due to the change of auxiliary heating and sawtooth model. A full investigation must be taken to resolve this issue. At this moment, we rather leave the issue for a future work.

The results of the ion pedestal temperature and the corresponding central ion temperature are summarized in table 1. It can be seen that the pedestal temperature ranges from 2.4 keV to 2.8 keV, where the central temperature ranges from 10.3 keV to 16.5 keV. The values of ion pedestal temperature are not much different, but the central ion temperature is significantly different among the simulations using the Mixed B/gB core transport and the MMM95 core transport models. It is worth noting that these pedestal temperature models were derived from different pedestal width scalings. The pedestal width constant in each models was chosen to minimize the RMS deviation with the experimental values obtained from the ITPA pedestal database [6].

Table. 1 Summary of pedestal values and central ion temperature in the simulations for different pedestal temperature models using either Mixed Bohm/gyro-Bohm or Multi-mode transport model

Pedestal model	Pedestal width	Mixed Bohm/gyro-Bohm model		Multi-mode model	
		$T_{i,ped}$ (keV)	$T_{i,0}$ (keV)	$T_{i,ped}$ (keV)	$T_{i,0}$ (keV)
1	$\Delta \propto \rho s^2$	2.6	10.8	2.7	16.3
2	$\Delta \propto \sqrt{\rho R q}$	2.4	10.3	2.4	16.1
3	$\Delta \propto R\sqrt{\beta_{\theta,ped}}$	2.8	11.5	2.9	16.5

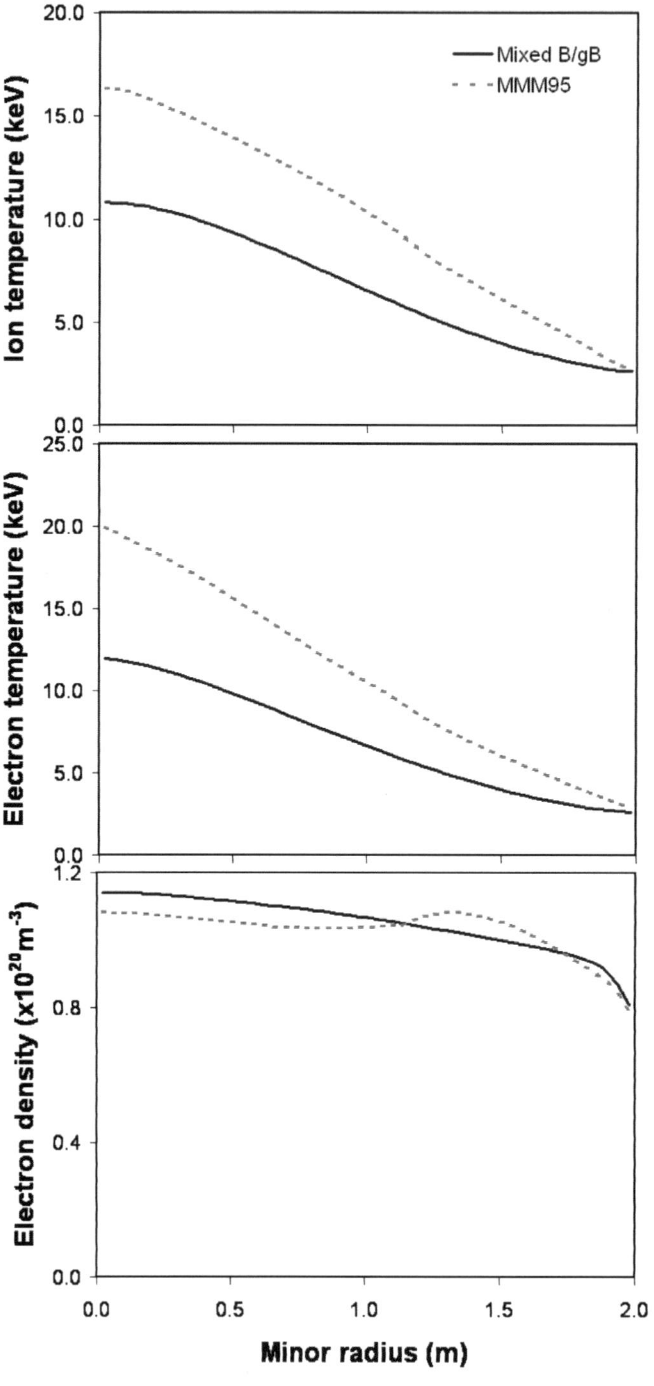

Figure 1. Profiles for ion temperature (top), electron temperature (middle) and electron density (bottom) are shown as a function of minor radius at a time of 300 sec. These BALDUR simulations are carried out using Mixed B/gB model (solid line) and MMM95 model (dotted line). The pedestal temperature model based on magnetic and flow shear stabilization is used to provide the temperature boundary conditions.

Figure 2 shows the ion thermal diffusivities as a function of minor radius from simulations using the Mixed B/gB transport model. Note that the ion temperature and density profiles for this simulation are shown in figure 1. The pedestal model based on magnetic shear and flow shear stabilization width model is used for predicting boundary conditions. The "effective" thermal diffusivity, for example, is defined as the heat flux divided by the density times temperature gradient—with no separate contribution for convection. The total thermal diffusivities shown in figure 2 are the contributions from the Bohm and gyro-Bohm terms in the Mixed B/gB model as well as neoclassical transport, which has gyro-Bohm scaling. It can be seen in figure 2 that the Bohm contribution to the Mixed B/gB transport model is the dominant contribution to the ion thermal diffusivities everywhere in the plasma, except in the region closed to the center of the plasma, where the neoclassical transport is the most dominant. It can be seen that the ion thermal diffusivity is less than 1 m^2/s in the region from the center of the plasma to the radius of 80% of minor radius. This result for the transport is observed in all simulations with Mixed B/gB in this work. In is worth mentioning that the dominance of the Bohm contribution is similar to those results with the Mixed B/gB transport model reported in references [9, 10].

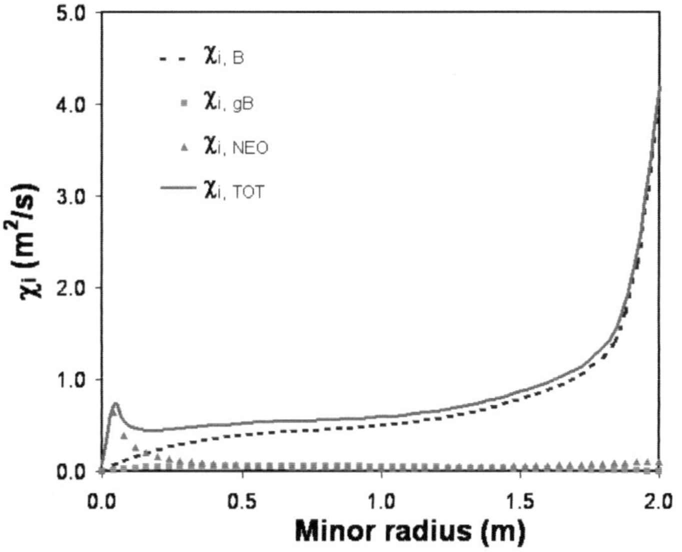

Figure 2. Ion diffusivities from the Mixed B/gB transport model are shown as a function of minor radius at a time of 300 sec. This simulation is carried out using the pedestal temperature model based on magnetic and flow shear stabilization width model. The blue line shows the Bohm contribution, the pink line shows the gyro-Bohm contribution, the red line shows the neoclassical contribution, and the green line shows the total ion thermal diffusivity.

Figure 3 shows the ion thermal diffusivities as a function of minor radius from a simulation using the MMM95 transport model for ITER. Note that the temperature and density profiles for this simulation are shown in figure 1. The pedestal model based on magnetic shear and flow shear stabilization width model is used for predicting boundary conditions. Note also that the Multi-mode ion thermal transport model consists of the ion temperature gradient and trapped electron modes, the drift-resistive ballooning modes, and the kinetic ballooning modes. It can be seen in figure 3 that the contribution from the ITG and TEM modes is the main contribution to most of the region of the plasma, except the region closed to the center of the plasma, where the neoclassical transport is the most dominant. The thermal transport from the kinetic ballooning mode plays a significant role in the region between normalized minor radius of 10% to 50%, whereas the resistive ballooning mode is small almost everywhere. This behavior of the kinetic ballooning mode is quite surprising since the

Journal of Physics: Conference Series **123** (2008) 012034 doi:10.1088/1742-6596/123/1/012034

contribution from this mode is normally small. It is worth mentioning that in the region from the center of the plasma to the edge of plasma, the ion thermal diffusivity is less than 1 m^2/s, except at the last grid point that the ion thermal is slightly above 1 m^2/s. Because of stronger thermal transport in the region near the edge of plasma when the Mixed B/gB transport model is used, the lower temperature profile is found. This result for the transport is observed in all simulations with MMM95 in this work. It is worth noting that there is a large drop in the kinetic ballooning mode transport and a notch in the ITG/TEM transport at the minor radius of about 1.1 m. This probably associates with the inverse density gradient because of the density hump at around that radius. This issue requires a further study. At this moment, we will leave this issue for future work.

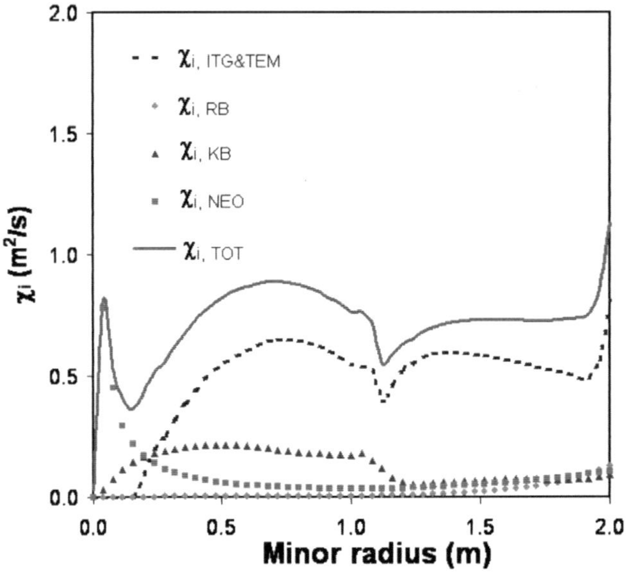

Figure 3. Ion diffusivities from the MMM95 transport model are shown as a function of minor radius at a time of 300 sec. This simulation is carried out using the pedestal temperature model based on magnetic and flow shear stabilization width model. The blue line shows the ITG and TEM contribution, the pink line shows the RB contribution, the violet line show the KB contribution, the red line shows the neoclassical contribution, and the green line shows the total ion thermal diffusivity.

There are two types of auxiliary heating used in the ITER simulation. The total amount of neutral beam injection heating power, P_{NBI}, is 33 MW. Another source of auxiliary heating is the RF heating. The total amount of RF heating power is 7 MW. For simplicity, the RF heating profiles are taken to be a parabolic shape, although it is recognized that the physics of RF heating might be more complicated in the ITER plasma. Note that Ohmic heating is small compared to other types of heating. The alpha heating power is also shown in figure 4. It is found that the alpha heating power is about 49.4 MW.

In figure 5, the deuterium and tritium density profiles are plotted as a function of minor radius at the time of 300 sec. In this simulation, the anomalous core transport is described by MMM95 model. It can be seen that the shape of both profiles is similar to the shape of electron density profile. There is a small humps in the region about 60% of the plasma and contain a smaller peak at the region close to

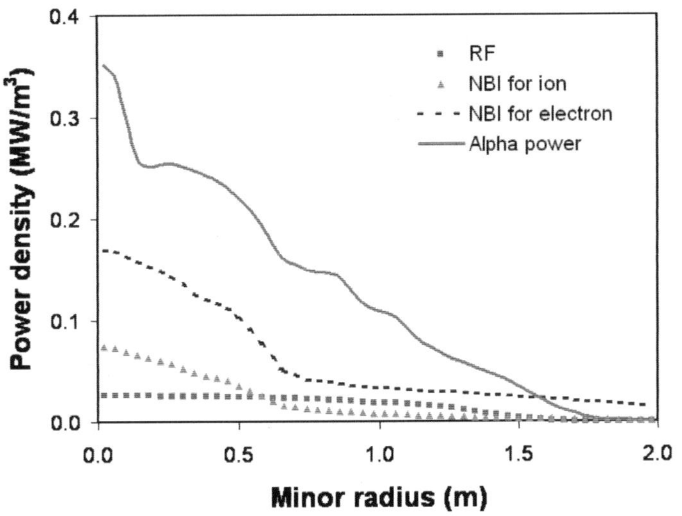

Figure 4. The heating power density profiles are plotted as a function of minor radius. In this simulation, 40 MW of auxiliary heating is used (33.0 MW of NBI heating power and 7.0 MW of RF heating power). The alpha heating power from the fusion reaction is about 49.4 MW. This simulation is carried out using MMM95 model and the pedestal temperature model based on magnetic and flow shear stabilization width model.

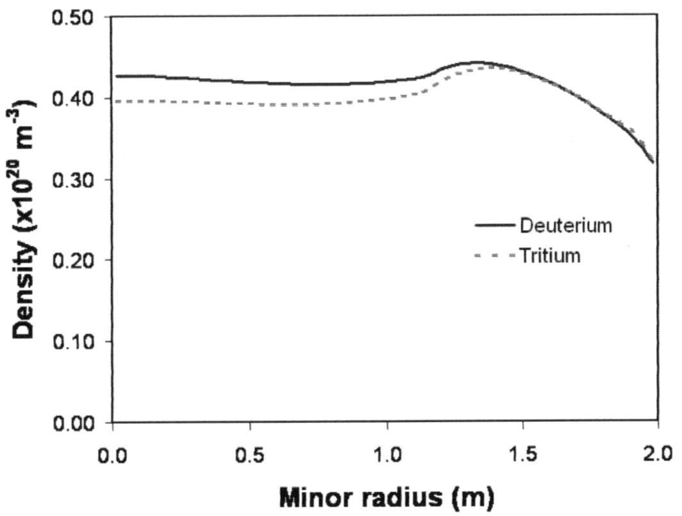

Figure 5. The deuterium and tritium density profiles are plotted as a function of minor radius. This simulation is carried out using MMM95 model and the pedestal temperature model based on magnetic and flow shear stabilization width model.

the center of the plasma. This type of density profiles can not yield good alpha power production. If this is a case, a consideration of density peaking might be required. It is also found that in the simulation with MMM95 model, the particle transport and impurity transport is the same up to the minor radius of 0.6 m. On the other hand, in the simulations with Mixed B/gB, the impurity transport is set to be the same with the particle transport.

In figure 6, the central ion temperature and alpha power is plotted as a function of time during time of 250 sec to 300 sec for the simulations using the Mixed B/gB and MMM95 models. Note that the pedestal model based on magnetic shear and flow shear stabilization width model is used for predicting boundary conditions in this simulation. It can be seen that the central ion temperature and alpha power production oscillate due to the effects of sawtooth crashes. It is found that during a sawtooth crash, the central ion temperature drops about 26% and 24% for the simulations using Mixed B/gB and MMM95 models, respectively. On the other hands, it is found that during a sawtooth crash,

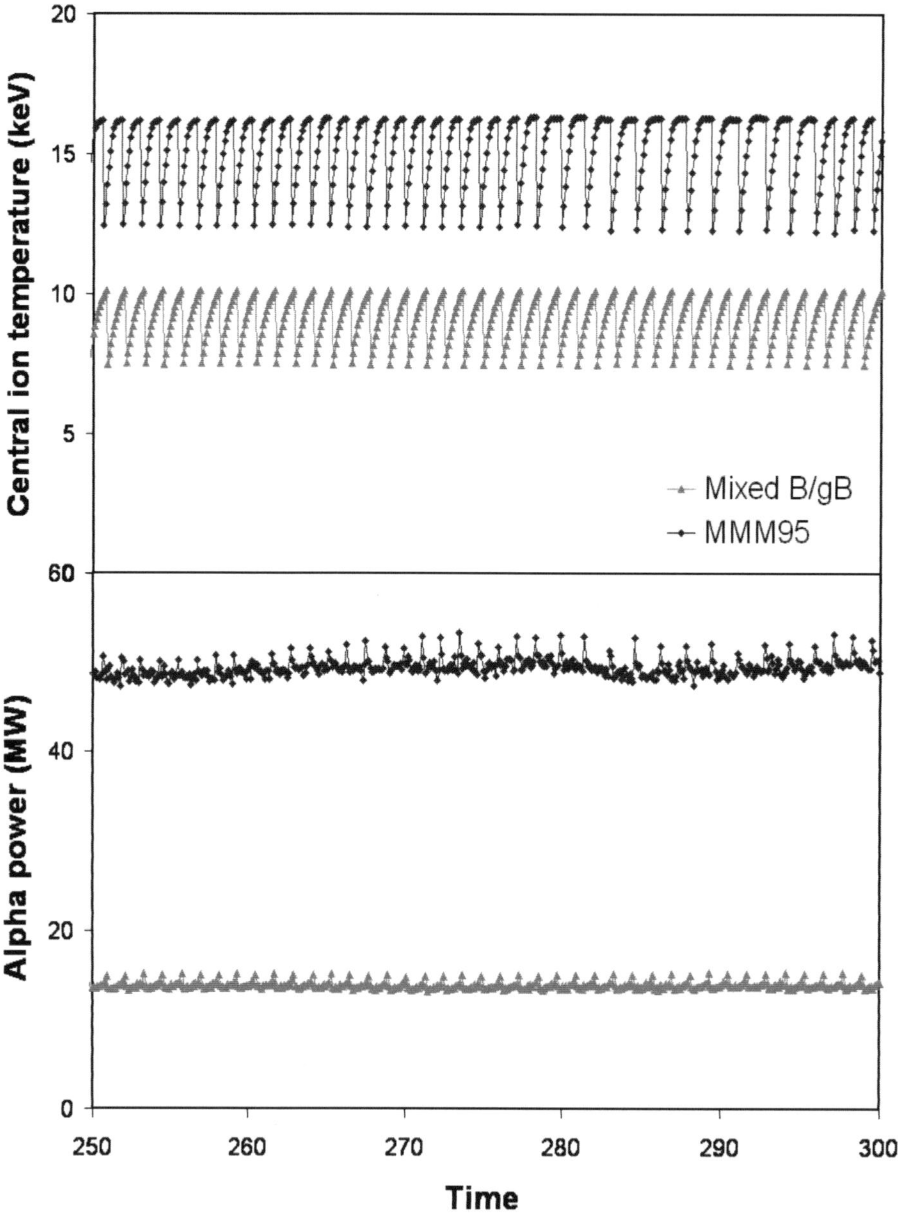

Figure 6. Central ion temperature (top panel) and alpha power (bottom panel) are plotted as a function of time. This simulation is carried out using the pedestal temperature model based on the magnetic and flow shear stabilization width model. The red line shows the simulation using Mixed B/gB and the blue line shows the simulation using MMM95.

the alpha power production suddenly increases about 7% in both simulations. The results was reported and explained in reference [2]. Note that the Porcelli model is used for calculating when sawtooth crash occurs and a modified Kadomtsev magnetic reconnection model is used to describe the effects of sawtooth crash. The details of the sawtooth triggering model and the Kadomtsev magnetic reconnection model can be found in references [14, 15]. It is found in all simulations that almost all of sawtooth crashes are triggered by the trapped of fast ion (equation 13 in reference [14]). It is found that the frequency of the sawtooth oscillation is almost the same for both transport model (0.8 Hz for Mixed B/gB and 0.7 Hz for MMM95 during the last 20 sec). The sawtooth mixing radius in the simulation using Mixed B/gB tends to be larger than that using MMM95 (116.1 cm for Mixed B/gB and 109.2 cm for MMM95 during the last 20 sec). It can be seen that the mixing radius of sawtooth oscillation in ITER is about half of the minor radius. This is an important issue for ITER and further investigation must be taken.

The fusion performance can be evaluated in term of Fusion Q, which can be calculated as

$$FusionQ = \frac{5 \times P_{\alpha.avg}}{P_{AUX}}, \tag{10}$$

where $P_{\alpha,avg}$ is an average alpha power and P_{AUX} is an auxiliary heating power (equal to 40 MW for these simulations). The results of average alpha power and fusion Q are summarized in table 2. The higher alpha power production and fusion Q in the simulation using MMM95 results from higher temperature prediction. Note that the average alpha power is taken during the time of 250 sec to 300 sec.

Table. 1 Summary of pedestal values and central ion temperature in the simulations for different pedestal temperature models using either Mixed Bohm/gyro-Bohm or Multi-mode transport model

Pedestal model	Pedestal width	Mixed Bohm/gyro-Bohm model		Multi-mode model	
		P_α (MW)	*Fusion Q*	P_α (MW)	*Fusion Q*
1	$\Delta \propto \rho s^2$	13.9	1.7	49.4	6.2
2	$\Delta \propto \sqrt{\rho R q}$	11.7	1.5	47.3	5.9
3	$\Delta \propto R\sqrt{\beta_{\theta,ped}}$	16.3	2.0	51.2	6.4

4. Conclusions

Self-consistent modeling of the International Thermonuclear Experimental Reactor (ITER) has been carried out using the 1.5D BALDUR integrated predictive modeling code. In these simulations, the boundary is taken to be at the top of the pedestal, where the pedestal values are described using the theoretical-based pedestal model. These pedestal temperature models are based on three different pedestal width scalings: magnetic and flow shear stabilization, flow shear stabilization, and normalized poloidal pressure. These pedestal width scalings are combined with a pedestal pressure gradient model based on ballooning mode limit to predict the pedestal temperature. The developed pedestal temperature models are used together with a core transport model, which is a combination of an anomalous transport and a neoclassical transport. An anomalous transport is calculated either using the Mixed Bohm/gyro-Bohm (Mixed B/gB) core transport model or Multimode (MMM95) core transport model, while a neoclassical transport is computed using the NCLASS model. At the reference design point, it is found that the pedestal temperatures in all simulations with different pedestal width scaling are nearly the same. As a result, the performances with the same transport model are similar. It is also found that the simulations using MMM95 yield more fusion power than those using Mixed B/gB. When the MMM95 is used, it appears that the ion temperature gradient

Journal of Physics: Conference Series **123** (2008) 012034 doi:10.1088/1742-6596/123/1/012034

(ITG) and trapped electron modes (TEM) are the most dominant modes. When the Mixed B/gB is used, it appears that the Bohm contribution is the most dominant term.

Acknowledgments
T. Onjun is grateful to Prof. Arnold H. Kritz and Dr. Glenn Bateman at Lehigh University for their generous supports. This work is supported by National Research Council of Thailand (NRCT), Thailand Toray Science Foundation, and Thammasat University Research Fund.

References
[1] Aymar R *et al.* 2002 *Plasma Phys. Control. Fusion* **44** 519
[2] Bateman G *et al.* 2003 *Plasma Phys. Control. Fusion* **45** 1939
[3] Singer C E *et al.* 1988 *Comput. Phys. Commun.* **49** 399
[4] Bateman G *et al.* 1998 *Phys. Plasmas* **5** 1793
[5] Houlberg W A *et al.* 1997 *Phys. Plasmas* **4** 3231
[6] Onjun T *et al.* 2002 *Phys. Plasmas* **9** 5018
[7] Onjun T *et al.* 2005 *Phys. Plasmas* **12** 082513
[8] Erba M *et al.* 1997 *Plasma Phys. Control. Fusion* **39** 261
[9] Hannum D *et al.* 2001 *Phys. Plasmas* **8** 964
[10] Onjun T *et al.* 2001 *Phys. Plasmas* **8** 975
[11] Sugihara M *et al.* 2000 *Nucl. Fusion* **40** 1743
[12] Osborne T H *et al.*1999 *J. Nucl. Mater.* **266-269** 131
[13] Hatae T *et al.* 2001 *Nucl. Fusion* **41** 285
[14] Porcelli F *et al.* 1996 *Plasma Phys. Control. Fusion* **38** 2163
[15] Bateman G *et al.* 2006 *Phys. Plasmas* **13** 072505

Journal of Physics: Conference Series **123** (2008) 012035 doi:10.1088/1742-6596/123/1/012035

H-Mode confinement properties close to the power threshold in ASDEX Upgrade

F Ryter, JC Fuchs, W Schneider, ACC Sips, A Stäbler, J Stober and ASDEX Upgrade Team

Max-Planck-Institut für Plasmaphysik, EURATOM Association, D-85748 Garching

E-mail: ryter@ipp.mpg.de

Abstract. Confinement properties close to the H-mode power threshold are studied in the ASDEX Upgrade tokamak. The results show that good confinement can be obtained close to the threshold with Type-I ELMs. The existence of Type-I ELMs does not necessarily require the heating power to be higher than the H-Mode power threshold, but it requires collisionality to be low enough. At higher collisionality Type-III ELMs replace the Type-I ELMs and confinement time is reduced by about 20%.

1. Introduction

The confinement time predictions for the ITER standard scenario, the ELMy H-mode, are based on scaling laws deduced from the ITPA multi-machine H-mode confinement database. The most widely used expression at present is the ITER98Y2 [1], from which one deduces the H-Mode factor $H98Y2 = \tau_{th}/\tau_{ITER98Y2}$, where τ_{th} is the thermal confinement time, i.e. without fast ion contribution. A new public version of this database has been recently released and its main properties are reported in [2].

It is well-known that the H-Mode is reached above a certain power threshold P_{thr} which is, in particular, studied using the ITPA multi-machine power threshold database [3, 4]. These studies yielded several scaling expressions for P_{thr}, the most recent one, recommended for ITER, is discussed in a companion paper at this conference [5]. It yields P_{thr} in MW and reads:

$$P_{thr} = 1.97 B_T^{0.77} n_{20}^{0.78} a^{0.98} R^{1.0} \qquad (1)$$

where B_T, n_{20}, R and a are respectively the magnetic field, line averaged density, minor and major radii, with the following units: T, $10^{20}\mathrm{m}^{-3}$, m. The heating power to be compared to P_{thr} is the usual net loss thermal power, P_{LTH}, also used in the ITPA databases and written as:

$$P_{LTH} = P_{Ohm} + P_{heat} - P_{losses} - dW/dt \qquad (2)$$

Here P_{Ohm} is the residual Ohmic power, P_{heat} the auxiliary heating power, whereas P_{losses} accounts for all losses, which are prompt fast ion losses in the case of NBI and RF absorption in the case of ICRF, W is the plasma energy content. In several present-day tokamaks the net heating power P_{LTH} can exceed P_{thr} by a large factor. Indeed, this is confirmed by the examination of the ITPA confinement database for the divertor tokamaks with ITER like shape, such as ASDEX Upgrade, Alcator C-Mod, DIII-D, JET and JT-60U for deuterium operation, as discussed in [5]. Except in Alcator C-Mod for which P_{LTH}/P_{thr} does not exceed 1.3, in the other devices this ratio can be high. In contrast to the general present situation, one expects P_{LTH} to be rather close to P_{thr} in ITER due to the limitation on the auxiliary heating power. It is therefore of great interest for predictions to investigate in detail the behaviour of confinement in H-Modes for heating powers close to the threshold power. This is the aim of this paper which deals with ASDEX Upgrade confinement data.

Journal of Physics: Conference Series **123** (2008) 012035 doi:10.1088/1742-6596/123/1/012035

2. ASDEX Upgrade data-set

ASDEX Upgrade is a divertor tokamak (R = 1.65m, a = 0.5m , $\kappa \approx 1.6$) from which about 1500 time slices are recorded in the ITPA confinement database. The device is equipped with NBI heating (8 beams, 2.5 MW each), delivering up to 20 MW for deuterium, and with ICRF power up to 6 MW, the scheme being here (H)D. Both methods have been used, either separately or combined, in the discharges included in the data-set analyzed here. For the present work we selected from the ASDEX Upgrade contribution to the ITPA database time slices fulfilling the condition $P_{LTH}/P_{thr} \leq 1.6$. This builds a data subset of about 270 records, to which we added data from other ASDEX Upgrade shots chosen on purpose for the present work. This yields finally a data-set of 360 time slices. These discharges were all run in the magnetic configuration in which the ion ∇B drift was directed towards the active X-point which provides the lowest LH power threshold. In addition, plasma and beam gases were deuterium and no time slice with high radiation losses has been included. The choice of the restriction $P_{LTH}/P_{thr} \leq 1.6$ is arbitrary and has been dictated by the wish to investigate the confinement behaviour close enough to the threshold while providing an amount of data which is, on one hand, sufficient to be representative and on the other hand limited enough to allow the assessment of the individual time points if necessary. The time slices have been chosen such that they are not taken during (3,2) or (2,1) NTM modes. They were chosen in quasi steady-state phases, in time intervals taken late enough after the last change of the heating power, such that the dW/dt contribution to P_{LTH} is negligible. The length of the time interval over which the data are taken and averaged is also long enough to guarantee reliable and representative values. This data-set contains data for time intervals with Type-I and Type-III ELMs, as well as intermediate cases for which Type-I and Type-III ELMs are mixed within the selected time interval. One can distinguish two types of "mixed ELMs": (i) regular Type-I ELMs each of them being followed by a train of 3-4 Type-III ELMs, often named compound ELMs, (ii) cases in which Type-I and Type-III ELMs occur independently and without any regular pattern. In the following we do not distinguish between these 2 cases which are both categorized as "mixed" and labeled Type-I/III. The database contains 218 time slices categorized as Type-I ELMs, 80 Type-I/III and 62 Type-III. Type-I ELMs are thought to be caused by an ideal MHD instability whereas Type-III ELMs are most probably linked with a resistive phenomenon [6, 7]. The ELM type can be determined from the time evolution of the discharges, in particular, the variation of the ELM frequency when power is varied is a precious indication of the ELM type. The ELM frequency increases with power for Type-I ELMs, but it decreases with power for Type-III ELMs [6]. The ELM type may be difficult to determine when the density is strongly increased at constant heating power and some uncertainty might exist in such cases.

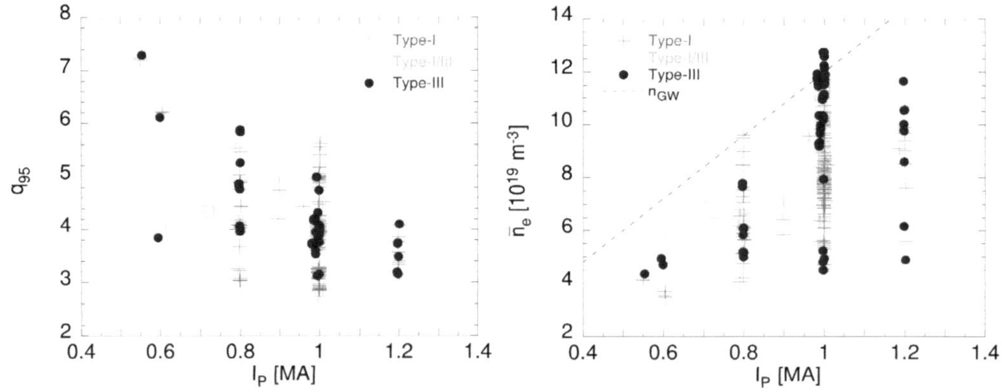

Figure 1. *Ranges of the ASDEX Upgrade data-set used in this work. The 3 different ELM types are indicated by the symbols. Left plot: q_{95} versus plasma current. Right plot: line-averaged density versus plasma current, the Greenwald limit is indicated by the dashed line.*

The ranges covered by the main control parameters are given in figures 1 and 2. They all correspond to the usual operational window of the device. For plasma current, I_p, the range extends from 0.6 to 1.2 MA, with more discharges at 1 MA, values used the most often. The range in safety factor q_{95} covers the usual range around 4 for all I_p values, with a much larger range for $I_p \leq 1$ MA. The different symbols indicate the three ELM types mentioned above and will be kept throughout the paper. The left plot of figure 1 indicates that all 3 types of ELMs can be obtained at any current. The right plot of figure 1 shows the line-averaged density range at different values of I_p. The Greenwald density, $n_{GW} = I_p/(\pi a^2)$ [8], is indicated by the dashed line. This line corresponds to an average value and variations of $\pm 15\%$ for individual time points can be caused by slight variations of a. The density has been explored over a large range up to n_{GW} for currents below 1.2 MA.

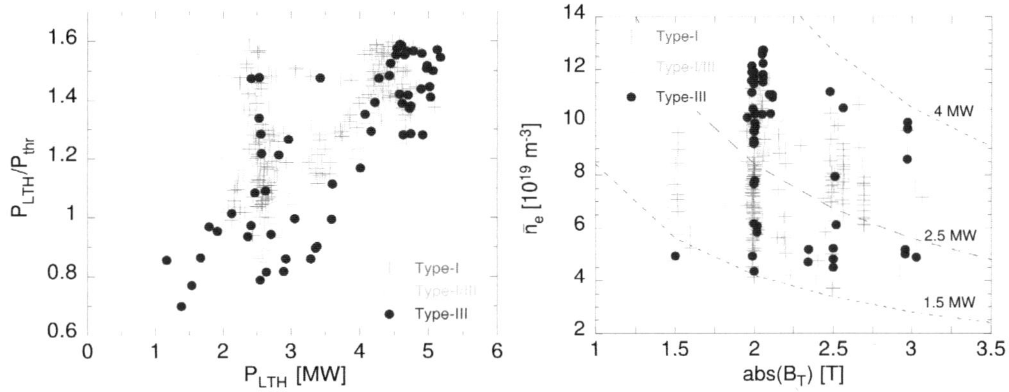

Figure 2. *Ranges of the ASDEX Upgrade data-set used in this work. Left plot: ratio P_{LTH}/P_{thr} versus net heating power. Right plot: line-averaged density versus magnetic field. As explained in the text, the dashed lines correspond to the H-Mode threshold at fixed value of P_{LTH} (values in MW in the graph).*

The left plot of figure 2 shows the ratio P_{LTH}/P_{thr} versus heating power. The range in P_{LTH} is limited to values below about 5 MW by the restriction $P_{LTH}/P_{thr} \leq 1.6$ and remains well below the machine possibility of 20 MW. The large number of points at $P_{LTH} \approx 2.5$ MW and $P_{LTH} \approx 5$ MW correspond respectively to one and two NBI beams. The points in-between correspond either to the use of ICRF power, or to reduced NBI power by modulation or reduced acceleration voltage. The values of P_{LTH}/P_{thr} range between ≈ 0.7 and 1.6, the imposed upper limit. The lower boundary of P_{LTH}/P_{thr} is determined by the H-to-L transition. The existence of H-Mode points for $P_{LTH}/P_{thr} < 1$ is due to the fact that the scaling of P_{thr} represents the L-to-H transition which is in general higher than the back transition due to the good confinement provided by the H-Mode, a phenomenon often called H-Mode hysteresis. At $P_{LTH} \approx 2.5$ MW the ratio P_{LTH}/P_{thr} varies by a factor of about 2, which is due to the variation of P_{thr} induced by different B_T and $\bar{n}_e$ values. For $P_{LTH} \approx 5$ MW, there are no points for $P_{LTH}/P_{thr} < 1.2$ which suggests that P_{thr} cannot not be pushed to extremely high values by increasing magnetic field or density. This is in agreement with previous studies in ASDEX Upgrade which indicated that, when approaching n_{GW}, a very high heating power is required to sustain the H-Mode and P_{LTH}/P_{thr} must be much larger than 1 [9, 10]. As P_{thr} depends on B_T and $\bar{n}_e$ we show the ranges covered by these two quantities in the right plot of figure 2. There is a factor of 2 range in B_T and the density has been explored over a significant range at each value of B_T, providing a rather balanced data-set. The larger number of points at 2 T corresponds to the discharges at 1 MA and $q_{95} \approx 3.5$. In this plot we also added lines which represent the values of B_T and $\bar{n}_e$ corresponding to the threshold scaling of equation 1 for three fixed values of P_{thr}: 1.5 MW, 2.5 MW and 4 MW. One sees that the lower points in this plot correspond to $P_{thr} \approx 1.5$ MW, whereas the higher points almost reach 4 MW. Therefore, in our data-set P_{thr} varies by a factor of about 2.5. The points with the highest values of P_{thr} in this plot correspond to the upper right corner of the left plot, $P_{LTH} \approx 5$ MW and $P_{LTH}/P_{thr} > 1.3$. They correspond, in particular, to the highest density values which can be reached, limited by n_{GW}. Triangularity, δ, has also been varied over the whole I_p range between 0.2 and 0.45. The analysis of our data-set shows that, as well-known [11, 12, 13], higher δ helps sustaining the H-mode with good confinement at high density. However, this parameter does not play a crucial role in the results presented here and we will not discuss its influence in detail.

3. Power threshold

Before investigating the confinement properties, it is important to verify that the scaling used for P_{thr} reproduces well the ASDEX Upgrade data. This is expected because the contribution of the ASDEX Upgrade data to the threshold database is significant and also because it is close to the center of gravity of the ITPA database and is therefore well fitted.

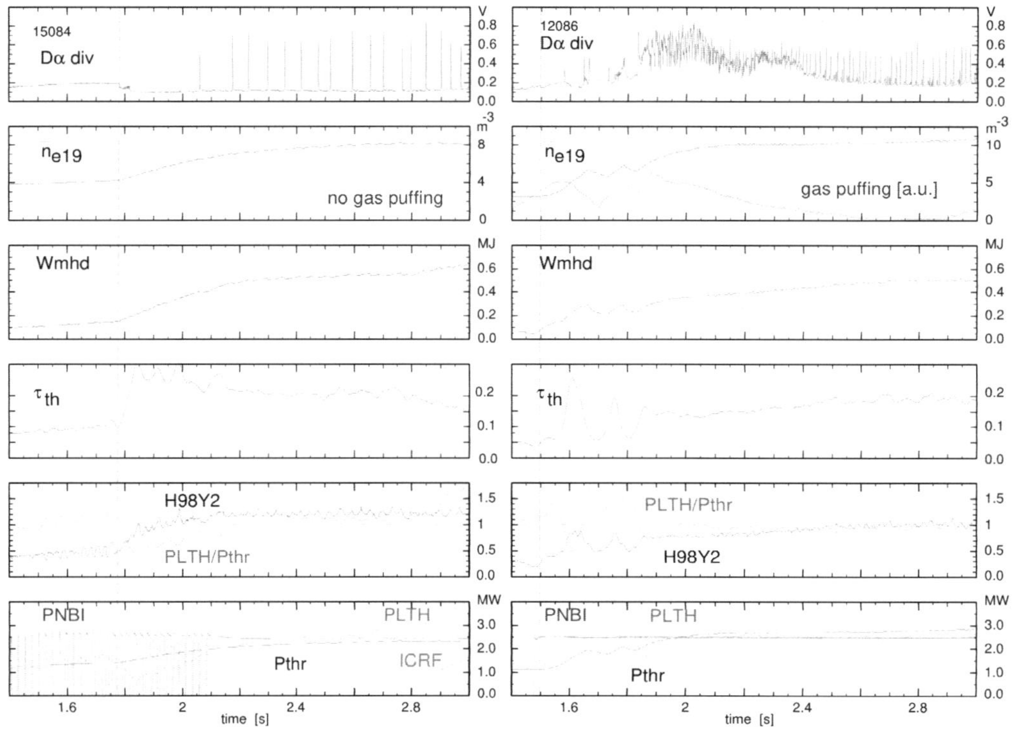

Figure 3. *Two examples of the time evolution of H-Mode discharges at low heating power, both at $I_p = 1$ MA, $B_T = 2$ T. The different signals are indicated in the corresponding boxes. Left plot: power ramp, no gas puffing. Right plot: heating power in one step, gas puffing to increase density.*

Figure 3 shows a selection of time traces for two representative discharges, both at $I_p = 1$ MA, $B_T = 2$ T, $q_{95} \approx 3.5$. The first example (left plot) has a slow heating power ramp obtained by NBI modulation with varying duty-cycle, as preferred for accurate threshold studies. The L-H transition occurs at about 1.77 s, as indicated by the vertical line. At this time point, the ratio P_{LTH}/P_{thr} is very close to 1 as revealed by the corresponding curve. This corresponds to $P_{LTH} \approx 1.6$ MW. After the L-H transition, as usual, plasma energy and therefore confinement time, represented by τ_{th} and $H98Y2$, increase by about a factor of 2. Density also starts to increase after the L-H transition. Note that P_{LTH} first quickly decreases after the L-H transition due to the dW/dt term. It should be underlined that the hysteresis allows this behaviour. As the power ramp continues up to 2.5 MW, P_{LTH} increases again, leading to a decrease of τ_{th}, whereas $H98Y2$ remains constant. The first Type-I ELM occurs at about 2.05 s followed at regular intervals by further ELMs. The ELM frequency is low, about 8 Hz, which is due to the low heating power. There was no gas puffing in this discharges but the low ELM frequency leads to a rather slow but large increase of density up to $8 10^{19} m^{-3}$. Consequently, P_{thr} increases and P_{LTH}/P_{thr} remains very close to 1.

The other discharge illustrated in the right plot is quite different. There is no power ramp, but 1 NBI beam is turned on, leading quickly to the L-H transition. Here also, P_{LTH}/P_{thr} is very close to 1 at the L-H transition. Gas puffing is applied to increase density, which indeed reaches $\approx 10^{20}m^{-3}$ later in the discharge. Due to this gas puffing, the discharge oscillates 2 times between ELM-free H-mode and L-mode in the time interval 1.5 - 1.9 s to finally remain in the H-mode with rather high frequency Type-III ELMs and a transition to slower Type-I ELMs at about 2.4 s. In this case the ratio P_{LTH}/P_{thr} is also close to one and confinement is good, as shown by $H98Y2 \approx 1$.
The good agreement of P_{LTH} with P_{thr} at the L-H transition illustrated by these cases is general for the data used here. This is also confirmed by previous L-H threshold power studies for ASDEX Upgrade data only, which yield a scaling very close to that used here [3].

4. Confinement behaviour

An overview of the confinement behaviour, given in figure 4, shows the H-Mode enhancement factor, H98Y2, as a function of P_{LTH}/P_{thr}. The question is often asked whether it is possible to define a minimum value of P_{LTH}/P_{thr} above which confinement is good, i.e. $H98Y2 \approx 1$. Figure 4 indicates that obviously good confinement can be achieved with high probability as soon as $P_{LTH}/P_{thr} \geq 1$.

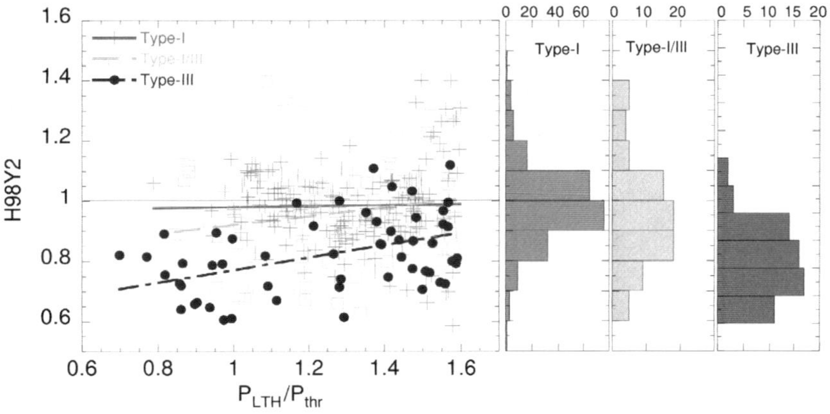

Figure 4. *Confinement properties illustrated by H98Y2. Left part: dependence upon P_{LTH}/P_{thr}. The line are linear fit for each type of ELM. Right part: the 3 statistical distributions of H98Y2 for each ELM type.*

The symbols for the different ELM types indicate that Type-III ELMs have systematically a somewhat lower confinement than the two other types. This is quantified by the statistical distribution plots for each ELM type in the right part of figure 4. For Type-I ELMs, $H98Y2$ is very close to unity whereas it is around 0.8 for Type-III ELMs. This is in agreement with the usual expectation that, in device of the size of ASDEX Upgrade at least, the confinement of H-Modes with Type-III ELMs is somewhat lower than that of Type-I ELMy H-Modes. This difference can be attributed to a lower pedestal pressure for Type-III ELMs, see e.g. [11]. Therefore good confinement is determined by the ELM type and not directly by the value of P_{LTH}/P_{thr}. In the scatter plot of figure 4 we added linear fits of $H98Y2$ versus P_{LTH}/P_{thr} for each ELM type. For Type-I ELMs the line is, as expected, very close to 1 and exhibits almost no dependence upon P_{LTH}/P_{thr}. For Type-III ELMs, the line is clearly below 1 but shows a significant increase with increasing P_{LTH}/P_{thr}. This might be caused by an increase of pedestal pressure and/or pressure gradient for Type-III, linked with the improved stability of these ELMs as collisionality decreases. The fit for Type-I/III ELMs lies in-between. In addition, we have verified that the ability to achieve good confinement at rather low values of P_{LTH}/P_{thr} is not restricted to specific values of control parameters such as I_p, B_T or q_{95}.
The left plot of figure 5 shows the relation between normalized β, β_N, and confinement factor. They are here closely related: The rather distinct two groups of points which built two parallel patterns increasing with $H98Y2$ are due to the discretisation of the NBI input power either 2.5 MW (1 beam)

Journal of Physics: Conference Series **123** (2008) 012035 doi:10.1088/1742-6596/123/1/012035

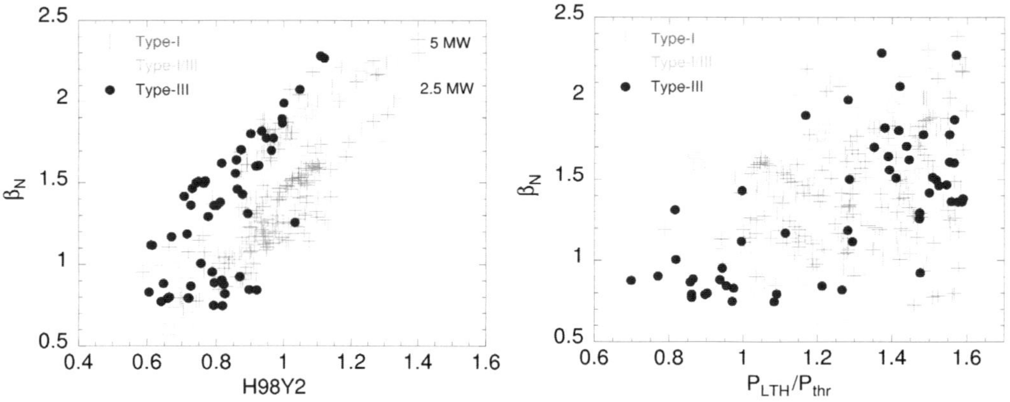

Figure 5. β *range and dependence Left plot:* β_N *versus normalized confinement. Right plot:* β_N *versus power normalized by threshold power.*

or 5 MW (2 beams). The excursion in $H98Y2$ is due to the fact that the confinement scaling cannot reproduce all dependencies with a very high accuracy. In particular, machine conditioning plays an important role through its influence on pedestal pressure, as investigated for ASDEX Upgrade in [14]. Triangularity also contributes to this variation, as discussed in [11, 12] for ASDEX Upgrade. This plot shows that $\beta_N \approx 2.4$ can be reached, despite the restriction in power imposed in the present study. The right plot indicates that reaching $\beta_N \approx 2$ only requires $P_{LTH}/P_{thr} \geq 1.2$. This is an important result for ITER for which the standard scenario does not require β_N above 2. It should be possible to reach this value with $H98Y2 \approx 1$ and for P_{LTH}/P_{thr} not significantly above 1. Moreover, it should be underlined that high values of β_N can be obtained with Type-III ELMs, most of them corresponding here to high density cases.

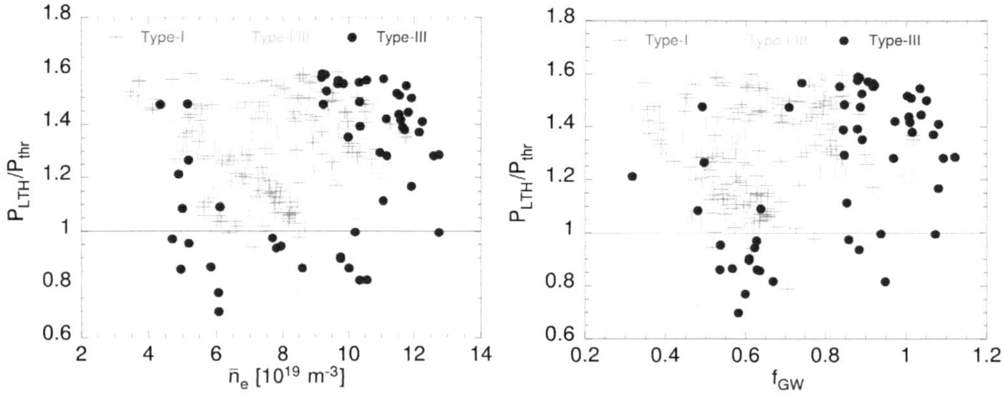

Figure 6. *Domains for each ELM type in the density-power plane. Left plot:* P_{LTH}/P_{thr} *versus line-averaged density. Right plot:* P_{LTH}/P_{thr} *versus Greenwald factor.*

The question of the minimum value of P_{LTH}/P_{thr} required to get a certain type of ELM, in particular Type-I, is often asked. Elements for the answer are provided by figure 6 in which P_{LTH}/P_{thr} is plotted versus line-averaged density or normalized density. The possible relation with physics quantities, such as resistivity or collisionality, is discussed below. Clearly, the ELM type is not fully dictated by P_{LTH}/P_{thr} but also, in particular by density if we consider control parameters. At medium densities there is a clear trend for Type-III ELMs to only occur at values of P_{LTH}/P_{thr} below 1, whereas at high density Type-III ELMs also occur frequently for P_{LTH}/P_{thr} values clearly above 1. This behavior does not seem to be more clearly related with f_{GW} than with $\bar{n}_e$. This is partly due to the fact that many of the high density points with f_{GW} close to 1 were obtained at the same plasma current of 1 MA in this data-set. There are a few Type-III ELM points, at rather low density and $P_{LTH}/P_{thr} > 1$ which seem to be in disagreement with the rest of the figure. These "low density outliers" have been verified carefully and are indeed Type-III ELMs. This might be attributed to the increase of P_{thr} towards low density, which is not taken into account in the scaling. However this would be somewhat unexpected at this density of about $5 10^{19} m^{-3}$ which is large compared to the minimum threshold density $\approx 2.5 10^{19} m^{-3}$ in ASDEX Upgrade [3]. This could also be due to machine conditions, but we have not found any concrete indication for this hypothesis. Apart from these few points, overall the two plots of figure 6 suggest that resistivity or collisionality may play a role. This would be in agreement with the hypothesis that Type-I and Type-III ELMs are respectively ideal or resistive MHD events. While resistivity, in general, formally separates ideal and resistive MHD, in the case of ELMs, collisionality also plays a role, due to the fact that at the plasma edge, the bootstrap current contribution, j_{BS}, to the local current is important, see e.g. [15, 16, 17, 18]. As j_{BS} strongly depends on normalized collisionality ν_e^* (and ν_i^*), [19], these quantities should also be considered when discussing ELM types. Indeed, ν_e^* has been shown to unify the dependence of ELM losses between devices and is widely used in studies dealing with ELMs. We show here diagrams for the ELM types in figure 7 using Spitzer resistivity $\eta_{Spitzer}$ and ν_e^*.

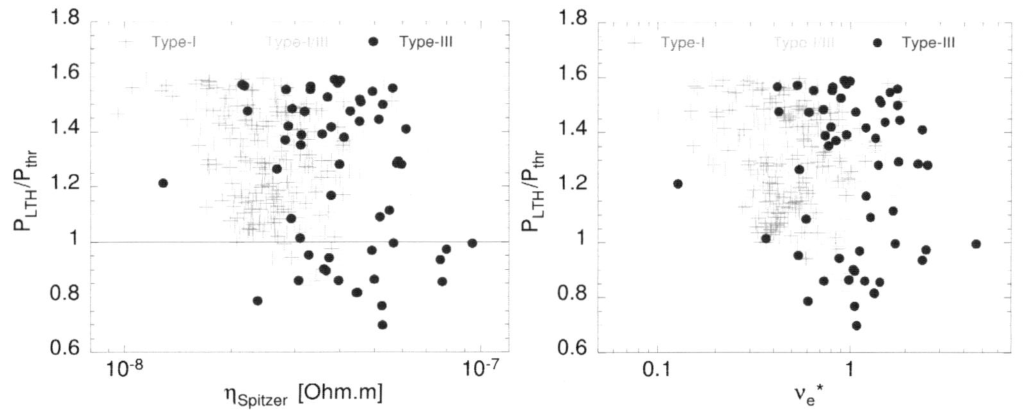

Figure 7. *ELM domain comparison when plotted versus resistivity or electron collisionality.*

These plots indeed show that Type-III ELMs are concentrated at higher values of $\eta_{Spitzer}$ or ν_e^*, whereas Type-I ELMs are rare in this region and populate in majority the low values. The difference between the two representations does not allow to state whether one or the other better reflects reality. It should be noted that, as both quantities depend on electron temperature, they are strongly correlated. There is one Type-III point at very low value of $\eta_{Spitzer}$ and ν_e^*. This is one of the low density outliers mentioned above and the only remaining outsider in the plots of figure 7. Indeed, the other few outliers have rather poor confinement and, despite low density, higher values of $\eta_{Spitzer}$ and ν_e^*. They are therefore in the bulk of the other Type-III points in figure 7. The reason for the poorer confinement of these points could not be identified.

Confinement reflected by $H98Y2$ is provided in the plots of figure 8 which shows a confinement degradation with increasing $\eta_{Spitzer}$ and ν_e^* , in concert with the occurrence of Type-III ELMs. It should be noted, that this figure is not plotted at constant ρ^* and β and cannot be interpreted as a general result on the dependence of confinement upon collisionality. It should be underlined that plotting $H98Y2$ versus density (not shown) does not exhibit a clear degradation for this data-set and does not order at all the ELM types.

5. Discussion and conclusion

The results presented here indicate that good confinement, $H98Y2 \geq 1$, can be achieved at heating powers very close to the H-mode power threshold. This requires, however, H-Modes plasmas with

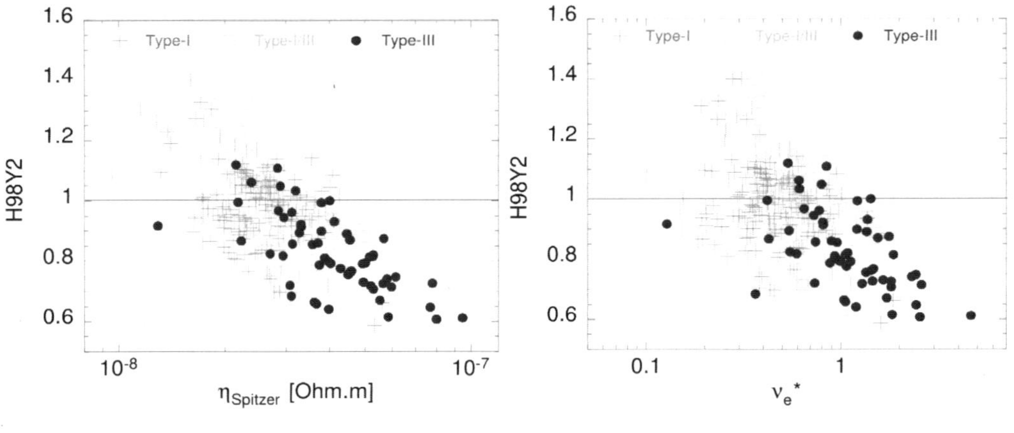

Figure 8. *Confinement and ELM type versus resistivity and electron collisionality.*

Type-I ELMs, probably not acceptable in ITER because of their high power load in the divertor. Type-III ELMs, which are more favorable from the point of view of divertor power load, exhibit values of $H98Y2$ which are 10 to 20 % lower than the Type-I ELM cases. In addition, our results indicate that values of β_N above 2 can be reached at low heating power and even with Type-III ELMs.

The confinement degradation linked with Type-III ELMs is a general behaviour in tokamaks. It occurs in particular at high density and is caused by a reduction of the pedestal height which propagates to the core by profile stiffness, see for instance Alcator C-Mod [20], ASDEX Upgrade [21], DIII-D [22], JET [13]. In some cases this degradation can be partly compensated by density peaking, as reported in ASDEX Upgrade [23, 12] and DIII-D [24]. However, high density peaking has deleterious effects on impurity transport and NTM stability. To avoid these different difficulties, several types of plasma scenarios with small ELMs, generally at high density, have been developed, see for instance review [25]. Their properties and ability to be extrapolated to ITER are still under assessment.

The majority of the Type-III ELMs included in the present work have been obtained at high density and rather low temperature. As a result, the occurrence of Type-I or Type-III ELMs is clearly linked with collisionality or resistivity. Figure 7 shows that, within our data, it is not possible to conclude whether one or the other quantity is more important to separate the ELM types. This is due to the fact that both strongly depend on temperature and possibly also because collisionality plays an important role in the ELM mechanism through the edge bootstrap current. Whether resistivity can play a major role in the ELM type may be clarified in the future in dedicated experiments in which density and temperature are varied adequately. Most of the recent analyses on the occurrence of Type-III ELMs, carried out in other devices, were focused, so far, on the effect of collisionality. In agreement with our observations, they also indicated that Type-III ELMs are obtained as density is sufficiently increased [26, 27, 28, 17]. However, high density or collisionality is not always required for Type-III ELMs which have also been observed at low collisionality in DIII-D [28] and JET [27]. Indeed, two branches for Type-III ELMs seem to exist: one at high density, with $\nu_e^* > 1.0$, where the ELMs seem to exist only below a critical temperature, independently of density, and a second one at low density, approximatively in the region $\nu_e^* < 0.3$ in which the Type-III ELM boundary seems to be linked with β, [28, 17, 27]. Whether resistivity plays a key role in such cases remains to be investigated. Our Type-III outlier at low ν_e^* in figure 7 could be such a low density case. These must be very rare in ASDEX Upgrade, because the lowest density in the H-mode is in general higher than that achieved in JET and DIII-D. This should be investigated in future ASDEX Upgrade experiments.

In ITER, the predicted values of ν_e^* at the pedestal lies between 0.03 and 0.07 [29, 30]. This shows that achieving high density Type-III ELMs comparable to those occurring in present devices with $\nu_e^* > 1.0$ would required a strong decrease of ν_e^* at the very edge by radiative cooling or other control method. This is a difficult task and this indicates that the low density Type-III ELMs seem to be a possible way for ITER which should be further explored.

In ASDEX Upgrade, Type-I ELMs do not necessarily require P_{LTH}/P_{thr} above unity. This is interpreted as being provided by the H-Mode hysteresis linked with the transport reduction in the H-Mode. In JET, the situation seems to be a bit different: Type-I ELMs require P_{LTH}/P_{thr} to be larger than 1.3 to 2.5, depending on the conditions [27]. This is somewhat surprising as ν_e^* in JET is expected to be in general lower than in ASDEX Upgrade which should favor the Type-I ELMs. In both devices, as power is increased, Type-III ELMs appear just above the threshold. However, once in the Type-I ELM regime with P_{LTH} significantly above P_{thr}, JET plasmas seem to be more sensitive to a fall-back into Type-III ELMs than in ASDEX Upgrade. This could be linked with the associated density decrease in such JET cases and to the existence of the low density Type-III ELMs in JET.
It should also be underlined that no physics reason links the ELM type to threshold power itself. Indeed, the L-H transition depends on edge parameters in the L-Mode just prior to the H-Mode, whereas ELMs are dictated by collisionality, pressure gradient and MHD stability in the pedestal region, when the H-Mode has developed. This is illustrated, for instance in ASDEX Upgrade, by plasmas in which the ion ∇B drift is directed away from the X-point, which is well-known to lead to a high power threshold. In this configuration, Type-III ELMs do not seem to exist. Another argument is provided by the fact that Type-III are the most usual ELMs in devices smaller than ASDEX Upgrade, for instance ASDEX [6], even when the heating power is well above the threshold. Thus, this observation is also in agreement with the fact that collisionality seems to play a key role for the ELM type. In view of ITER, these questions are being addressed in comparative studies between devices and will certainly improve our future understanding.

References

[1] ITER Physics Basis, Chap II, page 2208, Table 5 1999 *Nucl. Fusion* **39**
[2] McDonald D C, Cordey J G, Thomsen K, Kardaun O J W F, Snipes J A, Greenwald M, Sugiyama L, Ryter F, Kus A, Stober J *et al.* 2007 *Nucl. Fusion* **47** 147–174
[3] Ryter F and the H-mode Treshold Database Group 2002 *Plasma Phys. Controlled Fusion* **44** A415–A421
[4] Martin Y R and H-Mode Threshold Database Group 2005 *Proc. of the 20th IAEA Conference Fusion Energy (CD-Rom), Vilamoura, Portugal, November 2004* vol IAEA-CSP-25/CD (Vienna: IAEA) pp IAEA–CN–116/IT/P3–35
[5] Martin Y and ITPA CDBM H-Mode Group 2007 *This Conf, Paper P6-05*
[6] Zohm H, Osborne T, Burrell K, Chu M, Doyle E, Gohil P, Hill D, Lao L, Leonard A, Taylor T *et al.* 1995 *Nucl. Fusion* **35** 543–550
[7] Huysmans G T A 2005 *Plasma Phys. Controlled Fusion* **47** B165–B178
[8] Greenwald M, Terry J L, Wolfe S M, Ejima S, Bell M G, Kaye S M and Neilson G H 1988 *Nucl. Fusion* **28** 2199–2207
[9] Mertens V, Kaufmann M, Neuhauser J, Schweinzer J, Stober J, Büchl K, Gruber O, Haas G, Herrmann A, Kallenbach A *et al.* 1997 *Nucl. Fusion* **37** 1607–1614
[10] Mertens V, Borrass K, Gafert J, Laux M, Schweinzer J and ASDEX Upgrade Team 2000 *Nucl. Fusion* **40** 1839–1843
[11] Ryter F, Stober J, Stäbler A, Tardini G, Fahrbach H U, Gruber O, Herrmann A, Kallenbach A, Kaufmann M, Kurzan B *et al.* 2001 *Nucl. Fusion* **41** 537–550
[12] Stober J, Zohm H, Gruber O, Herrmann A, Kaufmann M, Maraschek M, Meo F, Sips A, Treutterer W and ASDEX Upgrade Team 2001 *Plasma Phys. Controlled Fusion* **43** A39–A53
[13] Saibene G, Sartori R, Loarte A, Campbell D J, Lomas P J, Parail V, Zastrow K D, Andrew Y, Sharapov S, Korotkov A *et al.* 2002 *Plasma Phys. Controlled Fusion* **44** 1769–1800
[14] Ryter F, Fahrbach H U, Gude A, Neu R, Rohde V, Stober J and ASDEX Upgrade Team 2002 *Plasma Phys. Controlled Fusion* **44** A407–A413
[15] Connor J W 1998 *Plasma Phys. Controlled Fusion* **40** 531–542
[16] Saarelma S, Günter S, Kurki-Suonio T and Zehrfeld H P 2000 *Plasma Phys. Controlled Fusion* **42** A139–A145
[17] Suttrop W 2000 *Plasma Phys. Controlled Fusion* **42** A1–A14
[18] Thomas D M, Leonard A W, Osborne T H, West R J G W P and Burrell K H 2006 *Plasma Phys. Controlled Fusion* **48** A183–A191
[19] Sauter O, Angioni C and Lin-Liu Y R 1999 *Phys. Plasmas* **6** 2834–2839
[20] Greenwald M, Boivin R L, Bombarda F, Bonoli P T, Fiore C L, Garnier D, Goetz J A, Golovato S N, Graf M A, Granetz R S *et al.* 1997 *Nucl. Fusion* **37** 793–807
[21] Suttrop W, Kaufmann M, de Blank H J, Brüsehaber B, Lackner K, Mertens V, Murmann H, Neuhauser J, Ryter F, Salzmann H *et al.* 1997 *Plasma Phys. Controlled Fusion* **39** 2051–2066
[22] Osborne T H, Ferron J R, Groebner R J, Lao L L, Leonard A W, Mahdavi M A, Maingi R, R L Miller A D T, Wade M and Watkins J 2000 *Plasma Phys. Controlled Fusion* **42** A175–A184
[23] Stober J, Fuchs C, Gruber O, Kaufmann M, Kurzan B, Meo F, Müller H W, Ryter F and ASDEX Upgrade Team 2001 *Nucl. Fusion* **41** 1535–1538

[24] Osborne T H, Leonard A W, Mahdavi M A, Chu M, Fenstermacher M E, Haye R L, McKee G, Petrie T W, Doyle E, Staebler G *et al.* (DIII-D Team) 2001 *Phys. Plasmas* **8** 2017–2022

[25] Oyama N, Gohil P, Horton L D, Hubbard A E, Hughes J W, Kamada Y, Kamiya K, Leonard A W, Loarte A, Maingi R *et al.* 2006 *Plasma Phys. Controlled Fusion* **48** A171–A182

[26] Hubbard A, Boivin R L, Drake J F, Greenwald M, In Y, Irby J H, Rogers B N and Snipes J A 1998 *Plasma Phys. Controlled Fusion* **40** 689–692

[27] Sartori R, Saibene G, Horton L D, Becoulet M, Budny R, Borba D, Chankin A, Conway G D, Cordey G, McDonald D *et al.* 2004 *Plasma Phys. Controlled Fusion* **46** 723–750

[28] Osborne T H, Groebner R J, Lao L L, Leonard A W, Maingi R, Miller R L, Porter G D, Thomas D M and Waltz R E 1997 *Europhysics Conference Abstracts (Proc. of the 24th EPS Conference on Controlled Fusion and Plasma Physics, Berchtesgaden, 1997)* vol 21A, part III ed Schittenhelm M, Bartiromo R and Wagner F (Petit-Lancy: EPS) pp 1101–1104

[29] Loarte A, Saibene G, Sartori R, Bécoulet M, Lomas P J, Ali-Arshad S, Alper B, Beurskens M, Kempenaars M, Koslowski R *et al.* 2003 *Europhysics Conference Abstracts (CD-ROM, Proc. of the 30th EPS Conference on Controlled Fusion and Plasma Physics, St. Petersburg, 2003)* vol 27A ed Koch R and Lebedev S (Geneva: EPS) pp P–1.96

[30] Saibene G, Lomas P, Sartori R, Loarte A, Stober J, Andrew Y, Arshad S, Conway G, de la Luna E, Günther K *et al.* 2005 *Nucl. Fusion* **45** 297–317

Journal of Physics: Conference Series **123** (2008) 012036 doi:10.1088/1742-6596/123/1/012036

H-mode plasma transport simulation in ITER with effect of neoclassical tearing mode

Y Takahashi, K Yamazaki, H Arimoto and T Shoji

Nagoya University, Chikusa-ku, Nagoya 464-8603, Japan

E-mail: yamazaki@ees.nagoya-u.ac.jp

Abstract. For the prediction of the ITER plasmas, the effect of the neoclassical tearing mode (NTM) on the plasma confinement has been calculated using the 1.5-dimensional equilibrium and transport simulation code TOTAL. The time evolution of the NTM magnetic island has been analyzed using the modified Rutherford equation for ITER normal shear plasmas. The anomalous transport model used here is GLF23. The saturated magnetic island widths are w/a ~ 0.048 at 3/2 mode and w/a~0.21 at 2/1 mode, and the reduction in fusion power output by NTM is 27% at the 3/2 mode, and 82% at the 2/1 mode. The stabilization effect of the electron cyclotron current drive (ECCD) with EC is also clarified. The threshold of ECCD power for the full stabilization at high beta NTM island formation is ~10 MW against the 3/2 mode, and ~23 MW against the 2/1 mode.

1. Introduction

The neoclassical tearing mode (NTM) might limit the plasma pressure and would lead to a disruption in future tokamak reactors. The NTM makes the temperature profile flat inside magnetic island, and the central plasma temperature is reduced. Therefore, when the magnetic island is formed in the plasma, the fusion output power is decreased. The analysis and control of NTMs are one of the crucial issues in tokamak reactor [1]. The NTM is caused by a lack of the bootstrap current inside the magnetic island where the pressure profile is flattened [2]. The electron cyclotron current drive (ECCD) gives the NTM stabilization to replace the missing bootstrap current. The effect of NTM on the ITER plasma should be investigated, and the stabilization method of NTM should be clarified.

For the prediction of ITER plasmas, the time evolution of neoclassical tearing modes has been calculated by using the 1.5-dimensional (1.5-D) equilibrium and transport simulation code (toroidal transport linkage code TOTAL [3,4]). The magnetic island width is evaluated using the modified Rutherford equation [5,6]. In the simulation code, we used the GLF23 anomalous transport model that can simulate H-mode plasmas [7].

The purpose of this paper is to clarify the effect of the NTM magnetic island formation on the plasma confinement and to demonstrate its stabilization by the electron cyclotron current drive (ECCD) in ITER. Especially, we focused on the ITER plasma performance with the GLF23 transport model and evaluated the decrease in the output fusion power and Q value due to NTMs, in addition to the ECCD feedback control as discussed in reference [6]. In the next section, a numerical model is described. The details of modified Rutherford equation, the ECCD current profile and the current drive efficiency are also shown in this section. In section 3, simulation results are shown. The summary is given in section 4.

Journal of Physics: Conference Series **123** (2008) 012036 doi:10.1088/1742-6596/123/1/012036

2. Numerical model

The time evolution of NTM has been calculated using 1.5-D equilibrium and transport code (toroidal transport linkage code TOTAL [3,4]). The plasma equilibrium is solved by the free-boundary Apollo code [8], and the plasma transport is evaluated including the impurity dynamics [9]. The anomalous transport model used here is the GLF23 [7] that can simulate H-mode plasmas.

2.1. Modified Rutherford equation

The time evolution of a NTM island width, W, is calculated according to the modified Ruther ford equation. Here, W is the normalized magnetic island width with respect to the minor plasma radius, a.

$$\frac{dW}{dt} = \Gamma_{\Delta'} + \Gamma_{BS} + \Gamma_{GGJ} + \Gamma_{pol} + \Gamma_{EC} \tag{1}$$

$$\Gamma_{\Delta'} = k_1 \frac{\eta}{\mu_0} \Delta'(W) \left\langle |\nabla\rho|^2 \right\rangle \tag{2}$$

$$\Gamma_{BS} = k_2 \eta L_q j_{BS} \left\langle \frac{|\nabla\rho|}{B_p} \right\rangle \frac{W}{W^2 + W_d^2} \tag{3}$$

$$\Gamma_{GGJ} = -k_3 \frac{\eta}{\mu_0} \varepsilon_s^2 \beta_{ps} \frac{L_q^2}{\rho_s L_p} \left(1 - \frac{1}{q_s^2}\right) \left\langle |\nabla\rho|^2 \right\rangle \frac{1}{W} \tag{4}$$

$$\Gamma_{pol} = -k_4 \frac{\eta}{\mu_0} g(\varepsilon_s, \nu_i) \beta_{ps} \left(\frac{\rho_{pi} L_q^2}{L_p}\right)^2 \left\langle |\nabla\rho|^2 \right\rangle \frac{1}{W^3} \tag{5}$$

$$\Gamma_{EC} = -k_5 \eta \frac{L_q}{\rho_s} \left\langle \frac{|\nabla\rho|}{B_p} \right\rangle f \eta_{EC} \frac{I_{EC}}{a^2} \frac{1}{W^2} \tag{6}$$

Here, $\Gamma_{\Delta'}$ is the classical stability index defined as the logarithmic jump of the radial magnetic field perturbation across the rational surface [10]. Γ_{BS}, Γ_{GGJ}, Γ_{pol} and Γ_{EC} are the perturbed bootstrap current, the stabilizing effect of the field line curvature [11], the ion polarization current and the EC current effect [12]. ρ is the coordinate of the normalized minor radius. η, ε_s, β_{ps}, ρ_{pi} and ρ_s are the neoclassical resistivity, the inverse aspect ratio, the local poloidal beta, the poloidal Larmor radius normalized by minor radius a and the rational surface position, respectively. The scale lengths, L_q and L_p, are defined as $L_q = q(dq/d\rho)^{-1}$ and $L_p = -p(dp/d\rho)^{-1}$.

2.2. EC current

In this paper, the EC current profile is modelled by a Gaussian distribution

$$j_{EC} = j_{EC0} \exp\left(-C\left(\frac{\rho - \rho_s}{W_{EC}}\right)^2\right), \tag{7}$$

as shown in figure 1. Here, $C = 4\ln 2$, and j_{EC0} is calculated from the total EC current I_{EC}. The value ρ_s is the position of the current density peak, and W_{EC} is the width of the localized EC current. For an ITER plasma given in Table 1, we assumed for simplicity that the value of I_{EC} is proportional to the EC power, P_{EC}, as I_{EC} [kA] = 4.35P_{EC} [MW] for the 3/2 mode and 4.15P_{EC} [MW] for the 2/1 mode [6].

Journal of Physics: Conference Series **123** (2008) 012036 doi:10.1088/1742-6596/123/1/012036

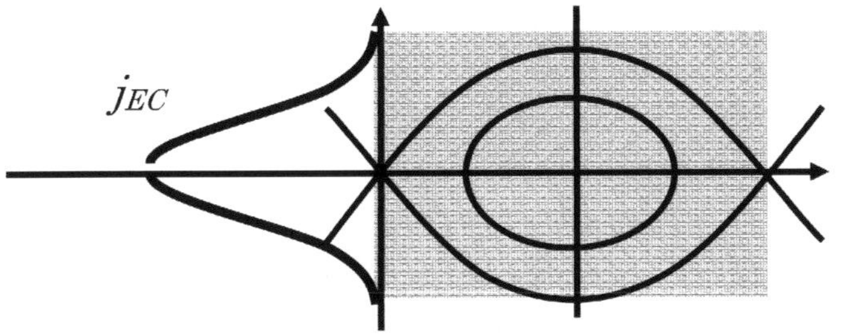

Figure 1. Model of EC current profile. The EC current density j_{EC} is modelled by the Gaussian distribution.

3. Numerical Results

Table 1 shows parameters of typical ITER plasma analysed in this paper. The local parameters at the rational surfaces of q = 3/2 and 2/1 are shown table 2, where ρ_s, β_{ps} and j_{BS} are the normalized rational surface position, the local poloidal beta, and the local bootstrap (BS) current density, respectively. Typical BS current density and the heating power profiles are given in figure 2. The coefficients of each term in the modified Rutherford equation used here are shown in table 3. Here, we used almost same parameters as those in the reference [6] based on the JT 60 experiments.

Table 1. Plasma parameters used here for ITER.

R_0 : major radius (m)	6.2
a : minor radius (m)	2.0
B_{t0} : troidal field at R_0 (T)	5.3
I_P : plasma current (MA)	15
$<n_e>$ (10^{20} m^{-3})	1.01
$<T_e>$ (keV)	10.9
$<T_i>$ (keV)	9.8
β_N	3.1

Table 2. Parameters relevant to the rational surface.

m/n	**3/2**	**2/1**
ρ_s	0.67	0.84
β_{ps}	0.65	0.46
j_{BS} (MA/m^2)	0.11	0.11

Table 3. Coefficients of each term in the modified Rutherford equation used here

k_1	1.0
k_2	10.0
k_3	1.0
k_4	1.0
k_5	5.0

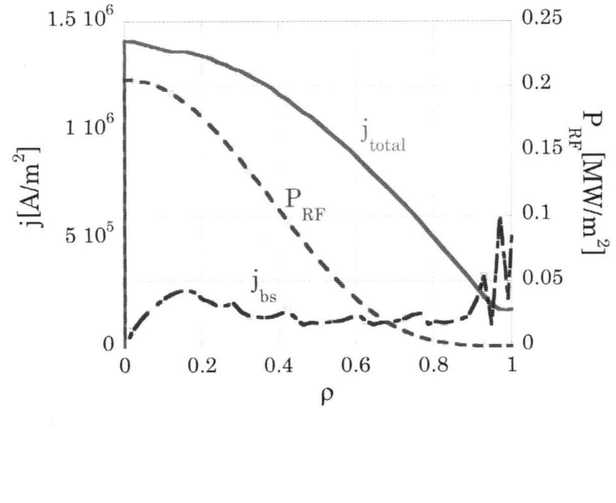

Figure 2. Plasma current density and input heating power P_{RF} profile as a function of normalized minor radius. The total input power is 40 MW, the average electron density is 1.0 $\times 10^{20}$ m^{-3} and the total current is 15 MA.

Journal of Physics: Conference Series **123** (2008) 012036 doi:10.1088/1742-6596/123/1/012036

3.1. Reduction in plasma temperature and fusion power by NTM

In an ITER plasma the NTM magnetic islands are saturated around ten seconds after introducing the seed island. Figure 3 shows the electron temperature profile and the q profile when the magnetic island is saturated. The 3/2 mode island (q = 1.5) exists at r/a = 0.67, and the 2/1 mode island (q = 2) exists at r/a = 0.84. We assumed the transport coefficient is quite large inside the magnetic island. According to figure 3, the electron temperature at plasma center decreases due to the magnetic island formation. That is, the fusion power decreases too due to the NTM magnetic island. Table 4 shows the plasma parameters when the magnetic island exists in the plasma. Here, $T_e(0)$, I_{BS}, and I_{TOTAL} are the central electron temperature, the total bootstrap current, and the total current. The Q value is defined by the ratio of the fusion output power to the input power. Here, we assumed inductively driven plasma current, and did not include the current drive power for evaluation the Q value. The alpha heating power can be evaluated by the Fokker-Planck code in the TOTAL code. Here we used a simple model of alpha particle power deposition to electrons and ions depending on the slowing-down critical energy. At 3/2 mode, the Q value decreases to 73% of no NTM case, and at 2/1 mode, the Q value decreases to 18% of no NTM case. To reduce the magnetic island width is important in order to raise the fusion power output.

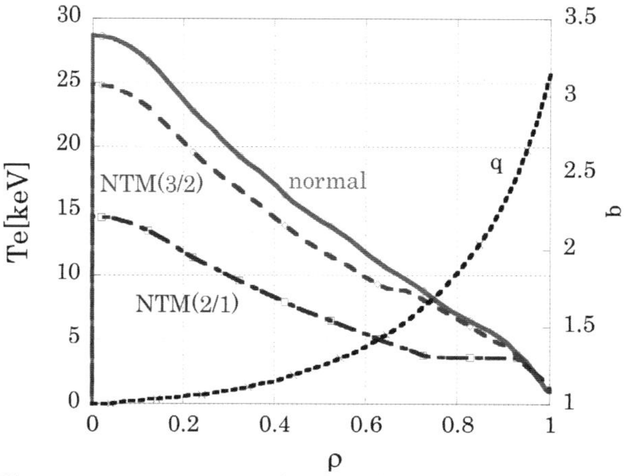

Figure 3. Electron temperature and safety factor q profile without and with 3/2 and 2/1 neoclassical tearing modes.

Table 4. Central temperature, bootstrap current fraction and Q value in three cases.

	$Te(0)$[keV]	I_{BS}/I_{total}(%)	Q
Normal	28.7	23.2	14.6
NTM(3/2)	24.9	20.3	10.7
NTM(2/1)	14.9	12.6	2.6

3.2. Time evolution and saturation of magnetic island width

The NTM magnetic island grows in time, and the island width is finally saturated. We assumed a seed island with w/a = 0.05 introduced at time = 10 s to compare the plasma performances with and without NTMs. When the NTM is introduced on the early stage, the time constant of the island evolution becomes shorter, but the final island saturation width is same in the case of no ECCD stabilization. Figure 4 shows the time evolution of the central electron temperature and the magnetic island width with 3/2 and 2/1 modes. The 3/2 mode island width is saturated at w/a = 0.048, and the 2/1 mode is saturated at w/a=0.21. The time constant for saturation is about 10 seconds. The central temperature decreases due to the magnetic island evolution, and the time constant for equilibrium saturation is about 10 seconds, too. Figure 5 shows every terms in the modified Rutherford equation at time = 10 s as a function of the seed island width *w/a*. The stabilization terms are $\Gamma_{\Delta'}$ (when *w/a* is large), Γ_{GGJ} and Γ_{pol}, and the destabilization terms are $\Gamma_{\Delta'}$ (when w/a is small) and Γ_{BS}. In both (a) 3/2 and (b) 2/1 cases in figure 5, the derivative *dw/dt* is negative when the normalized seed magnetic island width is smaller than 0.02. When *dw/dt* is negative, the magnetic island width is decreased. A magnetic island grows when the seed island is larger than 0.02. Figure 6 shows bootstrap current profile when the magnetic island width is saturated with 3/2 and 2/1 mode. The bootstrap current is decreased inside the magnetic island, and the local bootstrap current density at resonant surface is 5.1 kA/m^2 at the 3/2 mode and 3.6 kA/m^2 at the 2/1 mode. Figure 7 shows the time evolution of every term in the modified Rutherford equation. It should be noted that in equilibrium, the term $\Gamma_{\Delta'}$ changes from positive to negative. The island evolution in this case is determined initially by the delta-prime term and finally by the bootstrap current term as shown in figure 7. The time constant of the island saturation might be determined by the resistive time initially and the local transport time related to bootstrap current decay time finally.

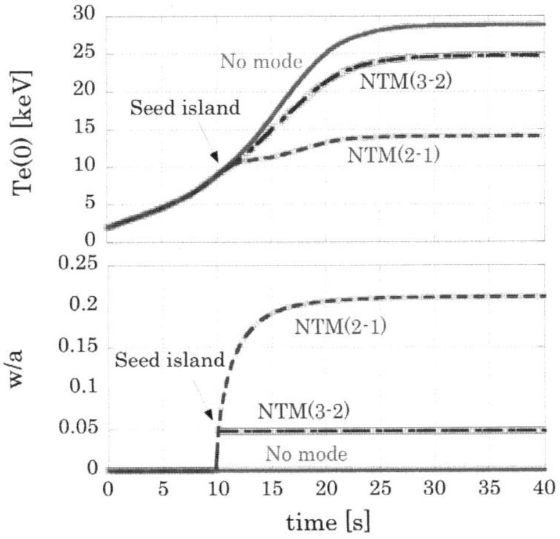

Figure 4. Time evolution of central temperature and normalized magnetic island width of each mode when a seed island is introduced at time = 10 s.

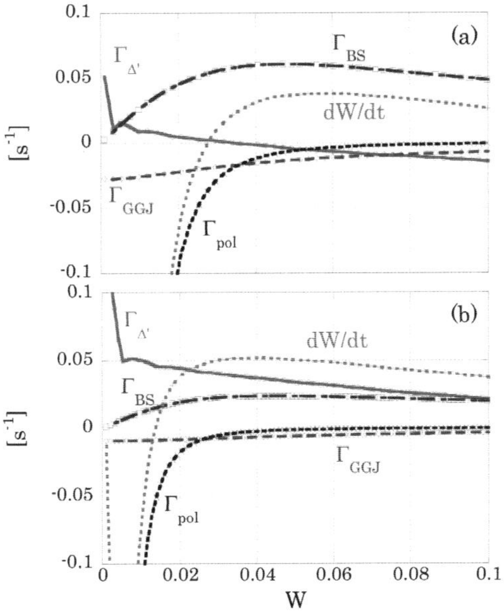

Figure 5. Every terms in the modified Rutherford equation at time = 10 s as a function of the island width *w/a* in (a) 3/2 mode and (b) 2/1 mode cases.

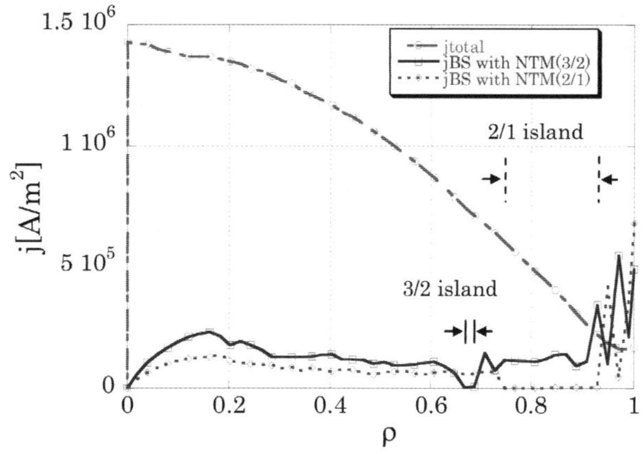

Figure 6. Total current and bootstrap current profile with 3/2 and 2/1 mode.

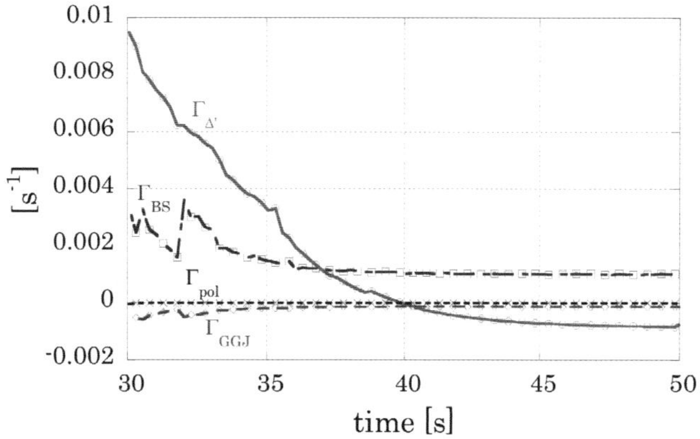

Figure 7. Time evolution of each terms in the modified Rutherford equation for m/n = 1/2.

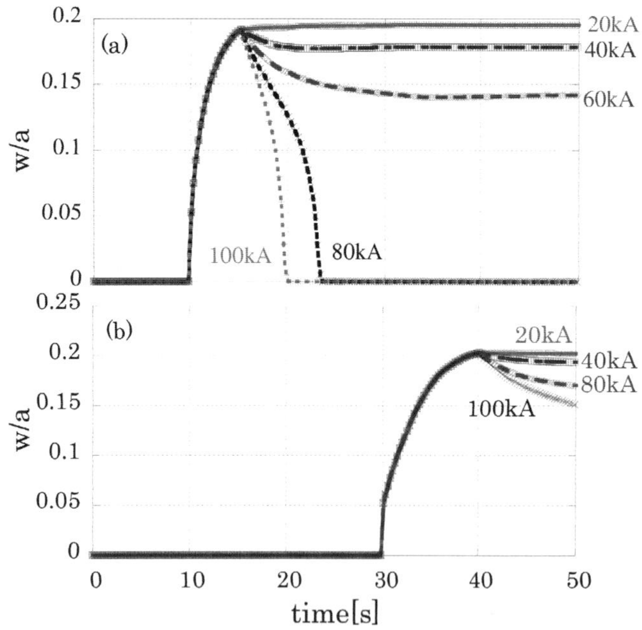

Figure 8. Time evolution of central temperature and magnetic island width at 2/1 modes. Magnetic seed island with w/a = 0.05 is introduced at (a) time = 10 s and (b) time = 30 s, and the EC current was injected from (a) 15 s and (b) 40 s with current width W_{EC} = 0.04 and with EC total current of 20, 40, 60, 80 and 100 kA.

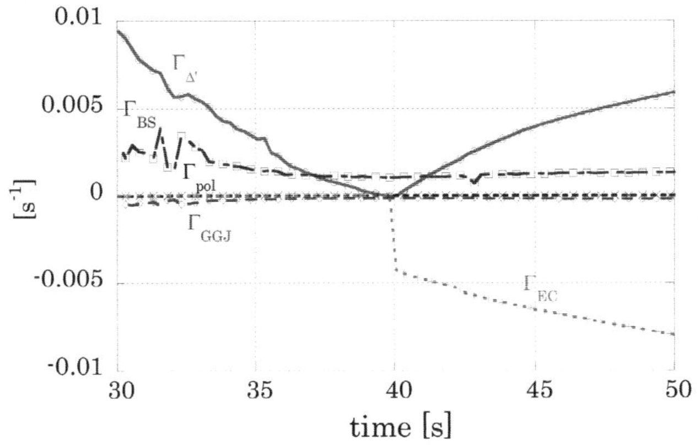

Figure 9. Time evolution of each term in the modified Rutherford equation at 1/2 by adding 100 kA ECCD power in the case of figure 8(b).

3.3. Stabilization of NTM by ECCD

We simulated the reduction in the magnetic island width by adding the electron cyclotron current drive (ECCD). Figure 8 shows the effect of the ECCD injected at 20 s in lower beta (time = 10 s) island growing phase and at 40 s for higher beta (time = 30 s) island-saturated phase using the model described in section 3.2. The magnetic island width is fully erased by adding the large ECCD power. In the earlier and lower-beta phase, the required ECCD power is lower. Figure 9 shows every terms of the modified Rutherford equation, when the magnetic island isn't erased by the ECCD. The stabilization of the ECCD power is balanced by the other terms. In the 2/1 mode case, the ECCD term balances with Γ_{BS} and $\Gamma_{\Delta'}$. We change the injection time of ECCD as shown in figure 10. Early injection of ECCD can easily reduce the magnetic island width. We need 10 MW ECCD power for 2/1 mode stabilization, and 23 MW for 3/2 mode stabilization, when the NTM is fully saturated at higher beta.

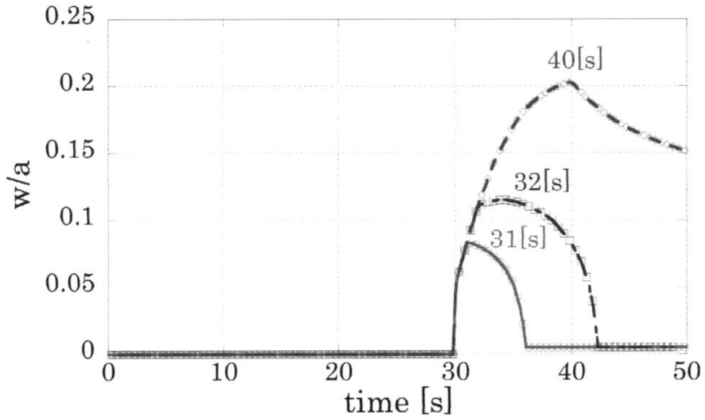

Figure 10. Time evolution of magnetic island of the seed island with *w/a* = 0.05 formed at time = 30 s. The ECCD power of 100 kA is injected at the time shown in the figure.

4. Summary and discussions

The ITER H-mode plasma with the neoclassical tearing mode (NTM) is simulated using integrated transport code TOTAL. The anomalous transport model used here is GLF23, and the magnetic island width of NTM is calculated by the modified Rutherford equation. The magnetic island formation reduces the plasma temperature and the fusion output power. The decrease in the Q value due to NTM is estimated. The reduction in fusion out power by NTM is 27% at the 3/2 mode, and 82% at the 2/1 mode. The saturated magnetic island widths of each modes are w/a ~ 0.048 at 3/2 mode and w/a ~ 0.21 at 2/1 mode.

The injection of ECCD is considered to stabilize the NTM and to recover the plasma confinement. We calculate the stabilization effect of ECCD with EC width $w_{EC}/a = 0.04$. The threshold power for the NTM full stabilization by ECCD in higher-beta island-saturation phase is ~10 MW against the 3/2 mode, and ~23 MW against the 2/1 mode.

For the reduction in ECCD power and the increase in the stabilization effects, the EC current width should be narrow. When the injection current width is half, the threshold power for the full stabilization is considered about half. The other method for ECCD power reduction is to modulate the EC current. The efficiency of ECCD can be raised by the current injection to the O point in magnetic island. The details of these analyses will be described somewhere in the future.

Acknowledgements

Authors wish to thank Dr. Jeronimo Garcia of Universitat Polite`cnica de Catalunya, Spain for his important collaborative work to introduce the GLF23 code to our TOTAL code.

References

[1] ITER Physics Basis Editors *et al.* 1999 *Nucl. Fusion* **39** 2137
[2] Hegna C C and Callen J D 1997 *Phys. Plasmas* **39** 2940
[3] Yamazaki K and Amano T 1992 *Nucl. Fusion* **32** 633
[4] Yamazaki K *et al.* 2006 *Fusion Engineering and Design* **81** 2743
[5] Sauter O *et al.* 1997 *Phys. Plasmas* **4** 1654
[6] Hayashi N, Ozeki T, Hamamatsu K and Takizuka T 2004 *Nucl. Fusion* **44** 477
[7] Waltz R E, Staebler G M, Dorland W, Hammett G W, Kotschenreuther M and Konings J A 1997 *Phys. Plasmas* **4** 2482
[8] Yamazaki K, Amano T, Naitou H, Hamada Y and Azumi M 1985 *Nucl. Fusion* **25** 1543
[9] Yamazaki K *et al.* 2006 *J. Plasma Fusion Res. SERIES,* **7** 102
[10] Furth H P, Rutherford P H and Selberg H 1973 *Phys. Fluids* **16** 1054
[11] Kotschenreuther M, Hazeltine R D and Morrison P J 1985 *Phys. Fluids* **28** 899
[12] Wilson H R *et al.* 1996 *Phys. Plasmas* **3** 248

Journal of Physics: Conference Series **123** (2008) 012037 doi:10.1088/1742-6596/123/1/012037

Effects of low central fuelling on density and ion temperature profiles in reversed shear plasmas on JT-60U

H Takenaga, S Ide, Y Sakamoto, T Fujita and the JT-60 Team

Japan Atomic Energy Agency, Naka Ibaraki 311-0193, Japan

E-mail: takenaga.hidenobu@jaea.go.jp

Abstract. Effects of low central fuelling on density and ion temperature profiles have been investigated using negative ion based neutral beam injection and electron cyclotron heating (ECH) in reversed shear plasmas on JT-60U. Strong internal transport barrier (ITB) was maintained in density and ion temperature profiles, when central fuelling was decreased by switching positive ion based neutral beam injection to ECH after the strong ITB formation. Similar density and ion temperature ITBs were formed for the low and high central fuelling cases during the plasma current ramp-up phase. Strong correlation between the density gradient and the ion temperature gradient was observed, indicating that particle transport and ion thermal transport are strongly coupled or the density gradient assists the ion temperature ITB formation through suppression of drift wave instabilities such as ion temperature gradient mode. These results support that the density and ion temperature ITBs can be formed under reactor relevant conditions.

1. Introduction

A density profile has large impacts on fusion performance in a fusion reactor. A peaked density profile can produce higher fusion output even with the low edge density. Since operation regime of the edge density is limited below the Greenwald density [1] in tokamak plasmas, a peaked density profile has an advantage for higher fusion output. In addition, a large density gradient can assist formation of the internal transport barrier (ITB) in an ion temperature profile through suppression of drift wave instabilities such as ion temperature gradient (ITG) mode [2]. On the other hand, a peaked density profile raises a concern of impurity accumulation, because neoclassical inward pinch velocity increases with increasing the bulk ion density gradient. Impurity transport analyses in JT-60U advanced tokamak plasmas with ITBs have shown that argon impurities are accumulated inside the ITB, but accumulation level is smaller than the neoclassical prediction [3]. At this impurity accumulation level, argon density profiles yield acceptable radiation profile even with a density ITB in a steady-state tokamak reactor [4]. In JT-60U reversed shear plasmas, high confinement has been achieved with high radiation loss fraction of ~0.9 [5]. In order to optimize fusion performance in a fusion reactor, it is important to understand physical mechanisms responsible for the density profile under reactor relevant conditions, such as low or no central fuelling, electron dominant heating and no Ware pinch velocity.

In ELMy H-mode plasmas, tokamak experiments have shown that a density peaking factor increases with decreasing effective collisionality due to anomalous inward pinch driven by drift wave instabilities such as ITG and/or trapped electron mode (TEM) [6, 7]. On the other hand, in advanced scenarios such as reversed shear (RS) plasmas, physics mechanisms for determining the density profile are not understood well. It is crucial issue for prediction of fusion performance in an advanced steady state tokamak reactor whether a density ITB can be formed with low central fuelling or not. It is expected that the anomalous inward pinch can play less important role in the ITB region, where anomalous transport is reduced by suppressing instabilities. Also, investigation of correlation between a density ITB and an ion temperature ITB is important, because the large density gradient can assist formation of an ion temperature ITB through suppression of ITG. It is crucial issue to understand how large density gradient is necessary for the formation of an ion temperature ITB.

In JT-60U, reversed shear plasmas have been optimized using positive ion based neutral beam injection (P-NBI) [8]. The P-NBI provides high central fuelling, which is different from a reactor relevant condition of low or no central fuelling. JT-60U also has negative ion based neutral beam injection (N-NBI) and electron cyclotron wave heating (ECH), which can provide the reactor relevant condition of low or no central fuelling and electron dominant heating. In this paper, effects of low central fuelling on density and ion temperature profiles were investigated using N-NBI and ECH in reversed shear plasmas on JT-60U.

This article is structured as follows: The NBI and ECH systems are described in section 2. Effects of low central fuelling after the strong ITB formation are discussed in section 3. In section 4, effects of low central fuelling during the ITB formation phase are discussed. In section 5, correlation between a density ITB and an ion temperature ITB during the ITB formation phase are discussed, followed by a summary in section 6.

2. NBI and ECH systems in JT-60U

JT-60U is a large tokamak device with a major radius (R) of ~3.4 m and an aspect ratio of ~4. The toroidal magnetic field (B_T) is 4 T at maximum and the plasma current (I_p) is limited below 3 MA. Input power up to 25 MW is available from the P-NBI system with beam energy of 85 keV. There are 7 units for perpendicular injection, 2 units for co-tangential injection and 2 units for counter-tangential injection. The Charge Exchange Recombination Spectroscopy (CXRS) measurement system uses 1 perpendicular injection unit and the Mortional Stark Effect (MSE) measurement system uses 0.5 or 1 unit of counter-tangential injection unit. For the N-NBI system, input power up to 5-6 MW is available with beam energy of 360-380 keV. The injection direction of N-NBI is co-tangential.

The ECH system with a frequency of 110 GHz also provides an input power of 3 MW by using 4 units of gyrotron. The poloidal injection angle can be scanned by steerable mirrors to change the deposition position. The toroidal injection angle is fixed for 3 units with current drive (CD) mode (+20 degree). It can be changed for 1 unit with co-ECCD mode (+20 degree), counter-ECCD mode (-20 degree) and pure heating mode (0 degree).

3. Effects of low central fuelling after the strong ITB formation

First, effects of low central fuelling and accompanying electron dominant heating on density and ion temperature profiles were investigated after the strong ITB formation using ECH. The central ECH was applied from $t = 5.6$ s with co-ECCD mode to the RS plasma at $I_p = 1.3$ MA and $B_T = 3.7$ T after the strong ITB was formed using P-NBI, as shown in figure 1 [9]. The P-NBI heating power was reduced at $t = 5.8$ s. In this discharge, D_2 gas-puffing was applied only in the early phase ($t < 4.4$ s). During the low central fuelling phase ($t > 5.8$ s), modulated Helium gas-puffing was applied to estimate helium transport coefficients [10]. The line averaged density ($\bar{n}_e$) continued to increase without gas-puffing after the P-NBI heating power was decreased. The stored energy also continued to increase and the central electron temperature (T_e) exceeds to the central ion temperature (T_i) under dominant electron heating. The electron heating power was estimated to be higher by a factor of 1.6 than the ion heating power at $t = 6.5$ s. The high confinement enhancement factor over IPB98(y,2) scaling of 2 was obtained with low central fuelling and central (T_e/T_i) of about 1.3.

Figure 2 shows the density, safety factor (q) and temperature profiles at $t = 6.5$ s in the discharge shown in figure 1. The strong electron density (n_e) ITB was maintained near the position of minimum safety factor with low central fuelling as shown in figure 2 (a). The strong T_i and T_e ITBs

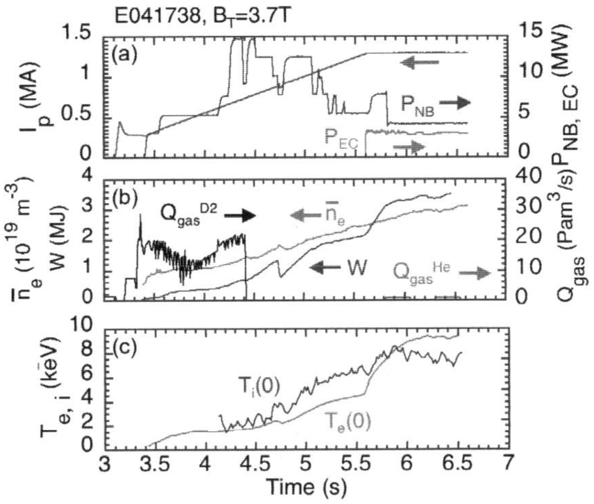

Figure 1. Waveforms of the RS plasma with low central fuelling after the strong ITB formation. (a) Plasma current (I_p), P-NBI heating power (P_{NB}) and ECH power (P_{EC}). (b) Line averaged electron density ($\bar{n}_e$), stored energy (W) and gas-puffing rates for D_2 ($Q_{gas}{}^{D2}$) and He ($Q_{gas}{}^{He}$). (c) Central ion and electron temperatures ($T_i(0)$ and $T_e(0)$).

were also maintained as shown in figure 2 (b). The effective electron diffusivity (D^{eff}) defined by considering only diffusion term was smaller in the low central fuelling case than those in the large central fuelling case [3] at the same range of the ion thermal diffusivity (χ_i) as shown in figure 3. The reduction of the central particle fuelling by a factor of 2 was compensated with the reduction of D^{eff} and the large density gradient was sustained in the ITB region. The value of D^{eff} is much smaller than the ion neoclassical particle diffusivity, indicating important role of inward convective flux. The outward diffusive flux could be almost balanced with the inward convective flux and small unbalance between these two fluxes corresponds to the flux produced by the central fuelling. Therefore, the central fuelling might have only small effects on the density profile. In the previous studies on effects of low central fuelling and electron heating in JT-60U using ion cyclotron heating (ICH) [11], the density gradient in the ITB region increased with central fuelling together with reduction of effective particle diffusivity. In the discharge shown in figure 1, the effective particle diffusivity did not increase even with low central fuelling as shown in figure 3. Different effects on the density profile between ECH and ICH should be investigated in future.

The helium diffusivity was estimated based on Helium gas-puffing modulation experiments separately from the convective term. Figure 4 shows relation between helium diffusivity and ion thermal diffusivity. Here, the diffusivities were normalized by their neoclassical values, in order to compare how large anomalous transport remains in particle and thermal transport channels. The

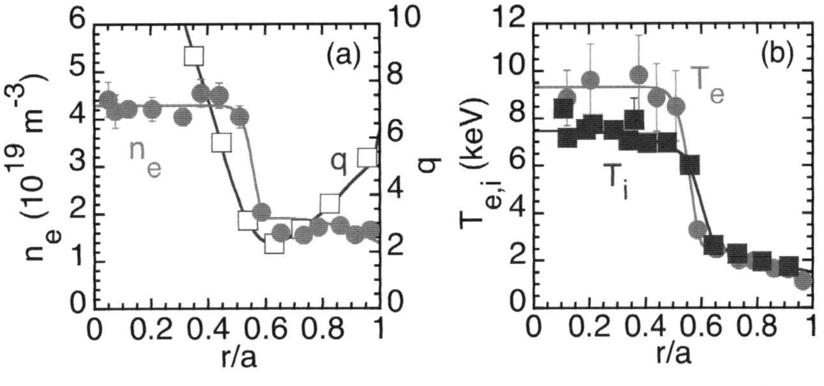

Figure 2. (a) electron density (circles) and safety factor (squares) profiles and (b) electron (circles) and ion (squares) temperature profiles under low central fuelling and dominant electron heating at t = 6.5 s in the discharge shown in figure 1.

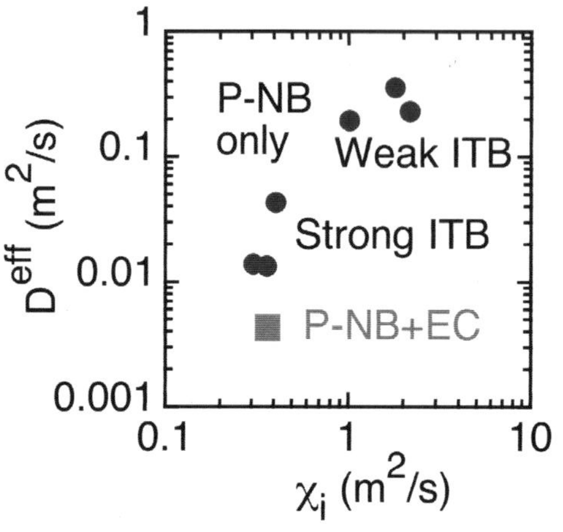

Figure 3. Relation between the effective electron diffusivity and the ion thermal diffusivity. Circles and square show the data in the high and low central fuelling cases, respectively.

helium diffusivity is similar to or even higher than those with high central fuelling at the same χ_i/χ_i^{NC} range. The convection velocity in the low central fuelling case is slightly negative (inward), however, its level is smaller than those in the high central fuelling case. The systematic helium transport study under the low central fuelling and the electron dominant heating is necessary in future.

The effects of low central fuelling on argon (Ar) accumulation were investigated in the similar discharge [3]. The central soft X-ray signal continued to increase after switch to the low central fuelling. The profile of the total Ar density summed over all ionization states was estimated using an impurity transport code, where the transport coefficient is determined by fitting the calculated soft x-ray profile to the measurement. The Ar radiation coefficient was taken from the ADAS database [12] considering the JT-60U diagnostic setup. The Ar density estimated from the soft X-ray intensity profile was more peaked by a factor of about 4 than the n_e profile as shown in figure 5. Here, 4 times larger diffusivity than the neoclassical value and the neoclassical inward pinch velocity were used in the ITB region, which suggests that the Ar accumulation is weaker than the neoclassical prediction. In the weak shear plasma, the flattening of the density profile was observed with the central ECH and the Ar accumulation was drastically reduced [3]. However, in the RS plasma, the strong n_e ITB was kept

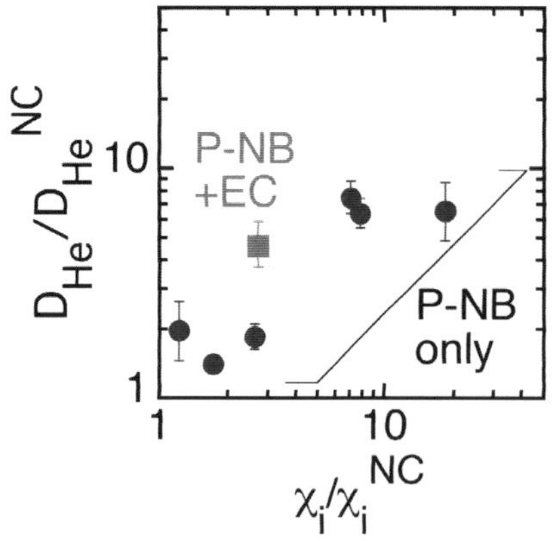

Figure 4. Relation between helium diffusivity and ion thermal diffusivity normalized by their neoclassical transport values. Circles and square show the data in the high and low central fuelling cases, respectively.

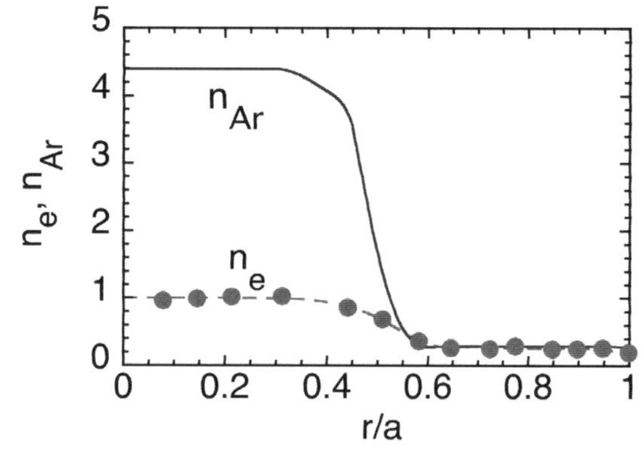

Figure 5. Electron density profile (circles) normalized at $r/a = 0$ and Ar density profile (solid line) adjusted to electron density outside the ITB.

and Ar was still accumulated inside the ITB even with the central ECH.

4. Effects of low central fuelling during the ITB formation phase

Next, effects of low central fuelling and electron dominant heating on n_e and T_i profiles were investigated during the ITB formation phase at I_p = 1 MA and B_T = 3.7 T. In the low central fuelling case, ECH (~ 2.5 MW), N-NBI (~ 4.6 MW) and P-NBI (~ 3.3 MW) were injected during I_p ramp-up phase as shown in figure 6 (a). The P-NBI injection included 1 unit of perpendicular beam and half unit of counter-tangential beam for diagnostics of CXRS and MSE. In the high central fuelling case, only P-NBI (~ 10.4 MW) was injected during I_p ramp-up phase as shown in figure 6 (b). In this case, 4 units of perpendicular beam and half unit of counter-tangential beam were used. The central fuelling rate was estimated to be 3.9×10^{20} /s in the low central fuelling case, which includes 3.2×10^{20} /s from P-NBI and 6.6×10^{19} /s from N-NBI. On the other hand, the central fuelling rate was estimated to be 1.0×10^{21} /s in the high central fuelling case. Central fuelling was smaller by a factor of 2.5 in the low central fuelling case than that in the high central fuelling case. In these discharges, density feedback control was applied during I_p ramp-up phase. The gas-puffing rate was higher in the low central fuelling case than that in the high central fuelling case. However, difference is not large compared with difference of the central fuelling rate. In the low central fuelling case, n_e and T_i at r/a = 0.4 continued to increase, while those at r/a = 0.6 stayed constant after t = 4.2-4.4 s, indicating the n_e and T_i ITB formation. The increasing rate for a number of electrons inside the separatrix (dN_e/dt) was

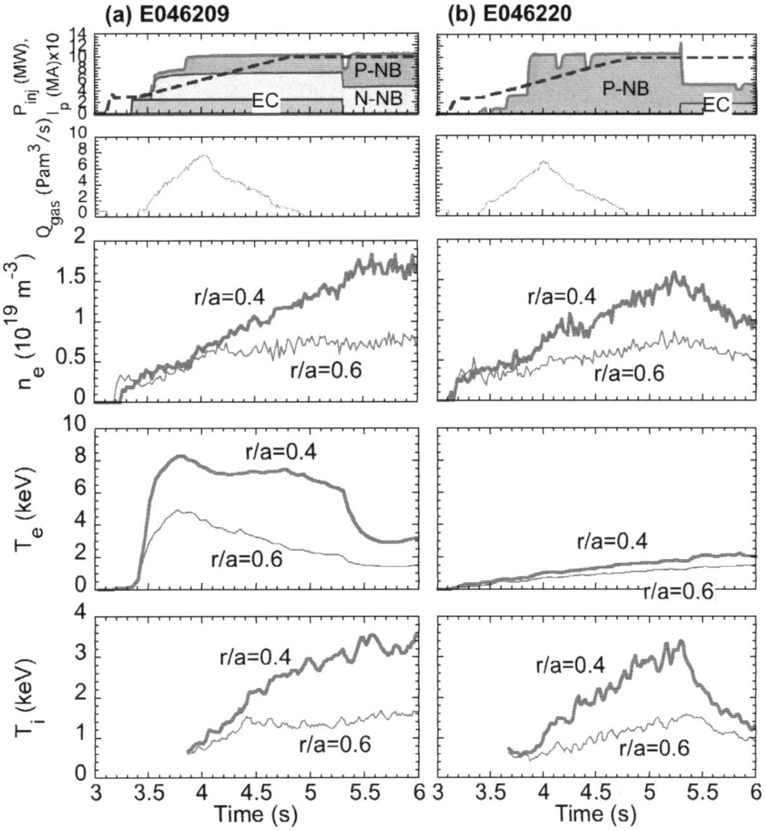

Figure 6. Waveforms of RS plasmas with (a) low and (b) high central fuelling during the ITB formation phase. First panel : plasma current, heating powers from P-NBI, N-NBI and ECH. Second panel : gas-puffing rate. Third panel : electron density. Fourth panel : electron temperature. Bottom panel : ion temperature. In the third, fourth and bottom panels, thick and thin lines show the data at r/a = 0.4 and 0.6, respectively.

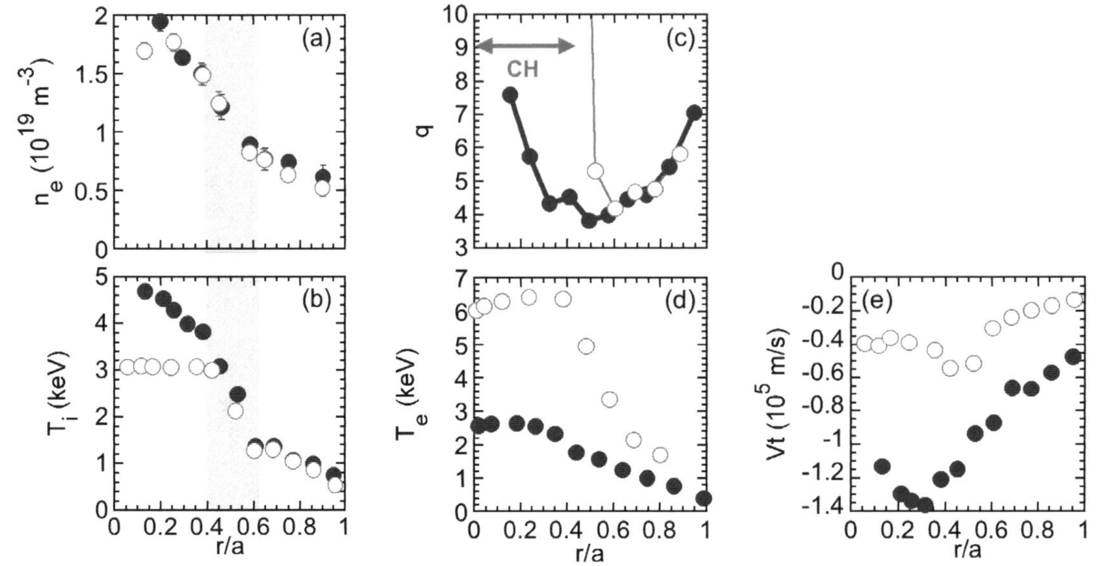

Figure 7. Profiles of (a) electron density, (b) ion temperature, (c) safety factor, (d) electron temperature and (e) toroidal rotation velocity. Open and closed symbols show the data in the low and high central fuelling case, respectively.

estimated to be 1.1×10^{20} /s for the low central fuelling case and to be 2.0×10^{20} /s for the high central fuelling rate. The central fuelling rate was larger by a factor of 4-5 than dN_e/dt for both cases. In the high central fuelling case, the formation of the n_e and T_i ITBs seems to occur earlier. However, the time evolutions of n_e and T_i were similar to those in the low central fuelling cases, while the time evolution of T_e was quite different for two cases. Note that electron heating was dominant in the low central fuelling case. In the low central fuelling case, the T_e ITB was formed independently from the n_e and T_i ITBs.

Figure 7 shows profiles of n_e, T_i, q, T_e and toroidal rotation velocity (V_t) at $t = 5.2$ s in the discharges shown in figure 6. It can be seen from figure 7 (a) that the density ITB can be formed even in the low central fuelling case. The effects of low central fuelling on the density ITB seems to be small, because n_e and T_i ITBs were similar in the region of $r/a = 0.4$-0.6 for the low and high central fuelling cases. In the low central fuelling case, wide current hole (CH) was produced due to higher T_e as shown in figure 7 (c). In the central region ($r/a \leq 0.4$), the T_i profile was flat in the low central fuelling case due to CH. On the other hand, the T_i profile was peaked in the high central fuelling case. The T_e profile was quite different for two cases due to the electron dominant heating in the low central fuelling case. The fluctuation dominating electron thermal transport could be different from that dominating density and ion thermal transport. Also, the toroidal rotation was different for two cases as shown in figure 7 (e). The counter V_t was larger in the high central fuelling case due to fast ion loss induced by the toroidal field ripple in the edge region. The notch structure was observed in the low central fuelling case. These results indicated that the central fuelling does not strongly affect the n_e and T_i ITB formation. The penetration of the recycling neutrals is very shallow [13] and the neutrals do not reach ITB region. Therefore, the edge particle source does not play important role for the ITB formation. This result suggests that a ratio of convection velocity and particle diffusivity does not strongly depends on q, T_e and V_t, although it is necessary to investigate the effects of different q and V_t profiles on the n_e and T_i ITBs more carefully.

5. Relation between density ITB and ion temperature ITB during the ITB formation phase

Finally, relation between n_e and T_i ITBs was investigated during the ITB formation phase. Here, two RS discharges with low central fuelling at $I_p = 1$ MA and $B_T = 3.7$ T shown in figure 8 were referred.

In the discharge shown in figure 8 (a), n_e was first increased by gas-puffing and then gas-puffing was stopped at $t = 4.4$ s. After gas-puffing was stopped, particle flux could be produced in the core region due to decrease in the edge density, resulting increase in the n_e gradient. The particle flux could give a similar effect to the central fuelling. In the other discharge shown in figure 8 (b), n_e was further increased by gas-puffing to get optimum density for the L-H transition and then gas-puffing was gradually decreased. The increase in the edge n_e induced by the L-H transition could decrease the n_e gradient in the ITB region.

In the discharge where gas-puffing was stopped at $t \sim 4.4$ s, the n_e and T_i ITBs were formed just after the gas-puffing stop as shown in figure 8 (a). The edge n_e decreased just after gas-puffing was stopped, while the central n_e continued to increase, resulting formation of the n_e ITB. At the same time, the central T_i largely increased, although the edge T_i stayed constant. The n_e and T_i profiles at $t = 5.4$ s were similar to the profiles shown in figure 7. In the discharge shown in figure 8 (b), the L-H transition occurred at around $t = 5$ s and ELMs appeared. After the L-H transition, the edge n_e increased and the n_e gradient decreased. In this phase, increase in the central T_i seems to be prevented. By decreasing the gas-puffing rate, the central T_i restarted to increase together with increase in the n_e gradient. These results indicated that the T_i ITB was changed when the n_e ITB was modified by gas-puffing and L-H transition.

Figure 9 shows relation between the n_e gradient and the T_i gradient around half the minor radius.

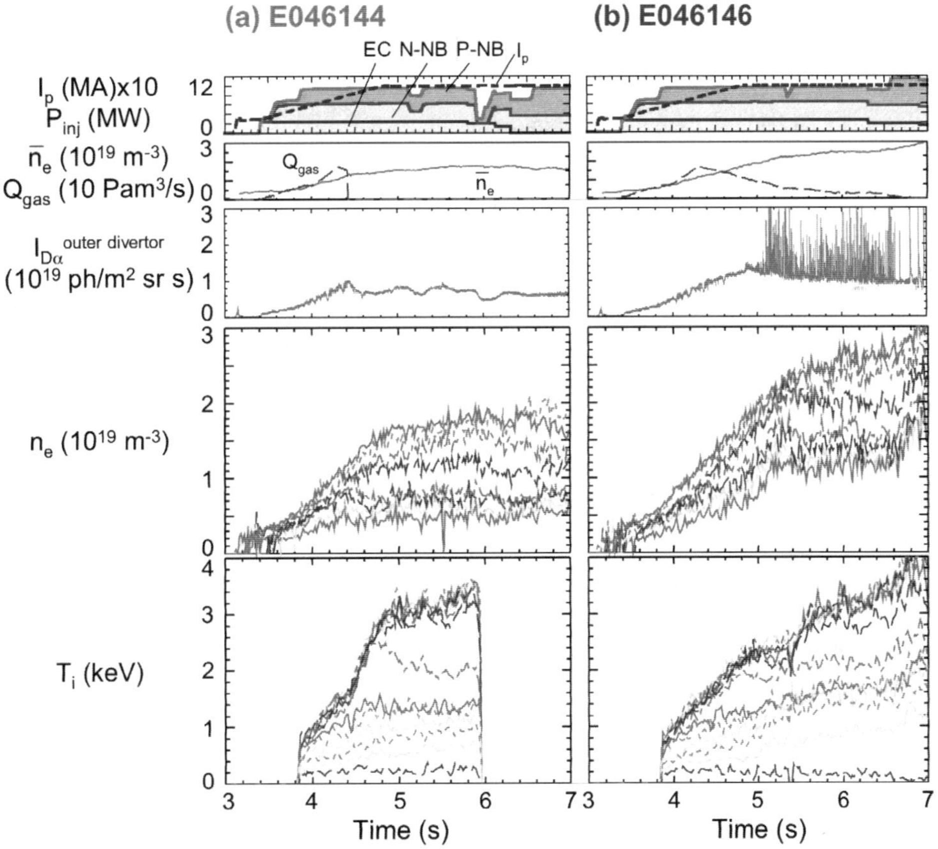

Figure 8. Waveforms of plasma current (I_p) and heating powers from ECH, N-NBI and P-NBI (first panel), line averaged electron density ($\bar{n}_e$) and gas-puffing rate (Q_{gas}) (second panel), electron densities (third panel) and ion temperatures (bottom panel). In the discharge shown in (a), gas-puffing was stopped at $t = 4.4$ s. In the discharge shown in (b), gas-puffing continued after $t = 4.4$ s and was gradually decreased.

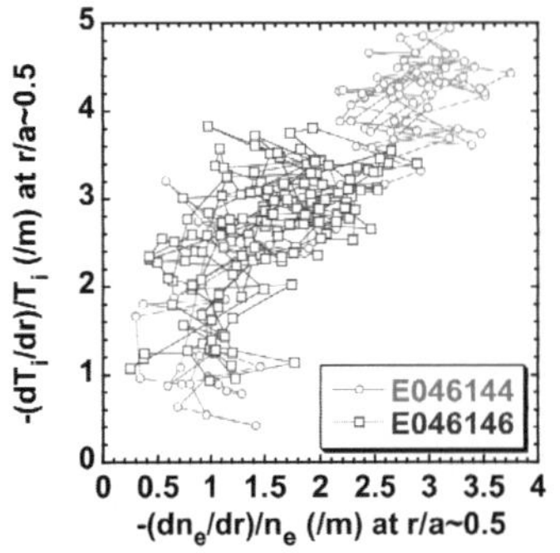

Figure 9. Relation between ion temperature gradient and electron density gradient. Circles and squares show the data in E046144 and E046146 shown in figure 8 (a) and (b), respectively.

The T_i gradient increased with the n_e gradient in these plasmas. The relation between the n_e gradient and the T_i gradient has been reported in [3], where the n_e gradient and the T_i gradient are correlated in the small value region ($-\nabla n_e / n_e$ and $-\nabla T_i / T_i < \sim 5$). In the higher T_i gradient region ($-\nabla T_i / T_i > \sim 5$), the n_e gradient was almost constant. The data points plotted in figure 9 are selected from this small region where the n_e gradient and T_i gradient are correlated. This correlation indicated that particle transport and ion heat transport might decrease simultaneously or the n_e ITB might assist the formation of the T_i ITB. In the latter case, there might be time delay of the increase in the T_i gradient from the increase in the n_e gradient, however it was not obvious from figure 9. In order to understand the causality, fast measurements are important. Recently, fast CXRS measurement was introduced in JT-60U. This measurement can provide precious information to understand the causality for the relation between n_e and T_i ITBs.

6. Summary

Effects of low central fuelling on the n_e and T_i ITBs were investigated in the RS plasmas on JT-60U. Strong n_e and T_i ITBs were maintained, when central fuelling was decreased after the strong ITB formation. Similar n_e and T_i ITBs were formed for the low and high central fuelling cases. An anomalous inward pinch could be still important with suppressed instabilities. Possibility for the formation of the strong n_e and T_i ITBs with ion thermal transport reduced to the neoclassical level and role of anomalous inward pinch in such strong ITB region are future issues. The T_i gradient increased with the n_e gradient, but their causality was not well understood yet. These results indicated that central fuelling does not largely affect the n_e and T_i profiles. Thus, it seems that the peaked density profiles with the ITB maybe possible in a fusion reactor.

References

[1] Greenwald M 2002 *Plasma Phys. Control. Fusion* **44** R27
[2] Rewoldt G and Tang W M 1990 *Phys. Fluids* B **2** 318
[3] Takenaga H *et al.* 2003 *Nucl. Fusion* **43** 1235
[4] Takenaga H, Kubo H, Kamada Y, Miura Y, Kishimoto Y and Ozeki T 2006 *Fusion Sci. Tech.* **50** 503
[5] Takenaga H *et al.* 2005 *Nucl. Fusion* **45** 1618
[6] Angioni C, Peeters A G, Pereverzev G V, Ryter F, Tardini G and ASDEX Upgrade Team 2003 *Phys. Rev. Lett.* **90** 205003
[7] Weisen H *et al* 2005 *Nucl. Fusion* **45** L1
[8] Takenaga H and the JT-60 Team 2007 *Nucl. Fusion* **47** S563
[9] Ide S, Suzuki T, Sakamoto Y, Takenaga H, Fujita T, Oyama N, Isayama A, Koide Y, Kamada Y and the JT-60 Team 2004 *Nucl. Fusion* **44** 87
[10] Takenaga H *et al.* 1999 *Nucl. Fusion* **39** 1917
[11] Iwase M, Koide Y, Tobita K, Moriyama S, Takenaga H, Fujita T, Shirai H, Kusama Y, Kramer G J and Kimura H 1999 *Plasma Phys. Control. Fusion* **41** 1189
[12] Summers H P 1994 *JET-IR 06* JET Joint Undertaking Culham
[13] Nakashima Y, Higashizono Y, Kawano H, Takenaga H, Asakura N, Oyama N, Kamada Y and Yatsu K submitted to *Journal of Physics: Conference Series*

The role of MHD in the sustainment of electron internal transport barriers and H-mode in TCV

G Turri , O Sauter, L Porte, S Alberti, E Asp, T P Goodman, Y R Martin, V S Udintsev and C Zucca

Ecole Polytechnique Fédérale de Lausanne (EPFL) Centre de Recherches en Physique des Plasmas, Association Euratom-Confederation Suisse CH-1015 Lausanne, Switzerland.

E-mail: `Gianpaolo.Turri@epfl.ch, Olivier.Sauter@epfl.ch`

Abstract. Advanced scenarios exhibit improved confinement properties, which make them attractive candidate for ITER. For these to be achieved, the sustainment of transport barriers and therefore high pressure gradients is inherent. Their stability properties both in the transient and steady state phases is a major issue [1], because of the relationship between high performances and proximity to a stability limit. Core MHD modes are one of the key issues in the development and sustainment of transport barriers, as they degrade the confinement properties and, in the worse case, disrupt the plasma. The understanding of the underlying physics can provide the means of finding regimes without modes. In TCV (Tokamak à Configuration Variable) H-mode and electron internal transport barriers (eITBs) have been obtained with different schemes, usually accompanied by various types of MHD phenomenon [2, 3, 4]. In this paper we focus on the low-shear Quasi-Stationary ELM free H-mode (QSEFHM) scenarios [4], which displays infrequent sawteeth and/or NTMs. In addition to that, high-performance eITBs shots are discussed, during which a variety of resistive to ideal modes are observed and ascribable to the infernal stability limit [3, 5]. Analysis of data from TCV highest performance discharges can clarify the potential threats of MHD modes in advanced scenarios. MHD core analysis of the QSEFHM [4], and of eITBs is presented, focusing on the existence of stability windows.

1. Introduction

In TCV (Tokamak à Configuration Variable) the ECH (Electron Cyclotron Heating) system allows to locally modify the electron density and temperature. The ECCD (Electron Cyclotron Current Drive) together with the neoclassical effects allow tailoring the current profile. Small and reverse magnetic shear plasma are routinely obtained, two instances of which are respectively the Quasi-Stationary ELM-Free H-Mode (QSEFHM, [4]) and high-performance eITBs (electron Internal Transport Barrier). These types of advanced scenarios, discussed in this paper, provide better confinement properties than conventional ones, mainly thanks to the reduced transport across flux surfaces, impeded by the arising of transport barriers.

The main characteristic of QSEFHM discharges is the transition to a quasi-stationary ELM-free H-mode phase which lasts for much longer than a confinement time, once the auxiliary heating is applied. This is not accompanied by density (and impurities) uncontrolled build-up often observed in ELM-free H-Modes, drawback of the conventional ELM-free scenarios. In addition to that, the robustness of the QSEFHM makes it a potential candidate for future exploitation, subject to further understanding and investigation of its properties.

The formation and sustainment of eITB in TCV [6, 7] can also be accompanied by the development of MHD instabilities. The character of the observed instabilities varies from disruptive (ideal infernal mode) to resistive (tearing mode developing magnetic island, classical or neoclassical). In addition, the MHD instabilities can cause periodic, spontaneous, slow, global oscillations of the plasma (electron temperature, density, plasma current, SXR radiation, etc). These are reminiscent of the Oscillatory Regime (O-regime [8]), which was first observed in the Tore Supra tokamak. In TCV the MHD activity and the O-regime are linked, with the first being the cause for the second [9].

MHD modes are detrimental for the confinement as they enhance perpendicular transport, and could limit the access to higher confinement, as possibly in the case of discharges not developing the QSEFHM. Therefore, the study of the interplay between advanced scenarios and MHD activity is of crucial importance, as well as the implementation of ways to avoid instabilities or to limit their effects.

2. The Quasi-Stationary ELM-Free H-Mode -*QSEFHM*-

In TCV, H-mode is obtained with and without auxiliary electron cyclotron heating. The QSEFHM has been achieved in TCV by employing three gyrotrons at full power (1.4 MW, 118 GHz) with the vertical launcher at the third harmonic, X-polarization (X3). This regime appears to be fusion relevant because it displays good confinement properties, absence of ELMs without density and impurity uncontrolled build-up. Furthermore, the scenario is obtained with

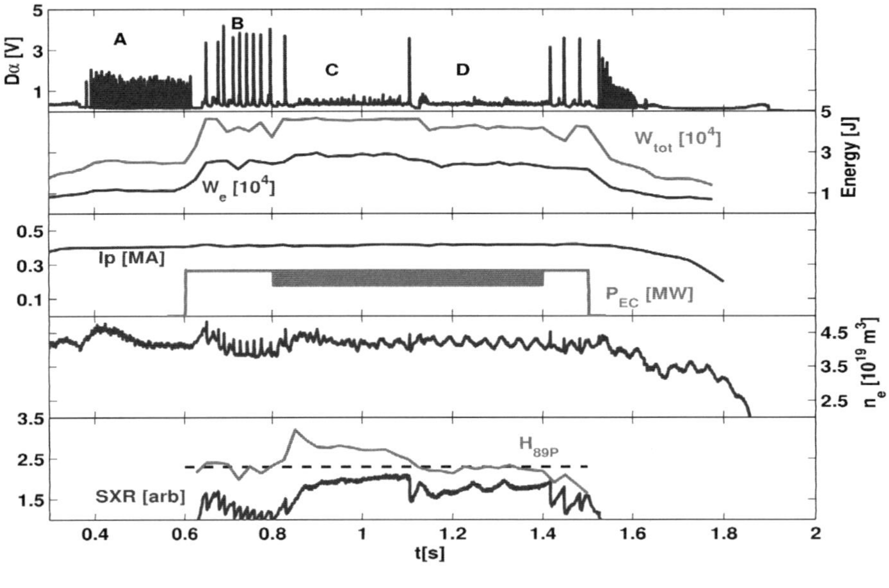

Figure 1. TCV Discharge #29892: overview of a discharge developing the quasi-stationary ELM-free H-mode. Transition A-B (top) occurs after the X3 heating is applied. At t=0.8 s, power is modulated (third plot) and soon after the plasma achieves the QSEFHM. The total and electron energy reach their maximum during the C-phase (second plot). The averaged density is almost constant (fourth plot) and the H_{89P} reaches the maximum value $\sim$ 3.2 (bottom plot). In phase D, a 3/2 NTM degrades slightly the performance.

no momentum input, and in target plasmas ($\beta_N \sim$ 2 and q_{95} = 2.5) that are ITER relevant. Finally, TCV is in the unique position to be able to achieve H-mode by heating solely the electrons, leaving the ions to be heated by ion-electron collisions, a scenario that mimics that

Journal of Physics: Conference Series **123** (2008) 012038 doi:10.1088/1742-6596/123/1/012038

of a burning plasma [4]. The QSEFHM (figure 1) is characterized by a H-mode without ELMs (top plot, phase C and D), high energy content (second plot), high constant plasma current (third plot, corresponding to $q_{95} \sim 2.5$), approximately constant electron line-averaged density n_{el} (fourth plot). The Ohmic H-mode phase (A) begins at t≃0.4s and is characterized by small and frequent ELMs. These ELMs do not display any particular MHD precursor. The sawtooth crashes and the ELMs are not coupled. The X3 heating starts at t=0.6s, which causes the transition between phases A and B in figure 1. The ELMs become larger, and the frequency drops from $230Hz$ to $60Hz$. At this time the increase in plasma energy can be observed from the energy trace (second plot, red trace). During the B phase, the three gyrotrons are used at full power (third plot, red trace). At $t = 0.8s$, one of the three gyrotrons is power modulated ($f = 240Hz$), slightly before the QSEFHM begins (Dα emission, top plot). At $t = 0.83s$, the last ELM marks the beginning of phase-C, the quasi-stationary phase (QSEFHM). This phase does not show variation in the plasma current and averaged electron density. The plasma total energy (black trace, second plot) and electron energy (blue trace) reach their maximum value, which is kept stationary during the C-phase. The H_{89P} scaling gives an average value 2.7 during this phase C underlying the improved confinement. At $t \simeq 1.1s$, a big ELM causes a large

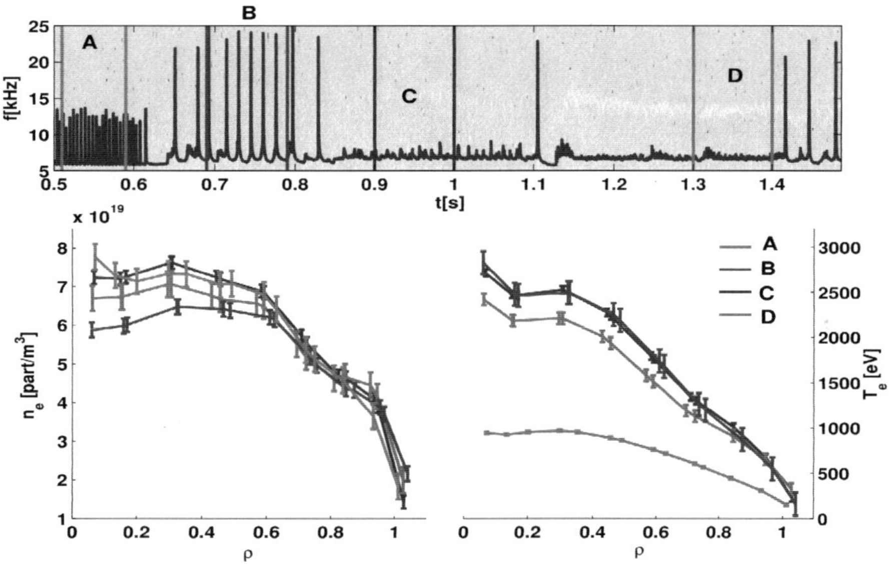

Figure 2. **TCV Discharge #29892:** the variation of the electron temperature and density profiles can be used to follow the different confinement properties of the four phases. The pedestal is almost constant during the H-mode, and the best confinement is achieved during the QSEFHM. The ELMy (X3) H-mode (green) and the second QSEFHM with an NTM (red) are similar in terms of confinement.

drop in the SXR emission and energy. Conversely, the ELM does not cause any change in the averaged density. This single ELM marks the transition between the first (C) and the second QSEFHM phase (D), with the latter displaying a lower energy content.

The various levels of confinement properties can be better appreciated by following the evolution of the electron temperature and density profiles throughout the four phases of the QSEFHM discharge (figure 2). The top plot shows the spectrogram of the central soft X-ray signal (digitized at 200kHz), with a mode starting at 1.1s whose onset is linked to the large ELM separating phase C and D (Dα trace, blue). Four time windows are chosen during the different phases, characterized by colored vertical lines. The violet profiles (electron density, bottom left; electron temperature, bottom right) represent the average Thomson Scattering (TS) profiles during the Ohmic ELMy H-mode. The density is as high as for the QSEFHM (no changes in the gas puffing are implemented throughout these phases), while the electron temperature is less than one third in the core. After the X3 heating is applied, the density slightly drops (-17%, larger ELMs) but the electron temperature strongly increases (green). The blue vertical lines delimit the QSEFHM; during this phase the confinement properties are the highest for both the electron temperature and density. The large ELM at $t \simeq 1.1s$ triggers a NTM mode at frequency $f \sim 15kHz$, which reduces the energy (figure 2, spectrogram plot, and also visible in figure 1, second plot) with the temperature and the density being lower when compared to C-phase. Their product is almost constant, suggesting that the effect on confinement of this NTM in the absence of ELMs is similar to the X3 heated, ELMy H-mode.

In the following subsection, a characterization of the MHD activity during the different phases is reported.

2.1. MHD characterization of the QSEFHM

The different phases (B to D, figure 1) show varying MHD activity, which plays an important role in defining β and W (the total thermal energy).

- **B**: In the X3 heated phase the ELMs are larger and their frequency lower. The core is stabilized against the sawtooth instability. The Soft X-Ray (SXR) radiation profiles in this case do not show any inversion radius, instead a global collapse of the radiation profile up to the center is observed at every ELM, with subsequent loss of the pedestal. Sometimes in between ELMs, a coherent m/n=2/1 oscillation is observed. This mode at f ~3kHz lasts for a few milliseconds after the crash, and it is then stabilized during the inter-ELM time, thus not having a significant impact on the confinement.
- **C**: During this phase, there is intermittent m/n=2/1 activity with f ~5kHz, which appears in bursts. The frequency is almost double than that of the 2/1 observed in phase-B, and the size smaller. Small SXR crashes are likely to be related to the m/n=2/1 mode, rather than being sawtooth crashes. This is because they appear to be linked to the mode, and the SXR profiles does not suggest sawteeth activity.
- **D**: The D-phase is characterized by the single ELM event which triggers an NTM mode with $f \sim 15kHz$, and main periodicity m/n=3/2. This mode has the same periodicity of a magnetic island developing in a similar discharge that will be discussed in 2.2. The NTM in this discharge has a much higher frequency and smaller effect on confinement. Indeed, the island full width is approximately $w \simeq 5cm$. This corresponds to an expected drop in energy of $\Delta W \sim 25\%$, consistent with the drop in electron energy from the reconstruction (~20%, figure 2). The resulting confinement level, from the electron energy calculations and the pressure profile from TS is comparable to that of the B-phase, i.e. that of an ELMy, X3-heated H-mode. The fact that the D-phase follows a major crash and the QSEFHM is recovered despite of the presence of an NTM mode is a sign of robustness of the scenario. The QSEFHM is therefore a viable scenario for ITER and burning plasmas. In order for it to be exploited the uncertainties regarding its attainment (in the following section a similar shot not developing the quasi-stationary regime is discussed), and the potentially lethal singular ELM events require further studies.

2.2. Comparison between ELMy X3 H-mode and QSEFHM

Discharge #29894 (figure 3) was obtained with the same settings as discharge #29892 (QSEFHM), except for the power modulation. The pre-X3 phase for the two discharges are the same, with similar plasma current, electron density, plasma shape (l_i, k, δ). As soon as the X3 power is switched on, a large m/n=3/2 mode appears at low frequency ($f \simeq 6kHz$). This is at a time before the power modulation for #29892, hence the absence of it cannot be the cause for the mode onset. In discharge #29894, the mode rotation speeds up during the ELM-free phase lasting for approximately a confinement time ($t \sim 50ms$), and then it slows down just before the first ELM. At the ELM event the frequency spectrum shows a steep deceleration of the mode, which is repeated for all the other ELMs (possibly suggesting a locking of the edge mode with the vessel, due to finite resistivity of the wall), all of which separated in time by approximately a confinement time. The $m/n = 3/2$ mode is non-linearly coupled with a

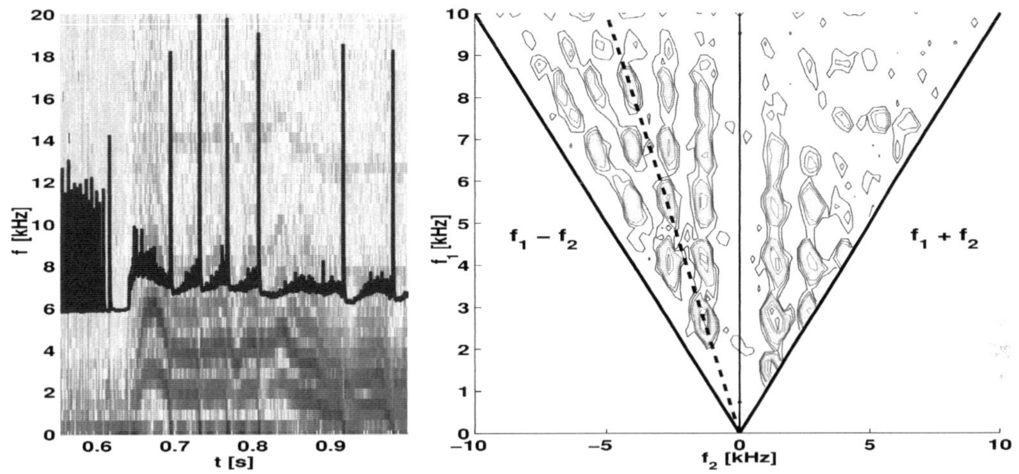

Figure 3. **TCV Discharge #29894:** the early MHD activity, m/n=3/2 coupled with 2/1 could prevent the access to the QSEFHM. The two modes are phase-locked, as obtained from the bi-coherence analysis, with the 3/2 being dominant. The bi-coherence is performed for the time interval [0.7s, 0.75s]

$m/n = 2/1$ at $f \simeq 3kHz$. The mode closer to the core, i.e. the one at the $q = 1.5$ resonant surface, is the dominant one. From a bi-coherence analysis (figure 3) the coupling (if any) of the two modes can be verified. The analysis consider the frequency relation between two coherent modes and the phase information within the Fourier transformation in the frequency space [10]. In figure 3 the correlation level between the two modes is approximately 0.7 at the frequency equal to the combinations of the two modes fundamental frequencies, *f1±f2*. This means that the two modes are phase-locked. The confirmation of the existence of two separate, phase-locked modes, is also obtained through magnetic island modeling [11], where current filaments are simulated at the flux surfaces reconstructed with the LIUQE [12] code. The perturbed magnetic field is then extrapolated to the poloidal magnetic coils placed around the vessel at different poloidal locations. The result gives the degree of confidence on the mode periodicity, which is mainly m/n=3/2 with non negligible 2/1 component. Furthermore, at a later time the lower frequency mode (m/n=2/1) is stabilized for 50ms after an ELM, whereas the dominant 3/2 remains. Finally, the bi-coherence plot during this time interval does not show any similarities with that obtained during the times when the two modes are coexisting.

Figure 4 shows the dual multiwire proportional x-ray (DMPX) detector array reconstruction of the modes [13], with the iso-surface of perturbed SXR emission for the m/n=3/2 mode. The black lines represent the maximum extent of the m/n=2/1 mode, for which the interpolation returns larger error bars. Hence, these are the extremes ρ_s occupied by the external island, not its real contours. This reconstruction is useful to add resonance points on the q-reconstruction by LIUQE (figure 4). The diamond represent the best fit for the DMPX signals, and the horizontal line is an error-bar ±7% of the central location, with this value chosen according to profile inversion simulation.

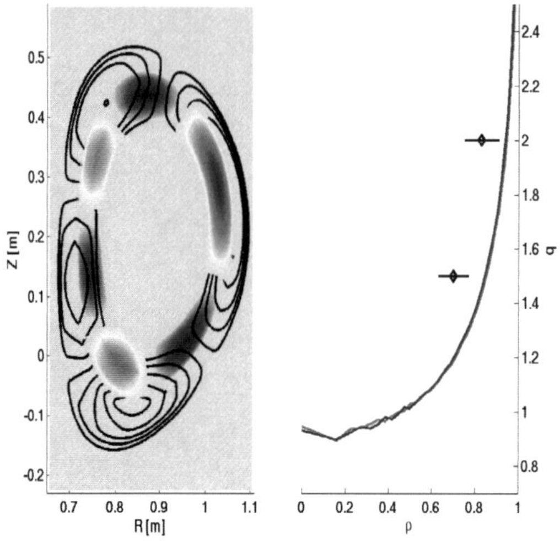

Figure 4. TCV Discharge #29894: the m/n=3/2 and 2/1 locked modes are reconstructed (left) with the high-spatial resolution SXR diagnostic. This allows to add constraints for the q-profile reconstruction (right).

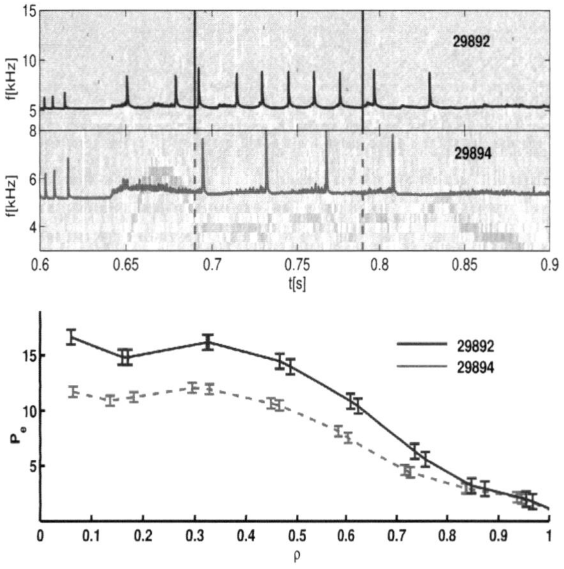

Figure 5. TCV Discharge #29894: comparison of the electron pressure profile between X3 heated ELMy H-mode phase for 29892 and 29894. It is evident the drop in confinement (red) due to the presence of the coupled modes.

This is useful to provide further constraints when running plasma transport simulations. The effect of the mode can be observed through the detailed comparison of the averaged TS profiles between the X3 heated phases of the two discharges (figure 5). In the Ohmic phase the two pressure profiles overlap. As soon as the X3 heating is started, discharge 29894 (second spectrogram from the top, figure 5) shows the coupled modes, together with ELMs. In discharge 29892, instead, only the ELMs and intermittent m/n=2/1 activity are observed. It is interesting to point out that discharge 29892 did not access QSEFHM during the full power phase, without an MHD mode, but only when the modulation had started. Because of the lack of available plasmas with QSEFHM and the apparent contradiction of this, further experiment will be run in the next campaign with and without modulation. It should be noted however that QSEFHM phases have been obtained without power modulation.

Figure 5 shows a comparison of the X3 heated ELMy H-mode phase of discharge 29892 (no modes) and 29894 (with 3/2 and 2/1 modes coupled). The effect of the 3/2 island ($w \simeq 7cm$), which corresponds to a loss in confinement of approximately 33% in agreement with the pressure drop between 29892 and 29894 (figure 5).

The development of the QSEFHM might require sufficient pressure gradients; it is possible that the 3/2 mode causing a significant drop in the confinement properties is preventing the appearance of the ELM-free quasi-stationary state. This statement requires further

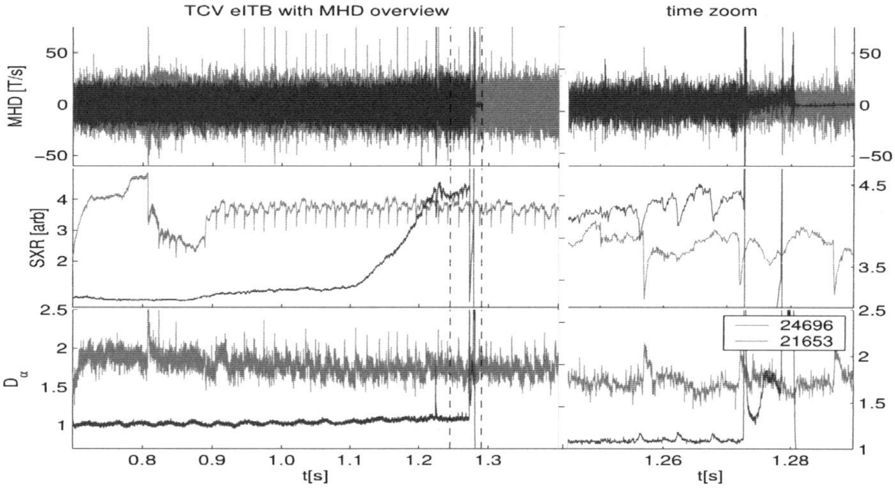

Figure 6. MHD overview for discharges #24696 (red) and #21653 (blue). From top to bottom: magnetic Mirnov raw signal, SXR emission from the core of the plasma, Dα emission. On the right-hand side, zooms of the time window delimited by the vertical dashed lines on the left.

investigation, and further experiments are planned for the next TCV campaign.

3. Electron Internal Transport Barriers (eITBs) and the effect of MHD

Scenarios involving MHD instabilities during the flat-top of eITBs are the interest of the second part of this paper. The character of the observed instabilities varies from disruptive (ideal infernal mode) to resistive (tearing mode developing magnetic island with or without a neoclassical contribution). Figure 6 shows two eITB experiments displaying ideal-like events: TCV discharge #21653 and #24696. The first one (blue) reaches a very high confinement factor (H_{RLW} up to 6) thanks to a significant reverse shear q-profile. This causes a fast approach to the ideal stability limit because of the improved confinement. At $t \simeq 1.28s$ a major disruption ideal event (ideal infernal mode [5]) terminates the discharge.

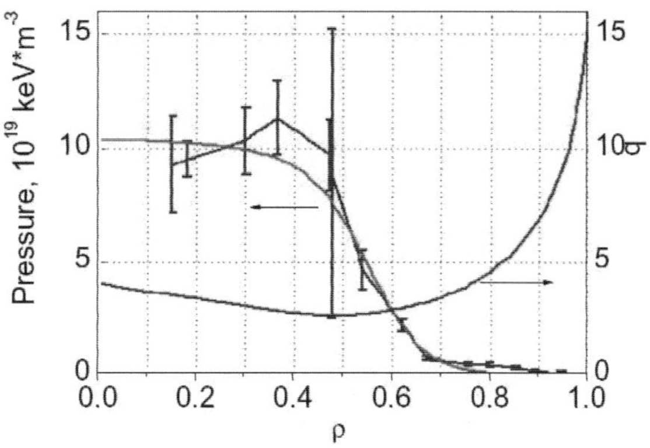

Figure 7. Thomson Scattering pressure profile from raw data (blue), with basic fit (red) and q-profile from CQL3D [14] for discharge TCV#21655. q-min∼2.7 at ρ=0.5 [5].

This is shown in details on the right side of figure 6, where the time zoom of the traces between the two vertical dashed lines is reported. For this discharge (figure 6) the non inductive current j_{CD} generated off-axis, in addition to the bootstrap current density j_{BS}, led to a hollow current density profile. The effect of the gyrotron on-axis added at $t = 1.1s$ is evident in the middle blue trace of figure 6, with the steep increase in SXR emission [7]. The current profile for a similar discharge #21655 but with slightly less counter-CD in the center has been reconstructed thanks to the Fokker-Planck code CQL3D (figure 7), resulting in a $q_{min} \approx 2.7$ at the radial location $\rho_\psi \simeq 0.5$, where the barrier is formed [5, 15]. Magnetic signals analysis reveals the presence of MHD modes with periodicity m/n=3/1 and 2/1 during the disruption of #21653, with a growth time $\tau_{MHD} \approx 20\mu s$, which is typical for ideal instabilities in TCV. For discharge #21655, analysis with KINX code reveals an unstable ideal infernal 3/1 mode with a significant 2/1 component, for β_N larger than 1. The β_N experimental value is close to unity, confirming that discharge #21655 is close to ideal stability threshold, with q_{min} near 3 (figure 8). The instability is caused by a high pressure gradient at $\rho_\psi \simeq 0.5$, where the eITB is formed in a low-shear region. This is the characteristic of infernal modes. Infernal mode appears in regions of low-shear [16]. In this region the development of low-n pressure-driven modes is possible, which was first described in [17] for reversed shear profiles. TCV plasmas are generally stable against the 1/1 mode, as q_{min} is generally above the values for which the internal kink is unstable. Infernal and external modes with m=2,3,4 can develop when q_{min} is close to those integer values [5].

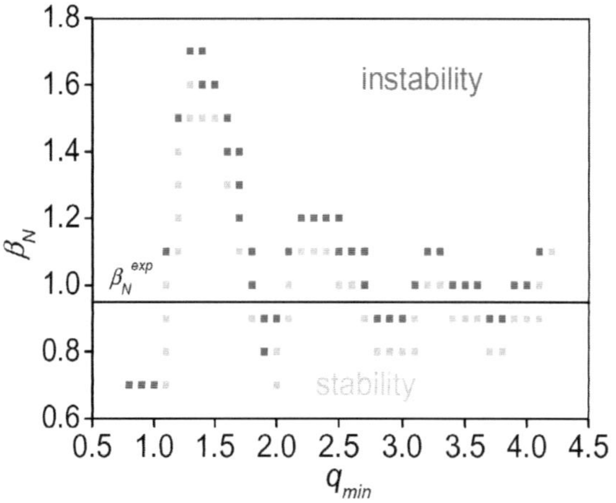

Figure 8. The n=1 mode stability boundary plotted in q_{min} space. Green and red squares correspond to ideally stable and unstable configurations respectively. β_N^{exp} is plotted for TCV discharge #21655

From the evolution of the Thomson Scattering profiles, these modes are clearly detrimental for the attainment of a steady state internal transport barrier. This is a major goal for a fusion reactor, as the confinement for this regime is highly enhanced compared to that of non-advanced scenarios. A way to avoid infernal modes is to tailor the q-profile through fine adjustments of the injected power in order to move the q_{min} away from the minimum values of 2 and 3, or to slightly reduce the local pressure gradient. For $q_{min} > 4$, the infernal mode becomes an external kink mode, which is less sensitive to the values of q_{min} [5].

The red trace in fig. 6, TCV discharge #24696, represents a different scenario involving ideal crashes with a slightly different character [18]. These ideal-like modes have the same detrimental effect on the confinement and they resemble the β-collapse observed in JT-60U [19]. These crashes have a *sawtooth-like* behavior of the electron temperature, with drops and subsequent rises (fig. 6, middle red trace, zoomed on the right side of the figure). It is found that each crash involves the region close to the $q = 2$ surface, i.e. the same region involved in the presented infernal modes. This type of mode has also been previously observed in JET [20]. During this discharge (figure 6, red trace) on-axis counter ECCD was preceded by off-axis ECH, which resulted in a broader electron temperature profile. A significant Ohmic current was also present [18]. Thus, current and pressure profiles are different from the experiments described above, and ideal periodic modes develop. The mode is ideal kink-like and it is dominated by high local pressure gradient at the barrier location (low shear) and global β limit. Therefore the same characteristics as the infernal mode limit discussed above for #21653. The pressure profile obtained from TS measurements together with the modelling of the q-profile through the Mirnov coils data and DMPX inversion analysis shows that the strong pressure gradient is localized in the region of the minimum of the safety factor, i.e. the region of low shear. This is destabilizing the infernal mode, according to the modeling reported in [16] and figure 8. A fast, large crash of type 2/1 at $t = 0.81s$ stops the initial fast growth of the eITB. The crash trigger a 2/1 resistive mode that is stabilized later.

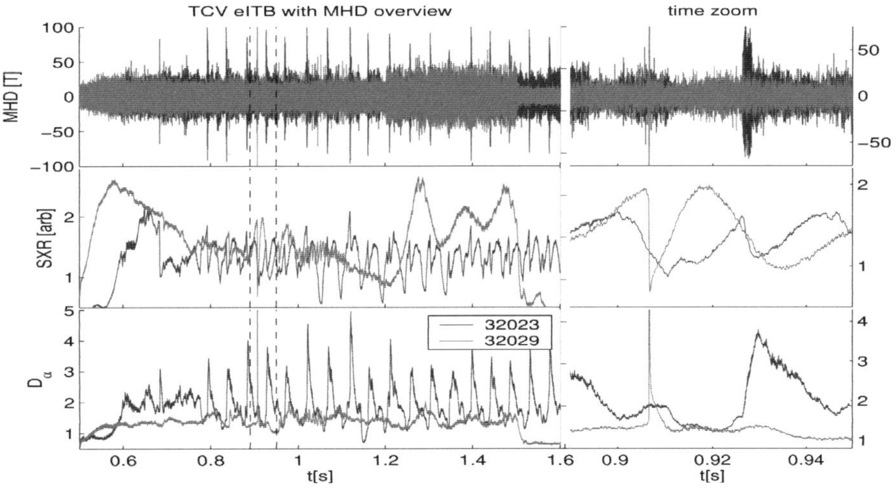

Figure 9. MHD overview for discharges #32029 (red) and #32023 (blue). From the top to the bottom: magnetic Mirnov raw signal, SXR emission from the core of the plasma, Dα emission. On the right side, zooms of the time window delimited by the vertical dashed lines.

At this time the confinement improves again until the pressure gradient is strong enough to trigger another infernal mode, with a period $\tau \simeq 16ms$, which is responsible for the absence of higher H_{RLW} for this discharge. The pressure peaking factor ($p_{e0}/ < p_e >$) reaches a very high value ($\simeq 15$) at $t = 0.81s$, and this causes the relaxation of the pressure profile and a subsequent decrease in β. The following smaller crashes develop in a region with $p_{e0}/ < p_e >\simeq 10$ and $\beta_N \simeq 0.75$.

Regarding experiments displaying a resistive character, figure 9 shows two discharges of global plasma oscillations. The blue traces refer to discharge #32023, on-axis counter-ECCD with Ohmic contribution. The oscillations are fast and dominated by an initial ideal phase which triggers the longer lasting ($\simeq 2 - 5ms$) resistive phase, where the topology of the plasma is changed by the appearance of a small magnetic island. This discharge is characterized by regular small infernal mode crashes with frequency $f \simeq 20Hz$. This type of regime, with the continuous bursts of ideal activity which are followed by a longer lasting resistive mode, causes electron temperature oscillations that are triangular-like due to the ideal character. Figure 10 shows a selection of plasma parameters evolving during four events. The spectrogram of magnetic signal (top trace) allows to appreciate the dual character of these oscillations. Every event is

Journal of Physics: Conference Series **123** (2008) 012038 doi:10.1088/1742-6596/123/1/012038

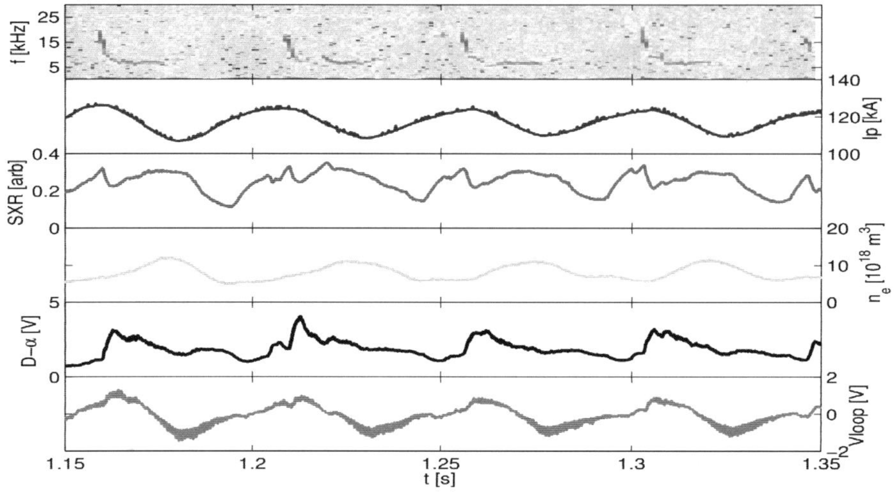

Figure 10. Overview of regular mixed ideal-type and resistive instabilities, discharge TCV #32023. From the top: magnetic spectrogram, indicating the fast ideal phase followed by a resistive mode. Plasma current (blue) which reaches the top at the onset of the infernal modes. SXR radiation from the core (red), showing the crashes at the ideal event. Plasma density (green). D-α light (black). V_{loop} (magenta) showing the continuous changes in the magnetic configuration

characterized by an ideal crash, which is responsible for the almost vertical signature indicating a large band of frequencies, from $15kHz$ to $5kHz$. The ideal phase can be observed also in the SXR trace (red, third from top) and Dα light (black, 5th from top). The drop in the SXR at the occurrence of the ideal mode indicates the fast collapse of the barrier, which is generally accompanied by the emission of light. These events happen at the top of the confinement phase (highest SXR core signal), where the plasma is unstable to the infernal mode. The secondary mode is different in character and separated by the first. It is resistive and lasts for approximately $\Delta t \simeq 20ms$. The magnetic island has a 2/1 periodicity, which is also the main component of the ideal event. The plasma current (blue, second trace) shows oscillations ($\Delta I \simeq 10kA$) with the top of the plasma current synchronous with the ideal mode. The change in the current is smoother than that of SXR radiation, as these crashes are small and the current diffusion time is much longer than the MHD time. The average density oscillates almost in phase with the SXR radiation, but with a smoother behavior. This is an indication of the different electron temperature and density dynamics. The I_p trace in particular shows that even cycles triggered by ideal modes can lead to oscillations similar to the so called *O-regime* [8]. The V_{loop} follow the general behavior of the other traces, indicating the continuous evolution of the magnetic configuration during this oscillatory regime. The drop in SXR emission during the small infernal crashes (proportional to the loss of confinement) is very limited (10% of the signal in the core). The plasma appears to shrink during the ideal mode, which probably leads to a removal of the barrier and the subsequent loss of confinement.

Discharge #32029, fully non inductive, is unique because it shows both type of modes in a distinctive way (figure 9, $0.8s \leq t \leq 1s$ for the infernal mode unstable plasma, $1.2s \leq t \leq 1.5s$ for the resistive mode). At the time when the barrier is formed ($t \leq 0.8s$) the electron temperature suddenly increases (fig. 9). We can infer the beginning of an extremely large global oscillation, with a full cycle completed before an ideal instability provokes the abrupt collapse of the temperature. The zoom on the right part of figure 9 shows the first big crash at $t \simeq 0.908s$, which alone does not stop the oscillations. The radiation profiles before and after the crash show the character of a sawtooth-like instability, with an inversion radius that is located in the proximity

Journal of Physics: Conference Series **123** (2008) 012038 doi:10.1088/1742-6596/123/1/012038

of $q = 2$. The drop in SXR radiation, proportional to the loss of confinement, is approximately 60% of the emission in the core. The gradients at the barrier are lost and expulsion of particles and heat is observed. Based on the pressure profile and the phenomenology of the mode, the crash appears to be the result of a major infernal mode, just not large enough to disrupt the plasma as in #21653. It could be called a minor disruption or a β-collapse. In effect, it is a crash triggered by an ideal $n = 1$ infernal mode. The applied heating, and the change in pressure profile, allows the quick recreation of the barrier. The second time the SXR trace reaches the top ($t \simeq 0.95s$), a smaller infernal mode is observed. This reduces the confinement and begins a second phase with frequent, small infernal modes ($1s \leq t \leq 1.2s$), which has a detrimental effect on the confinement and prevent the development of an eITB. These tinier collapses ($t \geq 0.95s$, red trace) are very frequent ($\Delta t \simeq 5ms$) and accompanied by MHD oscillations ($f \simeq 15kHz$). Later on the ideal modes are stabilized. At $t = 1.22s$ the eITB begins to reform (red middle trace, figure 9), which is accompanied by the development of a resistive tearing mode [9]. This mode involves the periodical growth and shrink of a magnetic island located near the foot of the barrier. At the top of the oscillations the island has the minimum width and begins to increase in size, very rapidly. This reduces the confinement factor until the bottom of the oscillation; at this time the island is essentially stabilized, the confinement improves, and the cycle starts over. The main periodicity of the mode, obtained by concurrent analysis with Magnetic coils, Soft X-ray tomography (SXR), singular value decomposition (SVD) and Fourier analysis reveals a 2/1 character [9].

4. Conclusions

The role of MHD stability and control is of crucial importance for the development of steady-state advanced scenarios in tokamak plasmas. Ohmic and ECH heated H-modes are routinely obtained on TCV, displaying type-I ELMs and other types of MHD instabilities. If X3 power is injected from the top, and the conditions in terms of plasma parameters and shape are met (together with the absence of large MHD modes) the plasma can access a quasi-stationary ELM-free H-mode (QSEFHM), with better confinement properties and absence of ELMs. The reasons for the attainment (or not) of this regime are still partly unknown, and further experiments are planned. The positive aspects are the absence of ELMs, the improved confinement, the robustness of the regime and the absence of uncontrolled density build-up. On the other hand, the regime cannot be easily reproduced and sometimes large ELMs are seen to interrupt the quasi-stationary phase, inducing a threat for the maximum thermal load and the possibility of triggering tearing modes, as in #29892. MHD stability plays a fundamental role, since discharges without large magnetic islands have confinement properties up to 30% higher than discharges that develop MHD modes. In addition, no discharge has been found (up to this time) that develops the QSEFHM when the X3 heating triggers a mode on the q=1.5-2 rational surfaces before entering the quasi-stationary phase. Understanding the interplay of core MHD with the scenario achievement (if any), and the stability properties of X3 heated ELMs is a key factor for the reproducibility and safe operation of the QSEFHM. The sizes of the tearing modes discussed in this paper are consistent with the observed loss of confinement, stressing the importance of MHD control for plasma performances.

Regarding electron internal transport barriers, values of H_{RLW} higher than 4 have been obtained during eITBs discharges in TCV, with and without the development of MHD instabilities. MHD modes can be of ideal type or resistive (tearing modes, NTMs), and all the intermediate states are possible. The origin of these modes is the unfavorable conjunction of large pressure gradient in a low shear region, i.e. they represent the infernal stability limit [5, 17, 16]. Since eITBs are essentially obtained in reverse shear plasmas and occur near q_{min}, large gradients in small shear regions are inherent of these scenarios. Therefore it is likely that the various regimes described in the literature, β-collapse, q=2 sawteeth, O-regime, minor and major disruptions in reverse shear are all related to the nearby stability limit of infernal modes, as observed in TCV. This is why they are sensitive to q_{min}, $p_0/<p>$, p' and magnetic shear [5].

The effect of all these modes is detrimental for the barrier. Avoidance of the ideal modes is necessary because of their fast growth and thus the impossibility of controlling them once developed. This can be achieved by changing the current density and pressure profiles to avoid the proximity to rational integers of the safety factor and/or to reduce the local pressure gradient. The effects of the resistive modes are less abrupt, as they slowly change the topology and act on the resistive time. Nevertheless, the integral effect on the plasma confinement properties can be as much damaging as that of the ideal mode. Favorable for these modes is the fact that they can be detected by the plasma control system, which can react in order to stabilize them, by changing the profiles for example with local ECCD. Experiments with Ohmic contribution and current injection to prove this principle have been successfully achieved [2, 9].

References

[1] Strait E J 1994 *Phys. Plasma* **1** 1415
[2] Udintsev V S *et al.* 2007 *Fusion Sci. Technol.* **52** 161
[3] Turri G *et al.* 2007 *IAEA Technical Meeting Steady State Operation, Daejeon - Korea*
[4] Porte L *et al.* 2007 *Nucl. Fusion* **47** 952
[5] Martynov A 2005 *PhD thesis, Ecole Polytechnique Fédérale de Lausanne* **3218**
[6] Sauter O *et al.* 2005 *Phys. Rev. Lett.* **94**
[7] Goodman T P *et al.* 2005 *Plasma Phys. Control. Fusion* **47** B107
[8] Giruzzi G *et al.* 2003 *Phys. Rev. Lett.* **91**
[9] Turri G *et al.* 2008 *Plasma Phys. Control. Fusion* **50** 065010
[10] Raju D *et al.* 2003 *Plasma Phys. Control. Fusion* **45** 369
[11] Reimerdes H 2001 *PhD thesis, Ecole Polytechnique Fédérale de Lausanne* **2399**
[12] Hofmann F and Tonetti G 1988 *Nucl. Fusion* **28** 1871
[13] Rossel J 2005 *LRP-report EPFL - CRPP*
[14] Parail V V 2002 *Plasma Phys. Control. Fusion* **44** A63
[15] Henderson M A *et al.* 2003 *Phys. Plasmas* **10** 1796
[16] Manickam J *et al.* 1987 *Nucl. Fusion* **27** 1461
[17] Ozeki T *et al.* 1993 *Nucl. Fusion* **33** 1025
[18] Zhuang G *et al.* 2004 *31st EPS Conference on Plasma Phys., London, 2004* **28G** P-2.143
[19] Pomphrey J *et al.* 1996 *Plasma Phys. Control. Fusion* **38** 1791
[20] Hender T C *et al.* 2002 *Plasma Phys. Control. Fusion* **44** 1143

Particle confinement of pellet-fuelled H-mode plasmas in the Mega Ampere Spherical Tokamak

M Valovič, K Axon, L Garzotti, S Saarelma, A Thyagaraja, R Akers, C Gurl, A Kirk, B Lloyd, G P Maddison, A Patel, S Shibaev, R Scannell, D Taylor, M Walsh and the MAST team

EURATOM/UKAEA Fusion Association, Culham Science Centre, Abingdon, Oxfordshire OX14 3DB, U. K

E-mail: martin.valovic@ukaea.org.uk

Abstract. This paper quantifies the particle confinement of pellet-fuelled plasmas in the Mega Ampere Spherical Tokamak (MAST). The dataset is restricted mostly to neutral beam heated plasmas and to shallow pellets launched from the high field side. It is shown that the pellet deposition can be explained only by invoking the ∇B drift of the pellet ablatant. The pellet creates a zone with positive density gradient and increased temperature gradient. Simulations show that these changes could increase the level of micro-turbulence and thus enhance further the penetration of pellet-deposited particles towards the core. Post–pellet dynamics of the density profile is characterised by the pellet retention time τ_{pel}. It is shown that τ_{pel} correlates with the status of the edge transport barrier (L-mode or H-mode) and decreases rapidly for pellet deposition radius r_{pel} approaching the plasma edge. For ELMy H-mode and ITER-like pellets, $r_{pel} \approx 0.8a$, the pellet retention time is about 20% of the energy confinement time. The fuelling requirement by the pellets for ITER is discussed.

1. Introduction

The present consensus is that in a reactor, high densities are difficult to be achieved by gas puffing due to the poor penetration of neutrals into the plasma. The most promising technique to control the density in reactor grade plasmas is the injection of cryogenic pellets. The price to pay for such deeper fuel deposition is a substantial perturbation to the plasma by the pellet itself. For example for ITER, local density increase up to 50% is expected in the pellet deposition zone located at the outer 15% of the minor radius [1, 2]. Pellet-fuelled plasmas are not stationary and the pellet deposition region is in a state of permanent response to perturbations. Pellets modify the confinement physics in the outer part of the plasma and to a large extent the pellet deposition zone takes over the role from the H-mode pedestal for setting up the boundary conditions for the plasma core.

Two practical parameters, the pellet deposition radius r_{pel} and the post-pellet particle confinement time τ_{pel}, provide the link between the confinement physics and the design of the pellet fuelling system. The particle throughput necessary to maintain the required plasma density is then determined by these two parameters as:

$$\Phi_{pel} \approx n_e S(a - r_{pel})/\tau_{pel} , \qquad (1)$$

where a is the plasma minor radius, S is the plasma surface area and n_e is the time averaged density (over pellets) at r_{pel} assuming flat profile in $r_{pel} < r < a$. In the absence of density peaking n_e is equal to the central density. The present paper reports on first measurement of pellet deposition radius and pellet retention time in MAST.

2. Experimental conditions

The targets for pellet injection were both L-mode and H-mode plasmas with plasma current $I_p = (0.66 - 0.76)MA$, geometric major radius $R_{geo} = (0.8 - 0.87)m$, minor radius $a = (0.55 - 0.60)m$, elongation $\kappa = 1.8 - 2.1$, vacuum toroidal field at the geometric radius $B = (0.47 - 0.50)T$, with double null divertor configuration and line averaged density $\bar{n}_e = (1.6 - 7.5)\times 10^{19} m^{-3}$. The working gas was deuterium and plasmas were heated with neutral beams with injected power of $P_{INJ} = (1.1 - 3.0)MW$ and with energy of $\le 65keV$. Pellets were launched from the top and they enter the plasma from the high field side of the machine as shown in figure 1a. The pellet injection system is described in more detail in [3]. Low field side launch is also possible in MAST, but was not used in this study. The nominal pellet sizes are $d_{pel} = (1.1; 1.35; 1.7)mm$ with $N_{pel} = (0.6; 1.2; 2.4)\times 10^{20}$ atoms and pellet velocities were in the range of 240-450 m/s. We selected data with only those pellets which cause relatively modest plasma perturbations approaching the conditions similar to those expected in ITER.

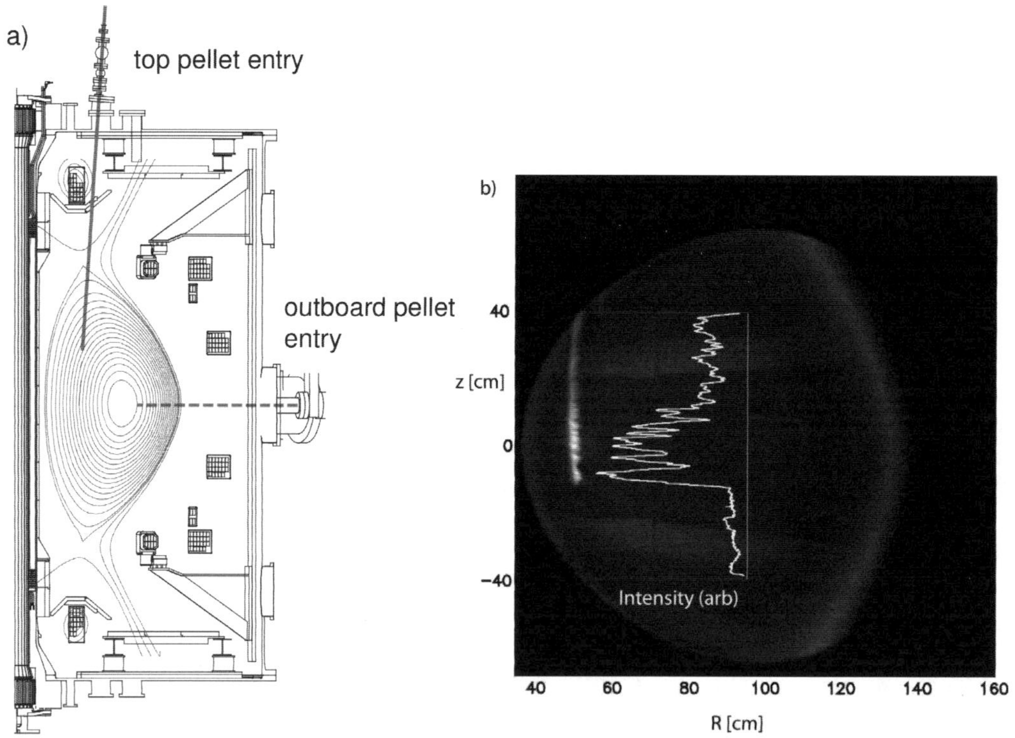

Figure 1. (a) left: The geometry of pellet injection. (b) right: The pellet image in narrow band ($\lambda_0 = 457.25nm$, $\Delta\lambda_{FWHM} = 2.43nm$) visible bremsstrahlung (exposure is 7ms). The insert shows the emission intensity along the pellet trajectory.

Journal of Physics: Conference Series **123** (2008) 012039 doi:10.1088/1742-6596/123/1/012039

3. Pellet deposition radius

The first parameter controlling the pellet particle throughput in equation (1) is the radius at which the majority of pellet particles are deposited. In MAST, the evaporation and deposition of a single pellet (pellet lifetime) lasts about 2 ms. The end of this process is clearly seen as a sharp inflexion point on the interferometer signal - an interesting observation as in neutral gas and plasma shielding models (NGPS) [4] the pellet evaporation rate decreases continuously with decreasing pellet size. The pellet evaporation and the deposition process is captured in the visible bremsstrahlung emission spectrum (figure 1b) revealing the clear burst-like structure (striations) of this process ([5] and references within). The image also confirms the sharp discontinuity in evaporation rate at the end pellet deposition. Absolute intensity of the emission indicates densities of the cold (1-2 eV [6]) ablatant up to $10^{23} m^{-3}$.

In order to measure the pellet deposition profiles without a distorting effect of post-pellet density evolution we triggered the high spatial resolution (~1cm) Thomson scattering system directly from pellet signal with controlled delay. The density profile just at the end of the pellet evaporation is shown in figure 2. It is seen that the the inner and outer density profiles are a relatively good function of the poloidal flux coordinate as determined from magnetic equilibrium reconstruction (EFIT). This shows that the plasma density quickly equilibrates along the flux surfaces already during the pellet life time. This significantly simplifies the post pellet transport analysis as one can assume that shortly after the pellet evaporation the density and temperature profiles are conventional and do not remember the 3D character of the particle source from the pellet. Evolution of the density profile during the pellet deposition cannot be explained by the simple NGPS model [4] as seen from figure 2. Agreement is achieved only if $\nabla \mathbf{B}$-drift is added which effectively spreads the pellet deposition inwards by $\sim 0.15 r/a$. The best fit is obtained with the drift distance in the form of $\lambda = c_s^2 \tau_d^2 / (2R)$, where the ion sound speed c_s corresponds to temperatures of 1-2 eV and duration of the drift is $\tau_d = 10 - 40 \mu s$ [7, 8, 9].

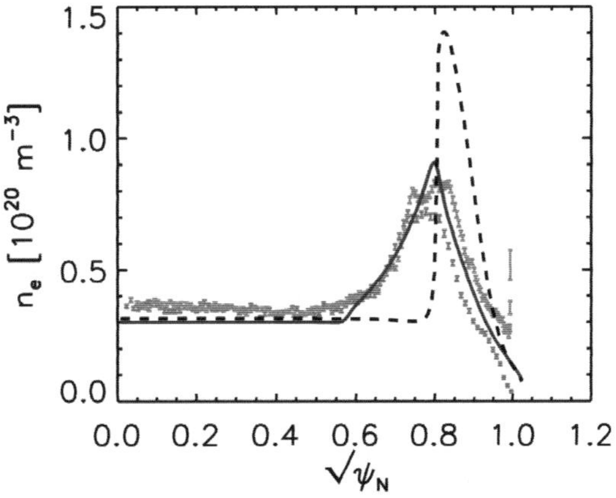

Figure 2. Density profile at the end of pellet deposition. Symbols: experimental data measured by pellet-triggered Thomson scattering. Inboard (magenta) and outboard (red) profiles measured at the same time are shown. Black dashed line: prediction by NGPS model. Blue line: prediction by NGPS model with ∇B-drift. ψ_N is the normalised poloidal flux.

The pellet deposition could be also affected by a temporary change in micro-turbulence induced by the pellet itself. Figure 3b shows two density profiles – one before and one just after the pellet deposition as captured by the 200 Hz Thomson scattering system. It is seen that the pellet creates a distinct zone with positive density gradient $\nabla n_e > 0$ and electron temperature gradient $a/L_{Te} \equiv -a\nabla \ln T_e$ doubled relative to its pre-pellet value (see figure 3a).

Calculations using the linear GS2 and TRANSP codes for the case in figures 3a and 3b show that at the radius of maximum increase of a/L_{Te} ($\sqrt{\psi_N} \sim 0.7$) the mixing length estimate of diffusivity γ/k_y^2 increases by a factor of 2 for modes with wavelengths $0.08 < k_y\rho_i < 0.8$ as seen in figure 3c. Here, γ is the linear growth rate, $\mathbf{k}_y$ is the poloidal wave vector perpendicular to total magnetic field, $\rho_i = \sqrt{T_i m_i}/(eB)$ is the ion Larmor radius where m_i in the ion mass, e is the electron charge. The ion temperature $T_i \approx T_e$, as measured by charge exchange spectroscopy and neutral particle analyser. The modes $0.08 < k_y\rho_i < 0.8$ are identified as micro-tearing and ion temperature gradient (ITG) instabilities. Stability of these modes strongly depends on the flow shearing rate $\omega_{E\times B}$ which is, however, not measured in the pellet case. In the non-pellet case, $\omega_{E\times B}$ is found to be of the same magnitude as that required to stabilise the long wavelength modes $\omega_{E\times B} \sim \gamma(k_y\rho_i < 1)$ [10].

The plasma in figures 3a and 3b has also been modelled by the global nonlinear fluid code CUTIE [11]. Figure 3d shows the relative amplitude of the density fluctuations $\tilde{n}/n$ in a simulation at the radius of maximum increase of a/L_{Te}. It is seen that the amplitude of fluctuations increases threefold for $0.07 < k_\theta\rho_s < 0.2$ modes and decreases for modes with $0.2 < k_\theta\rho_s < 0.4$, but the overall turbulent transport increases. Here, k_θ is the poloidal wave number and $\rho_s = \sqrt{(T_e + T_i)m_i}/(eB)$. Note that a similar enhancement of micro-turbulence in the pellet deposition zone has been also obtained by the CUTIE and TRB codes when simulating JET plasmas [12].

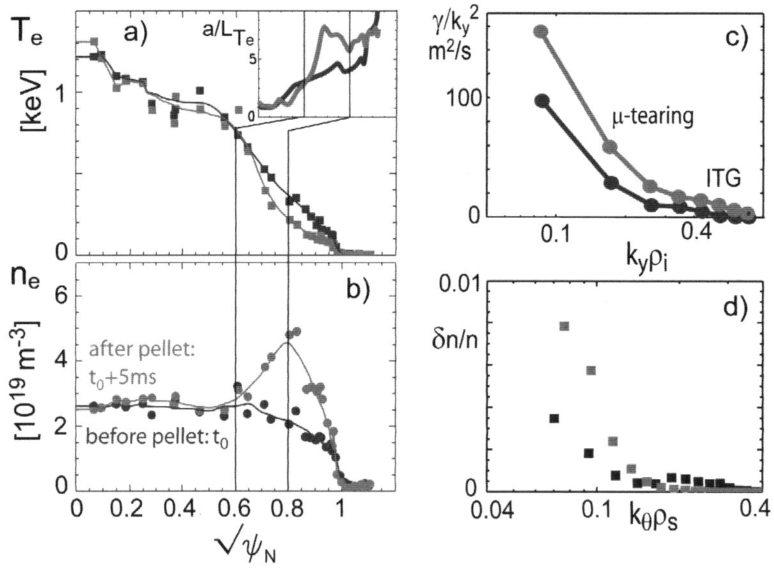

Figure 3. (a) Electron temperature, T_e, and (b) electron density, n_e, profiles before (blue) and after (red) the pellet injection as measured by 200Hz Thomson scattering. The insert panel shows the normalised electron temperature gradient a/L_{Te} over a small range of minor radius as indicated. The plasma is in ELMy H-mode, pulse number is 16335. (c) Normalised growth rates of micro-turbulence calculated by linear GS2 code without rotation shear. (d) Amplitudes of density fluctuations of micro-turbulence calculated by CUTIE code. Both GS2 and CUTIE results (c) and (d) refer to the normalised poloidal flux surface of $\sqrt{\psi_N} \sim 0.7$.

The temporary increase of particle transport in the zone with $\nabla n_e > 0$, as indicated above by both simulations, would result in further inward particle propagation in addition to the $\nabla\mathbf{B}$-drift and thus would be favourable for pellet deposition. Finally note that the distinct zone described above could takeover the role from the pedestal for setting-up the boundary condition for transport in the plasma core in general.

4. Pellet retention time

Deposition of pellet particles causes a significant perturbation to the plasma and therefore post-pellet plasma confinement can differ from that just before the pellet injection. In MAST, the response of the plasma to the pellet can be complex and this is illustrated in figure 4 which shows a full spectrum of observed cases. Figure 4c shows a common situation when the plasma density rapidly decays after the pellet deposition as a result of increased ELM frequency or even a short transition to L-mode. Figure 4b shows a case unique to MAST when the pellet is not followed by ELMs and plasma remains in ELM-free H-mode for very long time period until the first natural ELM occurs. Finally, the pellet can also induce a transition from L-mode to H-mode as shown in figure 4a. This full spectrum of post pellet plasma behaviour also demonstrates the potential of pellets as a control tool for particle transport in general.

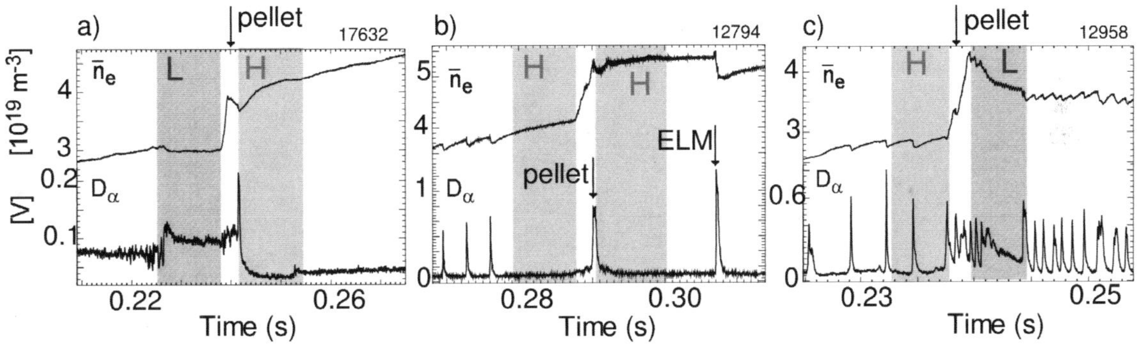

Figure 4. Different post-pellet behaviour in MAST. Top traces, $\bar{n}_e$, are the line averaged densities as measured by an interferometer. Lower traces show the emission of Balmer D_α line. Times of injection of pellets are indicated by arrows.

The post-pellet dynamics determines the second parameter which controls the fuelling efficiency in equation (1): the pellet retention time, τ_{pel}. In MAST this parameter can be measured directly with the help of the 200Hz Thomson scattering diagnostic. Here τ_{pel} is determined from the evolution of plasma density at fixed radial position:

$$n_e(t,\rho_{pel}) - n_{e,N}(t,\rho_{pel}) \propto \exp\left[-(t-t_{pel})/\tau_{pel}\right],$$

where, $\rho_{pel} = r_{pel}/a$ is the radius of maximum density perturbation due to the pellet and is referred to as the pellet deposition radius later in the text. t_{pel} is the time point just after the pellet deposition and $n_{e,N}(t,\rho_{pel})$ is the "natural density" due to the gas and beam fuelling and it is determined by a linear extrapolation from the pre-pellet trend. The parameter τ_{pel} is then calculated by a log-linear regression from discrete time slices.

It has to be noted that the time constant τ_{pel} encapsulates the very complex nature of post pellet transport. Firstly, the post pellet decay is not necessarily exponential and typically the time constant is shorter at the beginning of the decay than later on. Another complication is that the pellet retention time, as defined above, can depend on the fraction of gas and beam fuelling. In most of today's plasmas the contribution of gas and beam fuelling is much larger than expected in future devices.

Figure 5 shows the pellet retention time τ_{pel} as a function of pellet deposition radius. The error bars of τ_{pel} measure the goodness of exponential fit described above. The error bars of the pellet deposition radius, ρ_{pel}, are small as this is calculated as a position of the maximum of the fitted curve

Journal of Physics: Conference Series **123** (2008) 012039 doi:10.1088/1742-6596/123/1/012039

through the density profile. The width of the pellet deposition zone is, however, comparable to ρ_{pel} as seen on example in figure 3b.

Two types of correlations are evident from the data. Firstly, the values of τ_{pel} correlate with the status of the edge transport barrier. It is seen that for fixed pellet deposition radius ρ_{pel} L-modes have the poorer particle confinement while extremely good confinement is found in ELM-free H-modes. Secondly, the pellet retention time correlates with the pellet deposition radius. For a given group of data (L-mode, ELMy and ELM-free) the retention time τ_{pel} decreases for decreasing pellet penetration. In figure 5 this decreasing trend is compared with expectation due to diffusive loss if the effective diffusion coefficient, D_{eff} is independent of minor radius. It is seen that for L-mode, and to a lesser extent for ELMy H-mode, the pellet retention time decreases faster with decreasing pellet penetration than expected from the fixed value of D_{eff} indicating that effective diffusivity increases towards the edge.

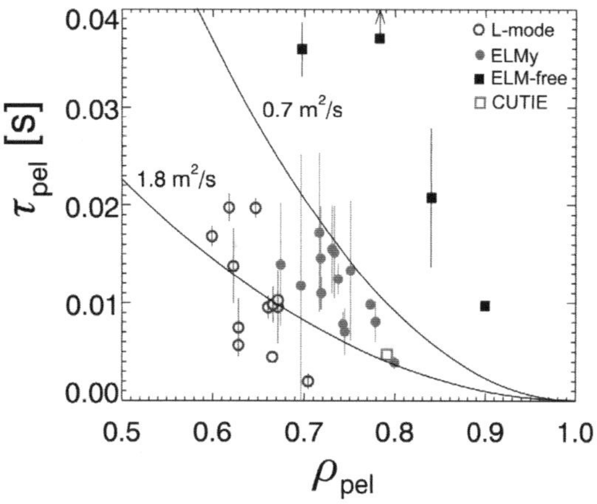

Figure 5. Pellet retention time τ_{pel} plotted against the normalised radius $\rho_{pel} = \sqrt{\psi_{N,pel}}$ taken at the maximum of the density perturbation by the pellet. The lines are the curves $(a - r_{pel})^2 / 2D_{eff}$ for two values of effective diffusivity D_{eff} with $a = 0.55m$.

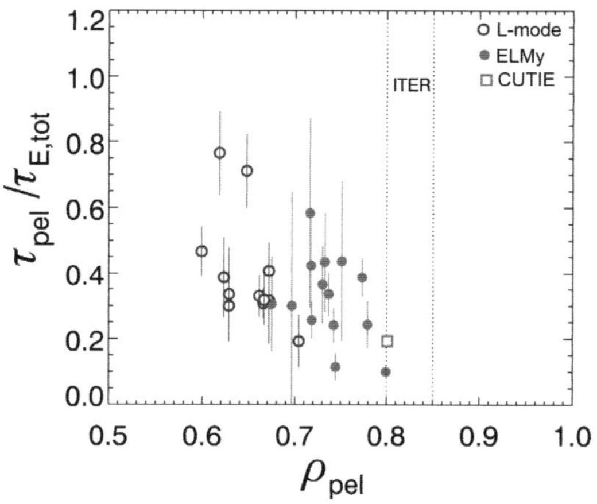

Figure 6. Pellet retention time normalised to total energy confinement time $\tau_{pel}/\tau_{E,tot}$ plotted against the pellet deposition radius as defined in figure 5.

Post pellet density evolution has been simulated by the CUTIE code. The starting profiles and plasma parameters were chosen to represent the discharge shown in figures 3a, 3b. The pellet retention time was found to be $\tau_{pel} \approx 4.5ms$ and as seen from figure 5 this value is in a good agreement with the data.

In order to obtain useful scalings the anomalous particle transport is usually normalised to the heat transport [13]. This is analogous to introduction of Prandtl-Peclet numbers in hydrodynamics and reflects the assumption that both the heat and particle fluxes are driven by the same turbulence. Figure 6 shows the pellet retention time τ_{pel} normalised to the global energy confinement time $\tau_{E,tot}$ for L-mode and ELMy H-mode plasmas. It is seen that similarly to τ_{pel} the ratio $\tau_{pel}/\tau_{E,tot}$ also decreases rapidly for shallower pellets. Further improvement of the scaling for τ_{pel} will require normalisation to the energy confinement time evaluated at the pellet deposition radius, analogous to the definition of pedestal energy confinement time in the two term energy scaling law [14].

5. Fuelling requirement for ITER

For pellet deposition expected in ITER, $r_{pel} \approx 0.80a$, the data in figure 6 extrapolate to the pellet retention time of $\tau_{pel}/\tau_{E,tot} \sim 0.2$. For this value the particle throughput calculated from equation (1) is $\Phi_{pel} = n_e S(a - r_{pel})/\tau_{pel} = 70 Pa m^3/s$. Here we used the parameters for a nominal ITER discharge [15] where the plasma density is $n_e = 10^{20} m^{-3}$, plasma surface $S = 680 m^2$ and the energy confinement time $\tau_E = 3.7 s$. Such value of pellet throughput is about 70% of the present ITER design value for steady state operation [15]. Choosing even the largest envisaged pellet (5mm) implies that the interval between pellets has to be much shorter than the pellet retention time: $f_{pel} = 6Hz \sim 4.4/\tau_{pel}$. Such conditions are rare in present pellet fuelled plasmas. Another uncertainty is due to the fact that the gas and beam fuelling in ITER will be small while this is not the case in present plasmas with pellet injection. These two effects could result in shorter pellet retention time in ITER than the value given above and thus the required particle throughput Φ_{pel} might be underestimated.

6. Conclusion

Two key parameters controlling the pellet fuelling efficiency, the pellet deposition radius and the pellet retention time, were measured in MAST using high field side pellet injection. The pellet deposition reveals the existence of $\nabla \mathbf{B}$-drift. The pellet creates a zone with positive density gradient and increased temperature gradient. Simulations show that these changes could increase the level of micro-turbulence and thus enhance further the penetration of pellet-deposited particles towards the core. In addition, this zone could modify the role of pedestal for setting-up the boundary condition for transport in the plasma core. The pellet retention time correlates with the status of the edge transport barrier and decreases rapidly for pellet deposition radius approaching the plasma edge. For shallow pellets, as expected in ITER, the data extrapolate to a pellet retention time to be about 20% of the confinement time. This would imply that the required pellet particle throughput is close to the design value for ITER for steady state operation. Large uncertainty of prediction in both parameters justifies further refinements of databases and models.

Acknowledgement

This work was funded jointly by the United Kingdom Engineering and Physical Sciences Research Council and by the European Communities under the contract of Association between EURATOM and UKAEA. The views and opinions expressed herein do not necessarily reflect those of the European Commission.

References

[1] Polevoi A R, Sugihara M, Takenaga H, Isayama A, Oyama N, Loarte A, Saibene G and Pereverzev G V *et al.* 2003 *Nucl. Fusion* **43** 1072
[2] Baylor L R, Parks P B, Jernigan T C, Caughman J B, Combs S K, Foust C R, Houlberg W A, Maruyama S and Rasmussen D A 2007 *Nucl. Fusion* **47** 443
[3] Axon K B, Maddison G P, Price M N, Scannell R, Shibaev S, Taylor D, Terranova D, Walsh M J 2004 *Proc. 31st EPS Conf. on Controlled Fusion and Plasma Physics (London, 2004)* ECA **28G** P4.197
[4] Garzotti L, Pégourié B, Geraud A, Frigione D and Baylor L R 1997 *Nucl. Fusion* **37** 1167
[5] Parks P B 1996 *Plasma Phys. Control. Fusion* **38** 571
[6] McNeill D H, Greene G J, Newburger J D and Owens D K 1991 *Phys. Fluids* **B 3** 1994
[7] Lengyel L L 1977 *Nucl. Fusion* **17** 805
[8] Rozhansky V A, Veselova I and Voskoboynikov S 1995 *Plasma Phys. Control. Fusion* **37** 399
[9] Pégourié B and Garzotti L 1997 *Proc. 24th EPS Conf. on Controlled Fusion and Plasma Physics (Berchtesgaden, 1997)* ECA **21A** 153
[10] Roach C M *et al.* 2005 *Plasma Phys. Control. Fusion* **47** B323
[11] Thyagaraja A A, Knight P J, de Baar M R, Hogeweij G M D and Min E 2005 *Phys. Plasmas* **12** 090907
[12] Garzotti L, Garbet X, Thyagaraja A, de Baar M R, Frigione D, Mantica P, Parail V, Pégourié B, Zabeo L and JET EFDA contributors 2006 *Nucl. Fusion* **46** 73

[13] Valovič M, Garzotti L, Voitsekhovitch I, Beurskens M, Garbet X, Giroud C, Keeling D and JET EFDA contributors 2007 *Nucl. Fusion* **47** 196
[14] Cordey J G and the ITPA H-mode Database Working Group, 2003 *Nucl. Fusion* **43** 670
[15] ITER Final Design Report 2001, http://www.iter.org/pdfs/PDD4.pdf

Journal of Physics: Conference Series **123** (2008) 012040 doi:10.1088/1742-6596/123/1/012040

β dependence of micro-instabilities using linear gyrokinetic simulations

L Vermare[1], C Angioni[1], A Bottino[1], A G Peeters[2] and ASDEX Upgrade Team

[1] Max-Planck-Institut für Plasmaphysik, EURATOM Association, Garching, Germany
[2] Centre for Fusion, Space and Astrophysics, University of Warwick, CV4 7AL Coventry, UK

E-mail: `laure.vermare@lptp.polytechnique.fr`

Abstract. The β scaling of micro-instabilities has been investigated by several authors using gyrofluid or gyrokinetic description in linear as well as in non-linear calculations. The typical picture well established now is that at low β, Ion Temperature Gradient (ITG) mode is stabilized with increasing β until Kinetic Ballooning modes become unstable. Such kind of numerical studies have been generally performed in plasma conditions close to the CYCLONE case, which means for ion temperature gradient length strongly above the ITG linear threshold and density gradient length around $R/L_n = 2.2$. However, in some standard H-mode plasmas, not only the ion temperature gradient length is closer to the ITG threshold but also the density gradient length can be rather small as compared to the CYCLONE case. In these conditions, the β scaling of micro-instabilities can significantly differ from the expected behavior described above. The goal of this paper is to investigate numerically the β scaling of micro-instabilities around such experimental conditions. Simulations have been performed using the gyrokinetic electromagnetic flux-tube code GS2 in its linear version and considering as reference, discharges from the dedicated β scaling experiments performed recently in ASDEX Upgrade. The effect of changing the gradient lengths (density and temperature) within experimental errors-bars is studied. It appears that for certain ranges of plasma parameters micro-tearing modes are the dominant micro-instability and coexist in the spectrum with ITG modes with a comparable growth rate. The effect of the plasma shape on the β scaling of micro-instabilities is also investigated. It is shown that the triangularity affects slightly the KBM branch which is destabilized at value of β clearly above the experimental range. However, it is found that at $\rho_t = 0.7$ Trapped Electron Modes are the most unstable instabilities and that such modes are destabilized with increasing β in agreement with results from dedicated β experiments performed in ASDEX Upgrade.

Journal of Physics: Conference Series **123** (2008) 012040 doi:10.1088/1742-6596/123/1/012040

1. Introduction

The investigation on electromagnetic effects on micro-turbulence and heat transport is very challenging, in both, theoretical and experimental studies but is of primary importance considering the interest in operating at high β in future devices.

Experimental studies on β scaling of heat transport did not lead yet to a clear conclusion. The empirical scaling law IPB98(y,2) [1] exhibits a strong β degradation of the thermal confinement time τ_{th} as $B\tau_{th} \propto \beta^{-0.9}$. The dimensionless parameters dependence of the global confinement time are obtained by replacing the engineering parameters in the empirical scaling law by dimensionless parameters. Therefore, a little change in the exponents of engineering parameters can lead to a significant change in the exponents of the dimensionless parameters. In particular, the β dependence is very sensitive to the exponent of the additional power which is by itself subject to large uncertainties. The alternative approach consists in measuring directly the β dependence of confinement by performing dedicated experiments in which β is varied keeping the other dimensionless parameters (normalized Larmor radius ρ^*, normalized collisionality ν^*, the plasma shape, etc ...) constant. Such kind of experiments have been performed in DIII-D and JET and no β dependence was found [2, 3]. In contrast, more recent experiments carried out in JT-60U [4] and ASDEX Upgrade [5] found a β degradation comparable to the dependence predicted in the scaling law.

From a theoretical point of view, a substantial effort to include electromagnetic effects in numerical simulations has been undertaken. The study of finite β effects on core turbulence has been investigated by several authors using gyrofluid [6, 7, 8, 9], as well as gyrokinetic [10, 11, 12, 13, 14] descriptions. The plasma conditions used for these studies were close to the CYCLONE case (the normalized density gradient lenght $R/L_n = R\nabla n/n = 2.2$ where R is the major plasma radius, the normalized temperature gradient length $R/L_T = R\nabla T/T = 6.9$, the magnetic shear $\hat{s} = 0.78$, the electron and ion temperature ratio $T_e/T_i = 1$ and the safety factor $q = 1.4$), for which at low β the most unstable modes are Ion Temperature Gradient (ITG) modes with the maximum of the growth rate around $k_\theta \rho_i \simeq 0.3$ (where k_θ is the poloidal wavenumber and ρ_i the ion Larmor radius). As a general result, linear simulations show that finite β effects on ITG modes are stabilizing. With increasing β, a transition from nearly electrostatic ion drift turbulence to electromagnetic modes as Kinetic Ballooning modes (KBM) is observed. The transition occurs when β is increased above a threshold value generally smaller than the ideal MHD ballooning limit. Non-linear simulations at low β give a β dependence of transport which agrees with the linear results : far from the ideal ballooning limit, the ion heat conductivity decreases with β but it starts to increase strongly when β approaches the ideal ballooning limit. However, in high collisionality H-mode plasmas, not only the ion gradient length at mid-radius is closer to the linear ITG threshold ($R/L_T < 5$) but also the density gradient length can be rather small (around $R/L_n \sim 1$) as compared to the CYCLONE case ($R/L_n = 2.2$). In these conditions, the behavior of the β scaling of micro-instabilities could be different from the expected picture described above. The goal of this paper is to study the β dependence of micro-instabilities in these particular experimental conditions which are close to the β scaling discharges performed in ASDEX Upgrade. The tool used for simulations is the gyrokinetic electromagnetic flux-tube code GS2 [15, 16] in its linear version.

The remainder of this paper is organized as follows: in section 2 a description of the ASDEX Upgrade discharges [5] used as reference for the numerical simulation parameters is given. In section 3, β scans considering different density and temperature gradient lengths, within the experimental error-bars, are presented. Spectra of micro-instabilities for low and high β cases are shown in section 4. Section 5 is devoted to a discussion on the plasma conditions for which the micro-tearing modes are the most unstable modes in the system. In section 6, the β scaling of micro-instabilities at $\rho_t = 0.7$ (ρ_t is the normalized minor radius defined by the toroidal flux) and the effect of the triangularity are presented. Finally, in section 8, a brief summary is given.

2. Description of the reference discharge

As mentioned above, the reference input parameters used for the following simulations are based on the best pair of discharges from the dedicated β scaling experiments recently performed in ASDEX Upgrade [5]. These discharges (♯21403-♯21426) are in standard H-mode type-I ELMs regime with pure NBI heating (all details can be found in the latter reference). During this β scan experiment, ρ^*, ν^*, β/B^4, the safety factor q and the plasma shape were kept constant while only β was varied. This means that temperature and density gradient lengths as well as the ratio T_e/T_i were equal in both discharges. Temperature, density and safety factor profiles for the low β shot (♯21403) are plotted in figure 1.

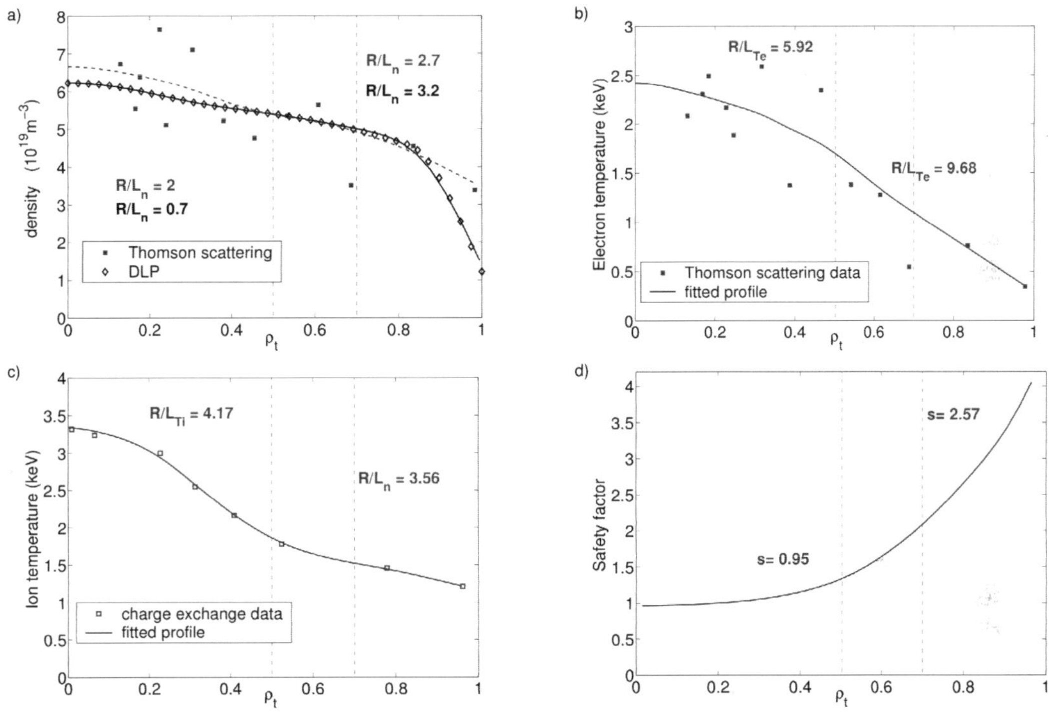

Figure 1. Toroidal radial profiles for the low β discharge ♯21403. Average from $t = [2.8 - 3.4]s$. a) Density profiles from interferometer data combined with lithium beam measurements (DLP) (full line) and from Thomson scattering measurements (dashed line). b) Electron temperature profile. c) Ion temperature profile. d) Safety factor profile.

The main input parameters for numerical simulations at $\rho_t = 0.5$ are : $T_e/T_i = 0.9$, $R/L_n = [0.7; 1.5; 2]$, $R/L_{Te} = 5.92$, $R/L_{Ti} = 4.17$, the electron collision frequency $\nu_e = 0.00279$ $log\Lambda n/T_e^{3/2} a\sqrt{m_i/T_i} = 0.16$ and the ion collision frequency $\nu_i = \nu_e \sqrt{m_e/m_i}\,(T_e/T_i)^{3/2}$ with the density in 10^{19} and the temperature in keV. Note that the experimental uncertainty on the gradient length is relatively large. In particular, the estimate of the density gradient length depends on the diagnostic used (cf. figure 1) and may vary significantly (from $R/L_n = 0.7$ to $R/L_n = 2$ at $\rho_t = 0.5$). The Thomson scattering profile is flatter at the edge and more peaked in the core than the profile obtained by combined measurements of interferometer and lithium beam diagnostics. For comparison, the scaling law density peaking from Reference [17] predicts for this discharge $R/L_n \simeq 1.5$. The experimental local-β value ranges between $\beta(\rho_t = 0.5) = 0.88\%$ (low β discharge) and $\beta(\rho_t = 0.5) = 1.48\%$ (high β discharge). All simulations presented in the following have been performed using the GS2 [15, 16] code which solves the gyro-averaged

Vlasov equation coupled to the Maxwell equations. Dynamic of electrons and ions, passing and trapped, is resolved and perturbations of the magnetic field in all directions (i.e., perturbations both parallel $\delta B_{||}$ and perpendicular $\delta B_{\perp}$ to the equilibrium field) are included. Particle collisions are modeled using a Lorentz collision operator. The GS2 code uses the ballooning representation in which the coordinates follow the magnetic field lines (the ballooning angle is noted θ_b).

To reproduce closely the experimental equilibrium, the Miller local equilibrium model [18] was used. The equilibrium input parameters are : the normalized half-width of the flux surface $\rho_c = 0.17$, the local safety factor $q = 1.3$, the local derivative of the Shafranov shift $\Delta' = -0.15$, the local elongation $\kappa = 1.377$ and its derivative $\kappa' = 0.58$, the local triangularity $\delta = 0.03$ and its derivative $\delta' = 0.33$, the magnetic shear $\hat{s} = (\rho/q) \times (dq/d\rho) = 0.95$. For the low β case, $\beta(\rho_t = 0.5) = 0.88\%$ and its local derivative $\beta' = \beta \sum[nT(R/L_T + R/L_n)] = [-0.096; -0.110; -0.118]$ depending on the density gradient length chosen. For the high β case, $\beta(\rho_t = 0.5) = 1.48\%$ and $\beta' = [-0.161; -0.183; -0.197]$.

All simulations have been performed including collisions and only two species were retained which means that the deuterium density is imposed to match the electron density and to verify the quasi-neutrality condition.

3. β scan close to experimental conditions

The β scans presented in this section were performed at $k_\theta \rho_i = 0.4$ where ITG modes, which are supposed to play the main role in heat transport in tokamak plasmas, are strongly unstable. In these simulations β and β' are varied consistently while all other input parameters were kept constant. In the local Miller equilibrium model [18] we used, the Shafranov shift is determined by β', therefore it has also been changed consistently with β. In term of plasma conditions, there are several main differences comparing the present β scan to previous studies. Firstly, the value of the ion temperature gradient length R/L_{Ti} is here taken very close to the ITG threshold while generally $R/L_{Ti} = [6.9-9]$. Secondly, the collisionality in the present simulations is rather high ($\nu^* = \nu_i/(\epsilon\ \omega_{bi}) = 0.05$, where ω_{bi} is the ion bounce frequency). And as third difference, the value of the density gradient length is small (generally $R/L_n = 2 - 2.2$ while here in most of the simulations $R/L_n = 0.7 - 1.5$).

The β scaling of micro-instabilities is very sensitive to the density gradient length while experimentally the uncertainty on this value is rather large. For these reasons, we have analyzed three different values of $R/L_n = [0.7, 1.5, 2]$. The evolutions of the growth rate and the real frequency as a function of β are plotted in figure 2.

For $R/L_n = 2$, ITG modes are the most unstable modes from $\beta = 0$ to $\beta = 4.5\%$. Between $\beta = 4\%$ and $\beta = 4.5\%$, the real frequency as well as the growth rate shows a discontinuity corresponding to the destabilization of KBM. In these specific conditions (R/L_{Ti} close to the ITG threshold), ITG modes are not stabilized with increasing β and the growth rate is very weakly affected by any increase of β. Note that in this scan, β and β' were varied simultaneously, indeed the stabilizing effect of β is balanced by the destabilizing effect of β' (at low β'), via the Shafranov shift. This has been proven by repeating the same scan keeping β' constant : in this case the ITG growth rate clearly decreases with increasing β. For $R/L_n = 1.5$, the behavior of the β scaling of micro-instabilities is similar to the $R/L_n = 2$ case with a slight shift of the KBM threshold toward higher β values (around $\beta = 5\%$).

For $R/L_n < 2$, the behaviour of the growth rate is significantly different from the expected theoretical picture of ITG stabilization with β up to the critical limit where the electromagnetic modes (KBM) become unstable. For $R/L_n = 0.7$, at low β ($\beta = [0 - 1\%]$), the most unstable modes are ITG modes which are stabilized with increasing β. For $\beta > 1\%$, the most unstable modes rotate in the electron direction. Looking at the eigenfunctions of such modes (figure 3), it appears that the electrostatic potential (Φ) is asymmetric with respect to the ballooning coordinate and the electromagnetic potential (A) is symmetric. This is not consistent with ITG,

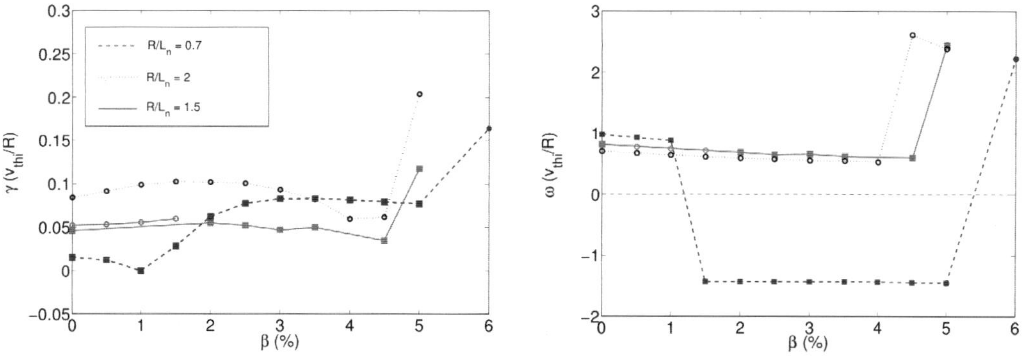

Figure 2. Growth rate and real frequency (normalized to v_{thi}/R where $v_{thi} = \sqrt{T_i/m_i}$ and R is the major radius) vs β for the experimental conditions of the reference shot ♯21403 at $\rho_t = 0.5$ for $R/L_n = 0.7$ (dashed line), $R/L_n = 1.5$ (full line) and $R/L_n = 2$ (dotted line). Simulations were performed on n = 15 toroidal turns (circle symbols) or on n = 79 toroidal turns (square symbols).

Trapped Electron Modes (TEM) or KBM instabilities for which the electrostatic potential (Φ) is symmetric and the electromagnetic potential (A) is asymmetric. This behaviour is the signature of micro-tearing modes which have been yet only detected, in experimental conditions, in high β simulations ($\beta = 10\%$) made for Spherical Tokamak plasmas [21, 22, 23, 24]. Micro-tearing modes are very extended in the ballooning coordinate (θ_b) meaning that they are extremely narrow in term of radial extension. Therefore, they required calculations on large boxes in term of θ_b (~79 toroidal turns) to be correctly resolved.

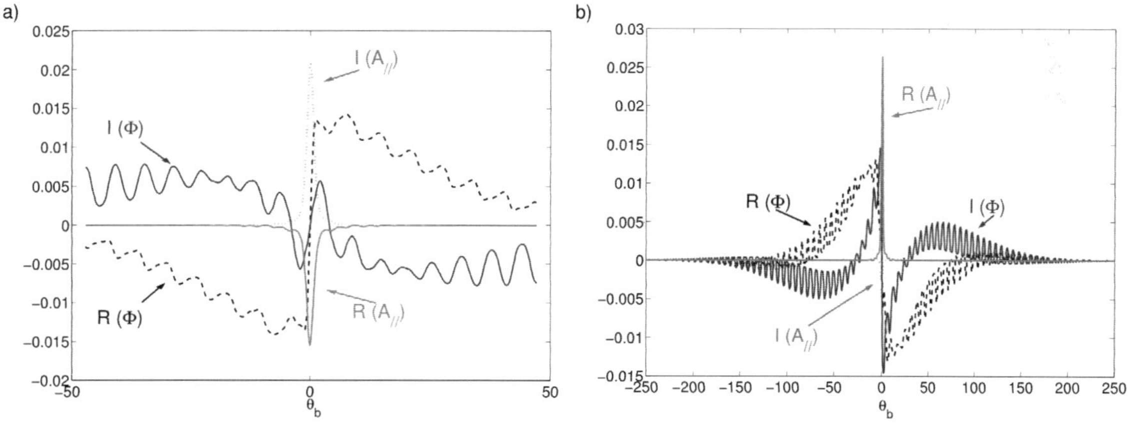

Figure 3. Eigenfunctions for $\beta = 2.5\%$ (a) for simulations performed on n = 15 toroidal turns and (b) for simulations performed on n = 79 toroidal turns as functions of the ballooning angle θ_b.

In calculations performed on narrow boxes, the growth rate of micro-tearing is overestimated and they can artificially appear as the most unstable modes in the system. Therefore, each time micro-tearing modes have been identified as the most unstable modes in a simulation, we have enlarged the box size until convergence was reached (θ_b grid extended up to a factor 5). In all the following figures, square symbols represent runs on large box while circle represent smaller box simulations.

Figure 2 shows that the growth rate of the micro-tearing modes, $\gamma_{\mu t}$, increases with increasing β until a saturation phase is reached, around $\beta \simeq 3\%$. The appearance of micro-tearing modes as most unstable modes is not only due to the ITG stabilization but also to the increase of the micro-tearing growth rate with β, for example, $\gamma_{\mu t}(\beta = 2\%) \gg \gamma_{ITG}(\beta = 0.5\%)$.

In this β scan, micro-tearing modes are the most unstable modes until $\beta = 6\%$ at which value a KBM is destabilized. However, a similar scan has been done with values of β' increased artificially by a factor 1.3 (not consistent with β and gradient length values) in which micro-tearing modes are the dominant micro-instabilities even at $\beta = 9\%$. KBMs have a stronger dependence on β than micro-tearing modes : when they become unstable, they become the most unstable modes as soon as β is slightly increased. This means that in such a case, where β' is artificially increased, KBM are still stable at $\beta = 9\%$.

The effect of changing the temperature gradient length within the experimental error-bars has been also investigated. Simulations of β scans at $k_\theta\rho_i = 0.4$ increasing R/L_{Ti} show that for $R/L_n = 1.5$, the expected β stabilizing effect on ITG is recovered for $R/L_{Ti} \geq 5$. For $R/L_n = 0.7$, the β range in which micro-tearing modes are the most unstable modes is reduced but still exists between $\beta = [2.5 - 4\%]$ even with $R/L_{Ti} = 7$. Similar simulations have been done to study the effect of the electron temperature gradient length R/L_{Te} on the β scaling of micro-instabilities. The results are presented in figure 4.

For $R/L_{Te} = 4$, the most unstable modes are ITG modes between $\beta = [0 - 2]\%$ for which the growth rate decreases with increasing β. For $\beta > 2\%$, the dominant instabilities become micro-tearing modes. Even at $\beta = 5\%$ no KBM is destabilized. Increasing slightly the electron temperature gradient length to $R/L_{Te} = 5$ has a destabilizing effect on ITG. The ITG growth rate increases slightly with increasing β and ITG modes stay the dominant instabilities between $\beta = [3.5 - 5]\%$ instead of micro-tearing modes. At $\beta = 5\%$, KBMs are destabilized. At low β, for $R/L_{Te} = 7$, simulations did not converge. For $\beta \geq 1.5\%$, the behavior of the dominant modes is similar to the β scan for the nominal value $R/L_{Te} = 5.9$ already discussed above. ITG modes which are the most unstable modes until $\beta = 4.5\%$ is very weakly affected by β.

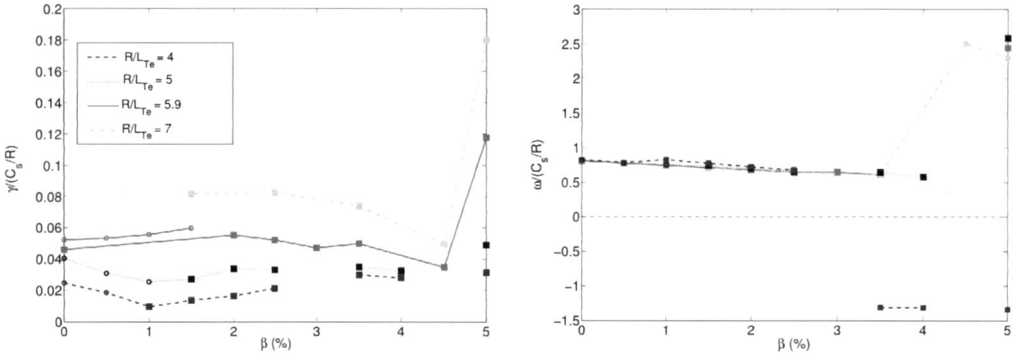

Figure 4. Growth rate and real frequency vs β for various values of R/L_{Te} for $R/L_n = 1.5$ and $k_\theta\rho_i = 0.4$.

From these simulations, the effect of R/L_{Te} on the β scaling of micro-instabilities is not very clear and surprising. In particular, ITG modes are found to be destabilized with increasing R/L_{te} which is not expected from collisionless simulations in which trapped electrons are affected by R/L_{Te} and lead to a stabilization of ITG with increasing R/L_{Te}.

Note that the AUG experiments, used as reference, cover the range $\beta(\rho_t = 0.5) = [0.88 - 1.48]\%$. Simulations show that in this range, the most unstable modes for $R/L_n = 1.5$ and $R/L_n = 2$ (considering nominal values for R/L_T) are ITG modes. The growth rate is slightly increasing from the low β to the high β (by about 9%). For $R/L_n = 0.7$, at $\beta(\rho_t = 0.5) = 0.88\%$ the most unstable modes are also ITG modes (with a growth rate very close to zero) but for the high β case, micro-tearing modes are the most unstable modes. The appearance of micro-tearing for such small values of β ($\beta < 2$) is rather unexpected especially at wavenumber $k_\theta\rho_i = 0.4$ for which the ITG growth rate is close to its maximum value.

4. Spectrum of micro-instabilities

The simulations discussed in the previous section showed the presence of micro-tearing modes for $R/L_n = 0.7$ at $k_\theta\rho_i = 0.4$. This value of R/L_n is at the lower limit of the experimental error-bars and may be considered too small even for a high collisionality H-mode plasma in ASDEX Upgrade. Therefore in this section, we discuss the spectrum of micro-instabilities for $R/L_n = 1.5$.

Figure 5 shows the linear spectrum of the most unstable modes in the range $k_\theta\rho_i = [0.1-0.6]$ for both low and high β cases. ITG modes appeared to be the most unstable modes for $k_\theta\rho_i = [0.2-0.5]$ with a maximum in the growth rate at $k_\theta\rho_i = 0.3$ for both values of β. Surprisingly, at lower wavenumber (between $k_\theta\rho_i = [0.1-0.2]$) micro-tearing are the most unstable modes. This result shows that even with a standard value of density gradient length, micro-tearing modes may be unstable in H-mode plasmas at low β values. Furthermore, the growth rate of micro-tearing and ITG modes are comparable.

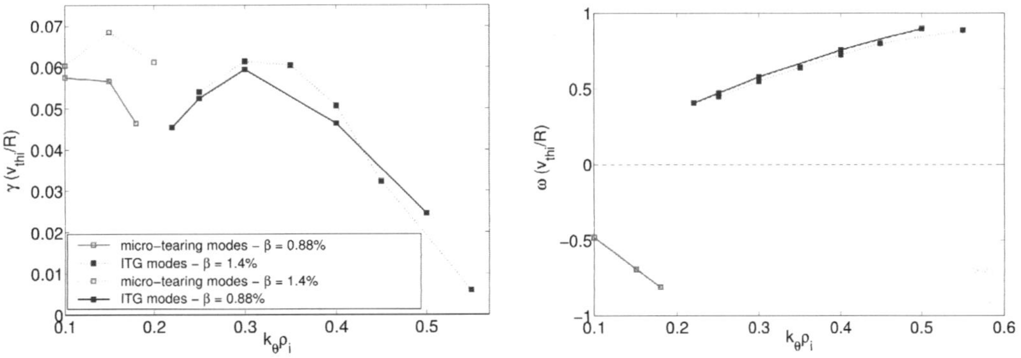

Figure 5. Spectrum for $R/L_n = 1.5$ for the low β case ($\beta = 0.88\%$) and for the high β case ($\beta = 1.4\%$).

However, it is important to underline that even if the poloidal wavenumber is smaller for micro-tearing than for ITG, the radial wavenumber of micro-tearing modes is much higher than the ITG one. This is due to the fact that micro-tearing have a very narrow radial extension. As a result, the average perpendicular wavenumber $k_\perp$, computed by GS2, is a factor of 5 higher for micro-tearing modes ($k_\perp\rho_i = 2.5$) as compared to ITG modes ($k_\perp\rho_i = 0.5$). According to the quasi-linear theory and the mixing length estimate ($\chi \propto \gamma/k_\perp^2$), ITG modes are therefore expected to play the dominant role in heat transport. However, it is well known that during the non-linear evolution, the spectrum of ITG turbulence is down shifted from $k_\theta\rho_i \simeq 0.3$ in the linear phase to $k_\theta\rho_i \simeq 0.15$ in the saturated non-linear phase [20]. In this spectral region, micro-tearing modes are linearly unstable and up to now the non-linear coupling of such modes with ITG modes is not completely understood. In reference [8], it has been shown when ITG modes are stabilized enough by the increase of β, the electron heat transport can become dominated by micro-tearing modes. In addition, it is apparently possible for micro-tearing to exhibit self-sustainment. On another hand, it is not certain that taking the perpendicular wavenumber of the linear modes as an average yields a reasonable estimate of the transport coefficients especially for electromagnetic modes such as micro-tearing.

In figure 5, it can also be observed that, for both branches, the increase of β has a weak effect on the growth rate and does not affect the real frequency.

The computation of the spectrum for $R/L_n = 0.7$ at $\beta = 0.88\%$ shows that ITG modes are weakly unstable, with a very low growth rate, as seen in figure 2 for $k_\theta \rho_i = 0.4$. Thus, at low density gradient, the micro-tearing branch extends to higher wavenumbers as compared to the $R/L_n = 1.5$ case (between $k_\perp \rho_i = 0.1 - 0.3$), with a comparable value of the growth rate.

5. Stability conditions of micro-tearing

Starting from the input parameters of the low β case with $R/L_n = 1.5$ for $k_\theta \rho_i = 0.15$, we have tried to define the domain in which micro-tearing modes are the most unstable instabilities. It has been found that, in AUG conditions, micro-tearing modes appear to be the most unstable modes when $R/L_{Te} \geq 3$. A minimum of $\beta = 0.4\%$ is required to obtain $\gamma_{\mu t} > \gamma_{ITG}$ and for $\beta > 6\%$ KBMs start to be dominant instabilities even at $k_\theta \rho_i = 0.15$. The micro-tearing threshold in term of collisionality has been found to be between $\nu_e = 0.066$ and $\nu_e = 0.1$.

Up to now, micro-tearing modes have been identified as the dominant micro-instabilities in gyrokinetic simulations applied to experimental discharges [21, 22, 23, 24] only for plasma conditions of the Spherical Tokamak (ST). The plasma parameters of ST discharges are rather different from AUG plasmas conditions. In particular, the plasma shape is different as compared to conventional tokamaks and very high β can be reached in ST (up to a factor 10 higher than in AUG plasmas). The number of parameters which may play a role in the stability of micro-tearing, ITG and TEMs is so large that it is very difficult to make a direct and complete comparison of the present study with MAST [23, 24] publications. However, we would like to underline some specific differences and agreements. The first difference is that, in AUG simulations, micro-tearing modes are the most unstable modes only for a limited part of the spectrum (around $k_\theta \rho_i = 0.15$) and coexist with ITG while on MAST they are dominant over the entire poloidal wavenumber range ($k_\theta \rho_i = [0 - 1]$) with a clear maximum of the growth rate localized around $k_\theta \rho_i = 0.8$. Secondly, the β scaling of the micro-tearing growth rate in MAST conditions exhibits a maximum at $\beta = 15\%$ [24] but these modes are almost completely stable at $\beta = 5\%$. The maximum of the growth rate of micro-tearing modes in MAST is around $\gamma = 7.56 \times 10^5 s^{-1}$ while in AUG simulation, $\gamma = 5.48 \times 10^5 s^{-1}$. These comparable amplitudes at very different β values show that the β scaling of micro-tearing is strongly affected by other plasma parameters. For example, micro-tearing stability is strongly dependent on collisionality [24] and in AUG simulations, ν_e is lower by about a factor of 3 as compared to the MAST case.

6. Effect of the triangularity at $\rho_t = 0.7$

Experimentally, the plasma shape is known to play a role in the quality of the confinement in tokamak plasmas. The increase of the elongation κ and of the triangularity δ is favorable in experiments. For example, it provides a significant increase in the β stability limit in DIII-D [25, 26] and the increase of triangularity changes the plasma edge dynamics (such as ELMs) and leads to an increased edge pressure gradient and consequently to an increase of the confinement time [27]. The effect of plasma shaping has been investigate numerically by several authors [28, 29, 30, 31]. As general results, it has been found that the elongation has a stabilizing effect on ITG mode and TEM while the effect of the triangularity seems to depend itself on the value of the elongation. In the electrostatic limit, δ has been found to be destabilizing at moderate κ and stabilizing at high κ [31]. In electromagnetic simulations, δ is also found to have a stabilizing effect on KBM by increasing the threshold of such modes.

The impact of increasing triangularity, in the conditions of the AUG β scaling experiments, on the micro-instabilities β scaling is explored in this section. As plasma shape affects more the edge, even if in this work we investigate core turbulence, triangularity effects have been study at $\rho_t = 0.7$ to use plasma parameters closer to the edge conditions. The main input parameters at $\rho_t = 0.7$ are : $T_e/T_i = 0.71$, $R/L_n = 3.26$, $R/L_{Te} = 9.68$, $R/L_{Ti} = 3.56$, $\nu_i = 0.0036$ and $\nu_e = 0.35$. For the low β case, $\beta(\rho_t = 0.7) = 0.68\%$ and its local derivative $\beta' = -0.109$. For the high β case, $\beta(\rho_t = 0.7) = 1.08\%$.

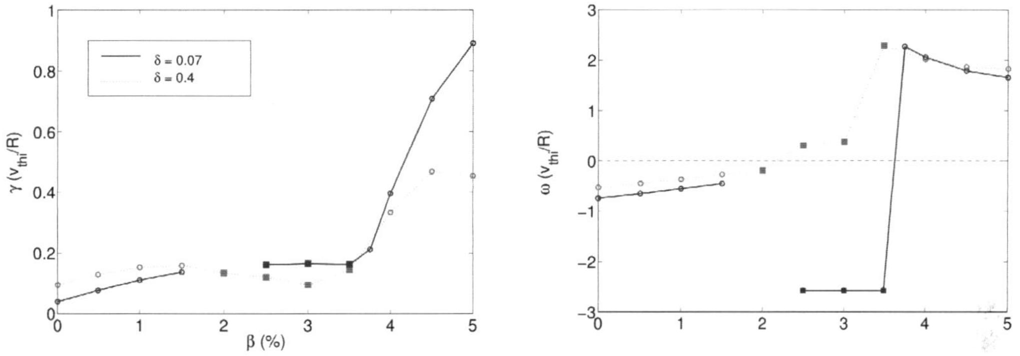

Figure 6. β scaling of the growth rate and the ral frequency at $\rho_t = 0.7$ for $\delta = 0.07$ (full line) and $\delta = 0.4$ (dotted line)

Figure 6 shows the growth rate and the real frequency of the most unstable modes at $\rho_t = 0.7$ for two values of triangularity ($\delta = 0.07$ which is the nominal value in the reference discharge and $\delta = 0.4$ corresponding to the experimental value of a high triangularity AUG plasma). Firstly, in the β scaling under experimental conditions at $\rho_t = 0.7$ ($\delta = 0.07$), it appears that TEMs are the most unstable modes from the electrostatic limit to $\beta = 1.5\%$ and that TEMs are destabilized with increasing β. The fact that TEMs are the most unstable modes for these conditions of low density and high density gradient is not unexpected. It has been found from linear gyrokinetic simulations that in ASDEX Upgrade discharges, with similar gradient lengths than in this present study, TEMs are the dominant instabilities [32]. However, the influence of β on TEMs are not well known and the destabilization of TEM with increasing β is an interesting result which requires deeper investigations.

Note that the experimental range of β values is in the regime of dominant TEM. The growth rate increases by about 30% from the low β ($\beta(\rho_t = 0.7) = 0.68\%$) to the high β case ($\beta(\rho_t = 0.7) = 1.08\%$) while the perpendicular wavenumber computed by GS2, $k_\perp$ decreases from $k_\perp = 5.37$ to $k_\perp = 5.03$. This behavior leads to an increase of the mixing length estimate $\chi \propto \gamma/k_\perp^2$ of about 50%. The simulation performed at $\beta = 2\%$ did not converge and we can observed a jump of the absolute value of the real frequency between $\beta = 1.5\%$ and $\beta = 2.5\%$ which means that the nature of the most unstable modes changed. At $\beta = 2.5\%$, micro-tearing modes are again the dominant instabilities (in these simulations, $k_\theta \rho_i = 0.4$). Another transition appears around $\beta = 4\%$, this time, the real frequency as well as the value of the growth rate jumps. The dominant modes, which rotate in the ion direction, are KBMs.

Now, increasing artificially the triangularity (keeping its derivative constant as compared to the low δ case), no micro-tearing modes appear as the most unstable modes at any value of β. TEMs are also the dominant micro-instabilities at low β and destabilized with increasing β. The β value for which the KBM becomes the dominant instability is not strongly affected by the change of the triangularity.

7. Summary

The β dependence of micro-instabilities have been studied using linear gyrokinetic simulations in experimental conditions of the ASDEX Upgrade dedicated β scan discharges. In such high density H-mode plasmas with low density gradient, it is found that when the ion temperature gradient length is close to the linear ITG threshold, β scaling of micro-instabilities is very sensitive to the density gradient length. The β scaling of micro-instabilities can significantly differ from the expected behaviour in which at low β, ITG modes are stabilized with increasing β until KBMs start to become unstable.

From linear gyrokinetic simulations, KBMs seem not to play a role in the β degradation observed in dedicated ASDEX Upgrade experiements. First, it is found that KBMs are destabilized at value of β clearly above the experimental β range. Even changing the density gradient length and the temperature gradient length within the error-bars do not change this result. The effect of the plasma triangularity on the KBM branch has been also investigated. It shows that at $\rho_t = 0.7$, the increase of the triangularity does not affect strongly the KBMs stability.

It is also shown that for low density gradient length, micro-tearing modes are the most unstable modes already for $\beta = 1\%$. Such modes are dominant at $k_\theta \rho_i = 0.4$ for the specific case of very low density gradient length ($R/L_n = 0.7$) but are also found to coexist with ITG modes for higher value of R/L_n ($R/L_n = 1.5$) at $k_\theta \rho_i = 0.15$ with similar value of the growth rate. These modes, which have a very narrow radial extension, are supposed not to play a crucial role in the turbulent heat transport observed in tokamak plasmas . However, non-linear evolution of such modes and their coupling with ITG modes are not completely understood. Therefore, their role in the β dependence of heat transport is not excluded and deeper investigations are required.

In addition to this, it has been found that, for values of R/L_{Ti} close to the linear ITG threshold, the β dependence of ITG modes is rather different from expected. In these conditions, β has a very weak effect on ITG stability. At low β ($\beta < 2\%$), ITG modes are slightly destabilized with increasing β while the stabilizing effect of β slowly starts for $\beta \geq 2\%$. Even if, in the experimental β range, ITG modes are destabilized with increasing β, the absolute effect is very weak and the perpendicular wavenumber decreases in the same time as the growth rate increases leading to a change in the mixing length estimate which is not in agreement with the observed experimental β degradation.

Finally, the study of the β scaling of micro-instabilities at $\rho_t = 0.7$ has shown that TEMs are the dominant modes and that β has a destabilizing effect on TEM. In this case, not only the growth rate increases with increasing β but also the perpendicular wavenumber decreases. In the experimental β range, the mixing length estimate $\chi \propto \gamma/k_\perp^2$ changes by about 50% from the low β to the high β case. Therefore, TEMs may play a role in the β degradation observed in ASDEX Upgrade.

Acknowledgments

The authors would like to acknowledge W. Dorland and M. Kotschenreuther for providing the gyrokinetic code GS2 and financial support for this work from the EURATOM programme of the European Community in the form of a Individual Fellowship, Contract No FU06-028309.

References

[1] ITER Physics Basis and ITER Physics Expert Groups 1999 *Nuclear Fusion* **39** 2175
[2] Petty C C, Luce T C, McDonald D C *et al.* 2004 *Phys. Plasmas* **11** 5
[3] McDonald D C *et al.* 2004 *Plasma Phys. and Control. Fusion* **46** A-215
[4] Urano H, Takizuaka T *et al.* 2006 *Nuclear Fusion* **46** 781
[5] Vermare L, Ryter F, Angioni C, Peeters A G *et al.* 2007 *Nuclear Fusion* **47** 490
[6] Scott B D 1997 *Plasma Phys. and Control. Fusion* **39** 1635

[7] Snyder P B and Hammett G W 2001 *Phys. Plasmas* **5** 3 744
[8] B.D. Scott 2003 *Plasma Phys. and Control. Fusion* **45** A385
[9] Scott B D 2006 *Plasma Phys. and Control. Fusion* **48** B277
[10] Jenko F and Dorland W 2001 *Plasma Phys. and Control. Fusion* **43** A141
[11] Falchetto G L, Vaclavik J and Villard L 2003 *Phys. Plasmas* **10** 5 1424
[12] Chen Y, Parker S E *et al.* 2003 *Nuclear Fusion* **43** 1121
[13] Parker S E, Chen Y *et al.* 2004 *Phys. Plasmas* **11** 5 2594
[14] Candy J 2005 *Physics of Plasmas* **12** 072307
[15] Kotschenreuther M, Rewoldt G and Tang W M 1995 *Comput. Phys. Commun* **88** 128
[16] Dorland W, Jenko F, Kotschenreuther M and Rogers B N 2000 *Phys. Rev. Lett.* **85** 5579
[17] Angioni C, Weisen H, Kardaun O J W F *et al.* 2007 *Nuclear Fusion* **47** 1326
[18] Miller R L, Chu M S, Greene J M, Lin-Liu Y R and Waltz R E 1998 *Phys. Plasmas* **5** 973
[19] Bourdelle C *et al.* 2003 *Phys. Plasmas* **10** 7 2881
[20] Lin Z and Hahm T S 2004 *Phys. Plasmas* **11** 1099
[21] Redi M H *et al.* 2003 *in Proceedings of 30th EPS Conference on Controlled Fusion and Plasma Physics, St Petersburg*, edited by R. Koch and S. Lebedev (Europoean Physical Society, Mulhouse), ECA Vol. 27A, P4.94
[22] Wilson H R *et al.* 2004 *Nuclear Fusion* **44** 917
[23] Applegate D J and Roach C M *et al.* 2004 *Physics of Plasmas* **11** 11 5085
[24] Applegate D J and Roach C M 2007 *Plasma Phys. and Control. Fusion* **49** 1113
[25] Lazarus E A, Chu M S, Ferron J R *et al.* 1991 *Phys. Fluids B* **3** 2220
[26] Ferron J R, Casper T A *et al.* 2005 *Physics of Plasmas* **13** 056126
[27] Suttrop W *et al.* 2000 *Plasma Phys. and Control. Fusion* **42** A97
[28] Rewoldt G, Tang W M, and Chance M S 1982 *Phys. Fluids* **25** 480
[29] Hua D D, Xu X Q, and Fowler T K 1992 *Phys. Fluids B* **4** 3216
[30] Waltz R E and Miller R L 1999 *Physics of Plasmas* **6** 4265
[31] Belli E A 2006 *PhD thesis*
[32] Peeters A G, Angioni C *et al.* 2005 *Physics of Plasmas* **12** 022505

AUTHOR INDEX

AUTHOR INDEX

AUTHOR INDEX

AUTHOR INDEX